# Die Grundlehren der mathematischen Wissenschaften

in Einzeldarstellungen
mit besonderer Berücksichtigung
der Anwendungsgebiete

Band 215

Helmut H. Schaefer

# Banach Lattices and Positive Operators

Springer-Verlag
Berlin Heidelberg New York 1974

Dr. rer. nat. Helmut H. Schaefer
Professor of Mathematics, University of Tübingen

AMS Subject Classification (1970):
06A65, 15A48, 15A51, 46–01, 46A35, 46A40, 46B99, 46E05,
47–01, 47B15, 47B55, 47B99, 47D05, 47D10, 47D15

ISBN 3–540–06936–4 Springer-Verlag Berlin Heidelberg New York
ISBN 0–387–06936–4 Springer-Verlag New York Heidelberg Berlin

Library of Congress Cataloging in Publication Data

Schaefer, Helmut H.
Banach lattices and positive operators.
(Die Grundlehren der mathematischen Wissenschaften in Einzeldarstellungen mit besonderer Berücksichtigung der Anwendungsgebiete; Bd. 215)
Bibliography: p.
Includes indexes.
1. Linear topological spaces. 2. Lattice theory. 3. Linear operators. I. Title. II. Series: Die Grundlehren der mathematischen Wissenschaften in Einzeldarstellungen; Bd. 215.
QA322.S27 515′.73 74–22499

 Typesetting and printing: Zechnersche Buchdruckerei, Speyer, Bookbinding: Konrad Triltsch, Würzburg.

*To Rhea, Christoph, and Mark*

# Preface

Vector lattices—also called *Riesz spaces*, *K-lineals*, or *linear lattices*—were first considered by F. Riesz, L. Kantorovič, and H. Freudenthal in the middle nineteen thirties; thus their early theory dates back almost as far as the beginning of the systematic investigation of Banach spaces. Schools of research on vector lattices were subsequently founded in the Soviet Union (Kantorovič, Judin, Pinsker, Vulikh) and in Japan (Nakano, Ogasawara, Yosida); other important contributions came from the United States (G. Birkhoff, Kakutani, M. H. Stone).

L. Kantorovič and his school first recognized the importance of studying vector lattices in connection with Banach's theory of normed vector spaces; they investigated normed vector lattices as well as order-related linear operators between such vector lattices. (Cf. Kantorovič-Vulikh-Pinsker [1950] and Vulikh [1967].) However, in the years following that early period, functional analysis and vector lattice theory began drifting more and more apart; it is my impression that "linear order theory" could not quite keep pace with the rapid development of general functional analysis and thus developed into a theory largely existing for its own sake, even though it had interesting and beautiful applications here and there.

When first confronted—almost a dozen years ago—with the task of writing a monograph on positive (linear) operators, I was painfully aware of this situation. I realized that the material on positive operators published by then, was not nearly sufficient to convince anyone of the close ties between positive operators and the mainstream of operator theory, at least as I anticipated them to exist. It was my good fortune that in the years from 1963 to 1973 several of my students contributed significantly, and perhaps even decisively, to bridge the aforementioned gap. Without the work of H. Lotz, R. Nagel, U. Schlotterbeck, and M. Wolff I could not have conceived this book in its present form; their results, along with those of others, encouraged me to offer this volume to the mathematical public.

The book contains five chapters, each of which begins with an introduction; thus a general introduction to the subject matter appears unnecessary. Suffice it to say here that Chapter I, dealing with (finite) positive square matrices, is intended to be both an introduction to and a motivation for the entire work; Chapter II and about one third of Chapter III are concerned with Banach lattices, thus laying the groundwork for the operator theory treated in the remainder of the book. The notes added to each chapter try to trace at least the more recent results to their origin, without any attempt at completeness or at determining priorities in each case. Concerning the exercises found at the end of each chapter (a total of 120), it should be said that the text would be far less complete, and possibly valuable, without them. Many of these exercises require advanced techniques which are generally indicated by hints, and some of them require auxiliary literature indicated by references. However, even though practically none of these exercises are required for an understanding of the text, every reader ought to take notice of them because of the supplementary information they contain—whether or not he can master them without undue effort.

The main objective of this book is to present the theory of Banach lattices and positive linear operators as an inseparable part of general Banach space and operator theory. Therefore, a prerequisite for an intelligent reading of the text and the numerous accompanying examples is familiarity with basic classical and functional analysis (including measure theory). Occasionally deeper results are needed; for these we refer uniformly to the books [DS], [H], and [S] by Dunford-Schwartz, Halmos, and the author, respectively (see Bibliography). Inclusion of these prerequisites would have enlarged the size of this volume beyond tolerable limits and, in many instances, it would have meant unnecessary duplication of easily accessible material.

A few words should be said about the reference system and the terminology employed. References are of the type $(u.v)$ (or $u.v$) within each chapter, where $u$ denotes the section number and $v$ the number of the proposition (definition, lemma, theorem) within Section $u$; references outside the chapter are by $(r.u.v)$ (or $r.u.v$), where the Roman numeral $r$ indicates the number of the chapter referred to. Definitions of considerable importance for the sequel (and other than standard definitions from auxiliary fields) have been numbered; frequent references (and, occasionally, repetitions) are made to facilitate the understanding of the text. Generally, the full hypothesis is included in the statement of each proposition and theorem to save the reader the trouble of locating implicit assumptions. For the definition of symbols, the sign $:=$ has been used where it promised to increase readability. For all these precautions, I realize that some sources of potential misunderstanding remain. For example, an *imbedding* is not meant to be necessarily injective; the term *unit ball* (of a normed space) refers to the closed unit ball throughout. Even though elsewhere these terms may be used in a slightly different meaning, it is hoped that the context will assist the reader in case of doubt. The terms *isomorphic* (sometimes abbreviated by $\cong$) and *isomorphism* are used to imply preservation of *all* structures referred to by a given concept; thus, an isomorphism of Banach spaces is a linear, surjective isometry. (Even so, the term *isometric* is sometimes added to preclude misunderstanding.) Unless explicitly stated to the contrary, the term *measure space* refers to a triple $(X, \Sigma, \mu)$ where $X$ is a non-void set, $\Sigma$ a $\sigma$-algebra ($\sigma$-field) of subsets of $X$, and $\mu$ a countably additive set function $\Sigma \to \overline{\mathbb{R}}_+$ (the extended positive reals). We point this out, because our usage departs from both [H] (who does not require $X \in \Sigma$) and from [DS] (who do not require $\mu$ to be positive). Concerning topological vector spaces, our terminology generally follows [S]. Otherwise, it is hoped that the terminology used is commonly known; thus, Halmos' *iff* (meaning if and only if) is used occasionally without being defined in the text.

Finally, I wish to make the following personal acknowledgments. M. Powell has contributed many critical remarks and helpful suggestions concerning Chapter II. R. Nagel has assisted me substantially in the preparation of Chapter III, while U. Schlotterbeck has done the same with respect to Chapter IV. C. Rall has supported me unfailingly in the preparation of the bibliography, the subject index, and the index of symbols; in addition, he has read the complete set of proofs. I express my sincere thanks to all of them, as I do to the publishers for their constant courtesy and ready cooperation.

Tübingen, September 1974

H. H. S.

# Table of Contents

Chapter I

# Positive Matrices

## Introduction

The purpose of this introductory chapter is twofold. First, it supplies a unified discussion of the most important operator theoretic properties of positive square matrices; it can be read independently of the remainder of the book, presupposing only familiarity with elementary polynomial algebra (cf. § 1). Second, and more important in our context, this first chapter is intended to serve as a motivation for the study of positive operators on Banach lattices—the central theme of this book. As in the later chapters, the ideal structure of a vector lattice (cf. Chapter II, §2) is used extensively and advantageously in Chapter I; even in the finite dimensional case discussed here, the judicious use of ideal theory turns out to be the key to many fruitful insights.

Section 1 reviews basic facts from the spectral theory of finite square matrices, considered as linear operators on $\mathbb{C}^n$. Section 2 discusses the peripheral spectrum of positive square matrices and related properties. Section 3 gives a survey of the asymptotic behavior of the powers of a matrix (or, in plusher terms, discusses the mean ergodic theorem for cyclic semi-groups of operators on $\mathbb{C}^n$) with an application to positive matrices, for which the theorem takes an interesting special form (3.4). Sections 4 and 5, concerned with stochastic and doubly stochastic matrices, respectively, derive the basic theory of these matrices; they supply, in fact, good examples for the concepts and methods developed so far.

The next three sections are grouped around the concept of operator-invariant ideal and the related notion of irreducibility. Section 6 derives the Perron-Frobenius theorem, without doubt the most striking single result on positive matrices, and discusses its ramifications; the special case of a primitive matrix is taken up in Section 7. Section 8 characterizes matrices, in various ways, that are direct sums of irreducible ones; it furnishes a combined application of the results obtained hitherto. At the same time, Section 8 is the key for the subsequent discussion (§ 9) of homogeneous Markov chains with finite state space; an application that illuminates the usefulness of the entire theory. Finally, Section 10 gives a few results on the localization of eigenvalues of positive matrices; it is mainly intended to exhibit some of the available methods.

Practically all topics of Chapter I are taken up later in this book for operators on Banach lattices. For example, operator-invariant ideals and mean ergodic theory are discussed in detail in Chapter III (§§ 7, 8); the spectral theory of positive (and certain non-positive) operators is discussed in Chapter V (§§ 4—7), and operators (important in ergodic theory) generalizing doubly stochastic matrices are investigated in Chapter V (§ 8). Since the methods and results applied in Chapter V presuppose much of the material supplied by Chapters II, III, and IV, it becomes apparent that the extension of the results of this first chapter to an infinite number of dimensions was, in many cases, not easy nor was the way to them always clear.

## § 1. Linear Operators on $\mathbb{C}^n$

Generally we denote by $\mathbb{N}, \mathbb{Z}$ the sets of integers $>0$ and of all integers, respectively, and by $\mathbb{Q}, \mathbb{R}, \mathbb{C}$ (in this order) the fields of all rational, real, and complex numbers. The field $\mathbb{C}$ is considered ordered by the relation "$\alpha \leqq \beta$ iff $\beta - \alpha$ is real and $\geqq 0$", so that $\mathbb{R}$ and $\mathbb{Q}$ can be considered to be totally ordered subfields of $\mathbb{C}$. The set of all real numbers $\geqq 0$ is denoted by $\mathbb{R}_+$.

Rather than considering abstract finite dimensional vector spaces over $\mathbb{C}$, in this chapter we shall restrict attention to the space $\mathbb{C}^n$ $(n \in \mathbb{N})$; for, it is well known that every vector space of dimension $n$ over $\mathbb{C}$ is isomorphic to $\mathbb{C}^n$. Elements (vectors) of $\mathbb{C}^n$ will be written $x = (\xi_i)_{1 \leqq i \leqq n}$ or briefly $x = (\xi_i)$; the vectors $e_j := (\delta_{ji})$ $(j = 1, \ldots, n)$ are said to constitute the *canonical basis* of $\mathbb{C}^n$. Under coordinatewise multiplication (which corresponds to the multiplication scheme $e_i e_j = \delta_{ij} e_i$, $i, j = 1, \ldots, n$), $\mathbb{C}^n$ is a commutative algebra with unit $e = (1, \ldots, 1)$; the ideals of this algebra are the vector subspaces of the form $J_{\mathsf{H}} := \{x \in \mathbb{C}^n : \xi_i = 0$ for $i \in \mathsf{H}\}$, where $\mathsf{H}$ is any subset of $\{1, \ldots, n\}$. If, as usual, $|x|$ denotes the modulus (or absolute value) $(|\xi_i|)$ of $x = (\xi_i)$ (§ 2), it is easy to verify that a vector subspace $J \subset \mathbb{C}^n$ is an ideal iff $x \in J$, $y \in \mathbb{C}^n$, and $|y| \leqq |x|$ implies $y \in J$. Thus the algebraic ideals of $\mathbb{C}^n$ are identical with the ideals of the complex vector lattice $\mathbb{C}^n$.[1]

The vector space of all linear forms on $\mathbb{C}^n$ is again $\mathbb{C}^n$ if we agree to identify a linear form $f$ with the vector $(f(e_i))$; thus if $x = (\xi_i)$ and $\eta'_i := f(e_i)$, then

$$f(x) = \langle x, f \rangle = \sum_{i=1}^n \xi_i \eta'_i .$$

In reality this identification amounts to selecting the linear forms $e'_i : x \mapsto \xi_i$ $(i = 1, \ldots, n)$ as a basis of the dual of $\mathbb{C}^n$, so that $f = \sum_{i=1}^n \eta'_i e'_i$. As a consequence of identifying $\mathbb{C}^n$ with its dual, we shall henceforth omit the primes; $\mathbb{C}^n$ is in duality with itself by virtue of

$$\langle x, y \rangle := \sum_{i=1}^n \xi_i \eta_i . \tag{1}$$

The bilinear form on $\mathbb{C}^n \times \mathbb{C}^n$ defined by (1) is called *canonical*; it should not be confused with the *canonical Hermitian form*

$$(x|y) := \sum_{i=1}^n \xi_i \eta_i^* \tag{2}$$

on $\mathbb{C}^n \times \mathbb{C}^n$, where $\eta^*$ denotes the complex conjugate of $\eta \in \mathbb{C}$.

---

[1] Cf. Chap. II, § 11 and Chap. III, § 1, Example 1.

Every linear map (operator) $A: \mathbb{C}^n \to \mathbb{C}^m$ is uniquely determined by the $m\,n$ complex numbers

$$\alpha_{ij} = \langle A e_j, e_i \rangle$$

$(i=1, \ldots, m;\ j=1, \ldots, n)$ so that $\alpha_{ij}$ is the $i$-th coordinate of the vector $A e_j \in \mathbb{C}^m$. The rectangular array $(\alpha_{ij})$ is called the matrix representing $A$ (with respect to the canonical bases of $\mathbb{C}^m$ and $\mathbb{C}^n$, respectively). It is easy to see that, as long as these bases are kept fixed, the mapping $A \mapsto (\alpha_{ij})$ is a bijection of the set of all linear maps $\mathbb{C}^n \to \mathbb{C}^m$ onto the set of all $m \times n$-matrices with complex entries. Thus we can and often shall identify $A$ with its matrix $(\alpha_{ij})$, which is a useful practice since the elementary operations of matrix algebra correspond (by definition) to the natural algebraic operations for linear mappings.

It is true that there is no a priori reason for fixing a particular basis of $\mathbb{C}^n$, and doing so would indeed be a handicap if we were to develop a general theory of linear mappings. But the order structure of the vector lattice $\mathbb{C}^n$ (and, similarly, the multiplicative structure of the algebra $\mathbb{C}^n$) is defined in terms of the canonical basis, and hence for linear operators related to these structures, representation by matrices with respect to the canonical basis is most adequate.

In the sequel we shall be exclusively concerned with endomorphisms of $\mathbb{C}^n$, canonically represented by matrices $(\alpha_{ij})$; no notational distinction will be made between an endomorphism $A$ and the matrix representing it. The identity map of $\mathbb{C}^n$ (unit matrix) is often denoted by $I$.

Let $A$ be an endomorphism of the vector space $\mathbb{C}^n$. The determinant $\Delta(\lambda) = \det(\lambda I - A)$ is called the *characteristic polynomial* of $A$; in particular, $\Delta(A) = 0$. The unique normalized polynomial of smallest degree that annihilates $A$, is called the *minimal polynomial* $m(\lambda)$ of $A$; $m(\lambda)$ always divides $\Delta(\lambda)$. The set of all zeros (in $\mathbb{C}$) of $\Delta(\lambda)$ is called the *spectrum* $\sigma(A)$ of $A$; equivalently, $\sigma(A)$ is the set of all $\lambda \in \mathbb{C}$ such that $\lambda I - A$ fails to be injective (and hence fails to be surjective). $A = (\alpha_{ij})$ and its *transpose (adjoint)* ${}^t A = (\alpha_{ji})$ have the same characteristic polynomial, the same minimal polynomial, and the same spectrum.

At this point we stress the difference between ${}^t A = (\alpha_{ji})$ and $A^* = (\alpha_{ji}^*)$, the conjugate transpose of $A$; we have $\sigma({}^t A) = \sigma(A)$ but $\sigma(A^*) = \sigma(A)^*$. In operator theoretic language, ${}^t A$ is related to $A$ by the identity $\langle A x, y \rangle = \langle x, {}^t A y \rangle$ but to $A^*$ by $(A x | y) = (x | A^* y)$ for all $x, y \in \mathbb{C}^n$ (see Formulae (1), (2) above).

The *resolvent* $R(\lambda)$ (or, if this distinction is necessary, $R(\lambda, A)$) is defined by $R(\lambda) := (\lambda I - A)^{-1}$ for all $\lambda \notin \sigma(A)$; note that $R(\lambda, {}^t A) = {}^t R(\lambda, A)$. $R(\lambda)$ is a rational function $B(\lambda)/\Delta(\lambda)$ (where $B(\lambda)$ and $\Delta(\lambda)$ possibly have common scalar divisors $q(\lambda)$, e.g. for $A = I$, $n > 1$). The numbers $\lambda \in \sigma(A)$ are called *characteristic values* (or *eigenvalues*) of $A$. There are three kinds of multiplicity attached to an eigenvalue $\lambda_0$ of $A$: The dimension of the characteristic space (eigenspace) $\{x : \lambda_0 x = A x\}$ is called the *geometric multiplicity* of $\lambda_0$. The order of $\lambda_0$ as a zero of $\Delta(\lambda)$ is called the *algebraic multiplicity* of $A$. Finally, there is the multiplicity of $\lambda_0$ as a pole of the resolvent, which agrees with the multiplicity of $\lambda_0$ as a zero of the minimal polynomial. The *spectral radius* $r$ (or $r(A)$) of $A$ is the largest modulus of any eigenvalue of $A$; it is the radius of the smallest circle of center 0 in $\mathbb{C}$ and containing $\sigma(A)$. The *spectral mapping theorem* of the operational calculus asserts that $\sigma(f(A))=f(\sigma(A))$ for each function $f$ locally holomorphic in a neighborhood of $\sigma(A)$.

Of the various standard notions of equivalence for $n\times n$-matrices, we shall mainly need that of similarity. Let us recall that two $n\times n$-matrices $A, B$ are called *similar* if there exists an invertible $U$ such that $B=U^{-1}AU$; in the special case where $U$ can be chosen to be unitary (that is, $U^*U=I$), $A$ and $B$ are called *unitarily equivalent*. It is easy to verify that similar matrices have identical characteristic and minimal polynomials, and the same spectrum. Moreover, the various multiplicities attached to an eigenvalue (see above) are invariant under similarity.

Let $F$ be a vector subspace of $\mathbb{C}^n$ invariant under $A$, that is, let $A(F)\subset F$, and denote by $p: F\to\mathbb{C}^n$ and $q: \mathbb{C}^n\to\mathbb{C}^n/F$ the canonical injection and quotient mappings, respectively. We denote by $A_{|F}$ the linear map induced by $A$ on $F$, and by $A_F$ the linear map induced by $A$ on $\mathbb{C}^n/F$; precisely, $p\circ A_{|F}=A\circ p$ [2] and $A_F\circ q=q\circ A$. If $\lambda\notin\sigma(A)$ then $(\lambda I-A)^{-1}(F)\subset F$ (Exerc. 1), thus $\sigma(A_{|F})\subset\sigma(A)$ and $\sigma(A_F)\subset\sigma(A)$; more precisely,

$$\sigma(A)=\sigma(A_{|F})\cup\sigma(A_F)\,. \tag{3}$$

Since vector subspaces and quotients cannot, in general, be conveniently represented in terms of the canonical basis of $\mathbb{C}^n$, the symbols $A_{|F}$ and $A_F$ cannot be interpreted as matrices until bases of $F$ and $\mathbb{C}^n/F$ have been specified. However, such an interpretation is possible without ambiguity when the $A$-invariant subspace $F$ is an ideal $J$ of $\mathbb{C}^n$ (see above). Thus if $J$ is the ideal $\{x\in\mathbb{C}^n: \xi_i=0$ for $i\in\mathsf{H}\}$ and $A$ is represented by the matrix $(\alpha_{ij})$, then $A_{|J}$ is canonically represented by the submatrix $(\alpha_{kl})$ where $(k,l)\in\mathsf{H}^c\times\mathsf{H}^c$ ($\mathsf{H}^c$ denoting the complement of $\mathsf{H}$ in $\{1,\dots,n\}$), and $A_J$ is canonically represented by the submatrix $(\alpha_{rs})$ where $(r,s)\in\mathsf{H}\times\mathsf{H}$. Since we shall primarily be concerned with $A$-invariant ideals and the respective quotients, the identification of operators and matrices agreed upon above will cause no difficulties.

We recall that a *norm* on $\mathbb{C}^n$ is a function $x\mapsto\|x\|$ of $\mathbb{C}^n$ into $\mathbb{R}_+$ satisfying these axioms:

$$\begin{gathered}\|x\|=0 \quad\text{implies}\quad x=0\\ \|\alpha x\|=|\alpha|\,\|x\|\\ \|x+y\|\leqq\|x\|+\|y\|\,.\end{gathered}\tag{4}$$

**Examples.** 1. The $l^\infty$-norm (or supremum norm) on $\mathbb{C}^n$ is defined by $\|x\|_\infty:=\sup_i|\xi_i|$.

2. The $l^p$-norm $(1\leqq p<+\infty)$ is defined by

$$\|x\|_p:=\left(\textstyle\sum_{i=1}^n|\xi_i|^p\right)^{1/p}.$$

The $l^2$-norm is often called the *Euclidean norm* of $\mathbb{C}^n$.

[2] $A_{|F}$ is often called the *restriction* of $A$ to $F$.

Every norm on $\mathbb{C}^n$ defines a norm on $L(\mathbb{C}^n)$, the vector space of all endomorphisms of $\mathbb{C}^n$, by means of the formula

$$\|A\| := \sup\{\|Ax\| : \|x\| \leqq 1\} . \tag{5}$$

In addition to (4) above, these norms satisfy

$$\begin{aligned} \|AB\| &\leqq \|A\| \, \|B\| \\ \|I\| &= 1 \end{aligned} \tag{6}$$

where juxtaposition denotes, as usual, the composite $A \circ B$. Thus (5) defines a norm on the algebra of $n \times n$-matrices which is called a *matrix norm*, since (in contrast with the most general norm on $L(\mathbb{C}^n)$) it satisfies (6). For the norm defined by (5) we shall occasionally write $\|A\|_p$ when the underlying norm on $\mathbb{C}^n$ is the $l^p$-norm $(1 \leqq p \leqq +\infty)$.

As an example, let us determine the $l^\infty$- and $l^1$-norms of an $n \times n$-matrix $A = (\alpha_{ij})$. $y = Ax$ or $\eta_i = \sum_{j=1}^n \alpha_{ij} \xi_j$ $(i = 1, \ldots, n)$ yields

$$\|y\|_\infty = \sup_i |\eta_i| \leqq (\sup_i \sum_{j=1}^n |\alpha_{ij}|) \, \|x\|_\infty .$$

Hence $\|A\|_\infty$ is not larger than the factor of $\|x\|_\infty$ on the right; suppose that the maximum of $\sum_{j=1}^n |\alpha_{ij}|$ is attained for $i = i_0$. Letting $x = (\xi_j)$ where $\xi_j = \operatorname{sgn} \alpha_{ij}^*$ for all $j$,[3] then $\|x\|_\infty = 1$ and $(Ax)_{i_0} = \sum_{j=1}^n |\alpha_{i_0 j}|$. Hence we obtain

$$\|A\|_\infty = \sup_i \sum_{j=1}^n |\alpha_{ij}| . \tag{7}$$

A similar procedure shows that the $l^1$-norm of $A$ is given by

$$\|A\|_1 = \sup_j \sum_{i=1}^n |\alpha_{ij}| . \tag{7'}$$

Thus the $l^\infty$- and $l^1$-norms are dual to each other, in the sense that $\|A\|_\infty = \|{}^t A\|_1$ for every $n \times n$-matrix $A$. The matrix norm $A \mapsto \|A\|_2$ is self-dual but, as the example of the unit matrix $(n \geqq 2)$ shows, it does not agree with the Euclidean norm of $\mathbb{C}^{n^2}$. The norms of $\mathbb{C}^n$ considered in Examples 1, 2 above have the special property

$$|x| \leqq |y| \;\Rightarrow\; \|x\| \leqq \|y\| . \tag{8}$$

A norm on $\mathbb{R}^n$ or $\mathbb{C}^n$ satisfying (8) is called a *lattice norm* (cf. Chap. II, § 5 and § 11). It follows from the characterization of ideals in $\mathbb{C}^n$ given above that if $J \subset \mathbb{C}^n$ is an ideal, then the quotient norm of a lattice norm is again a lattice norm on $\mathbb{C}^n/J$. On the other hand, the matrix norm (5) defined by a lattice norm of $\mathbb{C}^n$ is, in general, not a lattice norm of $\mathbb{C}^{n^2}$; for example, $A \mapsto \|A\|_2$ is not a lattice norm of $\mathbb{C}^{n^2}$.

Finally, we note that for any norm $A \mapsto \|A\|$ on $L(\mathbb{C}^n)$ we have the formula

$$r(A) = \lim_k \|A^k\|^{1/k} \tag{9}$$

for the spectral radius of $A$.

---

[3] For $\xi \in \mathbb{C}$, define $\operatorname{sgn} \xi = \xi/|\xi|$ if $\xi \neq 0$ and $\operatorname{sgn} 0 = 1$.

It is well known and easy to prove that any two norms $p, q$ on $\mathbb{C}^n$ are equivalent; that is, $p$ and $q$ generate the same topology, or equivalently, $c_1 p \leqq q \leqq c_2 p$ for suitable constants $c_1, c_2 > 0$. Thus all matrix norms are equivalent in this sense; yet it occurs that properties of $A \in L(\mathbb{C}^n)$ with respect to a particular norm imply algebraic properties of $A$ (cf. 2.8). This question seems to have been little investigated.

## § 2. Positive Matrices

We will be concerned with $\mathbb{C}^n$ not only as a vector space but chiefly as a complex vector lattice, that is, the complexification of the real vector lattice $\mathbb{R}^n$ (cf. Chap. II, § 11). The norms on $\mathbb{C}^n$ in which we shall be interested are lattice norms, in particular, the $l^1$- and $l^\infty$-norms (§ 1). The *absolute value* of $x = (\xi_i) \in \mathbb{C}^n$ is, of course, the real vector $|x| = (|\xi_i|)$; if $x \in \mathbb{R}^n$ then $x^+ := \sup(x, 0)$ and $x^- := \sup(-x, 0)$ are defined, whence $x = x^+ - x^-$ and $|x| = x^+ + x^-$. Loosely speaking, the lattice operation $x \mapsto |x|$ on $\mathbb{C}^n$ is defined coordinatewise. We will say that a vector $x = (\xi_i)$ is *strictly positive*, in symbols: $x \gg 0$, exactly when $\xi_i > 0$ for all $i$, $1 \leqq i \leqq n$. The ordering of $\mathbb{R}^n$ and $\mathbb{C}^n$ so defined is called *canonical*.

**2.1 Definition.** *A square matrix $A = (\alpha_{ij})$ of order n is called* positive, *in symbols: $A \geqq 0$, if $\alpha_{ij} \geqq 0$ for all i, j. A is called* fully positive, *in symbols: $A > 0$, if $\alpha_{ij} > 0$ for all i, j.*

If we identify, in the spirit of § 1, an endomorphism $A \in L(\mathbb{C}^n)$ with its matrix $(\alpha_{ij})$ with respect to the canonical basis of $\mathbb{C}^n$, then the ordering of matrices defined by (2.1) corresponds to the natural ordering of $L(\mathbb{C}^n)$, in which $A \geqq 0$ iff $Ax \geqq 0$ for all $x \geqq 0$. The latter is thus seen to be a lattice ordering which, by virtue of $A \mapsto (\alpha_{ij})$, corresponds to the canonical ordering of $\mathbb{C}^{n^2}$ (see above). In particular, $|A|$ is the matrix $(|\alpha_{ij}|)$; if $A$ is real then $A^+ = (\alpha_{ij}^+)$, etc. Also, if $A$ is positive and $J$ is an ideal of $\mathbb{C}^n$ (§ 1), the submatrices $A_{|J}$ and $A_J$ are again positive.

We point out that, in general, $A$ and $|A|$ do not commute; an example is furnished by the $2 \times 2$-matrix $\begin{pmatrix} 1 & -1 \\ 1 & 0 \end{pmatrix}$.

If $A$ is a positive matrix and $x \mapsto \|x\|$ is a lattice norm on $\mathbb{C}^n$, the corresponding matrix norm (§ 1, Formula (5)) is the same as the real norm $\|A\| = \sup\{\|Ax\| : \|x\| \leqq 1, x \in \mathbb{R}^n\}$ of $A$ (cf. Exerc. 2). Moreover, if $A \geqq 0$ an easy computation shows that even

$$\|A\| = \sup\{\|Ax\| : \|x\| \leqq 1,\ x \geqq 0\}. \tag{1}$$

In particular, for the $l^\infty$-norm of $A \geqq 0$ we obtain

$$\|A\|_\infty = \|Ae\|_\infty, \tag{2}$$

since $e = (1, \ldots, 1)$ is the largest element of the $l^\infty$-unit ball of $\mathbb{R}^n$.

We begin our discussion with the following simple but important result.

**2.2 Proposition.** *If $A \geqq 0$ then $R(\lambda) := (\lambda I - A)^{-1} \geqq 0$ for all $\lambda > r(A)$.*

In fact, this is an immediate consequence of the expansion of $R(\lambda)$ at infinity (the C. Neumann series)

$$R(\lambda) = \sum_{k=0}^{\infty} \lambda^{-(k+1)} A^k \tag{3}$$

which converges for $|\lambda| > r(A)$, and the obvious fact that $A^k \geqq 0$ for all $k \geqq 0$. (For a converse, see Exerc. 3.)

**2.3 Proposition.** *If $A \geqq 0$, the spectral radius $r = r(A)$ is an eigenvalue of $A$ with at least one eigenvector $x \geqq 0$.*

*Proof.* Suppose that for each $y \geqq 0$, $R(\lambda)y$ remains bounded as $\lambda \downarrow r$. If $x \in \mathbb{C}^n$ is arbitrary, then (3) yields

$$|R(\lambda)x| = |\sum_{n=0}^{\infty} \lambda^{-(n+1)} A^n x| \leqq \sum_{n=0}^{\infty} |\lambda|^{-(n+1)} A^n |x| = R(|\lambda|)|x|$$

for all $\lambda$, $|\lambda| > r$, whence it follows that $R(\lambda)x$ is uniformly bounded in the region $|\lambda| > r$; this is plainly impossible. Denote by $y_0$ a vector $\geqq 0$ for which $R(\lambda)y_0$ is unbounded as $\lambda \downarrow r$, and let $\| \ \|$ denote any lattice norm on $\mathbb{C}^n$. For $\lambda > r$, set $z(\lambda) = R(\lambda)y_0 / \|R(\lambda)y_0\|$; the denominator tends (monotonically) to $+\infty$. The set $S = \{x : \|x\| = 1 \text{ and } x \geqq 0\}$ is compact and $z(\lambda) \in S$. Thus $\{z(\lambda)\}_{\lambda > r}$ has a cluster point $x_0 \in S$ as $\lambda \downarrow r$. Since

$$(rI - A)z(\lambda) = (r - \lambda)z(\lambda) + y_0 / \|R(\lambda)y_0\|,$$

it follows that $rx_0 = Ax_0$. □

**Corollary.** *If $A \geqq 0$ and $\lambda z = Az$ for some $z \gg 0$, then $\lambda = r(A)$.*

*Proof.* $\lambda \neq r(A)$ implies $\langle z, y \rangle = 0$ for every eigenvector $y$ satisfying $r(A)y = {}^tAy$; since there exists a positive eigenvector $y$ of ${}^tA$ pertaining to $r(A)$, we must have $\lambda = r(A)$. □

**2.4 Proposition.** *Let $A$ be a matrix $\geqq 0$ and denote by $s$ and $t$ the smallest and greatest row sums of $A$, respectively. Then $s \leqq r(A) \leqq t$. Further, if $A \geqq 0$ and $B \geqq 0$ then $r(A+B) \geqq \sup(r(A), r(B))$.*

*Proof.* We employ the $l^\infty$-norm on $\mathbb{C}^n$. Then $\|A\|_\infty = \|Ae\|_\infty$ by Formula (2) above; moreover, $se \leqq Ae \leqq te$. Repeated application of $A$ to this inequality yields $s^k e \leqq A^k e \leqq t^k e$, hence on taking norms we obtain

$$s \leqq \|A^k\|_\infty^{1/k} \leqq t,$$

which implies the first assertion (§ 1, Formula (9)). The proof of the second assertion can be left to the reader. □

A somewhat deeper result than (2.3), equally valid for all matrices $A \geqq 0$, asserts that if $r\alpha \in \sigma(A)$, $|\alpha| = 1$, then $rH \subset \sigma(A)$ where $H$ is the cyclic subgroup, generated by $\alpha$, of the circle group $\Gamma$. (In particular, this implies that $\alpha$ is a root of unity unless $r=0$.) For convenience of expression, let us agree on the following terminology where for $x=(\xi_i)\in\mathbb{C}^n$, $\operatorname{sgn} x$ denotes the vector $(\operatorname{sgn}\xi_i)$.[1] We also write $xy$ for $(\xi_i\eta_i)$, in particular, $x^k$ for $(\xi_i^k)$.[2]

**2.5 Definition.** *If $A$ is a matrix with spectral radius $r$, the set of eigenvalues of modulus $r$ is called the* peripheral spectrum *of $A$. The latter is said to be* cyclic *if $r\alpha\in\sigma(A)$, $|\alpha|=1$, implies $r\alpha^k\in\sigma(A)$ for all $k\in\mathbb{Z}$; it is said to be* fully cyclic *if whenever $r\alpha x=Ax$ $(x\neq 0)$, then $|x|(\operatorname{sgn} x)^k$ is an eigenvector corresponding to $r\alpha^k$ for all $k\in\mathbb{Z}$.*

From the proof of (2.7) below we isolate the following lemma which embodies a conclusion typical for many proofs in the general theory of positive operators on Banach lattices (cf. V. 4.2).

**2.6 Lemma.** *Let $A\geqq 0$. If $\alpha x=Ax$, $|\alpha|=1$, and $|x|=A|x|$, then $\alpha^k|x|(\operatorname{sgn} x)^k=A(|x|(\operatorname{sgn} x)^k)$ for all $k\in\mathbb{Z}$.*

*Proof.* Let $\beta_i=\operatorname{sgn}\xi_i$ where $x=(\xi_i)$. Simultaneously we have $|\xi_i|\alpha\beta_i=\sum_{j=1}^n\alpha_{ij}|\xi_j|\beta_j$ and $|\xi_i|=\sum_{j=1}^n\alpha_{ij}|\xi_j|$ $(i=1,\dots,n)$, which implies $\alpha\beta_i=\beta_j$ for all pairs $(i,j)$ such that $\alpha_{ij}|\xi_j|>0$. Clearly, then, $\alpha^k\beta_i^k=\beta_j^k$ $(k\in\mathbb{Z})$ for these same $(i,j)$ which implies the desired result. □

**2.7 Theorem.** *The peripheral spectrum of every matrix $A\geqq 0$ is cyclic. Moreover, $\lambda=r(A)$ is a pole of $R(\lambda)$ of maximal order among the elements of the peripheral spectrum.*

*Proof.* Without loss of generality we assume that $r(A)=1$, since for $r(A)=0$ the result is trivial. Let $\alpha$ denote an element of the peripheral spectrum and set $\lambda=\rho\alpha$ where $\rho$ is any real number $>1$. Then from the C. Neumann series (3) we obtain

$$|(\lambda-\alpha)^k R(\lambda)|\leqq(\rho-1)^k R(\rho)$$

for each $k\in\mathbb{N}$, which proves the second assertion.

To prove the first assertion, let $\alpha x=Ax$ where $|\alpha|=1$ and $x\neq 0$; then $|x|=|\alpha x|=|Ax|\leqq A|x|$. Denote by $J$ the $A$-invariant ideal generated by $|x|$, that is, the smallest ideal of $\mathbb{C}^n$ (§ 1) containing $A^m|x|$ for each integer $m\geqq 0$. We suppose first that $J=\mathbb{C}^n$. By (2.3) there exists a vector $y_0>0$ satisfying $y_0={}^tAy_0$; we must have $\langle|x|,y_0\rangle>0$, since otherwise $\langle|x|,y_0\rangle=\langle A^m|x|,y_0\rangle=0$ for all $m$ which implies $y_0=0$, a contradiction. The ideal $K:=\{z:\langle|z|,y_0\rangle=0\}$ is $A$-invariant, and $A|x|-|x|\in K$ while $x\notin K$. Hence the canonical image $\hat{x}$ of $x$ in $\mathbb{C}^n/K$ is $\neq 0$, and we have $\alpha\hat{x}=A_K\hat{x}$, $|\hat{x}|=A_K|\hat{x}|$. Now from (2.6) it follows that $\alpha^k\in\sigma(A_K)$ for all $k\in\mathbb{Z}$, whence $\alpha^k\in\sigma(A)$ for all $k\in\mathbb{Z}$ (§ 1, Formula (3)). Second, if $J\neq\mathbb{C}^n$ the preceding argument applies to the restriction $A_{|J}$ (note that $r(A_{|J})=1$) and the conclusion remains unchanged. □

[1] $\operatorname{sgn}\xi=\xi/|\xi|$ if $\xi\neq 0$; for $\xi=0$, define $\operatorname{sgn}\xi=1$.

[2] This amounts to regarding $\mathbb{C}^n$ as an algebra (over $\mathbb{C}$) with $e_ie_j=\delta_{ij}e_j$ $(i,j=1,\dots,n)$.

The conclusion of (2.7) cannot be improved upon, as the following simple example shows. The matrix

$$\begin{pmatrix} 0 & 1 & 0 \\ 1 & 0 & 0 \\ 1 & 1 & 1 \end{pmatrix}$$

has the minimal polynomial $m(\lambda)=(\lambda-1)^2(\lambda+1)$ (equal to the characteristic polynomial), so that $\lambda=1$ is a pole of order 2 of the resolvent while $\lambda=-1$ is of order 1. $x_1=(0,0,1)$ and $x_{-1}=(1,-1,0)$ are essentially the only eigenvectors for 1 and $-1$, respectively. $|x_{-1}|<A|x_{-1}|$ shows that the peripheral spectrum is not fully cyclic.

In the subsequent sections we shall be concerned with situations where much stronger conclusions are valid. Let us conclude this section with a condition that bars both of the phenomena apparent in the preceding example. Recall that $\|A\|_1$ is the $l^1$-norm of $A$ (§ 1, Formula (7')).

**2.8 Proposition.** *Let $A\geqq 0$. If there exists $y_0\gg 0$ satisfying ${}^tAy_0\leqq r(A)y_0$ (in particular, if $\|A\|_1=r(A)$), then the peripheral spectrum of $A$ is fully cyclic and $\lambda=r(A)$ is a simple pole of the resolvent.*

*Proof.* Since for $A\geqq 0$, $\|A\|_1=r$ $(r=r(A))$ is equivalent with ${}^tAe\leqq re$, the second condition is included in the first. Further, the second paragraph in the proof of (2.7) carries over almost *verbatim* with $K=\{0\}$, so that the peripheral spectrum of $A$ is fully cyclic. Finally, ${}^tA^ky_0\leqq r^ky_0$ for all $k\geqq 1$ whence we conclude, using the C. Neumann series (3), that

$$(\lambda-r)[{}^tR(\lambda)]y_0\leqq y_0 \qquad (\lambda>r),$$

is bounded as $\lambda\downarrow r$. Since $y_0\gg 0$, it follows that $(\lambda-r)[{}^tR(\lambda)]$, and therefore $(\lambda-r)R(\lambda)$, is uniformly bounded for $\lambda>r$. Thus $\lambda=r$ is a simple pole of the resolvent. ∎

We remark that for a simple pole $\lambda_0$ of the resolvent of any matrix $A$, the geometric and algebraic multiplicities of $\lambda_0$ (§ 1) agree; in particular, such $\lambda_0$ is a simple root of the characteristic equation $\Delta(\lambda)=0$ iff the subspace (eigenspace) $\{x:\lambda_0x=Ax\}$ is one-dimensional. (Cf. Exerc. 6.)

## § 3. Mean Ergodicity

Let $A$ be a matrix (square matrix of order $n$). In many applications, for example in the theory of finite homogeneous Markov chains (§ 9), it is important to know the asymptotic behavior of the powers $A^k$ as $k\to+\infty$. The simplest case occurs when $A^k$ converges[1] to a matrix $P$; it is easily seen that $P$ is an idempotent

[1] For the Euclidean topology of $\mathbb{C}^{n^2}$, identical with the topology of elementwise convergence, or equivalently, with respect to any matrix norm.

($P^2=P$) satisfying $AP=PA=P$, that is, a projection commuting with $A$ of $\mathbb{C}^n$ onto the fixed space of $A$. A less restrictive property is *mean convergence*[2], that is, convergence of the averages

$$M_k = k^{-1}(I + A + \cdots + A^{k-1}) \qquad (k = 1, 2, \ldots) \tag{1}$$

as $k \to +\infty$. We note first that $r(A) < 1$ implies $A^k \to 0$ hence $M_k \to 0$; conversely, if $r(A) > 1$ then $(M_k)$ is necessarily an unbounded[1] sequence hence cannot be convergent. We begin with the following basic result (mean ergodic theorem) for arbitrary matrices $A$.

**3.1 Theorem.** *Let $A$ be a matrix having spectral radius $r(A)=1$. If the sequence $(A^k)_{k\in\mathbb{N}}$ is bounded then* $\lim_k M_k = P$ *where $P$ is a projection, commuting with $A$, onto the fixed space of $A$.*

*Proof.* Suppose $x \mapsto \|x\|$ is any norm on $\mathbb{C}^n$ and denote by $\|A\|$ the corresponding matrix norm (§ 1, Formula (5)). By hypothesis, there exists a constant $c>0$ such that $\|A^k\| < c$ for all $k \in \mathbb{N}$. An elementary computation shows that $AM_k - M_k = k^{-1}(A^k - I)$, whence by the triangle inequality

$$\|AM_k - M_k\| < k^{-1}(1+c)\,. \tag{2}$$

Now fix $x \in \mathbb{C}^n$ arbitrarily. The sequence $(M_k)$ is evidently bounded (namely, $\|M_k\| < 1+c$), hence the sequence $(M_k x)$ has a cluster point $y \in \mathbb{C}^n$. We show that $y = \lim_k M_k x$. In fact, if $z$ is any cluster point of $(M_k x)$ and $\varepsilon > 0$ is preassigned, there exist subscripts $l, m$ for which $\|M_l x - y\| < \varepsilon/2(1+c)$ and $\|M_m x - z\| < \varepsilon/2(1+c)$. Moreover, (2) implies that $Ay = y$ and $Az = z$. Thus $M_k y = y$ and $M_k z = z$ for all $k$ whence

$$y - z = M_l(M_m x - z) - M_m(M_l x - y)\,,$$

which implies

$$\|y - z\| \leqq \|M_l\|\,\|M_m x - z\| + \|M_m\|\,\|M_l x - y\|\,,$$

and hence $\|y - z\| \leqq \varepsilon$. Thus $y = z$ and, therefore, $(M_k x)$ converges for all $x \in \mathbb{C}^n$; that is, $(M_k)$ converges to some matrix $P$. By (2), $AP = PA = P$ hence $M_k P = P$ for all $k$ which shows that $P^2 = P$. Clearly, the range of $P$ consists of all vectors $x \in \mathbb{C}^n$ fixed under $A$. □

The reader may find it worthwhile to prove the result without any reference to vectors $x \in \mathbb{C}^n$. The preceding proof has the advantage of carrying over to any Banach space (in place of $\mathbb{C}^n$), provided $(A^k)$ is bounded and for each $x$, $(M_k x)$ has a weak cluster point. (Cf. Chap. III, § 7, Example 3.)

---

[2] This is the same as Cesaro convergence of the sequence $(A^k)$. $A^k \to P$ implies $M_k \to P$, but not conversely. Cf. (3.5) below.

From the C. Neumann series (§ 2, Formula (3)) it is quickly seen that boundedness of $\{A^k\}$ implies $r(A) \leqq 1$ and each unimodular eigenvalue (if there are any) is a simple pole of the resolvent. Moreover, $P \neq 0$ iff $1 \in \sigma(A)$, and in this case $P$ is the residuum of the Laurent expansion of $R(\lambda)$ at $\lambda = 1$. Finally, as a matrix equation $AP = P$ states that each column vector of $P$ is a fixed vector of $A$; any maximal linearly independent subset of these is a basis of the fixed space of $A$.

**3.2 Definition.** *Let $A$ be a matrix with spectral radius 1. The semi-group $\{A^k\}_{k \in \mathbb{N}}$ (or, briefly, $A$ itself) is said to be* mean ergodic *if* $\lim_k k^{-1}(I + A + \cdots + A^{k-1})$ *exists.*

Thus (3.1) asserts that $A$ is mean ergodic whenever $r(A) = 1$ and the semi-group $\{A^k\}$ is bounded; conversely, it can be shown (Exerc. 6) that mean ergodicity of $A$ implies the boundedness of $\{A^k\}$. In more detail: $A$ is mean ergodic iff $A = D + B$ where $D$ is similar to a diagonal matrix, $DB = BD = 0$, $r(D) = 1$, and $r(B) < 1$. Thus the existence of $\lim_{\lambda \to 1} (\lambda - 1) R(\lambda)$ does not, in general, imply the existence of $\lim_k k^{-1}(I + A + \cdots + A^{k-1})$ (Exerc. 6). For positive matrices is *does*; this is closely related to (2.7). For a proof not making use of the theory of elementary divisors, we need the following lemma.

**3.3 Lemma.** *Let $A$ be any matrix satisfying $r(A) = 1$. Suppose $\lambda = 1$ is a simple pole of the resolvent and the only eigenvalue of modulus 1. Then $A = P + B$ where $P = \lim_{\lambda \to 1} (\lambda - 1) R(\lambda)$ is a projection onto the fixed space of $A$, $PB = BP = 0$, and $r(B) < 1$.*

*Proof.* A simple computation (using Neumann's series) shows that $AP = PA = P$, whence $P^2 = \lim_{\lambda \to 1} PR(\lambda) = P$. Moreover, $Ax = x$ is equivalent with $Px = x$. Defining $B = A - P$ we see that $PB = BP = 0$. To show that all eigenvalues of $B$ have modulus $< 1$, suppose that $\alpha x = Bx$ where $|\alpha| \geqq 1$. Then $Px = 0$, since $\alpha Px = PBx = 0$, and this implies $\alpha x = Ax$. Here $\alpha \neq 1$ implies $x = 0$ by hypothesis, and $\alpha = 1$ implies $x = Px$ hence $x = 0$ as well. □

**3.4 Proposition.** *Let $A \geqq 0$ have spectral radius 1. The following assertions are equivalent:*

(a) *$\lambda = 1$ is a simple pole of $R(\lambda, A)$.*
(b) *The sequence $(A^k)_{k \in \mathbb{N}}$ is bounded.*
(c) *$A$ is mean ergodic.*

*Moreover,* $\lim_{\lambda \to 1} (\lambda - 1) R(\lambda) = \lim_k k^{-1}(I + A + \cdots + A^{k-1})$ *if either limit exists.*

*Proof.* (a) $\Rightarrow$ (b): Since by (2.7) the peripheral spectrum of $A$ consists only of roots of unity, there exists an integer $q > 0$ such that the peripheral spectrum of $A^q$ is $\{1\}$ (spectral mapping theorem, § 1). We show that $\mu = 1$ is a simple pole of $R(\mu, A^q)$. Putting $\mu = \lambda^q$ where $\lambda > 1$ we obtain, since $A \geqq 0$,

$$0 \leqq (\mu - 1) R(\mu, A^q) \leqq (\lambda^q - 1) R(\lambda, A) \leqq q 2^{q-1} (\lambda - 1) R(\lambda, A)$$

whenever $1<\lambda\leqq 2$. Thus, since $(\lambda-1)R(\lambda,A)$ is bounded as $\lambda\downarrow 1$ by hypothesis, it follows that $(\mu-1)R(\mu,A^q)$ is bounded as $\mu\downarrow 1$. Hence we can apply (3.3) to $A^q$, whence $A^q=S+B$ where $S^2=S$, $SB=BS=0$, and $r(B)<1$. $r(B)<1$ implies that $B^k\to 0$ for $k\to\infty$, hence it follows from

$$A^{qk}=S+B^k \qquad (k\geqq 1)$$

that the sequence $H=(A^{qk})_{k\in\mathbb{N}}$ is bounded (even convergent). Now the set $\{A^k\}_{k\in\mathbb{N}}$ is the union of the $q$ bounded sets $A^r H$ $(r=0,1,\dots,q-1)$, hence itself bounded.

(b) $\Rightarrow$ (c) is asserted by (3.1).

(c) $\Rightarrow$ (a): Denote by $P$ the projection $\lim_k k^{-1}(I+A+\cdots+A^{k-1})$. The coefficients of $(\lambda-1)^{-m}$ $(m=1,2,\dots)$ in the Laurent expansion of $R(\lambda)$ at $\lambda=1$ are given by $Q_m=(I-A)^{m-1}Q_1$, $Q_1$ being the residuum at $\lambda=1$. $PA=P$ implies $PQ_m=0$ whenever $m>1$. On the other hand, if $m_0$ is the order of the pole, $Q_{m_0}x$ is a fixed vector of $A$ for each $x\in\mathbb{C}^n$, whence $PQ_{m_0}=Q_{m_0}$. Thus $m_0=1$.

Finally, if either one of the limits mentioned in the theorem exists, by the preceding so does the other. Suppose so; we have $AP=PA=P$ and $AQ_1=Q_1A=Q_1$. Thus $Q_1P=\lim_k Q_1M_k=Q_1$ and $PQ_1=\lim_{\lambda\to 1}(\lambda-1)PR(\lambda)=P$. Since obviously $P$ and $Q_1$ commute, $Q_1=P$. □

From (3.3) we also obtain necessary and sufficient conditions for the existence of $\lim_k A^k$.

**3.5 Proposition.** *Let $A$ be any matrix. Then $\lim_k A^k=0$ is equivalent with $r(A)<1$. When $r(A)=1$, $\lim_k A^k$ exists if and only if $\lambda=1$ is a simple pole of the resolvent and the only eigenvalue of modulus* 1. *When $r(A)>1$, $\lim_k A^k$ does not exist.*

*Proof.* Suppose first that $r(A)=1$. If the hypothesis of (3.3) is satisfied, then $A^k=P+B^k$ $(k\geqq 1)$ where $B^k\to 0$ as $k\to+\infty$, hence $\lim_k A^k=P$. Conversely, suppose that $(A^k)$ converges. If $\alpha$ is an eigenvalue of modulus 1, then $\alpha x=Ax$ for some $x\neq 0$, and $A^k x=\alpha^k x$ for all $k\in\mathbb{N}$. Hence $(\alpha^k)$ converges which implies $\alpha=1$. Moreover, as a convergent sequence $(A^k)$ is bounded; a simple estimate shows that $(\lambda-1)R(\lambda)=(\lambda-1)\sum_{k=0}^{\infty}\lambda^{-(k+1)}A^k$ is bounded for $\lambda>1$, so the order of the pole $\lambda=1$ of $R(\lambda)$ is unity.

The preceding argument shows that $\lim_k A^k=0$ implies $r(A)<1$, the converse being obvious (for instance, from the convergence of Neumann's series for some $\lambda$, $0<\lambda<1$). For similar reasons, $r(A)>1$ implies that $(A^k)$ is not bounded, much less convergent. □

## § 4. Stochastic Matrices

In the preceding section we have seen that for positive matrices $A$ with spectral radius 1, the order of the pole $\lambda=1$ of the resolvent (equivalently, the multiplicity of $(\lambda-1)$ in the minimal polynomial of $A$) is of considerable importance, which is due to the fact that it dominates the respective orders of the remaining poles of $R(\lambda)$ on the spectral circle (see (2.7)). Stochastic matrices, to be defined instantly, form a class for which this order is always unity.

**4.1 Definition.** *A matrix* $A \geqq 0$ *is called* stochastic (*more precisely*, row stochastic) *if each row sum of* $A$ *is* 1, *or equivalently, if* $Ae = e$.[1]

A matrix $A \geqq 0$ is sometimes called *column stochastic* if its transpose ${}^tA$ is row stochastic (equivalently, if ${}^tAe = e$). The following elementary properties of stochastic matrices are easy consequences of our earlier results.

**4.2 Proposition.** *Every stochastic matrix* $A$ *has* $r(A) = 1$, *is mean ergodic, and the peripheral spectrum of* ${}^tA$ *is fully cyclic.*

*Proof.* $r(A) = 1$ follows from (2.4), since $s = t = 1$. The two remaining assertions result from (2.8) and (3.4), since $R(\lambda, A)$ and $R(\lambda, {}^tA)$ have poles of the same order (viz., unity) at $\lambda = 1$. ▯

This implies, in particular, that all unimodular eigenvalues of a stochastic matrix are simple poles of the resolvent (hence give rise only to linear elementary divisors).

The set $S_n$ of all stochastic matrices of order $n$ has some remarkable geometric and algebraic properties. First, $S_n$ is a convex subset of $\mathbb{R}^{n^2}$, namely, the intersection of the positive cone (orthant) of $\mathbb{R}^{n^2}$ with the $n$ hyperplanes defined by $\sum_{j=1}^{n} \alpha_{ij} = 1$ $(i = 1, \ldots, n)$; $S_n$ is obviously compact, since each $A \in S_n$ has $l^\infty$-norm $\|A\|_\infty = 1$ (§ 1). Thus $S_n$ is a convex polyhedron. The dimension of $S_n$ (that is, the dimension of the smallest affine[2] subspace of $\mathbb{R}^{n^2}$ containing $S_n$) is $n^2 - n = n(n-1)$. Secondly, since $ABe = BAe = e$ for all $A, B \in S_n$, $S_n$ is a compact (and, for $n > 1$, non-abelian) semi-group under composition.

Since $S_n$ is a convex polyhedron, the set of extreme points of $S_n$ is finite. Let us determine these extreme points explicitly. A $(0,1)$-*matrix* is a matrix $(\alpha_{ij})$ where for each pair $(i,j)$, either $\alpha_{ij} = 0$ or $\alpha_{ij} = 1$.

**4.3 Proposition.** *The* $n \times n$-*matrix* $A$ *is an extreme point of* $S_n$ *if and only if* $A$ *is a* $(0,1)$-*matrix with exactly one* 1 *in each row.*

*Proof.* Suppose $C = (\gamma_{ij})$ is a stochastic $(0,1)$-matrix, and let $j$ denote the unique index for which $\gamma_{1j} = 1$. Then if $C = \lambda A + (1-\lambda) B$ where $0 < \lambda < 1$ and $A, B$ are stochastic, $1 = \lambda \alpha_{1j} + (1-\lambda)\beta_{1j}$ implies $\alpha_{1j} = \beta_{1j} = 1$, since $0 \leqq \alpha_{1j} \leqq 1$, $0 \leqq \beta_{1j} \leqq 1$. Similarly, $\alpha_{1k} = \beta_{1k} = 0$ whenever $k \neq j$. This procedure obviously applies to each row, hence we have $A = B = C$ which shows $C$ to be an extreme point of $S_n$.

Conversely, let $C = (\gamma_{ij})$ be stochastic but not a $(0,1)$-matrix. By renumbering rows and columns, if necessary, we can arrange that $0 < \gamma_{11} < 1$, $0 < \gamma_{12} < 1$. Define $\varepsilon = \inf(\gamma_{11}, \gamma_{12})$ and $\alpha_{11} = \gamma_{11} - \varepsilon$, $\alpha_{12} = \gamma_{12} + \varepsilon$, $\beta_{11} = \gamma_{11} + \varepsilon$, $\beta_{12} = \gamma_{12} - \varepsilon$. Moreover, let $\alpha_{ij} = \beta_{ij} = \gamma_{ij}$ whenever $i \neq 1, j \neq 1, 2$. Then $A = (\alpha_{ij})$ and $B = (\beta_{ij})$ are stochastic, $C = \frac{1}{2}A + \frac{1}{2}B$, and $A \neq B$. Thus, $C$ is not an extreme point of $S_n$. ▯

---

[1] $e = (1, \ldots, 1)$.

[2] An *affine subspace* $H$ of $\mathbb{R}^m$ (or $\mathbb{C}^m$) is a translate, $H = x + L$, of a vector subspace $L$; $\dim H$ is defined to be $\dim L$.

It follows that $S_n$ has $n^n$ extreme points. Since $n^n > n(n-1)+1$ for $n>1$, $S_n$ is not a simplex when $n>1$. It is natural to ask whether the extreme points of $S_n$ can be characterized without reference to their special matrix form; this is indeed possible.

To this end we define a *(vector) lattice homomorphism* of $\mathbb{C}^n$ to be a linear map $A: \mathbb{C}^n \to \mathbb{C}^n$ such that $|Ax| = A|x|$ for all $x \in \mathbb{C}^n$ (cf. Chap. II, § 2 and § 11); $A$ is a *lattice isomorphism* if both $A$ and $A^{-1}$ are lattice homomorphisms.

**4.4 Proposition.** *A stochastic matrix A defines a lattice homomorphism of* $\mathbb{C}^n$ *if and only if A is an extreme point of* $S_n$.

*Proof.* If $A \in S_n$ is an extreme point then by (4.3) $A$ is a stochastic (0,1)-matrix, and it is readily verified that $|Ax| = A|x|$ for all $x \in \mathbb{C}^n$.

For the converse, we recall that the linear lattice homomorphisms $\mathbb{C}^n \to \mathbb{C}$ are exactly the linear forms $x \mapsto \alpha\langle x, e_i\rangle = \alpha\xi_i$ where $\alpha \geqq 0$ and $1 \leqq i \leqq n$. Suppose now that $A \in S_n$ defines a lattice homomorphism; then for each $i$, $x \mapsto f_i(x) = \langle Ax, e_i\rangle$ is a scalar lattice homomorphism satisfying $f_i(e) = 1$. So there exists $j = j(i)$ such that $\langle Ax, e_i\rangle = \langle x, e_j\rangle$ for all $x$. Substituting $x = e_k$, we find $\langle Ae_k, e_i\rangle = \langle e_k, e_j\rangle = \delta_{kj}$; that is, $A$ is a (0,1)-matrix. □

**Corollary.** *A stochastic matrix defines a lattice isomorphism if and only if A is a permutation matrix.*

*Proof.* Clearly every permutation matrix[3] defines a linear lattice isomorphism of $\mathbb{C}^n$. Conversely, if $A \in S_n$ defines a lattice isomorphism, then by (4.3) $A$ is a (0,1)-matrix; in this case, $A^{-1}$ exists iff $A$ is a permutation matrix. □

The method employed in the proof of (4.4) serves readily to determine all lattice homomorphisms of $\mathbb{C}^n$; in fact, if $A$ defines a lattice homomorphism then $A \geqq 0$ (since $|Ax| = Ax$ whenever $x \geqq 0$), and each row of $A$ contains at most one element $>0$; the converse is obvious. Similarly, $A$ defines a lattice isomorphism of $\mathbb{C}^n$ iff $A \geqq 0$ and replacement of each non-zero entry by 1 transforms $A$ into a permutation matrix (Exerc. 9). The spectrum of such isomorphisms does not show exceptional properties; by contrast, among stochastic matrices, permutation matrices are characterized by their spectrum in a simple way. By $\Gamma$ we denote the circle group $\{\lambda \in \mathbb{C} : |\lambda| = 1\}$.

**4.5 Theorem.** *Let A be stochastic. Then A is a permutation matrix if and only if* $\sigma(A) \subset \Gamma$.

*Proof.* The necessity of the condition is evident: In fact, if $P$ is a permutation matrix and $\pi = (c_1)\dots(c_m)$ the decomposition of the corresponding permutation $\pi$ into independent cycles of length $l_\mu$ $(\mu = 1, \dots, m)$, then $\sigma(A)$ is the union of the respective groups of $l_\mu$-th roots of unity $(\sum_1^m l_\mu = n)$.

The condition is sufficient. From (2.7) it follows that each $\alpha \in \sigma(A)$ is a root of unity, so there exists an integer $q \geqq 1$ for which $\sigma(A^q) = \{1\}$ (spectral mapping theorem, § 1). Since $A^q$ is stochastic, $R(\lambda, A^q)$ has a simple pole at $\lambda = 1$[4] (cf. remark after (4.1)). From Lemma (3.3) we conclude that $A^q = P + B$, where $P$ is

[3] $P$ is a *permutation matrix* if $Pe_i = e_k$ where $i \mapsto k$ is a permutation of $\{1, \dots, n\}$.

[4] This implies, in view of $\sigma(A^q) = \{1\}$, the minimal polynomial of $A^q$ to be $\lambda - 1$ whence $A^q = I$.

a projection, and $B=0$ in view of $\sigma(A^q)=\{1\}$; since $\sigma(P)=\{1\}$, it follows that $A^q=P=I$. Thus $A^{-1}=A^{q-1}$ is $\geqq 0$, hence $A$ a lattice isomorphism; by the corollary of (4.4), $A$ is a permutation matrix. ▯

**Corollary 1.** *The group of permutation matrices of order $n$ is the largest group with unit $I$ contained in the semi-group $S_n$.*

*Proof.* If $A \in S_n$ has $A^{-1} \in S_n$ then $A^{-1} \geqq 0$, hence $A$ is a lattice isomorphism. ▯

**Corollary 2.** *If $A$ is a stochastic lattice homomorphism then $\sigma(A) \subset \{0\} \cup \Gamma$.*

*Proof.* Let $J$ denote the set of all $x \in \mathbb{C}^n$ for which there exists $k \in \mathbb{N}$ such that $A^k|x|=0$. Then $J$ is an ideal $\neq \mathbb{C}^n$ invariant under $A$. The induced operator (submatrix) $A_J$ on $\mathbb{C}^n/J$ is again stochastic and a lattice homomorphism; moreover, by definition of $J$, $A_J$ is bijective. Hence, by (4.3) and (4.4), $A_J$ is a permutation matrix which implies $\sigma(A_J) \subset \Gamma$. Since $A$ is nilpotent on $J$, it follows that $\sigma(A) \subset \{0\} \cup \Gamma$ (§ 1, Formula (3)). ▯

The algebraic structure of the compact semi-group $S_n$ of stochastic matrices of order $n$ has been the object of numerous investigations (cf. Notes). In particular, the standard results on compact abelian semi-groups of operators (see Chap. III, § 7) can be applied to closed abelian subsemi-groups of $S_n$. Let us note here only a simple characterization of the minimal two-sided semi-group ideal $K_n$ of $S_n$. The tensor product $x \otimes y$ $(x=(\xi_i), y=(\eta_i) \in \mathbb{C}^n)$ denotes, as usual, the matrix $A=(\alpha_{ij})$ where $\alpha_{ij}=\eta_i \xi_j$.

**4.6 Proposition.** *The intersection of all two-sided semi-group ideals in $S_n$ is the ideal $K_n=\{x \otimes e : x \geqq 0, \langle x, e \rangle =1\}$.*

*Proof.* Clearly, $K_n \subset S_n$. If $x \otimes e \in K_n$ and $A \in S_n$ then $A(x \otimes e)=x \otimes Ae = x \otimes e$, and $(x \otimes e)A = {}^tAx \otimes e$ where $\langle {}^tAx, e \rangle = \langle x, e \rangle =1$, thus $K_n$ is a two-sided ideal. Secondly, if $H$ is any right ideal, $A \in H$ implies $A(x \otimes e)=x \otimes e \in H$ by the preceding, hence $K_n \subset H$. ▯

The elements $x \otimes e \in K_n$ are projections $P$ of rank 1, where in the matrix $P$ each row vector equals $x$. Note that the extreme points of $K_n$ are the stochastic lattice homomorphisms of rank 1.

## § 5. Doubly Stochastic Matrices

A more restricted but in many respects more interesting class of positive matrices is defined as follows.

**5.1 Definition.** *A square matrix $A \geqq 0$ is called* doubly stochastic (bistochastic) *if each row sum and each column sum of $A$ is 1, or equivalently, if $Ae=e$ and ${}^tAe=e$.*

For example, every permutation matrix is doubly stochastic. The $n \times n$-matrix $J_n$ each of whose entries equals $n^{-1}$, is doubly stochastic. In fact, $A \geqq 0$ is doubly stochastic iff $J_n A = A J_n = J_n$. From (4.2) we note that when $A$ is doubly stochastic, then $r(A)=1$ and the peripheral spectrum of both $A$ and ${}^tA$ is fully cyclic (cf. § 8). Moreover, the $l^\infty$- and $l^1$-norm of $A$ is $\|A\|_\infty=\|A\|_1=1$ (§ 1, Formulae (7), (7')).

Let us denote by $D_n$ the set of doubly stochastic matrices of order $n$. Clearly, $D_n$ is a closed convex subset of $S_n$ (the set of stochastic matrices of order $n$), and a closed subsemi-group of $S_n$. More precisely, $D_n$ is (like $S_n$) a convex polyhedron in $\mathbb{R}^{n^2}$, since it is the bounded intersection of the positive cone (orthant) with the $2n$ hyperplanes $H_i=\{A:f_i(A)=1\}$ and $K_j=\{A:g_j(A)=1\}$ $(i,j=1,\dots,n)$, where for $A=(\alpha_{ij})$, $f_i(A):=\sum_{j=1}^n \alpha_{ij}$ and $g_j(A):=\sum_{i=1}^n \alpha_{ij}$.

To determine the dimension of $D_n$, let us note first that exactly $k=2n-1$ of the forms $f_i, g_j$ are linearly independent on $\mathbb{R}^{n^2}$. Since $\sum_{i=1}^n f_i=\sum_{j=1}^n g_j$, we have $k\leqq 2n-1$. On the other hand, suppose that $k<2n-1$, then at least two forms are linear combinations of the remaining ones. If these are $f_i$ and $g_j$, say, then the matrix $A$ whose entries are $n^{-1}$ except for $\alpha_{ij}=2n^{-1}$, is not in $D_n$ but satisfies $f_k(A)=g_l(A)=1$ whenever $k\neq i$, $l\neq j$. Similarly, if the forms in question were $f_i$ and $f_k$ $(i\neq k)$, then defining $\alpha_{i1}=\alpha_{i2}=7/4n$, $\alpha_{k1}=\alpha_{k2}=1/4n$, and $\alpha_{hm}=n^{-1}$ for the remaining subscripts yields a matrix $A$ not in $D_n$ but satisfying the $2n-2$ conditions not involving $f_i$ and $f_k$. Likewise, no two distinct forms of the $g_j$ are linear combinations of the remaining $2n-2$ functionals so we conclude that $k=2n-1$. Therefore, the affine subspace $H$ of $\mathbb{R}^{n^2}$ which is the intersection of the $2n$ hyperplanes $H_i, K_j$ has dimension $n^2-(2n-1)=(n-1)^2$, and since $D_n$ has interior points in $H$ (for example, $J_n$) it follows that $D_n$ has dimension $(n-1)^2$.

It is clear that each permutation matrix is an extreme point of $D_n$, since by (4.3) it is even an extreme point of $S_n$. Conversely, each extreme point of $D_n$ is a permutation matrix. Our proof of this famous result follows Hoffman and Wielandt [1953].

**5.2 Theorem** (Birkhoff). *The permutation matrices of order $n$ are the extreme points of $D_n$.*

*Proof*. We have only to show that each extreme point $P=(p_{ij})$ of $D_n$ is a permutation matrix. The crux of the proof is the observation that if $P$ is extreme, there must exist at least one pair $(i,j)$ for which $p_{ij}=1$; in turn, this depends decisively on the fact that $\dim D_n=(n-1)^2$ (see above). In the $(n-1)^2$-dimensional affine subspace $H\subset\mathbb{R}^{n^2}$ which is the intersection of the hyperplanes $H_i, K_j$ $(i,j=1,\dots,n)$, $D_n$ is the intersection of the $n^2$ half-spaces defined by $\alpha_{ij}\geqq 0$. But since $D_n$ has dimension $(n-1)^2$, the extreme point $P$ must lie on the boundary of at least $(n-1)^2$ of these half-spaces, for $\{P\}$ has dimension 0. This implies that $P$ has at most $n^2-(n-1)^2=2n-1$ entries $p_{ij}>0$, so not each row can contain more than one entry $>0$. Thus $p_{ij}=1$ for some pair $(i,j)$.

The rest of the proof follows now by induction on $n$. In fact, the matrix $Q$ that emerges when the $i$-th row and $j$-th column are canceled from $P$, is clearly an element of $D_{n-1}$ $(n\geqq 2)$. But $Q$ must be an extreme point of $D_{n-1}$ or else $P$ would not be extreme in $D_n$. Since the assertion is clear for $n=1$, it follows that $P$ is a permutation matrix. ∎

Since every compact convex polyhedron is the convex hull of the set of its extreme points, (5.2) implies that each $A\in D_n$ can be expressed (in general, not uniquely) as a convex combination (barycenter) of permutation matrices. More precisely:

**5.3 Proposition.** *Every doubly stochastic matrix* $A \in D_n$ *can be expressed as a convex combination of at most* $(n-1)^2+1$ *permutation matrices.*

*Proof.* Since the dimension of $D_n$ is $(n-1)^2$, each of the simplexes[1] contained in $D_n$ has at most $(n-1)^2+1$ vertices, and each $A \in D_n$ must lie in some simplex contained in $D_n$. □

It is, moreover, clear that there exist $A \in D_n$ which cannot be the barycenter of fewer than $(n-1)^2+1$ permutation matrices (Exerc. 11). However, if $A \in D_n$ is irreducible (see § 6), and if $h$ is the number of points in the peripheral spectrum of $A$, then Marcus-Minc-Moyls [1961] have shown that

$$\nu(A) \leqq h\left(\frac{n}{h} - 1\right)^2 + 1,$$

where $\nu(A)$ is the minimum number of permutation matrices needed for a convex representation of $A$.

Birkhoff's Theorem (5.2) has the following interesting consequence for the spectrum $\sigma(A)$ of a doubly stochastic matrix $A$.

**5.4 Proposition.** *If* $A$ *is doubly stochastic of order* $n$, *then* $\sigma(A) \subset C_n$ *where* $C_n$ *is the convex hull of the set of all* $\nu$*-th roots of unity,* $1 \leqq \nu \leqq n$.

*Proof.* Every permutation matrix $P$ is normal, since ${}^tP = P^*$ (§ 1) and ${}^tPP = I$. For normal matrices $B$, it is well known that the numerical range $\{(Bx|x): \|x\|_2 = 1\}$, where $\|x\|_2 = \sqrt{(x|x)}$ is the Euclidean norm of $x \in \mathbb{C}^n$, equals the convex hull of $\sigma(B)$. But for each permutation matrix $P$ of order $n$, we have $\sigma(P) \subset C_n$, cf. (4.5); thus $(Px|x) \in C_n$ whenever $\|x\|_2 = 1$. Suppose $\lambda \in \sigma(A)$ and $\lambda x = Ax$, where $\|x\|_2 = 1$. By (5.3), $A$ is a convex combination: $A = \sum \mu_k P_k$ of certain permutation matrices $P_k$ of order $n$. Hence

$$\lambda = \lambda(x|x) = (Ax|x) = \sum \mu_k (P_k x|x)$$

which implies that $\lambda \in C_n$. □

Beside (5.2) another major (and much earlier) result has had much influence on the study of doubly stochastic matrices, particularly their applications in the theory of convexity (cf. Mirsky [1963]). To state it concisely, we introduce some additional notation.

If $x = (\xi_i) \in \mathbb{R}^n$, let $\bar{x} = (\bar{\xi}_i)$ where $(\bar{\xi}_i)$ is a rearrangement of the $n$ numbers $\xi_i$ into a non-increasing sequence:

$$\bar{\xi}_1 \geqq \bar{\xi}_2 \geqq \cdots \geqq \bar{\xi}_n .$$

[1] Recall that a *k-dimensional simplex* is the convex hull of $k+1$ points $x_0, x_1, \ldots, x_k$ where $x_1 - x_0, x_2 - x_0, \ldots, x_k - x_0$ are linearly independent ($k \geqq 1$). A 0-dimensional simplex is a singleton $\{x_0\}$.

Thus $\bar{\xi}_i = \xi_{\pi(i)}$ for a suitable permutation $\pi$ of $\{1, \ldots, n\}$. Moreover, for any such $\pi$ let us denote by $x_\pi$ the vector $(\xi_{\pi(i)})$ and by $\Pi(x)$ the convex hull of the set $\{x_\pi : \pi \in \mathsf{S}_n\}$ where $\mathsf{S}_n$ is the symmetric group on $n$ letters. Finally, for two vectors $x, y \in \mathbb{R}^n$ we write $y \prec x$ whenever

$$\bar{\eta}_1 + \bar{\eta}_2 + \cdots + \bar{\eta}_k \leqq \bar{\xi}_1 + \bar{\xi}_2 + \cdots + \bar{\xi}_k \quad (k = 1, \ldots, n) \tag{1}$$

with equality holding for $k = n$.

**5.5 Theorem** (Hardy-Littlewood-Polya). *The following assertions are equivalent for any pair of vectors $x, y \in \mathbb{R}^n$:*

(a) *There exists $A \in D_n$ such that $y = Ax$,*

(b) $y \prec x$,

(c) $y \in \Pi(x)$.

*Proof.* (a)⇒(b): If $y = Ax$, $A \in D_n$, there exist permutation matrices $P$ and $Q$ such that $\bar{y} = B\bar{x}$ where $B = PAQ \in D_n$, $B = (\beta_{ij})$. We note that for fixed $m$, and $1 \leqq i \leqq m$, we have

$$\bar{\eta}_i \leqq \textstyle\sum_{j=1}^m \beta_{ij} \bar{\xi}_j + (1 - \sum_{j=1}^m \beta_{ij}) \bar{\xi}_m .$$

Hence it follows that

$$\textstyle\sum_{i=1}^m \bar{\eta}_i \leqq \sum_{j=1}^m \bar{\xi}_j + \sum_{j=1}^m (\sum_{i=1}^m \beta_{ij} - 1) \bar{\xi}_j + \sum_{i=1}^m (1 - \sum_{j=1}^m \beta_{ij}) \bar{\xi}_m \leqq \sum_{i=1}^m \bar{\xi}_i .$$

Equality in (1) for $m = n$ is obvious.

(b)⇒(c): We begin by establishing the relation

$$\langle \bar{y}, \bar{z} \rangle = \max_{\pi \in \mathsf{S}_n} \langle y_\pi, z \rangle \tag{2}$$

for all vectors $y, z \in \mathbb{R}^n$. (2) is clear when $n = 2$; in the general case it follows by effecting successive transpositions. Secondly, if $y \prec x$ then

$$\langle \bar{y}, \bar{z} \rangle \leqq \langle \bar{x}, \bar{z} \rangle \tag{3}$$

for all $z \in \mathbb{R}^n$. For,

$$\begin{aligned} \langle \bar{y}, \bar{z} \rangle &= \textstyle\sum_{k=1}^{n-1} (\bar{\zeta}_k - \bar{\zeta}_{k+1})(\bar{\eta}_1 + \cdots + \bar{\eta}_k) + \bar{\zeta}_n (\bar{\eta}_1 + \cdots + \bar{\eta}_n) \\ &\leqq \textstyle\sum_{k=1}^{n-1} (\bar{\zeta}_k - \bar{\zeta}_{k+1})(\bar{\xi}_1 + \cdots + \bar{\xi}_k) + \bar{\zeta}_n (\bar{\xi}_1 + \cdots + \bar{\xi}_n) = \langle \bar{x}, \bar{z} \rangle \end{aligned}$$

by virtue of (1). Suppose now that $y \prec x$ but $y \notin \Pi(x)$. By the separation theorem for convex sets, there exists a hyperplane in $\mathbb{R}^n$ separating $\{y\}$ and $\Pi(x)$ strictly; equivalently, there exists a vector $z \in \mathbb{R}^n$ such that $\langle y, z \rangle > \langle u, z \rangle$ for all $u \in \Pi(x)$. In view of (2), this implies

$$\langle y, z \rangle > \max_{\pi \in \mathsf{S}_n} \langle x_\pi, z \rangle = \langle \bar{x}, \bar{z} \rangle .$$

But then by (2), $\langle \bar{y}, \bar{z} \rangle > \langle \bar{x}, \bar{z} \rangle$ which contradicts (3).

(c)⇒(a): $y \in \Pi(x)$ means that $y = \sum \lambda_\pi x_\pi$ for suitable numbers $\lambda_\pi \geqq 0$ satisfying $\sum \lambda_\pi = 1$. Now $x_\pi = P_\pi x$ where $P_\pi$ is the permutation matrix for which $P_\pi e_i = e_{\pi(i)}$ $(i = 1, \ldots, n)$. Thus $y = Ax$ for $A = \sum \lambda_\pi P_\pi \in D_n$. □

Without proof (cf. Exerc. 12) we mention the following theorem, which is closely related to (5.5) and also due to Hardy, Littlewood, and Polya:

*Let* $x = (\xi_i)$, $y = (\eta_i) \in \mathbb{R}^n$. *Then* $y \prec x$ *if and only if* $\sum_{i=1}^n F(\eta_i) \leqq \sum_{i=1}^n F(\xi_i)$ *for any convex, real-valued function* $F$ *on* $\mathbb{R}$.

Finally, concerning the semi-group $D_n$ similar remarks are valid, *mutatis mutandis*, as have been made at the end of Section 4 concerning the semi-group $S_n$ of stochastic matrices of order $n$. The minimal two-sided semi-group ideal of $D_n$ consists, trivially, of the single matrix $J_n = n^{-1}(e \otimes e)$; in fact, $J_n$ acts as the zero element of $D_n$ because of $J_n A = A J_n = J_n$ for all $A \in D_n$. Since $J_n$ defines a projection of $\mathbb{C}^n$ onto the vector subspace spanned by $\{e\}$, this implies that (up to scalar multiples) $e$ is the only fixed vector common to all $A \in D_n$.

## § 6. Irreducible Positive Matrices

Let $A$ be fully positive, $A \gg 0$. It was discovered by Perron [1907] that the spectral radius $r(A)$ (which is necessarily $>0$) is a simple eigenvalue of $A$ with a strictly positive eigenvector, and a simple root of the characteristic equation $\Delta(\lambda) = 0$. Little later Frobenius [1912] found what is perhaps the most important result on positive $n \times n$-matrices (see (6.5) below), a theorem that contains Perron's result as a special case. Frobenius' theorem rests on the concept of irreducibility, which is usually defined in matrix terms. But it is easy to see that this notion implicitly involves the ideals of the vector lattice (or equivalently, of the algebra) $\mathbb{C}^n$.

**6.1 Definition.** *A square matrix* $A \geqq 0$ *of order* $n$ *is called* irreducible *if there exists no permutation matrix* $P$ *such that*

$$P^{-1} A P = \begin{pmatrix} A_1 & 0 \\ B & A_2 \end{pmatrix}$$

*where* $A_i$ *is square of order* $m_i$ $(1 \leqq m_i < n)$, *or equivalently, if there exists no non-trivial ideal invariant under* $A$.[1]

It will be noted that Def. (6.1) continues to make sense when the restriction $A \geqq 0$ is omitted, but we wish to include this restriction for convenience of expression. On the other hand, if $n = 1$ the above definition implies that each $A \geqq 0$ is irreducible; again for reasons of convenience (cf. § 8), we do not want to exclude this trivial case.

---

[1] An ideal $J$ is called *trivial* if $J = \{0\}$ or $J = \mathbb{C}^n$. Some authors use the term *indecomposable* for irreducible. Also, matrices that are not irreducible are occasionally called *reducible* (or *decomposable*).

Every $A \gg 0$ is irreducible, and so is every full cycle permutation matrix $P$ ($P$ is a full cycle permutation matrix iff for each pair $(i,j)$, there exists $m=m(i,j)$ satisfying $e_i = P^m e_j$). Generally, if $n \geqq 2$ then $A$ is irreducible iff for any vector $y \geqq 0$ with $j$ coordinates $>0$ $(1 \leqq j < n)$, $(I+A)y$ has at least $j+1$ coordinates $>0$. (In fact, otherwise $y$ would generate an ideal $J(y) \neq \mathbb{C}^n$ and invariant under $A$.) Thus, whenever $A$ is irreducible and $B \geqq A$ then $B$ is irreducible.

We begin by establishing several properties of a positive matrix $A$ that are equivalent to irreducibility.

**6.2 Proposition.** *Let $A \geqq 0$ and let $r = r(A)$. The following assertions on $A$ are equivalent:*

(a) *$A$ is irreducible.*

(b) *${}^tA$ is irreducible.*

(c) *$r$ is a simple eigenvalue for both $A$ and ${}^tA$, and each of these matrices possesses a strictly positive eigenvector pertaining to $r$.*[2]

(d) *$(\lambda I - A)^{-1} x \gg 0$ for any $x > 0$ and any $\lambda > r$.*

*Note.* Condition (d) is equivalent with (d'): For each pair $(i,j)$ $(i,j=1,\dots,n)$ there exists an integer $k \geqq 1$ such that $A^k$ has an entry $>0$ in place $(i,j)$ (cf. Exerc. 13). Also, in (d) "any $\lambda > r$" can be replaced by "some $\lambda > r$" as the subsequent proof will show.

*Proof.* (a)⇔(b): For any ideal $J \subset \mathbb{C}^n$ and its polar $J^\circ = \{z : \langle x, z\rangle = 0$ for all $x \in J\}$ the relations $AJ \subset J$ and ${}^tAJ^\circ \subset J^\circ$ are equivalent; moreover, $J$ is a non-trivial ideal iff $J^\circ$ is.

(a)⇔(d): As usual, we set $R(\lambda) := (\lambda I - A)^{-1}$ $(\lambda \notin \sigma(A))$. Since $(\lambda I - A) R(\lambda) = I$, we have $A R(\lambda) \leqq \lambda R(\lambda)$ if $\lambda > r$. Thus for each $x > 0$, $R(\lambda)x$ generates a non-zero ideal invariant under $A$, which implies that $R(\lambda)x \gg 0$. Conversely, if (d) is satisfied and $J \neq \{0\}$ is an ideal invariant under $A$ we must have $J = \mathbb{C}^n$, since $0 \ll R(\lambda)x \in J$ for any $x$, $0 < x \in J$.

(a)⇒(c): By (2.3) there exists $x > 0$ satisfying $rx = Ax$. Since the ideal generated by $\{x\}$ is $A$-invariant, it follows that $x \gg 0$. Let $y$ be any other eigenvector belonging to $r$; without loss of generality we can assume $y \in \mathbb{R}^n$. There exists $\gamma \in \mathbb{R}$ such that $x - \gamma y \geqq 0$ but not $\gg 0$; since the ideal generated by $\{x - \gamma y\}$ is $A$-invariant, we have $x - \gamma y = 0$. By virtue of (a)⇔(b), the same conclusions hold for ${}^tA$.

(c)⇒(a): By hypothesis, there exists $y \gg 0$ satisfying $ry = {}^tAy$. Suppose $J$ is a non-trivial ideal invariant under $A$ and let $\rho = r(A_{|J})$.[3] By (2.3) there exists $z$, $0 < z \in J$, such that $\rho z = Az$. Now $\rho < r$, since by assumption $r$ is a simple eigenvalue with an eigenvector $x \gg 0$. But then $r\langle z, y\rangle = \langle z, {}^tAy\rangle = \langle Az, y\rangle = \rho\langle z, y\rangle$ implies that $\langle z, y\rangle = 0$ which conflicts with $z > 0$, $y \gg 0$. Thus $A$ is irreducible. ☐

---

[2] Since by the corollary of (2.3) any strictly positive eigenvector of a matrix $A \geqq 0$ must belong to the eigenvalue $r(A)$, the qualification "pertaining to $r$" can be omitted. Cf. (6.3).

[3] $r(A_{|J})$ denotes the spectral radius of the restriction (submatrix) $A_{|J}$ of $A$ to $J$.

**6.3 Proposition.** *Let $A$ be irreducible with spectral radius $r$, and suppose $n \geqq 2$. Then*

$$r > \max_{1 \leqq i \leqq n} \alpha_{ii}$$

*and $r$ is a simple root of the characteristic equation. Moreover, if $z > 0$ is an eigenvector of $A$ then $z \gg 0$ and $z$ belongs to $r$.*

*Proof.* Let $rx = Ax$ where $x = (\xi_i) \gg 0$ (Prop. 6.2). Then for each $i$, $1 \leqq i \leqq n$,

$$r\xi_i = \alpha_{ii}\xi_i + \sum_{j \neq i} \alpha_{ij}\xi_j .$$

But for any fixed $i$ we cannot have $\alpha_{ij} = 0$ for all $j \neq i$, since otherwise the ideal $J = \{x : \xi_i = 0\}$ were $A$-invariant. Thus $r\xi_i > \alpha_{ii}\xi_i$ hence, $r > \alpha_{ii}$ for all $i$. Further, $\lambda = r$ is a simple pole of the resolvent by (2.8) and a simple eigenvalue by (6.2), hence a simple root of $\Delta(\lambda) = 0$. Finally, if $z > 0$ satisfies $\rho z = Az$ then $\rho \neq r$ would imply $\langle z, y \rangle = 0$ for some $y = r^{-1}({}^t A) y \gg 0$ (cf. (6.2)(c)), which is impossible. Thus $\rho = r$, and again from (6.2) it follows that $z \gg 0$. □

**Corollary.** *If $A$ is irreducible and $n \geqq 2$, then $r > 0$ and $r^{-1}A$ is similar to a stochastic matrix.*

*Proof.* Let $rx = Ax$ where $x = (\xi_i) \gg 0$, and denote by $D_x$ the diagonal matrix with diagonal entries $\xi_i$. If $S = D_x^{-1}(r^{-1}A) D_x$, then $rx = Ax$ shows that $e = Se$ so $S$ is stochastic (§ 4). □

It is a rather trivial fact (and an immediate consequence of the expansion of the resolvent at infinity, cf. § 2, Formula (3)) that $r(B) \leqq r(A)$ whenever $A, B$ are $n \times n$-matrices satisfying $|B| \leqq A$. However, if in addition $A$ is irreducible then $r(B) = r(A)$ can occur only when $A$ and $B$ are very closely related. The precise relationship is as follows.

**6.4 Proposition** (Wielandt). *Let $A$ be an irreducible and $B$ any $n \times n$-matrix satisfying $|B| \leqq A$. If $r(B) = r(A)$ and $\lambda$ is any eigenvalue of $B$ of maximum modulus, there exists a unitary diagonal matrix $D_\lambda$ such that*

$$B = \lambda D_\lambda A D_\lambda^* . \tag{1}$$

*Note.* The subsequent proof will show that $D_\lambda$ is unique to within a unimodular scalar factor; the diagonal entries of $D_\lambda$ can be taken to be $\operatorname{sgn} \xi_i$ for any vector $x = (\xi_i) \neq 0$ satisfying $\lambda x = Bx$.

*Proof.* Without loss of generality we can suppose that $r(B) = r(A) = 1$. Suppose that $\alpha x = Bx$ where $x \neq 0$, $|\alpha| = 1$. Then $|x| \leqq |B|\,|x|$ and, by hypothesis, $|x| \leqq |B|\,|x| \leqq A|x|$. Since $y = {}^t A y$ for some $y \gg 0$ by (6.2), it follows that $|x| = |B|\,|x| = A|x|$. Hence $|x| \gg 0$, and this implies $|B| = A$ by virtue of $|B| \leqq A$. Now let $u = \operatorname{sgn} x$ so that $x = |x| u$, and denote by $D_u$ the diagonal matrix which has the $i$-th coordinate of $u$ in place $(i, i)$. From $\alpha x = Bx$ we conclude that $|x| = C|x|$, where $C = \alpha^* D_u^* B D_u$. But $|C| = |B| \leqq A$, hence $C_1 \leqq A$ if $C_1$ is the real part of $C$. Since $C_1 |x| = |x|$, we have $(A - C_1)|x| = 0$, thus $A = C_1$. This implies $C_1 \leqq |C| \leqq C_1$, hence $C_1 = C$ and so $A = C$. □

**Corollary.** *If $A$ is irreducible, $|B| \leqq A$, and $|\beta_{ij}| < \alpha_{ij}$ for at least one pair $(i,j)$, then $r(B) < r(A)$.*

Proposition (6.4) is the key to the celebrated theorem of Frobenius; recall that the *peripheral spectrum* of $A$ is the set of eigenvalues $\lambda$ satisfying $|\lambda| = r(A)$.

**6.5 Theorem** (Frobenius). *Let $A (\geqq 0)$ be irreducible with spectral radius $r$.*

(i) *The peripheral spectrum of $A$ is fully cyclic and of the form $rH$, where $H$ is the group of all $h$-th roots of unity for some $h \geqq 1$.*

(ii) *Each $\lambda \in rH$ is a simple root of the characteristic equation (hence a simple eigenvalue) of $A$.*

(iii) *The spectrum of $A$ is invariant under the group of rotations (of the complex plane) corresponding to $H$.*

(iv) *$\mathbb{C}^n$ is the direct sum of $h$ ideals $J_\nu$ such that $A(J_{\nu+1}) \subset J_\nu$, where $\nu = 0, \ldots, h-1$ and $J_h = J_0$.*

*Proof.* (i): We can suppose that $n \geqq 2$, and hence that $r = 1$. Let $\alpha, \beta$ be unimodular eigenvalues of $A$. Applying (6.4) to $B = A$ and $\lambda = \alpha, \lambda = \beta$, we obtain $A = \alpha D_\alpha A D_\alpha^*$ and $A = \beta D_\beta A D_\beta^* = \beta^* D_\beta^* A D_\beta$. So

$$A = \alpha \beta^* D_\alpha D_\beta^* A (D_\alpha D_\beta^*)^* . \tag{2}$$

This shows the peripheral spectrum of $A$ to be a finite subgroup of the circle group $\Gamma$, hence the group $H$ of all $h$-th roots of unity for some $h \geqq 1$. The remainder follows from (2.8) and (6.2).

(ii): Let $\varepsilon$ be a primitive $h$-th root of unity. Then for $\alpha = 1, \beta = \varepsilon^\nu$ $(\nu = 1, \ldots, h-1)$, (2) shows $A$ to be similar to $(\varepsilon^*)^\nu A$ whence $\Delta(\lambda) \equiv (\varepsilon^*)^{\nu n} \Delta(\varepsilon^\nu \lambda)$. Since $\lambda = 1$ is a simple root of $\Delta(\lambda) = 0$ by (6.3), so is each $\varepsilon^\nu$.

(iii): Since $A$ is similar to $\varepsilon A$, then $\sigma(A) = \sigma(\varepsilon A) = \varepsilon \sigma(A)$.

(iv): As before, denote by $\varepsilon$ a primitive $h$-th root of unity. Let $\varepsilon x = A x$ $(x \neq 0)$, then $|x| \leqq A|x|$ but actually $|x| = A|x|$, since ${}^t A$ has a strictly positive fixed vector by (6.2). Without loss of generality we can suppose that $\operatorname{sgn} x =: (\gamma_i)$ has at least one coordinate $= 1$. As in the proof of (2.6) (which the reader is asked to recall) it follows inductively (on $\nu$) that if $\gamma_i = \varepsilon^\nu$ then $\gamma_j = \varepsilon^{\nu+1}$ whenever $\alpha_{ij} > 0$; so the *set* $\{\gamma_i\}$ is exactly $H$. Now denote by $x_\nu$ the vector obtained from $|x| = (|\xi_i|)$ by replacing with 0 those $|\xi_i|$ for whose subscripts $i$ we have $\gamma_i \neq \varepsilon^\nu$ $(\nu = 0, \ldots, h-1)$. Then $|x| = \sum_{\nu=0}^{h-1} x_\nu$, and from the preceding it follows that $x_\nu = A x_{\nu+1}$ where $\nu = 0, \ldots, h-1$ and $x_h = x_0$. Hence if $J_\nu$ denotes the ideal generated by $x_\nu$, $A(J_{\nu+1}) \subset J_\nu$ and $\mathbb{C}^n$ is the direct sum of the $J_\nu$. □

If $Ae = e$ (that is, if $A$ is (irreducible and) stochastic; cf. (6.3) Corollary), then the preceding shows that the unimodular eigenvectors of $A$ pertaining to unimodular eigenvalues form a group $G$. The mapping $G \to H$ which maps each $u \in G$ onto the corresponding eigenvalue, is a homomorphism (character) with kernel $\{\alpha e : \alpha \in \Gamma\}$. Thus $H \cong G/\Gamma$. We also point out that for positive matrices $A$ with $r(A) = 1$, there exist weaker conditions than irreducibility implying that the peripheral spectrum is a group (Exerc. 8).

**Corollary 1** (Perron). *If* $A>0$, *then* $r(A)$ *is the only eigenvalue of modulus* $r(A)$ *and a simple root of the characteristic equation*[4].

*Proof.* Clearly, $A$ is irreducible. If $0<x\in J_0$, then $Ax\gg 0$. Hence $\mathbb{C}^n=J_1=J_0$ which shows that $h=1$. □

**Corollary 2.** *If* $A$ *is irreducible and possesses at least one diagonal element* $\alpha_{ii}>0$, *the conclusion of Cor. 1 remains valid.*

*Proof.* Let $\alpha_{ii}>0$. There exists $\nu$, $0\leqq\nu\leqq h-1$, such that the $i$-th coordinate of $x_{\nu+1}$ (notation as above) is $>0$. This implies $J_\nu\cap J_{\nu+1}\neq\{0\}$, whence $h=1$. □

For an irreducible matrix $A$ ($\geqq 0$), the number $h$ of points in the peripheral spectrum of $A$ is called the *index of imprimitivity* of $A$. (Cf. § 7.)

**Corollary 3.** *If* $A$ *is irreducible with index of imprimitivity* $h>1$, *there exists a permutation matrix* $P$ *such that*

$$P^{-1}AP=\begin{pmatrix} 0 & A_{12} & 0 & \dots 0 \\ 0 & 0 & A_{23} & \dots 0 \\ \dots & \dots & \dots & \dots \\ 0 & 0 & 0 & \dots A_{h-1,h} \\ A_{hl} & 0 & 0 & \dots 0 \end{pmatrix}$$

*where the zero blocks in the main diagonal are square.*

*Proof.* There exists a permutation $\pi$ of $\{1,\dots,n\}$ such that the non-zero coordinates of the vector $x_\nu$ (see proof of (6.5), Part (iv)) belong to the indices $i$ for which $k_\nu\leqq i\leqq k_{\nu+1}-1$, where $\nu=0,1,\dots,h-1$ and $1=k_0<k_1<\dots<k_h=n+1$. If $P$ is the permutation matrix corresponding to $\pi$, the relations $A(J_{\nu+1})\subset J_\nu$ show that $P^{-1}AP$ has exactly the form indicated. □

If the characteristic polynomial of $A$ is explicitly known, the index of imprimitivity can be computed as follows.

**6.6 Proposition.** *Let* $A$ *be irreducible with characteristic polynomial*

$$\Delta(\lambda)=\lambda^n+a_1\lambda^{n_1}+\dots+a_t\lambda^{n_t},$$

*where* $a_\tau\neq 0$ ($\tau=1,\dots,t$). *Then* $h$ *is the greatest common divisor of the differences* $n-n_1, n_1-n_2,\dots,n_{t-1}-n_t$.[5]

*Proof.* As before let $\varepsilon$ denote a primitive $h$-th root of unity. If $\lambda_0\neq 0$ is a root of $\Delta(\lambda)=0$, then $\lambda_0\varepsilon,\dots,\lambda_0\varepsilon^{h-1}$ are roots of the same multiplicity as $\lambda_0$, since $A$ is similar to $\varepsilon^\nu A$ ($\nu=1,\dots,h-1$) by virtue of (6.4), Formula (1), and since similar matrices have identical characteristic polynomials. Thus $\Delta(\lambda)\lambda^{-n_t}$ is a polynomial $q(\lambda^h)$, that is,

$$\Delta(\lambda)=q(\lambda^h)\lambda^{n_t}.$$

---

[4] In view of (2.3), it is clear that $A$ has an eigenvector $\gg 0$.

[5] Equivalently, $h$ is the g.c.d. of $n-n_t, n_1-n_t,\dots,n_{t-1}-n_t$. The trivial case $n=1$, $A=0$ is to be excluded.

It follows that $\lambda^h$ divides $\lambda^{n_\tau - n_t}$ $(\tau=0,\dots,t-1)$ where $n_0=n$, and hence $h$ divides all differences $n_\tau - n_t$. So $h$ divides the greatest common divisor $d$ of these differences. On the other hand, by definition of $d$ we have $\Delta(\lambda)=q_1(\lambda^d)\lambda^{n_t}$ for some non-constant polynomial $q_1$ [5] which shows that the peripheral spectrum of $A$ contains $d$ points. Thus $h=d$. □

It is evident that the set of all fully positive matrices of order $n$ forms a convex cone in $L(\mathbb{C}^n)$ which is dense in the cone $C_n$ of all positive matrices. *A fortiori* the set $K_n$ of irreducible $n\times n$-matrices is dense in $C_n$. More precisely, $K_n$ is a dense subcone (not containing its vertex 0 unless $n=1$) of $C_n$ which, in addition, satisfies $K_n + C_n \subset K_n$; in fact, we have observed earlier that $A \in K_n$ and $B \geqq A$ implies $B \in K_n$. On the other hand, if $A\in K_n$ and $B \in K_n$ then it can happen that $AB \notin K_n$; an example is furnished by $A=P$, $B=P^{-1}$ where $P$ is an irreducible permutation matrix. However, $K_n$ has the following multiplicative properties with respect to the cone of fully positive matrices of order $n$.

**6.7 Proposition.** *Let $A$ be an irreducible matrix and $B$ any fully positive matrix of order $n\geqq 2$. Then $AB > 0$ and $BA > 0$. Moreover, $(I+A)^{n-1} > 0$ and, if all diagonal elements of $A$ are $>0$, even $A^{n-1} > 0$.*

*Proof.* Since $A$ leaves no non-trivial ideal invariant, $z \gg 0$ implies $Az \gg 0$. Thus if $y>0$ and $B > 0$, we have $z=By\gg 0$ and so $ABy \gg 0$; that is, $AB > 0$. Similarly, $BA > 0$ follows from the fact that $y>0$ implies $Ay>0$ whenever $A$ is irreducible $(n\geqq 2)$. At the beginning of this section we have observed that whenever $A$ is irreducible $(n \geqq 2)$ and $y$ has $j$ non-zero $(>0)$ coordinates, where $1\leqq j<n$, then $(I+A)y$ has at least $j+1$ non-zero coordinates; thus $(I+A)^{n-1} > 0$. Finally, if all diagonal elements of $A$ are $>0$, then $kA > I+A$ for a suitable integer $k>0$; thus by the preceding, $k^{n-1}A^{n-1} > 0$ which implies $A^{n-1} > 0$. □

In conclusion, let us look briefly at the mean ergodic behavior of irreducible matrices $A$ (§ 3). By (6.3) there is no restriction of generality in supposing that $r(A)=1$. Then $\lambda=1$ is a simple root of the characteristic equation $\Delta(\lambda)=0$, and *a fortiori* a simple pole of the resolvent $R(\lambda)$. So (3.4) shows that

$$P=\lim_k k^{-1}(I+A+\cdots+A^{k-1})$$

is a projection onto the fixed space of $A$ which, as we know from (6.2), has dimension 1. Thus $P=z\otimes x$ where $x\gg 0$ is fixed under $A$, and where $z$ is the unique fixed vector of ${}^tA$ satisfying $\langle x,z\rangle=1$. Since ${}^tA$ is irreducible, $z\gg 0$. Thus $P > 0$. $P$ equals $J_n$ (all entries $=n^{-1}$) iff $A$ is doubly stochastic.

## § 7. Primitive Matrices

It is clear that any non-trivial ideal $J\subset\mathbb{C}^n$ which is invariant under some $A\geqq 0$, is invariant under each power $A^k$ $(k\geqq 1)$; that is, if $A$ is reducible so is each $A^k$. Conversely, if $A$ is irreducible then not each power of $A$ is necessarily irreducible, as the example of an irreducible permutation matrix shows. On the other hand, if $A > 0$ then each $A^k$ $(k\geqq 1)$ is $> 0$ hence irreducible. We shall see that the question of irreducibility for the powers of an irreducible matrix $A$ is directly connected with the number of points in the peripheral spectrum of $A$ (index of imprimitivity, cf. § 6).

**7.1 Definition.** *An irreducible matrix $A$ ($\geqq 0$) of order $n \geqq 2$ is called* primitive *if $r = r(A)$ is the only eigenvalue of $A$ having modulus $r$.*

We recall that the *trace* of an $n \times n$-matrix $A = (\alpha_{ij})$ is defined to be $\operatorname{tr} A := \sum_{i=1}^{n} \alpha_{ii}$; $\operatorname{tr} A$ is the negative coefficient of $\lambda^{n-1}$ in the characteristic polynomial $\Delta(\lambda)$.

**7.2 Proposition.** *If $A$ is irreducible and $B \geqq 0$ has $\operatorname{tr} B > 0$, then $A + B$ is primitive.*

*Proof.* It is clear that $A + B$ is irreducible, since it is $\geqq A$. Moreover $\operatorname{tr}(A+B) \geqq \operatorname{tr} B > 0$ so the coefficient of $\lambda^{n-1}$ in the characteristic polynomial of $A + B$ is $< 0$, whence it follows that $n - n_1 = 1$ (notation of (6.6)). Hence, $h = 1$ by (6.6). ∎

We note that an irreducible $A$ is primitive whenever $\operatorname{tr} A > 0$; this is, in fact, the essential assertion of Cor. 2 of (6.5). The following proposition illuminates the interplay between full positivity and irreducibility of the powers of $A$.

**7.3 Proposition.** *For an irreducible matrix $A$ (of order $n \geqq 2$) with spectral radius $r$, the following assertions are equivalent:*

(a) *$A$ is primitive.*

(b) *$\lim_{k \to \infty} (r^{-1} A)^k$ exists.*

(c) *There exists an integer $p$ such that $A^p \gg 0$.*

(d) *$A^k$ is irreducible for all $k \geqq 1$.*

*Proof.* (a) ⇒ (b): Clear from (3.5), since $\lambda = r$ ($> 0$) is a simple pole of the resolvent $R(\lambda, A)$.

(b) ⇒ (c): As has been observed at the end of Section 6, the projection $P = \lim_k (r^{-1} A)^k$ is fully positive, $P \gg 0$. Hence $A^p \gg 0$ for all sufficiently large $p$.

(c) ⇒ (d): If $A^k$ were reducible for some $k \in \mathbb{N}$, then so would be $A^{qk}$ for all $q \in \mathbb{N}$ which conflicts with $A^p \gg 0$ for large $p$.

(d) ⇒ (a): Suppose $A$ is not primitive. There exists an eigenvalue $r\alpha$ where $|\alpha| = 1$ but $\alpha \neq 1$. Since $\alpha^h = 1$ by (6.5), it follows that $A^h$ has at least two linearly independent fixed vectors which contradicts the irreducibility of $A^h$. ∎

For irreducible $A$ it follows from (6.7) by induction on $k$ that $A^p \gg 0$ implies $A^{p+k} \gg 0$ for all $k \in \mathbb{N}$. Hence, if $A$ is primitive there exists a unique $p \geqq 1$ such that $A^q \gg 0$ for all $q \geqq p$ but for no $q < p$ ($q \geqq 0$). Wielandt [1950] asserts (without proof) that $p \leqq (n-1)^2 + 1$. This bound is best possible, as is shown by the $n \times n$-matrix

$$\begin{pmatrix} 0 & 1 & 0 & \dots & 0 \\ 0 & 0 & 1 & \dots & 0 \\ \dots & \dots & \dots & \dots & \dots \\ 0 & 0 & 0 & \dots & 1 \\ 1 & 1 & 0 & \dots & 0 \end{pmatrix}.$$

Recalling that a square matrix $A$ of order $n$ is called the *direct sum* of the square matrices $A_\mu$ $(\mu=1,\dots,m)$ if

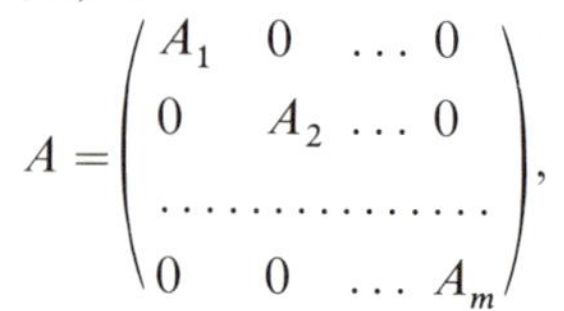

$$A=\begin{pmatrix} A_1 & 0 & \dots & 0 \\ 0 & A_2 & \dots & 0 \\ \dots & \dots & \dots & \dots \\ 0 & 0 & \dots & A_m \end{pmatrix},$$

we finally show that the index $h$ of imprimitivity of an irreducible matrix $A$ relates $A$ to primitive matrices as follows.

**7.4 Proposition.** *Let $A$ be irreducible with index of imprimitivity $h$. There exists a permutation matrix $P$ such that $P^{-1}A^hP$ is the direct sum of $h$ primitive matrices $A_\nu$ satisfying $r(A_\nu)=r(A)^h$ $(\nu=0,\dots,h-1)$.*

*Proof.* By (6.5), $\mathbb{C}^n$ is the direct sum of $h$ ideals $J_\nu$ such that $A(J_{\nu+1})\subset J_\nu$ $(J_h=J_0)$. Hence $A^h(J_\nu)\subset J_\nu$ for each $\nu=0,1,\dots,h-1$, and if we denote by $A_\nu$ the restriction of $A^h$ to $J_\nu$, it is immediately clear that for some permutation matrix $P$, $P^{-1}A^hP$ is the direct sum of the matrices $A_\nu$. (Explicitly this can be obtained by taking the $h$-th power of the matrix occurring in Corollary 3 of (6.5).) Since $A$ is irreducible, there exists a vector $z\gg 0$ satisfying $rz={}^tAz$, where $r=r(A)$. Clearly, $r^hz={}^t(A^h)z$ so each $A_\nu$ must have spectral radius $r(A_\nu)=r^h$ because of (2.3) Cor. and $z\gg 0$. It is also clear that no $A_\nu$ can have an eigenvalue of modulus $r^h$ and distinct from $r^h$; for, by (6.5) and the spectral mapping theorem (§ 1), $A^h$ has no such eigenvalues. It remains to show that each $A_\nu$ is irreducible. In fact, suppose there exists some $\nu$, $0\leqq\nu\leqq h-1$, and a non-trivial subideal $J\subset J_\nu$ such that $J$ is invariant under $A^h$. Since $A(J_{\nu+1})\subset J_\nu$ for all $\nu$, the ideal of $\mathbb{C}^n$ generated by $\{J, A(J),\dots, A^{h-1}(J)\}$ is non-trivial and invariant under $A$ which contradicts the irreducibility of $A$. Thus $A_\nu$ is irreducible $(\nu=0,\dots,h-1)$. □

## § 8. Invariant Ideals

The preceding sections leave no doubt as to the usefulness of the ideal concept for the study of positive matrices. We have seen that the most striking result (theorem of Frobenius, (6.5)) holds for matrices $A\geqq 0$ having no non-trivial invariant ideals; the natural next step is to investigate the class of those positive square matrices that are direct sums of irreducible matrices (cf. (7.4)). This is the purpose of the present section.

**8.1 Definition.** *If $A\geqq 0$ is an $n\times n$-matrix, an* $A$-ideal *is an ideal $J\subset\mathbb{C}^n$ satisfying $A(J)\subset J$. $J$ is called* maximal *if $J$ is maximal (with respect to $\subset$) among the $A$-ideals $\neq\mathbb{C}^n$; $J$ is called* minimal *if $J$ is minimal (with respect to $\subset$) among the $A$-ideals $\neq\{0\}$.*

We begin by observing that for any ideal $J$ and its annihilator $J^\circ=\{z:\langle x,z\rangle=0$ for all $x\in J\}$, the relations $A(J)\subset J$ and ${}^tA(J^\circ)\subset J^\circ$ are equivalent. Thus the family (lattice)[1] of $A$-ideals corresponds, for fixed $A$, to the family (lattice) of

[1] $\sup(J_1,J_2)=J_1+J_2$, $\inf(J_1,J_2)=J_1\cap J_2$.

${}^tA$-ideals by polarity; $J \mapsto J^\circ$ is an anti-isomorphism of the corresponding lattice structures. In particular, maximal (minimal) $A$-ideals correspond by polarity to minimal (maximal) ${}^tA$-ideals. Accordingly, if $J$ is a minimal $A$-ideal then the restriction of $A$ to $J$ is irreducible, and if $J$ is a maximal $A$-ideal then the operator induced on $\mathbb{C}^n/J$ (which can be identified with the transpose of the restriction of ${}^tA$ to $J^\circ$) is irreducible. Thus to every assertion concerning $A$-ideals there corresponds a dual assertion concerning ${}^tA$-ideals, and conversely.

It is easy to see that every minimal $A$-ideal is generated by an (essentially unique) positive eigenvector of $A$. In fact, if $J$ is $A$-minimal and the restriction of $A$ to $J$ has spectral radius $\rho$, then $\rho x = Ax$ for some $x$, $0 < x \in J$, by (2.3); since the ideal $J(x)$ of $\mathbb{C}^n$ generated by $\{x\}$ is $A$-invariant, we must have $J(x) = J$. (Of course, this can also be inferred from (6.2).) Thus as an example of the duality mentioned above, we obtain:

**8.2 Proposition.** *Every maximal $A$-ideal is of the form $J = \{x : \langle |x|, z \rangle = 0\}$ for an (essentially unique) positive eigenvector $z$ of ${}^tA$.*

*Proof.* Since $J^\circ$ is ${}^tA$-minimal, $J^\circ$ is generated by an (essentially unique) positive eigenvector $z$ of ${}^tA$ whence the assertion follows. □

Eigenvalues $\rho$ (necessarily $\geqq 0$) of a positive matrix $A$ with corresponding positive eigenvectors are thus closely related to $A$-invariant ideals. Let us agree on this definition.

**8.3 Definition.** *The eigenvalue $\rho$ $(\geqq 0)$ of $A \geqq 0$ is called* distinguished *if there exists an eigenvector $x > 0$ pertaining to $\rho$.*

**Examples.** 1. From (2.3) it follows that for every $A \geqq 0$, the spectral radius $r(A)$ is a distinguished eigenvalue of $A$ (hence also of ${}^tA$).

2. If $r = r(A)$ and if $rz = {}^tAz$ for some $z \gg 0$, then $r$ is the only distinguished eigenvalue of $A$. This is true, for instance, if $A$ is irreducible or if $A$ is column stochastic (in particular, if $A$ is doubly stochastic).

3. The respective sets of distinguished eigenvalues of $A$ and ${}^tA$ are, in general, distinct. Thus the stochastic matrix

$$A = \begin{pmatrix} 1 & 0 & 0 \\ \frac{1}{4} & \frac{3}{4} & 0 \\ \frac{1}{2} & 0 & \frac{1}{2} \end{pmatrix}$$

has the distinguished eigenvalues $1, \frac{3}{4}, \frac{1}{2}$ while 1 is the only distinguished eigenvalue of ${}^tA$. The minimal $A$-ideals are those generated by the eigenvectors $(0,1,0)$ and $(0,0,1)$, respectively.

Before attempting a characterization of direct sums of irreducible matrices, we shall first investigate the "local" behavior of a positive matrix "at" a distinguished eigenvalue $\rho$. For any such $\rho$, let us call the ideal $J_\rho$ generated by the eigenvectors $x > 0$ pertaining to $\rho$, the $A$-ideal *pertaining to $\rho$*. Clearly, $J_\rho$ is the ideal of $\mathbb{C}^n$ generated by $\{x_0\}$, where $x_0 > 0$ is an eigenvector for $\rho$ with a maximal number of coordinates $> 0$.

**8.4 Proposition.** *Let $\rho$ be a distinguished eigenvalue of $A$, let $J_\rho$ be the $A$-ideal pertaining to $\rho$, and let $\Phi_\rho=\{x\geqq 0: \rho x=Ax \text{ and } \|x\|_1=1\}$. The following assertions are equivalent:*

(a) *$J_\rho$ is the direct sum of minimal $A$-ideals.*

(b) *Each extreme point of $\Phi_\rho$ generates a minimal $A$-ideal.*

(c) *The transpose of $A_{|J_\rho}$ has a strictly positive eigenvector*[2].

(d) *For each minimal $A$-ideal $J\subset J_\rho$, $r(A_{|J})=\rho$.*

*Proof.* (a) $\Rightarrow$ (b): Let $J_\rho$ be the direct sum of the minimal $A$-ideals $I_\mu$ $(\mu=1,\dots,m)$. If $x$ is an extreme point of $\Phi_\rho$, then $x=\sum_{\mu=1}^m x_\mu$ $(x_\mu\in I_\mu)$ is possible only when $x=x_{\mu_0}$ for some $\mu_0$ while $x_\mu=0$ for all $\mu\neq\mu_0$. Thus $x\in I_{\mu_0}$ and, clearly, $I_{\mu_0}$ is generated by $\{x\}$.

(b) $\Rightarrow$ (c): Since any two distinct minimal $A$-ideals have intersection $\{0\}$, (b) implies first that the set $\mathsf{E}$ of extreme points of $\Phi_\rho$ is finite, $\mathsf{E}=\{x_\mu: \mu=1,\dots,m\}$ say. Since $J_\rho$ is the ideal of $\mathbb{C}^n$ generated by $\mathsf{E}$, it follows that $J_\rho$ is the direct sum of $m$ minimal ideals $I_\mu=J(x_\mu)$ (hence (a)); since $\rho x_\mu=Ax_\mu$ by definition of $\Phi_\rho$, we have $r(A_{|I_\mu})=\rho$ for all $\mu$. So ${}^tA_{|J_\rho}$ is the direct sum of the irreducible matrices ${}^tA_{|I_\mu}$ $(\mu=1,\dots,m)$, which each have spectral radius $\rho$, and it follows from (6.2) (c) that the transpose ${}^tA_{|J_\rho}$ has a strictly positive eigenvector (pertaining to $\rho$).

(c) $\Rightarrow$ (d): If $J$ is a minimal $A$-ideal contained in $J_\rho$ and $r(A_{|J})$ denotes the spectral radius of $A_{|J}$, then $r(A_{|J})=\rho$; for, we must have $r(A_{|J})\leqq\rho$, and $r(A_{|J})<\rho$ would imply that each eigenvector $x>0$ of $A_{|J}$ is orthogonal to each positive eigenvector of ${}^tA_{|J_\rho}$ pertaining to $\rho$, which conflicts with (c).

(d) $\Rightarrow$ (b): Suppose $x$ is an extreme point of $\Phi_\rho$. If the ideal $J$ generated by $\{x\}$ is not $A$-minimal, there exists a minimal $A$-ideal $I$ properly contained in $J$. By hypothesis, $r(A_{|I})=\rho$, so $\rho y=Ay$ for some $y$, $0<y\in I$, by (2.3). Since $y\in J$, there exists a real number $\varepsilon>0$ such that $z=x-\varepsilon y>0$; but $x=\varepsilon y+z$ (note that $\rho z=Az$) contradicts the extreme point property of $x$, since $y$ cannot by a scalar multiple of $x$. Hence $J$ is a minimal $A$-ideal.

Finally, the implication (b) $\Rightarrow$ (a) is contained in the proof of (b) $\Rightarrow$ (c) above. ▯

**Corollary.** *Suppose $A$ is column-stochastic and denote by $m$ the dimension of the fixed space $F=\{x: x=Ax\}$. There exist exactly $m$ (distinct) minimal $A$-ideals, and their direct sum is the ideal generated by $F$. Moreover, $\Phi_1$ is an $(m-1)$-dimensional simplex in $\mathbb{R}^n$ and $F$ is a vector sublattice*[3] *of $\mathbb{C}^n$.*

*Proof.* By hypothesis, ${}^tAe=e$ (cf. § 4) which implies that assertion (c) of (8.4) is valid with $\rho=1$. Hence $J_1$ is the direct sum of $k$ minimal ideals, say, each generated by an extreme point of $\Phi_1$. Hence these $k$ extreme points (constituting all extreme points) of $\Phi_1$ are linearly independent, whence it follows that $\Phi_1$ is a simplex of dimension $k-1$ (cf. § 5, Footnote[1]). On the other hand, $x=Ax$ implies $|x|=A|x|$ because of ${}^tAe=e$, hence $F$ is the vector sublattice of $\mathbb{C}^n$ generated by $\Phi_1$. Thus $k=m$ and $J_1$ is the ideal generated by $F$. ▯

---

[2] As before, $A_{|J}$ denotes the restriction (submatrix) of $A$ to $J$. If $\rho>0$, a condition equivalent with (c) is that $\rho^{-1}D^{-1}A_{|J_\rho}D$ be column-stochastic for some invertible diagonal matrix $D\geqq 0$.

[3] I.e., a conjugation invariant vector subspace $L$ of $\mathbb{C}^n$ such that $x\in L$ implies $|x|\in L$.

We are now prepared to characterize those matrices $A \geqq 0$ that are direct sums of irreducible components.

**8.5 Definition.** *An* $n \times n$*-matrix* $A \geqq 0$ *is said to be* completely reducible *if* $A$ *is the direct sum of irreducible matrices, or equivalently, if* $\mathbb{C}^n$ *is the direct sum of minimal* $A$*-ideals.*

Each doubly stochastic matrix (§ 5) is completely reducible; this follows quickly from the preceding corollary, since $Ae=e$. In particular, each permutation matrix is completely reducible; its minimal ideals are obviously determined by the independent cycles of the corresponding permutation. More generally, if both $A$ and ${}^tA$ have strictly positive eigenvectors (necessarily pertaining to $r(A)$), then $A$ is completely reducible; cf. assertion (c) of (6.2). These examples are easily seen to satisfy one or the other condition of the following theorem.

**8.6 Theorem.** *For any square matrix* $A \geqq 0$*, the following assertions are equivalent:*

(a) *$A$ is completely reducible.*

(b) *For each pair* $(i,j)$ $(i,j=1,\dots,n)$, $e_i \in J_A(e_j)$ *implies* $e_j \in J_A(e_i)$, *where* $J_A(e_i)$ *denotes the* $A$*-ideal generated by* $\{e_i\}$.[4]

(c) *The intersection of all maximal* $A$*-ideals is* $\{0\}$.

(d) *The respective sets of distinguished eigenvalues of* $A$ *and* ${}^tA$ *are identical,* $\{\rho_\mu : \mu=1,\dots,m\}$ *say, and there exist corresponding positive eigenvectors:* $\rho_\mu y_\mu = A y_\mu$ *and* $\rho_\mu z_\mu = {}^tA z_\mu$ $(\mu=1,\dots,m)$ *such that* $\sum_{\mu=1}^m y_\mu \gg 0$ *and* $\sum_{\mu=1}^m z_\mu \gg 0$.

*Proof.* (a)$\Leftrightarrow$(c): By the duality mentioned near the beginning of this section, the intersection of all maximal $A$-ideals is $\{0\}$ if and only if the direct sum of all minimal ${}^tA$-ideals is $\mathbb{C}^n$. But by Def. (8.5) (first part) $A$ is completely reducible if and only if ${}^tA$ is, since the transpose of an irreducible matrix is irreducible.

(a)$\Rightarrow$(b): Each $e_i$ $(i=1,\dots,n)$ belongs to some minimal $A$-ideal. Obviously, $e_i \in J_A(e_j)$ iff $e_i$ and $e_j$ belong to the same minimal $A$-ideal.

(b)$\Rightarrow$(a): We claim that each ideal $J_A(e_i)$ is a minimal $A$-ideal (of course, these $n$ ideals are not necessarily distinct). In fact, if some $J_A(e_i)$ were not $A$-minimal it would properly contain a minimal $A$-ideal $I$. Then $I$ contains some $e_j$ so $e_i \notin J_A(e_j)$, which is contradictory.

(a)$\Rightarrow$(d): If $A$ is completely reducible, then

$$A = \begin{pmatrix} A_1 & 0 & \dots & 0 \\ 0 & A_2 & \dots & 0 \\ \dots & \dots & \dots & \dots \\ 0 & 0 & \dots & A_m \end{pmatrix}$$

where each $A_\mu$ is irreducible $(\mu=1,\dots,m)$. Thus any eigenvalue of $A$ (respectively, of ${}^tA$) is an eigenvalue of some $A_\mu$ (respectively, of some ${}^tA_\mu$), and by (6.3) the

[4] $e_i = (0, \dots, 1, 0, \dots)$.

(not necessarily distinct) spectral radii $\rho_\mu$ ($\mu=1,\dots,m$) of $A_\mu$ are exactly the distinguished eigenvalues of both $A$ and ${}^tA$. The remainder is now obvious from (6.2).

(d) ⇒ (a): Let $y_\mu$, $z_\mu$ ($\mu=1,\dots,m$) be respective eigenvectors $>0$ of $A$ and ${}^tA$ satisfying (d). If $S_\mu \subset \{1,\dots,n\}$ denotes the support of $y_\mu$ and if $T_\mu$ denotes the support of $z_\mu$ then, since $\inf(y_\mu, z_\nu)=0$ whenever $\mu \neq \nu$, we have

$$S_\mu \cap \left(\bigcup_{\nu \neq \mu} T_\nu\right) = \emptyset$$

for all $\mu$. But $\sum_{\mu=1}^m z_\mu \gg 0$ implies that $\bigcup_{\mu=1}^m T_\mu = \{1,\dots,n\}$, hence $S_\mu \subset T_\mu$ and, by symmetry, $T_\mu \subset S_\mu$. Thus $S_\mu = T_\mu$ ($\mu=1,\dots,m$). Consequently, each of the ideals $J_\mu = \{x\colon \langle |x|, z_\nu\rangle = 0 \text{ for all } \nu \neq \mu\}$ is invariant under $A$ and ${}^tA$ so $A$ is the direct sum of the $m$ restrictions (submatrices) $A_\mu$ of $A$ to $J_\mu$ (note that $A_\mu$ is not necessarily irreducible). Moreover, each of the vectors $y_\mu$ and $z_\mu$ generates the ideal $J_\mu$. Therefore, application of (8.4) to each $A_\mu$ yields the desired conclusion. ☐

Condition (c) of (8.6) suggests the introduction of the $A$-*radical* $R$, to be defined as the intersection of all maximal $A$-ideals. Clearly, $R$ is an $A$-ideal, and it is easy to show that the operator induced by $A$ on $\mathbb{C}^n/R$ is completely reducible (Exerc. 17).

We have seen above that each doubly stochastic matrix is completely reducible; let us determine conditions under which a stochastic matrix (§ 4) is completely reducible.

**8.7 Proposition.** *For a stochastic matrix $A$, these assertions are equivalent:*

(a) *$A$ is completely reducible.*

(b) *${}^tA$ possesses a strictly positive eigenvector.*

(c) $\lim_{k\to\infty} A^k|x|=0$ *implies* $x=0$.

(d) $x>0$ *implies* $Px>0$, *where* $P=\lim_k k^{-1}(I+A\cdots+A^{k-1})$.[5]

*Proof.* (a) ⇔ (b): In view of $Ae=e$, $r(A)=1$ is the only distinguished eigenvalue of ${}^tA$; hence, since the ideal generated by $\{e\}$ is $\mathbb{C}^n$, the equivalence of (a) and (b) is immediate from (8.4).

(b) ⇒ (c): If $z={}^tAz$ for some $z \gg 0$, $\lim_k A^k|x|=0$ implies $\lim_k \langle A^k|x|, z\rangle = \langle |x|, z\rangle = 0$, thus $x=0$.

(c) ⇒ (d): By (2.7), there exists an integer $q \geqq 1$ such that 1 is the only eigenvalue of $A^q$ of modulus 1. Since $A^q$ is stochastic, it follows from (3.5) that $P_1 = \lim_k A^{qk}$ exists. We observe that $P_1|x|=0$ implies $x=0$; in fact, $A^{qk}|x| \to 0$ ($k\to\infty$) implies $A^k|x| \to 0$, hence $x=0$ by hypothesis. (Note that $A^l = A^{qk}A^r$ for any $l$, suitable $k$, and suitable $r$, $0 \leqq r < q$.) Since $P=\lim_k (qk)^{-1}(I+A+\cdots+A^{qk-1})$, an easy computation shows that for all $x$,

$$Px = q^{-1} P_1(x + Ax + \cdots + A^{q-1}x).$$

Thus if $x \geqq 0$ and $Px=0$ it follows that $P_1 x = 0$, since $A \geqq 0$ and $P_1 \geqq 0$. Therefore, as we have just proved, $x=0$.

(d) ⇒ (b): If $x>0$ implies $Px>0$, the matrix $P$ cannot have a zero column. Because of $AP=P$ each row vector of $P$ is a fixed vector of ${}^tA$, hence the sum of all row vectors of $P$ is a vector $z={}^tAz$ satisfying $z \gg 0$. ☐

[5] See (3.2) and (4.2).

**Corollary.** *A stochastic matrix is completely reducible if and only if the projection* $P=\lim_k k^{-1}(I+A+\cdots+A^{k-1})$ *is completely reducible.*

*Proof.* Since $P$ is stochastic, it suffices for the proof to use the equivalence of conditions (a) and (d) of (8.7). □

Using the concept of minimal $A$-ideal, it is easy to obtain a normal form for an arbitrary positive square matrix $A$ by applying a suitable simultaneous permutation of rows and columns. In fact, suppose $A\geqq 0$ is an $n\times n$-matrix having exactly $m$ $(1\leqq m\leqq n)$ minimal $A$-ideals $I_\mu$; clearly, $I_\mu\cap I_\nu=\{0\}$ whenever $\mu\neq\nu$. The annihilator of the direct sum $\bigoplus_{\mu=1}^m I_\mu$ is an ideal $I_0$ which is trivial $(I_0=\{0\})$ exactly when $A$ is completely reducible. By applying a suitable permutation of $\{1,\dots,n\}$ we can arrange that $I_\mu$ lives on the coordinates $i$ satisfying $k_{\mu-1}+1\leqq i\leqq k_\mu$ $(\mu=1,\dots,m)$ while $I_0$ lives on $\{i: 1\leqq i\leqq k_0\}$. It is then clear that $I_0$ contains no $A$-ideal $\neq\{0\}$; that is, for each $i$ satisfying $1\leqq i\leqq k_0$, there exists $j>k_0$ such that $\langle A^l e_i, e_j\rangle>0$ for some $l\geqq 1$. Thus we have obtained this result.

**8.8 Proposition.** *Suppose $A\geqq 0$ is an $n\times n$-matrix, and denote by $m$ the number of minimal $A$-invariant ideals $(1\leqq m\leqq n)$. For a suitable permutation matrix $P$,*

$$P^{-1}AP=\begin{pmatrix} A_0 & 0 & \dots 0 \\ B_1 & A_1 & \dots 0 \\ \dots & \dots & \dots \\ B_m & 0 & \dots A_m \end{pmatrix},$$

*where $A_\mu$ $(\mu=1,\dots,m)$ are irreducible*[6] *and where the $k_0\times k_0$-matrix $A_0$ is empty $(k_0=0)$ if and only if $A$ is completely reducible; if $k_0>0$, the $B_\mu$ are not all zero. This normal form is unique up to permutations of the coordinates within each diagonal block and up to the order of $A_1,\dots,A_m$.*

The proof is covered by the discussion preceding (8.8). Let us note that for $k>1$, $P^{-1}A^kP$ is of like form and that for each $j$, $1\leqq j\leqq k_0$, there exists $i>k_0$ and $l\in\mathbb{N}$ such that $P^{-1}A^lP$ has an entry $>0$ in place $(i,j)$. We also remark that the set formed by the respective spectral radii of $A_\mu$ $(\mu=1,\dots,m)$ is contained in the set of distinguished eigenvalues of $A$ (Exerc. 18). If $A_0$ is not irreducible, then $A_0$ can be rearranged into a matrix with irreducible square blocks down the main diagonal and with zero blocks above the latter (cf. Def. (6.1)). Finally, a different normal form for $A$ can be obtained by arranging ${}^tA$ into the normal form above and then transposing back (cf. Gantmacher [1970]).

## § 9. Markov Chains

The theory of positive, finite square matrices has well-known applications to probability theory: The theory of Markov processes with discrete time parameter *(Markov chains)* and finite state space. Even though this theory deals of necessity

---

[6] Of course, the $A_\mu$ $(\mu=1,\dots,m)$ are the restrictions of $A$ to the minimal $A$-ideals. Note that $A_\mu=0$ can occur when the corresponding minimal $A$-ideal is one-dimensional (cf. remark after (6.1)).

only with stochastic matrices (§ 4), practically all the concepts and major results developed so far have important probabilistic applications of great intuitive appeal. We shall, therefore, discuss some basic features of these Markov chains in the present section.

We consider a physical system (henceforth called "the system") capable of being in $n$ ($\geqq 2$) distinct states. The system is observed at times $k=1,2,\ldots$; if $X_k$ is the state of the system at time $k$, the sequence of random variables $\{X_k : k \in \mathbb{N}\}$ is a *stochastic process* reflecting the evolution of the system. Our interest centers on what happens to the system in the long run, that is, on the asymptotic properties of the sequence $\{X_k\}$. The stochastic process $\{X_k\}$ becomes a *Markov process* by the requirement that the conditional probability of the event $X_{k+1}=j$, given $X_1=j_1,\ldots,X_k=j_k$, does not depend (explicitly) on $j_1,\ldots,j_{k-1}$; in other words, the probability distribution of the random variable $X_{k+1}$ is completely determined by that of $X_k$, for all times $k$. Moreover, we assume the process to be *homogeneous* (or *stationary*), so that the relation between the distributions of $X_{k+1}$ and $X_k$ is independent of $k$. Together with an initial probability distribution (namely, that of $X_1$) the process is then completely characterized by the $n^2$ *transition probabilities*

$$\mathbb{P}\{X_{k+1}=j \mid X_k=i\} = p_{ij} \tag{1}$$

($i,j=1,\ldots,n$; $k \in \mathbb{N}$).[1] Thus $p_{ij}$ is the probability for the system to be in state $j$ at time $k+1$ if it was in state $i$ at time $k$. This implies $\sum_{j=1}^n p_{ij}=1$ ($1 \leqq i \leqq n$), so the $n \times n$-matrix $P=(p_{ij})$ is stochastic; $P$ is called the *transition matrix* of the process. Furthermore, the probability of transition $i \mapsto j$ in two steps is given by

$$p_{ij}^{(2)} = \sum_{l=1}^n p_{il} p_{lj},$$

since this transition can occur in $n$ mutually exclusive ways $i \mapsto l \mapsto j$ ($1 \leqq l \leqq n$). Thus the matrix $(p_{ij}^{(2)})$ is the square $P^2$ of $P$; more generally, the probability of passage $i \mapsto j$ in exactly $m$ steps is given by the entry $p_{ij}^{(m)}$ of $P^m$. Any particular distribution of probabilities is given by a vector $z=(\zeta_i) \geqq 0$, $\|z\|_1=1$ ($l^1$-norm one); it follows from the preceding that, if $X_1$ has the probability distribution $z_1=(\zeta_i^{(1)})$, then $X_2$ has the distribution $z_2=(\zeta_i^{(2)})$ where

$$\zeta_j^{(2)} = \sum_{i=1}^n p_{ij} \zeta_i^{(1)} \qquad (1 \leqq j \leqq n).$$

Hence, $z_2 = {}^tP z_1$ and, generally, $X_k$ has the distribution

$$z_k = {}^tP^k z_1 . \tag{2}$$

In contrast with the (relative) transition probabilities, the coordinates $\zeta_i \geqq 0$ of a distribution $z$ are called *absolute probabilities.*

The asymptotic properties of the system (as $k \to +\infty$) are thus reflected by the asymptotic behavior of $p_{ij}^{(k)}$; roughly speaking,

$$\lim_k p_{ij}^{(k)} = q_{ij}$$

---

[1] As usual, the symbol $\mathbb{P}$ stands for the probability of an event in some underlying probability space (which, for our purposes, may be taken to be $\{1,\ldots,n\}$). Some deviations from previous notation are made to conform with standard usage.

(if the limit exists) will be the approximate probability that the system, initially in state $i$, will be in state $j$ after a large number of transitions. Accordingly, if $z_1$ is the distribution of $X_1$ and if $\lim_k P^k = Q$ exists, then ${}^tQ z_1$ will be the approximate distribution of $X_k$ for large $k$.

A distribution $z$ is called *stationary* if $z = {}^tP z$. Hence, the stationary distributions are exactly the fixed vectors $z \geqq 0$ of ${}^tP$ satisfying $\|z\|_1 = 1$. Since 1 is the spectral radius of $P$ and ${}^tP$ (cf. (4.2)), it is already clear from (2.3) that stationary distributions always exist. In general, however, $\lim_k P^k$ does not exist but the following is true.

**9.1 Proposition.** *For any transition matrix $P$, the asymptotic mean transition probabilities*

$$q_{ij} = \lim_k k^{-1}(p_{ij} + p_{ij}^{(2)} + \cdots + p_{ij}^{(k)}) \tag{3}$$

*always exist. $Q = (q_{ij})$ is a transition matrix, and for each initial distribution $z_1$ the mean asymptotic distribution is*

$$z = {}^tQ z_1 .$$

*Proof.* The assertion is an immediate consequence of (4.2) (ergodicity of $P$ and ${}^tP$). Note that $Q$ is stochastic and that, for any matrix $A$,

$$\lim_k k^{-1}(I + A + \cdots + A^{k-1}) = \lim_k k^{-1}(A + A^2 + \cdots + A^k)$$

whenever either limit exists. □

The set $\{1, \ldots, n\}$ of all possible states of the system is now classified, for a fixed transition matrix $P$, by their mutual iterated passage probabilities $p_{ij}^{(k)}$.[2] A state $j$ is called a *consequent* of the state $i$ if there exists $k \in \mathbb{N}$ such that $p_{ij}^{(k)} > 0$. The state $i$ is called *transient* if $i$ is not a consequent of each of its consequents, otherwise *non-transient*. Thus if $i$ is a transient state and the system is initially in state $i$ then it will pass, with probability $>0$, through a state $j$ from which it can never return to $i$. On the other hand, if $i$ is non-transient then the set of all consequents of $i$ forms a subset $T$ of $\{1, \ldots, n\}$ with the following property: If the system is ever in a state $i \in T$, it will not leave $T$ thereafter with probability 1; moreover, $T$ is minimal with respect to this property. Such (non-empty) subsets $T \subset \{1, \ldots, n\}$ are called *ergodic classes* of states.

**9.2 Proposition.** *There exists at least one ergodic class. Moreover, the (possibly empty) set $T_0$ of transient states and the family of ergodic classes $T_\mu$ $(\mu = 1, \ldots, m)$ together form a partition of the set of all states.*

*Proof.* Let $\{J_\mu : \mu = 1, \ldots, m\}$ $(m \geqq 1)$ denote the set of all minimal ${}^tP$-ideals (Def. ((8.1)), and let $T_\mu = \{i : e_i \in J_\mu\}$. Then it is clear from the notion of minimal ${}^tP$-ideal that $J_\mu$ is the smallest ${}^tP$-ideal containing $e_i$, for all $i \in T_\mu$. But then by Formula (2) above, the states $i \in T_\mu$ are all consequents of each other, and each consequent of any $i \in T_\mu$ is in $T_\mu$. Thus each $T_\mu$ is an ergodic class and conversely, and these

[2] We adopt the terminology of Doob [1953]. In the terminology of Kolmogorov, transient states are called *inessential*, non-transient ones *essential*.

classes are mutually disjoint. Define $T_0=\{1,\dots,n\}\setminus\bigcup_1^m T_\mu$. If $i\in T_0$, then the ${}^tP$-ideal $J$ generated by $\{e_i\}$ must contain a vector $e_j$ contained in some $J_\mu$, since otherwise $J$ would contain a minimal ${}^tP$-ideal distinct from all $J_\mu$ $(\mu=1,\dots,m)$. This is clearly impossible; hence, $T_0$ is the set of all transient states. □

The preceding classification of states becomes evident when ${}^tP$ is brought into the canonical form given in (8.8) (which possibly necessitates a renumbering of states). We obtain

$$(4')\qquad {}^tP=\begin{pmatrix} {}^tR_0 & 0 & \dots 0 \\ {}^tS_1 & {}^tP_1 & \dots 0 \\ \dots & \dots & \dots \\ {}^tS_m & 0 & \dots {}^tP_m \end{pmatrix}$$

and, transposing,

$$(4)\qquad P=\begin{pmatrix} R_0 & S_1 & \dots S_m \\ 0 & P_1 & \dots 0 \\ \dots & \dots & \dots \\ 0 & 0 & \dots P_m \end{pmatrix}.$$

Here the matrices $P_\mu$ $(\mu=1,\dots,m)$ are irreducible stochastic, and each $P_\mu$ is the transition matrix of the process within the ergodic class $T_\mu$. It can happen that the set $T_0$ of transient states is empty (in which case $R_0$ is the empty matrix); this occurs, for example, if $P$ is symmetric or (more generally) doubly stochastic.

**9.3 Proposition.** *For $T_0=\emptyset$ it is necessary and sufficient that there exist a strictly positive stationary distribution. If $T_0\neq\emptyset$, then*

$$\lim_k p_{ij}^{(k)}=0$$

*whenever $i,j\in T_0$.*[3]

*Proof.* Since $T_0=\emptyset$ is equivalent with complete reducibility (§ 8) of $P$, the first assertion follows from (8.7). To prove the second assertion, by (3.5) we have to show that the square matrix $R_0$ in (4) has spectral radius $r(R_0)<1$. If not, by (2.3) there exists a vector $z=(\zeta_i)_{i\in T_0}$, $z>0$, satisfying $z=R_0z$. Since $z=R_0^kz$ for all $k\geqq 1$, it follows that $\sum_{j\in T_0}p_{ij}^{(k)}=1$ (all $k\geqq 1$) whenever $\zeta_i=\|z\|_\infty$; thus for at least one $i\in T_0$, the $i$-th row of each $P^k$ has only zero entries outside $T_0$. But this is impossible, since it implies that the system, once in state $i$, remains in $T_0$ with probability 1. □

The limit relation of (9.3) can be used for a characterization of transient states (Exerc. 19): A state $j$ is transient iff for each $\varepsilon>0$, there exists $k_0(\varepsilon)$ such that after $k\geqq k_0$ transitions, the system is in state $j$ with probability $<\varepsilon$ irregardless of its initial state.

To obtain further asymptotic properties of the transition matrix $P$, we consider several special cases increasing in generality.

[3] Note that by (4), $p_{ij}^{(k)}=0$ for all $k$ whenever $j\in T_0$, $i\notin T_0$.

(i) *$P$ is primitive.* There are no transient states, and there is a single ergodic class consisting of all states $i$, $1 \leqq i \leqq n$. By (7.3), $P^k > 0$ for large $k$ and $\lim_k P^k = Q$ exists. Since $P$ is irreducible, the projection $Q$ is fully positive and of rank 1 (cf. end of § 6); thus since $Q$ is stochastic, $Q$ has $n$ identical rows $(q_1, \ldots, q_n)$. Hence,

**9.4 Proposition.** *If the transition matrix is primitive, then*

$$\lim_k p_{ij}^{(k)} = q_j > 0 \qquad (i, j = 1, \ldots, n)$$

*exists and is independent of i. Moreover,* $z = (q_1, \ldots, q_n)$ *is the unique stationary probability distribution.*

It follows that the system behaves asymptotically like a system for which $p_{ij} = p_j$ is independent of $i$. After a large number of transitions it is in state $j$ with probability approximately $q_j > 0$, irregardless of the initial state. By Perron's theorem ((6.5) Cor. 1) this occurs whenever $p_{ij} > 0$ for all $(i,j)$. We note this special case:

*If P is primitive and doubly stochastic, then* $\lim_k p_{ij}^{(k)} = n^{-1}$ *for* $i, j = 1, \ldots, n$.

(ii) *$P$ is irreducible, but not primitive.* Again there are no transient states, and $\{1, \ldots, n\}$ is the only ergodic class. By renumbering the states if necessary, we can arrange that $P$ takes the form ((6.5) Cor. 3)

$$P = \begin{pmatrix} 0 & P_{12} & 0 & \ldots 0 \\ 0 & 0 & P_{23} & \ldots 0 \\ \multicolumn{4}{c}{\dots\dots\dots\dots\dots\dots} \\ 0 & 0 & 0 & \ldots P_{d-1,d} \\ P_{d1} & 0 & 0 & \ldots 0 \end{pmatrix}$$

where, according to standard probabilistic usage, we have employed the symbol $d$ to denote the index of imprimitivity of $P$ (§§ 6, 7). Let $C_1, C_2, \ldots, C_d$ denote the sets of states involved in the (not necessarily square) matrices $P_{12}, \ldots, P_{d1}$, respectively, as column indices. Then if at time $k$ the system is in some state $i \in C_\delta$, at time $k+1$ it will be in some state $j \in C_{\delta+1}$ ($\delta$ modulo $d$) with probability 1. Therefore, the sets $C_\delta$ are called *cyclically moving classes* of states with (common) *period d*. The transition probabilities in $d$ steps form the matrix

$$P^d = \begin{pmatrix} P_1^{(d)} & 0 & \ldots 0 \\ 0 & P_2^{(d)} & \ldots 0 \\ \multicolumn{3}{c}{\dots\dots\dots\dots\dots} \\ 0 & 0 & \ldots P_d^{(d)} \end{pmatrix}$$

where, by (7.4), each $P_\delta^{(d)}$ is primitive (and, of course, stochastic). Hence, from case (i) above we conclude

$$\lim_{k \to \infty} P^{kd} = Q^{(d)} = \begin{pmatrix} Q_1^{(d)} & 0 & \ldots 0 \\ 0 & Q_2^{(d)} & \ldots 0 \\ \multicolumn{3}{c}{\dots\dots\dots\dots\dots} \\ 0 & 0 & \ldots Q_d^{(d)} \end{pmatrix}$$

where each $Q_\delta^{(d)}$ is a fully positive stochastic projection matrix of rank 1. Hence, any row of $Q_\delta^{(d)}$ is of the form $(q_{\delta j})$ $(j \in C_\delta, \delta = 1, \ldots, d)$.

Now let $m$ be fixed, $1 \leqq m \leqq d$. We have, since $P^{kd+m} = P^m P^{kd}$,

$$p_{ij}^{(kd+m)} = \sum_{l=1}^n p_{il}^{(m)} p_{lj}^{(kd)}$$

and, as just has been proved,

$$\lim_{k \to \infty} p_{lj}^{(kd)} = \begin{cases} q_{\delta j} & \text{if } \; l, j \in C_\delta \\ 0 & \text{if } \; l \in C_\gamma, j \in C_\delta \, (\gamma \neq \delta). \end{cases}$$

If $i \in C_\gamma$, $j \in C_\delta$ and $\gamma + m \not\equiv \delta \bmod d$, then $p_{ij}^{(kd+m)} = 0$. On the other hand, if $\gamma + m \equiv \delta \bmod d$, then

$$p_{ij}^{(kd+m)} = \sum_{l \in C_\delta} p_{il}^{(m)} p_{lj}^{(kd)}$$

and

$$\lim_{k \to \infty} p_{ij}^{(kd+m)} = q_{\delta j} \sum_{l \in C_\delta} p_{il}^{(m)} = q_{\delta j}.$$

An easy computation now shows that, for all $i$,

$$\lim_k k^{-1}(p_{ij} + p_{ij}^{(2)} + \cdots + p_{ij}^{(k)}) = \frac{q_{\delta j}}{d} = q_j \tag{5}$$

$(j \in C_\delta, \delta = 1, \ldots, d)$. We have proved:

**9.5 Proposition.** *If the transition matrix $P$ is irreducible, the asymptotic mean transition probability $q_{ij}$ is independent of $i$, $q_{ij} = q_j > 0$, and $(q_1, \ldots, q_n)$ is the unique stationary probability distribution.*

(The uniqueness and strict positivity of $(q_1, \ldots, q_n)$ are immediate consequences of the irreducibility of $P$, cf. (6.2).)

(iii) *$P$ is completely reducible, but not irreducible.* No transient states exist, and there are $m \geqq 2$ ergodic classes of states $T_\mu$ $(\mu = 1, \ldots, m)$. There is no interchange between these ergodic classes, and within each ergodic class the behavior of the process is as described in the preceding case (ii). If the irreducible restriction $P_\mu$ of $P$ (Formula (4) above) to $T_\mu \times T_\mu$ has index of imprimitivity $d_\mu$, then $d_\mu$ is called the *period* of $T_\mu$ (or of the states $i \in T_\mu$). If $C_\delta^{(\mu)}$ $(\delta = 1, \ldots, d_\mu)$ denote the cyclically moving subclasses of the ergodic class $T_\mu$ $(\mu = 1, \ldots, m)$ then by (5) above, the asymptotic mean transition probability is given by

$$\lim_{k \to \infty} k^{-1}(p_{ij} + p_{ij}^{(2)} + \cdots + p_{ij}^{(k)}) = \begin{cases} 0 & \text{if } \; i \in T_\mu, j \in T_\nu \quad (\mu \neq \nu) \\ q_{\delta j}^{(\mu)}/d_\mu & \text{if } \; i \in T_\mu, j \in C_\delta^{(\mu)}. \end{cases}$$

There are exactly $m$ linearly independent stationary probability distributions (Cor. of (8.4)).

Two particular cases, already mentioned above, deserve attention. If $P$ is symmetric or doubly stochastic then $q_{\delta j}^{(\mu)}$ becomes independent of $j \in T_\mu$ also whence, in this case,

$$\lim_k k^{-1}(p_{ij} + p_{ij}^{(2)} + \cdots + p_{ij}^{(k)}) = \begin{cases} 0 & \text{if } \; i \in T_\mu, j \in T_\nu \quad (\mu \neq \nu) \\ 1/n_\mu & \text{if } \; i, j \in T_\mu \end{cases}$$

where $n_\mu$ is the number of states in $T_\mu$.

(iv) *P is not completely reducible.* From (9.3) we know that $\lim_k p_{ij}^{(k)}=0$ whenever $j\in T_0$ ($T_0\neq\emptyset$ in the present case). If $i,j$ both belong to (possibly distinct) ergodic classes, the asymptotic behavior of $p_{ij}^{(k)}$ for $k\to\infty$ is as described in Case (iii) above. The only novel feature in the present case is the existence of mean asymptotic transition probabilities $q_{ij}$ for $i\in T_0$, $j\notin T_0$. This is reflected in the matrix $Q$ of (9.1) which, if the states are numbered as in (4), has the form

$$Q=\begin{pmatrix} 0 & U_1\dots U_m \\ 0 & Q_1\dots 0 \\ \dots & \dots \\ 0 & 0 \quad \dots Q_m \end{pmatrix}. \tag{6}$$

Here the column index $j$ runs successively through $T_0, T_1,\dots, T_m$. If $n_\mu$ ($\mu=0,1,\dots,m$) is the number of states in $T_\mu$, then $U_\mu$ is a matrix with $n_0$ rows and $n_\mu$ columns ($\mu=1,\dots,m$). Each $Q_\mu$ is a fully positive stochastic projection matrix of order $n_\mu$, with $n_\mu$ identical rows $(q_1^{(\mu)},\dots,q_{n_\mu}^{(\mu)})$; $(q_1^{(\mu)},\dots,q_{n_\mu}^{(\mu)})$ is the unique stationary probability distribution supported by $T_\mu$, and each stationary distribution of the process is a (unique) convex combination of these $m$ distributions (cf. (8.4)). To determine the elements $q_{ij}$ of $U_\mu$, we observe first that

$$\textstyle\sum_{j\in T_\mu} p_{ij}^{(k)} \qquad (1\leqq i\leqq n,\ 1\leqq\mu\leqq m)$$

is the probability for the system, initially in state $i$, to enter $T_\mu$ at some time not later than $k$ (for, once in $T_\mu$ the system stays there). Since this probability cannot decrease as $k$ increases,

$$c_{i\mu}=\lim_k \textstyle\sum_{j\in T_\mu} p_{ij}^{(k)} \tag{7}$$

exists and is the probability that the system, initially in state $i$, ever enters $T_\mu$. Clearly, $c_{i\mu}=1$ if $i\in T_\mu$ and $c_{i\mu}=0$ if $i\in T_\nu$ ($\nu\neq 0,\mu$); if $i\in T_0$, then $0\leqq c_{i\mu}\leqq 1$.

Now suppose that the matrices $P_\mu$ ($\mu=1,\dots,m$) in (4) are all primitive, and let $i\in T_0$, $j\in T_\mu$ ($\mu\neq 0$).

$$p_{ij}^{(k+m)}=\textstyle\sum_{l\in T_\mu} p_{il}^{(k)}p_{lj}^{(m)}+\sum_{l\in T_0} p_{il}^{(k)}p_{lj}^{(m)}$$

for all $k$, $m\in\mathbb{N}$. The second term on the right tends to 0 for $k\to\infty$, since it is majorized by $\sum_{l\in T_0} p_{il}^{(k)}$ which has limit 0 by (9.3). Thus on letting $k\to\infty$, $m\to\infty$ we obtain, in view of $\lim_m p_{lj}^{(m)}=q_j^{(\mu)}$ ($l\in T_\mu$; see Case (iii)) and (7),

$$\lim_k p_{ij}^{(k)}=c_{i\mu}q_j^{(\mu)} \qquad (j\in T_\mu).\text{[4]} \tag{8}$$

Hence, *the i-th row of* $U_\mu$ ($i\in T_0$) *is proportional to the (identical) rows of* $Q_\mu$ ($\mu=1,\dots,m$), *with constant of proportionality* $c_{i\mu}$.

Finally, if the hypothesis that $P_\mu$ be primitive is dropped, it can be shown in a similar fashion that

$$q_{ij}=\lim_k k^{-1}(p_{ij}+\dots+p_{ij}^{(k)})=c_{i\mu}\frac{q_{\delta j}^{(\mu)}}{d_\mu} \qquad (j\in C_\delta^{(\mu)}). \tag{9}$$

[4] Note that (8) and (9) are valid for all $i$, $1\leqq i\leqq n$.

Here $d_\mu$ is the common period of the states $j \in T_\mu$, $C_\delta^{(\mu)}$ runs through the cyclically moving subclasses of $T_\mu$, and $q_{\delta j}^{(\mu)}/d_\mu$ is the asymptotic mean transition probability $q_{ij}$ for $i \in T_\mu$, $j \in C_\delta^{(\mu)}$ $(\delta = 1,\dots,d_\mu; \mu = 1,\dots,m)$. Thus the above assertion is equally true in the general case.

## § 10. Bounds for Eigenvalues

There is an extensive literature on the localization of eigenvalues for complex $n \times n$-matrices $A$ under various special assumptions on $A$ (cf. Marcus-Minc [1964, Chap. III]). The most important results on matrices $A \geqq 0$ are contained in the preceding sections of this chapter (in particular, §§ 2, 6) and are concerned with the peripheral spectrum. Beyond this, two problems appear to have received particular attention:

(i) To give bounds, in simple terms involving the entries of $A$, for the spectral radius $r = r(A)$, and (ii) whenever $r$ is the only eigenvalue of $A$ of modulus $r$, to give upper bounds for the moduli of all eigenvalues $\lambda \neq r$ of $A$. An answer to each of these questions is presented below; first let us record the application of a well-known proposition (Gershgorin discs; cf. Markus-Minc [1964]) to positive matrices.

**10.1 Proposition.** *Let $A = (\alpha_{ij})$ be $\geqq 0$ and define $p_i := \sum_{j \neq i} \alpha_{ij}$ $(i = 1, \dots, n)$. Then each eigenvalue of $A$ is contained in at least one of the $n$ circular discs*

$$\{\zeta : |\zeta - \alpha_{ii}| \leqq p_i\} \qquad (i = 1, \dots, n)$$

*of the complex plane.*

*Proof.* If $\lambda \in \sigma(A)$ then $\lambda x = Ax$ for some $x \neq 0$, $\|x\|_\infty = 1$. Let $i$, $1 \leqq i \leqq n$, be an integer for which $|\xi_i| = 1$; since $|\xi_j| \leqq 1$ for all $j$, it follows that $|\lambda - \alpha_{ii}| = |(\lambda - \alpha_{ii})\xi_i| \leqq \sum_{j \neq i} \alpha_{ij} |\xi_j| \leqq p_i$. □

Several improvements of this result have been obtained by A. Brauer (Exerc. 14).

In (2.4) above it has been proved that for any matrix $A \geqq 0$, the spectral radius $r = r(A)$ satisfies $s \leqq r \leqq t$, where $s$ and $t$ denote the smallest and largest row sums of $A$, respectively. (Since $A$ and ${}^tA$ have the same spectral radius it is clear that the same extimate holds for the smallest and largest column sums, from which an obvious improvement can be derived; cf. als Marcus-Minc [1964, III. 1.6].) For $A > 0$, we have the following result.

**10.2 Proposition.** *Let $A > 0$, and let $\mu$ denote the smallest entry of $A$. If $s, t$ denote the smallest and largest row sums of $A$ and if $\sigma^2 := (s - \mu)/(t - \mu)$, then*

$$s + \mu(\sigma^{-1} - 1) \leqq r(A) \leqq t - \mu(1 - \sigma).$$

*Proof.* By (6.2) $A$ possesses an (essentially unique) eigenvector $x = (\xi_i) \gg 0$, and we can arrange (by applying a suitable permutation to both rows and columns of $A$) that $1=\xi_1\geqq\xi_2\geqq\cdots\geqq\xi_n>0$. Letting $s_i:=\sum_{j=1}^n\alpha_{ij}$ $(i=1,\ldots,n)$, for any $i$ we have

$$r\xi_i\geqq\alpha_{i1}+(\textstyle\sum_{j=2}^n\alpha_{ij})\xi_n=\alpha_{i1}(1-\xi_n)+s_i\xi_n,$$

whence

$$r\geqq\frac{s_i\xi_n+(1-\xi_n)\mu}{\xi_i}. \tag{1}$$

Similarly, from

$$r\xi_j\leqq\textstyle\sum_{k=1}^{n-1}\alpha_{jk}+\alpha_{jn}\xi_n$$

it follows that

$$r\leqq\frac{s_j-\mu(1-\xi_n)}{\xi_j}. \tag{1'}$$

We wish to apply (1) to $i=n$ and (1′) to $j=1$; in addition, an estimate of $\xi_n$ is needed. To obtain the latter, we use (1) and (1′), respectively, for the subscripts $i$ and $j$ for which $s_i=t$ and $s_j=s$; this yields, since $\xi_n\leqq\xi_i\leqq1$,

$$r\geqq\frac{t\xi_n+(1-\xi_n)\mu}{\xi_i}\geqq(t-\mu)\xi_n+\mu \tag{2}$$

and

$$r\leqq\frac{(s-\mu)+\mu\xi_n}{\xi_j}\leqq\frac{s-\mu}{\xi_n}+\mu. \tag{2'}$$

(2) and (2′) combined yield $(t-\mu)\xi_n\leqq r-\mu\leqq(s-\mu)/\xi_n$ hence, in particular, $\xi_n^2\leqq(s-\mu)/(t-\mu)=\sigma^2$. Since $\xi_n\leqq\sigma$, an application of (1) to $i=n$ and of (1′) to $j=1$ and the remark that $s_n\geqq s$ and $t\geqq s_1$, finally gives the desired result. □

We now want to establish an upper bound for the moduli of eigenvalues $\lambda\neq r$; for certain matrices $A\geqq0$ (e.g., for the irreducible ones with trace $>0$, (6.5) Cor. 2) it is known that each such eigenvalue $\lambda$ satisfies $|\lambda|<r$. The following result does not require $A$ to be irreducible.

**10.3 Proposition.** *Let $A\geqq0$ have spectral radius $r$ and suppose ${}^tA$ possesses an eigenvector $y\gg0$. Let*

$$m=\inf_H\langle Ax,u\rangle,\qquad M=\sup_H\langle Ax,u\rangle$$

*where $H:=S_{n-1}\times T_{n-1}$ and $S_{n-1}:=\{x:x\geqq0,\langle x,y\rangle=1\}$, $T_{n-1}:=\{u:u\geqq0,\langle u,e\rangle=1\}$.*[1] *Then each eigenvalue $\lambda\neq r$ of $A$ satisfies*

$$|\lambda|\leqq\inf(M-r,r-m). \tag{3}$$

[1] As usual, $e=(1,1,\ldots,1)$.

*If* $m>0$ *(that is, if* $A \gg 0$*) then* (3) *implies*

$$(4) \qquad |\lambda| \leqq \frac{M-m}{M+m} r .$$

*Proof.* Consider the matrices $B=A-m(y \otimes e)$ and $C=M(y \otimes e)-A$; since $\langle Bx,u \rangle \geqq 0$ and $\langle Cx,u \rangle \geqq 0$ for all $(x,u) \in H$, it follows that $B \geqq 0$ and $C \geqq 0$. On the other hand, if $\lambda \neq r$ is an eigenvalue of $A$, then $\lambda z = Az$ for some $z \neq 0$, and $z$ satisfies $\langle z,y \rangle = 0$. This implies $Bz=Az=\lambda z$ and $Cz=-Az=-\lambda z$. Moreover, ${}^tCy=(M-r)y$ and ${}^tBy=(r-m)y$. Since $y \gg 0$, it follows from (2.3) Cor. that $M-r$ and $r-m$ are the spectral radii of $C$ and $B$, respectively. This proves (3).

We show now that (3) implies (4), supposing that $m>0$. We distinguish the cases $M-r \geqq r-m$ and $M-r<r-m$. In the first case, $M+m \geqq 2r$, we have by (3)

$$|\lambda| \leqq r-m = \tfrac{1}{2}(2r-m-M)+\tfrac{1}{2}(M-m)$$

or

$$|\lambda| \left(1+\frac{M+m-2r}{2|\lambda|}\right) \leqq \frac{1}{2}(M-m),$$

excluding the trivial case $\lambda=0$. Since $|\lambda|<r$, the second factor on the left is not smaller than $(1/2r)(M+m)$, which yields (4). In the second case, $M+m<2r$, we have by (3)

$$|\lambda| < \frac{M-r}{2}+\frac{r-m}{2}=\frac{1}{2}(M-m)=\frac{M-m}{2r} r .$$

But $2r>M+m$ by hypothesis, so again we obtain (4). (Note that in case $M+m=2r$, (3) and (4) are equivalent; if $M+m \neq 2r$, then (3) is strictly sharper than (4).) □

Let us determine the numbers $m, M$ of (10.3) explicitly. The sets $S_{n-1}$ and $T_{n-1}$ defining $H=S_{n-1} \times T_{n-1}$ are $(n-1)$-dimensional simplices. If $y=(\eta_i)$, the extreme points (vertices) of $S_{n-1}$ are the points $\eta_i^{-1} e_i$ $(i=1,\dots,n)$; the extreme points of $T_{n-1}$ are $e_i$. Since for each fixed $u \in T_{n-1}$, the linear form $x \mapsto \langle Ax,u \rangle$ attains its infimum at an extreme point of $S_{n-1}$ (and similarly for $x$ and $u$ interchanged), it follows that

$$(5) \qquad m=\inf_{i,j} \frac{\alpha_{ij}}{\eta_j}, \qquad M=\sup_{i,j} \frac{\alpha_{ij}}{\eta_j}.$$

Thus if $A$ is column stochastic (§ 4), ${}^tAe=e$, we can take $y=n^{-1}e$ and obtain $m=n\mu$, $M=n\mathsf{M}$ where $\mu, \mathsf{M}$ denote the smallest and largest entry of $A$, respectively. Thus we obtain this corollary.

**Corollary.** *If $A$ is a (row- or column-) stochastic $n \times n$-matrix, each eigenvalue $\lambda \neq 1$ of $A$ satisfies*

$$|\lambda| \leqq \inf(n\mathsf{M}-1, 1-n\mu) .$$

*If, in addition,* $\mu>0$ *then this implies*

$$(6)\qquad |\lambda| \leqq \frac{\mathsf{M}-\mu}{\mathsf{M}+\mu}.$$

(6) is a special case of an inequality obtained by E. Hopf [1963] for certain integral operators.

## Notes

§ 1: A review of standard material, especially concerning the spectral theory of finite square matrices, with no attempt at completeness. References for this material and generally for Chapter I: Gantmacher [1970], Marcus-Minc [1964].

§ 2: Improvement of (2.4) in the sense of giving upper and lower bounds for $r(A)$ is possible under additional hypotheses (see, e.g., (10.2) and Marcus-Minc [1964]). (2.7) is rarely mentioned explicitly in the literature; its first assertion can, of course, be obtained from Frobenius' theorem (6.5) by considering principal irreducible submatrices of $A$ and employing induction on $n$. (This method is, however, entirely unsuitable for generalization to positive linear operators on Banach lattices.)

§ 3: (3.1) is a very special case of the mean ergodic theorem (Chap. III, § 7). In matrix theory it is most frequently applied to stochastic matrices (cf. § 9). The point of (3.4) is that for positive matrices $A$ with spectral radius 1, the mean ergodic behavior of $A$ is determined by the order of the pole of $(\lambda-A)^{-1}$ at $\lambda=1$.

§ 4: The primary interest of stochastic matrices seems to stem from their occurence as matrices of transition probabilities (§ 9); their connection (4.4) with linear lattic homomorphisms seems to be less widely recognized. (4.5) appears to be due to Romanovsky [1936].

§ 5: Doubly stochastic matrices play an important role in the theory of convexity. While the Hardy-Littlewood-Polya theorem (5.5) goes back to 1929, it is surprising that Birkhoff's theorem (5.2) was not discovered until 1946. Various possibilities to generalize the notion doubly stochastic to operators on infinite dimensional vector lattices are discussed by Rota [1961], Mirsky [1963] and by Ryff [1963], [1965]; Ryff's approach appears to be most natural. Generalizations of (5.5) to measurable functions were also considered by W.A.J. Luxemburg [1967]. It seems to be an open question whether (5.4) extends to stochastic matrices (cf. Exerc. 14(b)). For properties of semi-groups of stochastic and doubly stochastic matrices, see also Caubet [1965] and Farahat [1966].

§ 6: Perhaps Frobenius' theorem (6.5) can be considered to be the most important single result on positive square matrices. In view of this it is striking that the notion of $A$-invariant ideal (Def. 8.1) does not appear explicitly in the literature until recently (cf. Schaefer [1963b]), even though it is the key for an extension of the theorem to abstract positive operators on Banach lattices

(Chap. V, § 5), and even though the structural importance of the ideal concept for lattices has long been recognized (Birkhoff [1967]). Note that, unless $A$ is primitive (§ 7), the inequality in (6.3) only asserts that $r>0$ when $n\geqq 2$. The key lemma (6.4), due to Wielandt [1950], is well suited for generalization (Chap. V, Prop. 5.1).

§ 8: This section attempts to establish the relationship between irreducible and general positive matrices. This relationship depends on the notion of a completely reducible matrix (Def. 8.5) and the concept of $A$-radical (Exerc. 17). (Note that irreducible matrices are special cases of completely reducible ones.) The results of this section have been given in Schaefer [1968] for the case of a weakly compact positive operator on a Banach lattice $C(X)$ ($X$ compact).

§ 9: The discussion of stationary Markov processes with finite state space given in this section is well suited to test and apply the ideal and spectral theory of positive square matrices developed in the preceding sections. The recognized concept of ergodic class serves particularly well to illuminate the notion of $P$-invariant ideal, a notion advantageously employed throughout this chapter. For a detailed discussion the reader is referred to Kemmeny-Snell-Knapp [1966], and to Doob's treatise [1953] for advanced study; see also Chung [1960].

§ 10: The literature on the localization of eigenvalues of a positive matrix is extensive, and no distantly representative account can be given here. We have selected only a very few results with proofs rather typical of the methods available for positive matrices. (10.2) is due to Ostrowski [1952], (10.3) to Schaefer [1970a]. For further results and references, see Markus-Minc [1964] and Schneider [1964] (particularly, the report by A. Brauer [1964]), F.L. Bauer [1965], and Bauer-Deutsch-Stoer [1969].

## Exercises

Throughout the following exercises, we denote by $A, B, \dots$ $n\times n$-matrices with real or complex entries and, simultaneously, the endomorphisms of $\mathbb{C}^n$ these matrices represent via the canonical basis of $\mathbb{C}^n$ (cf. § 1).

1. If $F\subset\mathbb{C}^n$ is a vector subspace invariant under $A$, then $\sigma(A)=\sigma(A_F)\cup\sigma(A_{|F})$, where $A_{|F}$ denotes the restriction of $A$ to $F$ and $A_F$ denotes the endomorphism of $\mathbb{C}^n/F$ induced by $A$. (Show that $(\lambda-A)^{-1}F\subset F$ whenever $(\lambda-A)^{-1}$ exists.)

2. Consider $\mathbb{C}^n$ as the complexification of $\mathbb{R}^n$, and suppose $x\mapsto\|x\|$ is any norm on $\mathbb{R}^n$. Extend this norm to $\mathbb{C}^n$ by

$$\|z\|_{\mathbb{C}}:=\sup_{0\leqq\theta<2\pi}\|(\cos\theta)x+(\sin\theta)y\|$$

where $z=x+iy$ $(x,y\in\mathbb{R}^n)$. Show that, if $A$ is real, then $\|A\|=\|A\|_{\mathbb{C}}$ where

$$\|A\| = \sup\{\|Ax\| : \|x\| \leqq 1, x\in\mathbb{R}^n\},$$
$$\|A\|_{\mathbb{C}} = \sup\{\|Az\|_{\mathbb{C}} : \|z\|_{\mathbb{C}} \leqq 1, z\in\mathbb{C}^n\}$$

and where, as usual, $Az:=Ax+iAy$ for $z=x+iy$ $(x,y\in\mathbb{R}^n)$.

3. If $A\geqq 0$, if $(\lambda-A)^{-1}$ exists and is $\geqq 0$, then $\lambda$ is real and $\lambda>r(A)$.

4. Suppose $A\geqq 0$ and $r(A)=1$. If both $A$ and ${}^tA$ possess strictly positive fixed vectors, then the peripheral spectrum of $A$ is fully cyclic and the linear hull of the corresponding eigenvectors forms a subalgebra of $\mathbb{C}^n$ (cf. Chap. III, Exerc. 28).

5. Show the following assertions to be equivalent ($B$ any square matrix):
(a) $B$ is similar to a diagonal matrix.
(b) The minimal polynomial of $B$ has simple roots only.
(c) The resolvent of $B$ has no singularities other than first order poles.

6. For any $A$ with spectral radius $r(A)=1$, these properties are equivalent:
(a) $A$ is mean ergodic (Def. 3.2).
(b) The sequence $(A^k)_{k\in\mathbb{N}}$ is bounded.
(c) Every unimodular eigenvalue of $A$ is a first order pole of the resolvent (or equivalently, a simple root of the minimal polynomial).
(d) There exists a decomposition $A=D+B$, where $D$ is diagonal with unimodular diagonal entries, $DB=BD=0$, and $r(B)<1$.
Give an example of an $n\times n$-matrix $A$ $(n\geqq 3)$ which is not mean ergodic but for which $\lim_{\lambda\to 1}(\lambda-1)(\lambda-A)^{-1}$ exists.

7. If $A\geqq 0$ is mean ergodic (Def. 3.2) and $\sigma(A)\subset\Gamma$ ($\Gamma$ the circle group), then $A$ is periodic; that is, there exists an integer $q\geqq 1$ such that $A^q=I$. If $q$ is minimal then $\lim_k M_k=q^{-1}(I+A+\cdots+A^{q-1})$.

8. Let $A\geqq 0$ satisfy $r(A)=1$.
(a) If the fixed space of $A$ is one dimensional and $A$ or ${}^tA$ has a strictly positive fixed vector, the peripheral spectrum of $A$ is a subgroup of the circle group and its elements are simple roots of the minimal polynomial.
(b) If $\Delta(\lambda)$ denotes the characteristic polynomial of $A$ and if $B(\lambda):=\Delta(\lambda)(\lambda-A)^{-1}$ then $B(\lambda)\geqq 0$ and $\dfrac{dB(\lambda)}{d\lambda}\geqq 0$ for $\lambda\geqq 1$.
Moreover, $A$ is irreducible (Def. 6.1) iff $B(1)>0$.

9. Let $A\geqq 0$.
(a) For $A$ and ${}^tA$ to be lattice homomorphisms of $\mathbb{R}^n$, it is necessary and sufficient that in each row and in each column, $A$ contains at most one entry $\neq 0$.
(b) If $A$ is stochastic and both $A$ and ${}^tA$ are lattice homomorphisms, then $A$ is a permutation matrix.
(c) Suppose $A^{-1}$ exists. If $A$ has more than $n$ entries $>0$, then $A^{-1}$ has entries $>0$ as well as $<0$.

(d) If $A^{-1}$ exists and is $\geqq 0$, then $A=DP$ where $P$ is a permutation matrix and $D$ a diagonal matrix with diagonal entries $>0$.

(e) If $A \gg 0$, there exists a unique stochastic matrix $S$ such that $S=DAD$, where $D$ is diagonal with diagonal entries $>0$.

10. Give examples of groups (under matrix multiplication) of doubly stochastic matrices with unit not the unit matrix.

11. Give an example of a doubly stochastic $n \times n$-matrix which is not the barycenter of fewer than $(n-1)^2+1$ permutation matrices.

12. A function $F: \mathbb{R} \to \mathbb{R}$ is called *convex* if for all $\xi, \eta \in \mathbb{R}$, $F(\lambda\xi+\mu\eta) \leqq \lambda F(\xi)+\mu F(\eta)$ whenever $\lambda, \mu>0$ and $\lambda+\mu=1$. Show that for vectors $x=(\xi_i)$ and $y=(\eta_i)$ in $\mathbb{R}^n$, $y \prec x$ (notation of 5.5) if and only if

$$\sum_{i=1}^{n} F(\eta_i) \leqq \sum_{i=1}^{n} F(\xi_i)$$

for every convex $F$.

13. Let $A \geqq 0$.

(a) $A$ is irreducible if and only if for each pair $(i,j), 1 \leqq i,j \leqq n$, there exists an integer $k$ not larger than the degree of the minimal polynomial of $A$, such that entry $(i,j)$ of $A^k$ is $>0$.

(b) A positive projection $P$ is irreducible if and only if $P \gg 0$.

(c) Suppose $A$ to be mean ergodic (Def. 3.2) with $r(A)=1$. Then $A$ is irreducible if, and only if, $\lim_k k^{-1}(I+A+\cdots+A^{k-1}) \gg 0$.

14. Let $A=(\alpha_{ij})$ be stochastic $(i,j=1,\ldots,n)$.

(a) If $m$ denotes the smallest entry of $A$ and if $\alpha_{ii}+\alpha_{jj}>1-(n-2)m$ for all pairs $(i,j)$, then $\det A>0$ and $\operatorname{Re}\lambda>0$ for all eigenvalues $\lambda$ of $A$.

(b) If $\lambda$ is an eigenvalue of $A$ whose argument satisfies $|\arg\lambda|<2\pi/n$, then $\lambda$ is contained in the convex hull $C_n$ of the set of all $v$-th roots of unity, $1 \leqq v \leqq n$.

(c) If there exists $i$ such that $\alpha_{ij}=\alpha_{lj}$ for all $l \neq i$, $j \neq i$, there exists an eigenvalue $\lambda_0 \neq 1$ of $A$ such that $|\lambda| \leqq \lambda_0$ for all eigenvalues $\lambda \neq 1$ of $A$. (For this and the following problem, cf. A. Brauer [1964].)

15. A real $n \times n$-matrix $A$ is called *power positive* (A. Brauer [1964]) if $A^k \gg 0$ for some $k \in \mathbb{N}$.

(a) If $A$ is power positive, then $r(A)>0$ and there exists a real eigenvalue $\lambda_0$ which possesses a strictly positive eigenvector, satisfies $|\lambda_0|=r(A)$, and is a simple root of the characteristic polynomial.

(b) If $A^k \gg 0$ for some odd $k$, or if $A$ has at least one row $\geqq 0$, then $\lambda_0>0$.

(c) If $A$ is power positive and $s,t$ denote the smallest and largest row sums of $A$, respectively, then $s \leqq \lambda_0 \leqq t$.

16. Let $A$ be real and suppose $r(|A|)=1$.

(a) If $\varepsilon$ is a unimodular eigenvalue of $A$, then $\varepsilon^{2k+1}$ is an eigenvalue of $A$ and $\varepsilon^{2k}$ is an eigenvalue of $|A|$, for each $k \in \mathbb{Z}$.

(b) Show that $\varepsilon^{2n}=1$ for each unimodular eigenvalue of $A$. Deduce from this that if $B$ is a real $n \times n$-matrix with a unimodular eigenvalue not a $2n$-th root of unity, then $r(|B|)>1$.

(c) If $A$ possesses no non-trivial invariant ideals in $\mathbb{R}^n$ and if $\varepsilon$ is any unimodular eigenvalue of $A$, then $\varepsilon^2 \in G_m$ and the unimodular spectrum of $A$ is the coset $\varepsilon G_m$, where $G_m$ is the group of all $m$-th roots of unity for some $m$, $1 \leqq m \leqq n$. (For (c), use (6.4) and (6.6); see also Schaefer [1971 b].)

17. Let $A \geqq 0$. The intersection $R$ of all maximal $A$-ideals is called the *radical* of $A$. $A$ is completely reducible (Def. 8.5) if and only if $R = \{0\}$, cf. (8.6).

(a) The endomorphism $A_R$ of $\mathbb{C}^n/R$ induced by $A$ (and representable by a submatrix of $A$, since $R$ is an ideal) is completely reducible.

(b) If $A$ is stochastic, the restriction of $A$ to $R$ has spectral radius $<1$.

(c) If $A$ is completely reducible and nilpotent, then $A=0$.

18. We use the notation of (8.8). A real number $\rho \geqq 0$ is a distinguished eigenvalue of $A$ if and only if ($\alpha$) $\rho$ equals the spectral radius $\rho_\mu := r(A_\mu)$ for some $\mu$ $(1 \leqq \mu \leqq m)$ or ($\beta$) $\rho$ is a distinguished eigenvalue of $A_0$ with an eigenvector $x_0 \geqq 0$, $\rho x_0 = A_0 x_0$, such that $B_\mu x_0 > 0$ implies $\rho > \rho_\mu$ $(\mu = 1, \ldots, m)$.

19. Let $P = (p_{ij})$ be the transition matrix of a stationary Markov process with state space $\{1, 2, \ldots, n\}$ (§9).

(a) $j$ is a transient state if and only if for each $\varepsilon > 0$, there exists $k_0(\varepsilon)$ such that $p_{ij}^{(k)} \leqq \varepsilon$ for all $k \geqq k_0$ and all $i$.

(b) If $\lim_k p_{ij}^{(k)}$ exists and is $>0$ for all pairs $(i,j)$ of states, then $P$ is primitive (§ 7) and the limit is independent of $i$. (Use Exerc. 13(c) to show that $P$ is irreducible, and apply (7.3).)

20. If $A \geqq 0$ is an irreducible $n \times n$-matrix and $n \geqq 2$, the spectral radius $r$ of $A$ is characterized by the following minimax property:

$$r = \max_{x \gg 0} \min_{1 \leqq i \leqq n} (Ax)_i/\xi_i = \min_{x \gg 0} \max_{1 \leqq i \leqq n} (Ax)_i/\xi_i,$$

where $x = (\xi_i) \in \mathbb{R}^n$ and $(Ax)_i$ denotes the $i$-th coordinate of $Ax$ (Wielandt [1950]). In particular, for each $x \gg 0$ one obtains the bounds

$$\min_i (Ax)_i/\xi_i \leqq r \leqq \max_i (Ax)_i/\xi_i$$

with equality holding only if $x$ is the (essentially) unique positive eigenvector of $A$.

Chapter II

# Banach Lattices

## Introduction

This chapter treats the basic theory of vector and Banach lattices; it lays the foundation for the operator theory to which most of Chapters III—V is devoted. Roughly, the chapter can be divided into three parts. The algebraic theory of vector lattices, including duality, is found in Sections 1—4. Sections 5 and 6 contain a general theory of Banach lattices with the exception of representation theory, which is postponed to Chap. III, §§ 4—6. The third part, consisting of Sections 7—10, deals with special Banach lattices; particularly, with abstract $M$- and $L$-spaces and their duality. It is here that classical analysis (especially measure theory) and central parts of functional analysis blend most conspicuously with Banach lattice theory; it can be said without exaggeration that the power and appeal of the general theory largely stem from this relationship with the core of analysis. (This is one of several respects in which the theory of Banach lattices resembles the theory of commutative Banach algebras.) For an illustration, the reader is referred to Chap. IV, §§ 5 or 9. Finally, Section 11 introduces complex vector and Banach lattices which are indispensable for satisfactory applications to spectral theory (Chap. V, §§ 3—7). A more detailed account follows.

Section 1 supplies the elementary algebraic theory of vector lattices, collecting the basic tools. In Section 2 a discussion of the central notions of ideal and band, including their immediate applications such as the Riesz decomposition theorem (2.10), are given and the relationship of Boolean algebras of band projections with the extreme boundary of certain order intervals is established. Vector lattices of finite dimension are investigated in detail in Section 3; this discussion is centered on the notions of maximal ideal and radical. Section 4 is devoted to the duality of vector lattices and leads up to Nakano's important theorem (4.12).

Section 5 begins with the elementary theory of normed vector lattices, then turns of the Šilov boundary (5.7) and an application to monotone convergence (5.9). The remainder of this section is concerned with various useful characterizations of Banach lattices with order continuous norm (particularly, of reflexive Banach lattices), both through intrinsic properties relating order and topology and through imbedding properties with respect to $c_0$, $l^1$, and $l^\infty$ (which are the simplest infinite dimensional representatives of $AL$-spaces and order complete $AM$-spaces without and with unit, respectively). Section 6 contains a detailed discussion of the properties of quasi-interior positive elements (perhaps more suggestively called topological units), a notion playing an important role throughout the remainder of the book.

Spaces of all continuous real functions on a compact space furnish a very important (probably the most important) class of Banach lattices, both in their own right (cf. Semadeni [1971]) and because of their constant occurence as principal ideals of a general Banach lattice; they are considered in detail, and from several angles, in Section 7. Not surprisingly, the situation is similar for spaces of integrable functions (Section 8). They are central objects of measure theory (cases in point: the theorems of Vitali-Hahn-Saks and of Lebesgue-Radon-Nikodym), as well as intimately related to abstract $M$-spaces by their mutual duality. This duality and its most fascinating aspects are investigated in Sections 9 and 10. Next to Hilbert space, $AM$- and $AL$-spaces seem to furnish (among Banach spaces) the most fertile ground for the development of operator theory—an opinion which the author believes to be substantiated by many results derived in the remaining three chapters of this book.

Finally, the complexification of vector and Banach lattices studied in Section 11 is generally useful, and indispensable for the applications to spectral theory made in Chap. V, §§ 3—7.

## § 1. Vector Lattices Over the Real Field

An *ordered set* is a set $A$ endowed with a binary relation, usually denoted by "$\leqq$", which is supposed to be *transitive* ($x\leqq y \,\&\, y\leqq z \Rightarrow x\leqq z$), *reflexive* ($x\leqq x$, all $x\in A$), and *anti-symmetric* ($x\leqq y \,\&\, y\leqq x \Rightarrow x=y$). (If the last axiom is not postulated, the relation is called a *preorder* on $A$ and denoted by some other symbol such as $\prec$.)

Let $(A,\leqq)$ be an ordered set. We write $y\geqq x$ to mean $x\leqq y$, and $x<y$ to express that ($x\leqq y \,\&\, x\neq y$); similarly for $y>x$. A subset $B$ of $A$ is called *majorized (minorized)* if there exists $a_0\in A$ such that $b\leqq a_0$ for all $b\in B$ (respectively, $a_0\leqq b$ for all $b\in B$); $a_0$ is called a *majorant* or *upper bound* (respectively, a *minorant* or *lower bound*) of $B$ in $A$. An *order interval* $[x,y]$, where $x,\, y\in A$, is the set of all $z\in A$ satisfying $x\leqq z\leqq y$; a subset $C\subset A$ contained in some order interval of $A$, is called *order bounded.* If $B$ is a majorized subset of $A$ and if there exists a majorant of $B$ that minorizes all majorants of $B$ (in $A$), such an element is unique, called the *supremum* or *least upper bound* of $B$, and denoted by $\sup B$; similarly for $\inf B$ (the *infimum* or *greatest lower bound* of $B$). (Note that these concepts depend on the set $A$ of which $B$ is thought to be a subset; hence, sometimes the notation $\sup_A B$, $\inf_A B$ is used.) The set $A$ is called *totally ordered* if for each pair $(x,y)\in A\times A$, at least one of the relations $x\leqq y$ and $y\leqq x$ holds.

A non-void subset $D$ of an ordered (or preordered) set is called *directed* ($\leqq$) or *directed upward* if for each pair $(x,y)\in D\times D$ there exists $z\in D$ satisfying $z\geqq x$ and $z\geqq y$; subsets $D$ directed downward are defined analogously. If $D$ is a directed ($\leqq$) set, each subset $D_x:=\{z\in D: z\geqq x\}$ is called a *section* of $D$; the family $(D_x)_{x\in D}$ is a base of a filter on $D$ called the *section filter* of $D$.

**1.1 Definition.** *An ordered set* $(L, \leqq)$ *is called a* lattice *if for each pair* $(x,y) \in L \times L$, *the elements* $x \vee y := \sup\{x,y\}$ *and* $x \wedge y := \inf\{x,y\}$ *exist in* $L$. *If, in addition, the distributive law*

$$\text{(LD)} \qquad (x \vee y) \wedge z = (x \wedge z) \vee (y \wedge z)$$

*is satisfied for all* $x, y, z \in L$, $L$ *is called a* distributive lattice. *A distributive lattice is called a* Boolean algebra *if* $L$ *has a smallest element* $0 := \inf L$ *and a largest element* $1 := \sup L$, *and if for each* $x \in L$ *there exists a (unique) element* $x^c$, *called the* complement *of* $x$ *in* $L$, *satisfying* $x \wedge x^c = 0$, $x \vee x^c = 1$.

We point out (cf. Birkhoff [1967]) that Axiom (LD) is equivalent to the dual distributive law:

$$\text{(LD)}' \qquad (x \wedge y) \vee z = (x \vee z) \wedge (y \vee z).$$

If $(L, \leqq)$ is a lattice, the mappings $(x,y) \mapsto x \vee y$ and $(x,y) \mapsto x \wedge y$ are usually called the *lattice operations*. As laws of composition they are idempotent, associative, commutative, and satisfy $x \wedge (x \vee y) = x$ as well as $x \vee (x \wedge y) = x$; conversely, it is not difficult to verify that if a non-void set $L$ is endowed with two laws of composition having these properties, then "$x \leqq y$ *iff* $x \vee y = y$" defines an ordering under which $L$ is a lattice in the sense of Def. 1.1. Recall also that a lattice $L$ is called *(countably) complete* if every (countable) subset of $L$ possesses a least upper bound and a greatest lower bound.

If $L$ is a lattice, a subset $L_0$ closed under the lattice operations is called a *sublattice* of $L$; $L_0$ is called a *(countably) complete sublattice* of $L$ if $L_0$ is closed under the formation of arbitrary (countable) infima and suprema. However, it should be noted that a subset $L_0$ of a (countably complete, complete) lattice $L$ can be a (countably complete, complete) lattice under the ordering induced by that of $L$, without being a (countably complete, complete) sublattice of $L$; in other words, infima and suprema of subsets of $L_0$ (whenever they exist) need not be the same in $L_0$ and in $L$ (see following examples).

**Examples.** 1. Let $X$ be a topological space. The sets $\boldsymbol{O}$ and $\boldsymbol{C}$ of all open and of all closed subsets of $X$, respectively, are distributive lattices under set inclusion $\subset$, and anti-isomorphic under complementation $M \mapsto M^c$ in $X$. $\boldsymbol{O}$ and $\boldsymbol{C}$ are sublattices of the power set $2^X$ and complete lattices but, in general, not complete sublattices of $2^X$.

The set $\boldsymbol{Q} = \boldsymbol{O} \cap \boldsymbol{C}$ of all open-and-closed subsets of $X$ is a sublattice of $\boldsymbol{O}$ and of $\boldsymbol{C}$ and a Boolean algebra. But even if $\boldsymbol{Q}$ is (countably) complete it is not, in general, a (countably) complete subalgebra of the complete Boolean algebra $2^X$. In this context, we recall the representation theorem of M. H. Stone (Exerc. 1):

*Every Boolean algebra* $A$ *is isomorphic with the Boolean algebra of all open-and-closed subsets of a totally disconnected compact space* $K_A$. $K_A$ *is unique (to within homeomorphism) and called the Stone representation space of* $A$.

(Note that the term compact is employed for Hausdorff (or $T_2$) spaces only. The term "Stone representation space" should not be confused with the concepts of Stonian and quasi-Stonian space (§ 7).)

2. If $G$ is a group, the set $\boldsymbol{G}$ of all subgroups is a lattice under set inclusion but, in general, not a sublattice of $2^G$. Similarly, if $V$ is a vector space the set $\boldsymbol{V}$ of all vector subspaces is a lattice but, in general, not a sublattice of $2^V$; except for trivial cases, the lattices $\boldsymbol{G}$ and $\boldsymbol{V}$ are not distributive. Moreover, if $G$ is a topological group or $V$ a topological vector space, the respective sets of closed subgroups or closed vector subspaces are lattices (under set inclusion) but, in general, not sublattices of $\boldsymbol{G}$ or $\boldsymbol{V}$.

3. If $M$ is a non-void set and $(L, \leqq)$ is a lattice, the set $L^M$ of all mappings $f: M \to L$ is a lattice under the *canonical ordering* defined by "$f \leqq g$ *iff* $f(t) \leqq g(t)$ *for all* $t \in M$". It is frequently a difficult problem to determine under what conditions on $K$ a subset $K \subset L^M$ is a lattice for the induced ordering; for example, this problem arises when $L$ is a vector lattice (see below) and $K$ a vector subspace of $L^M$. We will be extensively concerned with this latter problem in Chapter IV.

We wish to apply the order and lattice concepts discussed so far, to vector spaces $E$ over the real field $\mathbb{R}$. If a vector space $E$ is endowed with an order relation $\leqq$, useful results can be expected only if certain compatibility axioms are met. These axioms require the ordering to be invariant under translations $x \mapsto x + x_0$, and under homothetic maps $x \mapsto \lambda x$ with ratio $\lambda > 0$.

**1.2 Definition.** *A vector space $E$ over $\mathbb{R}$, endowed with an order relation $\leqq$, is called an* ordered vector space *if these axioms are satisfied:*

$(\mathrm{LO})_1$ $\quad x \leqq y \Rightarrow x+z \leqq y+z$ *for all* $x, y, z \in E$,

$(\mathrm{LO})_2$ $\quad x \leqq y \Rightarrow \lambda x \leqq \lambda y$ *for all* $x, y \in E$ *and* $\lambda \in \mathbb{R}_+$.

*A* vector lattice *is an ordered vector space such that* $x \vee y := \sup\{x, y\}$ *and* $x \wedge y := \inf\{x, y\}$ *exist for all* $x, y \in E$.

Vector lattices are also called *Riesz spaces* or *linear lattices* (cf. Vulikh [1967], Luxemburg-Zaanen [1971]). If $(E, \leqq)$ is an ordered vector space, the subset $E_+ := \{x \in E : x \geqq 0\}$ is called the *positive cone* of $E$; elements $x \in E_+$ are called *positive*[1].

In conformity with the usage of the term "isomorphism" (without qualifier) to indicate preservation of all structures implied by a given concept, we understand by an *isomorphism* of an ordered vector space $E_1$ onto an ordered vector space $E_2$, a linear bijection $T: E_1 \to E_2$ such that $x \leqq y$ in $E_1$ iff $Tx \leqq Ty$ in $E_2$. It is clear that if $E_1$, $E_2$ are isomorphic ordered vector spaces then $E_1$ is a vector lattice iff $E_2$ is; however, the range of an order isomorphism of a vector lattice $E_1$ *into* a vector lattice $E_2$ need not be a sublattice of $E_2$ (cf. remarks preceding Example 1).

**Examples.** 4. Clearly the one-dimensional vector space $\mathbb{R}_0$ over $\mathbb{R}$ is a vector lattice under its usual ordering; if $M$ is any set, $\mathbb{R}_0^M$ is a vector lattice under

---

[1] To avoid confusion with the terminology introduced in Def. 2.4 (cf. Chap. I, Def. 2.1), non-zero positive elements of $E$ will not be termed "strictly positive".

its canonical ordering (Example 3 above; for $M=\emptyset$, the usual convention is $\mathbb{R}_0^{\emptyset}=\{0\}$). Many of the vector lattices occurring in analysis are vector sublattices of $\mathbb{R}_0^M$, and quotients of such sublattices by ideals (§ 2).

5. Let $A$ be a Boolean algebra. A real function $\mu\colon A\to\mathbb{R}$ is called (finitely) *additive* on $A$ if $\mu(x\vee y)=\mu(x)+\mu(y)$ whenever $x\wedge y=0$. Under the vector and order structures induced by $\mathbb{R}^A$, the set of all bounded additive functions on $A$ is a vector lattice (though not a sublattice of $\mathbb{R}^A$, in general). It is not difficult to verify that $\mu_1\vee\mu_2$ and $\mu_1\wedge\mu_2$ are given by

$$\mu_1\vee\mu_2(x)=\sup_{y\leqq x}\{\mu_1(y)+\mu_2(x\wedge y^c)\}$$

$$\mu_1\wedge\mu_2(x)=\inf_{y\leqq x}\{\mu_1(y)+\mu_2(x\wedge y^c)\}$$

for all $x\in A$. The example can be extended to Boolean rings, a *Boolean ring* being defined as a distributive lattice $L$ such that $0:=\inf L$ exists and such that for each pair $(x,y)\in L\times L$, there exists $z\in L$ satisfying $x\wedge z=0$, $x\vee z=x\vee y$ (relative complementation).

6. If $(E_\alpha)_{\alpha\in\mathrm{A}}$ is a family of vector lattices, the Cartesian product $\prod_\alpha E_\alpha$ is a vector lattice if the vector and lattice operations are defined "coordinatewise". The ordering of $\prod_\alpha E_\alpha$ so obtained is called *canonical* and can be viewed as a generalization of the construction given in Example 3. The direct sum $\bigoplus_\alpha E_\alpha$ of the family $(E_\alpha)_{\alpha\in\mathrm{A}}$ is understood to be the vector sublattice of $\prod_\alpha E_\alpha$ containing precisely all finitely non-zero families $(x_\alpha)_{\alpha\in\mathrm{A}}$.

We turn to the basic arithmetical relations valid in any vector lattice $E$. (It should be noted that all results based exclusively on the translation axiom $(\mathrm{LO})_1$ are valid in commutative lattice ordered groups; for further extensions, see G. Birkhoff [1967].) It is readily seen from $(\mathrm{LO})_1$ that whenever $\sup A$ exists for a non-void subset $A\subset E$ then $\inf(-A)$ exists (and conversely); moreover, for each $x\in E$ we have the following:

$$x+\sup A=\sup(x+A), \tag{1}$$

$$x+\inf A=\inf(x+A), \tag{1'}$$

$$\sup A=-\inf(-A). \tag{2}$$

**1.3 Definition.** *Let $E$ be a vector lattice. For all $x\in E$, we define $x^+:=x\vee 0$, $x^-:=(-x)\vee 0$, $|x|:=x\vee(-x)$. $x^+$, $x^-$ and $|x|$ are called the* positive part, *the* negative part, *and the* modulus *(or* absolute value*) of $x$, respectively. Two elements $x,y\in E$ are called* orthogonal (disjoint, *or* lattice disjoint) *if $|x|\wedge|y|=0$; this is denoted by $x\perp y$. Two subsets $A,B$ of $E$ are called* orthogonal (lattice disjoint) *if $x\perp y$ for each pair $(x,y)\in A\times B$. For $A\subset E$, $A^\perp$ denotes the set of all $x\in E$ disjoint from each $y\in A$.*

*A subset $S\subset E_+$ is called an* orthogonal system *if $0\notin S$ and if $u\wedge v=0$ for each pair of distinct elements of $S$.*

We now prove some fundamental properties of the absolute value function.

**1.4 Proposition.** *Let $E$ be a vector lattice. For all $x, y, x_1, y_1 \in E$ and all $\lambda \in \mathbb{R}$, the following relations are valid.*

(3) $$x = x^+ - x^-$$

(3′) $$|x| = x^+ + x^-.$$

(4) $$|x| = 0 \Leftrightarrow x = 0; \quad |\lambda x| = |\lambda|\,|x|, \quad |x+y| \leqq |x| + |y|.$$

(5) $$x+y = x \vee y + x \wedge y$$

(5′) $$|x-y| = x \vee y - x \wedge y.$$

(6) $$|x \vee y - x_1 \vee y_1| \leqq |x-x_1| + |y-y_1|$$

(6′) $$|x \wedge y - x_1 \wedge y_1| \leqq |x-x_1| + |y-y_1|.$$

*Moreover,* (3) *is the unique representation of $x$ as a difference of disjoint positive elements of $E$.*

*Proof.* From (1) and (2) we obtain the identity

$$x_1 - (x \wedge y) + y_1 = (x_1 - x + y_1) \vee (x_1 - y + y_1).$$

Letting $x = x_1$ and $y = y_1$ we obtain (5) which, in turn, yields (3) for $y=0$. The first two assertions of (4) being easy consequences of the anti-symmetry of the ordering and of $(\mathrm{LO})_2$, respectively, the triangle inequality follows from the obvious relations $\pm x \leqq |x|$, $\pm y \leqq |y|$ which imply $\pm(x+y) \leqq |x|+|y|$. Now $x^+ + x^- = x + 2x^- = x + (-2x) \vee 0 = (-x) \vee x$ verifies (3′). Further, if $x = y - z$ where $y \wedge z = 0$ then $y \geqq x^+$, $z \geqq x^-$ and $y - x^+ \perp z - x^-$. But $y - x^+ = z - x^-$ so $y = x^+$, $z = x^-$.

To obtain (5′), observe that translation invariance (1) implies $x \vee y = x + (y-x)^+$ and $x \wedge y = x + (y-x) \wedge 0 = x - (x-y)^+$; subtraction yields $x \vee y - x \wedge y = (y-x)^+ + (x-y)^+ = (x-y)^+ + (x-y)^- = |x-y|$, by virtue of (3′).

To verify (6), write $x \vee y - x_1 \vee y_1 = (x \vee y - x_1 \vee y) + (x_1 \vee y - x_1 \vee y_1)$. Again by translation invariance, $x \vee y - x_1 \vee y = (x-y)^+ - (x_1 - y)^+$; letting $z = x - y$, $z_1 = x_1 - y$ we have to prove $z^+ - z_1^+ \leqq |z - z_1|$, which is evident because of $z^+ \leqq z_1^+ + |z - z_1|$. Thus we have $(x \vee y - x_1 \vee y)^+ \leqq |x - x_1|$ and interchanging $x$ with $x_1$ we obtain $|x \vee y - x_1 \vee y| \leqq |x - x_1|$. Since the operation $\vee$ is commutative it follows that $|x_1 \vee y - x_1 \vee y_1| \leqq |y - y_1|$ and hence, the triangle inequality (4) implies (6). (6′) is readily obtained from (6) replacing $x, x_1, y, y_1$ by $-x, -x_1, -y, -y_1$ respectively. □

*Remark.* If $E$ is an ordered vector space (not necessarily a vector lattice) and $x, y$ are given elements of $E$, then $x \vee y$ exists iff $x \wedge y$ exists and these elements are related by (5); a similar observation applies to Formulae (7), (7′) below.

From Prop. 1.4 we obtain these additional identities.

**Corollary 1.** *For arbitrary elements $x$, $y$ of a vector lattice $E$, one has the relations:*

(7) $$x \vee y = \tfrac{1}{2}(x+y+|x-y|),$$

(7′) $$x \wedge y = \tfrac{1}{2}(x+y-|x-y|),$$

(8) $$|x| \vee |y| = \tfrac{1}{2}(|x+y|+|x-y|),$$

(8′) $$|x| \wedge |y| = \tfrac{1}{2}\big||x+y|-|x-y|\big|.$$

*In particular, $x \perp y$ if and only if $|x+y|=|x-y|$.*

*Proof.* (7) and (7′) are readily obtained from (5) and (5′) by addition and subtraction, respectively. Using (7) and the commutativity and associativity of the lattice operations, we find

$$\begin{aligned}|x| \vee |y| &= (x \vee -x) \vee (y \vee -y) = (x \vee -y) \vee (y \vee -x)\\ &= \tfrac{1}{2}[(x-y+|x+y|) \vee (y-x+|x+y|)]\\ &= \tfrac{1}{2}|x+y| + \tfrac{1}{2}[(x-y) \vee (y-x)] = \tfrac{1}{2}(|x+y|+|x-y|),\end{aligned}$$

which is (8). To prove (8′) observe that by (5) and (8),

$$|x| \wedge |y| = |x|+|y|-|x| \vee |y| = |x|+|y|-\tfrac{1}{2}|x+y|-\tfrac{1}{2}|x-y|.$$

Substituting $x=u+v$, $y=u-v$ this yields $|x| \wedge |y|=|u+v|+|u-v|-|u|-|v|$ $=2(|u| \vee |v|)-|u|-|v|$ through application of (8) to the first two summands. Another application of (7) gives $2(|u| \vee |v|)-|u|-|v|=\big||u|-|v|\big|$ and hence, substituting back, we receive (8′). ☐

**Corollary 2.** *In every vector lattice $E$, the relation $x \leqq y$ is equivalent to $(x^+ \leqq y^+ \;\&\; y^- \leqq x^-)$, and $x \perp y$ is equivalent to $|x| \vee |y|=|x|+|y|$. Moreover, $x \perp y$ implies $(x+y)^+=x^+ + y^+$ and $|x+y|=|x|+|y|$.*

We omit the easy proof.

The concept of orthogonality (Def. 1.3) in a vector lattice $E$ is particularly useful because of the infinite distributivity of the lattice underlying $E$.

**1.5 Proposition** (Distributive Laws). *Suppose that $(x_\alpha)_{\alpha \in \mathsf{A}}$ and $(y_\alpha)_{\alpha \in \mathsf{A}}$ are families in $E$ for which $x := \sup_\alpha x_\alpha$ and $y := \inf_\alpha y_\alpha$ exist. Then for each $z \in E$ one has*

$$x \wedge z = \sup_\alpha (x_\alpha \wedge z),$$

$$y \vee z = \inf_\alpha (y_\alpha \vee z).$$

*Proof.* It is clear that $x \wedge z \geqq x_\alpha \wedge z$ for each $\alpha \in \mathsf{A}$. Suppose that $u \geqq x_\alpha \wedge z$ for all $\alpha$ then by (5), $u \geqq x_\alpha + z - x_\alpha \vee z$ and hence, $u + x \vee z \geqq x_\alpha + z$ for all $\alpha$. This implies $u + x \vee z \geqq x + z$ and hence, using (5) again, we obtain $u \geqq x \wedge z$ which shows that $x \wedge z = \sup_\alpha (x_\alpha \wedge z)$. The proof of the second distributive law is similar or can be reduced to the first using (2) and (5). ☐

**Corollary 1.** *Let $A$ be any subset of the vector lattice $E$. The subset $A^\perp$ orthogonal to $A$ is a vector subspace of $E$ which contains the suprema and infima (whenever they exist in $E$) of any of its subsets.*

*Proof.* Let $z \in E$ be fixed. If $x, y \in \{z\}^\perp$ and $\alpha, \beta \in \mathbb{R}$ then $|\alpha x + \beta y| \wedge |z| \leqq 2(|\alpha|\,|x| \vee |\beta|\,|y|) \wedge |z| \leqq 2(|\alpha|\,|x| \wedge |z|) \vee 2(|\beta|\,|y| \wedge |z|) = 0$, so $\{z\}^\perp$ is a vector subspace of $E$. Moreover, if $B \subset \{z\}^\perp$ and $\sup B$ (or $\inf B$) exists in $E$ it is contained in $\{z\}^\perp$ by (1.5) (in particular, $\{z\}^\perp$ is a vector sublattice of $E$). Now for any subset $A \subset E$, we have $A^\perp = \bigcap_{z \in A} \{z\}^\perp$ and the properties of $\{z\}^\perp$ are clearly intersection invariant. □

**Corollary 2.** *Every subset of E consisting of pairwise orthogonal non-zero elements, is linearly independent.*

*Proof.* Suppose $x_1, \ldots, x_n$ $(n \in \mathbb{N})$ are $\neq 0$ and pairwise orthogonal. If this finite set were linearly dependent, there would exist $i$ $(1 \leqq i \leqq n)$ and numbers $\alpha_j \in \mathbb{R}$ $(j \neq i)$ such that $x_i = \sum_{j \neq i} \alpha_j x_j$. By Cor. 1 this would imply $x_i \perp x_i$ and hence, $x_i = 0$ which is contradictory. □

*Remark.* In contrast with the distributive laws, infinite associative laws are valid in any lattice $L$. That is, if $A_\alpha$ are subsets of $L$ such that $x_\alpha := \sup A_\alpha$ exists for each $\alpha$, and if $x := \sup_\alpha x_\alpha$ exists, then $x = \sup \bigcup_\alpha A_\alpha$. Likewise for infima; the proof can be left to the reader.

Another constantly used property of vector lattices is the so-called decomposition (or interpolation) property.

**1.6 Proposition** (Decomposition Property). *If $x, y$ are positive elements of a vector lattice $E$, the relation*

(D) $$[0, x+y] = [0, x] + [0, y]$$

*holds for the corresponding order intervals. Equivalently, if $x_i$ $(i = 1, \ldots, m)$ and $y_j$ $(j = 1, \ldots, n)$ are positive elements satisfying $\sum_i x_i = \sum_j y_j$ there exist elements $z_{ij} \geqq 0$ in $E$ such that $x_i = \sum_j z_{ij}$ and $y_j = \sum_i z_{ij}$ for all $i$ and $j$.*

*Proof.* It is clear that $[0, x] + [0, y] \subset [0, x+y]$ whenever $x \geqq 0$, $y \geqq 0$. To prove the converse inclusion, let $z \in [0, x+y]$. We define $u, v$ by $u := x \wedge z$ and $v := z - u$. It remains to show that $v \in [0, y]$. But $v = z - x \wedge z = z + (-x \vee -z) = (z - x) \vee 0$ by (1) and (2) hence, $v \leqq (x + y - x) \vee 0 = y$. The second form of the decomposition property is proved by induction on $m$ and $n$; suffice it to show that it holds for $m = n = 2$. If $x_1 + x_2 = y_1 + y_2$ then $x_1 \in [0, y_1 + y_2]$ and it follows from (D) that $x_1 = z_{11} + z_{12}$ for suitable elements $z_{11} \in [0, y_1]$ and $z_{12} \in [0, y_2]$. Letting $z_{21} := y_1 - z_{11}$, $z_{22} := y_2 - z_{12}$ it is readily seen that $x_2 = z_{21} + z_{22}$, and the desired decomposition is obtained. Finally, it is clear that the second form of the decomposition property implies (D). □

**Corollary.** *If $x, y, z \in E_+$ then $(x+y) \wedge z \leqq x \wedge z + y \wedge z$.*

*Proof.* In fact, if $w := (x+y) \wedge z$ then $0 \leqq w \leqq x + y$ hence by (D), $w = w_1 + w_2$ where $w_1 \in [0, x]$ and $w_2 \in [0, y]$. But evidently $w_1 \leqq w \leqq z$ and $w_2 \leqq w \leqq z$ hence, $w_1 \leqq x \wedge z$ and $w_2 \leqq y \wedge z$ whence the assertion follows. □

As in the proof of the preceding corollary, the decomposition property is most frequently used in the form (D); that is, from $z \leqq x + y$ $(x, y, z \geqq 0)$ one concludes the existence of elements $z_1 \in [0, x]$ and $z_2 \in [0, y]$ satisfying $z = z_1 + z_2$.

We conclude this section with a brief discussion of the most elementary aspects of order convergence and order completeness. If $E$ is a vector lattice (or, more generally, an ordered set) and $B$ is a directed ($\leqq$) subset of $E$ for which $x_0 := \sup B$ exists, then the filter in $E$ generated by the section filter of $B$ contains the family of order intervals $([x, x_0])_{x \in B}$, and $\bigcap_{x \in B} [x, x_0] = \{x_0\}$. This motivates the following definition.

**1.7 Definition.** *Let $S$ be an ordered set.*

*A filter $F$ on $S$* order converges *to $x \in S$ if $F$ contains a family of order intervals with intersection $\{x\}$. A sequence in $S$* order converges *to $x \in S$ if its section filter contains a sequence of order intervals with intersection $\{x\}$.*

*A subset $S_0$ of $S$ is called* order dense *in $S$ if for each $x \in S$, there exists a filter on $S$ possessing a base in $S_0$ and order convergent to $x$.*

Thus the prime examples of order convergent filters are the section filters of upward (downward) directed families $(z_\alpha)_{\alpha \in \mathsf{A}}$ with supremum (infimum) $z$; for this type of order convergence, the notations $z_\alpha \uparrow z$ and $z_\alpha \downarrow z$ are used frequently in the literature.

If $E$ is a vector lattice and $F$ is a filter on $E$, then $F$ is order convergent to $x \in E$ iff there exists a family $(z_\alpha)_{\alpha \in \mathsf{A}}$ in $E$ such that $z_\alpha \downarrow 0$ and the sets $F_\alpha := \{z \in E : |z - x| \leqq z_\alpha\}$ $(\alpha \in \mathsf{A})$ belong to $F$. Similarly, a sequence $(x_n)_{n \in \mathbb{N}}$ in $E$ is order convergent to $x \in E$ iff there exists a sequence $z_n \downarrow 0$ such that $|x_m - x| \leqq z_n$ for all $m \geqq n$ and $n \in \mathbb{N}$ (cf. Exerc. 2). However, it should be noted that the topological associations evoked by the above terminology are only partly justified (Exerc. 2).

To verify the order convergence of a given sequence or filter, it is often necessary to form suprema and infima of certain order bounded sets; the existence of these suprema and infima is, however, not implied by the vector lattice axioms introduced so far (Def. 1.2). Therefore, we consider several additional axioms in increasing strength. The weakest of these, the axiom of Archimedes, amounts to requiring that for each positive element $x$ of a vector lattice $E$ the sequence $(n^{-1} x)_{n \in \mathbb{N}}$ have infimum 0.

**1.8 Definition.** *Let $E$ be a vector lattice. Consider these axioms:*

(A) *$x, y \in E$ and $nx \leqq y$ for all $n \in \mathbb{N}$ implies $x \leqq 0$.*

(OS) *$0 \leqq x_n \leqq \lambda_n x$, where $x_n, x \in E$ and $(\lambda_n) \in l^1$, implies the order convergence of $\sum_1^\infty x_n$.*[2]

(COC) *For each non-void countable majorized set $B \subset E$, $\sup B$ exists in $E$.*

(OC) *For each non-void majorized set $B \subset E$, $\sup B$ exists in $E$.*

*Then $E$ is said to be* Archimedean *if* (A) *is satisfied,* $l^1$-relatively complete *if* (OS) *is satisfied,* countably order complete *if* (COC) *is satisfied, and* order complete *if* (OC) *is satisfied.*

The validity of Axioms (COC) and (OC) is frequently expressed by saying that $E$ is Dedekind $\sigma$-complete and Dedekind complete, respectively. Each of the four axioms listed in Def. 1.8 implies the preceding, but not conversely

---

[2] I. e., the existence of $\sup_n \sum_{\nu=1}^n x_\nu$. Vector lattices satisfying (OS) are often called "relatively uniformly complete" (cf. 7.2).

(Exerc. 3); moreover, (A) is inherited by vector sublattices but none of the remaining three is. In some instances, reverse implications hold; for example, (COC) implies (OC) whenever each majorized orthogonal system in $E$ is countable (Exerc. 4). It is also clear from Formula (1) that the validity of (OC) (respectively, (COC)) implies the existence of $\inf B$ whenever $B \neq \emptyset$ is a minorized (respectively, countable minorized) subset of $E$.

**Examples.** 7. Let $\alpha$ be a fixed ordinal $>0$ and consider the vector space $E$ of all families $(x_\beta)_{\beta<\alpha}$ of real numbers indexed by the set of all ordinals $<\alpha$. Under the ordering defined by "$x \leqq y$ iff $x_\beta < y_\beta$ for the first ordinal $\beta < \alpha$ for which $x_\beta \neq y_\beta$" (lexicographic ordering) $E$ is a vector lattice which is not Archimedean unless $\alpha = 1$. Note that $E$ is totally ordered; it will be seen below (§ 3) that every totally ordered Archimedean vector lattice has linear dimension 0 or 1.

8. Let $K$ be a compact space and denote by $C(K)$ the vector lattice of all real-valued continuous functions on $K$ under its canonical ordering (Example 3). $C(K)$ satisfies (OS); every vector sublattice of $C(K)$ is Archimedean but not every vector sublattice satisfies (OS). (Consider the vector lattice of all piecewise linear, continuous real functions on $[0,1]$.) $C(K)$ satisfies (OC) (respectively, (COC)) iff the closure of every open subset (respectively, of every open $F_\sigma$) is open in $K$. For proofs, see § 7.

Let $E$ be a vector lattice. It is clear from Zorn's lemma that every orthogonal system of $E$ (Def. 1.3) is contained in a maximal orthogonal system. If $E_+$ contains an element $u$ such that the singleton $\{u\}$ is a maximal orthogonal system, then $u$ is called a *weak order unit (Freudenthal unit)* of $E$. However, these concepts are useful only in connection with the Archimedean axiom (A); here is an application.

**1.9 Proposition.** *Suppose $E$ to be Archimedean and let $S$ denote any maximal orthogonal system of $E$. For each $x \in E_+$ the directed family $(x_{n,\mathsf{H}})$, where $x_{n,\mathsf{H}} := \sum_{u \in \mathsf{H}} (x \wedge nu)$ for $n \in \mathbb{N}$ and $\mathsf{H}$ any finite subset of $S$, order converges to $x$; that is,*

$$x = \sup_{n,\mathsf{H}} x_{n,\mathsf{H}} \qquad (x \in E_+).$$

*Proof.* It is clear that $x \geqq x_{n,\mathsf{H}}$ for each $n \in \mathbb{N}$ and each finite $\mathsf{H} \subset S$. Suppose that $z \geqq x_{n,\mathsf{H}}$ for all $n$ and $\mathsf{H}$; we have to show that $z \geqq x$. Let $u \in S$ be fixed. Then $0 \leqq z - (x \wedge nu) = (z-x) \vee (z-nu)$ which implies $0 = [(z-x) \vee (z-nu)] \wedge 0 = (z-x)^- \wedge (z-nu)^-$ by (2) and distributivity (1.5), and so $(z-x)^- \wedge (u - n^{-1} z)^+ = 0$ for all $n \in \mathbb{N}$. Now $\sup_n (u - n^{-1} z) = u$, since $E$ is Archimedean and hence, $u$ being positive, $\sup_n (u - n^{-1} z)^+ = u$. Using the distributive law again we obtain

$$0 = \sup_n [(z-x)^- \wedge (u - n^{-1} z)^+] = (z-x)^- \wedge u .$$

Since $u \in S$ was arbitrary and $S$ is maximal, it follows that $(z-x)^- = 0$ or, equivalently, that $z \geqq x$. □

**Corollary.** *If $E$ is an Archimedean vector lattice containing a weak order unit $u$, then $x = \sup_n (x \wedge nu)$ for all $x \in E_+$.*

The classical method due to Dedekind of constructing the real numbers from the rationals is completion by cuts. This method[3] can be adapted to vector lattices $E$ if (and only if) $E$ is Archimedean; the procedure is, roughly, as follows. Define an ordered pair $(A', A'')$ of non-void subsets of $E$ to be a *cut* in $E$ if $A'$ is the set of all lower bounds of $A''$ and $A''$ is the set of all upper bounds of $A'$. The set $\overline{E}$ of all cuts in $E$, ordered by "$(A', A'') \leqq (B', B'')$ *iff* $A' \subset B'$", turns out to be a lattice in which each majorized subset $\neq \emptyset$ has a supremum and each minorized subset $\neq \emptyset$ has an infimum. Moreover, the mapping $x \mapsto (A'_x, A''_x)$, where $A'_x := \{y \in E: y \leqq x\}$ and $A''_x := \{y \in E: y \geqq x\}$, is an isomorphism of the lattice $E$ onto a sublattice $E_0$ of $\overline{E}$ such that for each $\overline{x} \in \overline{E}$, one has $\overline{x} = \sup\{z \in E_0 : z \leqq \overline{x}\} = \inf\{z \in E_0 : z \geqq \overline{x}\}$. The given vector structure of $E$ then suggests the definition of addition and multiplication by real scalars in $\overline{E}$ (define $(A', A'') + (B', B'') = (C', C'')$ so that $C''$ is the set of all upper bounds of $A' + B'$ in $E$); the Archimedean axiom (A) is then essentially needed to prove that $\overline{E}$ is an additive group. We summarize this in the following proposition.

**1.10 Proposition** (Dedekind Completion). *For every Archimedean vector lattice $E$, there exists an order complete vector lattice $\overline{E}$ containing $E$ as a vector sublattice and such that*

$$\overline{x} = \sup\{x \in E : x \leqq \overline{x}\} = \inf\{x \in E : x \geqq \overline{x}\}$$

*holds for each $\overline{x} \in \overline{E}$. By these properties, $\overline{E}$ is determined uniquely to within isomorphism.*

It follows, in particular, that $E$ is imbedded in its Dedekind completion with preservation of arbitrary infima and suprema. We shall, however, rarely need this result, leaving its detailed verification to the interested reader (cf. Vulikh [1967], Luxemburg-Zaanen [1971]).

## § 2. Ideals, Bands, and Projections

A subset $A$ of an ordered vector space $E$ is called *saturated* (with respect to the ordering or the positive cone of $E$) whenever $x, y \in A$ implies the order interval $[x, y]$ to be contained in $A$ (cf. [S, V. 3]). When $E$ is a vector lattice, the following stronger notion is frequently more useful.

**2.1 Definition.** *A subset $A$ of a vector lattice $E$ is called* solid *if $x \in A$, $y \in E$, and $|y| \leqq |x|$ implies $y \in A$. A solid vector subspace $I$ of $E$ is called an* ideal *(or* lattice ideal*) of $E$.*

It is readily verified that each ideal is a vector sublattice of $E$; conversely, each saturated vector sublattice is an ideal. The concepts "solid" and "ideal" are clearly intersection invariant; thus each subset $B \subset E$ is contained in a

[3] Completion by cuts serves to imbed any ordered set in a complete lattice with preservation of infima and suprema (cf. Birkhoff [1967]).

smallest solid subset $S(B)$ of $E$ containing $B$, called the *solid hull* or *solid cover* of $B$, and in a smallest ideal $I(B)$ of $E$ containing $B$, called the *ideal generated by* $B$.

The ideal of $E$ generated by a singleton $\{u\}$ (or less precisely, the ideal generated by $u \in E$) is called a *principal ideal* and denoted by $E_u$; obviously one can assume $u \in E_+$. An element $u \in E_+$ is called an *order unit* (or *strong order unit*) of $E$ if $E = E_u$.

**Example.** 1. Let $E$ be any vector lattice. If $x \in E_+$ then the symmetric order interval $[-x, x]$ is solid, and the vector subspace $\bigcup_1^\infty n[-x,x]$ is the principal ideal $E_x$. More generally, if $A$ is a directed ($\leq$) subset of $E_+$ then $\bigcup \{n[-x,x]: n \in \mathbb{N}, x \in A\}$ is the ideal generated by $A$. Any union of solid sets is solid, and $B \subset E$ is solid iff $B = \bigcup \{[-|x|, |x|]: x \in B\}$.

If $(E_\alpha)_{\alpha \in \mathrm{A}}$ is a family of vector lattices, each $E_\alpha$ can be identified with an ideal of $\prod_\alpha E_\alpha$ and of $\bigoplus_\alpha E_\alpha$ (§ 1, Example 6).

As is the case with the concepts "convex" and "circled" in vector spaces (recall that a subset $A$ of a real or complex vector space is called *circled* if $|\lambda| \leq 1$ implies $\lambda A \subset A$), the mapping $B \mapsto S(B)$ has certain typical properties; we mention monotoneity (with respect to set inclusion) and idempotency. Moreover, the following simple properties of solid sets are often needed.

**2.2 Proposition.** *Let $E$ be any vector lattice. For any $A \subset E$, we have $S(A) = \{y \in E$: there exists $x \in A$ such that $|y| \leq |x|\}$. If $A$ is solid then $A$ is circled, and the convex hull of $A$ is solid. Any finite sum $\sum_i A_i = \{\sum_i x_i: x_i \in A_i\}$ of solid sets $A_i$ $(i = 1, \ldots, n)$ is solid.*

*Proof.* If $A \subset E$, $A \subset B$ and $B$ is solid then clearly the set of all $y \in E$ such that $|y| \leq |x|$ for some $x \in A$, is contained in $B$; to see that this set is solid, note that if $z \in E$ and $|z| \leq |y|$ for some such $y$ then $|z| \leq |x|$ for some $x \in A$. The circledness of solid sets follows from (1.4), Formula (4). Next recall that the convex hull of $A \subset E$ is the set of all convex combinations $\sum_1^n \lambda_i a_i$ where $n \in \mathbb{N}$, $a_i \in A$, and where the numbers $\lambda_i > 0$ satisfy $\sum_1^n \lambda_i = 1$. Suppose $A$ is solid; if $y \in E$ and $|y| \leq |\sum_1^n \lambda_i a_i|$ then $y^+ \leq \sum_1^n \lambda_i |a_i|$ and $y^- \leq \sum_1^n \lambda_i |a_i|$. The decomposition property (1.6) implies that $y^+ = \sum_1^n \lambda_i b_i$, $y^- = \sum_1^n \lambda_i c_i$ for suitable elements $b_i \in [0, |a_i|]$ and $c_i \in [0, |a_i|]$ $(i = 1, \ldots, n)$. Since $A$ is solid, it follows that $b_i - c_i \in [-|a_i|, |a_i|] \subset A$ for all $i$; hence, $y = y^+ - y^- = \sum_1^n \lambda_i (b_i - c_i)$ is in the convex hull $\operatorname{co} A$ of $A$. Therefore, $\operatorname{co} A$ is solid.

The final assertion is an immediate consequence of (1.6). ☐

For any vector lattice $E$, let us denote by $\boldsymbol{I}(E)$ the set of all ideals of $E$. If $I, J \in \boldsymbol{I}(E)$, it is clear that $I \cap J$ is an ideal and it follows from (2.2) that $I + J \in \boldsymbol{I}(E)$. Thus under set inclusion $\boldsymbol{I}(E)$ is a lattice, the lattice operations being defined by $I \vee J := I + J$ and $I \wedge J := I \cap J$. It is also clear that $\boldsymbol{I}(E)$ is a complete lattice (with largest element $E$ and smallest element $\{0\}$) and, in fact, $\boldsymbol{I}(E)$ shares this property with the set of all vector subspaces of $E$; however, it is of great importance for the ideal theory of vector lattices (cf. Chap. III) that the lattice $\boldsymbol{I}(E)$ is distributive.

**2.3 Proposition.** *For any vector lattice $E$, the lattice $\boldsymbol{I}(E)$ is distributive.*

*Proof.* We have to prove the validity of Axiom (LD) (Def. 1.1); that is, for $I, J, K \in \boldsymbol{I}(E)$ we have to prove that $(I+J)\cap K=(I\cap K)+(J\cap K)$. The inclusion of the right hand member in the left is trivial. To prove the reverse inclusion, let $z\in(I+J)\cap K$, then $z=x+y$ where $x\in I$, $y\in J$. This implies $|z|\leqq|x|+|y|$ and hence, by (1.6), $|z|=u+v$ where $u\in[0,|x|]$, $v\in[0,|y|]$. Clearly then, $u\in I\cap K$ and $v\in J\cap K$. Thus $|z|\in I\cap K+J\cap K$ and hence, $z\in I\cap K+J\cap K$, since the latter sum is an ideal.

Since we have not proved the equivalence of Axioms (LD) and (LD)$'$, let us indicate that $(I\cap J)+K=(I+K)\cap(J+K)$ is valid also. Again, the inclusion $\subset$ is trivial. For the reverse inclusion let $z\in(I+K)\cap(J+K)$ then $z=x+z_1=y+z_2$ where $x\in I, y\in J$, and $z_1, z_2\in K$. We obtain $|z|\leqq|x|+|z_1|\vee|z_2|, |z|\leqq|y|+|z_1|\vee|z_2|$ which shows that $|z|\leqq|x|\wedge|y|+|z_1|\vee|z_2|$. Using (1.6) we have $|z|=u+v$ where $0\leqq u\leqq|x|\wedge|y|$, $0\leqq v\leqq|z_1|\vee|z_2|$. Thus $u\in I\cap J$ and $v\in K$ whence it follows that $|z|$, and therefore $z$, is an element of the ideal $(I\cap J)+K$. □

We turn to the discussion of linear maps.

**2.4 Definition.** *Let $E, F$ be ordered vector spaces and let $T: E\to F$ be a linear map (operator).*

*$T$ is called* positive *(in symbols: $T\geqq 0$) if $Tx\geqq 0$ for all $x\geqq 0$; $T$ is called* strictly positive *(in symbols: $T\gg 0$) if $Tx>0$ for all $x>0$.*

*If $E, F$ are vector lattices, $T$ is called a* lattice homomorphism (homomorphism of vector lattices) *if $T(x\vee y)=Tx\vee Ty$ and $T(x\wedge y)=Tx\wedge Ty$ for all $x, y\in E$.*

*$T$ is called* order continuous *if for each filter $\mathbf{F}$ on $E$ that order converges to $x\in E$, the filter with base $T(\mathbf{F})$ order converges to $Tx$ in $F$. $T$ is called* sequentially order continuous *if for each sequence $(x_n)$ in $E$ with order limit $x$, $(Tx_n)$ order converges to $Tx$ in $F$.*

If $E, F$ are ordered vector spaces, the set $K\subset L(E,F)$ of all positive linear maps satisfies $K+K\subset K$ and $\lambda K\subset K$ for all $\lambda\geqq 0$ (such a set is sometimes called a *wedge*). If $K\cap -K=\{0\}$, then $K$ is the positive cone for an ordering called the *canonical ordering* of $L(E,F)$ [S, Chap. V, § 5]; if $F\neq\{0\}$, $K\cap -K=\{0\}$ is equivalent to $E=E_+-E_+$ and holds, in particular, whenever $E$ is a vector lattice. Let us mention the following simple fact, often used without further comment: If $E, F$ are ordered vector spaces and $t: E_+\to F_+$ is an additive, positive homogeneous map so that $t(x+y)=tx+ty$ and $t(\lambda x)=\lambda tx$ for all $x, y\in E_+$ and $\lambda\geqq 0$, there exists a unique positive linear map $T:(E_+-E_+)\to F$ extending $t$. Indeed, if for $z=x-y$ $(x, y\in E_+)$ we define $Tz:=tx-ty$, then by the additivity of $t$ the value $Tz$ is independent of the particular representation of $z$ as the difference of positive elements, and it is equally straightforward to verify that $T$ is linear and positive.

Now suppose $E, F$ to be vector lattices. We observe that every lattice homomorphism $T: E\to F$ is positive, since for all $x\in E_+$ we have $Tx=Tx^+=(Tx)^+\geqq 0$. Moreover, if $T: E\to F$ is positive, then $|Tx|\leqq T|x|$ for all $x\in E$; in fact, $\pm x\leqq|x|$ implies $\pm Tx\leqq T|x|$.

**2.5 Proposition.** *Let $E, F$ be vector lattices. For any linear map $T: E \to F$, the following assertions are equivalent:*

(a) *$T$ is a lattice homomorphism.*

(b) *$|Tx| = T|x|$ for all $x \in E$.*

(c) *$Tx^+ \wedge Tx^- = 0$ for all $x \in E$.*

*Moreover, if $T$ is a lattice homomorphism mapping $E$ onto an ideal of $F$, then $T(A)$ is solid in $F$ for each solid subset $A$ of $E$.*

*Proof.* (a) $\Rightarrow$ (c): $Tx^+ \wedge Tx^- = T(x^+ \wedge x^-) = 0$, since $x^+ \wedge x^- = 0$ for all $x \in E$.

(c) $\Rightarrow$ (b): From $x = x^+ - x^-$ we have $Tx = Tx^+ - Tx^-$, where $Tx^+ \wedge Tx^- = 0$ by hypothesis. By (1.4), $Tx = (Tx)^+ - (Tx)^-$ is the unique representation of $Tx$ as the difference of orthogonal elements of $F_+$ and hence, $(Tx)^+ = Tx^+$ and $(Tx)^- = Tx^-$ for all $x \in E$. Therefore, $|Tx| = Tx^+ + Tx^- = T|x|$.

(b) $\Rightarrow$ (a): If $x \geqq 0$ then $x = |x|$ so $Tx = |Tx| \geqq 0$ and hence, $T \geqq 0$. This implies $(Tx)^+ \leqq Tx^+$ and $(Tx)^- \leqq Tx^-$; since $(Tx)^+ + (Tx)^- = Tx^+ + Tx^-$ by hypothesis, it follows that $(Tx)^+ = Tx^+$ and $(Tx)^- = Tx^-$ for all $x \in E$. So if $x, y \in E$ then $T(x \vee y) = Tx + T(y-x)^+ = Tx + (T(y-x))^+ = Tx \vee Ty$ and, similarly, $T(x \wedge y) = Tx \wedge Ty$.

To prove the last assertion, it suffices to show that $T(A)$ is solid in $F$ whenever $A$ is solid and $T$ is a surjective lattice homomorphism. We have to show that $x \in A$, $y \in E$ and $|Ty| \leqq |Tx|$ implies $Ty \in T(A)$. In fact, we have $(Ty)^+ = Ty^+ \leqq T|x|$ and $(Ty)^- = Ty^- \leqq T|x|$. It follows that $Ty^+ = T|x| \wedge Ty^+ = T(|x| \wedge y^+)$ and $Ty^- = T|x| \wedge Ty^- = T(|x| \wedge y^-)$; but $|x| \wedge y^+ \in A$ and $|x| \wedge y^- \in A$, since $A$ is solid, and so $z := |x| \wedge y^+ - |x| \wedge y^-$ is in $A$ because of $|z| \leqq |x| \wedge |y| \in A$. Therefore, $Ty = Tz \in T(A)$. ▯

If $E, F$ are vector lattices and $T: E \to F$ is a positive linear map, the set $\{x \in E: T|x| = 0\}$ is an ideal of $E$ called the *absolute kernel* (or the *null ideal*) of $T$. If $T$ is a lattice homomorphism then the kernel $T^{-1}(0)$ agrees with the absolute kernel of $T$ and hence is an ideal of $E$; conversely, each ideal of $E$ is the kernel of a lattice homomorphism, as the following result shows.

**2.6 Proposition.** *Let $E$ be a vector lattice, let $I$ be an ideal of $E$, and denote by $q$ the canonical map of $E$ onto the quotient vector space $E/I$. Under the finest ordering of $E/I$ for which $q$ is positive, $E/I$ is a vector lattice and $q$ a lattice homomorphism of $E$ onto $E/I$.*

*Proof.* The finest ordering of $E/I$ for which $q$ is positive, is obviously given by the relation "$q(x) \leqq q(y)$ iff there exist elements $x_1 \in x + I$ and $y_1 \in y + I$ satisfying $x_1 \leqq y_1$ in $E$". Evidently, the binary relation $\leqq$ thus defined on $E/I$ is reflexive and transitive; since $q$ is linear, it also satisfies Axioms $(\mathrm{LO})_1$ and $(\mathrm{LO})_2$ of Def. 1.1. To show the anti-symmetry of this preordering of $E/I$ (or equivalently, in view of $(\mathrm{LO})_1$, to prove that $q(E_+) \cap -q(E_+) = \{0\}$), we need that $I$ is an ideal. In fact, if $q(x) \geqq 0$ and $q(x) \leqq 0$ for some $x \in E$, there exist elements $x_1, x_2 \in E$ satisfying $0 \leqq x_1 \in x + I$ and $0 \leqq x_2 \in -x + I$. This implies $x_1 + x_2 \in I$ and hence $x_1 \in I$ (and $x_2 \in I$) and so, $q(x) = 0$.

To show that $E/I$ is a vector lattice and $q$ a lattice homomorphism, we prove that for given $x, y \in E$, $q(x \vee y)$ is the least upper bound of $\{q(x), q(y)\}$ in $E/I$.

Clearly, $q(x \vee y)$ is a majorant of $q(x)$ and $q(y)$, since $q$ is positive. Let $z \in E$ satisfy $q(z) \geqq q(x)$ and $q(z) \geqq q(y)$; there exist elements $z_1 \in z+I$ and $z_2 \in z+I$ such that $z_1 \geqq x$ and $z_2 \geqq y$. Now $z_1 - z_2 \in I$ and, since $I$ is a vector sublattice of $E$, $|z_1 - z_2| \in I$. Then $w := z_2 + |z_1 - z_2| \in z + I$ while, obviously, $w \geqq x \vee y$; it follows that $q(z) = q(w) \geqq q(x \vee y)$. Therefore, $q(x \vee y) = q(x) \vee q(y)$. The relation $q(x \wedge y) = q(x) \wedge q(y)$ now follows from (5) of (1.4), since $q$ is linear (cf. Remark after 1.4). □

**Corollary.** *If $E, F$ are vector lattices and $T: E \to F$ is a linear map satisfying $T(E_+) = F_+$, the following assertions are equivalent:*

(a) *$T$ is a lattice homomorphism.*

(b) *$T^{-1}(B)$ is solid for each solid set $B \subset F$.*

(c) *$T^{-1}(0)$ is an ideal in $E$.*

*Proof.* (a) ⇒ (b): If $x \in T^{-1}(B)$ and $y \in E$, $|y| \leqq |x|$ then $|Ty| = T|y| \leqq T|x| \in B$ hence $Ty \in B$ and $y \in T^{-1}(B)$.

(b) ⇒ (c) is trivial. (c) ⇒ (a): Let $I := T^{-1}(0)$ and let $q$ denote the quotient map $E \to E/I$. The bijection $T_0$ of $E/I$ onto $F$ defined by $T_0 \circ q = T$ is an isomorphism of ordered vector spaces and hence, of vector lattices. Since $q$ is a lattice homomorphism, so is $T$. □

If $I$ is an ideal of the vector lattice $E$, the order of $E/I$ defined in (2.6) is called *canonical*, and $q$ is called the *canonical lattice homomorphism* of $E$ onto $E/I$. We point out that $E/I$ does not inherit from $E$ any of the properties listed in Def. 1.8 (Exerc. 3) and that, in general, $q$ is not order continuous (see Example 3).

**Examples.** 2. Let $E$ be a vector lattice of real functions on a non-void set $X$ under its canonical order (§ 1, Example 3), and let $F := \mathbb{R}^{X_0}$ (canonical order) where $X_0$ is a non-void subset of $X$. The restriction map $f \mapsto f_0$, where $f \in E$ and $f_0 := f \mid X_0$, is a lattice homomorphism of $E$ into $F$ and, if $E = \mathbb{R}^X$, evidently order continuous. However, if $X$ is a normal topological space, $X_0$ a closed nowhere dense subset of $X$, and if $E$ is the vector lattice of all continuous real functions on $X$, then $f \mapsto f_0$ is not order continuous. (Consider the family of all $f \in E_+$ satisfying $f(X_0) = \{1\}$; this family is directed downward and order converges to 0 in $E$.) Thus, the restriction of an order continuous lattice homomorphism $E \to F$ to a vector sublattice of $E$ need not be order continuous.

3. We use the notation of the preceding example. Letting $I := \{f \in E: f(X_0) = \{0\}\}$, $I$ is an ideal in $E$ and $E/I$ can be identified with the vector sublattice of $F$ whose elements are restrictions to $X_0$ of functions $f \in E$. Under this identification, the canonical map $E \to E/I$ becomes the restriction map $f \mapsto f_0$. From Example 2 it follows now that $q: E \to E/I$ is not necessarily order continuous. Moreover, it occurs that $E$ is order complete while for certain ideals $I$, $E/I$ is not even Archimedean (Exerc. 3).

In (2.3) above it was shown that the lattice of ideals $\boldsymbol{I}(E)$ is distributive; however, in general $\boldsymbol{I}(E)$ is not complemented (Def. 1.1), that is, not a Boolean algebra (cf. Example 5 below). We turn to the investigation of the set of all complemented ideals of $E$.

**2.7 Proposition.** *Let $E$ be any vector lattice, and suppose $I$, $J$. to be complementary ideals of $E$ (that is, $I \cap J = \{0\}$ and $I + J = E$). The projection $E \to I$ with kernel $J$ is positive, and $I = J^\perp$ (hence, $I = I^{\perp\perp}$).*

*Proof.* By assumption, $E = I + J$ is a direct sum of vector spaces. Let $x = y + z$ denote the corresponding decomposition of $x \in E$. If $x \geqq 0$ we have $0 \leqq y^+ + z^+ - (y^- + z^-)$ or, equivalently, $0 \leqq y^- + z^- \leqq y^+ + z^+$; considering that $y^- \perp y^+ + z^+$ (because of $y^- \wedge y^+ = 0$ and $y^- \wedge z^+ \in I \cap J = \{0\}$) we obtain $y^- = 0$; that is, the projection $x \mapsto y$ is positive and, by symmetry, $x \mapsto z$ is positive.

From $I \cap J = \{0\}$ it is clear that $I \subset J^\perp$; let us show that $J^\perp \subset I$. In fact, if $x \in J^\perp$ then $|x| \in J^\perp$ and we have $|x| = u + v$, where $u \in I_+$ and $v \in J_+$ by the preceding. This implies $v \leqq |x| \in J^\perp$ and hence, $v \in J \cap J^\perp = \{0\}$. So $|x| = u \in I$ and hence, $x \in I$. Therefore, $I = J^\perp$; by symmetry we obtain $J = I^\perp$ and consequently, $I = I^{\perp\perp}$. □

Thus if $I$ is a complemented ideal of $E$, we have $I = I^{\perp\perp}$ and from (1.5) Cor. 1 it follows that $I$ is closed under the formation of arbitrary suprema and infima; this motivates the following definition.

**2.8 Definition.** *Let $E$ be any vector lattice. An ideal $I$ of $E$ is called a* band *if $A \subset I$ and $\sup A = x \in E$ implies $x \in I$. A complemented ideal $I$ of $E$ (that is, an ideal satisfying $I + I^\perp = E$) is called a* projection band; *the associated projection $E \to I$ with kernel $I^\perp$ is called a* band projection.

Since for each $A \subset E$, $A^\perp$ is clearly solid, it follows from (1.5) Cor. 1 that $A^\perp$ is a band. If $E$ is Archimedean, then every band $B$ is of this form, namely, $B = B^{\perp\perp}$; moreover, the property that $B \subset E$ is a band iff $B = B^{\perp\perp}$, characterizes Archimedean vector lattices (Exerc. 3).

A notion intermediate between those of ideal and band is that of a $\sigma$-ideal; an ideal $I \in \boldsymbol{I}(E)$ is called a *$\sigma$-ideal* if for each countable set $A \subset I$, $\sup A = x \in E$ implies $x \in I$.

For any vector lattice $E$, we denote by $\boldsymbol{B}(E)$ and $\boldsymbol{P}(E)$ the set of all projection bands and band projections of $E$, respectively; $1_E$ denotes the identity mapping of $E$.

**2.9 Theorem.** *Let $E$ be any vector lattice. The set $\boldsymbol{B}(E)$ is a Boolean algebra and a sublattice of $\boldsymbol{I}(E)$.*

*An idempotent endomorphism $P$ of the vector space $E$ is a band projection if and only if $0 \leqq P \leqq 1_E$; every band projection is an order continuous lattice homomorphism.*

*$\boldsymbol{P}(E)$ is a Boolean algebra under the operations*

$$P \vee Q := P + Q - PQ$$

$$P \wedge Q := PQ$$

*and isomorphic with $\boldsymbol{B}(E)$ under the mapping $P \mapsto PE$. In particular, every pair of band projections commutes.*

*Proof*. If $A, B \in \boldsymbol{B}(E)$ then $E = A + A^{\perp} = B + B^{\perp}$ (Def. 2.8); from $E = (A + A^{\perp}) \cap (B + B^{\perp})$ and (2.3) we obtain

$$(\star) \qquad E = A \cap B + (A \cap B^{\perp} + A^{\perp} \cap B + A^{\perp} \cap B^{\perp}),$$

$$(\star\star) \qquad E = (A \cap B + A \cap B^{\perp} + A^{\perp} \cap B) + A^{\perp} \cap B^{\perp}.$$

From ($\star$) we conclude, since the term in parentheses denotes an ideal orthogonal to $A \cap B$, that $A \cap B \in \boldsymbol{B}(E)$ (cf. 2.7). On the other hand, from $A = A \cap (B + B^{\perp}) = A \cap B + A \cap B^{\perp}$ and $B = (A + A^{\perp}) \cap B = A \cap B + A^{\perp} \cap B$ we see that the term in parentheses in ($\star\star$) denotes the ideal $A + B$, which is evidently orthogonal to $A^{\perp} \cap B^{\perp}$; therefore, $A + B \in \boldsymbol{B}(E)$. This shows $\boldsymbol{B}(E)$ to be a sublattice of $\boldsymbol{I}(E)$ and, obviously, to be a Boolean algebra.

If $A \in \boldsymbol{B}(E)$ and $P_A$ is the associated band projection (with kernel $A^{\perp}$) then from (2.7) we see that $P_A \geqq 0$ and $1_E - P_A \geqq 0$, so that $0 \leqq P_A \leqq 1_E$ (in the vector ordering of $L(E, E)$, see above). Conversely, let $P \in L(E, E)$ satisfy $P^2 = P$ and $0 \leqq P \leqq 1_E$. Then for $x \in E$ we have $0 \leqq P x^{+} \leqq x^{+}$ and $0 \leqq P x^{-} \leqq x^{-}$ and hence, $0 \leqq P x^{+} \wedge P x^{-} \leqq x^{+} \wedge x^{-} = 0$ which shows $P$ to be a lattice homomorphism by (2.5) (c). Thus $P^{-1}(0)$ is an ideal and similarly $(1_E - P)^{-1}(0)$ is an ideal; so by (2.7) $P$ is a band projection. Moreover, if $P$ is a band projection and $x = x_1 + x_2$, $y = y_1 + y_2$ are the corresponding decompositions of $x, y \in E$ into components in $PE$ and $P^{-1}(0)$, respectively, we have $[x, y] = [x_1, y_1] + [x_2, y_2]$ for the order interval with endpoints $x$, $y$; this shows $P$ to be order continuous (Def. 1.7, 2.4).

It is clear that $P \mapsto PE$ is a bijection of $\boldsymbol{P}(E)$ onto $\boldsymbol{B}(E)$; for the remaining assertions it suffices to show that, if again by $P_A$ we denote the band projection with range $A \in \boldsymbol{B}(E)$, $P_A \wedge P_B$ and $P_A \vee P_B$ are band projections with ranges $A \cap B$ and $A + B$, respectively. Since $P_A P_B$ vanishes on the ideal $A \cap B^{\perp} + A^{\perp} \cap B + A^{\perp} \cap B^{\perp}$ and leaves each element of $A \cap B$ fixed, it follows from ($\star$) that $P_A P_B$ is the band projection with range $A \cap B$; since the same reasoning applies to $P_B P_A$, we have $P_A P_B = P_B P_A$. Therefore, $P_A \vee P_B$ is a projection of $E$; since it is easy to see that $P_A \vee P_B$ leaves each element of $A + B = (A \cap B + A \cap B^{\perp} + A^{\perp} \cap B)$ fixed and vanishes on $A^{\perp} \cap B^{\perp}$, it follows from ($\star\star$) that $P_A \vee P_B$ is the band projection with range $A + B$. ∎

In general, not every band in a vector lattice $E$ is a projection band, even if $E$ is countably order complete (Exerc. 3); however, in order complete vector lattices the notions of band and projection band are coextensive *(Riesz decomposition theorem)*.

**2.10 Theorem.** *For any subset $A$ of an order complete vector lattice $E$, $E$ is the direct sum of the band generated by $A$ and of the band $A^{\perp}$. In particular, each band of $E$ is a projection band.*

*Proof*. Denote by $B_A$ the band generated by $A$. Then $B_A$ and $A^{\perp}$ are orthogonal bands, since $A \subset A^{\perp\perp}$ and $A^{\perp\perp}$ is a band. Since $E$ is order complete, $y := \sup [0, x] \cap B_A$ exists and is in $B_A$ for any given $x \in E_+$. Let $z := x - y$; the proof will be complete if we show that $z \in A^{\perp}$. Suppose $u \in A$ is arbitrary and let $w := z \wedge |u|$. Since $w \in B_A$ and $w + y \leqq z + y = x$, it follows from the definition of $y$ that $w + y \leqq y$ and hence, that $w = 0$. ∎

**Examples.** 4. Let $X$ denote an infinite set, and let $E$ be a vector sublattice of $\mathbb{R}^X$ containing the constant-one function and each function with countable support. If $\boldsymbol{F}$ is a filter on $X$ with empty intersection (for example, the filter of all subsets of $X$ having finite complement), then the set $\{f \in E$: there exists $F \in \boldsymbol{F}$ with $f(F)=\{0\}\}$ is an ideal in $E$ which is not a band. (If $X$ is uncountable and $\boldsymbol{F}$ the filter of all subsets of $X$ with countable complement, this construction yields a $\sigma$-ideal which is not a band.)

5. Let $X=[0,1]$ be the real unit interval under its standard topology, and let $E:=C(X)$ denote the vector lattice (canonical order) of all real valued continuous functions on $X$. For each real number $a$, $0<a\leqq 1$, the ideal $B_a:=\{f\in E: f(t)=0 \text{ for all } t\geqq a\}$ is a band of $E$ but not a projection band. In fact, if $B\in\boldsymbol{B}(E)$ and $e=e_1+e_2$ is the decomposition of the constant-one function $e$ according to $E=B+B^\perp$, then $e_i$ are continuous functions on $X$ only capable of the values 0 and 1 and hence, the characteristic functions of a pair of complementary open-and-closed subsets of $X$. Since $X$ is connected, it follows that $e_1=0$ or $e_1=e$; that is, we have $\boldsymbol{P}(E)=\{0,1_E\}$. (Cf. § 7.)

The proof of the Riesz theorem (2.10) suggests a characterization of those bands in a vector lattice $E$ that are projection bands.

**2.11 Proposition.** *If $A$ is any subset of the vector lattice $E$, the band $B_A$ generated by $A$ is a projection band if and only if*

$$x_A := \sup_{n,\mathrm{H}} \left(x \wedge n \textstyle\sum_{y\in\mathrm{H}} |y|\right)$$

*exists for each $x\in E_+$, where $n$ runs through $\mathbb{N}$ and $\mathrm{H}$ through all finite subsets of $A$. If so, $x\mapsto (x^+)_A-(x^-)_A$ $(x\in E)$ is the band projection associated with $B_A$.*

*Proof.* An inspection of the proof of (2.10) shows that $B_A$ is a projection band iff $P_A x:=\sup[0,x]\cap B_A$ exists for each $x\in E_+$, and if so then $x\mapsto P_A x^+ - P_A x^-$ is the band projection associated with $B_A$ (cf. discussion after Def. 2.4). Now denote by $I(A)$ the ideal generated by $A$; $I(A)$ contains precisely those $z\in E$ satisfying $|z|\leqq n\sum_{y\in\mathrm{H}}|y|$ for suitable $n\in\mathbb{N}$ and suitable finite $\mathrm{H}\subset A$. Furthermore, denote by $J$ the vector subspace of $E$ generated by all $z\in E_+$ which are of the form $z=\sup C$ where $C$ is a directed $(\leqq)$ subset of $I(A)_+$; we claim that $J=B_A$. Now each $z\in J$ is of the form $z=\sup C_1-\sup C_2$ for suitable directed $(\leqq)$ sets $C_i\subset I(A)_+$; using (1.5) and (1.6) one verifies that $J$ is an ideal. Moreover, if $D\subset J_+$ is a directed $(\leqq)$ set such that $w:=\sup D$ exists in $E$ then, since there exist directed $(\leqq)$ sets $C_z\subset I(A)_+$ such that $z=\sup C_z$ for each $z\in D$, it follows that $w=\sup\bigcup_{z\in D}C_z$ and hence, $w\in J$. Thus $J$ is a band in $E$ and, because of $A\subset J\subset B_A$, we have $J=B_A$. Thus $P_A x$ (whenever it exists) equals $\sup[0,x]\cap I(A)$ and conversely; this proves our assertion. □

Let $E$ be any vector lattice. The band $B(u)$ of $E$ generated by the singleton $\{u\}$, is called a *principal band* of $E$. If each principal band of $E$ is a projection band, $E$ is said to have the *principal projection property* (abbreviated (PPP); cf. Luxemburg-Zaanen [1971]); it is easy to show that (PPP) implies the validity of the Archimedian Axiom (A) (Exerc. 3). We obtain these corollaries of (2.11).

**Corollary 1.** *A principal band $B(u)\subset E$ is a projection band if and only if for each $x\in E_+$, $\sup_n(x\wedge n|u|)$ exists in $E$.*

**Corollary 2.** *$E$ has the principal projection property if and only if for each pair $x,y\in E_+$, $\sup_n(x\wedge ny)$ exists. In particular, every countably order complete vector lattice has the principal projection property.*

If by (PP) we denote the property of a vector lattice that each band be a projection band, the following logical scheme results ("Main Inclusion Theorem", Luxemburg-Zaanen [1971], Thm. 25.1):

$$(\mathrm{OC}) \begin{array}{c}\nearrow\\ \searrow\end{array} \begin{array}{c}(\mathrm{COC})\\ \\ (\mathrm{PP})\end{array} \begin{array}{c}\searrow\\ \nearrow\end{array} (\mathrm{PPP}) \Rightarrow (\mathrm{A}).$$

In this scheme, none of the implications can be reversed (Exerc. 3).

We conclude this section by a result which permits, under suitable assumptions, the representation of the Boolean algebra $\boldsymbol{P}(E)$ of band projections of a vector lattice $E$ by an ordered subset of $E$ itself. This result will be useful in the discussion of $AL$-spaces (§ 8).

**2.12 Proposition.** *Let $E$ be a vector lattice having the principal projection property and containing weak order units. For each weak order unit $u\in E$, the mapping $P\mapsto Pu$ is an isomorphism of the Boolean algebra $\boldsymbol{P}(E)$ onto the ordered subset of all elements $v$ satisfying $v\wedge(u-v)=0$.*

*Proof.* We observe first that if $u$ is a weak order unit of $E$ (§ 1) and $P\in\boldsymbol{P}(E)$, then from (1.9) Cor. and the order continuity of $P$ (2.9) it follows that $u_P:=Pu$ is a weak order unit of the band $PE$. Now let $u$ be a fixed weak order unit of $E$. It is clear that $0\leqq u_P\leqq u$ and, in fact, $u_P\wedge(u-u_P)=0$ for all $P\in\boldsymbol{P}(E)$. Obviously $P\leqq Q$ implies $u_P\leqq u_Q$ in $E$; conversely, if $u_P\leqq u_Q$ then $PE\subset QE$ by our first observation and hence, $P\leqq Q$. Thus $P\mapsto u_P$ maps $\boldsymbol{P}(E)$ order isomorphically onto a set of elements $v\in[0,u]$ satisfying $v\wedge(u-v)=0$; in particular, this set is a Boolean algebra. To show that the range of the map $P\mapsto u_P$ consists of all $v\in E$ satisfying $v\wedge(u-v)=0$, we need the hypothesis that $E$ have the principal projection property: Indeed, the principal band $B(v)$ is a projection band, and $u-v\in B(v)^\perp$; if $P_v$ is the associated band projection then the uniqueness of the decomposition $u=v+(u-v)$ (according to $E=B(v)+B(v)^\perp$) shows that $v=P_v u$. □

If $e$ is any positive element of a vector lattice $E$, the set $B_e:=\{x\in E: x\wedge(e-x)=0\}$ is sometimes called the set of *characteristic elements* of the order interval $[0,e]$ (cf. Vulikh [1967]); this terminology is evidently derived from the special case where $E$ is a vector lattice of real functions on a set $X$ under its canonical ordering (§ 1, Example 3) and $e$ is the constant-one function on $X$. In these circumstances, $x\in B_e$ iff $x$ is the characteristic function of some subset of $X$.

Thus (2.12) shows that whenever $E$ has the principal projection property and possesses weak order units $u$, then $B_u$ is isomorphic (as a Boolean algebra) to $\boldsymbol{P}(E)$ and hence, independent of the choice of $u$; in this case, $B_u$ is called a *base* of $E$ (cf. Exerc. 4). However, $B_e$ is always a Boolean algebra and identical with the extreme boundary (set of all extreme points) of $[0,e]$.

**2.13 Proposition.** *Let $E$ be any vector lattice and let $e \in E_+$. The subset $B_e := \{x \in E: x \wedge (e-x) = 0\}$ is a Boolean algebra under the induced ordering and identical with the extreme boundary of the order interval $[0,e]$.*

*Proof.* If $x \in B_e$ then, clearly, $e-x$ is the complement of $x$ in $B_e$ (cf. Def. 1.1). To show that $B_e$ is a Boolean algebra it suffices (since each vector lattice is distributive (1.5)) to prove that $x, y \in B_e$ implies $x \vee y \in B_e$ and $x \wedge y \in B_e$. Let $z := x \vee y$; we have $e - z = e + (-x) \wedge (-y) = (e-x) \wedge (e-y)$ and hence, $z \wedge (e-z) = (x \vee y) \wedge ((e-x) \vee (e-y)) = 0$ by distributivity. Similarly, $(x \wedge y) \wedge (e - x \wedge y) = (x \wedge y) \wedge ((e-x) \vee (e-y)) = 0$.

We next show that each $x \in B_e$ is an extreme point of $[0,e]$. Suppose $x = \lambda y + (1-\lambda) z$ where $y, z \in [0,e]$ and $0 < \lambda < 1$. From $x \wedge (e-x) = 0$ we obtain $y \wedge (e-x) = 0$ and $z \wedge (e-x) = 0$ hence, $(y \vee z) \wedge (e-x) = 0$; this implies $y \vee z + (e-x) = y \vee z \vee (e-x) \leqq e$ and so $y \vee z \leqq x$. Now $y < x$ would imply $x < \lambda x + (1-\lambda) z \leqq x$ which is contradictory; therefore, $y = x (= z)$ and hence, $x$ is extreme.

Finally, let $x$ be an extreme point of $[0,e]$ and set $z = x \wedge (e-x)$. Then $0 \leqq x - z \leqq e$ and $0 \leqq x + z \leqq x + (e-x) = e$. Thus $x = \frac{1}{2}(x+z) + \frac{1}{2}(x-z)$ is a convex combination in $[0,e]$, which can only hold for $z = 0$. □

In conclusion, we remark that whenever $E$ is a Banach lattice (§ 5) which is countably order complete, then $[0,e]$ is the convex closure of $B_e$ (Exerc. 4).

## § 3. Maximal and Minimal Ideals. Vector Lattices of Finite Dimension

Let $E$ be any vector lattice, let $I$ be fixed in $\boldsymbol{I}(E)$ (the lattice of all ideals of $E$), and let $q$ denote the canonical map $E \to E/I$. The following result is often useful in the ideal theory of vector lattices.

**3.1 Proposition.** *The mapping $J \mapsto q(J)$ is a homomorphism of the lattice $\boldsymbol{I}(E)$ onto $\boldsymbol{I}(E/I)$; its restriction to the sublattice of ideals of $E$ containing $I$, is an isomorphism onto $\boldsymbol{I}(E/I)$.*

*Proof.* If $J \in \boldsymbol{I}(E)$ then, clearly, $q(J)$ is a vector subspace of $E/I$, and (2.5) and (2.6) imply that $q(J) \in \boldsymbol{I}(E/I)$. To see that $J \mapsto q(J)$ is a homomorphism of lattices, let $J, K \in \boldsymbol{I}(E)$; it is clear that $q(J+K) = q(J) + q(K)$ and $q(J \cap K) \subset q(J) \cap q(K)$, so we have only to prove that $q(J) \cap q(K) \subset q(J \cap K)$. In fact, if $z \in q(J) \cap q(K)$ there exist $x \in J$, $y \in K$ satisfying $z = q(x) = q(y)$; it follows that $|z| = q(|x|) \wedge q(|y|) = q(|x| \wedge |y|)$ so that $|z|$, and therefore $z$, is in $q(J \cap K)$.

Thus $J \mapsto q(J)$ is a homomorphism of lattices as asserted; it maps $\boldsymbol{I}(E)$ onto $\boldsymbol{I}(E/I)$, since for any ideal $H$ of $E/I$, we have $q^{-1}(H) \in \boldsymbol{I}(E)$ and $H = q(q^{-1}(H))$. Moreover, if $J \supset I$ then $q^{-1}(q(J)) = J + I = J$ which proves the second assertion. □

**3.2 Definition.** *Let $E$ be any vector lattice and let $I \in \boldsymbol{I}(E)$. $I$ is called* maximal *if $E$ is the only ideal properly containing $I$; $I$ is called* minimal *if $\{0\}$ is the only ideal properly contained in $I$. The intersection of the set of all maximal ideals of $E$ is called the* radical *$R_E$ of $E$. $E$ is called* semi-simple *if $R_E = \{0\}$ and $E$ is called* simple *if $\boldsymbol{I}(E) = \{0, E\}$.*

In general, maximal or minimal ideals do not exist in a given vector lattice $E$ (Example 1 below). (Note that if no maximal ideals exist in $E$ then $R_E = E$, intersection of the empty subset of $2^E$.)

**3.3 Proposition.** *For any vector lattice $E$, the quotient $E/R_E$ is semi-simple and Archimedean.*

*Proof.* Since by virtue of (3.1), $H \mapsto q^{-1}(H)$ ($q$ the canonical map $E \to E/R_E$) maps the set of all maximal ideals of $E/R_E$ bijectively onto the set of all maximal ideals of $E$, it follows that $E/R_E$ has radical $\{0\}$. Now let $u, v$ be positive elements of $E/R_E$ satisfying $nv \leqq u$ for all $n \in \mathbb{N}$. If $H$ is any maximal ideal of $E/R_E$ and if $u \in H$ then, clearly, $v \in H$; if $u \notin H$ then the ideal generated by $H \cup \{v\}$ cannot contain $u$ either, and hence is proper; since $H$ is maximal, we must have $v \in H$. Thus $v$ is in the radical of $E/R_E$ and so $v = 0$; that is, $E/R_E$ is Archimedean. □

**Corollary.** *Every semi-simple vector lattice is Archimedean.*

We can now characterize vector lattices isomorphic to $\mathbb{R}$ (or, more precisely, to the one-dimensional vector lattice $\mathbb{R}_0$ underlying the real number field $\mathbb{R}$).

**3.4 Proposition.** *For any vector lattice $E \neq \{0\}$, the following assertions are equivalent:*

(a) *$E$ is isomorphic to $\mathbb{R}_0$.*
(b) *$E$ is simple.*
(c) *$E$ is totally ordered and Archimedean.*

*Proof.* (a) $\Rightarrow$ (b) is evident.

(b) $\Rightarrow$ (c): If $x \in E$ is given, we must have $x^+ = 0$ or $x^- = 0$, since $x^+ > 0$ and $x^- > 0$ would imply $\{0\} \neq E_{x^+} \neq E$ against the hypothesis. Thus $x \geqq 0$ or $x \leqq 0$ for each $x \in E$, which shows $E$ to be totally ordered; by the corollary of (3.3), $E$ is Archimedean.

(c) $\Rightarrow$ (a): Denote by $e$ any fixed element $> 0$ in $E$. For given $x \in E$, consider the sets $C_1 := \{\lambda \in \mathbb{R} : x \leqq \lambda e\}$ and $C_2 := \{\lambda \in \mathbb{R} : x \geqq \lambda e\}$. $C_1$ and $C_2$ are convex subsets of $\mathbb{R}$ such that $\mathbb{R} = C_1 \cup C_2$. A simple consideration using the Archimedean axiom shows that $C_1$ and $C_2$ are both non-empty and closed (in the standard topology of $\mathbb{R}$). Since $\mathbb{R}$ is connected, $C_1 \cap C_2$ cannot be void; on the other hand, $C_1 \cap C_2$ cannot contain more than one real number $\lambda$. Therefore, $x = \lambda e$ and it is clear that the mapping $x \mapsto \lambda$ of $E$ onto $\mathbb{R}_0$ is an isomorphism of vector lattices. □

**Corollary.** *If $I$ is a maximal (respectively, minimal) ideal of the vector lattice $E$, then $E/I$ (respectively, $I$) is isomorphic to $\mathbb{R}_0$.*

*Proof.* If $I \in \boldsymbol{I}(E)$ is maximal, then $E/I$ is simple by (3.1). It is clear that if $I$ is minimal then $I$ is simple. □

There is a transparent relationship between maximal and minimal ideals of an Archimedean vector lattice, as follows.

**3.5 Proposition.** *Let $E$ be Archimedean, and let $I \in \boldsymbol{I}(E)$. If $I$ is minimal, then $I^\perp$ is maximal and $E = I + I^\perp$. If $I$ is maximal, then $I^\perp$ is minimal if and only if $I$ is a projection band.*

*Proof.* Because $E$ is Archimedean, it is an easy consequence of (2.11) that each minimal ideal $I$ is a projection band, since $I$ is isomorphic to $\mathbb{R}_0$. Thus in this case, $E = I + I^\perp$ and, since $I$ has linear dimension 1, $I^\perp$ is maximal. If $I$ is maximal, then by definition $I$ is a projection band iff $E = I + I^\perp$; since $I^\perp$ has at most dimension 1 by the corollary of (3.4), we have $E = I + I^\perp$ iff $I^\perp$ is one-dimensional, that is, iff $I^\perp$ is minimal. □

**Examples.** 1. Let $E := L^1(\mu)$ (cf. [H, § 42] or [DS, IV, § 8]). If $\mu$ is a measure without atoms (diffuse measure), then $E$ has no maximal or minimal ideals. (Since maximal and minimal ideals are necessarily closed in the Banach lattice $E$, this can be seen from the characterization of closed ideals of $E$ given in Chap. III, § 1, Example 2.) An example is furnished by Haar measure $\mu$ on an infinite compact group; note that lattice ideals should not be confused with convolution ideals of the group algebra $L^1(\mu)$.

2. Let $K$ be compact, and let $E := C(K)$ (§ 1, Example 8). The maximal ideals of $E$ are of the form $I_t := \{f \in E : f(t) = 0\}$ for $t \in K$ (Chap. III, § 1, Example 1). $I_t$ is a projection band iff $t$ is an isolated point of $K$. Hence there exist minimal ideals of $E$ iff $K$ has isolated points, and these ideals are necessarily of the form $I_t^\perp$ for $t$ isolated in $K$.

3. The $n$-dimensional number space $\mathbb{R}^n$ is a totally ordered vector lattice under its lexicographic ordering (§ 1, Example 7). We recall that two elements $x = (\xi_0, \dots, \xi_{n-1})$ and $y = (\eta_0, \dots, \eta_{n-1})$ of $\mathbb{R}^n$ satisfy $x \leqq y$ under this ordering iff $\xi_i < \eta_i$ for the smallest index $i$ for which $\xi_i \neq \eta_i$. As in any totally ordered vector lattice, the lattice of ideals $\boldsymbol{I}(\mathbb{R}^n)$ is totally ordered (under inclusion; cf. III. 2.2); hence, there exists precisely one maximal ideal (namely, $I = \{x : \xi_0 = 0\}$) and precisely one minimal ideal (namely, $J = \{x : \xi_i = 0$ for $i = 0, \dots, n-2\}$). If $\mathbb{R}^\alpha$ is the corresponding vector lattice with elements $(\xi_\beta)_{\beta < \alpha}$ for any ordinal $\alpha > 0$, then $I = \{x : \xi_0 = 0\}$ is the unique maximal ideal, and there exist no minimal ideals iff $\alpha$ is a limit ordinal.

An element $a \neq 0$ of a vector lattice $E$ is called an *atom* if the principal ideal $E_a$ is totally ordered; this concept, which we mention in passing, is mainly useful for Archimedean vector lattices. If $E$ is Archimedean, these are equivalent assertions: (a) *$a$ is an atom of $E$*, (b) *$E_a$ is minimal*, (c) *$E_a^\perp$ is maximal* (Exerc. 7).

We now turn to the discussion of vector lattices of finite dimension. If $E$ is a vector lattice of finite dimension $n \geqq 1$, maximal and minimal ideals obviously exist in $E$; moreover, $\boldsymbol{I}(E)$ is finite (Exerc. 8). Special cases are the spaces $\mathbb{R}^n$ under their canonical and lexicographic orderings (Example 2 above yields $C(K) = \mathbb{R}^n$ under its canonical order for $K$ the discrete space $\{1, \dots, n\}$); there are other possibilities as we shall see. We begin the discussion with the characterization of totally ordered vector lattices of finite dimension.

One further remark: In a finite dimensional vector lattice $E$, every ideal is principal. In fact, if $H$ is a Hamel basis of the vector space underlying the ideal $I$ and if $e := \sum_{u \in H} |u|$, then $I = E_e$; in particular, $E$ itself is a principal ideal (that is, $E$ contains order units).

**3.6 Theorem.** *Every totally ordered vector lattice $E$ of finite dimension $n$ is isomorphic to the lexicographically ordered number space $\mathbb{R}^n$.*

*Proof.* Suppose $n \geqq 1$ and denote by $u$ an order unit of $E$. The order interval $[-u, u]$ is a convex, absorbing subset of $E$ and hence, a 0-neighborhood for the unique Hausdorff topology under which $E$ is a topological vector space [S, I.3.2]. The positive cone $E_+$ of $E$ contains interior points, since it contains $[0, 2u] = u + [-u, u]$. Now $E$, being totally ordered, is the disjoint union of $E_+$ and $-E_+ \setminus \{0\}$ so it follows from a well known separation theorem for convex sets [S, II.9.1] that there exists a hyperplane $H$ separating $E_+$ from $-E_+$, and strictly separating the interior of $E_+$ from the interior of $-E_+$. Thus $0 \in H$ and each $x \in E$ has a unique representation $x = \lambda u + y$ where $\lambda \in \mathbb{R}$, $y \in H$. Since the interior of $E_+$ agrees with the open semi-space determined by $H$ and containing $u$, it follows that $x \geqq 0$ iff either $\lambda > 0$, or $\lambda = 0$ *and* $y \geqq 0$.

The proof is now completed by induction on $n$. The theorem is evidently true for $n = 1$; if $n > 1$ then by the preceding $E$ is isomorphic to the Cartesian product $\mathbb{R} \times H$, ordered by the relation $(\lambda, y) \geqq 0$ iff $\lambda > 0$ or $\lambda = 0$ *and* $y \geqq 0$. Since $H$ is a totally ordered vector sublattice of $E$ with dimension $n-1$, the induction hypothesis shows $H$ to be isomorphic to the lexicographically ordered number space $\mathbb{R}^{n-1}$ and hence, the assertion is proved. □

The preceding proof suggests the following construction: If $M$ is any vector lattice, the real vector space $\mathbb{R} \times M$ can be made into a vector lattice by stipulating that $(\lambda, y) \geqq 0$ iff either $\lambda > 0$ or $\lambda = 0$ *and* $y \in M_+$; this vector lattice will be denoted by $\mathbb{R} \circ M$ and called the *lexicographic union of* $\mathbb{R}$ *with* $M$ (Birkhoff [1967]). It is clear that $M$ is a maximal ideal of $\mathbb{R} \circ M$. In fact, $M$ contains every proper ideal of $\mathbb{R} \circ M$, since any ideal not contained in $M$ must contain an element $(\lambda, y)$ with $\lambda > 0$ (hence, an order unit of $\mathbb{R} \circ M$) and thus cannot be proper. Indeed, this property characterizes the vector lattices $\mathbb{R} \circ M$ (cf. Birkhoff [1967], Chap. XV, § 2).

**3.7 Proposition.** *If $E$ is a vector lattice possessing a proper ideal $M$ which contains every proper ideal of $E$, then $E$ is isomorphic to $\mathbb{R} \circ M$ and contains no projection bands other than $\{0\}$, $E$.*

*Proof.* Suppose $I \in \boldsymbol{I}(E)$ is a projection band so that $E = I + I^\perp$. If $\{0\} \neq I \neq E$ then $\{0\} \neq I^\perp \neq E$ and it follows that $I \subset M$, $I^\perp \subset M$ and hence, $E \subset M$ which is contradictory. To show that $E$ is isomorphic with $\mathbb{R} \circ M$ fix $x \in E_+$, $x \notin M$. Since $M$ is necessarily maximal, $E = L(x) + M$ where $L(x)$ denotes the one-dimensional vector subspace of $E$ generated by $x$. It suffices to show that for each $z = \lambda x + y$, where $y \in M$, we have $z > 0$ whenever $\lambda > 0$. But $\lambda > 0$ implies that the ideal $I_z$ equals $E$, since $I_z \not\subset M$. Therefore, $E = I_{z^+} + I_{z^-}$ and these ideals are complementary (hence, projection bands); thus either $I_{z^+} = \{0\}$ or $I_{z^-} = \{0\}$ by our first remark. But $I_{z^+} = \{0\}$ is impossible, since $q(z^+) \geqq q(z) = \lambda q(x) > 0$

where $q$ denotes the canonical map $E \to E/M$. Thus $\lambda > 0$ implies $z^- = 0$ and hence, $z > 0$. ▯

From the proof of the main theorem (3.9) on finite dimensional vector lattices, we isolate this lemma.

**3.8 Lemma.** *Let $E$ be any vector lattice and let $I \in \boldsymbol{I}(E)$. For any finite orthogonal system $\{v_i : i = 1, \dots, n\}$ of $E/I$, there exists an orthogonal system $\{x_i : i = 1, \dots, n\}$ of $E$ such that $v_i = q(x_i)$ for all $i$, $q$ denoting the canonical map $E \to E/I$.*

*Proof.* The proof is by induction on $n$; clearly, the assertion is true for $n = 1$. Given the orthogonal system $\{v_i\}$ of $E/I$ $(i = 1, \dots, n;\ n > 1)$, by the induction hypothesis there exist orthogonal elements $x_i \in E_+$ satisfying $v_i = q(x_i) (1 \leqq i \leqq n-1)$. Let $x_n > 0$ be an element for which $q(x_n) = v_n$ and define

$$x_i' = x_i - x_i \wedge x_n \quad (i = 1, \dots, n-1),$$
$$x_n' = x_n - x_n \wedge (x_1 + \cdots + x_{n-1}).$$

From the hypothesis on $\{v_i\}$ it follows that $q(x_i \wedge x_n) = v_i \wedge v_n = 0$ and hence, that $x_i \wedge x_n \in I$ for $i = 1, \dots, n-1$. Similarly, it follows that $x_n \wedge (x_1 + \cdots + x_{n-1}) \in I$. Thus $q(x_i') = q(x_i) = v_i$ for $i = 1, \dots, n$, and it is readily seen that $\{x_i' : i = 1, \dots, n\}$ is an orthogonal system in $E$. ▯

**3.9 Theorem.** *Let $E$ be a vector lattice of finite dimension $n (\geqq 1)$, and let $r$ denote the dimension of the radical of $E$. Then $E$ is the sum of $k = n - r$ mutually orthogonal ideals $I_j$ $(j = 1, \dots, k)$, each of which possesses a unique maximal ideal $M_j$ and hence is of the form $I_j = \mathbb{R} \circ M_j$. This decomposition of $E$ is unique except for a permutation of indices.*

*Proof.* Denoting by $R$ the radical of $E$ (Def. 3.2), $E/R$ is a vector lattice of dimension $k = n - r$, and Archimedean by (3.3). Since each minimal ideal of $E/R$ is one-dimensional by (3.4) and a projection band by (3.5), an induction argument shows at once that $E/R$ is the sum of $k$ mutually orthogonal minimal ideals. Thus $E/R$ contains an orthogonal system $\{v_j : j = 1, \dots, k\}$ (of course, the $v_j$ are atoms of $E/R$).

By Lemma 3.8 there exists an orthogonal system $\{x_j : j = 1, \dots, k\}$ in $E$ with canonical image $\{v_j\}$ in $E/R$. Denote by $L$ the vector subspace of $E$ generated by the set $\{x_j : j = 1, \dots, k\}$. Since $\{v_j : j = 1, \dots, k\}$ is a Hamel basis of $E/R$, we have $E = L + R$. We claim that the ideal $I(L)$ generated by $L$ is all of $E$. If not then $I(L)$ is a proper ideal not containing $R$ and hence, $I(L)$ is not maximal. But in this case there exists a maximal ideal $J$ of $E$ containing $I(L)$; since $J$ is proper, $J$ cannot contain $R$ (because of $E = L + R$) which contradicts the definition of the radical $R$. Therefore, $I(L) = E$.

Thus if $I_j$ $(j = 1, \dots, k)$ denotes the ideal of $E$ generated by $x_j$, $E$ is the sum of the $k$ mutually orthogonal ideals $I_j$. Let $M_j := I_j \cap R$; we claim that $M_j$ is the unique maximal ideal of the vector lattice $I_j$. In fact, $M_j$ is maximal, since it has codimension 1 in $I_j$ (note that the restriction of the canonical map $E \to E/R$ to $I_j$ can be identified with the canonical map $I_j \to I_j/M_j$). Moreover, if $K$ is any maximal ideal of $I_j$ then the ideal $K' := K + \sum_{i \neq j} I_i$ is maximal in $E$, since it has codimension

1; this implies $K' \supset R$ hence, $K = K' \cap I_j \supset R \cap I_j = M_j$ and so $K = M_j$. Therefore, $M_j$ contains each proper ideal of $I_j$ and it follows from (3.7) that $I_j = \mathbb{R} \circ M_j$ $(j = 1, \ldots, k)$.

It remains to prove that there exists no other decomposition of $E$ of this kind, except for a permutation of indices. In other words, we have to show that if $E = J + J^\perp$ where $J$ is an ideal of the form $J = \mathbb{R} \circ N$, then $J = I_l$ for some $l$, $1 \leqq l \leqq k$. By the distributivity of $\boldsymbol{I}(E)$ (2.3), for each $j$ we have $I_j = (J + J^\perp) \cap I_j = J \cap I_j + J^\perp \cap I_j$ and from (3.7) it follows that either $J \cap I_j = \{0\}$ or $J \cap I_j = I_j$. Since $J \cap I_j = \{0\}$ cannot occur for all $j$ $(1 \leqq j \leqq k)$, we must have $J \cap I_l = I_l$ for at least one index $j = l$. But again by (3.7), $J \neq I_l$ is impossible, since this would imply that $I_l$ is a non-trivial projection band in $J$. Thus $J = I_l$ and the proof is complete. □

**Corollary 1.** *For a vector lattice $E$ of finite dimension $n$ $(\geqq 1)$, the following assertions are equivalent:*

(a) *$E$ is Archimedean.*
(b) *$E$ is semi-simple.*
(c) *$E$ is isomorphic to $\mathbb{R}^n$ (canonical order).*

*Proof.* (a)⇒(b): Suppose the radical $R$ of $E$ to be $\neq \{0\}$; then there exists a minimal ideal $I$ of $R$, and evidently $I$ is minimal in $E$. Since $E$ is Archimedean, by (3.5) $I^\perp$ is a maximal ideal of $E$ not containing $R$; this contradiction shows that $R = \{0\}$.

(b)⇒(c): By (3.9) $E$ is the sum of $n$ orthogonal one-dimensional ideals; so $E$ is isomorphic to the canonically ordered number space $\mathbb{R}^n$ (§ 1, Example 4). (c)⇒(a) is immediate, since $\mathbb{R}$ is Archimedean ordered. □

Thus the canonical and lexicographic orderings of $\mathbb{R}^n$ correspond to the extreme cases where the radical is $\{0\}$ or a totally ordered maximal ideal, respectively; more generally, a finite dimensional vector lattice $E$ is the direct sum of an Archimedean and a totally ordered ideal iff its radical $R_E$ is totally ordered (Exerc. 7).

**Corollary 2.** *For any finite dimensional vector lattice $E$, the Boolean algebra $\boldsymbol{B}(E)$ of projection bands is isomorphic with $\boldsymbol{I}(E/R_E)$.*

*Proof.* Let $E = \sum_{j=1}^k I_j$ according to (3.9). It is clear from (3.7) that each projection band $B \in \boldsymbol{B}(E)$ is a sum $\sum_{j \in H} I_j$ for some finite subset $H$ of $\{1, \ldots, k\}$, and conversely. Thus the mapping $B \mapsto q(B)$, where $q$ is the canonical map $E \to E/R_E$, establishes an isomorphism of $\boldsymbol{B}(E)$ onto $\boldsymbol{I}(E/R_E)$ (cf. 3.1). □

**Corollary 3.** *If a finite dimensional vector lattice $E \neq \{0\}$ contains no projection bands $\neq \{0\}$, $E$ then $E$ is the lexicographic union $\mathbb{R} \circ M$, $M$ denoting the unique maximal ideal of $E$.*

*Proof.* By Cor. 2 and (3.4), the dimension of $E/R_E$ is 1 so that $R_E$ is a maximal ideal of $E$, and obviously the only one. The remainder now follows from (3.7). □

It should be noted that this converse of (3.7) fails in the infinite dimensional case (cf. § 2, Example 5).

If $E$ is a vector lattice of infinite linear dimension, the existence of an order unit suffices to guarantee the existence of maximal ideals; it is also easy to characterize the radical of $E$. An element $v$ of a vector lattice $E$ is called *infinitely small* if there exists $u \in E$ such that $n|v| \leqq u$ for all $n \in \mathbb{N}$; it is immediate that the set of all infinitely small elements of $E$ is an ideal which is the zero ideal if and only if $E$ is Archimedean. If $E$ is finite dimensional then, in the notation of (3.9), the ideal of infinitely small elements is the ideal $\sum_{j=1}^{k} M_j$ and hence, agrees with the radical of $E$. More generally:

**3.10 Proposition.** *Let $E$ be a vector lattice with order unit. The radical $R_E$ agrees with the ideal of all infinitely small elements of $E$, and $E$ is Archimedean if and only if $E$ is semi-simple.*

*Proof.* Since $E$ contains an order unit $e$, a straightforward application of Zorn's lemma shows that each proper ideal of $E$ is contained in a maximal ideal. If $v \in E$ is infinitely small, then $v$ is contained in every maximal ideal of $E$ (see proof of 3.4). Conversely, suppose $v \in R_E$ is a positive element not infinitely small; there exists $n \in \mathbb{N}$ such that $nv$ is not majorized by $e$, that is, $w_1 := (e - nv)^- = (nv - e)^+ > 0$. But then $w_2 := (e - nv)^+$ is not an order unit of $E$, so there exists a maximal ideal $I$ containing $w_2$. Since $0 \leqq w_1 \leqq nv$ and $v \in I$, it follows that $w_2 - w_1 \in I$ and hence, $e \in I$ which is contradictory. Thus if $E$ is Archimedean, $E$ is semi-simple; the converse was proved above (3.4 Cor.). □

**Corollary.** *Every totally ordered vector lattice $E \neq \{0\}$ with order unit is the lexicographic union of $\mathbb{R}$ with the ideal of all infinitely small elements.*

*Proof.* Since the lattice of ideals $\boldsymbol{I}(E)$ is totally ordered, there exists precisely one maximal ideal of $E$ which by (3.10) must be the ideal of infinitely small elements; the remainder follows from (3.5). □

## § 4. Duality of Vector Lattices

If $E$ is a vector space, we denote by $E^*$ the algebraic dual of $E$, that is, the vector space of all linear forms on $E$.

**4.1 Definition.** *Let $E$ be any vector lattice. A linear form $f \in E^*$ is called* order bounded *if for each order interval $[x, y] \subset E$, the set $f([x, y])$ is order bounded in $\mathbb{R}$. An order bounded $f$ is called* order continuous (sequentially order continuous) *if $f$ converges to 0 along each filter (each sequence) that order converges to 0 in $E$.*

*The vector spaces of all order bounded, sequentially order continuous, and order continuous linear forms on $E$ are denoted by $E^\star$, $E_0^\star$, and $E_{00}^\star$ respectively; $E^\star$ is called the* order dual *of $E$.*

This definition is in accordance with Def. 2.4; it is clear that $E_{00}^\star \subset E_0^\star \subset E^\star$ but it should be noted that on a non-Archimedean vector lattice (such as lexicographic $\mathbb{R}^n$, $n > 1$) a linear form convergent on order convergent filters, need not be order bounded (cf. Example 1 below).

We begin with a discussion of the order structure of $E^\star$. The canonical order of $E^\star$ is (in accordance with Def. 2.4) the order determined by "$f \leqq g$ *iff* $f(x) \leqq g(x)$ *in* $\mathbb{R}$ *for all* $x \in E_+$"; since each $f \in E^\star$ is uniquely determined by its values on $E_+$ (see remarks after 2.4), $E^\star$ can be considered a subspace of $\mathbb{R}^{E_+}$ and, under this identification, the order of $E^\star$ just defined is induced by the canonical order of $\mathbb{R}^{E_+}$ (§ 1, Example 3).

**4.2 Proposition.** *Let $E$ be any vector lattice. Under its canonical order, the order dual $E^\star$ is an order complete (in particular, Archimedean) vector lattice in which the lattice operations are given by the formulae*

$$
\begin{aligned}
f \vee g(x) &= \sup\{f(y)+g(z): y \geqq 0,\ z \geqq 0,\ y+z=x\} \\
f \wedge g(x) &= \inf\ \{f(y)+g(z): y \geqq 0,\ z \geqq 0,\ y+z=x\}
\end{aligned}
\tag{1}
$$

*for $f, g \in E^\star$ and $x \in E_+$. Moreover, if $A \subset E^\star$ is a directed ($\leqq$) majorized set, then at $x \in E_+$ the supremum $f_0 := \sup A$ is given by*

$$
f_0(x) = \sup_{f \in A} f(x)\,. \tag{2}
$$

*Proof.* Define $h: E_+ \to \mathbb{R}$ by $h(x) := \sup\{f(y)+g(x-y): 0 \leqq y \leqq x\}$; it is clear that $h(x) \leqq k(x)$ for any $k \in E^\star$ majorizing $f$ and $g$. Thus by the remarks following (2.4), to verify (1) it suffices to show that $h$ is additive and positive homogeneous, and that the linear form $\bar{h}: E \to \mathbb{R}$ defined by $h$ is order bounded. Clearly, $h(\lambda x) = \lambda h(x)$ whenever $\lambda \in \mathbb{R}_+$. Let $x = x_1 + x_2$ where $x_1, x_2 \in E_+$; since $[0,x] = [0,x_1] + [0,x_2]$ by the decomposition property (1.6).

$$
h(x_1+x_2) = \sup\{f(y_1)+f(y_2)+g(x_1-y_1)+g(x_2-y_2)\} = h(x_1)+h(x_2)\,,
$$

the supremum being taken over all $y_1 \in [0,x_1]$ and all $y_2 \in [0,x_2]$. To see that $\bar{h}$ is order bounded we observe that it is enough that $h([0,x])$ be bounded for each $x \in E_+$, and this is evident from the definition of $h$.

Finally, if $A \subset E^\star$ is directed ($\leqq$) and majorized, the mapping $r: E_+ \to \mathbb{R}$ defined by $r(x) := \sup\{f(x): f \in A\}$ is obviously positive homogeneous, and it is additive, since $A$ is directed ($\leqq$); the linear form $f_0$ defined by $r$ is order bounded, since $f_0$ is majorized and minorized by elements of $E^\star$. Clearly, then, $f_0 = \sup A$ in $E^\star$. ∎

**Corollary 1.** *If $f, g \in E^\star$ and $x \in E_+$, then $f^+ = f \vee 0$ and $f^- = (-f) \vee 0$ and $|f| = f \vee (-f)$ are given by*

$$
\begin{aligned}
f^+(x) &= \sup\{f(y): 0 \leqq y \leqq x\} \\
f^-(x) &= -\inf\{f(y): 0 \leqq y \leqq x\} \\
|f|(x) &= \sup\{|f(z)|: 0 \leqq |z| \leqq x\}\,.
\end{aligned}
\tag{3}
$$

*Moreover, if $\{f_\alpha : \alpha \in \mathsf{A}\}$ is any majorized subset of $E^\star$ and $f_0 := \sup_\alpha f_\alpha$, then at $x \in E_+$ $f_0$ is given by*

$$
f_0(x) = \sup\{f_{\alpha_1}(x_1) + \cdots + f_{\alpha_n}(x_n)\}\,, \tag{4}
$$

*where* $\{\alpha_1, \ldots, \alpha_n\}$ *runs over all non-void finite subsets of* A *and where* $\{x_1, \ldots, x_n\}$ *runs over all finite decompositions* $x = x_1 + \cdots + x_n$ *of* $x$ *into positive summands.*

*Proof.* The first two formulae of (3) are specializations of (1) for $g=0$. To prove the third, observe that for $x \in E_+$, $\{u-v: u \geqq 0, v \geqq 0, u+v \leqq x\} = \{z: |z| \leqq x\}$ and hence,

$$|f|(x) = f \vee (-f)(x) = \sup\{f(z): |z| \leqq x\} = \sup\{|f(z)|: |z| \leqq x\}.$$

Finally, if $\{\alpha_1, \ldots, \alpha_n\}$ is any finite non-void subset of A, repeated application of (1) together with the associativity of the lattice operations in $E^\star$ shows that

$$(\sup_i f_{\alpha_i})(x) = \sup\{f_{\alpha_1}(x_1) + \cdots + f_{\alpha_n}(x_n)\}$$

where the right hand supremum is taken over all positive decompositions $x = x_1 + \cdots + x_n$ $(x \in E_+)$. But the set of all suprema of finite non-void subsets of $\{f_\alpha: \alpha \in \mathrm{A}\}$ is clearly directed $(\leqq)$ in $E^\star$ and hence, by virtue of (2), $f_0(x)$ is given by (4). □

**Corollary 2.** *A linear form* $f$ *on a vector lattice* $E$ *is order bounded if and only if* $f$ *is the difference of two positive linear forms.*

*Proof.* It is clear that each positive linear form on $E$ is order bounded and hence, so is $f = f_1 - f_2$ whenever $f_i \geqq 0$ $(i=1,2)$; conversely, if $f \in E^\star$ then $f = f^+ - f^-$. □

**Corollary 3.** *Two elements* $f, g \in E^\star$ *are lattice orthogonal* $(f \perp g)$ *if and only if for each real number* $\varepsilon > 0$ *and each* $x \in E_+$, *there exists a positive decomposition* $x = y + z$ *such that* $|f|(y) < \varepsilon$ *and* $|g|(z) < \varepsilon$.

This is clear from Def. 1.3 and Formula (1) above.

**4.3 Proposition.** *Let* $\mathfrak{F}$ *be a filter on* $E$ *possessing a base of symmetric order intervals* $\{[-x, x]: x \in D\}$. *The set* $B_{\mathfrak{F}}$ *of all* $f \in E^\star$ *satisfying* $\lim_{\mathfrak{F}} f(x) = 0$, *is a band in* $E^\star$.

*Proof.* It is clear that $B_{\mathfrak{F}}$ is a vector subspace of $E^\star$. Suppose $f \in B_{\mathfrak{F}}$ and $\varepsilon$, $0 < \varepsilon \in \mathbb{R}$, are given; there exists $x \in D$ such that $|f(z)| \leqq \varepsilon$ for all $z \in E$, $|z| \leqq x$. From (3) we obtain $||f|(z)| \leqq |f|(x) \leqq \varepsilon$ whenever $z \in [-x, x]$; so $|f| \in B_{\mathfrak{F}}$, and $B_{\mathfrak{F}}$ is a vector sublattice of $E^\star$. It is now immediate that $B_{\mathfrak{F}}$ is an ideal in $E^\star$.

To show that $B_{\mathfrak{F}}$ is a band, let $(f_\alpha)_{\alpha \in \mathrm{A}}$ denote a directed $(\leqq)$ family in $B_{\mathfrak{F}}$ such that $f := \sup_\alpha f_\alpha$ exists in $E^\star$. Given $x_0 \in D$ and $\varepsilon > 0$, there exists $\alpha_0 \in \mathrm{A}$ such that $(f - f_{\alpha_0})(x_0) \leqq \varepsilon$; moreover, there exists $x_1 \in D$ such that $x_1 \leqq x_0$ and $|f_{\alpha_0}|(x_1) \leqq \varepsilon$. We obtain

$$|f(z) - f_{\alpha_0}(z)| \leqq (f - f_{\alpha_0})(x_1) \leqq (f - f_{\alpha_0})(x_0) \leqq \varepsilon$$

for all $z \in [-x_1, x_1]$ and, therefore, $|f(z)| \leqq 2\varepsilon$ for all $z \in [-x_1, x_1]$. This shows that $f \in B_{\mathfrak{F}}$. □

**Corollary.** *If $E$ is a vector lattice with order dual $E^\star$, the sets of all order continuous and all sequentially order continuous linear forms, respectively, are projection bands in $E^\star$.*

*Proof.* If $\mathfrak{F}_{00}$ (respectively, $\mathfrak{F}_0$) denotes the family of all order convergent filters on $E$ (Def. 1.7) possessing a base (respectively, a countable base) of symmetric order intervals, it is clear that $E_0^\star = \bigcap_{F \in \mathfrak{F}_0} B_F$ and $E_{00}^\star = \bigcap_{F \in \mathfrak{F}_{00}} B_F$. Hence, the assertion follows from (4.3) and (2.10). □

Before considering examples, let us characterize the scalar valued lattice homomorphisms (Def. 2.4) of $E$.

**4.4 Proposition.** *Let $f$ be a non-zero linear form on a vector lattice $E$. These assertions are equivalent:*

(a) *$f$ is a lattice homomorphism of $E$ onto $\mathbb{R}$.*

(b) *$f(x^+) \wedge f(x^-) = 0$ for all $x \in E$.*

(c) *$f \geqq 0$ and $f^{-1}(0)$ is a maximal ideal of $E$.*

(d) *$f \geqq 0$ and the ideal of $E^\star$ generated by $f$ is minimal.*

*Proof.* The equivalences (a)⇔(b)⇔(c) are immediate consequences of (2.6) and its corollary.

(a)⇒(d): Let $g \in E^\star$ be an element of the ideal $E_f^\star$ generated by $f$ so that $|g| \leqq cf$ for some $c \in \mathbb{R}_+$. Then $x \in f^{-1}(0)$ implies $|g(x)| \leqq cf(|x|) = 0$, since $f^{-1}(0)$ is an ideal of $E$, whence it follows that $f^{-1}(0) \subset g^{-1}(0)$. Therefore, $g = 0$ or $g^{-1}(0) = f^{-1}(0)$ which shows that $g = \lambda f$ for some $\lambda \in \mathbb{R}$. Thus $E_f^\star$ is one dimensional and hence, minimal in $E^\star$.

(d)⇒(b): Suppose $x \in E$ is given and assume that $f(x^-) > 0$. Consider the subcone $P := \bigcup_{n=1}^\infty n[0, x^-]$ of $E_+$ and define a mapping $r: E_+ \to \mathbb{R}_+$ by

$$r(y) := \sup\{f(z) : z \in [0, y] \cap P\} \qquad (y \in E_+).$$

Since $[0, y_1 + y_2] \cap P = [0, y_1] \cap P + [0, y_2] \cap P$ by the decomposition property (1.6), $r$ is readily seen to be additive; obviously, $r$ is positive homogeneous. Thus $r$ extends to a linear form $g$ on $E$ satisfying $0 \leqq g \leqq f$; since $E_f^\star$ is minimal by hypothesis, $g$ is a scalar multiple of $f$, and from $g(x^-) = f(x^-) > 0$ we obtain $g = f$. On the other hand, it is clear that $r(x^+) = 0$ hence, $f(x^+) = 0$. Thus $f(x^+) \wedge f(x^-) = 0$ for each $x \in E$. □

From the characterization (d) of (4.4) we see that each lattice homomorphism $E \to \mathbb{R}$ is a positive atom of $E^\star$(§ 3), and conversely; the band $E_a^\star$ of $E^\star$ generated by the set of all atoms is called the *atomic part* of $E^\star$ (Exerc. 7).

**Examples.** 1. Let $E = \mathbb{R} \circ M$ be a vector lattice which is the lexicographic union of $\mathbb{R}$ with a vector lattice $M$ (§ 3); since $M$ is order bounded in $E$, it follows that $f(M) = 0$ for each $f \in E^\star$. Thus $E^\star$ is isomorphic with $(E/M)^\star$ and, since $E/M \cong \mathbb{R}$, it follows that $\dim E^\star = 1$.

If $E$ is an $n$-dimensional vector lattice ($n \in \mathbb{N}$) with radical $R_E$ of dimension $r$, it now follows from (3.9) that $E^\star$ is isomorphic to $(E/R_E)^\star$ and $\dim E^\star = n - r$; in particular, $E^\star$ is isomorphic to $E$ iff $E$ is Archimedean. Moreover, if $E = \sum_{j=1}^{n-r} I_j$ is the representation of $E$ according to (3.9) and $\{x_j : j = 1, \dots, n-r\}$

is an orthogonal system of $E$ such that each $x_j$ is an order unit of $I_j$, then the $n-r$ linear forms which are order bounded and satisfy $f_i(x_j)=\delta_{ij}$ for $i,j=1,\dots,n-r$, exhaust the set of non-zero scalar lattice homomorphisms of $E$ (to within some normalization).

2. Let $A$ be a Boolean algebra with Stone representation space $K_A$ (§ 1, Example 1), and denote by $E_A$ the vector lattice of real functions on $K_A$ which is the linear hull of the characteristic functions of all open-and-closed subsets of $K_A$. The vector lattice $\mathsf{M}_A$ of all bounded, finitely additive real functions on $A$ (§ 1, Example 5) can be identified with the order dual $E_A^\star$ and, in fact, with the order dual of $C(K_A)$ (every positive linear form on $E_A$ is continuous for the uniform norm (cf. 7.2) and hence, has a unique positive extension to $C(K_A)$). Thus $\mathsf{M}_A$ is isomorphic with the vector lattice of all real Radon measures on the compact space $K_A$.

In the special case where $A$ is a $\sigma$-algebra $\Sigma$ of subsets of a set $X$, the order dual $E_\Sigma^\star$ is often denoted by $ba(X,\Sigma)$ [DS, Chap. IV, § 5] while the band $(E_\Sigma)_0^\star$ is the vector lattice of all real valued (countably additive) measures on $\Sigma$ and denoted by $ca(X,\Sigma)$.

3. It was seen above (Example 1) that there exist non-Archimedean vector lattices $E$ of arbitrary linear dimension and such that $\dim E^\star=1$. But even if $E$ is Archimedean (and of infinite dimension), it can occur that $E^\star=\{0\}$. Well known examples for this situation are the vector lattices $L^p(\mu)$ where $0<p<1$ and $\mu$ is Lebesgue measure on $\mathbb{R}$, and the vector lattice $L$ of all finite Lebesgue measurable functions on $\mathbb{R}$ modulo $\mu$-null functions (Exerc. 10). Endowed with the metrics $\varrho(f,g)=\int|f-g|^p\,d\mu$ and $\varrho(f,g)=\int(|f-g|/(1+|f-g|))\,d\mu$, respectively, the spaces $L^p(\mu)$ and $L$ are complete, metrizable topological vector spaces.

4. If $E$ is any vector lattice then $E^\star$, $E_0^\star$, and $E_{00}^\star$ are all distinct in general. For example, if $E=C[0,1]$ then $E_0^\star=\{0\}$ and if $E=B[0,1]$ (bounded Borel functions on $[0,1]$) then $E^\star\neq E_0^\star\neq E_{00}^\star$ (Exerc. 10).

Moreover, the atomic part $E_a^\star$ (the band generated by all atoms) of $E^\star$ shows behavior quite independent of $E_0^\star$ and $E_{00}^\star$. Thus if $E=L^1(\mu)$ where $\mu$ is Lebesgue measure on $[0,1]$ then $E_a^\star=\{0\}$, because $E$ contains no maximal ideals (§ 3, Example 1) and hence, by (4.4), $E^\star$ contains no atoms; on the other hand, $E^\star=E_{00}^\star$ (cf. 8.3). By contrast, if $E=C[0,1]$ then $E_0^\star=\{0\}$ while $E_a^\star$, the band generated by all Dirac measures on $[0,1]$, separates points of $E$.

5. For $X$ a locally compact space let $\mathscr{K}(X)$ denote the vector lattice of all real valued continuous functions with compact support on $X$. The order dual $\mathscr{M}(X):=\mathscr{K}(X)^\star$ is an order complete vector lattice whose elements are called Radon measures on $X$ (Bourbaki [1965], Intégration Chap. II, § 1). $\mathscr{K}(X)$ and $\mathscr{M}(X)$ play significant roles in the representation of Banach lattices (Chap. III, § 5).

In the present context it is natural to ask if a linear form, defined on a vector sublattice or ideal of a vector lattice $E$, can be extended to $E$ with preservation of order related properties. A glance at the preceding examples shows that, in general, the answer is negative. For instance, if $E=\mathbb{R}\circ M$ (Example 1) then no order bounded linear form $\neq 0$ on $M$ has an order bounded extension to $E$. A less trivial example furnishes the principal ideal $L_e$ in $L$ (Example 3) generated by the constant-one function $e$. In fact, $L_e$ can be identified with $C(K_{\Sigma'})$, the vector lattice of continuous real functions on the Stone space $K_{\Sigma'}$ of the real Borel field (modulo Lebesgue null sets) $\Sigma'$ and hence, $L_e^\star$ is the vector lattice of Radon measures on $K_{\Sigma'}$ (Example 2). None of these measures (other than 0) has an order bounded extension to $L$.

There are a number of useful and important extension theorems, some of which apply to the more general problem of extending linear operators; this subject will be taken up in connection with the discussion of $AM$- and $AL$-spaces below (§§ 7, 8). Moreover, the reader is referred to Ando [1965] and Luxemburg-Zaanen [1963a]—[1965c]. We limit ourselves to proving a general extension theorem, due independently to H. Bauer [1957] and I. Namioka [1957]; in the vector lattice context, interesting applications are (4.6) and (5.6) below.

As usual, the positive cone of an ordered vector space $E$ (over $\mathbb{R}$) is denoted by $E_+$, and a linear form $f$ on $E$ is called *positive* if $f(x) \geqq 0$ whenever $x \in E_+$.

**4.5 Proposition.** *Let $M$ be a vector subspace of the ordered vector space $E$. For a linear form $f_0$ on $M$ to have a positive extension $f$ to $E$, it is necessary and sufficient that $f_0$ be bounded above on $M \cap (U - E_+)$ where $U$ is a suitable convex absorbing subset of $E$.*

*Proof.* If $f$ is a positive extension of $f_0$, then the set $U := \{x \in E : f(x) < 1\}$ satisfies the requirement.

Conversely, let $U$ be a subset of $E$ satisfying the requirement of the theorem, and let $\gamma \in \mathbb{R}$ be chosen so that $x \in M \cap (U - E_+)$ implies $f_0(x) < \gamma$. Then $\gamma > 0$ and the set $N := \{x \in M : f_0(x) = \gamma\}$ is a hyperplane in $M$ (provided $f_0 \neq 0$, which we evidently can assume) and an affine subspace of $E$ not intersecting $U - E_+$. But $U$, and hence $U - E_+$, is a convex 0-neighborhood for the finest locally convex topology on $E$; so from the geometric form of the Hahn-Banach theorem [S, II.3.1] it follows that there exists a hyperplane $H \subset E$ containing $N$ and not intersecting the interior of $U - E_+$. Thus $0 \notin H$ and $H$ is of the form $H = \{x \in E : f(x) = \gamma\}$ for a suitable linear form $f$ on $E$. But since $N$ is a hyperplane (that is, a maximal proper affine subspace) of $M$ and since $0 \notin H$, we must have $H \cap M = N$; that is, $f$ is an extension of $f_0$ (cf. [S, I.4.1]). Because $f(x) < \gamma$ for all $x \in U - E_+$ and, in particular, for all $x \in -E_+$ it follows that $f(x) \geqq 0$ whenever $x \in E_+$. ▯

**4.6 Proposition.** *Let $E$ be a vector lattice, and let $M$ be a vector sublattice such that the ideal generated by $M$ equals $E$. Then every order bounded linear form on $M$ has an order bounded extension to $E$.*

*Proof.* In view of (4.2), it suffices to show that each positive linear form $f_0$ on $M$ has a positive extension $f$ to $E$. Define $p : E \to \mathbb{R}_+$ by

$$p(x) := \inf\{f_0(y) : y \in M,\ y \geqq |x|\}.$$

Since the ideal generated by $M$ equals $E$, $p$ is well defined and it is straightforward to verify that $p$ is a semi-norm (even a lattice semi-norm, Def. 5.1) on $E$; moreover, $f_0(y) \leqq f_0(|y|) = p(y)$ for $y \in M$. Defining $U := \{x \in E : p(x) < 1\}$, we show that $f_0(y) < 1$ for all $y \in M \cap (U - E_+)$ so that (4.5) applies. In fact, if $y = u - v$ where

$p(u)<1, v\in E_+$ and $y\in M$, then $0\leqq y^+\leqq u+y^-$ and we obtain $f_0(y)=p(y^+)-p(y^-)\leqq p(u+y^-)-p(y^-)\leqq p(u)<1$. □

**Corollary.** *If $E$ is Archimedean then each $f\in E^\star$ has an order bounded extension to the Dedekind completion $\bar{E}$ of $E$.*

This is clear from the characterization of the Dedekind completion $\bar{E}$ of $E$ given in (1.10); moreover, it is readily seen that each order continuous linear form on $E$ has a unique order continuous extension to $\bar{E}$ (Exerc. 9).

We turn to the duality relations between a vector lattice $E$ and vector sublattices $F$ of $E^\star$. As usual, we shall write $\langle E,F\rangle$ for the pair $(E,F)$ endowed with the canonical bilinear form $(x,f)\mapsto\langle x,f\rangle:=f(x)$ on $E\times F$; unless the contrary is expressly stated, we do not assume that $F$ separate $E$. Recall [S, Chap. IV, § 1] that the *polar* of a subset $A\subset E$ with respect to this duality is the set $A^\circ=\{f\in F: \langle x,f\rangle\leqq 1,\ \textit{all}\ x\in A\}$. In particular, if $A$ is a vector subspace of $E$ then $A^\circ$ is a vector subspace of $F$ which is sometimes called the *annihilator* of $A$.

**4.7 Proposition.** *If $E$ is a vector lattice and $F$ a vector sublattice of $E^\star$, the polar of each solid subset $A\subset E$ is solid in $F$. In particular, the polar $I^\circ$ of an ideal $I\in\boldsymbol{I}(E)$ is an ideal in $F$.*

*Proof.* For each $x\in A$ and each $f\in A^\circ$ we have, by Formula (3) of (4.1) Cor. 1, $|f|(|x|)=\sup\{|f(z)|: |z|\leqq|x|\}\leqq 1$, since $A$ is solid; hence, $|f|\in A^\circ$. Thus if $f\in A^\circ$, $g\in F$ and $|g|\leqq|f|$, it follows that $|g(x)|\leqq|g|(|x|)\leqq|f|(|x|)\leqq 1$ for $x\in A$, whence $g\in A^\circ$. □

The roles of $E$ and $F$ cannot be interchanged in (4.7) without further precautions; simple examples show that the polar $J^\circ$ of an ideal $J$ in $F$ is not necessarily an ideal in $E$ (Exerc. 11). This deficiency is due to the fact that, although each element of $E$ defines an order bounded linear form on $F$, the evaluation map $E\to F^\star$ is not a lattice homomorphism in general. The situation improves when $F$ is assumed to be an ideal of $E^\star$.

**4.8 Proposition.** *Let $E$ be any vector lattice, and let $G$ be an ideal of $E^\star$. The evaluation map $x\mapsto\tilde{x}$, where $\tilde{x}(g):=\langle x,g\rangle$ $(x\in E, g\in G)$, is a lattice homomorphism of $E$ into $G^\star$.*

*Proof.* Since $x\mapsto\tilde{x}$ is evidently linear, it suffices to show that for each $x\in E$, the elements $(\tilde{x})^+$ and $(x^+)^\sim$ of $G^\star$ agree. It is clear that $(\tilde{x})^+\leqq(x^+)^\sim$. To prove the converse inequality we define, for each $f\in G_+$, a mapping $r_f: E_+\to\mathbb{R}_+$ by virtue of

$$r_f(y)=\sup\{f(z): z\in[0,y]\cap P\}\qquad(y\in E_+)$$

where $P:=\bigcup_{n=1}^\infty n[0,x^+]\subset E_+$. $r_f$ is positive homogeneous and additive (cf. proof of 4.4) and hence defines a positive linear form $g_f\in E^\star$. Since $0\leqq g_f\leqq f$, the assumption that $G$ be an ideal of $E^\star$ implies that $g_f\in G$. Also, $g_f(x^-)=0$ by the construction of $g_f$, and $g_f(x)=g_f(x^+)=f(x^+)$. We obtain

$$(x^+)^\sim(f)=f(x^+)=g_f(x)\leqq\sup_{0\leqq h\leqq f}h(x)=(\tilde{x})^+(f).$$

Since this holds for each $f \in G_+$, we obtain $(x^+)^\sim \leqq (\tilde{x})^+$ and hence, the desired equality. □

**Corollary.** *For any ideal $G$ of $E^\star$, the annihilator $G^\circ \subset E$ is an ideal of $E$ and $E/G^\circ$ can be identified with a vector sublattice of $G^\star$.*

This is clear from (4.8) and (2.6) Cor., since $G^\circ$ is the kernel of the evaluation map $E \to G^\star$.

The reader should be cautioned not to believe that the evaluation map $E \to G^\star$ is sequentially order continuous let alone order continuous, unless $G \subset E_{00}^\star$ (Exerc. 11). However, if $G$ is an ideal contained in $E_{00}^\star$ then (4.8) can be considerably strengthened (Thm. 4.12 below); this result, due to Nakano [1950], is our ultimate goal in this section. For its proof we need several preparatory results, which also are of independent interest.

**4.9 Proposition.** *Let $E$ be a countably order complete vector lattice and suppose there exists a strictly monotone mapping $\varphi: E_+ \to \mathbb{R}$. Then $E$ is order complete. Moreover, if $A$ is a non-void set in $E$ such that each countable subset of $A$ is majorized, then $\sup A$ exists and there is a countable subset $A_0$ of $A$ for which $\sup A_0 = \sup A$.*

*Proof.* By an increasing transfinite sequence in $E$ we understand a mapping $\alpha \mapsto x_\alpha$ defined on the set of all ordinals less than some ordinal $\beta$, into $E$ such that $\alpha < \alpha' < \beta$ implies $x_\alpha \leqq x_{\alpha'}$. Since there exist no uncountable subsets of $\mathbb{R}$ which are well ordered under the ordering induced by $\mathbb{R}$, the postulated existence of a strictly monotone mapping $\varphi: E_+ \to \mathbb{R}$ implies that every strictly increasing transfinite sequence in $E_+$ (and therefore, in $E$) is countable.

Suppose now $A$ to be a non-void subset of $E$ such that each countable subset $C$ of $A$ is majorized, let $x_C = \sup C$ $(C \neq \emptyset)$ and denote by $A'$ the set of all these elements $x_C$. By transfinite recursion we can construct an increasing transfinite sequence $(x_\alpha)_{\alpha < \varepsilon}$, indexed by the set of all countable ordinals $\alpha$, in such a way that $x_\alpha = x_{\alpha+1}$ if and only if $x_\alpha$ is the greatest element of $A'$. Since this sequence cannot be strictly increasing at each countable ordinal $\alpha$, $A'$ has a greatest element $x_C$ and, evidently, $x_C = \sup A$. □

An order complete vector lattice $E$ such that each majorized subset $A \neq \emptyset$ of $E$ contains a countable subset $A_0$ satisfying $\sup A_0 = \sup A$, is called *super Dedekind complete* (cf. Luxemburg-Zaanen [1971]). It is trivial to prove but important to observe that whenever $E$ is super Dedekind complete, then $E_0^\star = E_{00}^\star$. The following is an immediate corollary of (4.9).

**Corollary.** *Every countably order complete vector lattice on which there exists a strictly positive linear form, is super Dedekind complete.*

If $E$ is a vector lattice and $0 \leqq f \in E^\star$, the absolute kernel $N(f) := \{x \in E: f(|x|) = 0\}$ (§ 2) is an ideal of $E$; clearly, $N(f)$ is a band whenever $f \in E_{00}^\star$ (cf. Exerc. 11). We are interested in a condition on $E$ guaranteeing that $N(f)$ is a projection band.

**4.10 Proposition.** *Let $E$ be countably order complete, and let $0 \leqq f \in E_{00}^\star$. Then $E = N(f) + N(f)^\perp$, and $N(f)^\perp$ is super Dedekind complete.*

*Proof*. First we observe that $N(f)^\perp$ is a band in $E$ on which $f$ is strictly positive and hence, $N(f)^\perp$ is super Dedekind complete by the corollary of (4.9). Moreover, for each $x \in E_+$ the set $[0,x] \cap N(f)^\perp$ is a directed $(\leqq)$ set each of whose countable subsets is majorized in $N(f)^\perp$; from (4.9) it follows that $\sup [0,x] \cap N(f)^\perp$ exists in $N(f)^\perp$ and thus, by (2.11), $N(f)^\perp$ is a projection band in $E$. So $E = N(f)^{\perp\perp} + N(f)^\perp$; finally, $N(f) = N(f)^{\perp\perp}$, since $N(f)$ is a band and $E$ is Archimedean (Exerc. 3). □

For each $0 \leqq f \in E^\star$, the band $N(f)^\perp \subset E$ will be called the *band of strict positivity* of $f$, and be denoted by $P(f)$; $P(f)$ is the unique band $B \subset E$ such that $x \in B$ iff $0 < y \leqq |x|$ implies $f(y) > 0$.

**Corollary.** *Let $E$ be countably order complete. For each pair $f, g$ of positive order continuous linear forms on $E$, one has the relations: $P(f \vee g) = P(f) + P(g)$, $P(f \wedge g) = P(f) \cap P(g)$ and $N(f \vee g) = N(f) \cap N(g)$, $N(f \wedge g) = N(f) + N(g)$ where all of the occurring bands are projection bands of $E$.*

*Proof*. Since for each $h$, $0 \leqq h \in E^\star_{00}$, $E = P(h) + N(h)$ is a band decomposition of $E$ by (4.10), it suffices by (2.9) to prove the relations $N(f \vee g) = N(f) \cap N(g)$ and $N(f \wedge g) = N(f) + N(g)$. While the former of these is an easy consequence of (4.2), Formula (1), the latter is contained in the following more general result. □

**4.11 Proposition.** *Let $E$ be a countably order complete vector lattice, and let $0 \leqq f, g \in E^\star_0$. Then $N(f \wedge g) = N(f) + N(g)$; in particular, if $f \perp g$ then $E = N(f) + N(g)$.*

*Proof*. Suppose $0 \leqq x \in N(f \wedge g)$. By (1) of (4.2), for each $n \in \mathbb{N}$ there exists a decomposition $x = y_n + z_n$ $(y_n, z_n \in E_+)$ such that $f(y_n) + g(z_n) < 2^{-(n+1)}$. Let $\bar{z}_n := \sup_{k \geqq n} z_k$ and $\bar{y}_n := x - \bar{z}_n$. The hypothesis $g \in E^\star_0$ implies that

$$g(\bar{z}_n) \leqq \textstyle\sum_{k=n}^{\infty} g(z_k) \leqq 2^{-n}$$

and, since $f(y_k) < 2^{-(k+1)}$ and $\bar{y}_n = x - \bar{z}_n \leqq x - z_k \leqq y_k$ for all $k \geqq n$, we have $f(\bar{y}_n) = 0$ for all $n \in \mathbb{N}$. Now define $z := \inf_n \bar{z}_n$ and $y := \sup_n \bar{y}_n$; we obtain $g(z) = 0$ and, using the hypothesis $f \in E^\star_0$, $f(y) = 0$. On the other hand, $x = y + z$ so $x \in N(f) + N(g)$. Therefore, $N(f \wedge g) \subset N(f) + N(g)$ while the inclusion $N(f) + N(g) \subset N(f \wedge g)$ is obvious. □

**Example.** 6. Let $\mu$ be a signed measure on a measurable space $(X, \mathbf{S})$ in the sense of Halmos [H, § 28], and let $E$ denote a countably order complete vector lattice of $\mu$-integrable real functions on $X$. If $N$ is the ideal of $\mu$-null functions in $E$ then $E/N$ is super Dedekind complete by the corollary of (4.9), since $|\mu|$ induces a strictly positive linear form on $E/N$, and since $E/N$ is countably order complete (Exerc. 5).

If $\mu$ is order continuous on $E$ then $P(|\mu|) = N(|\mu|)^\perp$ provides a "*lifting*" of $E/N$ to a band in $E$; in general, however, $\mu \notin E^\star_{00}$ and the existence of a lifting becomes a serious problem (see A. and C. Ionescu-Tulcea [1969]).

Finally, suppose $\mu$ to be totally $\sigma$-finite. Applying (4.11) to the vector lattice $E$ of all $\mu$-integrable real functions with $f = \mu^+$ and $g = \mu^-$, we conclude that there exists a measurable decomposition $X = X_1 \cup X_2$ such that

$X_1 \cap X_2$ is $|\mu|$-null and the restriction of $\mu$ to $\{X_1 \cap S: S \in \mathbf{S}\}$ is positive, to $\{X_2 \cap S: S \in \mathbf{S}\}$ negative (*Hahn decomposition* of $X$ with respect to $\mu$; cf. [H, § 29] and [DS, II.IV.10]).

The following result is the announced refinement of (4.8); interesting applications will be considered below (§§ 5, 8).

**4.12 Theorem** (Nakano). *Let $E$ be an order complete vector lattice, let $G$ be an ideal in $E^\star_{00}$, and denote by $q$ the evaluation map $E \to G^\star$. Then $q(E)$ is an order dense ideal of $G^\star_{00}$.*

*Proof.* Without loss of generality we can suppose that $G$ separates $E$; otherwise, observing that the polar $G^\circ$ (with respect to $\langle E, E^\star \rangle$) is a band in $E$, we can consider the band $(G^\circ)^\perp$ of $E$ which by (2.10) is isomorphic to $E/G^\circ$. Hence, we can suppose $q$ injective and thus identify $E$ with the order complete sublattice $q(E)$ of $G^\star_{00}$.

(i) Let $E(x)$ denote the band of $E$ generated by $x \in E_+$, let $N(x) := \{f \in G: \langle x, |f| \rangle = 0\}$, and let $P(x) := N(x)^\perp$ in $G$. The decompositions

$$E = E(x) + E(x)^\perp,$$
$$G = P(x) + N(x)$$

are dual in the sense that $N(x) = E(x)^\circ$, $P(x) = [E(x)^\perp]^\circ$ (and hence, $E(x) = N(x)^\circ$, $E(x)^\perp = P(x)^\circ$). In fact, it is clear that $G$ decomposes into the direct sum of ideals $G = [E(x)^\perp]^\circ + E(x)^\circ$ and it is straightforward that $N(x) = E(x)^\circ$, since each $f \in G$ is order continuous; because $P(x)$ and $[E(x)^\perp]^\circ$ are both ideals complementary to $N(x)$ in $G$, they must agree (2.7).

(ii) Let $0 < u \in E$. We show that $u = u_1 + u_2$, where $u_1, u_2 \in G^\star_{00}$ and $u_1 \wedge u_2 = 0$, implies already that $u_1$ and $u_2$ belong to $E$. If $P(u_i) \subset G$ denotes the band of strict positivity of $u_i$, where $u_i$ $(i = 1, 2)$ is considered a (order continuous) linear form on $G$, then $P(u) = P(u_1) + P(u_2)$ by the corollary of (4.10). Now if $B(u), B(u_i)$ denote the bands of $G^\star_{00}$ generated by $u$ and $u_i$, respectively, and if $E(u)$ denotes the band of $E$ generated by $u$, then in view of (i) we have the dual decompositions

$$\begin{aligned} G^\star_{00} &= B(u_1) + B(u_2) + B(u)^\perp, \\ E &= E(u) \qquad\qquad\quad + E(u)^\perp, \\ G &= P(u_1) + P(u_2) + N(u). \end{aligned}$$

The point of this part of the proof now consists in showing that there exist bands $E_i$ of $E$ satisfying $E_i \subset B(u_i)$ $(i = 1, 2)$ and $E_1 + E_2 = E(u)$. To this end let $E_i$ denote the band of $E$ generated by all bands $N(f)^\perp \subset E$, where $0 < f \in P(u_i)$ and $N(f) := \{x \in E: \langle |x|, f \rangle = 0\}$. By (4.10) Cor. we have $E_1 \cap E_2 = \{0\}$ and since $P(u) = [E(u)^\perp]^\circ$ by (i), $E_i \subset E(u)$. Suppose now that $x \in E$ is orthogonal to $E_1 + E_2$; then $\langle |x|, f \rangle = 0$ for all $f$, $0 < f \in P(u_1) + P(u_2) = P(u)$; therefore by (i), $x \in E(u)^\perp$. Thus, since $E_1 + E_2$ is a band in $E$, it follows from (2.10) that $E_1 + E_2 = E(u)$. But by (i), $0 < f \in P(u_i)$ implies $N(f)^\perp \subset B(u_i)$ and so $E_i \subset B(u_i)$ $(i = 1, 2)$. Now if $u = v_1 + v_2$ where $v_i \in E_i \subset B(u_i)$ it is clear that we must have $v_i = u_i \in E$ $(i = 1, 2)$, since $B(u_1) \perp B(u_2)$.

(iii) Next we prove that $E$ is an ideal in $G_{00}^{\star}$. Let $0 \leqq v \leqq u$ be satisfied for elements $u \in E$, $v \in G_{00}^{\star}$. The supremum $w := \sup[0, v] \cap E$ (taken in $G_{00}^{\star}$) is an element of $E$, $E$ being identified with an order complete sublattice of $G_{00}^{\star}$. Let $y := v - w$; the assertion is proved if $y = 0$. If not, there exists $\lambda_0 > 0$ in $\mathbb{R}$ such that $(y - \lambda_0 u)^+ > 0$. Put $z := y - \lambda_0 u$ and denote by $p$ the band projection of $G_{00}^{\star}$ onto the band generated by $z^+$. Then $p(y - \lambda_0 u) = py - \lambda_0 pu = pz^+ = z^+ > 0$, so $pu > 0$ and $0 < \lambda_0 pu \leqq py \leqq y$ (cf. 2.9). But $u = pu + (1_{G_{00}^{\star}} - p)u$ is a decomposition of $u \in E$ into orthogonal summands and hence, $pu \in E$ by (ii). This contradicts the definition of $w$, since $w + \lambda_0 pu \leqq w + y = v$ and $w + \lambda_0 pu \in E$. Thus $y = 0$.

(iv) To show that $E$ is order dense in $G_{00}^{\star}$ (or equivalently, that $G_{00}^{\star}$ is the band in $G^{\star}$ generated by $E$), it suffices to show that $0 \leqq z \in G_{00}^{\star}$ and $z \perp E$ implies $z = 0$. But $z \perp E$ implies $P(z) \perp P(u)$ in $G$ for all $u \in E_+$ by the corollary of (4.10), whence $P(z) \subset N(u)$ for all $u \in E_+$. Therefore, $P(z) \subset E^{\circ} = \{0\}$ and hence, $z = 0$.

This completes the proof of (4.12). □

## § 5. Normed Vector Lattices

We define a *topological vector lattice* $E$ to be a vector lattice and a Hausdorff topological vector space over $\mathbb{R}$ which possesses a base of solid 0-neighborhoods (Def. 2.1); this compatibility requirement between the vector lattice and topological structures of $E$ seems to be the most natural and is equivalent to normality of the positive cone $E_+$ and (uniform) continuity of the lattice operations [S, V.7.1]. If the topology of $E$ is locally convex, $E$ is called a *locally convex vector lattice*; since the convex hull of a solid set is solid by (2.2), every locally convex vector lattice has a 0-neighborhood base of convex solid sets. The gauge function (Minkowski functional, [S, Chap. II, § 1]) of an absorbing, convex solid subset of a vector lattice is called a *lattice semi-norm*; thus the topology of a locally convex vector lattice can be defined by a family of lattice semi-norms. We shall primarily be concerned with the case where the topology of $E$ is given by a single lattice semi-norm which is a norm. (Recall that a function $p: E \to \mathbb{R}$ is a semi-norm iff (i) $p(x+y) \leqq p(x) + p(y)$ and (ii) $p(\lambda x) = |\lambda| p(x)$ for all $x, y \in E$ and $\lambda \in \mathbb{R}$; $p$ is a norm iff, in addition, $p(x) = 0$ implies $x = 0$.)

**5.1 Definition.** *Let $E$ be a vector lattice. A semi-norm (norm) $p$ on $E$ is called a* lattice semi-norm (lattice norm) *if $|x| \leqq |y|$ implies $p(x) \leqq p(y)$ for all $x, y \in E$. If $p$ is a lattice norm on $E$, the pair $(E, p)$ is called a* normed (vector) lattice; *if, in addition, $(E, p)$ is norm complete it is called a* Banach lattice.

The defining property of a lattice semi-norm $p$ on a vector lattice $E$ is equivalent to requiring that $p$ be a semi-norm satisfying $p(x) = p(|x|)$, and $p(y) \leqq p(x)$ whenever $0 \leqq y \leqq x$; for this reason, lattice (semi-)norms are sometimes called *monotone* (semi-)norms. We observe that a normed vector lattice is necessarily Archimedean, and that a vector lattice and normed vector space $E$ (over $\mathbb{R}$) is a normed vector lattice iff the unit ball of $E$ is solid. (For the equivalence of a given norm on a vector lattice $E$ with a suitable lattice norm, cf. Exerc. 13.)

**Examples.** 1. Let $E$ be any vector lattice and let $f$ denote a positive linear form on $E$. Then $x \mapsto p(x) := f(|x|)$ is a lattice semi-norm on $E$. More generally, if $\mathsf{M}$ is any non-void $\sigma(E^\star, E)$-bounded set of positive linear forms on $E$, then $x \mapsto p_{\mathsf{M}}(x) := \sup_{f \in \mathsf{M}} f(|x|)$ is a lattice semi-norm on $E$; in fact, it is easy to see from (4.7) and (4.8) that each lattice semi-norm $p$ on $E$ is of this form with $\mathsf{M}$ the positive part $P^\circ \cap E^\star_+$ of the polar $P^\circ$ of $P := \{x \in E : p(x) \leq 1\}$. Moreover, the kernel $J := \{x \in E : p(x) = 0\}$ of any lattice semi-norm $p$ is an ideal of $E$, and $p$ induces a lattice norm on $E/J$. (Cf. Exerc. 13(d).)

2. For a detailed discussion of the classical spaces $\mathscr{L}^p(\mu)$ and $L^p(\mu)$, where $(X, \Sigma, \mu)$ is a measure space, the reader is referred to [DS, III § 3 and IV § 7]. Generalizations of these spaces are the Banach function spaces studied by Luxemburg and Zaanen [1963a]—[1965c] as well as Orlicz spaces (cf. Zaanen [1967]). See also Exerc. 14.

A general construction of Banach lattices proceeds as follows. Let $(E, p)$ be a semi-normed vector lattice, let $(X, \Sigma, \mu)$ be a measure space, and denote by $\mathscr{M}(X, E)$ the vector space (pointwise addition and multiplication by scalars) of all maps $f: X \to E$ which are measurable with respect to $\Sigma$ and the Borel field of the topological space $E$. Then $\mathscr{M}(X, E)$ is a vector lattice under its canonical order (§ 1, Example 3). If $\mathscr{M}_\mu$ denotes the ideal of all $f \in \mathscr{M}(X, E)$ for which

$$p_\mu(f) := \int p \circ f(s)\, d\mu(s) < +\infty,$$

then $(\mathscr{M}_\mu, p_\mu)$ is a semi-normed vector lattice, from which a normed vector lattice can be obtained in the usual way (Example 1). Of course, in this construction $p_\mu$ can be replaced by the lattice semi-norm $p_{\mathsf{M}} = \sup_{\mu \in \mathsf{M}} p_\mu$, where $\mathsf{M}$ denotes an arbitrary family of (countably additive) measures on $(X, \Sigma)$, and each $p_\mu$ can be replaced by some $p_{\mu,\alpha}(f) := (\int [p(f)]^\alpha d\mu)^{1/\alpha}$ $(1 \leq \alpha < +\infty)$ or by $p_{\mu,\infty}(f) := \sup_{s \in X} \operatorname{ess} p \circ f(s)$.

A particular case, which is very useful in spectral theory (cf. Chap. V, § 1), arises for $X = \mathbb{N}$, $\Sigma$ the power set of $\mathbb{N}$, and $\mathsf{M}$ any set of positive real sequences.

3. The vector lattice $\mathsf{M}_A$ of all bounded, finitely additive real functions on a Boolean algebra $A$ (§ 1, Example 5 and § 4, Example 2) is a Banach lattice under the norm $\mu \mapsto \|\mu\| := \sup_{a \in A} |\mu|(a)$; $\|\mu\|$ is called the total variation of $\mu$. As is well known, $\mathsf{M}_A$ can also be viewed as the Banach lattice of all Radon measures on the Stone representation space $K_A$ of $A$ (or equivalently, as the dual of $C(K_A)$; cf. § 7, Example 4).

If $A$ is a $\sigma$-algebra, the set $\mathsf{N}_A$ of all countably additive $\mu \in \mathsf{M}_A$ is a band in $\mathsf{M}_A$ by (4.3); hence $\mathsf{N}_A$ is closed in $\mathsf{M}_A$ (see 5.2 below) and, therefore, a Banach lattice.

4. Let $[a, b]$ denote a compact interval of the real axis. The space $C[a, b]$ of continuous real functions on $[a, b]$, endowed with its canonical order and the supremum norm, is a Banach lattice with order unit $e$ (the constant-one

function). If $C^{(n)}[a,b]$ $(n \geqq 1)$ denotes the vector space of all real functions $n$ times continuously differentiable on $[a,b]$, the identity

$$f(s) = \sum_{\nu=0}^{n-1} \frac{1}{\nu!} f^{(\nu)}(a)(s-a)^\nu + \frac{1}{(n-1)!} \int_a^s (s-t)^{n-1} f^{(n)}(t)\,dt$$

shows the vector space $C^{(n)}[a,b]$ to be isomorphic to $\mathbb{R}^n \times C[a,b]$; in an obvious manner, this isomorphism can be used to provide $C^{(n)}[a,b]$ with the structure of a Banach lattice. (For example, for $n=2$ one obtains the Banach lattice $C^{(2)}[a,b]$ whose positive cone consists of all convex $f \in C^{(2)}[a,b]$ satisfying $f(a) \geqq 0$ and $f'(a) \geqq 0$.)

5. Topological vector lattices whose topology is defined by a countable family of lattice norms arise in various contexts of analysis; for example, in distribution theory. Thus the vector lattice (canonical order) of rapidly decreasing real sequences $a = (\alpha_1, \alpha_2, \ldots)$ (which is isomorphic to the space of rapidly decreasing, infinitely differentiable real functions on $\mathbb{R}$) is a complete, metrizable locally convex vector lattice under the topology defined by the lattice norms $p_k(a) := \sum_{n=1}^{\infty} n^k |\alpha_n|$ $(k = 1, 2, \ldots)$.

We turn to the discussion of the elementary properties of normed vector lattices; many of these results are valid, mutatis mutandis, in more general situations. However, as a rule such generalizations will not be mentioned.

**5.2 Proposition.** *Let $E$ denote a normed vector lattice.*

(i) *The mappings $x \mapsto x^+$, $x \mapsto x^-$, $x \mapsto |x|$ and $(x,y) \mapsto x \vee y$, $(x,y) \mapsto x \wedge y$ are uniformly continuous from $E$ (respectively, from $E \times E$) into $E$.*

(ii) *The positive cone $E_+$ is closed (in particular, $E$ is Archimedean).*

(iii) *The closure of a solid set (and hence, the weak closure of a convex solid set) is solid.*

(iv) *Each band (more generally, each $\sigma$-ideal) is closed in $E$.*

(v) *Each band projection is continuous.*

*Proof.* (i) The uniform continuity of $(x,y) \mapsto x \vee y$ and $(x,y) \mapsto x \wedge y$ results immediately from Formulae (6) and (6′) of (1.4) by virtue of the monotonicity of the norm of $E$ (Def. 5.1); taking $y=0$ yields the uniform continuity of $x \mapsto x^+$ and $x \mapsto x^-$, and hence of $x \mapsto |x| = x^+ + x^-$.

(ii) $E_+ = \{x \in E : x^- = 0\}$ is closed as the inverse image of the closed set $\{0\}$ under the continuous map $x \mapsto x^-$. If $x, y \in E$ and $nx \leqq y$ for all $n \in \mathbb{N}$, then $x - n^{-1} y \in -E_+$ for all $n$ which implies $x \in -E_+$ or $x \leqq 0$, since $-E_+$ is closed.

(iii) Let $A \subset E$ be solid and suppose that $|y| \leqq |x|$ for some $x \in \bar{A}$ and $y \in E$. There exists a sequence $(x_n)$ in $A$ converging to $x$. Since $|x_n| \in A$, defining $y_n^+ := y^+ \wedge |x_n|$ and $y_n^- := y^- \wedge |x_n|$ we have $y_n = y_n^+ - y_n^- \in A$, since $|y_n| \leqq |x_n|$ and since $A$ is solid, and we have $y = \lim_n y_n$ by (i); therefore, $y \in \bar{A}$. If, in addition, $A$ is convex then the weak closure of $A$ agrees with the norm closure and hence, is solid.

(iv) If $(x_n)$ is an increasing sequence convergent to $x \in E$, then $x = \sup_n x_n$; this is an easy consequence of $E_+$ being closed in $E$ (see 5.8 below). Suppose now that $I$ is a $\sigma$-ideal in $E$ (§ 2) and let $(z_n)$ be a sequence in $I$ convergent to $z \in E$.

Letting $y_n := |z_n| \wedge |z|$ we have $y_n \in I$ and by (i) the sequence $(y_n)$ converges to $|z|$. Now define $x_n := \sup_{1 \leqq \nu \leqq n} y_\nu$ $(n \in \mathbb{N})$; the elements $x_n$ form an increasing sequence in $I$ and satisfy $y_n \leqq x_n \leqq |z|$. Therefore, we obtain

$$\|x_n - |z|\| \leqq \|y_n - |z|\|$$

which shows $(x_n)$ to converge to $|z|$ in $E$. But then $|z| = \sup_n x_n$ by our introductory remark; since $I$ is a $\sigma$-ideal, we have $|z| \in I$ and hence, $z \in I$. Therefore, $I$ is closed.

(v) If $B$ is a projection band of $E$ and $x \mapsto Px$ the associated band projection (§ 2), then $|Px| \leqq P|x| \leqq |x|$ implies that $\|Px\| \leqq \|x\|$ for all $x \in E$; hence, $P$ is continuous (and of norm $\leqq 1$). ∎

**Corollary 1.** *The closure of a vector sublattice (respectively, ideal) of E is a vector sublattice (respectively, ideal) of E.*

*Proof.* Since the closure of a vector subspace is a vector subspace of $E$, the assertion follows from (5.2)(i) for vector sublattices and from (5.2)(iii) for ideals (Def. 2.1). ∎

**Corollary 2.** *With respect to the unique continuous extensions of the vector and lattice operations and the norm, the completion of (the uniform space associated with) a normed vector lattice is a Banach lattice.*

The easy verification, based on the uniform continuity of the lattice operations and the norm, is left to the reader, who will note that the positive cone of the completion $\tilde{E}$ is the closure in $\tilde{E}$ of $E_+$. It should be noted that norm completeness and order completeness of a normed vector lattice are independent properties (Exerc. 15).

**Corollary 3.** *Every real valued, sequentially order continuous lattice homomorphism of a normed vector lattice is (norm) continuous.*

*Proof.* It suffices to observe that the kernel $f^{-1}(0)$ of such a lattice homomorphism $f$ is a $\sigma$-ideal of $E$ hence closed by (5.2)(iv), which implies the continuity of $f$ (cf. [S, I.4.2]). ∎

Much stronger continuity assertions can be made when $E$ is supposed to be a Banach lattice.

**5.3 Theorem.** *Let E, F be normed vector lattices and suppose E to be norm complete. Every positive linear map $T: E \to F$ is continuous.*

*Proof.* Let $T: E \to F$ be linear and positive (Def. 2.4). If $T$ is not continuous then $T$ must be unbounded on the unit ball $U$ of $E$ and hence, since $U \subset U \cap E_+ - U \cap E_+$, on $U \cap E_+$. This implies the existence of a sequence $(x_n)$ in $U \cap E_+$ such that $\|Tx_n\| \geqq n^3$ for all $n \in \mathbb{N}$. On the other hand, $E$ being complete, $z := \sum_n n^{-2} x_n$ exists in $E$ and we have $z \geqq n^{-2} x_n$ for all $n$, since $E_+$ is closed by (5.2)(ii). But then $Tz \geqq n^{-2} T x_n \geqq 0$ which implies $\|Tz\| \geqq n^{-2} \|Tx_n\| \geqq n$ for all $n$ by the monotonicity of the norm of $F$, a contradiction. ∎

*Note.* For the conclusion of (5.3) it suffices to have $E_+$ norm complete which is, however, equivalent to norm completeness of $E$. On the other hand, barreledness of a normed vector lattice $E$ does not suffice (Exerc. 13).

In the following corollaries, $E$ is supposed to be a Banach lattice, $F$ a normed vector lattice. A linear map $S: E \to F$ is called *absolutely majorized* if there exists a positive linear map $T: E \to F$ such that $|Sx| \leqq Tx$ for all $x \in E_+$.

**Corollary 1.** *Every absolutely majorized linear map $S: E \to F$ is continuous.*

*Proof.* If $T \geqq 0$ majorizes $S$, then $S = T - (T - S)$ is the difference of two positive linear maps and hence continuous. □

**Corollary 2.** *Every positive linear form on $E$ is continuous.*

**Corollary 3.** *Every maximal ideal of $E$ is closed.*

*Proof.* By the corollary of (3.4), each maximal ideal of $E$ is the kernel of a positive linear form on $E$ and hence closed by Cor. 2. □

**Corollary 4.** *If a vector lattice $E$ is a Banach lattice for two distinct norms, these norms are equivalent*[1].

*Proof.* In fact, the identity map of $E$ is positive, hence continuous in both directions. □

**5.4 Proposition.** *For any closed ideal $I$ of the normed vector lattice $E$, the quotient $E/I$ is a normed vector lattice under its canonical order and norm; $E/I$ is norm complete whenever $E$ is.*

*Proof.* It is well known that the normed space $E/I$, whose norm is defined by $\|\hat{x}\| = \inf\{\|x\|: x \in \hat{x}\}$, is norm complete whenever $E$ is (cf. [S, I.6.3]); to see that the quotient norm is a lattice norm on $E/I$, it suffices to observe that $q(U)$ is solid in $E/I$ where $U$ is the unit ball of $E$ and $q: E \to E/I$ is the quotient map. This follows from (2.5), since $U$ is solid in $E$ and $q$ is a lattice homomorphism onto $E/I$ (2.6). □

Thus the formation of closed vector sublattices, the completion process, and the formation of quotients over closed ideals generates new normed vector lattices from given ones; of course, so does the formation of finite Cartesian products (and certain more general procedures, cf. Example 3 above). Another important instance is the formation of the Banach dual of a normed vector lattice; recall that $E^\star$ denotes the order dual of $E$ (Def. 4.1).

**5.5 Proposition.** *If $E$ is any normed vector lattice and $E'$ denotes the vector space of continuous linear forms on $E$, then $E' \subset E^\star$ and $E'$ is an order complete Banach lattice under its dual norm and the ordering induced by $E^\star$. $E'$ is an ideal in $E^\star$ and, if $E$ is barreled, a band; moreover, $E' = E^\star$ whenever $E$ is norm complete.*

*Proof.* Denote by $U$ the unit ball of $E$, $U := \{x \in E: \|x\| \leqq 1\}$. Since $U$ is solid, Formula (3) of (4.2) shows that whenever $f \in E'$ then $f^+ \in E'$ and $f^- \in E'$ and

---

[1] I. e., there exist constants $k_1, k_2 > 0$ such that $k_1 \|x\|_1 \leqq \|x\|_2 \leqq k_2 \|x\|_1$ for all $x \in E$.

hence, that $E'$ is a vector sublattice of $E^\star$. By (4.7) the polar $U^\circ$ of $U$ (with respect to the duality $\langle E, E^\star\rangle$) is solid in $E^\star$ which simultaneously proves that $E'$ is a Banach lattice under its dual norm and an ideal of $E^\star$. Therefore, since $E^\star$ is order complete (4.2), $E'$ is order complete.

Suppose now that $E$ is barreled and $(f_\alpha)$ is a directed $(\leqq)$ family in $E'$ such that $f_0 := \sup_\alpha f_\alpha$ exists in $E^\star$. By (2) of (4.2), we have $\lim_\alpha f_\alpha(x) = f_0(x)$ for each $x \in E$ so that by virtue of the Banach-Steinhaus theorem (cf. [S, III.4.2 and III. 4.6]) we have $f_0 \in E'$; thus $E'$ is a band in $E^\star$.

The final assertion is clear from (5.3) (Cor. 2). ∎

**Corollary 1.** *For each closed ideal $I$ of a normed vector lattice $E$, the polar $I^\circ$ (with respect to $\langle E, E'\rangle$) is a $\sigma(E', E)$-closed band of $E'$. Moreover, $I'$ can be identified with the Banach lattice $E'/I^\circ$ (and hence, with $(I^\circ)^\perp$ in $E'$), and $(E/I)'$ can be identified with the normed ideal $I^\circ$ of $E'$.*

*Proof.* It is clear that $I^\circ$ is $\sigma(E', E)$-closed, and by (4.7) $I^\circ$ is an ideal of $E'$; the fact that $I^\circ$ is a band (and hence a projection band, since $E'$ is order complete) now follows easily from Formula (2) following (4.2). For the remaining assertions it suffices to verify that the norm isomorphisms $I' \to E'/I^\circ$ and $(E/I)' \to I^\circ$ (cf. [S, IV, p. 161]) are order isomorphisms as well; this can be left to the reader. ∎

**Corollary 2.** *Under evaluation, every normed vector lattice is isomorphic to a normed vector sublattice of its bidual $E''$.*

*Proof.* By virtue of the Hahn-Banach theorem, the evaluation map $E \to E''$ is a norm isomorphism; hence the assertion follows from (4.8), since $E'$ is an ideal of $E^\star$. ∎

**5.6 Proposition.** *Let $N$ be a vector sublattice of the normed vector lattice $E$. Each continuous positive linear form on $N$ has a norm preserving positive (linear) extension to $E$.*

*Proof.* We apply (4.5) with $U$ the unit ball of $E$ and assume that $0 < f_0 \in N'$ has norm 1, so that $f_0(x) \leqq 1$ for $x \in U \cap N$. Let $x \in (U - E_+) \cap N$, then $x \leqq u$ for some $u \in U$ which implies $x^+ \leqq u^+ \in U$ and hence, $x^+ \in U \cap N$. Since $f_0$ is positive, we obtain $f_0(x) \leqq f_0(x^+) \leqq 1$. Thus by (4.5) $f_0$ has a positive extension $f$ to $E$ such that the hyperplane $\{x \in E : f(x) = 1\}$ does not intersect the interior of $U$; so $\|f\| \leqq 1$ and, since $f$ extends $f_0$, $\|f\| = \|f_0\|$. ∎

The following remark concerning (5.6) may not be amiss. If $j: N \to E$ denotes the canonical injection, its adjoint $j': E' \to N'$ is a strict metric homomorphism (i.e., maps the unit ball of $E'$ onto that of $N'$, by virtue of the Hahn-Banach theorem); now (5.6) shows that $j'(E'_+ \cap U^\circ) = N'_+ \cap V^\circ$, where $V = U \cap N$ is the unit ball of $N$ and $V^\circ$ is the polar of $V$ with respect to the duality $\langle N, N'\rangle$. Thus $j'$ maps the positive cone of $E'$ onto that of $N'$ but is not, in general, a lattice homomorphism. For $j'$ to be a lattice homomorphism it is necessary and sufficient that $\overline{N}$ be an ideal in $E$ (cf. Chap. III, Exerc. 24); in this (and only this) case $N^\circ$ is an ideal in $E'$ and the Banach lattices $E'/N^\circ$ and $N'$ can be identified under the bijective map induced by $j'$ (cf. 5.5).

Let $E$ be a normed vector lattice with unit ball $U$, and let $0<f\in E'$. Since $|f(x)|\leqq f(|x|)$ $(x\in E)$ and since $x\in U$ implies $|x|\in U$, the norm of $f$ is given by

$$\|f\|=\sup\{f(x)\colon x\in U\cap E_+\}. \tag{1}$$

(This implies, in particular, the following property of dual Banach lattices: If $(f_\alpha)$ is a directed $(\leqq)$ subset of $E'_+$ such that $f:=\sup_\alpha f_\alpha$ exists in $E'$, then $\|f\|=\sup_\alpha\|f_\alpha\|$. Cf. (4.2), Formula (2) and Exerc. 15.) Dually, if $E$ is identified with a normed vector sublattice of $E''$ by virtue of (5.5) Cor. 2, then the norm of each $x\in E_+$ is given by

$$\|x\|=\sup\{\langle x,x'\rangle\colon x'\in U^\circ\cap E'_+\}. \tag{2}$$

Thus $E_+$ can be viewed as a convex cone of continuous real functions on the $\sigma(E',E)$-compact space $X:=U^\circ\cap E'_+$. If there exists a unique smallest closed subset $P$ of $X$ such that each $x\in E_+$ attains its maximum on $P$, then $P$ is called the *Šilov boundary* of $E_+$. Since we will make essential use of the Šilov boundary later (Chap. III, § 5), we proceed to prove its existence for the positive cone of a normed lattice and identify it explicitly. (Under more general assumptions, the proof of the existence of a Šilov boundary is due to H. Bauer [1961]. See also Alfsen [1971], Chap. I, §§ 5, 6.) We recall that a *face* $F$ of a convex set $C$ is a convex subset $\neq\emptyset$ which is extreme in $C$ (i.e., such that $a\in F$, $b\in C$, $c\in C$ and $a=\lambda b+(1-\lambda)c$ implies $b\in F$, $c\in F$ whenever $0<\lambda<1$). Also, if $E$ is a normed vector lattice with (closed) unit ball $U$, we write $U^\circ_+$ for the positive part $U^\circ\cap E'_+$ of the dual unit ball. Since $U^\circ\subset U^\circ_+-U^\circ_+$, it is clear that each $x\in E$ is determined by its values on $U^\circ_+$.

**5.7 Theorem.** *Let $E$ be a normed vector lattice with unit ball $U$. Then the Šilov boundary of $E_+$, $E_+$ being considered as a cone of continuous positive real functions on the $\sigma(E',E)$-compact space $U^\circ_+$, exists and is identical with the $\sigma(E',E)$-closure $P$ of the set of all extreme points of $U^\circ_+$ which are maximal in $U^\circ_+$ (for the canonical order of $E'$).*

*Proof.* Recall that a *boundary* $Q$ of $E_+$ is a closed subset of $U^\circ_+$ such that each $x\in E_+$ attains its maximum (equivalently, its norm; see (2) above) on $Q$. The proof consists of two parts: First, we show that $P$ is a boundary for $E_+$ and second, we show that every boundary $Q$ contains $P$.

Let $x_0\in E_+$ be given; we show there exists a point $x'_0\in P$ such that $\langle x_0,x'_0\rangle=\sup\{\langle x_0,x'\rangle\colon x'\in U^\circ_+\}=\|x_0\|$. A face $F$ of $U^\circ_+$ is called hereditary $(\geqq)$ if $x'\in F$, $y'\geqq x'$ and $y'\in U^\circ_+$ implies $y'\in F$; the family of all closed hereditary faces of $U^\circ_+$ is evidently ordered inductively under downward inclusion and hence, by Zorn's lemma, each closed hereditary face $F$ of $U^\circ_+$ contains a minimal such face $F_0$. Now consider the set $F:=\{y'\in U^\circ_+\colon\langle x_0,y'\rangle=\|x_0\|\}$; it is readily seen that $F$ is a closed hereditary face and hence, $F$ contains a minimal $F_0$. Since $F_0$ is convex compact, $F_0$ contains an extreme point $x'_0$ which, because $F_0$ is a face of $U^\circ_+$, is extreme in $U^\circ_+$. If $x'_0$ were not maximal in $U^\circ_+$ then, since $F_0$ is hereditary, there would exist $y'\in F_0$ satisfying $y'>x'_0$. But since $E_+$ separates $E'$ this would imply the existence of some $y \in E_+$ such that $\langle y, y' \rangle > \langle y, x_0' \rangle$ and, as direct verification shows, the set

$$F_y := \{y' \in F_0 : \langle y, y' \rangle = \sup_{z' \in F_0} \langle y, z' \rangle\}$$

would be a closed hereditary face properly contained in $F_0$, contradicting the minimality of $F_0$. Therefore, $F_0 = \{x_0'\}$ and so $x_0' \in P$.

The fact that every boundary $Q$ of $E_+$ contains $P$ follows at once from the subsequent lemma when applied to $H = E_\sigma'$, $G = E$, $K = U_+^\circ$. (We feel the lemma becomes more transparent when presented in greater generality than is needed for our purposes.)

**Lemma.** *Let $H$ be an ordered vector space with $\sigma(H,G)$-closed positive cone $H_+$, where $G$ is a $H$-separating subspace of the algebraic dual $H^*$, let $K$ be a convex $\sigma(H,G)$-compact subset of $H$, and let $x_0$ be an extreme point of $K$ which is maximal in $K$ (with respect to the order of $H$).*

*For each $\sigma(H,G)$-neighborhood $U$ of $x_0$, there exists a positive linear form $f \in G$ such that $\sup_{x \in K \setminus U} f(x) < f(x_0)$.*

*Proof.* By definition of the weak topology $\sigma(H,G)$, there exist closed semi-spaces $H_i := \{x \in H : f_i(x) \leqq \alpha_i\}$, where $f_i \in G$ $(i = 1, \ldots, n)$, such that

$$x_0 \in H \setminus (H_1 \cup \cdots \cup H_n) \subset U .$$

Let $K_i := K \cap H_i$ and denote by $K_0$ the convex hull of $\bigcup_1^n K_i$; then $K_0$ is compact [S, II.10.2] but we must have $x_0 \notin K_0$, since otherwise $x_0 \in \bigcup_1^n K_i$ which is impossible. Since $x_0$ is maximal in $K$, it follows that $(x_0 + H_+) \cap K = \{x_0\}$ and so $(x_0 + H_+) \cap K_0 = \emptyset$. By a well-known separation theorem [S, II.9.2], there exists a $\sigma(H,G)$-closed real hyperplane $\{x \in H : f(x) = \alpha\}$ strictly separating $K_0$ from $x_0 + H_+$; without loss of generality we can assume that $\sup\{f(x) : x \in K_0\} < \alpha < f(x_0)$. Then $f$ is positive, since it is bounded below on $H_+$, and the assertion follows because of $K \setminus U \subset K_0$.

This completes the proof of (5.7). □

*Remark.* An inspection of the proof of (5.7) shows that the following considerably more general result is valid: *Let $E$ be an ordered topological vector space with total positive cone $E_+$, and let $K$ be a convex $\sigma(E',E)$-compact subset of $E'$. Then the Šilov boundary of $E_+$, $E_+$ considered as a cone of continuous real functions on $K$, is the $\sigma(E',E)$-closure of the set of extreme points of $K$ which are maximal in $K$* (under the canonical order of $E'$).

The imbedding of a normed vector lattice, by means of evaluation, in the space $C(U_+^\circ)$ of continuous functions on $U_+^\circ$ (positive part of the dual unit ball) leads to the following extension of Dini's classical convergence theorem. (For a direct proof under more general assumptions, cf. [S, V.4.3].) First, let us record this frequently needed lemma.

**5.8 Lemma.** *Let $E$ be an ordered vector space and a topological vector space such that the positive cone $E_+$ is closed. If $A$ is a directed* ($\leqq$) *subset of $E$ whose section filter converges to $x \in E$ then $x = \sup A$.*

*Proof.* Let $z \in A$ be fixed; for $y \in A$, $y \geqq z$ we have $y - z \in E_+$ and hence, $x - z \in E_+$, since $E_+$ is closed. Thus $x$ majorizes $A$. On the other hand, if $u$ majorizes $A$ then $u - y \in E_+$ for all $y \in A$ and hence, $u - x \in E_+$ or $u \geqq x$; this shows that $x = \sup A$. □

**5.9 Theorem.** *Let $E$ be a normed vector lattice, let $(x_\alpha)$ be a directed* ($\leqq$) *family in $E$, and suppose there exists $x \in E$ such that $\lim_\alpha \langle x_\alpha, x' \rangle = \langle x, x' \rangle$ for each $x'$ in the Šilov boundary $P$ of $E_+$. Then $x = \sup_\alpha x_\alpha$ and $\lim_\alpha \|x - x_\alpha\| = 0$.*

*Proof.* The functions $x' \mapsto \langle x_\alpha, x' \rangle$ ($x' \in P$, cf. 5.7) form a directed ($\leqq$) family of continuous real functions on the $\sigma(E', E)$-compact space $P$, converging pointwise to the continuous function $x' \mapsto \langle x, x' \rangle$. By Dini's classical theorem, the convergence is uniform on $P$ so (5.7) implies $\lim_\alpha \|x - x_\alpha\| = 0$. Since $E_+$ is closed in $E$ by (5.2), (5.8) shows that $x = \sup_\alpha x_\alpha$. □

**Corollary.** *Every weakly convergent directed family in a normed vector lattice is norm convergent.*

It is clear that an assertion corresponding to (5.9) holds for downward directed families; moreover, a family satisfying the hypotheses of (5.9) contains a countable subfamily having the same limit and supremum, by virtue of the metrizability of $E$. The reader should illustrate to himself the significance of (5.9) by applying it to specific normed vector lattices (see Examples 1—4 above). For a discussion of topological convergence vs. order convergence in topological vector lattices we refer to [S, Chap. V, § 7], particularly [S, V.7.5] and its corollaries.

We devote the remainder of this section to an investigation of those Banach lattices which are isomorphic, under evaluation, to an ideal of their norm (and order) bidual $E''$; in this situation the evaluation map $q: E \to E''$ preserves arbitrary infima and suprema, which is not true in general (cf. discussion immediately preceding 4.9).

**5.10 Theorem.** *For any Banach lattice $E$, the following assertions are equivalent:*

(a) *$E$ is order complete and each continuous linear form on $E$ is order continuous.*
(b) *Each directed* ($\leqq$) *majorized family in $E$ converges weakly.*
(c) *Each directed* ($\geqq$) *family in $E$ with infimum* 0, *norm converges to* 0.
(d) *$E$ is countably order complete and each decreasing sequence in $E$ with infimum* 0, *norm converges to* 0.
(e) *Evaluation $E \to E''$ maps $E$ onto an ideal of the Banach lattice $E''$.*
(f) *Each order interval of $E$ is $\sigma(E, E')$-compact.*

*Proof*. (a)$\Rightarrow$(b) is clear.

(b)$\Rightarrow$(c): Condition (b) is equivalent to the condition that each directed ($\geqq$) minorized family in $E$ be weakly (i.e., $\sigma(E,E')$-)convergent. Thus if $A\subset E$ is directed ($\geqq$) and if $\inf A=0$, it follows from (5.8) and (5.9) that $A$ norm converges to 0.

(c)$\Rightarrow$(d): If $A$ is a directed ($\leqq$) majorized subset of $E$ and $B$ denotes the set of all majorants of $A$, then $C:=B-A$ is directed downward with $\inf C=0$. Thus (c) implies that $C$, and hence $A$, is a Cauchy family in $E$ which shows $\sup A$ to exist (cf. 5.8); therefore, $E$ is order complete and a fortiori countably order complete. The second statement of (d) is trivially implied by (c).

(d)$\Rightarrow$(a): Suppose $A$ is a directed ($\leqq$) majorized subset of $E$; we can assume that $A$ contains the supremum of each of its countable subsets. As in the proof of (4.9) we consider increasing transfinite sequences and claim that for each strictly increasing transfinite sequence $(x_\alpha)_{\alpha<\beta}$ contained in $A$, the ordinal $\beta$ is countable. In fact, if not there would exist an increasing sequence $(\alpha_n)_{n\in\mathbb{N}}$ of ordinals $\alpha_n<\beta$ and a real number $\varepsilon>0$ such that $\|x_{\alpha_{n+1}}-x_{\alpha_n}\|>\varepsilon$ for all $n\in\mathbb{N}$. However, $(x_{\alpha_n})$ converges to its supremum in $A$ by hypothesis (d); that is, the sequence $(x_{\alpha_n})$ is a Cauchy sequence in $E$, a contradiction. Thus by transfinite recursion we can construct an increasing transfinite sequence $(x_\alpha)$ in $A$, indexed by all countable ordinals $\alpha$, such that $x_\alpha=x_{\alpha+1}$ iff $x_\alpha=\sup A$. It follows that $E$ is super Dedekind complete and hence, by hypothesis (d), that each $f\in E'$ is order continuous.

(a)$\Rightarrow$(e): Since $E'(=E^\star)=E^\star_{00}$, the implication results immediately from Theorem (4.12).

(e)$\Rightarrow$(f): If, under evaluation, $E$ is identified with a sublattice of $E''$, then (e) implies that each order interval $[x,y]_E=(x+E_+)\cap(y-E_+)$ agrees with the corresponding interval $[x,y]_{E''}=(x+E''_+)\cap(y-E''_+)$. Since the latter interval is clearly $\sigma(E'',E')$-compact, the assertion follows.

(f)$\Rightarrow$(a): Suppose $A$ is a directed ($\leqq$) subset of $E$ majorized by $z\in E$; without loss of generality we can suppose that $A\subset E_+$ and hence, $A\subset[0,z]$. Weak compactness of $[0,z]$ implies that $A$ converges weakly (and hence, by (5.9), in norm) to some $x\in[0,z]$; by (5.8), then, $x=\sup A$, since $E_+$ is (norm and weakly) closed in $E$. It is immediately clear from this argument that each $f\in E'$ is order continuous.

This completes the proof of (5.10). □

*Remark*. $A$ theorem analogous to (5.10) for normed (not necessarily norm complete) vector lattices results if to each of the assertions (a)—(d), the requirement is added that order intervals be norm complete, (e) and (f) being left unchanged.

**Corollary 1.** *A Banach lattice E satisfying any one of the assertions in* (5.10) *is super Dedekind complete, and the band of* $E''$ *generated by E is the band of all order continuous linear forms on* $E'$.

*Proof*. The first assertion results from (5.10)(c) (by virtue of (5.8) and the metrizability of $E$), while the second follows from (4.12). □

**Corollary 2.** *If a Banach lattice E satisfies any one of the assertions in* (5.10), *then every band projection in* $E'$ *is* $\sigma(E',E)$*-continuous (hence, each band in* $E'$ *is* $\sigma(E',E)$*-closed).*

*Proof.* Consider the locally convex topology on $E'$ generated by the semi-norms $f \mapsto p_x(f) := \langle x, |f| \rangle$, where $x \in E_+$; it is the topology of uniform convergence on order bounded subsets of $E$ and denoted by $o(E',E)$ (Exerc. 28). Since by (5.10)(f) each order interval of $E$ is $\sigma(E,E')$-compact, the Mackey-Arens theorem [S, IV.3.2] shows $o(E',E)$ to be consistent with the dual system $\langle E,E' \rangle$; therefore, the weak topology associated with $o(E',E)$ is $\sigma(E',E)$. On the other hand, if $P$ denotes any band projection in $E'$ then by (2.9) we have $|Pf| \leqq |f|$ for all $f \in E'$. So $p_x(Pf) \leqq p_x(f)$ for each $x \in E_+$, which implies $P$ to be $o(E',E)$-continuous and hence, $\sigma(E',E)$-continuous. Finally, since $E'$ is order complete each band of $E'$ is a projection band (2.10) and hence, each band is $\sigma(E',E)$-closed in $E'$. □

Before considering examples, we prove a characterization of reflexive Banach lattices due to Ogasawara [1948] (cf. Luxemburg-Zaanen [1964c], p. 530). Of course, we call a Banach lattice $E$ *reflexive* if the underlying Banach space is reflexive; that is, if the evaluation map $q: E \to E''$ is surjective.

**5.11 Theorem.** *A Banach lattice E is reflexive if and only if these two conditions are satisfied:*

(i) *Every norm bounded increasing sequence in E is norm convergent.*

(ii) *Every positive decreasing sequence in* $E'$ *is norm convergent.*

*Moreover, if E is reflexive, E is super Dedekind complete and conditions* (i) *and* (ii) *hold with sequences replaced by arbitrary directed families.*

*Proof.* Conditions (i) and (ii) are necessary. In fact, if $E$ is reflexive each norm bounded increasing sequence (which is evidently $\sigma(E,E')$-Cauchy) is weakly convergent, hence norm convergent by (5.9) Cor. On the other hand, every decreasing positive sequence in $E'$ is $\sigma(E',E)$-Cauchy and equicontinuous, hence $\sigma(E',E)$-convergent and, since $E$ is reflexive, norm convergent in $E'$ by (5.9) Cor.

Conditions (i) and (ii) are sufficient. Clearly, (i) implies that (d) of (5.10) is satisfied; if $q: E \to E''$ denotes the evaluation map then by (5.10)(a) and (4.12), $q(E)$ is an ideal in $(E')^\star_{00}$. But $E'$ is order complete by (5.5) and hence, (ii) implies by (5.10), (d)$\Rightarrow$(a) that $E'' = (E')^\star_{00}$ which shows that $E''$ is the band (in $E''$) generated by $q(E)$. We have to show that $q(E) = E''$. Let $0 < z \in E''$ then by (2.11) there exists a directed ($\leqq$) set $A \subset E_+$ such that $z = \sup q(A)$. To show that $z \in q(E)$, we argue as in step (d)$\Rightarrow$(a) of the proof of (5.10): Let $(x_\alpha)_{\alpha<\beta}$ be a maximal strictly increasing transfinite sequence in $A$; if $(x_\alpha)_{\alpha<\beta}$ is countable then, since $A$ is norm bounded in $E$, (i) implies that $x := \sup_{\alpha<\beta} x_\alpha$ exists in $E$, and evidently we have $z = q(x) = \sup q(A)$, since $q(E)$ is an ideal of $E''$. But every strictly increasing transfinite sequence $(x_\alpha)_{\alpha<\beta}$ must be countable (i.e., $\beta$ a countable ordinal). Otherwise there would exist a strictly increasing sequence $(\alpha_n)_{n\in\mathbb{N}}$ of ordinals $<\beta$ and a real number $\varepsilon > 0$ such that $\|x_{\alpha_n} - x_{\alpha_{n+1}}\| > \varepsilon$; this is impossible, because $(x_{\alpha_n})$ is a norm bounded increasing sequence in $E$ and hence convergent by (i).

The final assertion is clear from (5.10) and its corollary. □

**Examples.** 6. An important example of a Banach lattice satisfying the equivalent assertions of (5.10) is $c_0$, the vector lattice of real null sequences under the supremum norm (more generally $c_0(\mathsf{A})$, $\mathsf{A}$ any index set). In fact $c_0$ is an ideal in its bidual $l^\infty$; by contrast, $l^\infty$ is not an ideal in its bidual. Standard examples of reflexive Banach lattices are the spaces $L^p(\mu)$ $(1<p<+\infty)$.

7. An intermediate class between reflexive Banach lattices and those which are ideals in their biduals, are the Banach lattices satisfying only condition (i) of (5.11); an inspection of the proof of (5.11) shows condition (i) to be equivalent to the condition that (under evaluation) $E$ be isomorphic to a band in $E''$ (precisely, the band of all order continuous linear forms on $E'$). Banach lattices $E$ having this property are called *KB-spaces* by Vulikh [1967]. Examples of non-reflexive $KB$-spaces are furnished by $l^1$ and $L^1(\mu)$ ($\mu$ Lebesgue measure on $\mathbb{R}$) or, more generally, by all infinite dimensional $AL$-spaces (§ 8). It will be shown below (10.6) that *a Banach lattice E is a KB-space iff E is weakly sequentially complete.*

**5.12 Definition.** *A normed vector lattice E is said to have* order continuous norm *if every order convergent filter in E norm converges.*

Thus a Banach lattice $E$ has order continuous norm iff $E$ satisfies the equivalent assertions of (5.10); in particular, reflexive Banach lattices (see 5.11) and $KB$-spaces (Example 7) have order continuous norm. It is interesting and surprising that Banach lattices with order continuous norm as well as the two subclasses mentioned, can be characterized by the non-imbeddability in $E$ of one or several of the spaces $c_0, l^1, l^\infty$ as closed vector sublattices. We devote the remainder of this section to these characterizations, the key device of the proofs being the following lemma (Meyer-Nieberg [1973b]) on the existence of certain disjoint sequences. (See also Exerc. 17, 18.)

**5.13 Lemma.** *Let E denote a normed vector lattice, let $\delta>0$ and $c>0$ be real numbers, and let $(v_n)_{n\in\mathbb{N}}$ be a sequence in $E_+$ such that $\|v_n\|\geqq 1+\delta$ and $\|\sum_{\nu=1}^n v_\nu\|\leqq c$ for all n. Suppose in addition that at least one of the following conditions is satisfied:*

(i) *$(v_n)$ is majorized by some $x\in E_+$.*

(ii) *E is order complete with order continuous norm.*

*There exists a sequence $(k(n))_{n\in\mathbb{N}}$ of natural numbers and a disjoint sequence $(x_n)$ in $E_+$, $\|x_n\|\geqq 1$, such that $x_n\leqq v_{k(n)}$ for $n\in\mathbb{N}$.*

*Proof.* First note: For each $\alpha>0$ there exists a sequence $(k(n))$ in $\mathbb{N}$ such that

$$\|(\alpha v_{k(1)}-v_{k(n)})^-\|\geqq 1+\delta/2 \tag{1}$$

for all $n$. Otherwise, there must exist a subsequence $(w_n)$ of $(v_n)$ such that $\|(\alpha w_j - w_k)^-\|<1+\delta/2$ whenever $j<k$. This implies

$$\begin{aligned}\alpha c &\geqq \alpha\|w_1+\cdots+w_n\| = \|n w_{n+1}-(w_{n+1}-\alpha w_1)-\cdots-(w_{n+1}-\alpha w_n)\|\\ &\geqq \|n w_{n+1}-(w_{n+1}-\alpha w_1)^+-\cdots-(w_{n+1}-\alpha w_n)^+\|\\ &> n(1+\delta)-n(1+\delta/2)=n\delta/2\end{aligned}$$

for all $n\in\mathbb{N}$, which is contradictory.

Second note: There exists $p\in\mathbb{N}$ such that

$$\|(v_1-v_{p+1}-\cdots-v_{p+n})^+\|\geqq 1+\delta/2 \tag{2}$$

for all $n$. Otherwise, for each $p\in\mathbb{N}$ there must exist a least natural number $r(p)$ such that $\|(v_1-v_{p+1}-\cdots-v_{p+r(p)})^+\|<1+\delta/2$. Defining $p_1=1$ and $p_{i+1}=p_i+r(p_i)$ recursively, we obtain

$$\begin{aligned}c &\geqq \|v_2+\cdots+v_{p_{n+1}}\| = \|n v_1-(v_1-v_2-\cdots-v_{p_2})-\cdots-(v_1-v_{p_n+1}-\cdots-v_{p_{n+1}})\|\\ &\geqq \|n v_1-(v_1-v_2-\cdots-v_{p_2})^+-\cdots-(v_1-v_{p_n+1}-\cdots-v_{p_{n+1}})^+\|\\ &> n(1+\delta)-n(1+\delta/2)=n\delta/2\end{aligned}$$

for all $n\in\mathbb{N}$, which is contradictory.

To prove the assertion we now show the existence of a sequence $(x_1, y_2, y_3, \ldots)$ satisfying $0\leqq x_1\leqq v_{k(1)}$, $\|x_1\|\geqq 1$ and $0\leqq y_n\leqq v_{k(n)}$, $\|y_n\|\geqq 1+\delta/2$, $x_1\wedge y_n=0$ for all $n\geqq 2$, under each of the additional assumptions (i) and (ii); the proof is then completed by induction.

Assumption (i): By (1) there exists a sequence $(k(n))$ in $\mathbb{N}$ such that $\|(3\delta^{-1}\|x\|v_{k(1)}-v_{k(n)})^-\|>1+\delta/2$ whenever $n\geqq 2$. Define

$$x_1:=\left(v_{k(1)}-\frac{\delta}{3\|x\|}x\right)^+,\qquad y_n:=\left(v_{k(n)}-\frac{3\|x\|}{\delta}v_{k(1)}\right)^+$$

for $n\geqq 2$. By definition, $\|y_n\|\geqq 1+\delta/2$, and it is clear that $x_1\wedge y_n=0$ $(n\geqq 2)$. Moreover,

$$\begin{aligned}\|x_1\| &= \left\|\left(v_{k(1)}-\frac{\delta}{3\|x\|}x\right)+\left(v_{k(1)}-\frac{\delta}{3\|x\|}x\right)^-\right\|\\ &\geqq \|v_{k(1)}\|-\delta/3-\left\|\left(v_{k(1)}-\frac{\delta}{3\|x\|}x\right)^-\right\|\geqq 1+\delta/3\,.\end{aligned}$$

Assumption (ii): By (1) there exists a sequence $(k(n))$ in $\mathbb{N}$ such that $\|(v_{k(n)}-v_{k(1)})^+\|\geqq 1+\delta/2$ for all $n\geqq 2$. By (2) there exists $p\in\mathbb{N}$ such that $\|(v_{k(1)}-v_{k(p+1)}-\cdots-v_{k(p+n)})^+\|\geqq 1+\delta/2$ for all $n$. Define

$$\begin{aligned}x_1 &:= \inf\nolimits_n (v_{k(1)}-v_{k(p+1)}-\cdots-v_{k(p+n)})^+,\\ y_n &:= (v_{k(p+n-1)}-v_{k(1)})^+\end{aligned}$$

for $n \geqq 2$. Then it is clear that $\|x_1\| \geqq 1+\delta/2$, since the norm of $E$ is order continuous, and that $\|y_n\| \geqq 1+\delta/2$ $(n \geqq 2)$. Finally, we obtain

$$0 \leqq x_1 \wedge y_n \leqq (v_{k(1)} - v_{k(p+1)} - \cdots - v_{k(p+n-1)})^+ \wedge (v_{k(p+n-1)} - v_{k(1)})^+ \leqq y_n^- \wedge y_n^+ = 0$$

whenever $n \geqq 2$.

This completes the proof of (5.13). □

**5.14 Theorem.** *Let $E$ denote a countably order complete Banach lattice. The following are equivalent:*

(a) *$E$ has order continuous norm (thus, $E$ satisfies the equivalent assertions of (5.10)).*

(b) *No Banach sublattice of $E$ is vector lattice isomorphic to $l^\infty$.*

(c) *Every positive linear map $l^\infty \to E$ is weakly compact*[1].

(d) *Every closed ideal of $E$ is a band.*

*Proof.* (a)⇒(b): Since every vector lattice isomorphism of a Banach lattice onto a closed vector sublattice of $E$ is bicontinuous by (5.3) Cor. 4, the implication follows because $l^\infty$ does not have order continuous norm.

(a)⇒(c) and (a)⇒(d) are clear from (5.10).

(b)⇒(a): Suppose the norm of $E$ is not order continuous; then there exist $\varepsilon > 0$ and a decreasing sequence $(z_n)$ in $E_+$ such that $\|z_n - z_{n+1}\| > \varepsilon$ for all $n \in \mathbb{N}$. Choose any $\delta > 0$ and define $v_n := \varepsilon^{-1}(1+\delta)(z_n - z_{n+1})$; the sequence $(v_n)$ satisfies the hypothesis of (5.13). Thus there exists a disjoint normalized sequence $(x_n)$ in $E_+$ which is majorized by $x_0 := \varepsilon^{-1}(1+\delta)z_1$.

Clearly, the linear hull of the sequence $(x_n)$ is a vector sublattice of $E$; moreover, since $\sum_{n=1}^k x_n \leqq x_0$ for all $k \in \mathbb{N}$ and since $E$ is countably order complete, for each sequence $(\alpha_n) \in l^\infty$ the series $\sum_{n=1}^\infty \alpha_n x_n$ order converges to some $x \in E$. It is clear that the correspondence $(\alpha_n) \mapsto x$ defines an isomorphism of $l^\infty$ onto a vector sublattice $E_0$ of $E$, and the estimate

$$\sup_n |\alpha_n| \leqq \|\textstyle\sum_{n=1}^\infty \alpha_n x_n\| \leqq (\sup_n |\alpha_n|)\|x_0\|$$

shows $E_0$ to be closed in $E$ which contradicts (b).

(c)⇒(b): The negation of (b) (given (c) holds) would imply that the unit ball of $l^\infty$ is weakly compact, which is false.

(d)⇒(a): Suppose the norm of $E$ is not order continuous. Then, as in (b)⇒(a), we can prove the existence of a disjoint, normalized sequence $(x_n)$ in $E_+$ which is majorized. $E$ being countably order complete, let $x := \sup_n \sum_{\nu=1}^n x_\nu$. Evidently $x$ belongs to the band of $E$ generated by $(x_n)$. Denoting by $I$ the closed ideal generated by the sequence $(x_n)$, we will show that $x \notin I$ and thus establish a contradiction with (d).

Now the assumption $x \in I$ implies the existence of $z \in E$ such that $0 \leqq z \leqq x$, $\|x - z\| < \frac{1}{2}$ and $z \leqq \sum_{n=1}^k \varrho_n x_n$ for suitable $k \in \mathbb{N}$ and numbers $\varrho_n \in \mathbb{R}_+$. Thus if $P_n$ denotes the band projection of $E$ onto the principal band generated by $x_n$

[1] Cf. (9.4) et seq.

(cf. (2.11) Cor. 2) and if $y_n := P_n z$, then we have $0 \leqq y_n \leqq P_n x = x_n$ for all $n \in \mathbb{N}$, and $z = \sum_{n=1}^{k} y_n$. Consequently,

$$\tfrac{1}{2} > \|x - z\| \geqq \|\textstyle\sum_{n=1}^{k+1} x_n - \sum_{n=1}^{k} y_n\| \geqq \|x_{k+1}\|$$

which is contradictory, since $\|x_{k+1}\| = 1$. □

**Corollary.** *Every separable, countably order complete Banach lattice has order continuous norm.*

A similar characterization, with $l^\infty$ replaced by $c_0$, can now be proved for $KB$-spaces (Example 7 above).

**5.15 Proposition.** *For any Banach lattice E, the following assertions are equivalent:*
(a) *Under evaluation, E is isomorphic to a band in $E''$.*
(b) *Every norm bounded increasing sequence in E converges.*
(c) *No Banach sublattice of E is vector lattice isomorphic to $c_0$.*
(d) *Every positive linear map $c_0 \to E$ is weakly compact.*

*Proof.* The equivalence (a)$\Leftrightarrow$(b) follows from (5.10) and (5.10) Cor. 1.

(b)$\Rightarrow$(d): Let $T: c_0 \to E$ be linear and positive, and let $z := \sup_k \sum_{n=1}^{k} Te_n$ where $e_n := (\delta_{nm}) \in c_0$. It is clear that the unit ball of $c_0$ is mapped by $T$ into $[-z, z]$, and intervals are weakly compact in $E$ by (5.10).

(d)$\Rightarrow$(c): If there exists a closed vector sublattice of $E$ vector lattice isomorphic to $c_0$ then (d) implies the identity map of $c_0$ to be weakly compact, which is contradictory.

(c)$\Rightarrow$(b): First, from (5.13) we conclude that (c) implies $E$ to have order continuous norm, since otherwise $E_+$ would contain a majorized, orthogonal normalized sequence and hence, a Banach sublattice isomorphic to $c_0$. Hence by (5.10), $E$ is an ideal in $E''$.

Now suppose $(y_n)$ to be a norm bounded increasing sequence which is not a Cauchy sequence; we can assume $(y_n) \subset E_+$. Since $(y_n)$ is majorized in $E''$, there exists a constant $c \in \mathbb{R}_+$ and a subsequence $(y_{n(k)})_{k \in \mathbb{N}}$ such that $v_k := c(y_{n(k+1)} - y_{n(k)})$ is a sequence in $E$ which, when considered as a sequence in $E''$, satisfies the hypothesis of (5.13). Thus, $E$ being an ideal in $E''$, there exists a disjoint normalized sequence $(x_n)$ in $E_+$. But it is readily seen that the closed vector sublattice $E_0$ of $E$ generated by $(x_n)$, is vector lattice isomorphic to $c_0$. Therefore, (c) implies (b). □

Our final result is an analogous characterization of reflexive Banach lattices that supplements (5.11). In contrast with (5.14), $E$ is not assumed to be countably order complete.

**5.16 Theorem.** *For any Banach lattice E, the following are equivalent assertions:*
(a) *E is reflexive.*
(b) *$E'$ and $E''$ have order continuous norm.*
(c) *No Banach sublattice of E is vector lattice isomorphic to either $c_0$ or $l^1$.*
(d) *No Banach sublattice of $E'$ is vector lattice isomorphic to either $c_0$ or $l^1$.*
(e) *Each separable Banach sublattice of E is reflexive.*

*Proof*. (a) $\Rightarrow$ (b) is clear from (5.10).

(b) $\Rightarrow$ (a): By virtue of (5.11) the assumption implies that $E'$ is reflexive, hence so is $E$ [S, IV. 5.7].

(a) $\Rightarrow$ (c) and (a) $\Rightarrow$ (d) are clear, since the dual and each Banach subspace of a reflexive Banach space are reflexive.

(d) $\Rightarrow$ (a): Suppose $E$ is not reflexive, then at least one of $E'$, $E''$ does not have order continuous norm.

(i) Suppose $E'$ does not have order continuous norm, then by (5.13) there exists a majorized, disjoint, normalized sequence in $E'$; the closed vector sublattice generated by this sequence is vector lattice isomorphic to $c_0$ which contradicts (d).

(ii) Suppose $E''$ does not have order continuous norm while $E'$ does. Since $E''$ is order complete, (5.14) shows that $l^\infty$ is contained (up to norm equivalence) in $E''$ as a closed vector sublattice. Denote by $B_n$ the band of $E''$ generated by the unit vector $e_n := (\delta_{nm}) \in l^\infty$. By (5.10) Cor. 2, the corresponding band projection is $\sigma(E'', E')$-continuous and hence, the adjoint of a band projection $E' \to B_n'$. Let $\delta \in \mathbb{R}$ satisfy $0 < \delta < 1$. For each $n$, there exists a positive normalized $x_n \in B_n'$ such that $\langle x_n, e_n \rangle \geqq c(1-\delta)$, where $c := \inf_n \|e_n\| > 0$. Now if $x = \sum_{n=1}^k \alpha_n x_n$ is any element of the vector subspace (and vector sublattice) $E_0$ of $E'$ generated by the sequence $(x_n)$ and if $C := \sup_k \|\sum_{n=1}^k e_n\|$, then

$$c(1-\delta) \sum_{n=1}^k |\alpha_n| \leqq \langle \sum_{n=1}^k |\alpha_n| x_n, \sum_{n=1}^k e_n \rangle \leqq C \|\sum_{n=1}^k \alpha_n x_n\| \leqq C \sum_{n=1}^k |\alpha_n|$$

and this shows the (norm) closure $\bar{E}_0$ in $E'$ to be vector lattice isomorphic to $l^1$. Again, this contradicts (d).

(a) $\Leftrightarrow$ (c) is now clear, since the Banach space $E$ is reflexive if and only if $E'$ is. (a) $\Rightarrow$ (e) is trivial.

(e) $\Rightarrow$ (a): If $E$ is not reflexive, then by (a) $\Leftrightarrow$ (c) $E$ contains at least one closed vector sublattice isomorphic to $c_0$ or to $l^1$, neither of which is reflexive. □

For further equivalences and characterizations, the interested reader is referred to Exerc. 17, 18.

## § 6. Quasi-Interior Positive Elements

By the Corollary of (1.9) an element $u \geqq 0$ of an Archimedean vector lattice $E$ is a weak order unit of $E$ if and only if the principal ideal $E_u$ is order dense in $E$. However, in topological vector lattices order convergence does not, in general, imply topological convergence (cf. 5.10) and thus the concept of weak order unit is frequently not appropriate for considerations involving topology. Therefore, we introduce a topological analog which will turn out to be extremely useful.

**6.1 Definition.** *An element $u \geqq 0$ of a topological vector lattice $E$ is called a* quasi-interior point of $E_+$ *(or a* quasi-interior positive element of $E$*) if the principal ideal $E_u$ is dense in $E$.*

It is easy to verify that the (possibly void) set $W$ of all weak order units of a vector lattice $E$ is a lattice subcone of $E_+$ satisfying $W+E_+=W$. Analogous behavior is shown by the set $Q$ of all quasi-interior points of $E_+$ when $E$ is a topological vector lattice (recall that a topological vector lattice is Hausdorff by definition, § 5); in particular, $Q=\operatorname{int} E_+$ whenever $E_+$ has interior points, which is the reason for our terminology.

**6.2 Proposition.** *The set $Q$ of quasi-interior positive elements of a topological vector lattice $E$ has the following properties:*

(i) *$Q$ is a sublattice and convex subcone of $E_+$ such that $Q+E_+=Q$.*

(ii) *If $E$ is locally convex and $Q\neq\emptyset$ then $Q$ is dense in $E_+$.*

(iii) *If $E_+$ has non-empty interior then $Q=\operatorname{int} E_+$.*

(iv) *If $E$ is separable and complete metrizable (in particular, if $E$ is a separable Banach lattice) then $Q\neq\emptyset$.*

*Proof.* (i) If $x\in Q$, $\lambda>0$, and $y\in E_+$ then $E_{\lambda x+y}\supset E_x$ which shows that $\lambda x+y\in Q$; therefore, $Q$ is a convex subcone of $E_+$ (not containing 0 unless $E=\{0\}$) such that $Q+E_+=Q$. Let $x,y\in Q$. It is clear from the preceding that $x\vee y\in Q$. On the other hand, it is easy to see (cf. proof of (III. 1.1)) that the intersection of two dense ideals is a dense ideal in $E$; therefore, since $E_{x\wedge y}=E_x\cap E_y$, we have $x\wedge y\in Q$.

(ii) By (i) the closure $\bar{Q}$ is a convex subset of $E_+$, since $E_+$ is closed in $E$ (recall that $E_+=\{x\in E: x^-=0\}$). Suppose there exists $y\in E_+\setminus\bar{Q}$; clearly, then, $y\neq 0$. So by the second separation theorem [S, II. 9.2] there exists a continuous linear form $f\in E'$ satisfying $f(y)=-1$ and $f(z)\geqq 0$ for $z\in\bar{Q}$. Hence, for suitable $z_0\in Q$ we obtain $f(z_0+y)<0$ which is contradictory, since $z_0+y\in Q$ by (i). Therefore, $E_+=\bar{Q}$.

(iii) Suppose $\operatorname{int} E_+$ to be non-void. For each $x\in\operatorname{int} E_+$ the set $(-x+\operatorname{int} E_+)\cap(x-\operatorname{int} E_+)$ is a neighborhood of 0 (in $E$) contained in the order interval $[-x,x]$ and hence, in $E_x$; this shows $E_x=E$ and so $x\in Q$. Conversely, if $x\in Q$ then $E_x$, which is dense in $E$, must contain some element $y\in\operatorname{int} E_+$; it follows that $y\leqq nx$ for some $n\in\mathbb{N}$. Since $n^{-1}y\in\operatorname{int} E_+$, we obtain $x\in n^{-1}y+E_+\subset(\operatorname{int} E_+)+E_+=\operatorname{int} E_+$ and so $x\in\operatorname{int} E_+$. Therefore, $Q=\operatorname{int} E_+$.

(iv) Suppose $E$ separable and complete metrizable. It is clear (since $E$ is metrizable, or by continuity of the lattice operations) that $E_+$ is separable. Denote by $x_n$ ($n\in\mathbb{N}$) the elements of a countable dense subset $S\subset E_+$ and let $\varrho$ be a pseudo-norm on $E$ such that the metric $(x,y)\mapsto\varrho(x-y)$ generates the topology of $E$ (cf. [S, I. 6.1]). For each $n$ choose $\alpha_n$, $0<\alpha_n\in\mathbb{R}$, so that $\varrho(\alpha_n x_n)<2^{-n}$. Since $E$ is complete and $E_+$ closed, the series $\sum_n \alpha_n x_n$ converges to an element $x\in E_+$ such that $E_x$ contains $S-S$ and hence is dense in $E$. Therefore, $x\in Q$. □

Thus we have $\operatorname{int} E_+\subset Q\subset W$ in any topological vector lattice; notice that interior points of $E_+$ are (strong) order units of $E$ (§ 2). The examples below will show that, in general, both inclusions are proper; in particular, $\operatorname{int} E_+\neq Q$ necessitates $\operatorname{int} E_+=\emptyset$, which is precisely the case making the concept of quasi-interior point useful. We return now to the setting of normed vector lattices to which the results of this section will be applied almost exclusively in the sequel.

**6.3 Theorem.** *Let $E$ be a normed vector lattice. For any $u \in E_+$ these assertions are equivalent:*

(a) *$u$ is quasi-interior to $E_+$.*

(b) *For each $x \in E_+$ the sequence $(x_n)$, where $x_n := x \wedge nu$ $(n \in \mathbb{N})$, norm converges to $x$.*

(c) *For each continuous linear form $f > 0$ on $E$, one has $f(u) > 0$.*

*Proof.* (a) $\Rightarrow$ (b): Let $x \in E_+$ be given. Since the ideal $E_u$ is dense in $E$ by hypothesis, there exists a sequence $(y_k)$ in $E_u$ convergent to $x$; without loss of generality we can assume $0 \leqq y_k \leqq y_{k+1}$ for all $k$, because $E_u$ is an ideal. By definition of $E_u$, there exists an integer-valued function $k \mapsto n(k)$ on $\mathbb{N}$ such that $y_k \leqq n(k)u$; it follows that $y_k \leqq x \wedge n(k)u = x_{n(k)} \leqq x$. This implies $\lim_k \|x - x_{n(k)}\| = 0$ and hence $\lim_n \|x - x_n\| = 0$, because $(x_n)$ is an increasing sequence.

(b) $\Rightarrow$ (c): Let $0 \leqq f \in E'$. Evidently $f(u) = 0$ implies $f(x_n) = 0$ $(x_n = x \wedge nu)$ for each $n$ and each $x \in E_+$; since $f$ is continuous, it follows that $f(x) = 0$ for all $x \in E_+$ and hence, $f = 0$.

(c) $\Rightarrow$ (a): Suppose $u \in E_+$ and assume that $\bar{E}_u \neq E$. By the Hahn-Banach theorem, there exists an $f \in E'$, $f \neq 0$, such that $f(\bar{E}_u) = \{0\}$. Then $|f| \in E'$ is a non-zero continuous positive linear form which vanishes at $u$ (cf. (4.2) Cor. 1). ☐

**Corollary 1.** *An element $x \geqq 0$ of a normed vector lattice $E$ is contained in a closed hyperplane supporting*[1] *$E_+$ if and only if $x$ is not quasi-interior to $E_+$.*

*Proof.* If $H = \{x \in E : f(x) = \alpha\}$ $(f \in E')$ is a closed hyperplane supporting $E_+$ then we must have $\alpha = 0$ and can assume $f \geqq 0$. (In fact, we can assume that $f(x) \geqq \alpha$ for all $x \in E_+$ which implies $\alpha \leqq 0$; since $H$ contains a point of $E_+$ we must have $\alpha = 0$.) The assertion follows immediately. ☐

Thus if $S$ denotes the set of points of $E_+$ in which $E_+$ is supported by a closed hyperplane, then $E_+$ is the disjoint union of $S$ and its quasi-interior $Q$; this observation furnishes another proof that $Q = \operatorname{int} E_+$ whenever $\operatorname{int} E_+ \neq \emptyset$. Moreover, if $E$ is a Banach lattice the term "closed" can be omitted from Cor. 1, because a hyperplane supporting $E_+$ is necessarily closed by (5.3) Cor. 2.

**Corollary 2.** *A quasi-interior positive element $x$ of a normed vector lattice $E$ is quasi-interior to $F_+$ for each vector sublattice $F$ containing $x$.*

*Proof.* This follows at once from (6.3) (c), since by (5.6) each $f_0 \in F'_+$ has an extension $f \in E'_+$. ☐

**Examples.** 1. Let $(X, \Sigma, \mu)$ denote a measure space. If $(X, \Sigma, \mu)$ is (totally) $\sigma$-finite then each of the Banach lattices $L^p(\mu)$ $(1 \leqq p \leqq +\infty)$ possesses quasi-interior positive elements. If $p < +\infty$ each function $f \in L^p$ such that $f(t) > 0$ a.e. $(\mu)$ is quasi-interior to the positive cone, and this concept is coextensive with that of weak order unit (cf. 6.5). By contrast, if $p = +\infty$ then $f$ is a quasi-interior positive element iff the $\mu$-essential infimum of $f$ is $>0$ (i.e., iff there exists $\delta > 0$ such that $f(t) \geqq \delta$ a.e. $(\mu)$). Thus $Q = \operatorname{int} E_+$ for $E = L^\infty(\mu)$

[1] A (real) hyperplane $H$ is $E$ is said to support $E_+$ at $x$ if $x \in E_+ \cap H$ and $E_+$ is contained in one of the semi-spaces determined by $H$ (cf. [S, Chap. II, § 9]).

while the functions $f \in E$ satisfying $f(t) > 0$ a.e. $(\mu)$ are precisely the weak order units of $E$.

Unless it is finite dimensional, the Banach space $L^\infty(\mu)$ is not separable (since it evidently contains a copy of $l^\infty$); moreover, $L^p(\mu)$ $(p < +\infty)$ is not necessarily separable if $(X, \Sigma, \mu)$ is (totally) $\sigma$-finite. These examples show that separability of a Banach lattice $E$ is not a necessary condition for the existence of quasi-interior points of $E_+$, even if $E$ is reflexive.

Finally, if $(X, \Sigma, \mu)$ is not (totally) $\sigma$-finite then none of the spaces $L^p(\mu)$ $(1 \leqq p < +\infty)$ contains quasi-interior positive elements; in these cases, an appropriate generalization is the concept of topological orthogonal system (Chap. III, § 5).

2. Let $X$ be a completely regular topological space and denote by $C_b(X)$ the Banach lattice (supremum norm) of all real valued, bounded continuous functions on $X$. The interior (equivalently, quasi-interior) positive elements are the functions $f$ for which $\inf_{t \in X} f(t) > 0$, and the weak order units are the positive functions whose zero set $\{t \in X : f(t) = 0\}$ is rare.

3. Suppose $X$ to be a locally compact, non-compact space and let $E := C_0(X)$ denote the Banach lattice (supremum norm) of all real valued, continuous functions on $X$ vanishing at infinity. Then $E_+$ has void interior, and the quasi-interior $Q$ of $E_+$ is non-void iff $X$ is countable at infinity. In fact, for an $f \in E_+$ to generate a dense ideal it is necessary and sufficient that $f(t) > 0$ for all $t \in X$; on the other hand, it is easy to see that a strictly positive function exists in $C_0(X)$ iff $X$ is countable at infinity.

4. The strong (normed) dual $E'$ of a Banach lattice does not, in general, contain quasi-interior positive elements even if $E_+$ has non-void interior. For example, if $E = C(K)$ where $K$ is a compact space, the dual $M(K)$ (§ 4, Example 5) contains the projection band $M_a(K)$ generated by all Dirac measures on $K$, as its atomic part (§ 4, Example 4). But $M_a(K)$ is isomorphic to $l^1(K)$ for the discrete space (index set) $K$; hence if $M(K)_+$ contains quasi-interior points then so does $l^1(K)$ (by Cor. 2 of (6.4) below), and this implies $K$ to be countable (Example 1).

The property of a Banach lattice $E$ to contain quasi-interior positive elements is not inherited by closed ideals of $E$ (Examples 2, 3) nor by the dual $E'$ (Example 4); however, it can be shown that if $E$ is reflexive then $E_+$ has non-void quasi-interior if and only if $E'_+$ does (see 6.6 Cor.). Moreover, the following positive assertion can be made.

**6.4 Proposition.** *Let $E, F$ be normed vector lattices and let $T: E \to F$ be a continuous, positive linear map with range dense in $F$. For each quasi-interior point $x \in E_+$, $Tx$ is quasi-interior to $F_+$.*

*Proof.* If $x$ is quasi-interior to $E_+$ then $E_x$ is dense in $E$, and from the continuity of $T$ it follows that $T(E_x)$ is dense in $T(E)$ and hence in $F$. Since $E_x = \bigcup_{n \in \mathbb{N}} n[-x, x]$, we have

$$T(E_x) = T(\textstyle\bigcup_n n[-x, x]) = \bigcup_n n\,T[-x, x] \subset \bigcup_n n[-Tx, Tx] = F_{Tx},$$

using that $T \geqq 0$; this shows $Tx$ to be quasi-interior to $F_+$. □

**Corollary 1.** *If $I$ is a closed ideal of the normed vector lattice $E$ and $q: E \to E/I$ denotes the canonical map, then $q$ maps the quasi-interior of $E_+$ into the quasi-interior of $(E/I)_+$.*

This is clear from (6.4), since $q$ is positive and surjective.

**Corollary 2.** *If $E_+$ contains quasi-interior points and $N$ is a vector sublattice of $E$ which is the range of a continuous, positive projection, then $N_+$ contains quasi-interior points.*

The latter corollary applies, in particular, to each projection band of $E$ (see 5.2 (v)).

It will be seen below (§ 9) that, except in trivial cases, a Banach lattice whose positive cone has non-void interior does not possess the properties listed in (5.10), and conversely. Thus the following result is a counterpart to (6.2)(iii).

**6.5 Proposition.** *If $E$ is a normed vector lattice isomorphic to an ideal of $E''$ under evaluation, then $W = Q$; that is, the respective sets of weak order units of $E$ and quasi-interior points of $E_+$ are identical.*

*Proof.* Without loss of generality we can assume $E$ to be a Banach lattice; in fact, the completion $\tilde{E}$ can be identified with the closure of $E$ in $E''$ and $\tilde{E}$ is an ideal in $E''$ whenever $E$ is. Moreover, each weak order unit and each quasi-interior point of $E_+$ preserves its status under the transition from $E$ to $\tilde{E}$. Now if $u \in E_+$ is a weak order unit of $E$ then $x = \sup_n x \wedge nu$ for all $x \in E_+$, by the corollary of (1.9); by (5.10) this supremum is a norm limit so (6.3) shows that $u$ is quasi-interior to $E_+$. Thus $W \subset Q$, the reverse inclusion being trivial. ☐

**Example.** 5. An example of a Banach lattice to which (6.5) applies is furnished by each of the spaces $L^p(\mu)$ ($1 \leqq p < +\infty$; cf. Example 1) and, more generally, by each reflexive Banach lattice and each $AL$-space (§ 8). Another example, which is neither reflexive nor an $AL$-space, is the Banach lattice $c_0$.

$c_0$ can be considered as the Banach lattice $C_0(X)$ for $X$ the (locally compact) discrete space $\mathbb{N}$. It is interesting to observe that among the Banach lattices $C_0(X)$ (Example 3) those with $X$ discrete are the only ones which are ideals in their bidual (cf. 5.10).

In fact, if $C_0(X)$ is an ideal in its bidual then it is order complete, and it is not difficult to see that for any $f \in C_0(X)_+$ the principal band generated by $f$ is isomorphic (as a Banach lattice) to $C_0(Y)$ for a suitable open-and-closed subspace $Y$ (of $X$) which is countable at infinity. Now for $C_0(Y)$ to be an ideal in its bidual, it is necessary by (6.5) that each $g \in C_0(Y)_+$ be either strictly positive or vanish on an open subset $\neq \emptyset$ of $Y$ (Examples 2, 3); this shows $Y$ to be discrete and hence, $X$ must be discrete.

Our final result concerns the existence of weak order units in the dual of a Banach lattice.

**6.6 Theorem.** *Let $E$ denote a countably order complete Banach lattice, and consider these assertions:*

(a) *$E$ has order continuous norm and $E_+$ has non-void quasi-interior.*
(b) *$E'$ possesses weak order units.*
(c) *$E$ has order continuous norm.*
*Then* (a) $\Rightarrow$ (b) $\Rightarrow$ (c).

*Proof.* (a) $\Rightarrow$ (b): Let $(v'_\alpha)_{\alpha \in \mathsf{A}}$ denote a maximal orthogonal system of $E'$; since $E'$ is a Banach lattice, it suffices to show that $(v'_\alpha)$ is necessarily countable. Now if $B'_\alpha$ denotes the band $\{v'_\alpha\}^{\perp}$ in $E'$, by (5.10) Cor. 2 there exists a band $B_\alpha$ in $E$ with polar $B_\alpha^\circ = B'_\alpha$; since $\bigcap_\alpha B_\alpha^\circ = \{0\}$, the bipolar theorem [S, IV. 1.5 Cor. 2] shows that the convex closure of $\bigcup_\alpha B_\alpha$ equals $E$; in particular, $E$ is the band generated by $\bigcup_\alpha B_\alpha$. Therefore, if $u$ is quasi-interior to $E_+$ and if $u_\alpha$ is the projection of $u$ in $B_\alpha$ $(\alpha \in \mathsf{A})$, we have $u = \sup_\alpha u_\alpha$ and $(u_\alpha)_{\alpha \in \mathsf{A}}$ is a maximal orthogonal system of $E$. But $E$ is super Dedekind complete by (5.10) Cor. 1, so $(u_\alpha)$ and hence $(v'_\alpha)$ is countable.

(b) $\Rightarrow$ (c): Suppose $E$ does not have order continuous norm. Then by (5.14) there exists a vector lattice isomorphism $i: l^\infty \to E$ with range a closed vector sublattice of $E$. Since order convergent filters in $E'$ are $\sigma(E', E)$-convergent, the adjoint $i'$ is an order continuous positive surjection (cf. discussion following (5.6)) of $E'$ onto $(l^\infty)'$; hence, if $w$ is a weak order unit of $E'$ then $i'(w)$ is a weak order unit of $(l^\infty)'$. But it is easy to see that the dual of $l^\infty$ contains no weak order units (in fact, such an element would have to be contained in the band $l^1 \subset (l^\infty)'$), and this contradiction proves the desired implication. ∎

Since the dual of a reflexive Banach lattice is reflexive, the following corollary results at once from (6.5).

**Corollary.** *Let $E$ be a reflexive Banach lattice. Then $E_+$ has non-void quasi-interior if and only if $E'_+$ does.*

## § 7. Abstract $M$-Spaces

Of particular importance both in the theory of Banach spaces and the theory of Banach lattices are the spaces $C(K)$ of continuous (real or complex valued) functions on a compact (Hausdorff) space $K$. An extensive investigation of these spaces has been made by Z. Semadeni [1971], to which the interested reader is referred for further study; in the present section we present their basic theory, which will be fundamental for the remainder of this book.

**7.1 Definition.** *A lattice norm $x \mapsto \|x\|$ on a vector lattice $E$ is called an* $M$-norm *if it satisfies the axiom*

(M) $$\|x \vee y\| = \|x\| \vee \|y\| \qquad (x, y \in E_+).$$

*$(E, \| \ \|)$ is called an* $M$-normed space, *and an $M$-normed Banach lattice is called an* abstract $M$-space (*briefly*, $AM$-space). *If the unit ball of an $M$-normed space contains a largest element $e$, $e$ is called the* unit *of $E$.*

Thus an $M$-normed space $E$ is a normed vector lattice (Def. 5.1) on whose positive cone $E_+$ the norm commutes with the formation of finite suprema; if

$E$ has a unit $e$, then the order interval $[-e,e]$ is the unit ball of $E$. This latter assertion has a converse, as follows.

**7.2 Proposition.** *Let $E$ be an Archimedean vector lattice possessing an order unit $e$. The gauge function of $[-e,e]$, given by*

$$p_e(x) := \inf\{\lambda \in \mathbb{R}: -\lambda e \leqq x \leqq \lambda e\} \quad (x \in E),$$

*is an $M$-norm on $E$. $(E,p_e)$ is an $AM$-space (with unit $e$) if and only if $E$ satisfies Axiom* (OS) (Def. 1.8).

*Proof.* It is at once clear that $p_e$ is a lattice semi-norm (Def. 5.1); moreover, $p_e(x)=0$ implies $x=0$, since $E$ is Archimedean. The Archimedean property of $E$ also implies that $|x| \leqq p_e(x)e$ for each $x \in E$; thus if $x, y \geqq 0$ then $x \vee y \leqq (p_e(x) \vee p_e(y))e$, and so $p_e(x \vee y) \leqq p_e(x) \vee p_e(y)$. Conversely, if $x, y \geqq 0$ and $x \vee y \leqq \lambda e$ then $\lambda \geqq p_e(x) \vee p_e(y)$ which shows that $p_e(x \vee y) \geqq p_e(x) \vee p_e(y)$. Therefore, $p_e$ is an $M$-norm.

If $(E,p_e)$ is an $AM$-space (i.e., norm complete) then clearly Axiom (OS) is satisfied. Conversely, suppose (OS) holds in $E$. To show that $(E,p_e)$ is norm complete it suffices to prove that, given a Cauchy sequence in $(E,p_e)$, there exists a subsequence $(x_n)$ such that $\sum_n (x_{n+1}-x_n)$ converges. But each Cauchy sequence in $(E,p_e)$ contains a subsequence $(x_n)$ such that $|x_{n+1}-x_n| \leqq \lambda_n e$ for all $n$, where $(\lambda_n)$ is a given sequence in $l^1$. Let such a sequence be chosen (for example, $\lambda_n = 2^{-n}$) and define $u_n := (x_{n+1}-x_n)^+$, $v_n := (x_{n+1}-x_n)^-$; we have $0 \leqq u_n \vee v_n \leqq \lambda_n e$ and it is enough to show that $\sum_n u_n$ converges. By Axiom (OS), $\sup_m \sum_{n=1}^m u_n =: u$ exists in $E$ and we obtain

$$u - \sum_{n=1}^m u_n = \sup_k \sum_{n=m+1}^{m+k} u_n \leqq \left(\sum_{n=m+1}^{\infty} \lambda_n\right) e,$$

whence it follows that $p_e(u - \sum_{n=1}^m u_n) \leqq \sum_{n=m+1}^{\infty} \lambda_n$ for all $m$. □

The following corollary explains the particular role played by $AM$-spaces in the theory of Banach lattices.

**Corollary.** *Let $E$ be a Banach lattice. For each $x \in E_+$, the principal ideal $E_x$ is an $AM$-space (under the norm whose unit ball is the order interval $[-x,x]$) with unit $x$, and the canonical imbedding $E_x \to E$ is continuous.*

This is clear from (7.2) and the obvious facts that each Banach lattice satisfies (OS) and that order intervals are bounded. It will be seen later (III. 4.2) that for any pair $x, y$ of quasi-interior points of $E_+$ (Def. 6.1), the $AM$-spaces $E_x$ and $E_y$ are canonically isomorphic.

**Examples.** 1. The Banach space $E$ (sometimes denoted by $m(E_n)$) of all bounded sequences $x=(x_n)$, where $x_n \in E_n$ and $E_n$ is an $AM$-space with unit $e_n$ $(n \in \mathbb{N})$, endowed with the norm $x \mapsto \|x\| := \sup_n \|x_n\|$ is an $AM$-space with unit $e=(e_n)$ under its canonical (coordinatewise) ordering. If $E_n = \mathbb{R}$ for all $n$, $E$ is usually denoted by $l^\infty$ (or $m$). $l^\infty$ is not separable (the characteristic functions of arbitrary subsets of $\mathbb{N}$ form an uncountable subset of $l^\infty$ whose elements have mutual distance 1); the closed vector sublattice $c$ of all convergent real sequences is a separable $AM$-space with unit $e=(1,1,\ldots)$, and the closed vector sublattice $c_0$ of real null sequences is an $AM$-space without unit.

2. Let $X$ be any topological space. The vector space $C_b(X)$ of all bounded, real valued continuous functions on $X$, endowed with its canonical order (§ 1, Example 3), is an $AM$-space with unit $e$ (where $e(s)=1$, all $s\in X$); for each topology $\boldsymbol{T}$ of $X$ coarser than the given one, $C_b(X)$ contains the $\boldsymbol{T}$-continuous, bounded real functions as a closed vector sublattice. (There is, however, no bijective correspondence between these topologies on $X$ and $AM$-subspaces with unit of $C_b(X)$, as the example $X=[0,1]$ (discrete topology) and the sublattice $B[0,1]$ of bounded Borel functions on $[0,1]$ (usual topology) shows. Cf. Example 4 below.)

Another important example is obtained as follows. Let $\varphi: X\to \mathbb{R}$ be a strictly positive function. A continuous real function $f$ on $X$ is called $\varphi$-dominated if for every real $\varepsilon>0$, there exists a compact subset $K$ of $X$ such that $|f(s)|\leqq\varepsilon\varphi(s)$ whenever $s\in X\setminus K$. Under the norm $\|f\|_\varphi:=\inf\{\lambda: |f|\leqq\lambda\varphi\}$, the vector lattice of $\varphi$-dominated functions is an $M$-normed space denoted by $D_\varphi(X)$; $D_\varphi(X)$ is an $AM$-space whenever $\varphi$ is upper semi-continuous. The spaces $D_\varphi(X)$ (with $X$ completely regular and $\varphi$ lower semi-continuous) play an important role for the representation of $AM$-spaces without unit (Goullet de Rugy [1972]).

3. Let $(X,\Sigma,\mu)$ be a measure space. (Recall that we require $\mu\geqq 0$ to be countably additive and $\Sigma$ to be a $\sigma$-algebra of subsets of $X$.) The vector space $\mathscr{L}^\infty(\mu)$ of all bounded, $\Sigma$-measurable real functions, endowed with its canonical order and the supremum norm, is an $AM$-space with unit $e$ $(e(s)=1$, all $s\in X)$. The subset $\mathscr{N}$ of all $\mu$-null functions is a $\sigma$-ideal (hence closed, (5.2) (iv)) in $\mathscr{L}^\infty(\mu)$, and the quotient $L^\infty(\mu):=\mathscr{L}^\infty(\mu)/\mathscr{N}$ is again an $AM$-space with unit.

The elementary permanence properties of $AM$-spaces are simple to describe: Each closed vector sublattice of an $AM$-space is an $AM$-space, the quotient of an $AM$-space (with unit) over a closed ideal is an $AM$-space (with unit), and the completion of an $M$-normed space (with unit) is an $AM$-space (with unit); the easy verification is omitted (Exerc. 19).

The prime example of an $AM$-space with unit is contained in Example 2 above; in fact, each $AM$-space with unit is isomorphic (as a Banach lattice) to $C(K)$ for a suitable compact space $K$. To prove this famous result (independently obtained by Kakutani [1941b] and M. and S. Krein [1940]) we need an equally important approximation theorem (Stone-Weierstrass theorem, lattice version) due to M. H. Stone [1937].

**7.3 Theorem.** *Let $K$ be a compact space, and let $F$ be a vector sublattice of $C(K)$ which contains the constant-one function $e$. If $F$ separates the points of $K$ then $F$ is dense in $C(K)$.*

*Proof.* Let $s$, $t$ be given points of $K$ and $\alpha$, $\beta$ given real numbers such that $\alpha=\beta$ if $s=t$, and suppose $F$ separates $K$. There exists $f\in F$ such that $f(s)=\alpha$ and $f(t)=\beta$. This is clear if $s=t$, since $e\in F$; if $s\neq t$, there exists $g\in F$ such that $g(s)\neq g(t)$, and a suitable linear combination of $g$ and $e$ will satisfy the requirement.

Now let $h \in C(K)$ and $\varepsilon > 0$ be preassigned; the lattice property of $F$ and the compactness of $K$, along with the foregoing remark, will permit the construction of $g \in F$ such that $\|h-g\| < \varepsilon$. First, let $s$ be any fixed element of $K$. For each $t \in K$ there exists $f_t \in F$ such that $f_t(s) = h(s)$ and $f_t(t) = h(t)$. The set $U_t := \{r \in K : f_t(r) > h(r) - \varepsilon\}$ is open and contains $t$; hence $K = \bigcup_{t \in K} U_t$ and the compactness of $K$ implies the existence of a finite set $\{t_1, \ldots, t_n\}$ such that $K = \bigcup_{v=1}^{n} U_{t_v}$. Since $F$ is a lattice, the function $g_s := \sup_v f_{t_v}$ is in $F$; it is clear that $g_s(t) > h(t) - \varepsilon$ for all $t \in K$, and $g_s(s) = h(s)$.

Now suppose $g_s \in F$ with these properties chosen for each $s \in K$. The set $V_s := \{r \in K : g_s(r) < h(r) + \varepsilon\}$ is open and contains $s$. Since $K = \bigcup_{s \in K} V_s$, there exists a finite subfamily $\{V_{s_\mu} : \mu = 1, \ldots, m\}$ covering $K$. Using the lattice property of $F$ again we obtain $g := \inf_\mu g_{s_\mu} \in F$, and it is clear that $h(r) - \varepsilon < g(r) < h(r) + \varepsilon$ for all $r \in K$; therefore, $\|h-g\| < \varepsilon$. □

The preceding theorem (as well as its algebraic version) have extensions to complex spaces $C(K)$ (cf. [S, V. 8.3]).

**7.4 Theorem.** *Let $E$ be an AM-space with unit and denote by $K$ the $\sigma(E', E)$-compact set of real valued lattice homomorphisms of norm 1 on $E$. The evaluation map $x \mapsto f$ (where $f(t) = \langle x, t \rangle$, $t \in K$) is an isomorphism of $E$ onto $C(K)$.*

*Proof.* It is readily seen that the set $X$ of all positive linear forms of norm 1 on $E$ is identical with $E'_+ \cap H$, where $H$ denotes the $\sigma(E', E)$-closed hyperplane $\{x' : \langle e, x' \rangle = 1\} \subset E'$ and $e$ denotes the unit of $E$ (cf. § 5, Formula (2)). It is also immediately verified that $x' \in X$ is an extreme point of $X$ iff $0 \leqq y' \leqq x'$, $y' \in E'$ implies that $y'$ is a scalar multiple of $x'$; that is, $x' \in X$ is extreme iff $x'$ generates a minimal (i. e., one dimensional) ideal of $E'$, and we conclude from (4.4) that the set of all extreme points of $X$ is identical with $K$. Since $K$ is the set of real lattice homomorphisms $x'$ satisfying $\langle e, x' \rangle = 1$, $K$ is closed for $\sigma(E', E)$ and hence, compact; in fact, (5.7) shows $K$ to be the Šilov boundary of $E_+$. This (or direct verification, since the convex closure of $K \cup \{0\}$ for $\sigma(E', E)$ is the positive part of the dual unit ball) shows that for each $x \in E$, we have $\|x\| = \||x|\| = \sup\{\langle |x|, t \rangle : t \in K\} = \sup\{|f(t)| : t \in K\}$, keeping in mind that each $t \in K$ is a lattice homomorphism. Thus $x \mapsto f$ is an isometric lattice isomorphism of $E$ into $C(K)$; since $K$ separates the points of $E$ and since the unit $e \in E$ is mapped onto the constant-one function, it follows from (7.3) that the isomorphism is surjective. □

**Corollary 1.** *Every AM-space $E$ with unit is isomorphic to $C(K)$, where $K$ is the Šilov boundary of $E_+$.*

Conversely, if $X$ is a compact space then $C(X)$ (Example 2) is an $AM$-space with unit the constant-one function, and the mapping $t \mapsto \delta_t$ ($\delta_t$ Dirac measure at $t \in X$) is a homeomorphism of $X$ onto the space $K$ defined in (7.4). In fact, the following stronger assertion is valid.

**Corollary 2.** *Let $K_1$, $K_2$ be compact spaces. If $C(K_1)$ and $C(K_2)$ are isomorphic as vector lattices, then $K_1$ and $K_2$ are homeomorphic.*

*Proof.* We can visualize $C(K_1)$ and $C(K_2)$ as the same vector lattice $E$ endowed with $M$-norms defined by order units $e_1, e_2$ of $E$ (cf. 7.2), respectively; these norms are necessarily equivalent. If $X_1 := \{x' \in E'_+ : \langle e_1, x' \rangle = 1\}$ and $X_2 := \{x' \in E'_+ : \langle e_2, x' \rangle = 1\}$, then $\varphi : x' \mapsto \langle e_2, x' \rangle^{-1} x'$ is a bijection of $X_1$ onto $X_2$ which is continuous for the topologies induced by $\sigma(E', E)$. Since the extreme points of $X_i$ are those generating minimal ideals in $E'$, $i = 1, 2$, it follows that $\varphi$ maps $K_1$ homeomorphically onto $K_2$. (This is a special case of Thm. III. 4.1.) □

*Remark.* The preceding corollary should be compared with the well-known theorem of Banach-Stone, which asserts that $K_1$ and $K_2$ are homeomorphic whenever $C(K_1)$ and $C(K_2)$ are linearly isometric, and whose conclusion fails if the isometry condition is omitted (Exerc. 19). Also, there is a close relationship between (7.3), (7.4) and their analogs for commutative $B^*$-algebras; the parallels in the respective proofs are particularly well exhibited by Z. Semadeni [1971], Chap. IV, § 13.

Thus, from the viewpoint of Banach lattices, $AM$-spaces with unit and spaces $C(K)$ ($K$ compact) are the same thing; consequently, in the remainder of this section we will formulate our results generally for spaces $C(K)$. A representation theory for $AM$-spaces without unit, related to the theory presented below (Chap. III, §§ 4—6), has been given by Goullet de Rugy [1972] (cf. Exerc. 19).

**7.5 Proposition.** *The Banach space $C(K)$ ($K$ compact) is separable if and only if $K$ is metrizable.*

*Proof.* Let $E = C(K)$ be separable, and denote by $S$ a countable total subset of $E$. The weak topology $\sigma(E', S)$ is metrizable (because it is generated by countably many semi-norms) and agrees on the dual unit ball $U^\circ$ with $\sigma(E', E)$ (because it is a Hausdorff topology coarser than $\sigma(E', E)$); thus $K$, which can be identified with a subspace of $U^\circ$, is metrizable.

Conversely, suppose $K$ metrizable; then $K \times K$ is compact metrizable and hence, the diagonal $\Delta := \{(t, t) : t \in K\}$ of $K \times K$ has a countable base $\{V_n\}$ of neighborhoods. Let $\{U_i^{(n)} : i = 1, \dots, k_n\}$ be a finite open cover of $K$ such that $\bigcup_{i=1}^{k_n} U_i^{(n)} \times U_i^{(n)} \subset V_n$, for each $n \in \mathbb{N}$; and let $\{f_i^{(n)} : i = 1, \dots, k_n\}$ be a partition of unity on $K$ subordinated to $\{U_i^{(n)}\}$. The set $F_n$ of all linear combinations of the functions $f_i^{(n)}$ ($i = 1, \dots, k_n$) with rational coefficients is countable, and so is $F := \bigcup_n F_n$; we claim that $F$ is dense in $C(K)$. In fact, given $f \in C(K)$ and $\varepsilon \in \mathbb{R}$, $\varepsilon > 0$, there exists $n \in \mathbb{N}$ such that $V_n \subset \{(s, t) : |f(s) - f(t)| < \varepsilon/2\}$; now for each $i$, $1 \leqq i \leqq k_n$, we choose $s_i \in U_i^{(n)}$ and a rational number $\alpha_i$ such that $|f(s_i) - \alpha_i| < \varepsilon/2$. By definition of a partition of unity, we have $f_i^{(n)} \geqq 0$, $\sum_{i=1}^{k_n} f_i^{(n)}(s) = 1$ identically in $K$ and $f_i^{(n)}(s) = 0$ if $s \notin U_i^{(n)}$. Therefore,

$$|f(s) - \sum_{i=1}^{k_n} \alpha_i f_i^{(n)}(s)| \leqq \sum_{i=1}^{k_n} |f(s) - \alpha_i| \, f_i^{(n)}(s) < \varepsilon$$

for all $s \in K$. This shows $F$ to be dense in $C(K)$. □

In the following examples we consider $AM$-spaces $E$ with unit which are not a priori given in the form $C(K)$ and try to determine the corresponding compact space $K$ of (7.4) from the defining properties of $E$.

**Examples.** 4. Let $X$ be a non-void set, and denote by $E$ a vector lattice of bounded, real-valued functions on $X$ which contains the constant-one function $e$ and is complete under the supremum norm. Then $K$ (the Šilov boundary of $E'_+$) can be viewed as the completion of the Hausdorff uniform space associated with $(X,\boldsymbol{U})$, where $\boldsymbol{U}$ is the coarsest uniformity on $X$ for which each $f\in E$ is uniformly continuous (Exerc. 20).

Special cases of this situation abound in analysis:

(a) If $E$ is the space of all bounded continuous functions on the completely regular topological space $X$, then $K$ is (homeomorphic with) the Stone-Čech compactification of $X$.

(b) If $X$ is a topological group and $E$ the space of bounded, continuous right almost periodic functions on $X$, then $K$ is the Bohr compactification of $X$. (For details and references, see Semadeni [1971], Chap. IV, § 14.)

(c) Let $X$ be a locally compact, non-compact space and let $E$ be the space of continuous real functions $f$ on $X$ for which $\lim_{\boldsymbol{F}} f(s)$ exists, where $\boldsymbol{F}$ denotes the filter of subsets of $X$ having relatively compact complement. Then $K$ is the one-point (or Alexandrov) compactification of $X$.

(d) If $\Sigma$ is a $\sigma$-algebra of subsets of $X$ and $E$ the space of all bounded, $\Sigma$-measurable real functions, then $K$ is the Stone representation space of $\Sigma$ (§ 1, Example 1).

5. The Banach lattice $L^\infty(\mu)$ over a measure space $(X,\Sigma,\mu)$ (Example 3) is isomorphic with $C(K)$, where $K$ is the Stone representation space of the Boolean algebra $\Sigma/N$ ($N$ the ideal of $\mu$-null sets in $\Sigma$) (Exerc. 20).

Let $K$ be a compact space. We recall that a bounded subset $A$ of $C(K)$ is relatively compact iff $A$ is equicontinuous, that is, iff for each $\varepsilon>0$ there exists a finite open cover $\{U_i: i=1,\dots,n\}$ of $K$ such that $(s,t)\in U_i\times U_i$ $(i=1,\dots,n)$ implies $|f(s)-f(t)|<\varepsilon$ for all $f\in A$. Now if $F$ is a non-void finite subset of $A$ and $g:=\sup F$, then the obvious relation

$$|g(s)-g(t)|\leq \sup_{f\in F}|f(s)-f(t)| \qquad (s,t\in K)$$

shows the set $A'$ of all suprema of finite subsets of $A$ to be equicontinuous whenever $A$ is. Thus if $A$ is a non-void relatively compact subset of $C(K)$, it follows that $\sup A$ (and $\inf A$) exist (cf. 5.8). We also recall that a bounded subset $A\subset C(K)$ is relatively weakly compact iff each sequence in $A$ contains a subsequence converging pointwise to a function in $C(K)$ [DS, IV. 6.4 and IV. 6.7].

From these well-known facts we derive the following result valid for $AM$-spaces, borrowing from (9.1) that each $AM$-space can be identified with a closed vector sublattice of an $AM$-space with unit (namely, its bidual).

**7.6 Proposition.** *Let $E$ be an $AM$-space, let $A$ denote a non-void bounded subset of $E$, and let $|A|:=\{|x|: x\in A\}$. If $A$ is relatively compact, then so is $|A|$, and* $\sup A$ *as well as* $\inf A$ *exist in E. If $A$ is relatively weakly compact, then so is $|A|$.*

*Moreover, the mappings $x\mapsto x^+$, $x\mapsto x^-$, $x\mapsto |x|$ are weakly sequentially continuous.*

We next turn to a characterization of order completeness properties of the Banach lattices $C_b(X)$ (Example 2 above) in terms of topological properties of $X$. Recall that a subset of a topological space is called an $F_\sigma$ (respectively, $G_\delta$) set if it is a countable union of closed subsets of $X$ (respectively, a countable intersection of open subsets of $X$). Recall also that a Hausdorff topological space $X$ is called *normal* if for any pair $A, B$ of disjoint closed subsets of $X$, there exists a continuous function $f: X \to [0,1]$ such that $f(A) \subset \{1\}$, $f(B) \subset \{0\}$. Every compact and every metrizable space is normal (Urysohn).

**7.7 Proposition.** *Let $X$ be a topological space and denote by $C_b(X)$ the AM-space of all bounded, continuous functions $X \to \mathbb{R}$. Consider these assertions on $X$:*

(a) *Each open subset of $X$ has open closure.*

(a′) *Each open $F_\sigma$-subset of $X$ has open closure.*

(b) *$C_b(X)$ is order complete.*

(b′) *$C_b(X)$ is countably order complete.*

*Then* (a) $\Rightarrow$ (b) *and* (a′) $\Rightarrow$ (b′); *if $X$ is normal, then also* (b) $\Rightarrow$ (a) *and* (b′) $\Rightarrow$ (a′).

*Proof.* For mappings $h: X \to \mathbb{R}$, we use the notations $[h > \alpha] := \{s \in X : h(s) > \alpha\}$ and, similarly, $[h \geqq \alpha]$, $[h < \alpha]$, $[h \leqq \alpha]$ $(\alpha \in \mathbb{R})$. Recall that $h$ is called *lower semi-continuous* if $[h > \alpha]$ is open (equivalently, $[h \leqq \alpha]$ closed) for all $\alpha \in \mathbb{R}$, *upper semi-continuous* if $[h < \alpha]$ is open (equivalently, $[h \geqq \alpha]$ closed) for all $\alpha \in \mathbb{R}$. The *upper limit function* $h_u$ of a bounded real function $h$ on $X$ is defined by

$$h_u(s) := \inf_{U(s)} \sup_{t \in U(s)} h(t)$$

$(s \in X)$, where $U(s)$ runs through a neighborhood base of $s$.

(a) $\Rightarrow$ (b): Let $A$ be a directed $(\leqq)$ majorized subset of $C_b(X)$ and let $g$ denote the numerical supremum of $A$, that is, define $g: X \to \mathbb{R}$ by $g(s) := \sup_{f \in A} f(s)$ for all $s \in X$. If $k \in C_b(X)$ is any majorant of $A$ then clearly $f \leqq g \leqq k$, and even $f \leqq g_u \leqq k$, for all $f \in A$. We show that (a) implies the continuity of $g_u$, whence it will follow that $g_u = \sup A$.

Observe first that $g_u$ is upper semi-continuous (regardless of (a)); in fact, for each $\alpha \in \mathbb{R}$ $[g_u < \alpha]$ is open because whenever $g(s_0) < \alpha$, there exists $\varepsilon > 0$ and a neighborhood $U$ of $s_0$ such that $\sup_{t \in U} g(t) \leqq g_u(s_0) + \varepsilon < \alpha$.

To see that $g_u$ is lower semi-continuous if (a) holds, we note first that for all $\alpha \in \mathbb{R}$,

$$[g > \alpha] = \textstyle\bigcup_{f \in A} [f > \alpha] = \bigcup_{f \in A, n \in \mathbb{N}} [f \geqq \alpha + n^{-1}]$$

is open (respectively, an open $F_\sigma$ whenever $A$ is countable). Moreover, by the definition of $g_u$ we have, for all $\alpha \in \mathbb{R}$,

$$(\star) \quad [g_u > \alpha] = \textstyle\bigcup_{n \in \mathbb{N}} [g_u \geqq \alpha + n^{-1}] = \bigcup_{n \in \mathbb{N}} [g \geqq \alpha + n^{-1}]^- = \bigcup_{n \in \mathbb{N}} [g > \alpha + n^{-1}]^-,$$

the last equality holding because of $[g > \alpha + n^{-1}] \subset [g \geqq \alpha + n^{-1}] \subset [g > \alpha + m^{-1}]$ for $m > n$. Thus (a) (respectively, (a′) if $A$ is countable) implies $[g_u > \alpha]$ to be open for all $\alpha \in \mathbb{R}$ and hence, $g_u$ to be continuous.

(a′) ⇒ (b′) is covered by the preceding.

Now suppose $X$ to be normal. (b′) ⇒ (a′): If $V \neq X$ is an open non-void $F_\sigma$-subset of $X$, $V = \bigcup_n F_n$ say where $F_n$ $(n \in \mathbb{N})$ is closed, let $f_n \colon X \to [0,1]$ denote a continuous function such that $f_n(F_n) = \{1\}$, $f_n(X \setminus V) = \{0\}$. $A := \{f_n : n \in \mathbb{N}\}$ is a countable majorized subset of $C_b(X)$ so by hypothesis, $f := \sup A$ exists in $C_b(X)$. It is clear that $f(V) = \{1\}$ and $f(X \setminus \overline{V}) = \{0\}$; since $f$ is continuous, $f(\overline{V}) = \{1\}$ which shows $f$ to be the characteristic function of $\overline{V}$. Hence, $\overline{V}$ is open.

Finally, the proof of (b) ⇒ (a) is similar. □

It is customary to call a completely regular topological space $X$ *Stonian* if $X$ satisfies (a) of (7.7), *quasi-Stonian* if $X$ satisfies (a′) of (7.7). We obtain this corollary.

**Corollary.** *For $K$ a compact space, the AM-space $C(K)$ is order complete (countably order complete) if and only if $K$ is Stonian (quasi-Stonian).*

*Note.* Stonian spaces are also called *extremally disconnected.* The terminology "Stonian space" should not be confused with "Stone representation space" (of a Boolean algebra). The Stone representation space of a Boolean algebra (Exerc. 1) is always totally disconnected (i. e., possesses a base of open-and-closed subsets), but not necessarily extremally disconnected. For example, the Stone representation space $K_\Sigma$ of the Boolean algebra $\Sigma$ of real Borel sets is quasi-Stonian but not Stonian (because $\Sigma$ is countably complete but not complete, [H, § 16]); by contrast, $K_{\Sigma'}$ $(\Sigma' = \Sigma/N$, $N$ the ideal of Lebesgue null sets) is a Stonian space.

We stop to consider a peculiarity of quasi-Stonian and Stonian spaces.

**7.8 Proposition.** *Let $X$ be a Stonian (respectively, quasi-Stonian) topological space. If $G$ is a dense open set (respectively, a dense open $F_\sigma$-set) in $X$, then each bounded, continuous function $g \colon G \to \mathbb{R}$ has a continuous extension $\overline{g} \colon X \to \mathbb{R}$.*

*Proof.* We consider the slightly more delicate case where $X$ is quasi-Stonian and $G$ a dense open $F_\sigma$-subset, and we use the notation of the proof of (7.7). For bounded, continuous $g \colon G \to \mathbb{R}$ we define the *lower limit function* $g_l \colon X \to \mathbb{R}$ by

$$g_l(s) := \sup_{U(s)} \inf_{t \in U(s) \cap G} g(t),$$

where $U(s)$ runs through a neighborhood base of $s \in X$. It is not difficult to verify that $g_l$ is lower semi-continuous on $X$; of course, $g_l$ agrees with $g$ on $G$. We prove that $g_l$ is upper semi-continuous as well. To see this we observe that for each $\alpha \in \mathbb{R}$, the relations

$$(**) \qquad [g_l < \alpha] = \bigcup_{n \in \mathbb{N}} [g_l \leqq \alpha - n^{-1}] = \bigcup_{n \in \mathbb{N}} [g \leqq \alpha - n^{-1}]^- = \bigcup_{n \in \mathbb{N}} [g < \alpha - n^{-1}]^-$$

are valid for entirely similar reasons as the corresponding relations (*) in the proof of (7.7); the bars denote closures in $X$. To prove that $[g_l < \alpha]$ is open in $X$, it suffices by (**) to show that each of the sets $[g < \alpha - n^{-1}]$ (which are subsets of $G$) is an open $F_\sigma$ in $X$. But clearly these sets are open $F_\sigma$-sets in $G$;

since $G$ is an open $F_\sigma$ in $X$, it follows that each set $[g<\alpha-n^{-1}]$ is an open $F_\sigma$ in $X$. This proves $g_l$ to be continuous on $X$ and hence the assertion. ☐

Since the Stone-Čech compactification $\beta X$ of a completely regular space $X$ (cf. Example 4 above) is a compact space densely containing $X$ characterized (up to homeomorphism) by the property that each bounded, continuous $g: X \to \mathbb{R}$ has a continuous extension $\bar{g}: \beta X \to \mathbb{R}$, (7.8) contains this result.

**Corollary 1.** *Let $K$ be a Stonian (quasi-Stonian) compact space, and let $G$ be an open (an open $F_\sigma$) subspace of $K$. The closure $\bar{G}$ in $K$ is homeomorphic with the Stone-Čech compactification of $G$.*

There is another, lattice theoretic application of (7.8). Consider the two point compactification $\bar{\mathbb{R}}:=[-\infty,+\infty]$ of $\mathbb{R}$; $\bar{\mathbb{R}}$ is a complete lattice if the subset $\mathbb{R}$ is ordered as usual and it is agreed that $-\infty<\alpha<+\infty$ for each $\alpha\in\mathbb{R}$ (lattice completion of $\mathbb{R}$, cf. Chap. III, § 3). Suppose $K$ to be a quasi-Stonian space; since the real interval $[-1,1]$ is homeomorphic to $\bar{\mathbb{R}}$, it follows from (7.8) that each continuous $g: G\to\mathbb{R}$, where $G$ is a dense open $F_\sigma$ in $K$, has (a unique) continuous extension $\bar{g}: K\to\bar{\mathbb{R}}$. Now if $C_\infty(K)$ denotes the set of all continuous functions $K\to\bar{\mathbb{R}}$ which are finite except on rare sets, this extension theorem enables us to introduce a natural vector structure on $C_\infty(K)$. In fact, if $f_i\in C_\infty(K)$, $\alpha_i\in\mathbb{R}$ $(i=1,2)$ and $G:=\{t\in K: |f_1(t)|+|f_2(t)|<+\infty\}$, then $G$ is a dense open $F_\sigma$-set, and the function $g: G\to\mathbb{R}$ defined by $g(t):=\alpha_1 f_1(t)+\alpha_2 f_2(t)$ has a continuous extension $K\to\bar{\mathbb{R}}$ which serves to define $\alpha_1 f_1+\alpha_2 f_2$. It is easy to see that the vector space $C_\infty(K)$ is a vector lattice under its natural (pointwise defined) lattice structure, with these properties (cf. 7.7 Cor.).

**Corollary 2.** *If $K$ is a quasi-Stonian (Stonian) compact space, then $C_\infty(K)$ is a countably order complete (order complete) vector lattice containing $C(K)$ as an order dense ideal.*

It will be seen later (Chap. III, § 4) that every countably order complete Banach lattice $E$ possessing quasi-interior positive elements (§ 6), is isomorphic with a vector sublattice of $C_\infty(K)$ (III. 4.5). For further properties of $C_\infty(K)$ related to the representation of vector lattices, see Vulikh [1967], Chap. V and Exerc. 21.

The remainder of this section is concerned with a property of order complete spaces $C(K)$ that characterizes these spaces in the category of Banach spaces. If $G$ is a real vector space and $E$ is a vector lattice, a function $p: G\to E$ is called *sublinear* if (i) $p(x+y)\leqq p(x)+p(y)$ $(x, y\in G)$ and (ii) $p(\lambda x)=\lambda p(x)$ $(x\in G, \lambda\in\mathbb{R}_+)$. The following extension of the classical Hahn-Banach theorem, which emerges for $E=\mathbb{R}$, holds by virtue of a re-interpretation of the standard proof.

**7.9** (Generalized Hahn-Banach Theorem). *Let $G$ be a vector space over $\mathbb{R}$, $E$ an order complete vector lattice, and let $p: G\to E$ be sublinear. If $G_0$ is a vector subspace of $G$ and $f: G_0\to E$ is a linear map satisfying $f(x)\leqq p(x)$ for all $x\in G_0$, there exists a linear extension $F: G\to E$ of $f$ satisfying $F(x)\leqq p(x)$ for all $x\in G$.*

*Proof.* Supposing that $G_0\neq G$, let $x_0\in G\setminus G_0$; we show that $f$ possesses a linear extension to the linear span $G_1$ of $G_0\cup\{x_0\}$, $f_1: G_1\to E$ such that $f_1$ is dominated by $p$. For any $y, y'\in G_0$ we have

$$f(y)-f(y')=f((y+x_0)+(-y'-x_0))\leqq p(y+x_0)+p(-y'-x_0)$$

hence

$$(\star)\qquad f(y)-p(y+x_0)\leqq f(y')+p(-y'-x_0).$$

Because $y, y'$ are independent, $(\star)$ shows the set of all elements $f(y)-p(y+x_0)$ $(y\in G_0)$ to be majorized in $E$; since $E$ is order complete,

$$:=\sup_{y\in G_0}(f(y)-p(y+x_0))$$

exists in $E$, and we have $u\leqq f(y')+p(-y'-x_0)$ for all $y'\in G_0$. Now we define $f_1\colon G_1\to E$ to be the extension of $f$ for which $f_1(x_0)=-u$. To show that $p$ dominates $f_1$ we distinguish two cases.

Let $z=x+\lambda x_0$ where $x\in G_0$ and $\lambda>0$. Since $y:=\lambda^{-1}x\in G_0$, $(\star)$ and the definition of $u$ shows that $f(\lambda^{-1}x)-u\leqq p(\lambda^{-1}x+x_0)$, whence $f_1(z)\leqq p(z)$.

Let $z=x+\lambda x_0$ where $x\in G_0$ and $\lambda<0$. Since $y':=\lambda^{-1}x\in G_0$, $(\star)$ and the definition of $u$ shows that $-f(\lambda^{-1}x)+u\leqq p(-\lambda^{-1}x-x_0)$, whence $f_1(z)\leqq p(z)$.

Now a routine argument involving Zorn's lemma completes the proof. □

The preceding, purely formal extension of the Hahn-Banach theorem has this surprising consequence.

**7.10 Theorem.** *Let $G$ be a real Banach space, let $G_0$ be a normed vector subspace of $G$, and let $K$ be any compact Stonian space. Every continuous linear map $T_0\colon G_0\to C(K)$ has a norm preserving linear extension $T\colon G\to C(K)$.*

*Proof.* If $e$ denotes the unit (constant-one function) of $C(K)$, then $p(x):=\|x\|e$ $(x\in G)$ defines a sublinear function $G\to C(K)$. Assuming without loss of generality that $\|T_0\|\leqq 1$, we observe that $T_0x\leqq\|x\|e=p(x)$ for all $x\in G_0$ and hence, $T_0$ satisfies the hypothesis of (7.9), $E=C(K)$ being order complete (7.7 Cor.). Therefore, there exists a linear extension $T\colon G\to C(K)$ of $T_0$ satisfying $Tx\leqq\|x\|e$ $(x\in G)$ which implies that $\|T\|\leqq 1$. □

The property of order complete $AM$-spaces with unit expressed by (7.10) is characteristic of these spaces among Banach spaces (Exerc. 22); also, under suitable assumptions (on $G$ and $G_0$) (7.10) can be adapted to the case of complex scalars (Exerc. 22). We record the following supplements of (7.10); recall that a continuous linear map $T\colon G\to H$ between normed spaces is called a *contraction* if $\|T\|\leqq 1$.

**Corollary 1.** *Let $G$ be a real Banach space. Every Banach subspace $G_0$ which is (isometrically) isomorphic to some order complete $C(K)$, is the range of a contractive projection.*

*Proof.* Identifying $G_0$ with $C(K)$, we conclude from (7.10) that the identity map $1_{G_0}$ has a norm preserving extension to $G$ with values in $G_0$. □

**Corollary 2.** *Let $E$ be an order complete $AM$-space with unit $e$. If $E_0$ is an order complete vector lattice which is a closed vector sublattice of $E$ containing $e$, then $E_0$ is the range of a positive contractive projection.*

*Proof*. The existence of a contractive projection $P: E \to E_0$ is immediately clear from Cor. 1; we show that in the present circumstances, $P$ is necessarily positive. In fact, one has $z \in E_+$, $\|z\| \leqq 1$ if and only if $z = e - x$ for some $x \in E$, $\|x\| \leqq 1$. Moreover, $Pe = e$ since $e \in E_0$; thus $z \in E_+$, $\|z\| \leqq 1$ implies $Pz = P(e-x) = e - Px \geqq 0$, because of $\|Px\| \leqq \|x\| \leqq 1$. □

We point out that Cor. 2 applies, in particular, to the bidual $E''$ of an order complete $AM$-space $E$ with unit $e$, since the unit ball of $E''$ is the set $(-e + E_+^{\circ\circ}) \cap (e - E_+^{\circ\circ})$ ($E_+^{\circ\circ}$ denoting the bipolar of $E_+$ with respect to $\langle E'', E' \rangle$) and hence, $e$ is also the unit of $E''$ (cf. 9.1).

Finally, Cor. 2 permits us to prove this refinement of (7.10).

**Corollary 3.** *Let $G$ be a Banach lattice, let $G_0$ be a Banach sublattice, and denote by $E$ an order complete $AM$-space with unit. Every positive linear map $T_0: G_0 \to E$ has a norm preserving, positive linear extension $T: G \to E$.*

*Proof*. Note first that by (5.3), $T_0$ is necessarily continuous. Now we may and shall identify $E$ with $C(K)$ ($K$ compact Stonian) and consider $C(K)$ contained in $l^\infty(K)$ as required in the hypothesis of Cor. 2; thus, there exists a positive contractive projection $P: l^\infty(K) \to C(K)$.

Now for each $t \in K$, the mapping $x \mapsto Tx(t)$ is a positive linear form of norm $\leqq \|T_0\|$ on $G_0$ and hence, by (5.6), possesses a positive norm preserving extension $\varphi_t \in G'$. It is clear that the mapping $x \mapsto (\varphi_t(x))_{t \in K}$ is a positive linear map $\tilde{T}: G \to l^\infty(K)$ which extends $T_0: G_0 \to C(K)$, and that $\|\tilde{T}\| = \|T_0\|$. Therefore, the composition $T := P \circ \tilde{T}$ is a norm preserving positive linear extension of $T_0$ with values in $E = C(K)$. □

We point out to the interested reader that the preceding extension theorems can be dualized to yield lifting theorems involving $AL$-spaces (Exerc. 22).

We conclude this section with a result that will be instrumental in proving Theorem 8.9, the analog for $AL$-spaces of Cor. 3 above.

**7.11 Proposition.** *Let $K_1, K_2$ denote compact spaces and suppose $K_2$ to be totally disconnected. Let $T: C(K_1) \to C(K_2)$ be a positive linear map satisfying these conditions, where $e_i$ denotes the unit of $C(K_i)$ $(i = 1, 2)$:*

(i) *$Te_1 = e_2$, and $T[0, f] = [0, Tf]$ for all $f \in C(K_1)_+$.*

(ii) *The absolute kernel $\{f: T|f| = 0\}$ of $T$ is a projection band of $C(K_1)$.*

*There exists an isometric lattice homomorphism $S$ of $C(K_2)$ into $C(K_1)$ such that $T \circ S = 1_{C(K_2)}$.*

*Proof*. By (2.13) the Boolean algebras $B_1 := \{f \in C(K_1): f \wedge (e_1 - f) = 0\}$ and $B_2 := \{g \in C(K_2): g \wedge (e_2 - g) = 0\}$ agree with the respective extreme boundaries of $[0, e_1]$ and $[0, e_2]$. Clearly, $B_i$ can be identified with the Boolean algebra of all open-and-closed subsets of $K_i$ $(i = 1, 2)$.

First, we suppose $T$ to be strictly positive (Def. 2.4) so that (ii) is trivially satisfied. We claim that $T$ maps a Boolean subalgebra $B_1'$ of $B_1$ isomorphically onto $B_2$. Condition (i) implies that for each $g \in B_2$ there exists $f \in [0, e_1]$ satisfying $Tf = g$. Now $T(f \wedge (e_1 - f)) \leqq g \wedge (e_2 - g) = 0$ shows that $f \in B_1$, since $T$ is strictly positive; moreover, $f \in B_1$ is determined uniquely by $g$. In fact, if $Tf_1 = Tf_2 = g$ (where $g \in B_2$ and $f_1, f_2 \in [0, e_1]$) then $g = Th$ for

$h=\frac{1}{2}(f_1 \vee f_2)+\frac{1}{2}(f_1 \wedge f_2)$; but this implies $h \in B_1$ so $h$ is extreme in $[0,e_1]$ by (2.13), and it follows that $f_1=f_2$.

Define $B_1':=T^{-1}(B_2)\cap[0,e_1]$. To see that $B_1'$ is a Boolean subalgebra of $B_1$ and the restriction $\tau:=T|B_1'$ a Boolean isomorphism onto $B_2$ it suffices to show, since $T$ is linear, that $Tf_1 \wedge Tf_2=0$ implies $f_1 \wedge f_2=0$ whenever $f_1, f_2 \in B_1'$. But this is clear from the relation $0 \leqq T(f_1 \wedge f_2) \leqq Tf_1 \wedge Tf_2$ and our present assumption that $T$ be strictly positive.

Therefore, the inverse map $\tau^{-1}$ is a Boolean isomorphism of $B_2$ onto $B_1'$. Since $K_2$ is totally disconnected, it follows from (7.3) that the linear hull $L_2$ of $B_2$ in $C(K_2)$ is a dense vector sublattice of $C(K_2)$. On the other hand, it is readily verified that $\tau^{-1}$ induces an isometric vector lattice isomorphism $S_0\colon L_2 \to C(K_1)$ such that $T \circ S_0 g=g$ for all $g \in L_2$; since $L_2$ is dense in $C(K_2)$, the unique continuous extension $S\colon C(K_2) \to C(K_1)$ of $S_0$ is an isometric lattice homomorphism satisfying $T \circ S=1_{C(K_2)}$.

We now remove the assumption that $T$ be strictly positive, supposing instead that (ii) holds. If $A:=\{f\colon T|f|=0\}$ is a projection band of $C(K_1)$, then $A$ and $A^\perp$ are $AM$-spaces with unit and the restriction $T_0\colon T|A^\perp$ is strictly positive while it still satisfies condition (i). Applying the preceding construction to $T_0$, we obtain a lattice isometry $S\colon C(K_2) \to A^\perp$ satisfying $T_0 \circ S=1_{C(K_2)}$. Now if $j\colon A^\perp \to C(K_1)$ denotes the canonical injection, the mapping $j \circ S$ provides the desired right inverse of $T$. □

**Corollary.** *Let $K_1, K_2, T$ satisfy the hypotheses of* (7.11). *For every Banach space E and every (linear) contraction $R\colon E \to C(K_2)$, there exists a contractive (linear) lifting $\hat{R}\colon E \to C(K_1)$ satisfying $R=T \circ \hat{R}$. If, in addition, E is a Banach lattice and if $R \geqq 0$, then $\hat{R}$ can be chosen to be positive.*

It is clear that the desired lifting is given by $\hat{R}:=S \circ R$ (cf. 7.11).

## § 8. Abstract $L$-Spaces

For the theory of Banach lattices, the spaces next in importance to abstract $M$-spaces (§ 7) are spaces of integrable functions axiomatically termed $AL$-spaces (abstract Lebegue spaces), which exhibit properties contrasting strongly with and dual to those of $AM$-spaces. The most important of the many interesting theorems resulting from this duality will be discussed in Sections 9, 10 below; the present section is concerned with the basic properties of $AL$-spaces and a first representation theorem (8.5) as well as an injective property (8.9).

**8.1 Definition.** *A lattice norm $x \mapsto \|x\|$ on a vector lattice E is called an* $L$-norm *if it satisfies the axiom*

(L) $$\|x+y\| = \|x\|+\|y\| \qquad (x, y \in E_+).$$

*$(E, \|\ \|)$ is called an* $L$-normed space, *and an $L$-normed Banach lattice is called an* abstract $L$-space (*briefly,* $AL$-space).

Thus an $L$-normed space is a normed lattice whose norm is additive on the positive cone. It is clear that a (closed) vector sublattice of an $L$-normed space

($AL$-space) is an $L$-normed space ($AL$-space), and that the completion of an $L$-normed space is an $AL$-space. In view of (8.3)(iii) below, the quotient of an $L$-normed space over a closed ideal is again $L$-normed.

**8.2 Proposition.** *In an $L$-normed space $E$, every directed ($\leqq$) norm bounded family is a Cauchy family. For $E$ to be an $AL$-space, it is necessary and sufficient that each directed ($\leqq$) norm bounded family have a supremum.*

*Proof.* Suppose $A$ to be a directed ($\leqq$) family in $E$ bounded in norm by a constant $k$, say. If $A$ were not Cauchy, there would exist a number $\varepsilon > 0$ and an infinite sequence $(x_n)_{n \in \mathbb{N}}$ in $A$ such that $x_{n+1} \geqq x_n$ and $\|x_{n+1} - x_n\| > \varepsilon$ for all $n \in \mathbb{N}$. Axiom (L) implies that

$$n\varepsilon \leqq \sum_{k=2}^{n+1} \|x_k - x_{k-1}\| = \|x_{n+1} - x_1\| \leqq 2k$$

for all $n$, which is contradictory.

Concerning the second assertion, it is clear that the condition is necessary (cf. 5.8). Conversely, suppose that $E$ is an $L$-normed space in which each directed ($\leqq$) norm bounded family has a supremum. Given a Cauchy sequence in $E$, it suffices to show that some subsequence converges. Let $(x_n)$ denote a subsequence of a given Cauchy sequence satisfying $\|x_{n+1} - x_n\| < 2^{-2(n+1)}$. Letting $u_k := (x_{k+1} - x_k)^+$ for $k \in \mathbb{N}$, it will be enough to show that $\sum_k u_k$ converges in $E$. Defining $v_k := 2^k u_k$ $(k \in \mathbb{N})$, the choice of $(x_n)$ and the hypothesis imply that the suprema $u := \sup_n \sum_{k=1}^n u_k$, $v := \sup_n \sum_{k=1}^n v_k$ exist in $E$. Now we have $0 \leqq u - \sum_{k=1}^n u_k \leqq 2^{-n} v$ which implies $\|u - \sum_{k=1}^n u_k\| \leqq 2^{-n}\|v\|$ $(n \in \mathbb{N})$, and the proof is complete. □

**8.3 Proposition.** *Every $AL$-space $E$ has the following properties:*

(i) *Each directed ($\leqq$) norm bounded family in $E$ converges.*

(ii) *$E$ is order complete and $E' = E^\star = E^\star_{00}$ (that is, continuity, order continuity, and order boundedness of linear forms on $E$ are equivalent).*

(iii) *Each closed ideal of $E$ is a projection band.*

(iv) *Each weak order unit of $E$ is quasi-interior to $E_+$.*

(v) *Under evaluation, $E$ is isomorphic to the band of all order continuous linear forms on $E'$.*

*Proof.* (i) is an immediate consequence of (8.2). (i) implies (ii) by (5.10). Furthermore, (i) implies that each closed ideal of $E$ is a band and hence, (iii) follows from (ii) and (2.10). (iv) follows from (i), (1.9) Cor., and (6.3). Finally, by (ii) and (4.12) $q(E)$ (where $q\colon E \to E''$ denotes the evaluation map) is an order dense ideal of $(E')^\star_{00}$, and it follows from (i) that the range of $q$ is all of $(E')^\star_{00}$. □

**Examples.** 1. Let $E$ be any vector lattice and let $x'$ denote a positive linear form on $E$. If $N$ denotes the absolute kernel $\{x : \langle |x|, x' \rangle = 0\}$ of $x'$, then $E/N$ is an $L$-normed space with respect to $\hat{x} \mapsto p_{x'}(\hat{x}) := \langle |x|, x' \rangle$ $(x \in \hat{x} \in E/N)$. (In a trivial sense, every $L$-normed space $E$ is of this form with $x'$ the strictly positive linear form defined by $\langle x, x' \rangle = \|x^+\| - \|x^-\|$ and called the *norm functional* of $E$.) The completion $(E/N, p_{x'})^{\tilde{}}$ is an $AL$-space which will be denoted by $(E, x')$ (cf. Chap. IV, § 3).

2. For $X$ a locally compact space, let $E := \mathscr{K}(X)$ denote the vector lattice of continuous functions $X \to \mathbb{R}$ with compact support, and let $0 \leqq \mu \leqq E^\star$. The $AL$-space $(E, \mu)$ (Example 1) is isomorphic to $L^1(\mu)$. It will be seen below (Thm. 8.5) that up to isomorphism, these spaces $L^1(\mu)$ are the most general $AL$-spaces.

3. Let $X$ be a set and let $\Sigma$ denote a Boolean $\sigma$-algebra of subsets of $X$. The space $\mathsf{M}(\Sigma)$ (or $ba(X, \Sigma)$ in the notation of [DS], cf. § 1, Example 5) is an $AL$-space with respect to the total variation norm (§ 5, Example 3). The vector subspace $M(X, \Sigma)$ (or $ca(X, \Sigma)$ in the notation of [DS]) of all countably additive set functions $\mu \in \mathsf{M}(\Sigma)$ is a band in $\mathsf{M}(\Sigma)$. For any fixed $\mu \in M(X, \Sigma)$, the principal band generated by $\{\mu\}$ agrees with the set of all $\mu$-continuous measures [DS, III. 10] and is isomorphic with $L^1(|\mu|)$ (Radon-Nikodym Theorem, cf. (8.7) Cor.).

In the special case where $X = \mathbb{N}$ and where $\Sigma = 2^{\mathbb{N}}$, $M(X, \Sigma)$ is isomorphic to $l^1$ while $\mathsf{M}(\Sigma)$ is isomorphic to the bidual of $l^1$ (cf. § 9).

Let $E$ denote a vector lattice having the principal projection property; from (2.12) we recall that for each weak order unit $u$ of $E$, the Boolean algebra $B_u := \{v \in E : v \wedge (u - v) = 0\}$ is isomorphic to the Boolean algebra of projection bands of $E$. Recall also that $E_u$ denotes the $AM$-space with (order) unit $u$ (7.2 Cor.).

**8.4 Lemma.** *Let $E$ denote a Banach lattice having the principal projection property and possessing a weak order unit $u$. The representation of $B_u$ by its Stone representation space $K_u$ induces an isomorphism of the $AM$-space $E_u$ onto $C(K_u)$.*

*Proof.* By (7.4) $E_u$ is isomorphic to $C(K)$ for some compact space $K$; since $E$ has the principal projection property, so does the ideal $E_u$ which clearly implies $K$ to be totally disconnected (cf. Exerc. 1). On the other hand, the isomorphism $E_u \to C(K)$ evidently maps the Boolean algebra $B_u$ isomorphically onto the Boolean algebra of characteristic functions of all open-and-closed subsets of $K$; since $K$ is totally disconnected, these sets form a base of the topology of $K$ and the unicity of the Stone space $K_u$ (Exerc. 1) shows $K$ to be homeomorphic to $K_u$. □

We now prove a first representation theorem for $AL$-spaces due to Kakutani [1941a]; the concept of Radon measure employed is that of Bourbaki [1965].

**8.5 Theorem.** *For every $AL$-space $E$, there exists a locally compact space $X$ and a strictly positive Radon measure $\mu$ on $X$ such that $E$ is isomorphic with $L^1(\mu)$.*

*If (and only if) $E$ possesses a weak order unit $u$, $X$ can be chosen to be compact and such that the isomorphism $E \to L^1(\mu)$ maps $E_u$ onto $L^\infty(\mu)$.*

*Proof.* (a) First we consider the case where $E$ has a weak order unit. Clearly, if $E$ is isomorphic to $L^1(\mu)$ for $\mu$ a Radon measure on the compact space $X$ then the element $e \in E$ corresponding to the constant-one function on $X$ is a weak order unit of $E$. Conversely, if $u$ is a weak order unit of $E$ then by (8.3) (iv) $E_u$ is a dense ideal of $E$, and by (8.4) $E_u$ can be identified with $C(K_u)$. Moreover, by virtue of $\mu(x) := \|x^+\| - \|x^-\|$ the norm of $E$ defines a strictly positive linear form $\mu$ on $E$ (cf. Example 1) for which $\|x\| = \mu(|x|)$, and the restriction of $\mu$ to $E_u$ defines a strictly positive Radon measure on $K_u$ (which we again denote by $\mu$).

Since $C(K_u)$ is dense in $L^1(\mu)$, it is clear that the isomorphism $E_u \to C(K_u)$ extends to an isomorphism of Banach lattices $E \to L^1(\mu)$; moreover, since $E_u$ is an ideal of $E$, $C(K_u)$ is the ideal of $L^1(\mu)$ generated by the constant-one function and hence, can be identified with $L^\infty(\mu)$.

(b) Suppose that $E$ possesses no weak order unit and denote by $S:=\{u_\alpha: \alpha \in \mathsf{A}\}$ a maximal orthogonal system of $E$. By (1.9) and (8.3) (i), the ideal $I$ generated by $S$ is dense in $E$. On the other hand, $I$ is the algebraic direct sum $\bigoplus_\alpha E_{u_\alpha}$ and the principal ideal $E_{u_\alpha}$ is isomorphic to $C(K_\alpha)$ (7.2 Cor.), where by (8.4) $K_\alpha$ can be identified with the Stone representation space of the Boolean algebra $B_\alpha:=\{v \in E: v \wedge (u_\alpha - v)=0\}$. Thus $I$ can be identified with $\mathscr{K}(X)$ (Example 2), where the locally compact space $X$ is the direct topological sum of the family of compact spaces $(K_\alpha)_{\alpha \in \mathsf{A}}$. Clearly, by virtue of $\mu(x):=\|x^+\|-\|x^-\|$ $(x \in I)$ the norm of $E$ defines a strictly positive Radon measure $\mu$ on $X$ such that $\|x\|=\int |x|\,d\mu$ $(x \in I)$, and again the isomorphism $I \to \mathscr{K}(X)$ extends to an isomorphism of Banach lattices $E \to L^1(\mu)$. □

Since each $AL$-space $E$ is order complete, it follows from (8.4) that for an $AL$-space $E$ with weak order unit $u$, the base $B_u=\{v \in E: v \wedge (u-v)=0\}$ is a total subset of $E$. In view of (6.2) (iv), we obtain this corollary.

**Corollary.** *For any $AL$-space $E$, these assertions are equivalent:*

(a) *$E$ is separable.*

(b) *$E$ has a weak order unit $u$ and the base $B_u$ (which is a complete metric subspace of $E$) is separable.*

*Remarks.* 1. The proof of (8.5) given above makes full use of the approach to measure theory chosen by Bourbaki [1965]; alternatively one can proceed as follows (Kakutani [1941 a]). Case (a): The norm of $E$, being additive on $E_+$, defines a set function $\mu_0$ on the Boolean algebra ($\cong B_u$) of all open-and-closed subsets $U$ of $K_u$, by virtue of $\mu_0(U)=\|\chi_U\|$. $\mu_0$ is (trivially) countably additive, and its Carathéodory extension $\mu$ is a Baire measure vanishing on meager Baire subsets of $K_u$. The linear hull $L \subset E$ of $B_u$, consisting of the functions in $C(K_u)$ with finite range, is dense in $E$ (even in $E_u$) and for $x \in L$ we have $\|x\|=\int |x|\,d\mu$. From standard measure theoretic arguments (see III. 9.4 and [H, § 55]), it follows that $L$ is dense in $L^1(\mu)$ and thus $E \cong L^1(\mu)$. Extension to case (b) then proceeds as above.

2. We consider again case (a) (existence of a weak order unit $u \in E$). Using that $E_u \cong C(K_u)$ is a dense ideal of $E$ and employing Egorov's theorem [H, § 21], we conclude that each class $\in L^1(\mu, K_u)$ contains a function which is continuous and finite on a dense open subset of $K_u$; thus, by (7.7) Cor. 1, each class of $L^1(\mu, K_u)$ contains a *continuous*, extended real function $f$ which is unique by strict positivity of $\mu$. Therefore, $E$ is isomorphic to the vector sublattice of $C_\infty(K_u)$ (cf. 7.7 Cor. 2) whose elements are $\mu$-integrable. Extension to the general case (b) is clear.

3. Representation by Banach lattices of numerical (i. e., extended real valued) *continuous* functions will be discussed in Chap. III, §§ 4—6 for a wide class of Banach lattices. The question of unicity of the locally compact space $X$ will also be settled later (Thm. III. 5.3).

In connection with Kakutani's representation theorem (8.5), two basic tools for the study of $AL$-spaces are the Radon-Nikodym theorem ([H, § 31], [DS, III. 10.2], and (8.7) Cor. below) and the Vitali-Hahn-Saks theorem (cf. [DS, III. 7.2]). Following essentially the presentation in [DS, l.c.] we prove a version of the latter theorem adapted to our general setting.

If $E$ is a Banach lattice and if $e \in E_+$, the Boolean algebra $B_e$ (cf. 2.13) is obviously a closed subset of $E$ and hence, a complete metric space for the metric $\varrho(u,v) := \|u-v\|$ induced by $E$. Application of the Baire category theorem to $B_e$ yields the theorem of Vitali-Hahn-Saks. Recall that a function $\mu: B \to \mathbb{R}$ on a Boolean algebra $B$ is called *additive* if $\mu(u \vee v) + \mu(u \wedge v) = \mu(u) + \mu(v)$ for all $u, v \in B$, and if $\mu(0) = 0$ (§ 1, Example 5).

**8.6 Theorem.** *Let $E$ denote a Banach lattice, let $e \in E_+$, and suppose $(\mu_n)_{n \in \mathbb{N}}$ is a sequence of continuous, additive real functions on the Boolean algebra $B_e = \{u \in E: u \wedge (e-u) = 0\}$ such that $(\mu_n(u))$ converges in $\mathbb{R}$ for each $u \in B_e$. Then $(\mu_n)$ is uniformly equicontinuous and hence, converges to a continuous, additive function $\mu: B_e \to \mathbb{R}$; in particular, $\lim_{\|u\| \to 0} \mu_n(u) = 0$ uniformly for $n$ in $\mathbb{N}$.*

*Proof.* Let $\varepsilon > 0$ denote a preassigned number and define

$$B_{nm} := \{u \in B_e : |\mu_n(u) - \mu_m(u)| \leqq \varepsilon\}$$

for $n, m \in \mathbb{N}$. By the continuity of $\mu_n$, $B_{nm}$ is a closed subset of $B_e$; the same is then true of $B_p := \bigcap_{m,n \geqq p} B_{nm}$. The hypothesis implies that $B_e = \bigcup_1^\infty B_p$; since $B_e$ is a complete metric space, the Baire category theorem shows that some $B_p$, $B_q$ say, contains interior points. That is, there exists $u_0 \in B_q$ and $\delta > 0$ such that

$$(\star) \qquad |\mu_n(u) - \mu_m(u)| \leqq \varepsilon$$

whenever $\|u - u_0\| \leqq \delta$ and $n, m \geqq q$. Noting that for $v \in B_e$ we have $u_0 - u_0 \wedge v = (e - u_0 \wedge v) \wedge u_0 \in B_e$, we conclude from the additivity of $\mu_n$ that

$$\mu_n(v) = \mu_n(u_0 \vee v) - \mu_n(u_0 - u_0 \wedge v)$$

for all $v \in B_e$ and $n \in \mathbb{N}$, and, hence,

$$\mu_n(v) = \mu_q(v) + [\mu_n(u_0 \vee v) - \mu_q(u_0 \vee v)] - [\mu_n(u_0 - u_0 \wedge v) - \mu_q(u_0 - u_0 \wedge v)].$$

Since $\|v\| \leqq \delta$ implies $\|u_0 \vee v - u_0\| \leqq \delta$ and $\|u_0 \wedge v\| \leqq \delta$, from $(\star)$ we obtain $|\mu_n(v) - \mu_q(v)| \leqq 2\varepsilon$ whenever $\|v\| \leqq \delta$ and $n \geqq q$. Thus the sequence $(\mu_n)$ is equicontinuous at $0 \in B_e$. If, as usual, $u \Delta u_1$ denotes the symmetric difference $u \vee u_1 - u \wedge u_1 \in B_e$, then from the disjoint decompositions $u = u \wedge u_1 + v$, $u_1 = u \wedge u_1 + v_1$ we obtain $|\mu_n(u) - \mu_n(u_1)| \leqq |\mu_n(v)| + |\mu_n(v_1)|$ and hence,

$$|\mu_n(u) - \mu_n(u_1)| \leqq 2 \sup_{v \leqq u \Delta u_1} |\mu_n(v)|.$$

By the preceding this implies that

$$|\mu_n(u)-\mu_n(u_1)| \leqq 2\varepsilon + \sup_{v \leqq u \Delta u_1} |\mu_q(v)|$$

whenever $n \geqq q$ and $\|u-u_1\| \leqq \delta$ $(u, u_1 \in B_e)$. Therefore, the sequence $(\mu_n)$ is uniformly equicontinuous which implies the continuity of the limit function $u \mapsto \mu(u) := \lim_n \mu_n(u)$. The additivity of $\mu$ is clear. □

**Corollary** (Vitali-Hahn-Saks). *Let $(X, \Sigma, \mu)$ be a $\sigma$-finite measure space, and let $(\lambda_n)$ be a sequence of finite (countably additive) measures on $\Sigma$ which are absolutely continuous with respect to $\mu$. If $\lim_n \lambda_n(S) =: \lambda(S)$ exists in $\mathbb{R}$ for each $S \in \Sigma$, then $\lambda$ is a measure absolutely continuous with respect to $\mu$ and the countable additivity of the measures $\lambda_n$ is uniform with respect to $n \in \mathbb{N}$.*

*Proof*. We observe first that $E := L^1(\mu)$ contains a weak order unit $u$ (§ 6, Example 1), and that the Boolean algebra $B_u$ is isomorphic with $\Sigma/N$ where $N$ denotes the ideal of $\mu$-null sets in $\Sigma$. Now if $h: \Sigma \to \Sigma/N$ is the canonical map, then $\lambda \mapsto \lambda' = \lambda \circ h$ defines a bijection of the additive real functions on $B_u$ onto the set of additive real functions on $\Sigma$ vanishing on each $S \in N$. Moreover, it is easy to see that $\lambda$ is continuous on $B_u$ (for the topology induced by $E$) if and only if $\lambda'$ is countably additive on $\Sigma$, and the proof of (8.6) shows that a family $\Lambda$ of such functions on $B_u$ is uniformly equicontinuous if and only if the countable additivity of $\lambda' = \lambda \circ h$ is uniform for $\lambda$ in $\Lambda$. □

For a $\sigma$-finite measure space $(X, \Sigma, \mu)$, we now identify the dual of $L^1(\mu)$ as the Banach lattice $L^\infty(\mu)$ (cf. Exerc. 23) and give a characterization of order continuous forms on $L^\infty(\mu)$ which implies the well-known theorem of Radon-Nikodym.

**8.7 Theorem.** *Let $(X, \Sigma, \mu)$ be a $\sigma$-finite measure space. The mapping $g \mapsto \varphi_g$, where*

$$\varphi_g(f) := \int f g \, d\mu \qquad (f \in L^1(\mu)),$$

*is an isomorphism of the Banach lattice $L^\infty(\mu)$ onto the dual of $L^1(\mu)$. Moreover, if $u$ is any weak order unit of $L^1(\mu)$ and $\lambda$ an order bounded linear form on $L^\infty(\mu)$, the following assertions are equivalent:*

(a) *$\lambda$ is order continuous.*

(b) *$\lambda$ is sequentially order continuous.*

(c) *The restriction of $\lambda$ to the unit ball of $L^\infty(\mu)$ is continuous for the topology of uniform convergence on $[-u, u] \subset L^1(\mu)$.*

(d) *$\lambda$ is $\sigma(L^\infty, L^1)$-continuous.*

*Proof*. Considering, if necessary, the measure $u.\mu$ in place of $\mu$, we can assume $\mu$ to be finite and $u$ to be the constant-one function $e$ on $X$. Then the Hilbert space $L^2(\mu)$ can be viewed as a dense ideal of $E := L^1(\mu)$ and hence, $E'$ can be identified with a dense ideal of the dual $L^2(\mu)'$ which, in turn, is $L^2(\mu)$, the canonical bilinear form being given by the inner product $(f, g) \mapsto \int f g \, d\mu$. Under this identification, the unit ball $V$ of $E'$ is clearly the order interval $[-e, e]$ which shows $E'$ to be isomorphic to $L^\infty(\mu)$.

(a) $\Rightarrow$ (d): Since $L^1(\mu)$ can be identified (under evaluation) with the band of all order continuous linear forms on $L^\infty(\mu)$ (8.3) (v), (a) implies (d).

(d) $\Rightarrow$ (c): On the unit ball $V=[-e,e]$ of $L^\infty(\mu)$, the weak topology generated by the finite subsets of $[-e,e]\subset L^1(\mu)$ agrees with $\sigma(L^\infty,L^1)$ (because the former topology is a coarser Hausdorff topology, and $V$ is $\sigma(L^\infty,L^1)$-compact), thus (d) implies (c).

(c) $\Rightarrow$ (a): The topology $\boldsymbol{T}$ of uniform convergence is none other than the topology of $L^\infty(\mu)$ induced by $L^1$ if, as usual, $L^\infty$ is considered a subspace of $L^1$; since by (8.3) (i) every order convergent filter on $V$ converges for $\boldsymbol{T}$, the implication follows.

(a) $\Leftrightarrow$ (b): Since $\mu$ defines a strictly positive linear form on $L^\infty(\mu)$, $L^\infty(\mu)$ is super Dedekind complete by the corollary of (4.9). □

**Corollary** (Radon-Nikodym). *Let $(X,\Sigma,\mu)$ be a $\sigma$-finite measure space and let $\lambda$ be a finite measure on $(X,\Sigma)$ absolutely continuous with respect to $\mu$. There exists a unique class $f\in L^1(\mu)$ such that for all $S\in\Sigma$,*

$$\lambda(S)=\int_S f\,d\mu.$$

*Proof.* The absolute $\mu$-continuity of $\lambda$ (i. e., the assumption that $\mu(S)=0$ implies $|\lambda|(S)=0$, $S\in\Sigma$) entails that $\lambda$ defines an order bounded linear form on $L^\infty(\mu)$. Since a sequence in $L^\infty(\mu)$ order converges iff it is bounded and converges a.e. $(\mu)$, it follows that $\lambda$ is sequentially order continuous on $L^\infty(\mu)$. □

The following properties of spaces $L^1(\mu)$ are striking consequences of the Vitali-Hahn-Saks theorem.

**8.8 Proposition.** *Let $(X,\Sigma,\mu)$ be a $\sigma$-finite measure space, and let $E:=L^1(\mu)$. Then $E$ is weakly sequentially complete, and for a subset $A\subset E$ to be relatively weakly compact, it is necessary and sufficient that $A$ be bounded and that the measures $v_f\colon S\mapsto\int_S f\,d\mu$ $(S\in\Sigma)$ be uniformly countably additive for $f$ in $A$. Moreover, if $A\subset E$ is relatively weakly compact then so is the solid hull $\bigcup_{f\in A}[-|f|,|f|]$.*

*Proof.* Let $(f_n)$ denote a weak Cauchy sequence in $E$ and consider the sequence $(v_n)$ of finite measures on $\Sigma$ defined by $v_n(S):=\int_S f_n\,d\mu$; clearly, each $v_n$ is absolutely continuous with respect to $\mu$. Since $v_n(S)=\langle f_n,\chi_S\rangle$ $(S\in\Sigma)$ where $\chi_S\in L^\infty(\mu)$ is the characteristic function of $S$, the Vitali-Hahn-Saks theorem (8.6 Cor.) shows that $S\mapsto v(S):=\lim_n v_n(S)$ is a measure absolutely continuous with respect to $\mu$ and hence, by the Radon-Nikodym theorem (8.7 Cor.), we have $v(S)=\int_S f\,d\mu$ for a unique $f\in L^1(\mu)$. But $(f_n)$ defines a bounded sequence of linear forms on $L^\infty(\mu)$ convergent to $f$ on the total subset $\{\chi_S\colon S\in\Sigma\}$; therefore, $\lim_n f_n=f$ weakly in $E$.

Next, let $A$ be bounded in $E$. If $A$ is relatively weakly compact, then by Eberlein's theorem [S, IV. 11.1 Cor. 2] each sequence in $A$ contains a weakly convergent subsequence, so if the countable additivity of $v_f$ were not uniform in $A$ then it would not be uniform on some weakly convergent sequence in $A$ which contradicts the corollary of (8.6). Conversely, suppose the measures $v_f$ to be uniformly countably additive for $f$ in $A$. Since $A$ is bounded it is relatively $\sigma(E'',E')$-compact as a subset of $E''$. Let $h$ be any point in the $\sigma(E'',E')$-closure $\bar{A}$.

Then $v_h(S):=\langle h,\chi_S\rangle$ $(S\in\Sigma)$ defines a measure on $\Sigma$ which is countably additive because of the uniform countable additivity of $v_f$ $(f\in A)$. Since $v_h$ is absolutely continuous with respect to $\mu$, it follows from (8.7) Cor. that $h\in L^1(\mu)$, hence $\bar{A}\subset E$ and thus $A$ is relatively $\sigma(E,E')$-compact.

Finally, suppose $A\subset E$ to be relatively $\sigma(E,E')$-compact. If $|A|:=\{|f|:f\in A\}$ were not relatively $\sigma(E,E')$-compact, by what has been shown there would exist a sequence $(f_n)$ in $A$ for which the sequence of measures $S\mapsto\int_S|f_n|\,d\mu$ is not uniformly countably additive; that is, there would exist a decreasing sequence $(S_n)$ in $\Sigma$ with void intersection and an $\varepsilon>0$ such that $\int_{S_n}|f_n|\,d\mu\geqq\varepsilon$ for infinitely many $n\in\mathbb{N}$. Define $S_n^+:=\{t\in S_n: f_n(t)\geqq 0\}$ and $S_n^-:=\{t\in S_n: f_n(t)<0\}$; then one of $(\int_{S_n^+} f_n\,d\mu)$ and $(\int_{S_n^-} f_n\,d\mu)$ cannot be a null sequence, in fact, we can assume that $\int_{S_n^+} f_n\,d\mu\geqq\varepsilon/2$ for all $n\in\mathbb{N}$. Now $\lim_n\mu(S_n)=0$ implies $\lim_n\mu(S_n^+)=0$ (supposing, without loss of generality, $\mu(S_1)$ finite) and this contradicts the uniform countable additivity of the measures $v_{f_n}$, $(f_n)$ being relatively $\sigma(E,E')$-compact. Therefore, $|A|$ is relatively $\sigma(E,E')$-compact, and it is now clear that the same is true of the solid hull of $A$. □

**Corollary.** *Every $AL$-space $E$ is weakly sequentially complete. Moreover, if $A\subset E$ is relatively weakly compact then so is the solid hull $\bigcup_{x\in A}[-|x|,|x|]$.*

*Proof.* By Eberlein's theorem [S, IV. 11.1 Cor. 2], weak compactness of subsets of $E$ is equivalent to weak sequential compactness. But each sequence $(x_n)$ in $E$ is contained in some $AL$-subspace possessing a weak order unit (for example, in $\bar{E}_x$ where $x:=\Sigma|x_n|/2^n\|x_n\|$, the sum being extended over the non-zero terms of the sequence). Thus (8.5) reduces the proof to the case $E=L^1(\mu)$, $\mu$ being a Radon measure on a compact space, and the assertion is immediate from (8.8). □

From (8.8) a compactness criterion for subsets of $L^1(\mu)$ ($\mu$ $\sigma$-finite) can be derived: A bounded subset $A\subset L^1(\mu)$ is relatively compact iff the countable additivity of the measures $S\mapsto\int_S f\,d\mu$ $(f\in A)$ is uniform and each sequence $(f_n)$ in $A$ contains a subsequence convergent in measure (Exerc. 23).

**Examples.** 4. If $X$ is a locally compact space and $\mu$ a Radon measure on $X$, the representation space of $E:=L^1(\mu)$ obtained in (8.5) is, in general, not homeomorphic to $X$. (Consider Lebesgue measure on the real line and compare Examples 5 and 7 of Chap. III, § 5.) For the $AL$-space $l^1(\mathbb{N})$, the representation space employed in the proof of (8.5)(a) is the Stone-Čech compactification $\beta\mathbb{N}$ of the discrete space $\mathbb{N}$.

5. Suppose $(X,\Sigma,\mu)$ is a finite measure space. If $N$ denotes the ideal of $\mu$-null sets in $\Sigma$, the Boolean algebra $\Sigma/N$ is a complete metric space under the metric $\varrho$ defined by

$$\varrho(S_1,S_2):=\mu(S_1\triangle S_2)\qquad(S_i\in\Sigma),$$

cf. [DS, II. 7.1]. If $u$ is any weak order unit of $E=L^1(\mu)$, the metric space $(\Sigma/N,\varrho)$ is uniformly isomorphic to the base $B_u=\{v\in E: v\wedge(u-v)=0\}\subset E$. Thus, by the corollary of (8.5), $E$ is separable iff $(\Sigma/N,\varrho)$ is separable. This should not be confused with separability of $C(K_u)$ (see 8.4); unless $E$ is finite dimensional, $C(K_u)$ is not separable (since it contains a Banach sublattice isomorphic to $l^\infty$).

6. While each order interval of an $AL$-space $E$ is weakly compact, even norm compact subsets of $E$ are not order bounded, in general. For example, if $E=L^1(\mu)$ with $\mu$ Lebesgue measure on $\mathbb{R}$ and if $(B_n)$ is a sequence of disjoint real Borel sets satisfying $\mu(B_n)=n^{-1}$ $(n\in\mathbb{N})$, then $(\chi_{B_n})$ is a null sequence not order bounded in $E$. It will be shown later (IV. 2.8) that in every Banach lattice which cannot be renormed into an $AM$-space, there exist null sequences which are not order bounded.

We finally prove that $AL$-spaces have the same injectivity property as has been established for order complete $AM$-spaces in (7.10) Cor. 3; this very recent result is due to Lotz [1974].

**8.9 Theorem.** *Let $G$ be a Banach lattice, let $G_0$ be a Banach sublattice, and denote by $E$ an $AL$-space. Every positive linear map $T_0\colon G_0\to E$ has a norm preserving, positive linear extension $T\colon G\to E$.*

*Proof*. First we note that $E'$ is an order complete $AM$-space with unit $e$, where $e\in E'_+$ is given by $e(x)=\|x^+\|-\|x^-\|$ (see Example 1 above, and 9.1); in fact, $[-e,e]$ is the unit ball of $E'$. Thus the adjoint map $T_0'\colon E'\to G_0'$ carries $E'$ into the ideal $(G_0')_{e_2}$, where $e_2:=T'e$ and $\|e_2\|=\|T'e\|=\|T'\|=\|T\|$ (§ 5, Formula (2)). By (5.6), $e_2$ has an extension $e_1\in G'_+$ such that $\|e_1\|=\|e_2\|$.

Now consider the adjoint $j'$ of the canonical injection $j\colon G_0\to G$. $j'$ is a positive linear map of $G'$ onto $G_0'$ (cf. the discussion following 5.6) and has the property that $j'[0,x']=[0,j'x']$ for each $x'\in G'_+$. (In fact, let $y'\in G_0'$ satisfy $0\leqq y'\leqq j'(x')$ and consider $G_0$ (modulo the null ideal of $x'$) as a vector sublattice of the $AL$-space $(G,x')$ (Example 1); by (5.6), $y'$ has a norm preserving extension $w\in(G,x')'_+$, and if $z':=w_{|G}$ then $z'\in[0,x']$ and $j'(z')=y'$.) Moreover, $j'$ is order continuous; since every directed ($\leqq$) majorized subset $A$ of $G'$ converges to $\sup A$ for $\sigma(G',G)$ and $j'$ is continuous for $\sigma(G',G)$ and $\sigma(G_0',G_0)$, we have $j'(\sup A)=\sup j'(A)$.

The ideals $(G')_{e_1}$ and $(G_0')_{e_2}$ are order complete $AM$-spaces with respective units $e_1$ and $e_2$, and thus can be identified with $C(K_1)$ and $C(K_2)$, respectively, for suitable compact Stonian spaces $K_1$ and $K_2$ (7.7 Cor.). Now the restriction $j_0'$ of $j'$ to $(G')_{e_1}\cong C(K_1)$ satisfies the hypotheses of (7.11), because the order continuity of $j'$ implies the absolute kernel of $j_0'$ to be a projection band. Consequently, by (7.11) there exists an isometric lattice homomorphism $S\colon(G_0')_{e_2}\to(G')_{e_1}$ such that $Se_2=e_1$ and $j_0'\circ S$ is the identity map of $(G_0')_{e_2}$. The composition $S\circ T_0'$ maps $E'$ into $G'$, is positive, and has norm $\|S\circ T_0'\|=\|T_0'\|$, moreover, $j'\circ S\circ T_0'=T_0'$. This implies $T_0''=T_0''\circ S'\circ j''$, and if $P$ denotes the band projection $E''\to E$ (cf. 8.3 (v)), then the restriction $T$ of $P\circ T_0''\circ S'$ to $G$ is a positive, linear, norm preserving extension of $T_0$. □

**Corollary 1.** *Let $G$ be a Banach lattice. If $G_0$ is a Banach sublattice of $G$ isomorphic to an $AL$-space, then $G_0$ is the range of a positive, contractive projection.*

In fact, every norm preserving positive linear extension of the identity map of $G_0$ is a projection of the desired kind. In particular, every closed vector sublattice of an $AL$-space is the range of such a projection (Dean [1965], cf. also (III. 11.4)). Moreover, the existence of such projections implies the following.

**Corollary 2.** *Let $G$ be a Banach lattice, and let $G_0$ be a Banach sublattice isomorphic to an $AL$-space. Every continuous linear map $T_0: G_0 \to H$ into a Banach space $H$ has a norm preserving extension $T: G \to H$.*

Clearly, if $H$ is ordered and $T_0 \geqq 0$, then $T$ can be chosen positive. Finally, the preceding extension theorems can be dualized to yield lifting theorems involving $AM$-spaces (Exerc. 22).

## § 9. Duality of $AM$- and $AL$-Spaces. The Dunford-Pettis Property

We recall from (5.5) that the dual of any normed vector lattice is a Banach lattice under its canonical norm and order. $AM$- and $AL$-spaces are dual to each other in the following precise sense.

**9.1 Proposition.** *The dual of each $M$-normed space is an $AL$-space, and the dual of each $L$-normed space is an $AM$-space with unit.*

*Proof.* Suppose first that $E$ is an $M$-normed space (Def. 7.1); in view of (5.5), it suffices to show that $\|x'+y'\| \geqq \|x'\|+\|y'\|$ whenever $x', y' \in E'_+$. Given $\varepsilon>0$. by § 5, Formula (1) there exist elements $x, y \in E_+$ such that $\|x\|=\|y\| \leqq 1$ and $\langle x, x'\rangle > \|x'\|-\varepsilon/2$, $\langle y, y'\rangle > \|y'\|-\varepsilon/2$. If $z:=x \vee y$ then $\|z\| \leqq 1$, since $E$ is $M$-normed, and this implies $\|x'+y'\| \geqq \langle z, x'+y'\rangle \geqq \langle x, x'\rangle + \langle y, y'\rangle > \|x'\|+\|y'\|-\varepsilon$. $\varepsilon$ being arbitrary, this shows (in view of the triangle inequality) the norm of $E'$ to be additive on $E'_+$.

On the other hand, if $E$ is an $L$-normed space (Def. 8.1) and if $z=x_1-y_1$ $=x_2-y_2$ are any two decompositions of $z \in E$ into differences of elements $x_i, y_i \geqq 0$ $(i=1,2)$ then, since the norm of $E$ is additive on $E_+$, the definition $e(z):=\|x_1\|-\|y_1\|$ $(=\|x_2\|-\|y_2\|)$ yields a positive linear form $e$ on $E$ for which $e(|z|)=\|z\|$ $(z \in E)$. Evidently, $e$ is continuous and every linear form $f \in E'$ of norm $\leqq 1$ satisfies $|f(z)| \leqq e(|z|)$ or, equivalently, $-e \leqq f \leqq e$. It now follows from (7.2) that $E'$ is an $AM$-space with unit $e$. ☐

Thus if $E$ is an $AL$-space then $E'$ is an order complete $AM$-space with unit and hence, by (7.4) and (7.7) Cor., isomorphic to some $C(K)$ for $K$ a compact Stonian space. Moreover, by evaluation $E \to E''$ each element of $E$ defines an order continuous Radon measure $\mu$ on $K$. (It is not difficult to verify (Exerc. 24) that order continuity of $\mu \in M(K)$ is equivalent to requiring that the regular Borel measure $\bar{\mu}$ extending $\mu$ vanish on each rare Borel subset of $K$.) It is clear from (4.3) that the set of all order continuous Radon measures (sometimes called *normal measures*) on $K$ is a band in $M(K)$. The compact Stonian space $K$ is called *hyperstonian* if the band $N$ of normal Radon measures on $K$ separates $C(K)$ (or equivalently, if the union of the supports of all $v \in N$ is dense in $K$).

These considerations lead to a second representation theorem for $AL$-spaces.

**9.2 Theorem.** *For every AL-space E there exists a (compact) hyperstonian space K, unique to within homeomorphism, such that E can be identified with the AL-space of all order continuous Radon measures on K.*

*Proof.* By (9.1) $E'$ can be identified with $C(K)$ for some compact Stonian space $K$, and from (8.3) (v) it follows that evaluation maps $E$ onto the band of all order continuous Radon measures on $K$. Now let $K'$ be any hyperstonian space with the property that there exists an isomorphism $i: E \to N(K')$ onto the $AL$-space $N(K')$ of order continuous Radon measures on $K'$. Borrowing from the proof of (9.3) the fact that $N(K')$ is a Banach lattice predual to $C(K')$, we see that the adjoint $i'$ is an isomorphism $C(K') \to C(K) \cong E'$ which by (7.4) Cor. 2 shows $K'$ to be homeomorphic with $K$. □

The reader will notice the connection between the first representation theorem for $AL$-spaces (8.5), and the second (9.2); the former can be derived from the latter via the theorem of Radon-Nikodym (8.7) Cor. Let us note this corollary of (9.2).

**Corollary.** *In the dual of an AL-space E, each band is $\sigma(E', E)$-closed.*

*Proof.* Let $B$ be a band of $E'$. Since $E' = C(K)$ is order complete, $B$ is a projection band (2.10). If $P_B$ denotes the associated band projection and $e$ denotes the unit of $E'$ (constant-one function on $K$), then $B = P_B(E)$ and $B$ is an $AM$-space with unit $e_1 := P_B e$, $e_1$ being the characteristic function of an open-and-closed subset of $K$. The adjoint $P_B'$: $M(K) \to M(K)$ of $P_B$ is the mapping $\mu \mapsto e_1 . \mu$ and, since $e_1 . \mu$ is an order continuous measure on $K$ whenever $\mu$ is, leaves $E$ invariant. Therefore, $P_B$ is $\sigma(E', E)$-continuous and $B$ is $\sigma(E', E)$-closed. □

**9.3 Theorem.** *Let K be compact. For the Banach lattice $C(K)$ to be isomorphic to a dual Banach lattice (and hence, to the dual of an AL-space), it is necessary and sufficient that K be hyperstonian.*

*Proof.* The condition is necessary. In fact, if $C(K)$ is isomorphic to the Banach lattice $E'$ for some Banach lattice $E$, then (since $E'$ is order complete) $K$ is Stonian by (7.7) Cor. 2 and $E$ is isomorphic to an $AL$-space by (9.1) and (5.5) Cor. 2. Clearly, then, $K$ is hyperstonian, since $E$ separates $E'$ and each element of $E$ defines an order continuous Radon measure on $K$.

The condition is sufficient. Suppose $K$ to be hyperstonian and denote by $N$ the band of order continuous Radon measures on $K$. By hypothesis, $\langle C(K), N\rangle$ is a separated duality; we will show that the unit ball $[-e, e]$ of $C(K)$ is $\sigma(C(K), N)$-compact. By the Mackey-Arens theorem [S, IV. 3.2] this will imply that $C(K)$ can be viewed as the dual of the Banach sublattice $N$ of $M(K)$.

To this end, we consider on $C(K)$ the topology $o(C(K), N)$ generated by the family of semi-norms

$$f \mapsto \nu(|f|) = \int |f| \, d\nu \qquad (\nu \in N_+)$$

(in the sequel briefly called the $\nu$-topology). Since $N$ is an ideal in $M(K)$, each interval $[-\nu, \nu]$ $(\nu \in N_+)$ is $\sigma(N, C(K))$-compact and hence, the $\nu$-topology is consistent with the duality $\langle C(K), N\rangle$; thus it suffices to show that the unit ball $[-e,e]$ is weakly compact with respect to the $\nu$-topology. On the other hand, the canonical imbeddings $C(K)\to L^1(\nu)$ $(\nu\in N_+)$ define an injection $C(K)\to\prod_{\nu\in N_+}L^1(\nu)$ which is an isomorphism for the $\nu$-topology and the product topology, respectively; if we can show the image of $[-e,e]$ in $\prod_{\nu\in N_+}L^1(\nu)$ to be closed then $[-e,e]$ is $\sigma(C(K),N)$-compact, because $[-e,e]$ is convex and its image is contained in the weakly compact product $\prod_{\nu\in N_+}[-e_\nu,e_\nu]$, $e_\nu$ denoting the canonical image of $e$ in $L^1(\nu)$. (Note that by (8.3) and (5.10), $[-e_\nu,e_\nu]$ is weakly compact in $L^1(\nu)$, and that the weak topology of $\prod_{\nu\in N_+}L^1(\nu)$ is the product of the respective weak topologies of the factors, [S, IV. 4.3].) Thus the proof is reduced to the following lemma.

**Lemma.** *The unit ball* $[-e,e]$ *of* $C(K)$ *is complete for* $o(C(K),N)$.

Let $\boldsymbol{F}$ denote a filter on $U:=[-e,e]$ which is a Cauchy filter for the $\nu$-topology. Let $\nu\in N_+$ be fixed and denote by $H_\nu$ the set of all limit points of $\boldsymbol{F}$ for the (non-Hausdorff, in general) topology generated by the single semi-norm $f\mapsto\nu(|f|)$. $H_\nu$ is non-void: If $(F_n)$ is a decreasing sequence of elements of $\boldsymbol{F}$ such that $\nu(|h-h'|)<2^{-n-1}$ for all $h,h'\in F_n$, if $h_n\in F_n$ is chosen arbitrarily, and if $f_n:=\sup_{m\geqq n}h_m$ ($U$ is order complete), then we have $f_n=h_n+\sup_p\sum_{m=n+1}^{n+p}(h_m-h_{m-1})$ and the order continuity of $\nu$ implies

$$\nu(|f_n-h_n|)\leqq\sum_{m=n+1}^{\infty}\nu(|h_m-h_{m-1}|)\leqq 2^{-n}.$$

It is now clear that $f:=\inf_n f_n$ is an element of $H_\nu$.

Next we observe that $H_\nu$ is lattice ordered. In fact, let $\varepsilon>0$ be given and choose $F_\varepsilon\in\boldsymbol{F}$ so that $\nu(|h-h'|)\leqq\varepsilon/2$ whenever $h,h'\in F_\varepsilon$; then we have $\nu(|f-h|)\leqq\varepsilon/2$ for all $f\in H_\nu$, $h\in F_\varepsilon$. Now for $f,g\in H_\nu$ and $h\in F_\varepsilon$ the identity $f\vee g-h=f\vee g-h\vee g+h\vee g-h\vee h$ implies $|f\vee g-h|\leqq|f-h|+|g-h|$, and so $\nu(|f\vee g-h|)\leqq\varepsilon$; this shows $f\vee g\in H_\nu$. Similarly, one obtains $f\wedge g\in H_\nu$.

Dropping the temporary assumption that $\nu$ be fixed, we define $f_\nu:=\sup H_\nu$ for all $\nu\in N_+$. Evidently, we have $f_\nu\in H_\nu$. Moreover, $0\leqq\nu'\leqq\nu\in N_+$ implies $H_\nu\subset H_{\nu'}$ which shows the family $(f_\nu)_{\nu\in N_+}$ to be directed downward. $f:=\inf_\nu f_\nu$ exists and the directed set $(f_\nu)$ converges to $f$ for the $\nu$-topology. Thus $f\in\bigcap_{\nu\in N_+}H_\nu$ which shows that $f=\lim\boldsymbol{F}$ and hence, that $U$ is complete for the $\nu$-topology.

This ends the proof of (9.3). □

**Examples.** 1. If $(X,\Sigma,\mu)$ is a $\sigma$-finite measure space, the $AM$-space $L^\infty(\mu)$ is the dual of $L^1(\mu)$ (8.7) and isomorphic to $C(K_{\Sigma'})$, where $K_{\Sigma'}$ is the Stone representation space of $\Sigma':=\Sigma/N$ ($N$ the ideal of $\mu$-null sets, cf. § 7, Examples 3, 5). Thus $K_{\Sigma'}$ is hyperstonian.

2. There exist Stonian spaces which are not hyperstonian. For example, let $E$ denote the Dedekind completion of $\mathbf{C}[0,1]$; since there exist no order continuous linear forms $\neq 0$ on $C[0,1]$ (Exerc. 10), we also have $E^\star_{00}=\{0\}$. On the other hand, $E$ is an order complete $AM$-space with unit and thus isomorphic to some $C(K)$, where $K$ is a compact Stonian space which is not hyperstonian.

Theorem (9.3) can be strengthened as follows: If $C(K)$ ($K$ compact) is (isometrically) isomorphic to some dual Banach space, then $K$ is hyperstonian (Exerc. 24). By contrast, the problem of characterizing $AL$-spaces which are dual Banach spaces, is quite complex (cf. Lindenstrauss [1970]).

We now turn to an investigation of weakly compact linear maps and their relationship with the duality of $AM$- and $AL$-spaces (cf. Grothendieck [1953]). For an intelligent reading of this discussion, familiarity with the duality of locally convex spaces (see, e. g., [S, Chap. IV, §§ 1—5]) is required. We use the standard notations $\beta(G,H)$, $\tau(G,H)$, and $\sigma(G,H)$ for the strong, Mackey, and weak topologies, respectively, which are defined by a separated duality $\langle G,H\rangle$. If $E,F$ denote Banach spaces, a continuous linear map $T: E\to F$ is called *weakly compact* if $T(B)$ is relatively $\sigma(F,F')$-compact for every bounded subset $B\subset E$.

In the following we consider the dualities $\langle E'',E'\rangle$ and $\langle F'',F'\rangle$, identifying (under evaluation) $E$ and $F$ with subspaces of $E''$ and $F''$, respectively.

**9.4 Proposition.** *Let $E,F$ denote Banach spaces, and let $T: E\to F$ be a continuous linear map with adjoint $T'$ and second adjoint $T''$, respectively. The following are equivalent:*

(a) *$T$ is weakly compact.*
(b) *$T''(E'')\subset F$.*
(c) *$T'$ is continuous from $(F',\tau(F',F))$ into $(E',\beta(E',E))$.*
(d) *$T'(V^\circ)$ is $\sigma(E',E'')$-compact, $V$ denoting the unit ball of $F$.*

*Proof.* (a) $\Rightarrow$ (b): Since $T''$ is the continuous extension of $T$ to $(E'',\sigma(E'',E'))$ with values in $(F'',\sigma(F'',F'))$, and since the unit ball $U^{\circ\circ}$ of $E''$ (bipolar with respect to $\langle E'',E'\rangle$ of the unit ball $U\subset E$) is the $\sigma(E'',E')$-closure in $E''$ of $U$, relative $\sigma(F,F')$-compactness of $T(U)$ implies $T''(U^{\circ\circ})\subset F$ and hence, $T''(E'')\subset F$.

(b) $\Rightarrow$ (c): $T''(E'')\subset F$ implies that $T''(U^{\circ\circ})$ is a $\sigma(F,F')$-compact subset $C$ of $F$. So $T'(C^\circ)\subset U^\circ$ shows that $T'$ is continuous for the topologies in question, since $C^\circ$ is a 0-neighborhood in $F$ for the Mackey topology $\tau(F',F)$.

(c) $\Rightarrow$ (d): The hypothesis implies that $T'$ is continuous for the respective weak topologies, that is, for $\sigma(F',F)$ and $\sigma(E',E'')$. But $V^\circ$ is $\sigma(F',F)$-compact hence, $T'(V^\circ)$ is $\sigma(E',E'')$-compact.

(d) $\Rightarrow$ (a): If $T'(V^\circ)$ is $\sigma(E',E'')$-compact then $T': F'\to E'$ is weakly compact and hence, by the preceding implication (a) $\Rightarrow$ (d), $T''(U^{\circ\circ})$ is $\sigma(F'',F''')$-compact. Since $F$ is a $\sigma(F'',F''')$-closed subspace of $F''$, $T''(U^{\circ\circ})\cap F$ is a $\sigma(F,F')$-compact set containing $T(U)$ and so $T$ is weakly compact. ▯

**Corollary.** *If $E,F$ are Banach spaces with respective Banach duals $E',F'$, then $T: E\to F$ is weakly compact if and only if $T': F'\to E'$ is weakly compact.*

The property of a Banach space characterized in the following proposition is possessed by all $AM$- and $AL$-spaces (see below). We recall that in a Banach space the closed convex circled hull of each weakly compact subset is weakly compact (Krein's theorem, [S, IV. 11.4]); this makes it unnecessary to deal primarily with convex circled sets.

**9.5 Proposition.** *Let $E$ be a Banach space and denote by $\mathbf{W}$ the family of all weakly compact subsets of $E$. These assertions are equivalent:*

(a) *For every Banach space $F$ and each weakly compact $T: E \to F$, $T(W)$ is compact in $F$ whenever $W \in \boldsymbol{W}$.*
(b) *Each $W \in \boldsymbol{W}$ is compact for $\tau(E'', E')$.*
(c) *Each $\sigma(E', E'')$-compact subset of $E'$ is relatively $\tau(E', E)$-compact.*

*Proof.* (a) $\Rightarrow$ (b): As every locally convex space, $(E'', \tau(E'', E'))$ is isomorphic to a subspace of a suitable product, $\prod_\alpha F_\alpha$ say, of Banach spaces [S, II. 5.4 Cor. 2]. Let $i$ denote an isomorphism $E'' \to \prod_\alpha F_\alpha$ and let $p_\alpha$ denote the projection map $\prod_\alpha F_\alpha \to F_\alpha$. The map $\bar{U}_\alpha := p_\alpha \circ i$ is continuous and hence, maps the unit ball $U^{\circ\circ}$ of $E''$ (which is $\sigma(E'', E')$-compact) onto a weakly compact subset of $F_\alpha$; thus the restriction $U_\alpha := \bar{U}_{\alpha|E}$ is a weakly compact map $E \to F_\alpha$. By (a) this implies that $U_\alpha(W)$ is compact in $F_\alpha$ for all $W \in \boldsymbol{W}$; but $i(W) \subset \prod_\alpha U_\alpha(W)$, and this shows $W$ to be precompact for $\tau(E'', E')$. But $W$, being weakly compact, is complete for $\tau(E'', E')$; thus $W$ is $\tau(E'', E')$-compact.

(b) $\Rightarrow$ (a): If $T: E \to F$, $F$ any Banach space, is weakly compact then so is $T'$ by the corollary of (9.4); applying (9.4) to $T'$ and using that $T''(E'') \subset F$, it follows that $T''$ is continuous from $(E'', \tau(E'', E'))$ into $F$. Hence $T(W) \subset F$ is compact whenever $W \in \boldsymbol{W}$.

(b) $\Rightarrow$ (c): If $A'$ is a convex circled $\sigma(E', E'')$-compact subset of $E'$, its polar $A'^\circ$ (in $E''$) is a 0-neighborhood for $\tau(E'', E')$. Thus, since each $W \in \boldsymbol{W}$ is $\tau(E'', E')$-compact by hypothesis, for a given $W$ there exists a finite set $\Phi \subset W$ such that $W \subset \Phi + A'^\circ$. Also, since $A'$ is $\sigma(E', E)$-bounded there exists a finite subset $\Phi' \subset A'$ with this property: Whenever $x' \in A'$, there exists $y' \in \Phi'$ such that $|\langle z, x' - y' \rangle| \leq 1$ for all $z \in \Phi$. Let $W \in \boldsymbol{W}$ and $x \in W$ be fixed. Considering the identity

$$\langle x, x' - y' \rangle = \langle x - z, x' - y' \rangle + \langle z, x' - y' \rangle,$$

for each $x' \in A'$ there exists $y' \in \Phi'$ such that $|\langle z, x' - y' \rangle| \leq 1$ for all $z \in \Phi$; and for each such $y' \in \Phi'$ there exists $z \in \Phi$ such that $|\langle x - z, x' - y' \rangle| \leq 2$. In all, this implies $A' \subset \Phi' + 3W^\circ$ and hence, since $W \in \boldsymbol{W}$ was arbitrary, $A'$ is precompact for $\tau(E', E)$. But then $A'$ is relatively $\tau(E', E)$-compact, because $E'$ is quasi-complete (boundedly complete) for $\tau(E', E)$.

The proof of (c) $\Rightarrow$ (b) is similar, and will be omitted. ▯

**Corollary.** *If the dual $E'$ of a Banach space $E$ satisfies the equivalent assertions of* (9.5), *then so does $E$.*

*Proof.* By hypothesis, the Banach space $E'$ satisfies (9.5) (c). Now if $W \subset E$ is $\sigma(E, E')$-compact then, since $\sigma(E, E')$ is induced on $E$ by $\sigma(E'', E''')$, (9.5) (c) implies that $W$ is compact for $\tau(E'', E')$. This shows $E$ to satisfy (9.5) (b). ▯

It was first shown by Dunford and Pettis [1940] that every space $L^1(\mu)$ satisfies (a) of (9.5). Following Grothendieck [1953] we adopt this definition.

**9.6 Definition.** *A Banach space $E$ is said to have the* Dunford-Pettis property *if $E$ satisfies the three equivalent assertions of* (9.5).

We give another characterization of these spaces.

**9.7 Theorem.** *For any Banach space $E$, the following are equivalent assertions:*

(a) *$E$ has the Dunford-Pettis property.*

(b) *Every weakly compact linear map $E \to F$, $F$ any Banach space, transforms weakly convergent sequences into convergent sequences.*

(c) *If $F$ is any Banach space, $b$ any continuous bilinear form on $E \times F$, and if $\lim_n x_n = x$ weakly in $E$ and $\lim_n y_n = y$ weakly in $F$, then $\lim_n b(x_n, y_n) = b(x, y)$.*

*Proof.* (a) $\Rightarrow$ (b) is trivial in view of (9.5)(a).

(b) $\Rightarrow$ (a): Let $T: E \to F$ be weakly compact and let $W \subset E$ be $\sigma(E, E')$-compact. Given any sequence $y_n = Tx_n$ in $T(W)$ $(x_n \in W)$, by Eberlein's theorem [S, IV. 11.1 Cor. 2] $(x_n)$ contains a weakly convergent subsequence; thus by hypothesis, $(y_n)$ contains a convergent subsequence. Since $T(W)$ is weakly compact (hence norm closed), $T(W)$ is compact in $F$.

(a) $\Rightarrow$ (c): If $F$ is any Banach space and $b$ any continuous bilinear form on $E \times F$, there exists a (unique) $S \in \mathscr{L}(F, E')$ such that $b(x,y) = \langle x, Sy \rangle$ identically on $E \times F$. Let $(x_n)$, $(y_n)$ be weakly convergent sequences with limits $x \in E$, $y \in F$, respectively. Since $S$ is weakly continuous, $(Sy_n)$ is $\sigma(E', E'')$-convergent and hence, by (9.5)(c), its range is relatively $\tau(E', E)$-compact. But this implies the convergence of $(Sy_n)$ to $Sy$ to be uniform on weakly compact subsets of $E$, in particular, on the range of $(x_n)$; thus $\lim_n b(x_n, y_n) = \lim_n \langle x_n, Sy_n \rangle = \langle x, Sy \rangle = b(x, y)$.

(c) $\Rightarrow$ (a): Using (9.5)(c) we show that the existence of a $\sigma(E', E'')$-compact set $A' \subset E'$ which is not $\tau(E', E)$-precompact, is contradictory. Suppose $A'$ is such a set; then there exists a weakly compact subset $W \subset E$ such that $A' \not\subset \Phi + W^\circ$ for any finite $\Phi \subset A'$. Inductively we can thus construct a sequence $(x'_n)$ in $A'$ such that $\sup_{x \in W} |\langle x, x'_n - x'_m \rangle| \geqq 1$ whenever $m \neq n$; clearly, $(x'_n)$ contains no subsequence uniformly convergent on $W$. Since $A'$ is $\sigma(E', E'')$-compact, by Eberlein's theorem we can choose a subsequence (again denoted by $(x'_n)$) which is weakly convergent to some $x' \in A'$; letting $z'_n := x'_n - x'$, $(z'_n)$ does not converge uniformly on $W$. So there exist $\varepsilon > 0$ and a sequence $(x_n)$ in $W$ such that $|\langle x_n, z'_n \rangle| \geqq \varepsilon$ for infinitely many $n$; moreover, we can assume $(x_n)$ to converge weakly to some $x \in W$. But this contradicts (c), applied to the canonical bilinear form on $E \times E'$. ☐

To prove that every $AM$- and every $AL$-space has the Dunford-Pettis property, we will use the following compactness criterion (Grothendieck [1953]); this criterion is of independent interest and will also be needed later (§ 10). The reader will notice the resemblance of (9.8) with (8.8) on which the proof is largely based.

**9.8 Proposition.** *Let $K$ be any compact space and denote by $M(K)$ the $AL$-space of all Radon measures on $K$, considered to be the strong dual of $C(K)$. For a bounded subset $A \subset M(K)$ the following properties are equivalent:*

(a) *$A$ is relatively weakly compact.*

(b) *For each weak null sequence $(f_n)$ in $C(K)$, $\lim_n \mu(f_n) = 0$ uniformly for $\mu$ in $A$.*

(c) *For each bounded sequence $(f_n)$ in $C(K)$ satisfying $f_m \wedge f_n = 0$ whenever $m \neq n$, $\lim_n \mu(f_n) = 0$ uniformly for $\mu$ in $A$.*

(d) *For each sequence* $(U_n)$ *of disjoint open subsets of* $K$, $\lim_n \mu(U_n)=0$ *uniformly for* $\mu$ *in* $A$.

*Proof.* (a) $\Rightarrow$ (b): Since, by virtue of the dominated convergence theorem, weak convergence of sequences in $C(K)$ is equivalent to uniform boundedness and pointwise convergence, if follows that $f_n \to 0$ weakly iff $|f_n| \to 0$ weakly (cf. 7.6). Now if $f_n \to 0$ weakly and $g_m$ $(m \in \mathbb{N})$ denotes the Baire function defined by $g_m(t) := \sup_{n \geq m} |f_n(t)|$ $(t \in K)$, $(g_m)$ is a monotone sequence converging to 0 pointwise; thus $\lim_m v(g_m)=0$ for each $v \in M(K)$. By (8.8) Cor. the set $|A| := \{|\mu| : \mu \in A\}$ is relatively weakly compact; considering $(g_m)$ as a sequence of continuous functions on the weak closure of $|A|$ in $M(K)$, it follows from Dini's theorem that $|\mu|(g_m) \to 0$ uniformly for $\mu \in A$ and hence, $\mu(f_n) \to 0$ uniformly for $\mu$ in $A$.

(b) $\Rightarrow$ (c) is trivial, since a sequence $(f_n)$ satisfying the hypothesis of (c) is evidently a weak null sequence in $C(K)$.

(c) $\Rightarrow$ (d): If for some disjoint sequence $(U_n)$ of open sets the convergence $\mu(U_n) \to 0$ is not uniform for $\mu$ in $A$, then (by regularity of the Borel extensions of $\mu \in A$) there exists a sequence $(C_n)$ of compact sets, $C_n \subset U_n$, for which the convergence $\mu(C_n) \to 0$ is not uniform in $A$; by an application of Urysohn's theorem this is readily seen to contradict (c).

(d) $\Rightarrow$ (a): To show that $A$ is relatively weakly compact it is enough to show that each sequence in $A$ contains a weakly convergent subsequence (Eberlein's theorem, [S, IV. 11.1 Cor. 2]). Let $(\mu_j)$ be any sequence in $A$ and define $\mu := \sum_j 2^{-j} |\mu_j|$; then the range of $(\mu_j)$ is contained in the principal band of $M(K)$ generated by $\mu$, and the latter is isomorphic to $L^1(\mu, K)$ (§ 8, Example 3). Thus, considering the $\mu_j$ as elements of $L^1(\mu)$, it suffices by (8.8) to show that the countable additivity of the (regular) Borel measures $B \mapsto \mu_j(B)$ is uniform with respect to $j \in \mathbb{N}$.

Now assume the countable additivity of the sequence $(\mu_j)$ is not uniform. There exists $\varepsilon > 0$, a disjoint sequence $(B_j)$ of Borel sets, and a subsequence $(v_j)$ of $(\mu_j)$ such that $|v_j(B_j)| > \varepsilon$ for all $j$; by regularity of $v_j$, we can and will assume that each $B_j$ is compact. By the lemma stated separately below, we can find open sets $U_j \supset B_j$ such that $|v_j|(\bar{U}_k \setminus B_k) \leq \varepsilon/2^{k+1}$ for all $j, k \in \mathbb{N}$. Defining $V_k := U_k \setminus \bigcup_{j=1}^{k-1} \bar{U}_j$ we obtain

$$V_k \triangle B_k \subset (U_k \setminus B_k) \cap [\bigcup_{j=1}^{k-1} (\bar{U}_j \setminus B_j)]$$

and hence, $|v_k|(V_k \triangle B_k) \leq \varepsilon/2$ for all $k \in \mathbb{N}$. It follows that $|v_k(V_k)| > \varepsilon/2$ for all $k$; since $(V_k)$ is a sequence of disjoint open sets, we arrive at a contradiction with (d). It remains to prove this lemma.

**Lemma.** *Let* $A = \{\mu_j : j \in \mathbb{N}\}$ *be a countable subset (not necessarily bounded) of* $M(K)$ *satisfying condition* (d) *of* (9.8). *Then for each compact* $B \subset K$ *and* $\eta > 0$, *there exists an open* $U \supset B$ *such that* $|\mu_j|(\bar{U} \setminus B) < \eta$ *for all* $j \in \mathbb{N}$.

*Proof.* Given $\eta > 0$ and a compact $B \subset K$, by the regularity of $\mu_j$ there exists a sequence $(U_n)$ of open sets satisfying $U_n \supset \bar{U}_{n+1} \supset B$ and such that $|\mu_j|(\bar{U}_n \setminus B) < n^{-1}$ for $j = 1, \ldots, n$ and all $n \in \mathbb{N}$. Thus $\lim_n |\mu_j|(U_n) = |\mu_j|(B)$ for each $j$; if the limit were not uniform for $j \in \mathbb{N}$, there would exist $\delta > 0$ and subsequences $(v_j)$ of $(\mu_j)$ and $(V_j)$ of $(U_j)$ satisfying $|v_j|(V_j \setminus \bar{V}_{j+1}) > \delta$. But then for each $j$ there would exist an open set $W_j \subset V_j \setminus \bar{V}_{j+1}$ satisfying $|v_j(W_j)| > \delta/4$ which is incompatible with (d).

(Note that condition (d) of (9.8) does not imply $A$ to be bounded; cf. Exerc. 27 (a).)

This ends the proof of (9.8). □

**9.9 Theorem.** *Every AM-space and every AL-space possesses the Dunford-Pettis property.*

*Proof.* By (9.1) and the corollary of (9.5), it suffices to prove that every space $C(K)$ ($K$ compact) satisfies (9.7) (b). Let $F$ denote any Banach space, let $T: E \to F$ be weakly compact where $E := C(K)$, and suppose $(f_n)$ to be a $\sigma(E, E')$-null sequence. If $V^\circ$ denotes the unit ball of $F'$ then by (9.4) $T'(V^\circ)$ is $\sigma(E', E'')$-compact; hence, from (9.8) (b) it follows that $\lim_n \langle f_n, T'y' \rangle = 0$ uniformly for $y'$ in $V^\circ$. Evidently, this is equivalent to $\lim_n \|T f_n\| = 0$. □

**Corollary 1.** *If $E$ is an AM- or AL-space and $T: E \to E$ is weakly compact, then $T^2$ is compact; more generally, the product of two weakly compact endomorphisms of $E$ is compact.*

**Corollary 2.** *An AM- or AL-space which is reflexive, is finite dimensional.*

*Proof.* Since the identity map $1_E$ of a reflexive Banach space $E$ is weakly compact, Cor. 1 implies that $1_E$ is compact for $E$ an $AM$- or $AL$-space and hence, that $E$ is finite dimensional [S, I. 3.6]. □

In the following corollary and its proof, the term "isomorphic" refers only to the structure of a topological vector space.

**Corollary 3.** *An infinite dimensional AM-space can never be isomorphic to a subspace of an AL-space, and an infinite dimensional AL-space can never be isomorphic to a quotient of an AM-space.*

*Proof.* If an $AM$-space $E$ is isomorphic to a subspace of an $AL$-space, then by (8.8) $E$ is weakly sequentially complete and, since its dual is an $AL$-space, reflexive by (5.11); Cor. 2 implies that $E$ is finite dimensional. If an $AL$-space $E$ is isomorphic to a quotient of an $AM$-space $F$, then $E'$ is an $AM$-space isomorphic to a subspace of the $AL$-space $F'$; by the preceding, $E'$ is finite dimensional and so is $E$. □

**Examples.** 3. As is well known, every Banach space $G$ with unit ball $W$ can be identified with a Banach subspace of $C(W^\circ)$. If $G$ is infinite dimensional and reflexive, then by (9.9) Cor. 1 the corresponding subspace of $C(W^\circ)$ is non-complemented.

Similarly, the closed vector subspace $F$ of $L^1(\mu)$ ($\mu$ Lebesgue measure on $[0,1]$) generated by the Rademacher functions is non-complemented, since $F$ is isomorphic to $l^2$ (cf. Chap. IV, Exerc. 14).

4. Let $G_1$ and $G_2$ be Banach spaces. The existence of isometric (linear) injections $G_1 \to G_2$ and $G_2 \to G_1$ does not imply $G_1$ and $G_2$ to be isomorphic.

In fact, this situation is at hand for $G_1 = C[0,1]$ and $G_2 = C[0,1] \times l^2$ but $G_1$ and $G_2$ are not isomorphic as topological vector spaces, since $G_2$ does not have the Dunford-Pettis property.

5. If $\mu$ is a positive Radon measure on a compact space $K$, then by (5.10) the canonical maps $C(K) \to L^1(\mu)$ and $L^\infty(\mu) \to L^1(\mu)$ are weakly compact; thus, weakly convergent sequences in $C(K)$ or $L^\infty(\mu)$ norm converge in $L^1(\mu)$ (and, in fact, this is true for weak Cauchy sequences, Exerc. 26). More generally, *every continuous linear map of an AM-space into a weakly sequentially complete Banach space is weakly compact* (Exerc. 27) and hence, maps weakly convergent sequences onto norm convergent sequences.

In conclusion, we remark that every $AM$-space also possesses a reciprocal Dunford-Pettis property (Exerc. 25).

## § 10. Weak Convergence of Measures

Let $E$ be a vector lattice and let $E^\star$ denote its order dual (§ 4). Under suitable assumptions on $E$, what properties does $E^\star$ have with respect to $\sigma(E^\star, E)$? An interesting example is furnished by (5.10) Cor. 2; there are some further remarkable results related to weak convergence of sequences to which this section is devoted. Generally, the difficulty in discovering lattice-related properties of the weak topology lies in the fact that as a rule, the lattice operations are not weakly continuous; thus, if $E$ is an infinite dimensional Banach lattice, the lattice operations in $E$ are never $\sigma(E, E')$-continuous and the lattice operations in $E'$ are never $\sigma(E', E)$-continuous (Exerc. 28).

We begin with an elementary result.

**10.1 Proposition.** *Let $E$ be a vector lattice satisfying Axiom* (OS) (Def. 1.8); *then $E^\star$ is $\sigma(E^\star, E)$-sequentially complete.*

*Proof.* By (7.2) each of the principal ideals $E_x$ $(x \in E_+)$ is an $AM$-space with unit. If $(f_n)$ is a $\sigma(E^\star, E)$-Cauchy sequence then by (5.5), the restrictions $f_n | E_x$ form $\sigma(E'_x, E_x)$-Cauchy sequences in the respective duals of the $AM$-spaces $E_x$ and hence, the Banach-Steinhaus theorem shows that $f := \lim_n f_n$ is a linear form on $E$ whose restriction to each $E_x$ is continuous; therefore, $f \in E^\star$. □

Borrowing two results on ideals from Chapter III, we can establish this dual characterization of Banach lattices with order continuous norm.

**10.2 Theorem.** *The following assertions on a Banach lattice $E$ are equivalent:*

(a) *$E$ has order continuous norm.*

(b) *Every band projection in $E'$ is $\sigma(E', E)$-continuous.*

(c) *Every band in $E'$ is $\sigma(E', E)$-closed.*

(d) *Every closed ideal of $E$ is a projection band.*

*Proof.* (a) ⇒ (b) follows from (5.10) Cor. 2. (b) ⇒ (c) is clear.

(c) ⇒ (d): Let $I$ denote a closed ideal of $E$ then $B := I^\circ$ is a band in $E'$, and $I = B^\circ$ (bipolar theorem, [S, IV.1.5]). Since $B^\perp$ is $\sigma(E', E)$-closed by hypothesis, we have $B=J^\circ$ for the closed ideal $J:=(B^\perp)^\circ\subset E$; again by the bipolar theorem, $I^\circ\cap J^\circ=\{0\}$ implies that $I+J$ is dense in $E$. Thus, by (III.1.2), $I+J=E$ and, since $I\cap J=\{0\}$, $J=I^\perp$ and $I$ is a projection band.

(d)$\Rightarrow$(a): This follows at once from (III.11.1). □

**Examples.** 1. If $K$ is a compact hyperstonian space (§ 9) then, by (9.3), $C(K)$ is the dual of the $AL$-space $N$ of order continuous Radon measures on $K$. Since $N$ has order continuous norm, an ideal $I$ of $C(K)$ is a band iff $I$ is $\sigma(C(K),N)$-closed.

2. An instructive example is furnished by the Banach lattice $c_0$ and its first three duals. It is well known and easy to prove that the dual of $c_0$ can be identified with $l^1$, and the dual of $l^1$ with $l^\infty$ (cf. 8.7); the dual of $l^\infty$ can be identified with the Banach lattice $ba(\mathbb{N})$ [DS, IV.2.15] of all bounded, (finitely) additive real functions on the Boolean algebra $2^{\mathbb{N}}$ (§ 5, Example 3).

Since $c_0$ has order continuous norm, each band in $l^1$ is $\sigma(l^1,c_0)$-closed and each band projection is $\sigma(l^1,c_0)$-continuous. By contrast, bands in $ba(\mathbb{N})$ are not necessarily $\sigma(ba(\mathbb{N}),l^\infty)$-closed as the band $l^1\subset ba(\mathbb{N})$ of all order continuous linear forms on $l^\infty$ shows, which is $\sigma(ba(\mathbb{N}),l^\infty)$-dense.

In view of the preceding example, the following fundamental lemma due to Phillips [1940] is utterly important; it will be the key for all of the remaining results in this section. We consider the duality $\langle l^\infty, ba(\mathbb{N})\rangle$ (Example 2) and note that the adjoint of the canonical injection (evaluation map) $c_0\to l^\infty$ is the band projection $P$ of $ba(\mathbb{N})$ onto $l^1$; thus for each $\lambda\in ba(\mathbb{N})$, $P\lambda$ is the countably additive component of $\lambda$ with norm $\|P\lambda\|=\sum_{s\in\mathbb{N}}|\lambda(s)|$. (Throughout, we write $\lambda(s)$ in place of $\lambda(\{s\})$.) We reproduce the proof given in Day [1973] but give the result in slightly sharper form.

**10.3 Lemma.** *Let $(\lambda_n)_{n\in\mathbb{N}}$ be a sequence in $ba(\mathbb{N})$ such that $\lim_n\lambda_n(J)=0$ for each subset $J\subset\mathbb{N}$. Then*

$$\lim_n \textstyle\sum_{s\in\mathbb{N}}|\lambda_n(s)|=0\,.$$

*Proof.* Supposing the assertion $\lim_n\|P\lambda_n\|=0$ to be false we will derive a contradiction. First, there exists $\varepsilon>0$ such that $\|P\lambda_n\|>\varepsilon$ for infinitely many $n\in\mathbb{N}$. Thus, in view of the hypothesis, there exists a subsequence $(\mu_k)_{k\in\mathbb{N}}$ of $(\lambda_n)$ and a disjoint sequence of finite subsets $J_k\subset\mathbb{N}$ $(k\in\mathbb{N})$ for which

$$\textstyle\sum_{s\in J_k}|\mu_k(s)|\geqq\|P\mu_k\|-\varepsilon/10\,. \tag{1}$$

We now determine further subsequences $(v_k)$ of $(\mu_k)$ and $(H_k)$ of $(J_k)$ satisfying

$$|v_i|\big(\textstyle\bigcup_{j>i}H_j\big)<\varepsilon/10 \tag{2}$$

for all $i\in\mathbb{N}$. This can be done as follows. We let $H_1:=J_1$ and $v_1:=\mu_1$; then we consider a partition of the set $\{J_k: k\geqq 2\}$ into a finite number, greater than $10\|v_1\|/\varepsilon$, of infinite subsets; at least one of these subsets coincides with the range of a subsequence $(J_{n(k)})_{k\in\mathbb{N}}$ of $(J_k)_{k\geqq 2}$ for which $|v_1|\big(\bigcup_{k\in\mathbb{N}}J_{n(k)}\big)<\varepsilon/10$. We now let $H_2 := J_{n(1)}$ and $v_2 := \mu_{n(1)}$ and continue this process with a partition of the set $\{J_{n(k)}: k \geqq 2\}$ into a finite number, greater than $10\|v_2\|/\varepsilon$, of infinite subsets; the process is completed by induction.

Next we construct a sequence $(x_i)_{i \in \mathbb{N}}$ of vectors $x_i$ by defining $(s \in \mathbb{N})$

$$x_i(s) = \begin{cases} (-1)^i \operatorname{sgn} v_i(s) & \text{for } s \in H_i \\ 0 & \text{for } s \notin H_i . \end{cases}$$

Since the $H_i$ are pairwise disjoint, $x := \sum_{i \in \mathbb{N}} x_i \in l^\infty$ is well defined. We have $|\langle x_j, v_i \rangle| \leqq |v_i|(H_j)$ for $i, j \in \mathbb{N}$ and hence, for fixed $i$,

$$|\langle \textstyle\sum_{j \neq i} x_j, v_i \rangle| \leqq \sum_{j<i} |v_i|(H_j) + |v_i|(\bigcup_{j>i} H_j) .$$

Here the first right hand term is $< \varepsilon/10$ by (1), because $\bigcup_{j<i} H_j$ is a finite set disjoint from $H_i$, and the second term is $< \varepsilon/10$ by (2). Hence we have, for each $i \in \mathbb{N}$,

$$(3) \qquad |\langle x, v_i \rangle - \langle x_i, v_i \rangle| = |\langle \textstyle\sum_{j \neq i} x_j, v_i \rangle| < 2\varepsilon/10 .$$

But by (1), $\langle x_i, v_i \rangle = (-1)^i \sum_{s \in H_i} |v_i(s)|$ oscillates from above $9\varepsilon/10$ to below $-9\varepsilon/10$, and so (3) is incompatible with the convergence of the sequence $(\langle x, v_i \rangle)_{i \in \mathbb{N}}$. On the other hand, the function $s \mapsto x(s)$ takes only the values $0, \pm 1$ and so $x \in l^\infty$ is a linear combination of the characteristic functions of three disjoint subsets $J \subset \mathbb{N}$; thus the hypothesis implies $\lim_i \langle x, v_i \rangle = 0$. ☐

**Corollary.** *Let* A *be any index set. Then weakly compact subsets of* $l^1(\mathsf{A})$ *are compact, and bounded subsets of* $l^\infty(\mathsf{A})$ *are* $\tau(l^\infty, l^1)$*-relatively compact.*

*Proof.* To prove the first assertion it suffices (by Eberlein's theorem [S, IV.11.1 Cor. 2]) to show that a weakly convergent sequence in $l^1(\mathsf{A})$ norm converges. But since such a sequence is contained in a band of $l^1(\mathsf{A})$ isomorphic to $l^1(\mathbb{N})$ (we suppose A infinite) the assertion follows from (10.3), because weak convergence is $\sigma(l^1, l^\infty)$-convergence. The second assertion follows from the theorem of Banach-Dieudonné [S, IV.6.3], since the Mackey topology $\tau(l^\infty, l^1)$ is the topology of compact convergence by the preceding, and since bounded subsets of $l^\infty \cong (l^1)'$ are equicontinuous. ☐

The following theorem, perhaps the most important consequence of (10.3), was first proved by Grothendieck [1953] for Stonian spaces $K$; later it was extended by G.L. Seever [1968] to a class of spaces somewhat more general than quasi-Stonian compact spaces; see also Semadeni [1964] and Schaefer [1971 a].

**10.4 Theorem.** *Let* $E := C(K)$ *where* $K$ *is compact quasi-Stonian. Every* $\sigma(E', E)$*-convergent sequence in* $E' = M(K)$ *is* $\sigma(E', E'')$*-convergent and, moreover, uniformly convergent on each weakly compact subset of* $E$.

*Proof.* Let $(\mu_n)$ be a sequence in $E'$ convergent for $\sigma(E', E)$ to $\mu \in E'$; by considering the sequence $(\mu_n - \mu)$ we can assume that $(\mu_n)$ converges to 0. Clearly,

$(\mu_n)$ is bounded in $M(K)$, and to prove the first assertion it suffices to show that the set $\{\mu_n : n \in \mathbb{N}\}$ is $\sigma(E', E'')$-relatively compact. Suppose not; then by (9.8)(c) there exists a bounded sequence $(f_n)$ in $C(K)$ satisfying $f_n \wedge f_m = 0$ for $n \neq m$ and such that $\lim_n \mu_k(f_n) = 0$ does not hold uniformly for $k$ in $\mathbb{N}$. That is, there exists $\varepsilon > 0$ and subsequences $(v_n)$ of $(\mu_n)$ and $(g_n)$ of $(f_n)$ such that $|v_n(g_n)| > \varepsilon$ for all $n \in \mathbb{N}$. We define $g_\emptyset = 0$ and $g_J = \sup_{n \in J} g_n$ for each non-void subset $J \subset \mathbb{N}$; these suprema exist in $C(K)$, because $C(K)$ is countably order complete (7.7 Cor.). It is now readily verified that

$$\lambda_n(J) := v_n(g_J) \qquad (J \subset \mathbb{N}, n \in \mathbb{N})$$

defines a bounded additive set function $\lambda_n \in ba(\mathbb{N})$, and the hypothesis on $(\mu_n)$ made above implies that $\lim_n \lambda_n(J) = 0$ for each $J \subset \mathbb{N}$. From (10.3) it now follows that $\lim_n |\lambda_n(n)| = \lim_n |v_n(g_n)| = 0$ which is absurd.

By virtue of (9.5)(c), the second assertion is clear from (9.9). □

**Corollary 1.** *If $K$ is compact quasi-Stonian, every continuous linear map of $C(K)$ into a separable Banach space $F$ is weakly compact (and, in particular, maps weakly compact subsets of $C(K)$ onto compact subsets of $F$).*

*Proof.* If $F$ is a separable Banach space, the dual unit ball $V^\circ$ is metrizable (and compact) for $\sigma(F', F)$ [S, IV.1.5]. Thus if $E := C(K)$ and $T: E \to F$ is linear and continuous, then each sequence in $T'(V^\circ)$ contains a $\sigma(E', E)$-convergent subsequence, which by (10.4) actually converges for $\sigma(E', E'')$. Therefore, by Eberlein's theorem $T'(V^\circ)$ is $\sigma(E', E'')$-compact and hence, $T$ is weakly compact (9.4). The assertion in parentheses now follows from (9.9). □

**Corollary 2.** *If $K$ is compact quasi-Stonian then every separable quotient of $C(K)$ is reflexive and every separable complemented subspace of $C(K)$ is finite dimensional.*

These assertions are clear from Cor. 1 and (9.9) Cor. 1.

**Examples.** 3. The assertion of (10.4) does not hold for all $C(K)$ spaces, and not even for order complete $AM$-spaces without unit. Thus if $K = [0, 1]$, Lebesgue measure $\mu$ is the $\sigma(E', E)$-limit of a sequence of atomic measures which can obviously not converge for $\sigma(E', E'')$, since the band of atomic measures is $\sigma(E', E'')$-closed. Also, (10.4) does not hold for $E = c_0$, as the sequence of unit vectors $e_n := (\delta_{nm})_{m \in \mathbb{N}}$ in $l^1$ shows which converges to 0 for $\sigma(E', E)$.

4. Corollary 2 above contains the classical result that $c_0$ is not complemented (i.e., not the range of a continuous projection) in $l^\infty$. More generally, if $K$ is an infinite, metrizable compact space (hence, $C(K)$ separable (7.5)) then $C(K)$ is not complemented in its bidual or even in the $AM$-space $B(K)$ of bounded Borel functions on $K$ ($B(K)$ is countably order complete). On the other hand, $B(K)$ is never separable (cf. (5.14) Cor.), and an infinite quasi-Stonian compact space is never metrizable.

5. Let $A$ denote a $\sigma$-complete Boolean algebra. If $(\mu_n)$ is a sequence of bounded, additive real functions on $A$ (§1, Example 5) such that $\mu(a) := \lim_n \mu_n(a)$

exists (in $\mathbb{R}$) for each $a \in A$, then $\mu$ is a bounded additive function on $A$ (Ando [1961]). In fact, it can be shown (Exerc. 29) that $(\mu_n)$ is uniformly bounded; considering $(\mu_n)$ as a sequence in the dual of $E := C(K_A)$ ($K_A$ the Stone representation space of $A$), it follows that $(\mu_n)$ converges for $\sigma(E', E)$, and even uniformly on each weak null sequence of $E$ by (10.4).

We conclude this section with two further applications of (10.4). As before, we denote by $o(E', E)$ the topology on $E'$ of uniform convergence on the order intervals of $E$.

**10.5 Proposition.** *Let $E$ denote a countably order complete vector lattice with order dual $E^\star$, and let $G$ denote the ideal of $E^{\star\star}$ generated by the canonical image of $E$. Then each $\sigma(E^\star, E)$-convergent sequence in $E^\star$ converges for $\sigma(E^\star, G)$. In particular, each band projection in $E^\star$ is $\sigma(E^\star, E)$-sequentially continuous and each band in $E^\star$ is $\sigma(E^\star, E)$-sequentially complete.*

*Proof.* Since $E$ satisfies Axiom (OS) (Def. 1.8), each of the principal ideals $E_x$ ($x \in E_+$) is an $AM$-space with unit $x$ (7.2); hence, by (7.4) $E_x = C(K_x)$ where the compact space $K_x$ is quasi-Stonian, $E$ being countably order complete (7.7) Cor. If $M(K_x)$ denotes the strong dual of $C(K_x)$ it follows that under $o(E^\star, E)$, $E^\star$ can be identified with a vector subspace of the topological product $F := \prod_{x \in E_+} M(K_x)$; since the dual of $(E^\star, o(E^\star, E))$ is $G$, the weak topology $\sigma(E^\star, G)$ is induced by the product of the respective weak topologies of $M(K_x)$ [S, IV.4.1 Cor. 2]. Thus if a sequence in $E^\star$ converges for $\sigma(E^\star, E)$, by (10.4) it necessarily converges for $\sigma(E^\star, G)$. On the other hand, each band in $E^\star$ is closed and each band projection continuous for $o(E^\star, E)$ (cf. proof of (5.10) Cor. 2) and hence, for $\sigma(E^\star, G)$; in view of (10.1), this proves the remaining assertions. □

We finally establish the following additional characterization of Banach lattices which are bands in their respective biduals ($KB$-spaces; cf. (5.15) and § 5, Example 7).

**10.6 Proposition.** *The following are equivalent properties of a Banach lattice $E$:*

(a) *$E$ is weakly sequentially complete.*

(b) *Under evaluation, $E$ is isomorphic to a band of $E''$.*

*Proof.* (a)⇒(b): Clearly, if $E$ is weakly sequentially complete then $E$ is countably order complete, and so the implication follows from (5.9) Cor. and (5.15).

(b)⇒(a): It suffices to apply (10.5) to $E'$. □

## § 11. Complexification

It is often necessary in functional analysis (for instance, in the theory of group representations and in spectral theory) to consider vector spaces and algebras over the complex field. It is thus desirable to have available the notion of a complex vector lattice and, in particular, of a complex Banach lattice. In many concrete cases (as in the case of the Banach lattices $C(K)$, $M(K)$, and $L^p(\mu)$) the extension of the modulus to complex scalars is immediate; essentially through

Kakutani's representation of $AM$-spaces with unit (7.4), we will show in this section that for a wide class of vector lattices, the basic concepts discussed in this chapter carry over to complex scalars. Since the transition is surprisingly smooth, a somewhat informal presentation appears appropriate.

*Throughout this section, vector lattices $E$ (over $\mathbb{R}$) are required to satisfy Axiom* (OS) (Def. 1.8); this implies, in particular, $E$ to be Archimedean and each principal ideal $E_x$ ($x \in E_+$) to be an $AM$-space with unit (7.2).

(i) *Complex Vector Lattices.* If $G$ is a real vector space, we recall (cf. [S, I.7]) that the complexification $G_{\mathbb{C}}$ of $G$ is the additive group $G \times G$ with scalar multiplication defined by $(\alpha,\beta)(x,y) := (\alpha x - \beta y, \beta x + \alpha y)$ for $(\alpha,\beta) = \alpha + i\beta \in \mathbb{C}$; identifying each $x \in G$ with $(x,0) \in G \times G$, each $z \in G \times G$ is uniquely represented as $z = x + iy$, where $x, y \in G$. Thus $G_{\mathbb{C}}$ is frequently written as $G + iG$, with the understanding that $G_{\mathbb{C}}$ is the direct sum over $\mathbb{R}$ of the real vector subspaces $G$ and $iG$; in particular, $x + iy \mapsto x$ and $x + iy \mapsto y$ are $\mathbb{R}$-linear projections of $G_{\mathbb{C}}$ onto $G$.

Now let $E$ denote any vector lattice satisfying Axiom (OS) (Def. 1.8), and let $E_{\mathbb{C}} = E + iE$ denote the complexification of the underlying (real) vector space of $E$. For $z = x + iy$ a fixed element of $E_{\mathbb{C}}$ ($x, y \in E$), we consider the principal ideal $E_{|x|+|y|}$ of $E$ which by (7.4) is isomorphic to some $C(K_z)$ ($K_z$ compact). It is then clear that the space $C_{\mathbb{C}}(K_z)$ can be identified with a vector subspace of $E + iE$; precisely, with $E_{|x|+|y|} + iE_{|x|+|y|}$. But in $C_{\mathbb{C}}(K_z)$ the modulus function $f \mapsto |f|$ is well defined by $|f|(s) := |f(s)|$ ($s \in K_z$) and, if $f = g + ih$ for $g, h \in C(K_z)$, then $|f| = \sup_{0 \leq \theta < 2\pi} |(\cos\theta)g + (\sin\theta)h|$. This shows the supremum

$$|z| := \sup_{0 \leq \theta < 2\pi} |(\cos\theta)x + (\sin\theta)y| \tag{1}$$

to exist in $E$ for each $z = x + iy \in E_{\mathbb{C}}$. Moreover, it is easy to verify that the mapping $z \mapsto |z|$ of $E_{\mathbb{C}}$ into $E$ satisfies the relations

$$\begin{aligned} &|z| = 0 \quad \text{if and only if} \quad z = 0 \\ &|\alpha z| = |\alpha|\,|z| \quad \text{for all} \quad \alpha \in \mathbb{C}, \quad z \in E_{\mathbb{C}} \\ &|z_1 + z_2| \leq |z_1| + |z_2| \quad \text{for all} \quad z_1, z_2 \in E_{\mathbb{C}}. \end{aligned} \tag{2}$$

**11.1 Definition.** *A* complex vector lattice $(E_{\mathbb{C}}, |\;|)$ *is defined to be the complexification of a vector lattice $E$ (satisfying Axiom* (OS)*) and endowed with the modulus function $z \mapsto |z|$ given by* (1).

By a *vector sublattice* of $E_{\mathbb{C}}$ we understand a conjugation invariant vector subspace $F$ such that $x \in F$ implies $|x| \in F$.

(ii) *Ideals, Bands, Projections.* A subset $A$ of a complex vector lattice $E_{\mathbb{C}}$ is called *solid* if $z_1 \in A$, $z_2 \in E_{\mathbb{C}}$ and $|z_2| \leq |z_1|$ imply $z_2 \in A$. It is clear that a solid vector subspace $J$ of $E_{\mathbb{C}}$ intersects $E$ in an ideal $J_0$; moreover, since by (1) we have $|x + iy| = |x - iy|$ for all $x, y \in E$, it follows that $J$ is conjugation invariant and hence, $J = J_0 + iJ_0$.

We therefore define an *ideal* $J$ of $E_{\mathbb{C}}$ to be a solid vector subspace of $E_{\mathbb{C}}$ (or equivalently, as the complexification of an ideal $J_0 \subset E$); an ideal $J \subset E_{\mathbb{C}}$ is called a *band (projection band)* if $J \cap E$ is a band (projection band) of $E$ (Def. 2.8). For example, it is easy to show that an ideal $J \subset E_{\mathbb{C}}$ is a projection band iff $E_{\mathbb{C}} = J + J^{\perp}$, where $J^{\perp} := \{u \in E_{\mathbb{C}}: |u| \wedge |z| = 0$ for all $z \in J\}$. Thus an ideal $J \subset E_{\mathbb{C}}$ is a projection band iff there exists a projection $P: E_{\mathbb{C}} \to J$ with kernel $J^{\perp}$ and leaving $E$ invariant. Also, it is natural to call $E_{\mathbb{C}}$ *(countably) order complete* if $E$ is (countably) order complete.

The decomposition property (1.6) carries over to complex vector lattices, as follows.

**11.2 Proposition.** *Let $z, z_1, z_2 \in E_{\mathbb{C}}$ satisfy $|z| \leqq |z_1| + |z_2|$. There exist elements $u_1, u_2 \in E_{\mathbb{C}}$ such that $z = u_1 + u_2$ and $|u_1| \leqq |z_1|$, $|u_2| \leqq |z_2|$.*

*Proof.* If as before we identify $E_{|z_1|+|z_2|}$ with an $AM$-space $C(K)$ then $z, z_1, z_2$ are functions in $C_{\mathbb{C}}(K)$ and hence, can be written as $z = |z|h$, $z_1 = |z_1|h_1$, $z_2 = |z_2|h_2$ for suitable, unimodular complex functions $h, h_1, h_2$ on $K$. By (1.6) we have $|z| = g_1 + g_2$ for functions $g_i \in C(K)$ satisfying $0 \leqq g_i \leqq |z_i|$ $(i = 1, 2)$; clearly, the elements $u_i := g_i h$ have the desired property. ▯

**Corollary.** *In a complex vector lattice, the sum $A_1 + A_2$ of any pair $A_1, A_2$ of solid subsets is solid. Moreover, the sum of any pair of ideals (bands, projection bands) is an ideal (band, projection band).*

*Proof.* The assertion concerning solid sets (in particular, ideals) is readily verified through (11.2); the assertion concerning bands and projection bands follows from the remarks above and from (2.9). ▯

(iii) *Linear Mappings.* Let $E, F$ be real vector spaces and denote by $E_{\mathbb{C}}, F_{\mathbb{C}}$ their complexifications. Every ($\mathbb{R}$-)linear map $T: E \to F$ has a unique ($\mathbb{C}$-)linear extension $\tilde{T}$ given by

$$\tilde{T}z := Tx + iTy \tag{3}$$

for all $z = x + iy \in E_{\mathbb{C}}$ $(x, y \in E)$ which is called *canonical* (and usually again denoted by $T$). Thus the vector space $L(E, F)$ of ($\mathbb{R}$-)linear maps $E \to F$ can be viewed as a real vector subspace of $L(E_{\mathbb{C}}, F_{\mathbb{C}})$; the elements $T$ of this subspace are characterized by the property that $T(E) \subset F$ ($E, F$ being considered subsets of $E_{\mathbb{C}}, F_{\mathbb{C}}$ respectively) and are called *real linear maps* of $E_{\mathbb{C}}$ into $F_{\mathbb{C}}$. We observe that each $T \in L(E_{\mathbb{C}}, F_{\mathbb{C}})$ has a unique decomposition $T = T_1 + iT_2$ where $T_j$ $(j = 1, 2)$ are real linear maps $E_{\mathbb{C}} \to F_{\mathbb{C}}$ defined by the requirement that $Tx = T_1 x + iT_2 x$ and $T_j x \in F$ $(j = 1, 2)$ for all $x \in E$. Thus $Tz = T_1 z + iT_2 z$ for all $z \in E_{\mathbb{C}}$, that is,

$$T = T_1 + iT_2 \qquad (T \in L(E_{\mathbb{C}}, F_{\mathbb{C}}))\,. \tag{3'}$$

In particular, $T: E_{\mathbb{C}} \to F_{\mathbb{C}}$ is real if and only if $T_1 = T$, $T_2 = 0$. In other words, *$L(E_{\mathbb{C}}, F_{\mathbb{C}})$ is isomorphic to the complexification of $L(E, F)$.*

Now let $E_{\mathbb{C}}, F_{\mathbb{C}}$ be complex vector lattices. A linear map $T: E_{\mathbb{C}} \to F_{\mathbb{C}}$ is *(strictly) positive* if $T$ is real and its restriction $T_0: E \to F$ is (strictly) positive;

$T$ is a *lattice homomorphism* if $T$ is real and $T_0$ is a lattice homomorphism. For example, it is not difficult to prove that $T$ is a lattice homomorphism if and only if $|Tz| = T|z|$ for all $z \in E_{\mathbb{C}}$.

As an application of these concepts, the reader may wish to verify that whenever $J$ is an ideal of $E_{\mathbb{C}}$ and $q: E_{\mathbb{C}} \to E_{\mathbb{C}}/J$ is the canonical map, then under the finest ordering for which $q$ is positive $E_{\mathbb{C}}/J$ is a complex vector lattice isomorphic to the complexification of $E/(J \cap E)$, provided only that the latter quotient satisfies (OS); in these circumstances, $q$ is a lattice homomorphism (cf. 2.6). Moreover, every surjective lattice homomorphism $E_{\mathbb{C}} \to F_{\mathbb{C}}$ maps solid sets onto solid sets.

Finally, a linear map $T: E_{\mathbb{C}} \to F_{\mathbb{C}}$ is called *order continuous* if its real and imaginary parts $T_1, T_2$ (see (3')) are both order continuous (when restricted to $E$) (Def. 2.4).

(iv) *Duality*. The preceding considerations apply, in particular, to the case where $F_{\mathbb{C}}$ is the field $\mathbb{C}$ (viewed as a complex vector lattice). If $E$ is a real vector space and $f$ is a ($\mathbb{C}$-)linear form on $E_{\mathbb{C}}$ then by (3'), $f = f_1 + i f_2$ where $f_1, f_2$ are (the canonical extensions of) linear forms on $E$. Thus if $E$ is a vector lattice (satisfying (OS)) it is natural to define the *order dual* $E_{\mathbb{C}}^\star$ of $E_{\mathbb{C}}$ to be the vector space of all linear forms $f$ on $E_{\mathbb{C}}$ for which $f_1$ and $f_2$ are order bounded (Def. 4.1); $E_{\mathbb{C}}^\star$ is a complex vector lattice with respect to the modulus

$$|f| := \sup_{0 \leqq \theta < 2\pi} |(\cos\theta) f_1 + (\sin\theta) f_2| \in E^\star \quad \text{(see (1))}.$$

The reader may find it worthwhile to convince himself that for $f \in E_{\mathbb{C}}^\star$, the modulus $|f|$ satisfies

$$(4) \qquad |f|(x) = \sup_{|z| \leqq x} |f(z)| \quad (x \in E_+)$$

(cf. Chap. IV, § 1). Hence, $E_{\mathbb{C}}^\star$ can equivalently be defined as the set of all "order bounded" linear forms on $E_{\mathbb{C}}$; that is, linear forms bounded on each "interval" $\{z: |z| \leqq x\}$ of $E_{\mathbb{C}}$.

If $f \in E_{\mathbb{C}}^\star$ has the canonical decomposition $f = f_1 + i f_2$ and *(sequential) order continuity* of $f$ is defined to mean (sequential) order continuity of the restriction to $E$ of both $f_1$ and $f_2$, it is easy to see that the respective sets of sequentially order continuous and of order continuous linear forms on $E_{\mathbb{C}}$ are projection bands of $E_{\mathbb{C}}^\star$ (cf. 4.3 Cor.).

From (4) it follows, in particular, that with respect to the duality $\langle E_{\mathbb{C}}, E_{\mathbb{C}}^\star \rangle$, polars of solid sets are solid (cf. 4.7).

(v) *Normed Vector Lattices*. If $E_{\mathbb{C}}$ is a complex vector lattice and $x \mapsto \|x\|$ is a lattice norm on $E$ (Def. 5.1), the modulus function (1) suggests the extension to $E_{\mathbb{C}}$ of the given lattice norm by

$$(5) \qquad \|z\| := \big\| |z| \big\| \quad (z \in E_{\mathbb{C}}).$$

Since $|x| \vee |y| \leqq |x + iy| \leqq |x| + |y|$ for all $x, y \in E$ by (1), the space $E_{\mathbb{C}}$ normed by (5) is, as a topological vector space, $\mathbb{R}$-linearly homeomorphic to the Cartesian product $E \times E$ and contains $E$ as a normed, real vector subspace; in particular, $E_{\mathbb{C}}$ is a complex Banach space if and only if $E$ is norm complete. Moreover, the norm (5) satisfies

$$(5') \qquad |z_1| \leqq |z_2| \Rightarrow \|z_1\| \leqq \|z_2\| .$$

**11.3 Definition.** *By a* complex Banach lattice *we understand the complexification $E_{\mathbb{C}}$ of a Banach lattice $E$, endowed with the modulus function* (1) *and the norm* (5).

It is clear that an analogous definition can be agreed upon for a normed vector lattice $E$, provided $E$ satisfies Axiom (OS) (Def. 1.8). Since the triangle inequality $|z_1 + z_2| \leqq |z_1| + |z_2|$ (see (2)) implies $\big||z_1| - |z_2|\big| \leqq |z_1 - z_2|$, it follows from (5′) that in every complex Banach lattice, the modulus function $z \mapsto |z|$ is uniformly continuous. For later reference, let us record the following fact.

**11.4 Proposition.** *For every closed ideal $J$ of the complex Banach lattice $E_{\mathbb{C}}$, $E_{\mathbb{C}}/J$ is a complex Banach lattice under the quotient norm and the modulus $|q(z)| := q(|z|)$ ($q$ the canonical map, $z \in E_{\mathbb{C}}$).*

In fact, $E_{\mathbb{C}}/J$ is isomorphic, under the quotient norm and the modulus indicated, to the complexification $(E/(J \cap E))_{\mathbb{C}}$. It is clear now that the majority of results proved in this chapter can be reformulated for complex vector lattices; the method exhibited so far shows how the transition to complex scalars can be made when needed. We confine ourselves to proving, as another application, the following approximation lemma (Krengel [1964b]).

**11.5 Approximation Lemma.** *Let $E$ be a countably order complete vector lattice and denote by $E_{\mathbb{C}}$ its complexification. For given $z \in E_{\mathbb{C}}$ and real $\varepsilon > 0$, there exists a finite set $\{x_j : j = 1, \dots, n\}$ of mutually disjoint elements of $[0, |z|]$ and a set $\{\alpha_j : j = 1, \dots, n\}$ of complex scalars satisfying $|\alpha_j| \leqq 1$, with the property that $|z - \sum_{j=1}^n \alpha_j x_j| \leqq \varepsilon |z|$.*

*Proof.* By (7.4) the ideal $E_{|z|}$ of $E$ can be identified with a Banach lattice $C(K)$, where $K$ is compact quasi-Stonian (7.6), in such a way that $|z|$ becomes the constant-one function on $K$. Evidently we have $z \in E_{|z|} + i E_{|z|}$, and the latter ideal of $E_{\mathbb{C}}$ can be identified with $C_{\mathbb{C}}(K)$. Since $K$ is totally disconnected, there exists a finite cover $\{U_j : j = 1, \dots, n\}$ of $K$ of mutually disjoint, open-and-closed subsets such that on each $U_j$, the variation of $z \in C_{\mathbb{C}}(K)$ is $\leqq \varepsilon$. Thus if $\alpha_j \in z(U_j)$ $(j = 1, \dots, n)$ then $|\alpha_j| \leqq 1$, and if $x_j \in E_{|z|}$ denotes the characteristic function of $U_j$ we obviously have $p(z - \sum_{j=1}^n \alpha_j x_j) \leqq \varepsilon$, where $p$ denotes the uniform norm of $C_{\mathbb{C}}(K)$. Translating back into $E_{\mathbb{C}}$, we obtain $|z - \sum_{j=1}^n \alpha_j x_j| \leqq \varepsilon |z|$ as desired. □

The reader will have noticed that (in addition to Axiom (OS) which is assumed throughout this section), it suffices for the proof to have $K$ totally disconnected and hence, that $E$ have the principal projection property; however, the latter property and Axiom (OS) together imply $E$ to be countably order complete (Exerc. 3). Let us also note this consequence of (11.5).

**Corollary.** *If $E$ is a real or complex, countably order complete Banach lattice, then the family of linear combinations of mutually disjoint positive elements forms a dense vector subspace of $E$.*

(vi) *Weak and Strong Order Units.* If $E_{\mathbb{C}}$ is a complex vector lattice and $e \in E_+$, then $e$ is a weak order unit of $E_{\mathbb{C}}$ if $e \wedge |z| = 0$ implies $z = 0$; $e$ is a (strong) order unit if the ideal generated by $e$ is all of $E_{\mathbb{C}}$. Accordingly, if $E_{\mathbb{C}}$ is normed then $u \in E_+$ is quasi-interior to $E_+$ if the ideal generated by $u$ is dense in $E_{\mathbb{C}}$; similarly for topological orthogonal systems (Chap. III, § 5).

(vii) *Complex AM- and AL-spaces.* A complex $AM$-space ($AL$-space) is the complexification (Def. 11.3) of an $AM$-space ($AL$-space). Clearly, complex analogs of (7.4), (8.5), (9.3) and other results are valid.

## Notes

§ 1: Exposition of the elementary algebraic theory of vector lattices. Early book references include Kantorovič-Vulikh-Pinsker [1950], Nakano [1955]; see also Nakano [1966], Birkhoff [1967], and Vulikh [1967]. The most recent extensive account is given by Luxemburg-Zaanen [1971 b]. The definition (1.7) of order convergence is designed to include the standard definition of an order convergent sequence (cf. Exerc. 2). $l^1$-relative completeness (Axiom (OS), Def. 1.8) is usually termed "relative uniform completeness", an unfortunate choice because of the established meaning of the term "uniform completeness" in the theory of uniformities.

§ 2: The situation concerning the material of § 2 is similar; most of the results in this section are standard and well known, even though some of them (such as the last assertions of (2.5) and (2.13)) are rarely found in the literature.

§ 3: Maximal and minimal ideals of vector lattices are widely used and their characterizations well known. The characterization of finite dimensional Archimedean vector lattices (Cor. 1 of Thm. 3.9) is due to Judin [1939]; the result that every finite dimensional vector lattice can be built up from $\mathbb{R}$ by "direct and lexicographical union" is found in Birkhoff [1967]. On the other hand, the introduction of the radical and its systematic use, culminating in (3.9), appear to be new.

§ 4: Systematic treatments of the duality of vector lattices are contained in Vulikh [1967], Luxemburg-Zaanen [1963 a]—[1965 c], Luxemburg [1965 a] to [1965 c], and Nakano [1966], to which we refer for references and further details. The extension theorem (4.5) is independently due to Bauer [1957] and Namioka [1957]. Simultaneous extensions of families of linear forms on Riesz spaces are studied by Ando [1965]. (4.9) seems to be new in the version given here. The proof of Nakano's fundamental theorem (4.12) (Nakano [1950]) given in the text has been adapted from Vulikh [1967].

§ 5: As pointed out in the text many of the results through (5.11) are valid, with suitable modifications, under more general assumptions (in particular, for locally convex vector lattices); relevant references are [S], Peressini [1967], Wong-Ng [1973], and Fremlin [1974 a]. The transparent and comparatively simple proof of (5.7) is due to M. Powell (oral communication). The systematic investigation of Banach lattices with order continuous norm (Def. 5.12) seems to have begun by Luxemburg-Zaanen [1963a]—[1965c], even though several important results, as incorporated in (5.10) and (5.11), were known much earlier to the Soviet and Japanese schools (cf. Kantorovič-Vulikh-Pinsker [1950] and Nakano [1950]); for example, the proof of (5.10) makes essential use of Nakano's theorem (4.12). The latter part of the section (from (5.13) on) is of more recent origin; (5.14) through (5.16) are essentially due to Lozanovskii [1967], [1968] (see also Tzafriri [1972], Lotz [1973a], Meyer-Nieberg [1973a]), and Exerc. 17, 18). The very useful lemma (5.13) of Meyer-Nieberg [1973a] provides the key to the proofs given here.

§6: The concept of quasi-interior point (of the positive cone of an ordered topological vector space) was introduced independently by Fullerton [1957] (cf. Fullerton-Braunschweiger [1966]) and by Schaefer [1960b]. In the case of a topological vector lattice, the concept is the precise topological analog of the notion of weak order unit (Freudenthal unit). (6.6) is due to Lozanovskii [1970]; see also Nagel [1973b]. The remaining results of this section are due to the author (cf. Schaefer [1960b]).

§7: The richest source of information and references concerning *AM*-spaces with unit is Semadeni [1971]; we limit ourselves to pointing out the pioneering work of M. and S. Krein [1940], Kakutani [1941a], [1941b], and Stone [1937], [1941]; cf. also Nakano [1941]. The role of these spaces in the representation theory of vector lattices is central; see Vulikh [1967], Luxemburg-Zaanen [1971], and Exerc. 21. A report on *AM*-spaces, including the generalized Hahn-Banach theorem (7.9), is found in Day [1973]. A completely new approach revealing an unexpectedly rich variety of *AM*-spaces without unit, has been made by Goullet de Rugy [1972]. (7.11) is due to Lotz [1974], and instrumental in the proof of (8.9).

§8: (8.5) is due to Kakutani [1941a]. In certain situations other representations of *AL*-spaces (such as (9.2)) are more useful; a survey of the literature concerned with variants of these representation theorems is given in Chapter VII of Semadeni [1971]. (8.6) and (8.7), of which the classical theorems of Vitali-Hahn-Saks and Lebesgue-Radon-Nikodym now appear as corollaries, attempt to place these latter theorems in a functional analytic context so that their close relationship with (8.8) be properly set in evidence. (8.9) is due to Lotz [1974], while (8.9) Cor. 1 was first proved by Dean [1965]; cf. also the Notes concerning Chap. III, §11. Finally, the interrelations of $L^1$-spaces and measure theory with topological vector lattices are well exhibited in a recent monograph by Fremlin [1974a].

§9: The concept of hyperstonian space is due to Dixmier [1951]. Thm. 9.3 (more precisely, the version given in Exerc. 24(c)) was proved by Dixmier [1951]; the present proof is due to the author. The converse problem of finding preduals of *AL*-spaces is more involved (see Lindenstrauss-Wulbert [1969]). A survey of this latter problem and information on its ramifications can be found in Lindenstrauss [1970] and Lindenstrauss-Tzafriri [1973].

The term "Dunford-Pettis Property" was introduced by Grothendieck [1953] with apparent reference to Dunford-Pettis [1940]. The results in this part of the section are selected and adapted from Grothendieck [1953]; independently of Grothendieck, Brace [1953] proved the implication (c)⇒(b) of (9.7).

§10: The dual characterization (10.2) of Banach lattices with order continuous norm hinges on the implication (d)⇒(a) which is due to Ando [1969]. The central result of this section is (10.4), which was first proved by Grothendieck [1953] for Stonian spaces $K$, then extended to quasi-Stonian spaces by several authors (cf. Ando [1961], Seever [1968], Semadeni [1964]). The most extensive study and the strongest results in this context are due to Seever [1968]; the application (10.5) of (10.4) to general vector lattices is due to Schaefer [1971a].

§11: Ordered vector spaces and vector lattices over the complex field have been considered by various authors (cf. Schaefer [1966a]); Lotz [1968] seems to be the first who recognized the possibility of defining complex vector lattices in the spirit of this section. Further recent references include Ito [1967], Luxemburg-Zaanen [1971a], and Mittelmeyer-Wolff [1974]. In the latter paper, the concept of a vector lattice (over the real or complex field) is axiomatized.

## Exercises

1. (Boolean Algebras). Let $A$ be a Boolean algebra (Def. 1.1).

(a) Show that $A$ is a commutative ring with unit if addition and multiplication in $A$ are defined by $x+y := (x \wedge y^c) \vee (y \wedge x^c)$ and $xy := x \wedge y$, respectively. Moreover, $x+x=0$ and $x^2=x$ for each $x \in A$, and $A$ can be considered to be an algebra (in the usual sense) over the field $\{0,1\}$.

(b) There exists a compact (Hausdorff) space $K_A$ which is totally disconnected (i.e., each point possesses a base of open-and-closed neighborhoods), such that $A$ is isomorphic to the set of all open-and-closed subsets of $K_A$ ordered by inclusion (§1, Example 1). (Consider the set $K_A$ of all homomorphisms $\neq 0$ of $A$ into the Boolean algebra $\{0,1\}$ and show that the family $(K_x)_{x \in A}$, where $K_x := \{h \in K_A : h(x)=1\}$, is a base for a compact (Hausdorff) topology on $K_A$ whose open-and-closed subsets are precisely the sets $K_x$. Cf. [DS, I.12.1]; see also Semadeni [1971, IV.16.2] where the result is proved via the representation theorem (7.4).)

(c) The space $K_A$ of (b) is determined uniquely (to within homeomorphism) by its defining properties and called the Stone representation space of $A$. (If $K, K'$ are totally disconnected compact spaces and $i$ is a Boolean isomorphism between the respective Boolean algebras of open-and-closed subsets of $K$ and $K'$, then $i$ induces an isomorphism of $C(K)$ onto $C(K')$ (use 7.3) and hence, $K$ is homeomorphic to $K'$ (7.4) Cor. 1.)

(d) The representation space $K_A$ is Stonian (quasi-Stonian) if and only if $A$ is complete ($\sigma$-complete).

2. Let $E$ denote a vector lattice.

(a) A net $(x_\alpha)_{\alpha \in \mathsf{A}}$ in $E$ (i.e., a non-void family in $E$ indexed by a pre-ordered, directed set $\mathsf{A}$) is frequently called *order convergent* to $x \in E$ if there exists a directed ($\geqq$) family $(u_\alpha)_{\alpha\in A}$ in $E$ satisfying $\inf_\alpha u_\alpha=0$ and such that $|x_\alpha-x|\leqq u_\alpha$ for $\alpha\in A$ (cf. Peressini [1967], Luxemburg-Zaanen [1971]). If $E$ is order complete, a net in $E$ order converges iff its section filter order converges (Def. 1.7). However, if $E$ is not order complete this assertion can fail (D.H. Fremlin). (Consider the one-point compactification $K$ of an uncountable discrete space, and let $E:=C(K)$. If $(f_n)_{n\in\mathbb{N}}$ denotes the characteristic functions of a sequence of singletons in $K$, the sequence $(f_n)$ is not order convergent but its section filter order converges to 0.)

(b) Denote by $\boldsymbol{T}_0$ the finest topology on $E$ such that each order convergent filter on $E$ $\boldsymbol{T}_0$-converges (to its order limit). If $(E,\boldsymbol{T})$ is a complete metrizable topological vector space in which $E_+$ is closed and such that $x=\sup A$ or $x=\inf A$ implies $x\in\bar{A}$ for each directed ($\leqq$ or $\geqq$) subset $A$ of $E$, then $\boldsymbol{T}=\boldsymbol{T}_0$; examples are the spaces $L^p[0,1]$ $(0\leqq p<+\infty)$. In general, however, $(E,\boldsymbol{T}_0)$ is not a topological vector space; an example is $L^\infty[0,1]$, where $\boldsymbol{T}_0$ is the finest topology agreeing on each order interval with the topology induced by $L^1[0,1]$ (cf. 8.7). (These results are due to D.H. Fremlin.)

For a discussion of order convergence vs. topological convergence, see also Peressini [1967] and [S, Chap. V, § 7].

(c) Let $(X,\Sigma,\mu)$ denote a measure space and let $E$ be $L^0(\mu)$ (the vector lattice of all $\Sigma$-measurable, $\mu$-almost everywhere finite ($\bar{\mathbb{R}}$-valued) functions modulo $\mu$-null functions). A sequence $(f_n)$ in $E$ order converges to $f\in E$ iff $\lim_n f_n(s)=f(s)$ a.e. $(\mu)$. $(f_n)$ converges to $f$ in measure (i.e., $\lim_n \mu\{s:|f_n(s)-f(s)|\geqq\varepsilon\}=0$ for each $\varepsilon>0$) iff each subsequence contains an order convergent subsequence (*order $\star$-convergence*; Birkhoff [1967], Peressini [1967]).

(d) A sequence $(x_n)$ in $E$ is *relatively uniformly (r.u.) convergent* to $x\in E$ if there exists $u\in E_+$ and a real sequence $\varepsilon_n\downarrow 0$ such that $|x_n-x|\leqq\varepsilon_n u$ for all $n\in\mathbb{N}$. Axiom (OS) (Def. 1.8) is equivalent to r.u. completeness. If $E$ is a Banach lattice, a sequence $(x_n)$ converges in norm to $x\in E$ iff every subsequence of $(x_n)$ contains a subsequence r.u. convergent to $x$ *(r.u. $\star$-convergence)*. (Concerning r.u. convergence, see Luxemburg-Zaanen [1971], Birkhoff [1967], Peressini [1967].)

3. Let $E$ be a vector lattice.

(a) The axioms of Def. 1.8 are logically dependent, as follows: (OC) $\Rightarrow$ (COC) $\Rightarrow$ (OS) $\Rightarrow$ (A), and none of these implications can be reversed. (The vector lattice of piecewise affine continuous functions on $[0,1]$ has (A) but not (OS), $C[0,1]$ has (OS) but not (COC); the vector lattice of bounded Borel functions $B[0,1]$ has (COC) but not (OC).)

(b) Prove the logical scheme stated after (2.11) Cor. 2. (For a complete discussion, see Luxemburg-Zaanen [1971], Thm. 25.1.)

(c) None of the properties expressed by the axioms of Def. 1.8 is, in general, inherited by quotients over ideals. (Consider an infinite compact Stonian space $K$ and let $E:=C_\infty(K)$ (see 7.8 Cor. 2). If $s_0\in K$ is not isolated and $I:=\{f\in C_\infty(K): f(s_0)=0\}$, then $I$ is a prime ideal not of codimension 1. Hence $E/I$ is not Archimedean (cf. III.2.2), while $E$ is order complete.)

(d) $E$ is Archimedean iff $A=A^{\perp\perp}$ for each band $A$ of $E$. (Suppose $E$ Archimedean, let $A$ be any band in $E$, and let $S$ be a maximal orthogonal system of $A$.

There exists a maximal orthogonal system $S_1 \supset S$ of $E$; then $S_1 \setminus S \subset A^{\perp}$. Now if $x \in A^{\perp\perp}$, (1.9) implies $x \in A$. Conversely, if $E$ is not Archimedean there exist $x, y > 0$ in $E$ satisfying $ny \leqq x$ for all $n \in \mathbb{N}$. Letting $A := E_0 + E_0^{\perp}$ where $E_0$ is the band generated by $\{y\}$, we have $E = A^{\perp\perp}$ so $x \in A^{\perp\perp}$; however, $x \in A$ would be contradictory.)

(e) Even if $E$ is countably order complete, a band in $E$ is not necessarily a projection band. (Consider the vector lattice $E$ of all real functions on $[0,1]$ which are constant except on a countable subset of $[0,1]$; the band $\{f \in E : f(t) = 0$ for $0 \leqq t \leqq \frac{1}{2}\}$ is not a projection band.)

(f) Let $K$ be compact, and let $E := C(K)$. $E$ has the principal projection property iff $E$ is countably order complete; $E$ has the projection property (every band is a projection band) iff $E$ is order complete. Considering the principal ideals $E_x$ $(x \in E_+)$ (cf. 7.2), show that these equivalences are true for any vector lattice $E$ satisfying Axiom (OS).

4. A vector lattice $E$ is said to be of *countable type* if $E$ is countably order complete and each majorized set of mutually orthogonal positive elements is countable; a $\sigma$-complete Boolean algebra $A$ is said to be of *countable type* if every orthogonal subset of $A$ is countable.

(a) Every Boolean algebra of countable type is complete. (For a given subset $A_0$ of $A$, consider the family $\boldsymbol{C}$ of orthogonal sets $C \subset A$ such that each $c \in C$ is majorized by some $a \in A_0$. There exists a maximal element $C_0 \in \boldsymbol{C}$ (Zorn's lemma); one must have $\sup C_0 = \sup A_0$.)

(b) Every vector lattice $E$ of countable type is super Dedekind complete. (Reduce the problem to the case where $E$ has a weak order unit $e$, and apply (a) to the Boolean algebra (base) $B_e := \{x \in E : x \wedge (e - x) = 0\}$ (cf. 2.13).)

(c) If $E$ is of countable type and $(x_\alpha)_{\alpha \in \mathsf{A}}$ is a net in $E$ order convergent to $x$ (Exerc. 2), there exists a sequence $(\alpha_n)$ in $\mathsf{A}$ such that the sequence $(x_{\alpha_n})_{n \in \mathbb{N}}$ order converges to $x$. (Use (b).)

(d) If $E$ is an order complete vector lattice with weak order unit $e$, then $E$ is of countable type iff the Boolean algebra $B_e$ (see (b)) is of countable type.

(e) Every Banach lattice with order continuous norm (Def. 5.12) (in particular, every reflexive Banach lattice) is of countable type. (Use 5.10.)

(f) Suppose $E$ to be a countably order complete Banach lattice. For each $x \in E_+$, the order interval $[0, x]$ is the convex closure of its extreme boundary $B_x := \{y \in E : y \wedge (x - y) = 0\}$. (Use (2.13) and (7.4).)

5. (a) If $(E_\alpha)_{\alpha \in \mathsf{A}}$ is a family of vector lattices and if $E := \prod_\alpha E_\alpha$ (§ 1, Example 6), then $E$ satisfies any one of the four axioms of Def. 1.8 iff each $E_\alpha$ does $(\alpha \in \mathsf{A})$. The same is true of the direct sum $\bigoplus_\alpha E_\alpha$ (which is an ideal of $E$).

(b) Let $E$ be a vector lattice, $I$ an ideal of $E$, and let $q$ denote the canonical map $E \to E/I$. If $E$ is countably order complete and $I$ is a $\sigma$-ideal, then $E/I$ is countably order complete and $q$ is sequentially order continuous. Similarly, if $E$ is order complete and $I$ is a band, then $E/I$ is order complete and $q$ is order continuous.

(c) Every order complete vector lattice $E$ is isomorphic to an order dense ideal of an order complete vector lattice $F$ possessing a weak order unit. (Consider a maximal orthogonal system $(x_\alpha)_{\alpha\in A}$ of least cardinality in $E$ and let $F:=\prod_\alpha E_\alpha$, where $E_\alpha$ denotes the principal band of $E$ generated by $x_\alpha$.)

(d) Let $E$ be a vector lattice. For each subset $A\subset E$, denote by $\bigvee A$ ($\bigwedge A$) the set of suprema (infima) of all non-empty finite subsets of $A$. Then $\bigvee(\bigwedge A)=\bigwedge(\bigvee A)$ is the smallest subset of $E$ containing $A$ and closed under $\vee$, $\wedge$ (i.e., the sublattice generated by $A$); in particular, if $A$ is a vector subspace of $E$ then $\bigvee(\bigwedge A)$ is a vector sublattice of $E$. (See Jameson [1970].)

(e) Show that in a topological vector lattice $E$ (§5), the vector sublattice generated by a separable subset $A\subset E$ is separable. (Use (d).)

6. (Spectral Representation. Freudenthal [1936]). Let $E$ denote a countably order complete vector lattice with weak order unit $u$. For fixed $x\in E$ and $\lambda\in\mathbb{R}$, denote by $P(x,\lambda)$ the band projection of $E$ onto the principal band generated by $(\lambda u-x)^+$, and let $p(x,\lambda):=P(x,\lambda)u$; then $p(x,\lambda)$ is an element of the base $B_u:=\{v\in E: v\wedge(u-v)=0\}$ (cf. 2.12, 2.13).

(a) The function $\lambda\mapsto p(x,\lambda)$ is increasing, left continuous in the sense that $p(x,\lambda)=\sup_{\mu<\lambda} p(x,\mu)$ for $\lambda\in\mathbb{R}$, and satisfies $\inf_{\lambda\in\mathbb{R}} p(x,\lambda)=0$, $\sup_{\lambda\in\mathbb{R}} p(x,\lambda)=u$.

(b) $\lambda\leqq\mu$ implies

$$\lambda(p(x,\mu)-p(x,\lambda))\leqq(P(x,\mu)-P(x,\lambda))x\leqq\mu(p(x,\mu)-p(x,\lambda)).$$

(c) One has $x=\int_{-\infty}^{+\infty}\lambda\,dp(x,\lambda)$ in the sense that for each partition of $\mathbb{R}$ into consecutive intervals $(\lambda_n,\lambda_{n+1}]$ ($n\in\mathbb{Z}$) with uniformly bounded length $|\lambda_{n+1}-\lambda_n|\leqq\delta$, and for any choice of $l_n\in[\lambda_n,\lambda_{n+1}]$, it is true that

$$\left|x-\sum_{n=-\infty}^{+\infty} l_n(p(x,\lambda_{n+1})-p(x,\lambda_n))\right|\leqq\delta u.$$

(Show that $\sum_{\nu=-n}^{n}(P(x,\lambda_{\nu+1})-P(x,\lambda_\nu))x$ order converges to $x$ for $n\to\infty$, and use (b).)

For a detailed discussion, see Vulikh [1967] or Luxemburg-Zaanen [1971]. A representation theory for countably order complete vector lattices which places the Freudenthal theorem in its proper context, was given by Hackenbroch [1972].

7. Let $E$ be an Archimedean vector lattice. An element $a\in E_+$ is called an *atom* (§3) if the principal ideal $E_a$ is one dimensional.

(a) $a\in E$ is an atom $\Leftrightarrow E_a$ is a minimal ideal $\Leftrightarrow E_a^\perp$ is a maximal ideal.

(b) $E$ is called *atomic* if there exists a maximal orthogonal system $(x_\alpha)_{\alpha\in A}$ of $E$ consisting of atoms. Show that if $E$ is atomic, each $x\in E$ has a unique representation $x=\sum_\alpha\xi_\alpha x_\alpha$ (the sum being the order limit of its finite subsums); if, in addition, $E$ is order complete then the mapping $x\mapsto(\xi_\alpha)_{\alpha\in A}$ defines an isomorphism of $E$ onto an ideal of $\mathbb{R}^A$ (canonical order).

(c) The band $E_a$ of $E$ generated by the set of all atoms of $E$ is an atomic vector lattice (called the *atomic part* of $E$).

(d) If $E$ is order complete, then $E$ is atomic iff the set of all real valued, order continuous lattice homomorphisms of $E$ separates $E$ (M. Wolff [1971]).

8. (a) A vector lattice $E$ is finite dimensional if and only if the lattice of ideals $I(E)$ is finite.

(b) If a vector lattice $E$ of finite dimension $n$ is the direct sum of $k=n-r$ ideals $I_j$, where $r$ is the dimension of the radical of $E$ (Def. 3.2), then $I_j$ $(j=1,\dots,k)$ must be the ideals of (3.9) (except for numeration).

(c) An $n$-dimensional ($n\in\mathbb{N}$) vector lattice $E$ is isomorphic to the ordered product of $\mathbb{R}^{n_1}$ (canonical order) and $\mathbb{R}^{n_2}$ (lexicographic order) with $n=n_1+n_2$, if and only if the radical $R_E$ is totally ordered; in this case, $n_1=k-1$ and $n_2=r+1$ in the notation of (3.9). (Observe that iff $R_E$ is totally ordered, the ideals $I_j$ of (3.9) are one dimensional with at most one exception.) This corrects [S, Chap. V, Exerc. 21(b)].

9. (Dedekind Completion. Cf. Exerc. 16.) (a) Supply a detailed proof of (1.10). Observe that completion by cuts can be applied to any positively generated vector space satisfying Axiom (A) (Def. 1.8).

(b) If $E$ is an Archimedean vector lattice, each order continuous linear form $f$ on $E$ possesses a unique order continuous extension $\bar{f}$ to the Dedekind completion $\bar{E}$ of $E$, and the map $f\mapsto\bar{f}$ is an isomorphism of $E^\star_{00}$ onto $\bar{E}^\star_{00}$.

(c) Let $C(K)$ denote the vector lattice ($AM$-space) of continuous real functions on the compact space $K$, and let $D(K)$ denote the vector lattice of bounded real functions $f$ on $K$ continuous at the points of the complement of some meager set $N_f\subset K$. If $N$ denotes the ideal of all $f\in D(K)$ for which $\{s: f(s)\neq 0\}$ is meager, then $D(K)/N$ is isomorphic to the Dedekind completion of $C(K)$.

10. (Order Continuous Linear Forms.)

(a) If $E:=C[0,1]$, then $E^\star_0=\{0\}$ (and, a fortiori, $E^\star_{00}=\{0\}$; for notation, see Def. 4.1). (Show that by sequential order continuity each $\mu\in E^\star_0$, considered as a bounded Baire measure on $[0,1]$, must vanish on each closed nowhere dense subset of $[0,1]$. Thus if $\mu\in E^\star_0$ then $\|\mu\|=|\mu|(U)$ for each open subset $U$ of $[0,1]$ containing the set $Q$ of rational numbers in $[0,1]$; since $|\mu|(Q)=0$, then $\mu=0$ by regularity.)

(b) If $E:=B[0,1]$ (bounded Baire, or since $[0,1]$ is metric, Borel functions), $E^\star_0$ is the space $M[0,1]$ of bounded Baire measures while $E^\star_{00}$ is the space of all atomic bounded measures (the atomic part of $M[0,1]$, cf. Exerc. 7). $E^\star$ is the space of all bounded, additive real functions on the Boolean algebra of Baire subsets of $[0,1]$. (§ 4, Example 5).

(c) If $E:=F/N$ where $F$ denotes the vector lattice of all finite Baire functions on $[0,1]$ and $N$ denotes the ideal of Lebesgue null functions, then $E^\star=\{0\}$. (Show that $F^\star$ contains no elements other than linear combinations of one point (Dirac) measures. Alternatively: Under the metric

$$\varrho(f,g)=\int_0^1 \frac{|f-g|}{1+|f-g|}\,dm$$

($m$ Lebesgue measure), $E$ is a topological vector lattice with topological dual $\{0\}$. (For each $\varepsilon>0$, the convex hull of the 0-neighborhood $\{f:\varrho(f,0)<\varepsilon\}$ is all of $E$.) Using [S, V.5.5] conclude that $E^\star=\{0\}$.)

11. Let $E$ denote a vector lattice with order dual $E^\star$.

(a) If $F$ is a vector sublattice of $E^\star$, its polar $F^\circ$ (with respect to the duality $\langle E, E^\star \rangle$) is not necessarily an ideal of $E$.

(b) Suppose $G$ is an ideal of $E^\star$ and let $q: E \to G^\star$ denote the evaluation map. If $q(E)$ is an ideal in $G^\star$ then $q$ is order continuous. Does the converse hold? By letting $E := l^\infty$, $G = (l^\infty)'$ show that $q$ need not be sequentially order continuous.

(c) Let $E$ be countably order complete, and let $0 < f \in E_0^\star$. The absolute kernel $N(f) := \{x : f(|x|) = 0\}$ is a band in $E$ iff $f \in E_{00}^\star$. (Use (4.9) to show that $N(f)^\perp$ is a projection band.)

(d) If $0 < f \in E^\star$, show that $N(f)^\perp$ is the set of all $x \in E$ such that $0 < y \leqq |x|$ implies $f(y) > 0$.

12. (a) Generalize (6.6) (a) $\Rightarrow$ (b) as follows: If $E$ is a super Dedekind complete Banach lattice containing a weak order unit and if $E_{00}^\star \neq \{0\}$, then $E_{00}^\star$ contains weak order units. (Show that every orthogonal system of $E$ is countable, and conclude from (4.10) Cor. that the same is true for $E_{00}^\star$.) In particular, if $E_{00}^\star$ separates $E$, there exists a strictly positive order continuous linear form on $E$.

(b) There exist order complete $AM$-spaces with unit on which no linear form is strictly positive. (Consider the Dedekind completion $\bar{E}$ of $E := B[0,1]$ (Exerc. 10(b)); the existence of a strictly positive linear form on $\bar{E}$ would imply $E$ to be super Dedekind complete (4.9) Cor.)

13. Let $E$ be a vector lattice.

(a) For a given norm $q$ on (the underlying vector space of) $E$, these assertions are equivalent:

($\alpha$) There exists a lattice norm $p$ on $E$ equivalent to $q$.

($\beta$) The set of all $y \in E$ such that $|y| \leqq |x|$ for some $x, q(x) \leqq 1$, is bounded in $(E, q)$.

($\gamma$) Every sequence $(y_n)$ in $E$ such that $0 \leqq y_n \leqq |x_n|$ for some null sequence $(x_n)$ in $(E, q)$, is a null sequence in $(E, q)$.

(b) If $E$ is a Banach lattice and if $q$ is a semi-norm on $E$ such that $0 \leqq y \leqq x$ implies $q(y) \leqq c\, q(x)$ for fixed $c > 0$, then $q$ is continuous. (Argue as in the proof of (5.3).) In particular, every lattice semi-norm on a Banach lattice is continuous.

(c) Infer from (b) that the topology of a Banach lattice is the finest locally convex topology for which order intervals are bounded (cf. [S, Chap. V, § 6]).

(d) Every (Hausdorff) locally convex vector lattice is isomorphic with a topological vector sublattice of a product of Banach lattices. (Adapt the proof of [S, II.5.4], considering a 0-neighborhood base in $E$ consisting of convex solid sets.)

(e) A positive linear form on a barreled normed vector lattice $E$ is not necessarily continuous (T. Ando). (Let $F$ denote the space of Lebesgue integrable functions over $[0,1]$, and represent $F$ as the space of all $\overline{\mathbb{R}}$-valued continuous $\mu$-integrable functions on the Stonian space $K_u$ ($u$ a weak order unit of $F$; see Remark 2 after (8.5)). Let $(U_n)$ be a sequence of disjoint, non-void open-and-closed subsets of $K_u$, let $s_n \in U_n$, and let $s$ be an accumulation point of $\{s_n\}$. Let $E := \{f \in F : |f(s)| < +\infty\}$ then, since $F$ admits no real lattice morphisms, the linear form $f \mapsto f(s)$ is well defined on $E$ but not continuous; the essential fact to be proved is that $E$ is a (dense) barreled normed vector sublattice of $F$.)

14. (Köthe Function Spaces.) Let $(X,\Sigma,\mu)$ denote a $\sigma$-finite measure space, denote by $L$ the vector lattice of all real $\Sigma$-measurable functions on $X$, and let $\varrho: L_+ \to \bar{\mathbb{R}}_+$ be a function satisfying $\varrho(f+g) \leqq \varrho(f)+\varrho(g)$, $\varrho(\lambda f)=\lambda\varrho(f)$ $(\lambda\in\mathbb{R}_+)$, and $\varrho(f)=0$ iff $f(s)=0$ a.e. $(\mu)$. (For the relevant operations in $\bar{\mathbb{R}}_+$ see Chap. III, § 3.) The set $L(\varrho):=\{f\in L: \varrho(|f|)<+\infty\}$ is a vector sublattice of $L$ and its quotient over the ideal of $\mu$-null functions, endowed with the lattice norm induced by $\varrho$ (§ 5, Example 1), is a normed vector lattice denoted by $L_\varrho$ and called a *Köthe function space.*

(a) $\varrho$ is said to have the Riesz-Fischer property (RFP) if $\sum_1^\infty \varrho(|f_n|)<+\infty$ implies $\sum_1^\infty |f_n|\in L_\varrho$. If $\varrho$ has RFP then $L_\varrho$ is a Banach lattice (Luxemburg, Halperin).

(b) $\varrho$ is said to have the *Fatou property* if $\lim_n \varrho(f_n)=\varrho(f)$ for every increasing sequence $(f_n)$ in $L_+$ with supremum $f$. The Fatou property implies RFP.

(c) If, for $f\in L_+$, $\varrho'(f):=\sup\{\int fg\,d\mu: g\in L_+, \varrho(g)\leqq 1\}$ then $\varrho'$ is called the *associate function norm* of $\varrho$. $\varrho'$ always has the Fatou property.

(d) $L_{\varrho'}$ can be identified with the band $(L_\varrho)_0^\star$ of all sequentially order continuous linear forms on $L_\varrho$.

(In connection with the preceding, see Luxemburg-Zaanen [1963a]—[1964c], [1971].)

15. (a) Show that for normed vector lattices the properties of being norm complete and order complete, respectively, are independent.

(b) A lattice norm $p$ on a vector lattice $E$ is said to have the *Fatou property* if whenever $(x_n)$ is an increasing sequence in $E_+$ with supremum $x$, then $p(x)=\sup_n p(x_n)$. Show that for any Banach lattice $E$, the following are equivalent:

($\alpha$) Every lattice norm on $E$ has the Fatou property.

($\beta$) $E'=E_0^\star$ (i.e., every $f\in E'$ is sequentially order continuous).

(Observe that every lattice norm $p$ on $E$ is given by $p(x)=\sup_{x'\in M}\langle |x|,x'\rangle$ for a suitable $\sigma(E',E)$-compact set $M\subset E'_+$. Cf. Exerc. 13(b) and § 5, Example 1.)

16. Denote by $E$ an Archimedean vector lattice with Dedekind completion $\bar{E}$.

(a) For any lattice norm $p$ on $E$, $\bar{p}(z):=\inf\{p(x): |z|\leqq x,\ x\in E\}$ defines a lattice norm $\bar{p}$ on $\bar{E}$ extending $p$; if $p_1$ is any lattice norm on $\bar{E}$ extending $p$, then $\bar{p}(z)\leqq p_1(z)$ for all $z\in\bar{E}$.

(b) Each of the following conditions implies that $\bar{p}$ (see (a)) is the unique extension to $\bar{E}$ of the lattice norm $p$ on $E$:

(i) $\bar{E}$ is of countable type (Exerc. 4) and $x_n\downarrow 0$ in $E$ implies $\lim_n p(x_n)=0$. (Show that $\bar{E}$ can be identified with the ideal generated by $E$ in the norm completion $(E,p)\tilde{}$. Cf. (5.10).)

(ii) $(E,p)$ is an $M$-normed space with unit (Def. 7.1).

(c) If $\bar{E}$ is of countable type (Exerc. 4) and $p$ is a lattice norm on $E$ such that each increasing $p$-bounded sequence $(x_n)$ in $E_+$ is majorized, then all lattice norm extensions of $p$ to $\bar{E}$ are equivalent. (Show that for each extension $p_1$ of $p$ to $\bar{E}$, $(\bar{E},p_1)$ inherits the properties assumed of $(E,p)$ and hence, is a Banach lattice; apply (5.3) Cor. 4. Cf. (5.15).)

The preceding results (b), (c) are due to Soloviev [1966].

17. Let $E$ denote any Banach lattice.

(a) $E$ is countably order complete (respectively, order complete) iff every principal band (respectively, every band) of $E$ is a projection band. (Use Exerc. 3(f).)

(b) $E$ is isomorphic (under evaluation) to a band in its bidual $E''$ iff $E$ contains no Banach subspace isomorphic (as a topological vector space) to $c_0$. (Use (10.6) and (5.15).)

(c) $E$ is reflexive iff $E$ contains no Banach subspace isomorphic to either $c_0$ or $l^1$. (Use 5.16.)

For these and related results, see G. Ja. Lozanovskii [1967], [1968]; Tzafriri [1969] and Meyer-Nieberg [1973a].

18. Let $E$ be a Banach lattice with dual $E' = E^\star$.

(a) Let $E$ be countably order complete. For each real $\varepsilon > 0$ and each orthogonal normalized sequence $(x'_n)$ in $(E_0^\star)_+$ there exists an orthogonal normalized sequence $(x_n)$ in $E_+$ such that $\langle x_m, x'_n \rangle = 0$ for $m \neq n$ and $\langle x_n, x'_n \rangle > 1 - \varepsilon$. (Cf. 4.11.)

(b) For each real $\varepsilon > 0$ and each orthogonal normalized sequence $(x_n)$ in $E_+$ there exists an orthogonal normalized sequence $(x'_n)$ in $E'_+$ such that $\langle x_m, x'_n \rangle = 0$ for $m \neq n$ and $\langle x_n, x'_n \rangle > 1 - \varepsilon$. (Use 4.10.)

(c) Let $E$ have order continuous norm. Prove the implications $l^1 \subset E \Leftrightarrow c_0 \subset E' \Leftrightarrow l^\infty \subset E'$ and $c_0 \subset E \Rightarrow l^1 \subset E'$, where "$\subset$" is to be understood to indicate the existence of a vector lattice isomorphism onto a closed vector sublattice. (Use (a), (b).)

(d) Under evaluation, $E$ is isomorphic to an ideal of $E''$ iff every closed vector sublattice of $E$ which is isomorphic to $c_0$, is the range of a positive projection. (Cf. Meyer-Nieberg [1973a].)

19. ($AM$-Spaces.)

(a) If $E$ is an $M$-normed space, the same is true of each vector sublattice and each quotient over a closed ideal; the completion of an $M$-normed space is an $AM$-space. Analogous results hold for $L$-normed spaces.

(b) *Let $K_1, K_2$ be compact spaces and suppose the Banach spaces $C(K_1)$ and $C(K_2)$ to be (isometrically) isomorphic; then $K_1$ and $K_2$ are homeomorphic* (Theorem of Banach-Stone). (Observe that each extreme point of the dual unit ball of $C(K)$ is a Dirac measure $\pm\delta_s$ $(s \in K)$ and conversely, and that the adjoint of an isomorphism $C(K_1) \to C(K_2)$ induces an affine weak* homeomorphism between the dual unit balls.) Show by an example that the assertion fails if the isometry condition is omitted.

(c) Let $E$ be an $AM$-space possessing a quasi-interior positive element $u$ (Def. 6.1). If $X$ denotes the completely regular topological space given by the set of extreme points of $\{x' \in E'_+ : \langle u, x' \rangle = 1\}$ under $\sigma(E', E)$ and if $\varphi$ denotes the restriction to $X$ of the norm of $E'$, then $E$ is isomorphic (as a Banach lattice) to $D_\varphi(X)$ (§ 7, Example 2). Moreover, the structure space of $E$ (in the sense of Def. III.4.2) is homeomorphic to the Stone-Čech compactification of $X$. (Goullet de Rugy [1972].)

(d) An $AM$-space $E$ is isomorphic to $C_0(X)$ (continuous real functions vanishing at infinity, uniform norm) on some locally compact space $X$, iff the norm of $E'$ when restricted to the extreme rays of $E'_+ \setminus \{0\}$ is continuous for $\sigma(E', E)$.

For an elaborate ideal and representation theory of $AM$-spaces without unit (in particular, for proofs of (c) and (d) above) we refer to Goullet de Rugy [1972].

20. Let $X$ denote a non-void set, and let $E$ be a vector lattice of bounded real functions on $X$ containing the constant functions and separating the points of $X$.

(a) If $\boldsymbol{U}$ denotes the coarsest (weakest) uniformity on $X$ for which each $f \in E$ is uniformly continuous, then $(X, \boldsymbol{U})$ is precompact. If $K$ denotes the (compact) completion of $(X, \boldsymbol{U})$ then under the uniform norm, $E$ is isomorphic to a dense normed vector sublattice of $C(K)$. (Show that the unique continuous extensions to $K$ of the functions in $E$ separate the points of $K$, and apply (7.3).) $E$ is isomorphic to $C(K)$ iff $E$ satisfies Axiom (OS) (Def. 1.8.) (Use 7.2.)

(b) If $\Sigma$ is a point separating algebra of subsets of $X$ and $E$ the vector lattice of all bounded $\Sigma$-measurable functions, then the space $K$ constructed in (a) is the Stone representation space of $\Sigma$ (cf. Exerc. 1).

(c) Suppose $X$ to be a completely regular topological space, and let $E$ be all bounded continuous functions on $X$. If $E$ is separable (under the uniform norm) then $X$ is metrizable. Conversely, if $X$ is metrizable then $E$ is not necessarily separable. (For the first assertion, use (a) and (7.5). For the second, consider $X = \mathbb{N}$ under the discrete topology.)

(d) An infinite, quasi-Stonian compact space is never metrizable. (Apply (7.5) and (10.4) Cor. 2.)

21. (Representation of Order Complete Vector Lattices. Cf. Chap. III, Exerc. 14.)

(a) Every countably order complete vector lattice $E$ with weak order unit is isomorphic to an order dense ideal, containing $C(K)$, of $C_\infty(K)$ (cf. (7.8) Cor. 2) where $K$ is the Stone representation space of $\boldsymbol{B}(E)$ (see 2.9). (Arguing as in the proof of (8.4), show that for any weak order unit $u$ of $E$, $E_u$ is isomorphic to $C(K)$; by means of the relation $x = \sup_n (x \wedge nu)$ valid for $x \in E_+$ (1.9) Cor., this isomorphism extends to an isomorphism of $E$ into $C_\infty(K)$.)

(b) Every order complete vector lattice is isomorphic to an order dense ideal of $C_\infty(K)$, where $K$ is the Stone representation space of $\boldsymbol{B}(E)$. (Use (a) and Exerc. 5(c).)

(c) The compact spaces $K$ constructed in (a), (b) are uniquely determined by the requirement that $E$ be isomorphic to an order dense ideal in $C_\infty(K)$.

(d) Along the lines of Chap. III, § 5, a representation theory extending (a) can be developed for arbitrary countably order complete vector lattices (Hackenbroch [1972]).

22. (a) Let $G$ be a (real) Banach space with this property: Whenever $F$ is a Banach space and $F_0$ a closed vector subspace isomorphic to $G$, then $F_0$ is the range of a contractive projection. Then $G$ is isomorphic to $C(K)$ for some compact Stonian space $K$. (See, e.g., Day [1973].)

(b) A Banach lattice $E$ is isomorphic to some $C(K)$, $K$ compact Stonian, if and only if $E$ is an $AM$-space and there exists a positive, contractive projection of $E''$ onto $E$.

(c) Using the method of complexification discussed in § 11, state and prove complex analogs of (7.10) and (8.9).

(d) Derive lifting theorems (such as the corollary of (7.11)) for $AL$- and $AM$-spaces by dualizing (7.10) and (8.9).

23. (Spaces $L^p(\mu)$.)

(a) The norm of a Banach lattice $E$ is said to be *p-additive* (where $1 \leq p < +\infty$) if $\|x+y\|^p = \|x\|^p + \|y\|^p$ for every orthogonal pair $x, y \in E$. Show that a $p$-additive norm is necessarily order continuous. (Use (5.13).) Following the pattern of the proof of (8.5), prove that every Banach lattice with $p$-additive norm is isomorphic to $L^p(\mu, X)$ for a strictly positive (even order continuous) Radon measure $\mu$ on a suitable locally compact space $X$. More precisely, (8.5) extends to the case $1 < p (< +\infty)$ and the remarks following (8.5) apply accordingly. Cf. also (IV. 6.7).

(b) If a Banach lattice $E$ has $p$-additive norm, $1 < p < +\infty$, its dual $E'$ has $q$-additive norm where $q$ is the exponent conjugate to $p$, $q = p/(p-1)$. (If $x', y'$ are orthogonal elements of $E'_+$, show that there exist sequences $(x_n)$, $(y_n)$ in $E_+$ such that $x_n \perp y_n$ and $\|x_n\| = \|y_n\| = 1$ $(n \in \mathbb{N})$ for which $\lim_n \langle x_n, x' \rangle = \|x'\|$ and $\lim_n \langle y_n, y' \rangle = \|y'\|$. (Use (a) and (4.10) Cor.). Since $x', y'$ are orthogonal, conclude that one has $\|x' + y'\| = \lambda \|x'\| + \mu \|y'\|$ where $\lambda, \mu \in \mathbb{R}_+$ are determined by the equations $\lambda^p + \mu^p = 1$, $(\lambda/\mu)^{p-1} = \|x'\|/\|y'\|$. Deduce that $\|x' + y'\|^q = \|x'\|^q + \|y'\|^q$.)

(c) Let $(X, \Sigma, \mu)$ be an arbitrary measure space. Show that for $E := L^p(\mu)$, where $1 < p < +\infty$, the dual $E'$ is isomorphic to $L^q(\mu)$ for $q = p/(p-1)$, the canonical bilinear form on $E \times E'$ being given by $\langle f, g \rangle = \int f g \, d\mu$. (Noting that, by (a) and (b), both $E$ and $E'$ have order continuous norm (hence are super Dedekind complete), show that for each $x' \in E'_+$ the band of strict positivity is a principal band of $E$; thus reduce the problem to the case where $\mu$ is finite. Under this assumption, show that $L^\infty(\mu) \subset E' \subset L^1(\mu)$ (inclusion of dense ideals; cf. (8.7)) and, using (b), that the norm induced on $L^\infty(\mu)$ by $E'$ is the norm of $L^q(\mu)$.)

(d) Let $(X, \Sigma, \mu)$ be an arbitrary measure space and let $L^\infty(\mu)$ denote the quotient of the $AM$-space of bounded measurable real functions on $X$ over the ideal of functions vanishing locally almost everywhere $(\mu)$. (For the terminology used, cf. Chap. III, § 1, Example 2.) Prove that there exists a natural injection $i: L^\infty(\mu) \to L^1(\mu)'$ onto a Banach sublattice, and show the following to be equivalent:

($\alpha$) $(X, \Sigma, \mu)$ is localizable.

($\beta$) $i$ is surjective.

($\gamma$) $L^\infty(\mu)$ is order complete.

(e) A subset $A \subset L^1(\mu)$, $(X, \Sigma, \mu)$ any measure space, is relatively compact iff $A$ satisfies these conditions: (i) $A$ is bounded. (ii) The countable additivity of the set functions $S \mapsto \int_S f \, d\mu$ is uniform for $f$ in $A$. (iii) Each sequence in $A$ contains a subsequence convergent in measure on every subset $S \in \Sigma$ such that $\mu(S) < +\infty$. (By considering sequences in $A$, reduce the problem to the case where $\mu$ is $\sigma$-finite, and employ (8.8).)

24. Let $K$ be a compact space.

(a) If $\mu$ denotes a Radon measure on $K$ and $\bar{\mu}$ denotes its regular Borel extension [H, § 56], these assertions are equivalent:

($\alpha$) $\bar{\mu}(A) = 0$ for each closed nowhere dense subset $A$ of $K$.

($\beta$) $\bar{\mu}(A) = 0$ for each meager (first category) Borel subset $A$ of $K$.

($\gamma$) Whenever $f = \sup_\alpha f_\alpha$ (in $C(K)$) for a directed ($\leqq$) family $(f_\alpha)$, then $\mu(f) = \lim_\alpha \mu(f_\alpha)$.

(b) If $K$ is Stonian, $K$ is the union of disjoint, open-and-closed subsets $K_1, K_2$ such that $K_1$ is hyperstonian while $K_2$ supports no order continuous Radon measures (see (a)) other than 0. In general, however, $K_1 = \emptyset$ (§ 9, Example 2). (Show that each order continuous Radon measure on $K$ has open-and-closed support.)

(c) If the Banach space $C(K)$ is (isometrically) isomorphic to the Banach dual $G'$ of a Banach space $G$, then $K$ is hyperstonian and $G$ isomorphic to an $AL$-space. (First observe that, identifying $G'$ with $C(K)$, the unit ball $[-e, e]$ of $C(K)$ is $\sigma(G', G)$-compact; hence, by the Krein-Šmulian theorem [S, IV.6.4], the positive cone $C(K)_+$ is $\sigma(G', G)$-closed. Thus, each directed ($\leqq$) majorized family $(f_\alpha)$ in $C(K)$ $\sigma(G', G)$-converges to $f = \sup_\alpha f_\alpha$ by (5.8); in particular, $C(K)$ is order complete and each $x \in G$ defines an order continuous Radon measure on $K$, and $K$ is hyperstonian. Use (9.3) to conclude that $G \subset C(K)'$ agrees with the $AL$-space of order continuous measures on $K$.)

25. (Reciprocal Dunford-Pettis Property.)

(a) For a Banach space $E$, these assertions are equivalent:

($\alpha$) Given any Banach space $F$ and any continuous linear map $T: E \to F$, $T$ is weakly compact whenever $T(W)$ is compact for each weakly compact $W \subset E$.

($\beta$) Each $\tau(E', E)$-compact subset of $E'$ is $\sigma(E', E'')$-compact. (Reverse the pattern of the proof of (9.5).)

A Banach space $E$ satisfying ($\alpha$) is said to have the *reciprocal Dunford-Pettis property* (Grothendieck [1953]).

(b) Every Banach lattice $C(K)$ ($K$ compact) and every closed ideal of $C(K)$ has the reciprocal Dunford-Pettis property. (Use (a) and (9.8), and observe that each closed ideal in $C(K)$ can be identified with $C_0(X)$ for a suitable locally compact space $X$ (cf. Chap. III, § 1, Example 1).)

(c) Can an infinite dimensional $AL$-space have the reciprocal Dunford-Pettis property?

26. (Strict Dunford-Pettis Property.)

(a) A Banach space $E$ is said to have the *strict Dunford-Pettis property* (Grothendieck [1953]) if it satisfies the following equivalent assertions:

($\alpha$) Every weakly compact linear map $T: E \to F$, where $F$ is any Banach space, maps weak Cauchy sequences in $E$ onto (norm) convergent sequences in $F$.

($\beta$) Every weak Cauchy sequence in $E$ is Cauchy for the topology of uniform convergence on the $\sigma(E', E'')$-compact subsets of $E'$.

($\gamma$) Every $\sigma(E', E'')$-compact subset of $E'$ is compact for the topology of uniform convergence on the (ranges of the) weak Cauchy sequences in $E$.

(Proceed as in (9.5) to prove the equivalence of ($\alpha$)—($\gamma$).)

(b) Every $AM$-space and every $AL$-space has the strict Dunford-Pettis property.

27. *Every continuous linear map of an $AM$-space into a weakly sequentially complete Banach space is weakly compact* (Grothendieck [1953]). Prove this theorem by successively establishing the following results. We let $E := C(K)$, $E' = M(K)$, $K$ compact.

(a) Let $(\mu_j)$ be a sequence in $M(K)$ satisfying condition (d) of (9.8). Using the lemma at the end of the proof (9.8), show that $(\mu_j)$ is bounded in $M(K)$ iff $(\mu_j(\{s\}))$ is bounded for each $s \in K$.

(b) A sequence $(\mu_j)$ in $M(K)$ is $\sigma(E', E'')$-convergent whenever $(\mu_j(U))$ converges (in $\mathbb{R}$) for each open $U \subset K$. (Dieudonné [1951], Grothendieck [1953].) (For a given sequence $(U_n)$ of disjoint open subsets of $K$, define a mapping $M(K) \to ba(\mathbb{N})$ (cf. 10.3) by letting $\lambda(J) := \mu(\bigcup_{n \in J} U_n)$; with the aid of (10.3), show that the hypothesis on $(\mu_j)$ implies condition (d) of (9.8). Using (a) and (9.8), conclude that $(\mu_j)$ is relatively $\sigma(E', E'')$-compact.)

(c) Suppose $K$ metrizable and denote by $B$ the linear hull in $E''$ of the characteristic functions $\chi_C$ of all closed subsets $C$ of $K$. Every $\sigma(E', B)$-compact subset $A$ of $E'$ is $\sigma(E', E'')$-compact. (By Eberlein's theorem it suffices to show that each sequence $(\mu_j)$ in $A$ contains a $\sigma(E', E'')$-convergent subsequence. Since $C(K)$ is separable (7.5), $A$ is $\sigma(E', E)$-metrizable [S, IV.1.7] hence, $(\mu_j)$ contains a $\sigma(E', E)$-convergent subsequence; this subsequence converges (in $A$) for $\sigma(E', B)$ by hypothesis and thus $\sigma(E', E'')$-converges by (b).)

(d) If $K$ is metrizable and $F$ is any $\sigma(F, F')$-sequentially complete Banach space, then every continuous linear map $T: C(K) \to F$ is weakly compact. (Consider $T': F' \to E'$ and $T'': E'' \to F''$. Since every closed $C \subset K$ is a $G_\delta$-set, its characteristic function $\chi_C$ is the $\sigma(E'', E')$-limit of a sequence in $E$; thus $T''(\chi_C) \in F$. Hence $T''(B) \subset F$, which implies $T'(V^\circ)$ to be $\sigma(E', B)$-compact ($V^\circ$ denoting the unit ball of $F'$). By (b) $T'(V^\circ)$ is $\sigma(E', E'')$-compact. Apply (9.4).)

(e) By considering the second adjoint, if necessary, the main theorem can be reduced to the case $E = C(K)$ ($K$ compact); the restriction of (d) that $K$ be metrizable, is lifted as follows: Given $T: E \to F$ it is enough to show that for each bounded sequence $(f_n)$ in $E$, $(Tf_n)$ contains a weakly convergent subsequence. But the closed subalgebra $E_0$ (in the complex case, the closed conjugation invariant subalgebra) generated by $(f_n)$ and the constant-one function $e$ is a separable $AM$-space, hence isomorphic to $C(K_0)$ for $K_0$ compact metrizable. (In fact, $K_0$ is the quotient of $K$ over the equivalence relation $s \sim t$ defined by $f_n(s) = f_n(t)$, all $n \in \mathbb{N}$.)

28. Let $E$ denote a vector lattice with order dual $E^\star$, and let $G$ denote an ideal of $E^\star$. By $o(E, G)$ (respectively, $o(G, E)$) we denote the topology of uniform convergence on the order intervals of $G$ (respectively, of $E$); $o(E, G)$ is the topology of $E$ generated by the lattice semi-norms $p_f$ ($f \in G_+$) where $p_f(x) := \langle |x|, f \rangle$ for $x \in E$. (Cf. [S, Chap. V, Exerc. 20] and Peressini [1967, Chap. 3, § 2].)

(a) Let $E$ be a normed vector lattice with dual $E'$ (cf. 5.5). Then $o(E, E')$ is the coarsest locally convex topology of $E$ finer than $\sigma(E, E')$ and for which the lattice operations are continuous; similarly, $o(E', E)$ is the coarsest locally convex topology of $E'$ finer than $\sigma(E', E)$ and for which the lattice operations are continuous. (Show that for a vector topology of $E$ for which the lattice operations are continuous, the solid hull of each equicontinuous subset of $E^\star$ is equicontinuous.)

(b) For the topology of a Banach lattice $E$ to be $o(E, E')$, it is necessary and sufficient that $E$ be isomorphic (as a topological vector lattice) to an $AL$-space.

Similarly, the topology $o(E',E)$ is the norm topology of $E'$ if and only if $E$ is isomorphic (as a topological vector lattice) to an $AM$-space with unit.

If $E$ denotes an $M$-normed space, the completion of $(E,o(E,E'))$ can be identified with $E''$. (Observe that order intervals in $E'$ are $\sigma(E',E'')$-compact.)

(c) Let $E$ be a normed vector lattice. If the lattice operations in $E$ are $\sigma(E,E')$-continuous, or if the lattice operations in $E'$ are $\sigma(E',E)$-continuous, then $E$ is finite dimensional. (Use (a), observing that each of the assumptions implies order intervals of $E'$ to be finite dimensional, i.e., contained in finite dimensional vector subspaces of $E'$.)

(d) If $E$ is an atomic Banach lattice (cf. Exerc. 7) with order continuous norm, then the lattice operations are weakly sequentially continuous in $E$. (Peressini [1967].)

29. Let $A$ be a $\sigma$-complete Boolean algebra and let $\mathsf{M}_A$ denote the $AL$-space of bounded, (finitely) additive real functions on $A$ under the total variation norm (§ 5, Example 3).

(a) $\mathsf{M}_A$ *is sequentially complete for the topology of simple convergence* (Ando [1961]). (The essential fact to be proved is that each $\sigma(\mathsf{M}_A, A)$-Cauchy sequence is bounded; for this, see Ando (op. cit.).)

(b) Considering the Stone representation space $K_A$ of $A$, infer from (10.4) that for every $\sigma(\mathsf{M}_A, A)$-Cauchy sequence $(\mu_n)$ with limit $\mu \in \mathsf{M}_A$ and every band projection $P$ in $\mathsf{M}_A$, $\lim_n P\mu_n = P\mu$ holds uniformly on every decreasing sequence $(a_n)$ in $A$. (Remark that $K_A$ is quasi-Stonian; Exerc. 1(d).)

30. (Weak Sequential Precompactness. Cf. Meyer-Nieberg [1973b].) A subset $A$ of a Banach space is called *weakly sequentially precompact* (w.s.p.) if every sequence in $A$ contains a subsequence which is weakly Cauchy.

Let $E$ denote a Banach lattice. For a linear form $\varphi \in E'_+$ with absolute kernel $N_\varphi$, denote by $(E,\varphi)$ the $AL$-space which is the completion of $E/N_\varphi$ for the $L$-norm induced by $\varphi$ (§ 8, Example 1). Moreover, for a bounded set $A \subset E$ denote by $\varrho_A$ the lattice semi-norm on $E'$ defined by $\varrho_A(\varphi) := \sup_{x \in A} \langle |x|, |\varphi| \rangle$ (that is, let $\varrho_A$ be the gauge function of the polar of the solid hull of $A$).

(a) If $E$ is order complete and if $\varphi \in E'_+$ is order continuous, the canonical image of $E$ in $(E,\varphi)$ is an ideal. (Use 4.12.)

(b) Let $A \subset E$ be bounded and consider these assertions:

($\alpha$) $A$ is w.s.p.

($\beta$) Each monotone ($\leqq$) majorized sequence in $E'$ is $\varrho_A$-Cauchy.

($\gamma$) For every majorized disjoint sequence $(\varphi_n)$ in $E'_+$, $\lim_n \varrho_A(\varphi_n) = 0$.

Then $(\alpha) \Rightarrow (\beta) \Leftrightarrow (\gamma)$ and, if $E$ has order continuous norm, $(\beta) \Rightarrow (\alpha)$. (To prove the equivalence of ($\beta$) and ($\gamma$), employ (5.13). For $(\alpha) \Rightarrow (\gamma)$, apply (8.8) to $(E,\varphi)$ where $\varphi \in E'_+$ majorizes the sequence $(\varphi_n)$. Assuming $E$ to have order continuous norm, prove $(\beta) \Rightarrow (\alpha)$ by reducing the problem to separable $E$ and showing that a fixed sequence in $A$ contains a subsequence converging weakly in every $(E,\varphi)$ where $\varphi$ is any strictly positive linear form on $E$.)

Note that the equivalence of ($\alpha$) and ($\gamma$) can be viewed as an extension of (8.8) to Banach lattices with order continuous norm.

(c) Let $E$ have order continuous norm. Equivalent are:
($\alpha$) The unit ball of $E$ is w.s.p.
($\beta$) $l^1$ is not contained in $E$ as a topological vector subspace.
($\gamma$) $l^1$ is not contained in $E$ as a topological vector sublattice.
($\delta$) $c_0$ is not contained in $E'$ as a topological vector sublattice.
($\varepsilon$) $E'$ has order continuous norm.
(Prove the equivalences with the aid of (b), (5.14), Exerc. 18.)

(d) Let $E$ have order continuous norm. If $A \subset E$ is w.s.p. and $B$ denotes the closed convex solid hull of $A$, then $B$ is w.s.p. (Observe that $\varrho_B = \varrho_A$ and use (b).) Infer that if $E$ is a $KB$-space (§ 5, Example 7), the closed convex solid hull of every weakly compact subset is weakly compact. (Use (10.6).)

(e) Unless $E$ is weakly sequentially complete, the property "w.s.p." in (d) cannot generally be replaced by "relatively weakly compact". Let $F := L^1(\mu)$, $F' = L^\infty(\mu)$ where $\mu$ is Lebesgue measure on $[0,1]$, and let $f_n$ $(n = 0, 1, 2, \ldots)$ denote the $n$-th Rademacher function. (If $[0,1]$ is divided into $2^n$ disjoint subintervals of equal length, $f_n$ is consecutively assigned the values $+1, -1, +1, \ldots$ on their interior, 0 at endpoints.) Now let $E$ be the Banach lattice $c_0(F)$ of all $F$-valued null sequences; it is easy to see that $E$ has order continuous norm and that $E'$ can be identified with $l^1(F')$. Consider the sequence $(x_n)$ in $E$, where $x_n := (h_{nm})_{m \in \mathbb{N}}$ and $h_{nm} := f_n$ for $m \leqq n$, $h_{nm} := 0$ for $m > n$, for each fixed $n \in \mathbb{N}$. Then $(x_n)$ is a weak null sequence in $E$ while $(|x_n|)$ is an increasing weak Cauchy sequence which is not convergent (and hence, contains no weakly convergent subsequence).

In particular, the example (Meyer-Nieberg [1973b]) shows that if $A \subset E$ is relatively weakly compact then, in general, the set of moduli $|A|$ is not, even if $E$ has order continuous norm.

Chapter III

# Ideal and Operator Theory

## Introduction

Ideals are the prime structural components of a vector lattice; accordingly, closed ideals are bound to play a decisive role in the theory of Banach lattices, and their study is the principal concern of this chapter. Its first part (Section 1—6) is concerned with the lattice of closed ideals of a Banach lattice, and with a representation theory for a wide class of Banach lattices (notably those containing quasi-interior positive elements) that ensues naturally. The second part of the chapter (Sections 7—11) then turns to various operator theoretic applications (mainly, mean ergodic theory), employing systematically the concept of operator-invariant ideal. A brief survey follows.

Section 1 begins with some important properties of the lattices of closed (respectively, of weak* closed) ideals; strangely, these properties seem to have gone unnoticed until recently (cf. Notes). Just as for commutative Banach algebras, a representation theory for Banach lattices would most conveniently be built on the concept of maximal ideal; unfortunately, maximal ideals do not exist in Banach lattices in general (striking example: the $L^1$-spaces of Lebesgue measure; see § 1, Example 2). The appropriate substitute are certain maximal families (precisely, ultracoprefilters) of closed ideals, which are discussed in Section 2. In the presence of quasi-interior positive elements (more generally, of a topological orthogonal system; Def. 5.1), these families are advantageously described by valuations that take the place of scalar valued lattice homomorphisms (Thm. 3.6). The class of Banach lattices for which this approach proves fruitful consists at least of those containing a topological orthogonal system; in particular, it contains all separable Banach lattices and all Banach lattices with order continuous norm. The resulting representation theory (Sections 4—6) is, even though not applicable to all Banach lattices, satisfactory in several respects: Among them is the essential uniqueness of the representation space (see Thm. 5.3), the fact that the classical representation theorems for $AM$-spaces with unit (II.7.4) and for $AL$-spaces (II.8.5) are immediately contained as special cases, and the fact that no assumptions on order completeness have to be made. We finally point out that a full analog of the Stone-Weierstrass approximation theorem is available (see end of Section 6).

The second part of the chapter begins with a presentation of basic mean ergodic theory in arbitrary Banach spaces (Section 7). Even though order structures do not enter this discussion, it appeared appropriate to include that section because of the subsequent (Section 8) applications made to semi-groups of positive operators in Banach lattices. (Further contributions to mean ergodic theory are found in Chap. V, § 8). The Banach spaces occurring most frequently in ergodic theory and topological dynamics are Banach lattices, and the theory of irreducible semi-groups of operators (Def. 8.1) has its most appealing applications in these areas. The discussion of lattice homomorphisms given in Section 9 is interwoven with, and its direction determined by, the representation theory mentioned above. Section 10 is concerned with the representation of compact groups of positive operators on a general Banach lattice; its main result (10.4) contains the well known Halmos-von Neumann theorem (10.5) in an appropriate setting. Sections 8—10 are another example for the close contact between Banach lattice theory and areas of classical and functional analysis of established interest.

Finally, Section 11 considers positive projections. Even though this topic is less intimately related to ideal theory it has been placed at the end of this chapter, particularly because of its relationship with the mean ergodic theory of semi-groups of positive operators.

## § 1. The Lattice of Closed Ideals

Throughout this section, $E$ will denote a normed vector lattice over $\mathbb{R}$ (Chap. II, § 5). As usual, $E'_\sigma$ will denote the topological dual $E'$ of $E$ under the weak topology $\sigma(E',E)$ [S, Chap. II, § 5]; we recall that $E'$ agrees with the order dual $E^\star$ whenever $E$ is norm complete (II.5.5). It is trivial to prove but important to observe that the closure of a vector sublattice of $E$ is a vector sublattice, and the closure of an ideal is an ideal (Chap. II, § 5). Further, the polar $I^\circ$ with respect to the duality $\langle E, E'\rangle$ of an ideal $I \subset E$ is solid in $E'$, hence an ideal in $E'$. Moreover, since each order interval in $E'$ is norm bounded, hence an equicontinuous subset of $E'$, it follows that $E'$ is order complete and that for each ideal $I \subset E$, $I^\circ$ is a band in $E'$ (for, each directed ($\leq$) majorized subset of $E'$ is relatively $\sigma(E',E)$-compact), cf. (II.5.5) Cor. 1.

**1.1 Proposition.** *If $I,J$ are ideals in $E$ then $\overline{I \cap J} = \bar{I} \cap \bar{J}$.*

*Proof.* We need only prove that $\bar{I} \cap \bar{J} \subset \overline{I \cap J}$. If $w \in \bar{I} \cap \bar{J}$ then $|w| \in \bar{I} \cap \bar{J}$ and there exist sequences $(x_n)$ in $I$ and $(y_n)$ in $J$, respectively, each of them positive and convergent to $|w|$. Obviously we have $w_n := \inf(x_n, y_n) \in I \cap J$ and $\lim_n w_n = |w|$. Hence, $|w| \in \overline{I \cap J}$ and since the intersection is an ideal, $w \in \overline{I \cap J}$. □

The dual assertion for sums is true but requires norm completeness of $E$ (Exerc. 1); surprisingly, it extends to weakly closed bands in $E'$.

**1.2 Theorem.** *Let $I,J$ be ideals in the Banach lattice $E$. Then $\bar{I} + \bar{J} = \overline{I + J}$ and $I^\circ + J^\circ = (I \cap J)^\circ$. That is, the sum of two closed ideals in $E$ is closed, and the sum of two $\sigma(E',E)$-closed bands in $E'$ is $\sigma(E',E)$-closed.*

*Proof.* For the first assertion it suffices to prove that $K + H$ is closed in $E$ whenever $K, H$ are closed ideals. Assume first that $K, H$ are orthogonal, that is, $K \cap H = \{0\}$.

Let $(z_n)$ be a Cauchy sequence in $K+H$; without loss of generality we can assume $0 \leqq z_n \leqq z_{n+1}$ for all $n$. (For, if $(z_n)$ does not have this property, $y_n := \sup\{|z_\nu|: 1 \leqq \nu \leqq n\}$ defines an increasing Cauchy sequence in $K+H$ whose limit in $E$ majorizes the absolute of $\lim_n z_n$.) Since $K \cap H = \{0\}$ and $z_n \geqq 0$, there exists a unique decomposition $z_n = x_n + y_n$ where $0 \leqq x_n \in K$, $0 \leqq y_n \in H$ (cf. II.2.7). This implies $z_{n+p} - z_n = (x_{n+p} - x_n) + (y_{n+p} - y_n)$, which is again a decomposition into positive components so that, in particular, $0 \leqq x_{n+p} - x_n \leqq z_{n+p} - z_n$. It follows that

$$\|x_{n+p} - x_n\| \leqq \|z_{n+p} - z_n\|$$

for all $n, p \in \mathbb{N}$ hence, $(x_n)$ is a Cauchy sequence in $K$ and, since $K$ is complete, convergent to $x \in K$. Clearly $y := \lim_n y_n$ exists and belongs to $H$, whence $z = x + y \in K + H$. If $K$ and $H$ are not orthogonal, we consider the closed ideal $J := K \cap H$. From what we have shown it follows that the canonical image $q(K) + q(H)$ ($q$ the quotient map $E \to E/J$) is closed in the Banach lattice $E/J$; thus $K + H = q^{-1}(q(K) + q(H))$ is closed in $E$.

The proof of the second assertion is slightly more delicate and uses the theorem of Krein-Smulian [S, IV.6.4 Cor.]: To show that $I^\circ + J^\circ$ is closed in $E'_\sigma$, it is enough to show that $(I^\circ + J^\circ) \cap U^\circ$ is $\sigma(E', E)$-closed where $U$ denotes the unit ball of $E$. To this end we consider the sets $A := (I^\circ + J^\circ) \cap U^\circ$, $B := I^\circ \cap U^\circ + J^\circ \cap U^\circ$, and $C := I^\circ + J^\circ$ in $E'$. The decomposition property (Chap. II, § 1) shows that $A \subset B$: In fact, if $f = g + h$ where $g \in I^\circ$, $h \in J^\circ$, and $f \in U^\circ$ then $|f| \leqq |g| + |h|$ hence, $|f| = g_1 + h_1$ where $0 \leqq g_1 \leqq |g|$, $0 \leqq h_1 \leqq |h|$. It follows that $|f|, g_1, h_1$ are all in $U^\circ$, since $U^\circ$ is solid (II.4.7). So $|f| \in B$ which implies $f \in B$, since $B$ is solid.

Therefore, we have $A \subset B \subset C$. Relative to $\sigma(E', E)$, $A$ is closed in $C$ and $B$ is compact hence, $A$ is closed in $B$ and thus compact. Thus $A$ is closed in $E'_\sigma$. □

**Corollary.** *If $E$ is a Banach lattice, the family $\boldsymbol{A}(E)$ of closed ideals is a distributive lattice; the same is true of the family $\boldsymbol{W}(E')$ of $\sigma(E', E)$-closed bands in $E'$.*

*Proof.* By (II.2.3) the respective families $\boldsymbol{I}(E)$ and $\boldsymbol{I}(E')$ of all ideals in $E$ and $E'$ are distributive lattices, and (1.2) shows that $\boldsymbol{A}(E)$ and $\boldsymbol{W}(E')$ are sublattices of these lattices. □

Thus if $J, H, K \in \boldsymbol{A}(E)$, then $J \cap (H + K) = J \cap H + J \cap K$ is closed, and similarly for $\boldsymbol{W}(E')$. It is, moreover, easy to see that $\boldsymbol{A}(E)$ and $\boldsymbol{W}(E')$ are anti-isomorphic lattices by virtue of $I \mapsto I^\circ$.

Another consequence of (1.2) is the following relationship between $\boldsymbol{A}(E)$ and $\boldsymbol{A}(E/I)$, where $I$ is a fixed, closed ideal in $E$ and $q$ denotes the canonical (quotient) map $E \to E/I$.

**1.3 Proposition.** *If $E$ is a Banach lattice and $I \in \boldsymbol{A}(E)$, the mapping $J \mapsto q(J)$ is an epimorphism of the lattice $\boldsymbol{A}(E)$ onto $\boldsymbol{A}(E/I)$, and an isomorphism of the lattice of closed ideals containing $I$, onto $\boldsymbol{A}(E/I)$.*

*Proof.* Since by (1.2) $q^{-1}(q(J)) = J + I$ is closed in $E$, it follows from the definition of the quotient topology that $q(J)$ is closed in $E/I$. On the other hand, if $H \in \boldsymbol{A}(E/I)$ then $H = q(J)$ where $J := q^{-1}(H)$. Clearly, $q(J_1 + J_2) = q(J_1) + q(J_2)$; furthermore $q(J_1 \cap J_2) = q(J_1) \cap q(J_2)$, since $q$ is a lattice homomorphism. Finally, it is now evident that the map $H \mapsto q^{-1}(H)$ is an isomorphism (with inverse $q$) of $\boldsymbol{A}(E/I)$ onto the lattice of closed ideals of $E$ containing $I$. □

We recall (Chap. II, § 5) that for each $I \in \boldsymbol{A}(E)$, the dual Banach lattices $I'$ and $(E/I)'$ are canonically isomorphic with $E'/I^\circ$ and $I^\circ$ $(\subset E')$, respectively. From (1.2) it follows now that for $I, J \in \boldsymbol{A}(E)$ the dual of the Banach lattice $I+J$ can be identified with the (norm closed) band $I^{\circ\perp} + J^{\circ\perp}$ in $E'$, and that the dual of $E/(I \cap J)$ can be identified with $I^\circ + J^\circ$. (For, $I^{\circ\perp} + J^{\circ\perp} = (I^\circ \cap J^\circ)^\perp$ which in turn is isomorphic to $E'/(I^\circ \cap J^\circ)$ by the Riesz decomposition theorem II.2.10).

Now if $I, J \in \boldsymbol{A}(E)$, to each equivalence class $[x] \in (I+J)/I$ there corresponds, in a natural way, a unique equivalence class $[y] \in J/(I \cap J)$; the (canonical) map $\varphi$ so defined is an isomorphism of vector lattices. By the remarks of the preceding paragraph, the dual of $(I+J)/I$ can be identified with the band $(I^\circ \cap J^\circ)^\perp \cap I^\circ$ in $E'$; an easy computation using the distributivity of $\boldsymbol{A}(E')$ shows this band to be $J^{\circ\perp} \cap I^\circ$. Likewise, the dual of $J/(I \cap J)$ can be identified with $J^{\circ\perp} \cap (I \cap J)^\circ$ $= J^{\circ\perp} \cap (I^\circ + J^\circ) = J^{\circ\perp} \cap I^\circ$. Under these identifications (all corresponding to isomorphisms of Banach lattices, in particular, to norm isomorphisms) the adjoint $\varphi'$ of $\varphi$ becomes the identity map, so $\varphi$ is a norm isomorphism also. We have proved:

**1.4 Proposition.** *If $I, J$ are closed ideals in a Banach lattice $E$, the canonical map $(I+J)/I \to J/(I \cap J)$ is an isomorphism of Banach lattices.*

**Examples.** We determine the lattices of closed ideals for two principal examples of Banach lattices: the spaces $C(K)$, $K \neq \emptyset$ compact, and $L^p(\mu)$ $(1 \leqq p < +\infty)$.

1. Let $E := C(K)$, $K$ compact, and denote by $\mathsf{P}$ the set of all probability measures on $K$, so $\mathsf{P} = \{v \in E' : v \geqq 0,\ \|v\| = 1\}$. Suppose $I$ to be an arbitrary closed ideal in $E$. $I^\circ \cap \mathsf{P}$ is convex and $\sigma(E', E)$-compact and it is easy to see that (if Extr $A$ denotes the set of extreme points of $A$) one has $\mathrm{Extr}(I^\circ \cap \mathsf{P})$ $= I^\circ \cap \mathrm{Extr}\,\mathsf{P}$; in other words, $I^\circ \cap \mathsf{P}$ is a face of $\mathsf{P}$. Since $I$ is closed, the bipolar theorem [S, IV.1.5 Cor. 2] implies $I = (I^\circ \cap \mathrm{Extr}\,\mathsf{P})^\circ$ for, $\mathsf{P}^\circ = \{0\}$. Now $I^\circ \cap \mathrm{Extr}\,\mathsf{P}$ is a weakly closed set of Dirac measures on $K$ and hence can be identified (under the evaluation map) with a closed subset $A$ of $K$ such that $I = \{f \in E : f(A) \subset \{0\}\}$.

On the other hand, it is clear that each closed subset of $K$ determines a closed ideal of $E$ in this manner. Therefore, *the lattice $\boldsymbol{A}(E)$ is anti-isomorphic to the lattice of all closed subsets of $K$*, the latter being ordered by set inclusion. In particular, since any maximal ideal in $E$ is closed (II.5.3, Cor. 3), the maximal ideals of $E$ correspond precisely to the points $s \in K$ by virtue of $s \mapsto I_s$ $= \{f : f(s) = 0\}$. At the same time it follows that the Dirac measures on $K$ are the only (normalized) scalar lattice homomorphisms of $E$.

2. Let $(X, \boldsymbol{S}, \mu)$ be a measure space in the sense of Halmos [H, § 17]; in addition we require that $\boldsymbol{S}$ contain no atoms of infinite measure. Denote by $\boldsymbol{S}_0$ the ring of integrable sets in $\boldsymbol{S}$ (sets of finite measure), and by $\boldsymbol{S}_1$ the $\sigma$-ring of locally measurable sets (sets which intersect each $S \in \boldsymbol{S}_0$ in a set of $\boldsymbol{S}$); finally, let $\boldsymbol{N}$ stand for the $\sigma$-ring of locally negligible sets (sets which intersect each $S \in \boldsymbol{S}_0$ in a set of measure zero). $X$ is called *localizable* if for each family $(f_S)_{S \in \boldsymbol{S}_0}$,

where $f_S$ is defined and measurable on $S$ and such that $f_{S_1}(t)=f_{S_2}(t)$ a.e. $(\mu)$ in $S_1 \cap S_2$ for all $S_1, S_2 \in \boldsymbol{S}_0$; there exists a locally measurable function $f$ which agrees with $f_S$ a.e. on $S$ for each $S \in \boldsymbol{S}_0$.

Let $E := L^p(\mu)$ where $1 \leqq p < +\infty$. *If $X$ is localizable, then each closed ideal in $E$ corresponds to an equivalence class of locally measurable sets modulo local null sets;* briefly, $S \mapsto I_S := \{f \in E : f$ is locally null on $S\}$ defines an anti-isomorphism of the lattice $\boldsymbol{S}_1/\boldsymbol{N}$ onto $\boldsymbol{A}(E)$.

In fact, let $I$ be a closed ideal in $E$. Since each directed $(\leqq)$ majorized family converges in $E$ to its least upper bound, $I$ is a projection band so $E = I + I^{\perp}$. Let $\{f_\alpha : \alpha \in \mathsf{A}\}$ denote a maximal orthogonal subset of $I^{\perp}$. If $S \in \boldsymbol{S}_0$ is fixed then $\int_S |f_\alpha| \, d\mu \neq 0$ for $\alpha$ in a subset $\mathsf{A}_S$ of $\mathsf{A}$ which is at most countably infinite; we define $f_S$ on $S$ by $f_S(t)=1$ if $f_\alpha(t) \neq 0$ for some $\alpha \in \mathsf{A}_S$, and by $f_S(t)=0$ otherwise. It is easy to verify that the family $(f_S)_{S \in \boldsymbol{S}_0}$ so defined satisfies the requirement of the localizability hypothesis and thus there exists a locally measurable function $f$ which agrees with $f_S$ a.e. $(\mu)$ on each $S \in \boldsymbol{S}_0$. $S_I := \{t \in X : f(t) \neq 0\}$ is a locally measurable set. Since $I = I^{\perp\perp}$, $g \in I$ is equivalent to $|g| \wedge |f_\alpha| = 0$ for all $\alpha \in \mathsf{A}$ and, in turn, to $|g| \wedge |f| = 0$. Hence, $g \in I$ if and only if $g(t) = 0$ locally a.e. on $S_I$. It is now easy to see that $S_I$ is unique to within a locally negligible set and that, conversely, each class of locally measurable sets determines a closed ideal of $E$.

We point out that, in the same fashion, each locally measurable subset of $X$ determines a closed ideal of $L^\infty(\mu)$, the space of locally measurable functions which are uniformly bounded locally almost everywhere $(\mu)$. However, this process does not exhaust the supply of closed ideals of $L^\infty(\mu)$ in general. For example, if $\mu$ is Lebesgue measure on $\mathbb{R}$, the set of function classes whose representatives satisfy $\lim_n \sup \mathrm{ess}_{|t| \geqq n} |f(t)| = 0$, is a closed ideal of $L^\infty(\mu)$ not determined by a measurable set as above. The reason becomes clear when $L^\infty(\mu)$ is represented as the space $C(K)$ of continuous real functions on the Stone representation space $K$ of the Boolean algebra of real Borel sets: The ideals of $L^\infty(\mu)$ originating from measurable sets as above, correspond precisely to those ideals in $C(K)$ that live on open-and-closed subsets of $K$. This example also shows that a space $L^p(\mu)$, $1 \leqq p < +\infty$, possesses maximal ideals if and only if $\boldsymbol{S}_0$ contains atoms.

## § 2. Prime Ideals

The preceding examples show that, in analogy to the Gelfand representation theory for commutative Banach algebras, a complete characterization of a Banach lattice through its maximal ideal space is possible only for $AM$-spaces with unit, that is, spaces $C(K)$ (Chap. II, § 7); in general, the space of maximal ideals of a Banach lattice $E$ is too small for a useful representation, and quite possibly empty (§ 1, Example 2). It will be shown in the subsequent sections that a representation of $E$ by continuous functions on a compact or locally compact space can be obtained under assumptions much less restrictive than the existence of an order unit. The essential idea is to replace maximal ideals by certain maximal families (ultra-coprefilters) of closed ideals. An interesting algebraic notion, which generalizes the concept of maximal ideal, intervenes here and will be discussed first.

**2.1 Definition.** *A proper ideal $I$ of a vector lattice $E$ is called* prime *if $x \in E$, $y \in E$ and $x \wedge y \in I$ implies $x \in I$ or $y \in I$.*

Generally an element $a$ of an abstract lattice is called *prime* if $b \wedge c \leqq a$ implies $b \leqq a$ or $c \leqq a$. It will be seen instantly that a proper ideal $I \subset E$ is prime iff $I$ is a prime element of the lattice $\boldsymbol{I}(E)$ of all ideals in $E$; that is, $I$ is prime iff $J, K \in \boldsymbol{I}(E)$ and $J \cap K \subset I$ implies $J \subset I$ *or* $K \subset I$. The following characterization of prime ideals is given by Johnson-Kist [1962].

**2.2 Theorem.** *Let $I \neq E$ be an ideal of the vector lattice $E$. The following assertions are equivalent:*

(a) *$I$ is a prime ideal.*
(b) *$x, y \in E$ and $x \wedge y = 0$ implies $x \in I$ or $y \in I$.*
(c) *$E/I$ is totally ordered.*
(d) *$\boldsymbol{I}(E/I)$ is totally ordered.*
(e) *$I$ is a prime element of $\boldsymbol{I}(E)$.*

*Proof.* (a)⇒(b) follows trivially from (2.1).

(b)⇒(c): Since the canonical map $q: E \to E/I$ is a lattice homomorphism (Chap. II, § 2), it follows from $x^+ \wedge x^- = 0$ and (b) that for each $x \in E$, either $q(x^+) = q(x)^+ = 0$ or $q(x^-) = q(x)^- = 0$ hence, that $E/I$ is totally ordered.

(c)⇒(d): Suppose $K_1, K_2$ are distinct ideals in $E/I$ such that for some $z \in E/I$, we have $0 < z \in K_2 \setminus K_1$, say. Then (c) implies that $|y| \leqq z$ for each $y \in K_1$ hence, that $K_1 \subset K_2$.

(d)⇒(e): If $J, K \in \boldsymbol{I}(E)$ satisfy $J \cap K \subset I$ then $\{0\} = q(J \cap K) = q((J+I) \cap (K+I)) = q(J) \cap q(K)$ whence, $q(J) = \{0\}$ or $q(K) = \{0\}$ and so $J \subset I$ or $K \subset I$.

(e)⇒(a): Assume that $x \wedge y \in I$ and define $u := x - x \wedge y$, $v := y - x \wedge y$. Then $u \perp v$ and the principal ideals $E_u, E_v$ generated by $u, v$, respectively, satisfy $E_u \cap E_v = \{0\} \subset I$. Therefore, $E_u \subset I$ or $E_v \subset I$ by hypothesis; since $x = u + x \wedge y$ and $y = v + x \wedge y$, it follows that $x \in I$ or $y \in I$. □

**Corollary.** *Each proper ideal containing a prime ideal, is itself prime; the family of all ideals in $E$ containing a given prime ideal, is totally ordered (under inclusion).*

The proof is immediate from the equivalence of (a) and (c) (or (d), respectively) of (2.2).

**Examples.** 1. Every proper ideal of a totally ordered vector lattice is prime.

2. In any vector lattice, each maximal ideal (Chap. II, § 3) is prime.

3. An instructive example of an Archimedean vector lattice containing prime ideals which are not maximal, can be obtained as follows. Denote by $E$ the set of numerical (i.e., extended real) functions on $[0,1]$ which are continuous into the two point compactification of $\mathbb{R}$, and finite with the exception

of a finite number (depending on $f$) of "first order poles" $\{t_\nu : \nu \in \mathsf{H}\}$, so that for $t \neq t_\nu$

$$f(t) = g(t) + \sum_{\nu \in \mathsf{H}} \frac{a_\nu}{|t - t_\nu|}$$

with $g \in C[0,1]$, and where $f(t_\nu) = +\infty$ or $-\infty$ accordingly as $a_\nu > 0$ or $a_\nu < 0$. $E$ is readily seen to be an Archimedean vector lattice, with the vector and lattice operations performed pointwise on sets of finiteness, the resulting functions having obvious and unique extensions to $[0,1]$ which are members of $E$.

Let $t_0 \in [0,1]$ be fixed and consider the ideal $J = \{f \in E : f(t_0) = 0\}$. Now each $f \in E$ is locally near $t_0$ (that is, for $t$ in a neighborhood $U_f(t_0)$ dependent on $f$) of the form

$$f(t) = \frac{a_0}{|t - t_0|} + b(t),$$

where $b$ is real valued and continuous. Evidently the equivalence class $[f] \in E/J$ is uniquely determined by the ordered pair $(a_0, b_0)$ where $b_0 := b(t_0)$; it is now readily seen that $E/J$ is isomorphic to the lexicographically ordered plane $\mathbb{R}^2$. Thus $E/J$ is totally ordered and $J$ is prime; the only proper ideal containing $J$ is the maximal ideal of all $f \in E$ which are finite at $t_0$.

By considering functions with a finite number of poles of arbitrary order, an Archimedean vector lattice is obtained which contains countably infinite chains of prime ideals but no maximal ideals (Exerc. 6).

The relationship between prime ideals and maximal ideals is illuminated by the following result.

**2.3 Proposition.** *Let $E$ be any vector lattice. For an ideal $I \subset E$, these assertions are equivalent:*

(a) *$I$ is maximal.*

(b) *$I$ is prime and the order of $E/I$ is Archimedean.*

(c) *$E/I$ is isomorphic to $\mathbb{R}$.*

*Proof.* (a)⇒(c) is clear from (II.3.4) Cor. and (c)⇒(b) is clear from (2.2) (c). (b)⇒(a): If $I$ (which is proper by Def. 2.1) is not maximal then $E/I$ contains a proper ideal $K \neq \{0\}$. By condition (c) of (2.2), any $0 < z \in E/I$ not contained in $K$ would majorize $K$ so that $ny \leqq z$ for all $n \in \mathbb{N}$ and each $y$, $0 < y \in K$. This contradicts the hypothesis that $E/I$ be Archimedean ordered. □

For the remainder of this section, we suppose $E$ to be a Banach lattice. First, let us note this corollary of (2.3).

**Corollary.** *If $I$ is a prime ideal in the Banach lattice $E$, then either $\bar{I} = E$ or $\bar{I}$ is maximal. In particular, a prime ideal in $E$ is closed iff it is a maximal ideal.*

If $I$ is any proper ideal in $E$, the set of all closed ideals contained in $I$ has (as a subset of $2^E$) properties dual to those defining a filter base. These properties are axiomatized in the following definition.

**2.4 Definition.** *Let $E$ be a Banach lattice. A set $V \subset \boldsymbol{A}(E)$ is called a* coprefilter *of closed ideals if $V$ satisfies these axioms:*

($C_1$) $E \notin V$.
($C_2$) $I, J \in V \Rightarrow I + J \in V$.
($C_3$) $(K \in \boldsymbol{A}(E),\ K \subset J, J \in V) \Rightarrow K \in V$.

*A coprefilter which is maximal (with respect to set inclusion) is called an* ultra-coprefilter.

It is easy to see that the set of all coprefilters in $\boldsymbol{A}(E)$ is inductively ordered by set inclusion; so by Zorn's lemma, each coprefilter is contained in an ultra-coprefilter of closed ideals. An example of an ultra-coprefilter is given by the set of closed ideals contained in a maximal ideal of $E$. But whether there exist maximal ideals in $E$ or not, each coprefilter possesses a maximal refinement, and this is precisely the reason why ultra-coprefilters (or equivalently, valuations; cf. § 3 and Exerc. 8) are an appropriate substitute for maximal ideals in representation theory. Note also that for each ultra-coprefilter $\boldsymbol{U}$, $\bigcup\{I : I \in \boldsymbol{U}\}$ is a prime ideal in $E$.

We conclude this section with a simple maximality criterion whose proof can be left to the reader.

**2.5 Proposition.** *A coprefilter $V \subset \boldsymbol{A}(E)$, $E$ a Banach lattice, is maximal if and only if for each closed ideal $J \notin V$, there exists $I \in V$ such that $I + J = E$.*

## § 3. Valuations

The purpose of this section is to lay the groundwork for the representation of a class of Banach lattices (precisely, those possessing a topological orthogonal system; Def. 5.1) as Banach lattices of continuous numerical functions on a locally compact space. We are guided by the representation of $AM$-spaces with unit $e$ as spaces $C(K)$, $K$ compact (Chap. II, § 7); in this special case the elements of $K$ are the maximal ideals of $E$, or equivalently, the scalar lattice homomorphisms $f$ of $E$ normalized to satisfy $f(e) = 1$. Since, in general, a sufficient number of such homomorphisms does not exist, we consider more general homomorphisms called valuations; these, in turn, are closely related to the coprefilters of closed ideals introduced in the preceding section.

The first part of this section, through Proposition 3.4, is algebraic in nature and applies to any vector lattice $E$.

$\overline{\mathbb{R}}_+$, the lattice completion of $\mathbb{R}_+$, is made into an additive and multiplicative semi-group by these additional conventions: $a + \infty = \infty + a = \infty$ for all $a \in \overline{\mathbb{R}}_+$, $0 \cdot \infty = \infty \cdot 0 = 0$, and $a \cdot \infty = \infty \cdot a = \infty$ for $a > 0$. (We write $\infty$ synonymously for $+\infty$). According conventions are made for addition and multiplication in $\overline{\mathbb{R}}$, the lattice completion of $\mathbb{R}$, with the understanding that $\infty - \infty$ remains undefined.

**3.1 Definition.** *A mapping $\varphi$ of the vector lattice $E$ into $\overline{\mathbb{R}}_+$ is called a* valuation *of $E$ if $\varphi$ satisfies the axioms:*

($V_1$) $\varphi(x+y)=\varphi(x)+\varphi(y)$ *for all* $x\geqq 0,\ y\geqq 0$.
($V_2$) $\varphi(x\wedge y)=\varphi(x)\wedge\varphi(y)$ *for all* $x\geqq 0,\ y\geqq 0$.
($V_3$) $\varphi(\lambda x)\ =|\lambda|\,\varphi(|x|)$ *for all* $x\in E,\ \lambda\in\mathbb{R}$.
*A valuation is called* non-trivial *if it assumes at least three distinct values.*

Thus a valuation is completely determined by its values on the positive cone $E_+$ of $E$, and its restriction to $E_+$ is a homomorphism of $E_+$ into the abstract cone $\overline{\mathbb{R}}_+$ (cf. Exerc. 7). Each valuation of $E$ has the following additional properties, which are easy consequences of ($V_1$)–($V_3$) and the basic vector lattice identity $x+y=x\vee y+x\wedge y$ (Chap. II, § 1):

$$(\star)\qquad \varphi(0)=0\,,\quad \varphi(x)\geqq\varphi(y)\geqq 0 \quad \textit{whenever}\quad x\geqq y\geqq 0\,,$$
$$\textit{and}\quad \varphi(x\vee y)=\varphi(x)\vee\varphi(y)\quad \textit{whenever}\quad x\geqq 0\,,y\geqq 0\,.$$

Finally, a valuation $\varphi$ of $E$ is non-trivial iff there exists $x\in E$ such that $0<\varphi(x)<+\infty$.

**Examples.** 1. If $f: E\to\mathbb{R}$ is a (non-trivial) lattice homomorphism then $\varphi: x\mapsto f(|x|)$ is a (non-trivial) valuation of $E$ which is finite throughout. The corollary of (3.3) below shows that each finite valuation of $E$ originates in this way.

2. Denote by $E$ the vector lattice constructed in § 2, Example 3. Then $\varphi(f):=|f(t_0)|$ defines, for fixed $t_0\in[0,1]$, a non-trivial valuation of $E$ which is not finite. $\varphi^{-1}(0)$ is a prime ideal of $E$ which is maximal in $\varphi^{-1}(\mathbb{R})$.

**3.2 Proposition.** *If $\varphi$ is a non-trivial valuation of $E$, then $\varphi^{-1}(0)$ and (unless $\varphi$ is finite) $\varphi^{-1}(\mathbb{R})$ are prime ideals, and $\varphi^{-1}(0)$ is a maximal ideal in $\varphi^{-1}(\mathbb{R})$.*

*Proof*. That $\varphi^{-1}(0)$ is a prime ideal is immediately clear from ($V_2$) and condition (b) of (2.2), and the corollary of (2.2) shows that $\varphi^{-1}(\mathbb{R})$ is a prime ideal as well. To prove that $\varphi^{-1}(0)$ is maximal in $\varphi^{-1}(\mathbb{R})$, let $J$ be an ideal satisfying $\varphi^{-1}(0)\subset J\subset\varphi^{-1}(\mathbb{R})$ and suppose that $x\in E$ satisfies $0<x\in\varphi^{-1}(\mathbb{R})\setminus J$. Then whatever $z\in J$ we must have $\varphi(x)\geqq\varphi(z)$; if not, we would have $x-z\wedge x\in\varphi^{-1}(0)\subset J$ for some $z\in J$ and this would imply $x\in J$, contrary to the assumption. Hence, $\varphi(z)\leqq\varphi(x)<+\infty$ for all $z\in J$. Clearly, by ($V_3$) this implies $\varphi(J)=\{0\}$, that is, $J=\varphi^{-1}(0)$. □

**3.3 Proposition.** *A prime ideal $I\subset E$ is the kernel of a non-trivial valuation if and only if $E/I$ possesses a minimal ideal. Each non-trivial valuation of $E$ is determined by its kernel to within a scalar factor $>0$.*

*Proof*. The part of the first statement asserting necessity is contained in (3.2). Conversely, if $I$ is prime and $J$ is a minimal ideal in $E/I$ then $\{0\}$ is a maximal ideal in $J$, that is, $J$ is isomorphic to $\mathbb{R}$ (Chap. II, § 3; see also (2.3)). Denote by $f$ a (vector lattice) isomorphism $J\to\mathbb{R}$ and by $q$ the canonical map $E\to E/I$. We define $\varphi$ by $\varphi(x):=f\circ q(|x|)$ for $x\in q^{-1}(J)$ and by $\varphi(x)=+\infty$ for $x\in E\setminus q^{-1}(J)$. Then (since $I$ is prime) $\varphi$ is a non-trivial valuation of $E$, and $I=\varphi^{-1}(0)$.

For the proof of the second assertion suppose that $\varphi,\psi$ are non-trivial valuations of $E$ with identical kernel $I$. Then $q(\varphi^{-1}(\mathbb{R}))$ and $q(\psi^{-1}(\mathbb{R}))$ agree in $E/I$, since both are minimal ideals and since $E/I$ is totally ordered (2.2). It follows that $\varphi = c\psi$, $0 < c < +\infty$. ☐

**Corollary.** *For a non-trivial valuation $\varphi$ of $E$, these assertions are equivalent:*

(a) *$\varphi$ is finite valued.*

(b) *The kernel $\varphi^{-1}(0)$ is a maximal ideal in $E$.*

(c) *$E/\varphi^{-1}(0)$ is Archimedean ordered.*

*Proof.* (a)⇒(b) is immediate from (3.2), since $\varphi^{-1}(\mathbb{R}) = E$.

(b)⇒(c): If $\varphi^{-1}(0)$ is maximal then $E/\varphi^{-1}(0)$ is isomorphic to $\mathbb{R}$ hence, Archimedean ordered.

(c)⇒(a): Since $\varphi^{-1}(0)$ is prime, $E/\varphi^{-1}(0)$ is totally ordered. But a totally ordered Archimedean vector lattice is isomorphic to $\mathbb{R}$ (Chap. II, § 3) and the only minimal ideal in $\mathbb{R}$ is $\mathbb{R}$ itself, whence it follows that $\varphi^{-1}(\mathbb{R}) = E$. ☐

Proposition (3.3) shows that a non-trivial valuation $\varphi$ of a vector lattice $E$ is uniquely determined by its kernel and by its value at some element $u \in E$ at which $\varphi$ is finite non-zero (by ($V_3$) of Def. 3.1, $u$ can be assumed to be $>0$). If $u$ is such an element, then $\varphi$ is finite on the principal ideal $E_u$ generated by $\{u\}$ and the preceding corollary shows that $\varphi$ originates from a vector lattice homomorphism $E_u \to \mathbb{R}$ as in Example 1 above. The following important extension theorem (Lotz [1969]) shows that, conversely, every finite non-trivial valuation defined on some principal ideal of $E$, can be extended to a valuation of $E$; moreover, this extension is unique.

**3.4 Theorem.** *Let $E$ be any vector lattice, and suppose that $0 < u \in E$. If $\varphi$ is a valuation of $E_u$ satisfying $\varphi(u) = 1$, there exists a unique valuation $\psi$ of $E$ extending $\varphi$.*

*Proof.* We define $\psi$ by setting

$$\psi(x) := \sup\{\varphi(y) : y \in [0, |x|] \cap E_u\}$$

for all $x \in E$; certainly $\psi$ is an extension of $\varphi$ (recall that $[x, y]$ always denotes the order interval $\{z : x \leqq z \leqq y\}$). To verify ($V_1$) of Def. 3.1, let $x = y + z$ where $y, z \in E_+$. A simple computation using the decomposition property (Chap. II, § 1) shows that

$$[0, y+z] \cap E_u = [0, y] \cap E_u + [0, z] \cap E_u .$$

This implies $\psi(y+z) = \psi(y) + \psi(z)$ hence, the validity of ($V_1$). In view of this, to prove ($V_2$) it suffices to show that $y \wedge z = 0$ implies $\psi(y) \wedge \psi(z) = 0$ for arbitrary $y, z \in E_+$. Consider the ideals $I := E_y \cap E_u$ and $J := E_z \cap E_u$ as ideals in $E_u$. $y \wedge z = 0$ implies $I \cap J = \{0\}$ and it follows, since the kernel of $\varphi$ is a prime ideal in $E_u$ by (3.2), that either $\varphi(I) = \{0\}$ or $\varphi(J) = \{0\}$. In turn this implies $\psi(y) \wedge \psi(z) = 0$. ($V_3$) is trivially satisfied. It remains to show that $\psi$ is unique: Suppose $\chi$ is another extension of $\varphi$. If $\psi$ and $\chi$ are distinct, their respective kernels are distinct by (3.3), and there exists $x \in E_+$ such that $\psi(x) = 0$ and $\chi(x) > 0$, say. If $z := x \wedge u$ then $\psi(z) = 0$ and $\chi(z) > 0$ which is contradictory, since $z \in E_u$. ☐

**Corollary.** *Every non-trivial valuation $\varphi$ of a vector lattice $E$ is uniquely determined by its values on the ideal $E_u$, for any $u$ such that $0<\varphi(u)<+\infty$.*

The properties of valuations considered so far were strictly algebraic; we now consider valuations on a Banach lattice $E$, relating them to sets of closed ideals in $E$. Recall that $u\in E_+$ is called a *quasi-interior* element of $E_+$ if the closed ideal generated by $\{u\}$ equals $E$, that is, if $E=\overline{E}_u$ (Chap. II, § 6).

**3.5 Proposition.** *Let $\varphi$ be a valuation on the Banach lattice $E$ such that $\varphi(u)=1$ for some quasi-interior $u\in E_+$. Then the set of all closed ideals contained in the kernel of $\varphi$ is an ultra-coprefilter.*

*Proof.* It is clear that $\boldsymbol{U}_\varphi:=\{I\in\boldsymbol{A}(E): I\subset\varphi^{-1}(0)\}$ is a coprefilter (Def. 2.4). By (2.5) we have to show that for each closed ideal $J\notin\boldsymbol{U}_\varphi$, there exists $I\in\boldsymbol{U}_\varphi$ satisfying $J+I=E$. Suppose $J\notin\boldsymbol{U}_\varphi$; there exists $z$, $0<z\in J$, such that $\varphi(z)\geqq 1$. Let $x:=u\wedge z$ and $y:=u-x$, then $\varphi(x)=1$, $\varphi(y)=0$. For $w:=x-x\wedge y$ and $v:=y-x\wedge y$ we have $w\perp v$ and still $\varphi(w)=1$, $\varphi(v)=0$. It follows that $\overline{E}_v\in\boldsymbol{U}_\varphi$, since $\varphi(w)=1$ and $\overline{E}_v$ is orthogonal to $w$. Now from $x\in J$, $y\in J+\overline{E}_v$ it follows that $u=x+y\in J+\overline{E}_v$, hence that $J+\overline{E}_v=E$ because $J+\overline{E}_v$ is closed by (1.2) and $u$ is quasi-interior to $E_+$. □

In the following theorem, for simplicity of notation we write $F(x,\lambda)$ for the closed ideal generated by $(x-\lambda u)^+$ and $G(x,\lambda)$ for the closed ideal generated by $(x-\lambda u)^-$, where $x\in E$, $\lambda\in\mathbb{R}$ and $u$ is a fixed quasi-interior point of $E_+$.

**3.6 Theorem.** *Let $E$ be a Banach lattice containing an element $u$ quasi-interior to $E_+$. There exists a bijective correspondence $\varphi\leftrightarrow\boldsymbol{U}$ between the set of all valuations of $E$ satisfying $\varphi(u)=1$, and the set of all ultra-coprefilters of closed ideals in $E$. This correspondence is given by the formula*

$$(\star)\qquad \begin{aligned}\varphi(x)&=\inf\{\lambda\in\mathbb{R}: F(|x|,\lambda)\in\boldsymbol{U}\}\\ &=\sup\{\lambda\in\mathbb{R}: G(|x|,\lambda)\in\boldsymbol{U}\}^1\quad (x\in E).\end{aligned}$$

*Proof.* First we show that for a given $\boldsymbol{U}$, $(\star)$ defines a valuation of $E$ satisfying $\varphi(u)=1$; obviously it suffices to consider the case $x\geqq 0$, $\lambda\geqq 0$. Since $u$ is quasi-interior to $E_+$, it is clear that $\lambda<1$ implies $F(u,\lambda)=E\notin\boldsymbol{U}$ and that $\lambda>1$ implies $F(u,\lambda)=\{0\}\in\boldsymbol{U}$; so $\varphi(u)=1$. To verify $(\mathrm{V}_1)$ of Def. 3.1, choose $x,y\in E_+$ and suppose that $\varphi(x)\leqq\lambda$, $\varphi(y)\leqq\mu$. Since

$$(x+y-(\lambda+\mu)u)^+\leqq(x-\lambda u)^+ +(y-\mu u)^+,$$

we have $F(x+y,\ \lambda+\mu)\subset F(x,\lambda)+F(y,\mu)$ hence, $\varphi(x+y)\leqq\varphi(x)+\varphi(y)$. On the other hand, if $\lambda,\mu\in\mathbb{R}$ satisfy $\varphi(x+y)<\lambda+\mu$ then $G(x+y,\ \lambda+\mu)\notin\boldsymbol{U}$ whence $G(x,\lambda)\notin\boldsymbol{U}$ or else $G(y,\mu)\notin\boldsymbol{U}$. This shows that $\varphi(x)\leqq\lambda$ or else $\varphi(y)\leqq\mu$, so $\varphi(x+y)\geqq\varphi(x)+\varphi(y)$.

To verify axiom $(\mathrm{V}_2)$ we observe that, since $\boldsymbol{U}$ is an ultra-coprefilter, $I\cap J\in\boldsymbol{U}$ implies $I\in\boldsymbol{U}$ or else $J\in\boldsymbol{U}$ for each pair $I,J$ of closed ideals in $E$. (Cf. Exerc. 9.)

[1] It is understood that $\inf\emptyset=\sup\mathbb{R}=+\infty$.

Now if $x \wedge y = 0$ and $\lambda \geqq 0$ it is clear that $(x-\lambda u)^+ \wedge (y-\lambda u)^+ = 0$ hence, that $F(x,\lambda) \cap F(y,\lambda) = \{0\}$. Thus for each $\lambda \geqq 0$, either $F(x,\lambda) \in \boldsymbol{U}$ or $F(y,\lambda) \in \boldsymbol{U}$ which implies $\varphi(x) \wedge \varphi(y) = 0$. In view of $(V_1)$ this suffices to establish $(V_2)$; verification of $(V_3)$ is trivial. The remark just made (with reference to Exerc. 9) also serves to establish equality between the two right hand members of (*), since $F(x,\lambda) \cap G(x,\lambda) = \{0\}$ for all $x \in E$, $\lambda \in \mathbb{R}$.

Conversely, (3.5) shows that for each valuation $\varphi$ of $E$ satisfying $\varphi(u) = 1$, $\boldsymbol{U}_\varphi := \{I \in \boldsymbol{A}(E) : I \subset \varphi^{-1}(0)\}$ is an ultra-coprefilter; it remains to show that, given such $\varphi$, $\varphi$ satisfies (*) for $\boldsymbol{U} = \boldsymbol{U}_\varphi$. Let $x \in E_+$ be arbitrary but fixed. If $F(x,\lambda) \in \boldsymbol{U}_\varphi$, then $\varphi[(x-\lambda u)^+] = 0$ and from $0 \leqq x \leqq \lambda u + (x-\lambda u)^+$ it follows that $\varphi(x) \leqq \lambda$. If, on the other hand, $F(x,\lambda) \notin \boldsymbol{U}_\varphi$ then $G(x,\lambda) \in \boldsymbol{U}_\varphi$, since $F(x,\lambda) \cap G(x,\lambda) = \{0\}$. So $\varphi[(x-\lambda u)^-] = 0$, and $\lambda u \leqq x + (x-\lambda u)^-$ implies that $\lambda \leqq \varphi(x)$. This completes the proof. □

For an algebraic analogue of (3.6), see Exerc. 7.

## § 4. Compact Spaces of Valuations

If the lattice completion $\bar{\mathbb{R}}_+$ of $\mathbb{R}_+$ introduced above is considered as the one point compactification of $\mathbb{R}_+$, the set $\bar{\mathbb{R}}_+^E$ of all mappings $E \to \bar{\mathbb{R}}_+$ ($E$ any vector lattice) is, by Tychonov's theorem, a compact space under the product topology. Now let $u \in E_+$ be fixed; the set of all valuations $\varphi$ of $E$ (Def. 3.1) satisfying $\varphi(u) = 1$ ($u$-normalized valuations) is a closed, hence compact subspace of $\bar{\mathbb{R}}_+^E$ to be denoted by $K_u$ (or by $K_u(E)$ if this distinction is desirable). Evidently the topology of $K_u$ thus defined is the coarsest topology for which the mappings $\varphi \mapsto \varphi(x)$ $(x \in E)$ of $K_u$ into $\bar{\mathbb{R}}_+$ are continuous.

By Theorem 3.4 $K_u$ can be identified with the set of all $u$-normalized valuations of $E_u$ (or equivalently, with the set of all $u$-normalized, lattice preserving linear forms on $E_u$). If $E$ is Archimedean, $E_u$ is (under the norm whose unit ball is the order interval $[-u,u]$) an $M$-normed space with unit $u$ (Chap. II, § 7) which is norm complete if and only if $E_u$ satisfies Axiom (OS) (II.7.2); this axiom is, in particular, satisfied whenever $E$ is a Banach lattice (II.7.2 Cor.). Thus if $E$ is a Banach lattice and $u \in E_+$, $E_u$ can be identified with $C(K_u)$ by (II.7.4); this identification will often be made tacitly in the sequel. Also, it is easy to see that when $K_u$ is viewed as a subset of the dual $E_u'$, the topology induced by $\sigma(E_u', E_u)$ agrees with the topology of $K_u$ defined above.

**4.1 Theorem.** *Let $E$ be a Banach lattice. For any two quasi-interior elements $u, v$ of $E_+$, the spaces $K_u$ and $K_v$ are homeomorphic.*

*Proof.* Let $u, v \in E_+$ be quasi-interior positive elements; then $U := \{\varphi \in K_v : 0 < \varphi(u) < +\infty\}$ is a dense open subset of $K_v$. In somewhat more general form, we shall prove this assertion below as a lemma.

We now define a mapping $h_{u,v}$ of $U \subset K_v$ into $K_u$ by $h_{u,v}(\varphi) := [\varphi(u)]^{-1}\varphi$, which is evidently continuous. It is easy to see that the range of this map is the set $V := \{\psi \in K_u : 0 < \psi(v) < +\infty\}$, and that $h_{v,u} \circ h_{u,v}$ is the identity map of $U$. Again,

$V$ is a dense open subset of $K_u$; if we can show that $h_{u,v}$ has a continuous extension $K_v \to K_u$ then the same argument will apply to $h_{v,u}$, and it will follow that this extension is a homeomorphism of $K_v$ onto $K_u$.

To show that such an extension exists, let $\varphi_0 \in K_v \setminus U$ and denote by $\boldsymbol{F}_0$ the trace on $U$ of the neighborhood filter of $\varphi_0$. Since $K_u$ is compact, the filter base $h_{u,v}(\boldsymbol{F}_0)$ has an adherent point $\chi \in K_u$. If $\varphi_0(u)=0$ then $\varphi_0(x)>0$ $(x \in E)$ implies $\chi(x)=+\infty$, so $\chi^{-1}(0) \subset \varphi_0^{-1}(0)$. If $\varphi_0(u)=+\infty$, then $\varphi_0(x)=0$ implies $\chi(x)=0$, so $\varphi_0^{-1}(0) \subset \chi^{-1}(0)$. Since the sets $\boldsymbol{U}_{\varphi_0}$ and $\boldsymbol{U}_\chi$ of closed ideals contained in the kernels of $\varphi_0$ and $\chi$, respectively, are ultra-coprefilters by (3.5), it follows that $\boldsymbol{U}_{\varphi_0}=\boldsymbol{U}_\chi$. But $\chi(u)=1$, and so it follows from (3.6) that the adherent point $\chi$ is unique. By a well-known result on compact spaces, this implies the convergence of $h_{u,v}(\boldsymbol{F}_0)$ to $\chi$; moreover, it is easy to verify that the extension of $h_{u,v}$ to $K_v$ thus defined, is continuous.

We conclude the proof with this lemma:

*Let $v$ be a weak order unit of the Banach lattice $E$. Then for each weak order unit $u \in E_+$, $H_1 := \{\varphi \in K_v : \varphi(u)=0\}$ is a rare subset of $K_v$ and for each $x \in E_+$, $H_2 := \{\varphi \in K_v : \varphi(x)=+\infty\}$ is a rare subset of $K_v$.*

*Proof.* We note first that $w := u \wedge v$ is a weak order unit of $E$, hence of $E_v$. Since $E_v$ can be identified with $C(K_v)$ it follows that $\{\varphi \in K_v : \varphi(w)=0\}$ is rare in $K_v$, and a fortiori $H_1$ is rare in $K_v$. (Clearly, $H_1$ and $H_2$ are closed.) If $H_2$ has interior points then by Urysohn's theorem there exists an $f \in C(K_v)$, $0<f \leqq 1$, having its support in $H_2$ and hence, there exists an element $y \in E_v$, $0<y \leqq v$, such that $n\varphi(y)=\varphi(ny) \leqq \varphi(x)$ for all $n \in \mathbb{N}$ and all $\varphi \in K_v$. Therefore, $\varphi(ny)=\varphi(x \wedge ny)$ for all $\varphi \in K_v$ which shows that $ny = x \wedge ny$ for each $n$, since $K_v$ separates the elements of $E_v \cong C(K_v)$. Thus $ny \leqq x$ for all $n$ which implies $y=0$, since the order of $E$ is Archimedean; this contradiction shows $H_2$ to be rare in $K_v$. □

The homeomorphism $K_v \to K_u$ established in the proof of (4.1) will again be denoted by $h_{u,v}$ and will be called *canonical*; it is clear that $h_{u,v}=h_{u,w} \circ h_{w,v}$ for any triple $u,w,v$ of quasi-interior positive elements of $E$. By virtue of the (canonical) isomorphism $E_u \cong C(K_u)$, we obtain this corollary of (4.1).

**Corollary.** *For each pair $u,v$ of quasi-interior positive elements of the Banach lattice $E$, the Banach lattices $E_u$ and $E_v$ are (canonically) isomorphic.*

We have seen above (Theorem 3.6) that for each quasi-interior $u \in E_+$, $K_u$ can be identified with the set $K$ of all ultra-coprefilters of closed ideals in $E$; it is not difficult to see that under this correspondence $K_u \to K$, the canonical homeomorphism $K_v \to K_u$ induces the identity map of $K$. This motivates the following definition.

**4.2 Definition.** *Let $E$ be a Banach lattice with quasi-interior positive elements. The compact space $K$ of all ultra-coprefilters in $\boldsymbol{A}(E)$ (topologized by the correspondence $K_u \to K$ of (3.6)) is called the* structure space *of $E$.*

The structure space $K$ (or $K(E)$) of $E$ is thus a direct generalization of the maximal ideal space of an $AM$-space with unit; the maximal ideals of $E$ (if any) can be identified with those ultra-coprefilters that are generated by a single element of

$A(E)$. It is possible to define the topology of $K$ without reference to valuations (Exerc. 9); however, we will find it more convenient to employ valuations and to identify $K$ with $K_u$ for suitable quasi-interior points $u \in E_+$. This applies, in particular, to the two following propositions concerned with the structure spaces of quotients and certain sublattices.

**4.3 Proposition.** *If $E$ is a Banach lattice with quasi-interior positive elements and $I \in A(E)$, then the structure space $K(E/I)$ can be identified with a closed subspace of $K(E)$.*

*Proof.* If $u$ is quasi-interior to $E_+$, then $v := q(u)$ is quasi-interior to $(E/I)_+$ by (II.6.4) Cor. 1. It is readily seen that the map $\psi \mapsto \varphi := \psi \circ q$ ($q$ the canonical map $E \to E/I$) of $K_v(E/I)$ onto the subspace $\{\varphi \in K_u(E): \varphi(I) = \{0\}\} =: K_I$ of $K_u(E)$ is a continuous bijection. (Note that $K_I$ corresponds precisely to those ultra-coprefilters in $A(E)$ that contain $I$.) Since $K_I$ is closed in $K_u(E)$, this map is the desired homeomorphism. □

*Remark.* The reader should be cautioned that the mapping $I \mapsto K_I$, which obviously is an anti-isomorphism of $A(E)$ onto a lattice of closed subsets of $K$, does not exhaust the closed subsets of $K$ in general (Exerc. 10).

**4.4 Proposition.** *Let $E$ be a Banach lattice and let $H$ be a closed vector sublattice containing some quasi-interior element $v \in E_+$. Then $K(H)$ is a continuous image of $K(E)$. More precisely: Restricting $v$-normalized valuations of $E$ to $H$ defines a continuous surjection $K_v(E) \to K_v(H)$.*

*Proof.* Let $j: H \to E$ denote the canonical injection. It is clear that $\varphi \mapsto \varphi \circ j$ defines a continuous map $\tau$ of $K_v(E)$ into $K_v(H)$; we have to show that this map is surjective. If not, by Urysohn's theorem there exists a continuous function $f > 0$ in $C(K_v(H))$ vanishing on $\tau(K_v(E))$; since $C(K_v(H))$ can be identified with $H_v$, it follows that there exists $x > 0$ in $H_v$ such that $\varphi(x) = 0$ for all $\varphi \in K_v(E)$. Since $x \in E_v$, it follows that $x = 0$ which is contradictory. □

**Corollary.** *If the mapping $K_v(E) \to K_v(H)$ is injective, then $H = E$.*

*Proof.* The assumption implies that the restriction map $K_v(E) \to K_v(H)$ is a homeomorphism, hence that $H_v = E_v$ and so $H = E$. □

The preceding corollary is a generalization to Banach lattices with quasi-interior positive elements, of the classical Stone approximation theorem for $AM$-spaces with unit (II.7.3); we will return to this circle of ideas in § 6 below. It should also be noted that for sublattices $H$ of $E$ which contain quasi-interior points of $H_+$ but not of $E_+$, no simple relation between the structure spaces of $E$ and $H$ exists (Exerc. 11).

Like the maximal ideal space of a commutative Banach algebra with unit, the structure space of a Banach lattice $E$ with quasi-interior positive elements appears to be the appropriate tool for a representation of $E$ by means of continuous functions. But in contrast with the situation encountered in the Gelfand theory, the function space $\hat{E}$ representing $E$ *contains* $C(K(E))$ as a dense ideal; hence it becomes plausible that the general case (absence of quasi-interior points of $E_+$) cannot be reduced to the present case by a simple adjunction process.

**4.5 Theorem.** *Every Banach lattice $E$ with quasi-interior positive elements is isomorphic (as a vector lattice) to a vector lattice $\hat{E}$ of continuous numerical functions*[1] *on the structure space $K(E)$, each of which is infinite only on some rare subset of $K(E)$. $\hat{E}$ contains $C(K(E))$ as a dense ideal*[2], *and this property determines $K(E)$ to within homeomorphism.*

*Proof.* Denote by $u$ any fixed quasi-interior point of $E_+$. We identify $K(E)$ (Def. 4.2) with the compact space $K_u$ defined at the beginning of this section, so that the ideal $E_u \subset E$ is isomorphic to $C(K_u)$ by virtue of the evaluation map $x \mapsto \hat{x}$, where $\hat{x}(\varphi) := \varphi(x)$ for $\varphi \in K_u$. We extend this mapping by defining $\hat{x}(\varphi) := \varphi(x^+) - \varphi(x^-)$ for arbitrary $x \in E$ and $\varphi \in K_u$. By $(V_2)$ of Def. 3.1 we have $\varphi(x^+) \wedge \varphi(x^-) = 0$ for all $x$ and $\varphi$, hence $\hat{x}$ is a well defined and evidently continuous function of $K_u$ into $\overline{\mathbb{R}}$. Moreover, from the italicized lemma in the proof of (4.1) it follows that $\hat{x}$ is finite except on some rare subset of $K_u$.

Let us denote the range of $E$ under $x \mapsto \hat{x}$ by $\hat{E}$. Again from axiom $(V_2)$ we see that $x \mapsto \hat{x}$ preserves the lattice operations, $\hat{E}$ being considered a sublattice of $\overline{\mathbb{R}}^{K_u}$. To show that the map is injective, it suffices to prove that whenever $x_1 \geqq 0$, $x_2 \geqq 0$ and $\varphi(x_1) = \varphi(x_2)$ for all $\varphi \in K_u$, then $x_1 = x_2$. The assumption and $(V_2)$ implies that $\varphi(x_1 \wedge nu) = \varphi(x_2 \wedge nu)$ for all $\varphi \in K_u$ hence, since $x_1 \wedge nu$, $x_2 \wedge nu \in E_u \cong C(K_u)$, that $x_1 \wedge nu = x_2 \wedge nu$ for all $n \in \mathbb{N}$. Since $u \in E_+$ is quasi-interior, this implies $x_1 = x_2$ (II.6.3).

We next show that pointwise addition and scalar multiplication on sets of finiteness defines a vector space structure on $\hat{E}$ such that $E \to \hat{E}$ is a linear isomorphism. In fact, given $x, y \in E$ there exists a dense open subset $U \subset K_u$ on which both $\hat{x}$ and $\hat{y}$ are finite. It is clear from $(V_1)$ of Def. 3.1 that $\hat{x}(\varphi) + \hat{y}(\varphi) = (x+y)\hat{}\,(\varphi)$ whenever $\varphi \in U$, hence the sum function $\hat{x} + \hat{y}$, defined on $U$, possesses a (necessarily unique) continuous extension to $K_u$ with values in $\overline{\mathbb{R}}$ which is given by $(x+y)\hat{}$. A similar remark applies to scalar multiplication, and thus $x \mapsto \hat{x}$ is a vector lattice isomorphism of $E$ to $\hat{E}$.

Finally, suppose that for some compact space $K$ there exists an isomorphism of $C(K)$ onto a dense ideal in $E$. If $v$ is the image of the unit of $C(K)$ under this isomorphism, then $E_v$ is isomorphic to $C(K)$ and $v$ is quasi-interior to $E_+$. By (4.1), $K$ is homeomorphic to $K_v$ and hence to $K(E)$. □

The preceding theorem extends the Kakutani representation theorems for $AM$-spaces with unit and for $AL$-spaces with weak order unit (Chap. II, §§ 7,8), to arbitrary Banach lattices possessing quasi-interior positive elements. But the theorem gives only the first half of an intrinsic representation, as it were, for it fails to characterize the normed structure of $E$. Also, the requirement that $E_+$ contain quasi-interior points, even though necessary to obtain a compact representation space (cf. Def. 5.3), is unnecessarily restrictive. An extension of these results taking both aspects into account will be given in the following section.

---

[1] Recall that a continuous *numerical* function on $K$ is a continuous map $K \to \overline{\mathbb{R}}$, and that *rare* means having nowhere dense closure. It is understood that the lattice operations in $\hat{E}$ agree with the pointwise operations and that the algebraic operations agree with the pointwise operations on sets of finiteness.

[2] For the topology transferred from $E$.

## § 5. Representation by Continuous Functions

Recall (Chap. II, § 1) that a maximal orthogonal system of a vector lattice $E$ is a subset $S \subset E_+$ of mutually orthogonal, non-zero positive elements such that $x \in E$ and $|x| \wedge u = 0$ for all $u \in S$, implies $x = 0$. Just as the concept of quasi-interior positive element in a Banach lattice (Chap. II, § 6) is the topological version of the concept of weak order unit, the following definition gives the topological version of the concept of maximal orthogonal system.

**5.1 Definition.** *An orthogonal set $S$ of non-zero positive elements in a Banach lattice $E$ is called a* topological orthogonal system (t. o. s.) *of $E$ if the ideal generated by $S$ is dense in $E$.*

It is immediately clear that any t.o.s. of a Banach lattice $E$ is a maximal orthogonal system of $E$; the example $E = C[0,1]$, $S = \{u\}$ where $u(t) = t$ for all $t$, shows that the converse is false. Now let $E$ be any Banach lattice and let $S = \{u_\alpha : \alpha \in \mathsf{A}\}$ denote a maximal orthogonal system of $E$. For each $n \in \mathbb{N}$, each finite set $\mathsf{H} \subset \mathsf{A}$, and $x \in E_+$, we define

$$x_{n,\mathsf{H}} := \textstyle\sum_{\alpha \in \mathsf{H}} (x \wedge n u_\alpha) .$$

If $x \in E_+$ is fixed and $\boldsymbol{F}(\mathsf{A})$ denotes the set of finite subsets of $\mathsf{A}$ ordered by inclusion, it is easy to see that the family $(x_{n,\mathsf{H}})_{n \in \mathbb{N}, \mathsf{H} \in \boldsymbol{F}(\mathsf{A})}$ is directed ($\leqq$) with respect to the natural directions of $\mathbb{N}$ and $\boldsymbol{F}(\mathsf{A})$, and majorized by $x$; moreover,

$$x = \sup_{n,\mathsf{H}} x_{n,\mathsf{H}} \tag{1}$$

by (II.1.9). We claim: *$S$ is a t.o.s. of $E$ if and only if*

$$x = \lim_{n,\mathsf{H}} x_{n,\mathsf{H}} \tag{2}$$

*for each $x \in E_+$.* In fact, if $I_\alpha$ denotes the ideal in $E$ generated by $u_\alpha$ then the direct sum $I := \bigoplus_\alpha I_\alpha$ is the ideal generated by $S$; clearly, $x_{n,\mathsf{H}} \in I$ for all $n \in \mathbb{N}$, $\mathsf{H} \in \boldsymbol{F}(\mathsf{A})$. Thus if (2) holds, $I$ is dense in $E$. Conversely, suppose that $I$ is dense in $E$ and let $x \in E_+$, $\varepsilon > 0$ be preassigned. There exists $x_\varepsilon \in I$ such that $\|x - x_\varepsilon\| < \varepsilon$; since $I$ is an ideal, we can assume that $0 \leqq x_\varepsilon \leqq x$. Now $x_\varepsilon \in I$ implies the existence of a finite subset $\mathsf{H}$ of $\mathsf{A}$ and integers $k_\alpha > 0$ ($\alpha \in \mathsf{H}$) for which $x_\varepsilon \leqq \sum_{\alpha \in \mathsf{H}} k_\alpha u_\alpha$. Letting $n := \sup_{\alpha \in \mathsf{H}} k_\alpha$ and $z := (\sum_{\alpha \in \mathsf{H}} n u_\alpha) \wedge x$, the decomposition property (Chap. II, § 1) shows that $z = \sum_{\alpha \in \mathsf{H}} y_\alpha$ where $0 \leqq y_\alpha \leqq n u_\alpha$. Since $y_\alpha \leqq x$ it follows that $y_\alpha \leqq (n u_\alpha \wedge x)$ hence, $z \leqq \sum_{\alpha \in \mathsf{H}} (n u_\alpha \wedge x) = x_{n,\mathsf{H}}$. But $x_\varepsilon \leqq z$, so we obtain $0 \leqq x_\varepsilon \leqq x_{n,\mathsf{H}} \leqq x$ and, therefore,

$$\|x - x_{n,\mathsf{H}}\| \leqq \|x - x_\varepsilon\| < \varepsilon .$$

**Examples.** 1. If $u$ is quasi-interior to $E_+$, $\{u\}$ is a t.o.s. of $E$. But even if $E_+$ contains quasi-interior points, $E$ may possess (countably) infinite t.o.s. (Exerc. 12).

2. Suppose $E$ is a Banach lattice in which each order convergent filter converges (equivalently, such that for each directed ($\leqq$) subset $A$ of $E_+$, $\sup A = x$ implies $\lim A = x$; Chap. II, § 5). Then each maximal orthogonal system of $E$ is topological; such is the case, for instance, in $L^p(\mu)$ $(1 \leqq p < +\infty)$.

3. If $E$ is weakly sequentially complete and so by (II.5.10) Cor. 1 each majorized orthogonal subset of $E_+$ is countable, then each maximal orthogonal system of $E$ is topological. In fact, under these assumptions the family $(x_{n,\mathsf{H}})$ is a countable weak Cauchy family and hence weakly convergent to $x \in E_+$; Dini's theorem (II.5.9) then shows the validity of (2). (This is actually a special case of Example 2, since under the present assumptions $E$ is even a band in its bidual (II.10.6).)

4. There exist Banach lattices not containing any t.o.s. For example, let $X$ denote a connected locally compact space not countable at infinity (e.g., the Stone-Čech compactification of $\mathbb{R}$ less some point of $\beta\mathbb{R} \setminus \mathbb{R}$). The Banach lattice (sup-norm) $C_0(X)$ of continuous real functions on $X$ vanishing at infinity, does not contain a function strictly positive throughout $X$ (Exerc. 13). On the other hand, the connectedness of $X$ implies that any t.o.s. of $C_0(X)$ would necessarily be a singleton $\{u\}$ (cf. (5.2) below); but since $u$ has zeros in $X$, the principal ideal generated by $u$ cannot be dense.

**5.2 Proposition.** *Let $\{u_\alpha : \alpha \in \mathsf{A}\}$ be a t.o.s. of the Banach lattice E and denote by $E_\alpha$ the closed ideal generated by $u_\alpha$. Then each $E_\alpha$ is a projection band, and if $x \mapsto x_\alpha$ denotes the associated band projection,*

$$x = \sum_{\alpha \in \mathsf{A}} x_\alpha \qquad (x \in E) \tag{3}$$

*is unconditionally convergent in E. Moreover, if $\{v_\beta : \beta \in \mathsf{B}\}$ is any other t.o.s. of E, the non-zero members of the family $u_\alpha \wedge v_\beta$, $(\alpha, \beta) \in \mathsf{A} \times \mathsf{B}$, constitute a t.o.s. of E.*

*Proof.* To show that $E_\alpha$ is a projection band, by (II.2.11) Cor. 1 we must prove the existence of $x_\alpha := \sup_n (x \wedge n u_\alpha)$ for each $x \in E_+$ and each $\alpha \in \mathsf{A}$. For this it is evidently sufficient (and in fact necessary, since $u_\alpha$ is quasi-interior to $(E_\alpha)_+$, cf. (II.6.3)) that the sequence $(x_{n,\alpha})_{n \in \mathbb{N}}$, where $x_{n,\alpha} := x \wedge n u_\alpha$, be a Cauchy sequence in $E$. From (2) it follows that for given $\varepsilon > 0$, there exists $n_0 \in \mathbb{N}$ and a finite $\mathsf{H}_0 \subset \mathsf{A}$ such that $\|x_{n_1,\mathsf{H}_1} - x_{n_2,\mathsf{H}_2}\| < \varepsilon$ whenever $n_i \geqq n_0$ and $\mathsf{H}_i \supset \mathsf{H}_0$ $(i = 1, 2)$. Fixing $\alpha$ and letting $\mathsf{H}_1 = \mathsf{H}_2 = \mathsf{H}_0 \cup \{\alpha\}$, we have (using the mutual orthogonality of the elements $u_\alpha$)

$$|x_{n_1,\alpha} - x_{n_2,\alpha}| \leqq |x_{n_1,\mathsf{H}_1} - x_{n_2,\mathsf{H}_2}|$$

and hence, $\|x_{n_1,\alpha} - x_{n_2,\alpha}\| < \varepsilon$ whenever $n_1, n_2 \geqq n_0$. This shows $E_\alpha$ to be a projection band.

Similarly we have $\|\sum_{\alpha \in \mathsf{H}} x_{m,\alpha}\| < \varepsilon$ whenever $\mathsf{H} \cap \mathsf{H}_0 = \emptyset$ and $m \geqq n_0$. Thus we have

$$\|\sum_{\alpha \in \mathsf{H}} x_\alpha\| \leqq \varepsilon$$

whenever $\mathsf{H} \cap \mathsf{H}_0 = \emptyset$. This shows the unconditional convergence of $\sum_{\alpha \in \mathsf{A}} x_\alpha$;

it is clear from (2) that the sum equals $x$ ($x \in E_+$). Since the projection bands are mutually orthogonal, it is now easy to see that the expansion (3) is valid for arbitrary $x \in E$.

Writing down the expansion (3) of $x = u_\alpha$ ($\alpha$ fixed in A) with respect to the t.o.s. $\{v_\beta : \beta \in \mathsf{B}\}$ we obtain, in obvious notation,

(4) $$u_\alpha = \sum_{\beta \in \mathsf{B}} (u_\alpha)_\beta .$$

Now the relation $u_\alpha \wedge n v_\beta \leqq n(u_\alpha \wedge v_\beta)$ ($n \in \mathbb{N}$) shows that $(u_\alpha)_\beta$ is contained in the closed ideal $G_{\alpha\beta}$ generated by $u_\alpha \wedge v_\beta$; hence by (4), $u_\alpha$ is contained in the closed ideal generated by the family $(u_\alpha \wedge v_\beta)_{\beta \in \mathsf{B}}$. Since this is true for each $\alpha \in \mathsf{A}$, it follows that the non-zero members of $(u_\alpha \wedge v_\beta)_{(\alpha,\beta) \in \mathsf{A} \times \mathsf{B}}$ constitute a t.o.s. of $E$ as asserted. □

Before proving the main representation theorem we recall a few elementary facts on Radon measures. For $X$ a locally compact space, we denote by $\mathscr{K}(X)$ the vector lattice of all real-valued continuous functions on $X$ having compact support. An element $\mu$ of the order dual $\mathscr{M}(X) := \mathscr{K}(X)^\star$ is called a *Radon measure* on $X$. $\mathscr{M}(X)$ is an order complete vector lattice (Prop. II.4.2) and the topological dual of $\mathscr{K}(X)$ for the order topology, under which $\mathscr{K}(X)$ is barreled. (The order topology is the finest locally convex topology for which each order interval is bounded, or equivalently, the topology of the inductive limit with respect to the canonical injections $\mathscr{K}_K(X) \to \mathscr{K}(X)$, $K$ compact $\subset X$. Cf. Bourbaki [1965], Intégration, Chap. III, § 2, and [S, (V.6.3) Cor. 1].) This implies that boundedness and relative compactness are equivalent for the *vague topology* $\sigma(\mathscr{M}(X), \mathscr{K}(X))$.

**5.3 Theorem.** *Let $E$ be a Banach lattice possessing a topological orthogonal system.*

(i) *There exists a pair $(X, \mathsf{M})$, where $X$ is a direct sum of compact spaces and $\mathsf{M}$ is a minimal vaguely compact set of positive Radon measures on $X$, with this property: $E$ is isomorphic to the completion $\hat{E}$ of $(\mathscr{K}(X), p_\mathsf{M})$ with respect to the norm*

(5) $$p_\mathsf{M}(f) := \sup_{\mu \in \mathsf{M}} \int |f| \, d\mu ,$$

*and this completion can be identified*[1] *with the set of all continuous functions $X \to \overline{\mathbb{R}}$ which can be approximated, with respect to (5), by elements of $\mathscr{K}(X)$.*

(ii) *Each pair $(Y, \mathsf{N})$ having the properties enumerated in* (i), *is equivalent to $(X, \mathsf{M})$ in the following sense: There exists a homeomorphism of a dense open subspace $X_0 \subset X$ onto a like subspace $Y_0 \subset Y$, an isomorphism of vector lattices $\mathscr{K}(Y_0) \to \mathscr{K}(X_0)$ whose adjoint carries $\mathsf{M}_0 := \mathsf{M} | X_0$ to $\mathsf{N}_0 := \mathsf{N} | Y_0$, and the pair $(X_0, \mathsf{M}_0)$ satisfies* (i).

*Note.* If $F_\mathsf{M}$ denotes the set of all continuous functions $f: X \to \overline{\mathbb{R}}$ for which $p_\mathsf{M}(f) < +\infty$ then, since $p_\mathsf{M}$ is a norm on $\mathscr{K}(X)$, each $f \in F_\mathsf{M}$ is finite except on a rare subset of $X$ which is $\mu$-negligible for each $\mu \in \mathsf{M}$. Thus if $f_1, f_2 \in F_\mathsf{M}$

[1] See the following note. As before, $\overline{\mathbb{R}}$ denotes the two point compactification (and lattice completion) of $\mathbb{R}$.

then $f_1 - f_2$ is defined a.e. for each $\mu \in \mathsf{M}$ and $\rho(f_1, f_2) := p_{\mathsf{M}}(f_1 - f_2)$ defines a metric on $F_{\mathsf{M}}$. Then (i) asserts that the closure of $\mathscr{K}(X)$ in $(F_{\mathsf{M}}, \rho)$ is a Banach lattice $\hat{E}$ isomorphic to $E$, where it is understood that the lattice and vector operations in $\hat{E}$ agree with the corresponding pointwise operations on sets of finiteness. The implication that $\hat{E}$ is closed under these operations and a Banach lattice under the norm $p_{\mathsf{M}}$, is an essential part of the assertion. Finally, it is important to observe that $\mathscr{K}(X)$ is an ideal in $\hat{E}$ (cf. Example 7 below).

*Proof* of (5.3). (i) Using the notation of (5.2), we denote by $S = \{u_\alpha : \alpha \in \mathsf{A}\}$ a t.o.s. of $E$. Consider the set $V_S$ of all valuations $\varphi$ of $E$ such that $\varphi(u_\alpha) = 1$ for some $\alpha \in \mathsf{A}$. Under the coarsest topology making each of the maps $\varphi \mapsto \varphi(x)$ $(x \in E)$ of $V_S$ into $\overline{\mathbb{R}}$ continuous, $V_S$ is easily seen to be the direct topological sum of the compact subspaces $K_\alpha := \{\varphi \in V_S : \varphi(u_\alpha) = 1\}$ $(\alpha \in \mathsf{A})$. Moreover, since $\varphi(E_\alpha^\perp) = \{0\}$ for all $\varphi \in K_\alpha$ and since $E = E_\alpha + E_\alpha^\perp$ by (5.2), $K_\alpha$ can be identified with the structure space of $E_\alpha$ (Def. 4.2). This identification defines an isomorphism of vector lattices of $C(K_\alpha)$ onto the dense ideal $E_{u_\alpha}$ of $E_\alpha$ (cf. discussion preceding 4.1). On the other hand, $\mathscr{K}(V_S) = \bigoplus_{\alpha \in \mathsf{A}} C(K_\alpha)$ and it follows now from (5.2) that $\mathscr{K}(V_S)$ can be identified with the ideal $I(S) \subset E$ generated by $S$.

Thus, letting $X := V_S$, we have an isomorphism $h: \mathscr{K}(X) \to E$ with range $I(S)$; clearly, $h$ is continuous with respect to the order topology of $\mathscr{K}(X)$ and the norm topology of $E$. Since $h$ has dense range, its adjoint $h'$ is a bijection of $E'$ onto a vaguely dense subspace (in fact, an ideal) of $\mathscr{M}(X)$. By Bauer's theorem (II.5.7) there exists a unique minimal $\sigma(E', E)$-closed subset $P$ of $\{x' \in (E')_+ : \|x'\| \leq 1\}$ which determines the norm of $E$; that is, for which

$$\|x\| = \sup_{x' \in P} \langle |x|, x' \rangle \qquad (x \in E).$$

Its image $\mathsf{M} := h'(P)$ is vaguely compact, since $h'$ is continuous for the corresponding weak topologies. We define $p_{\mathsf{M}}$ by

$$p_{\mathsf{M}}(f) = \sup_{\mu \in \mathsf{M}} \int |f| \, d\mu \qquad (f \in \mathscr{K}(X));$$

trivially, then, $h$ is an isomorphism of $(\mathscr{K}(X), p_{\mathsf{M}})$ onto the normed vector lattice $I(S) \subset E$. Evidently, $\mathsf{M}$ is the unique minimal, vaguely compact subset of $\mathscr{M}(X)_+$ for which $h$ has this property. It is also clear that $h$ extends to an isomorphism $\overline{h}$ of the completion $\hat{E}$ of $(\mathscr{K}(X), p_{\mathsf{M}})$ onto $E$, and that $\mathscr{K}(X)$ is an ideal in the Banach lattice $\hat{E}$.

To verify the final statement of (i) we note that by (4.5), $\overline{h}^{-1}(E_\alpha)$ can be identified (as explained in the note above) with a vector lattice $\hat{E}_\alpha$ of continuous numerical functions on $K_\alpha$, in which $C(K_\alpha)$ is dense. Hence, a continuous function $f_\alpha : K_\alpha \to \overline{\mathbb{R}}$ is in $\hat{E}_\alpha$ if and only if for each $\varepsilon > 0$, there exists $g \in C(K_\alpha)$ satisfying $p_{\mathsf{M}}(f_\alpha - g) < \varepsilon$. Equation (3) of (5.2) now shows that $\hat{E} = \overline{h}^{-1}(E)$ can be identified with the vector lattice of all continuous functions $f: X \to \overline{\mathbb{R}}$ which can be approximated, in the same sense, by elements of $\mathscr{K}(X)$.

(ii) Suppose now that $(Y, \mathsf{N})$ satisfies part (i) of (5.3). In particular, then, $Y$ is a direct sum of compact spaces, $Y = \sum_{\beta \in \mathsf{B}} L_\beta$ say, and there exists an isomorphism $k$ of the normed vector lattice $(\mathscr{K}(Y), p_{\mathsf{N}})$ onto a dense ideal $J \subset E$. If $f_\beta \in \mathscr{K}(Y)$ denotes the characteristic function of $L_\beta \subset Y$ and if $v_\beta := k(f_\beta)$, it is clear that

$T:=\{v_\beta : \beta \in \mathsf{B}\}$ is a t.o.s. of $E$. Moreover, the non-zero members of the family $(u_\alpha \wedge v_\beta)_{(\alpha,\beta)\in \mathsf{A}\times\mathsf{B}}$ constitute a t.o.s. $R$ of $E$ by (5.2).

Denote by $F_\beta$ the closed ideal in $E$ generated by $v_\beta$ ($\beta \in \mathsf{B}$) and define $G_{\alpha\beta}:=E_\alpha \cap F_\beta$; then by (1.1), $G_{\alpha\beta}$ is the closed ideal (possibly $\{0\}$) generated by $u_\alpha \wedge v_\beta$. Let $\alpha \in \mathsf{A}$ be fixed. By (5.2) $E_\alpha$ and $F_\beta$ are projection bands in $E$ and hence, so is $G_{\alpha\beta}$; in particular, $G_{\alpha\beta}$ is a projection band in $E_\alpha$. Thus $E_\alpha = G_{\alpha\beta} + G_{\alpha\beta}^\perp$, and by virtue of (4.3) the respective structure spaces $M_{\alpha\beta}$ and $N_{\alpha\beta}$ of $G_{\alpha\beta}$ and $G_{\alpha\beta}^\perp$ can be identified with complementary (hence, open-and-closed) subspaces of $K_\alpha$ (cf. Exerc. 10). This implies that, for each $\alpha \in \mathsf{A}$, the direct sum $\sum_{\beta\in\mathsf{B}} M_{\alpha\beta}$ is homeomorphic with an open subspace of $K_\alpha$; this subspace is dense in $K_\alpha$, since the ideal generated by the family $(u_\alpha \wedge v_\beta)_{\beta\in\mathsf{B}}$ is dense in $E_\alpha$ (Formula (4) above). Accordingly, the direct topological sum $Z:=\sum_{\alpha,\beta} M_{\alpha\beta}$ is homeomorphic with a dense open subspace $X_0$ of $X$. By symmetry, $Z$ is also homeomorphic with a dense open subspace $Y_0$ of $Y$.

Now denote by $\tau_1 : Z \to X_0$ and by $\tau_2 : Z \to Y_0$ the homeomorphisms just constructed. Thus we obtain a homeomorphism $\tau := \tau_2 \circ \tau_1^{-1}$ of $X_0$ onto $Y_0$. By definition of $X_0, Y_0$ there exist vector lattice isomorphisms $i: \mathscr{K}(X_0) \to I(R)$ and $j: \mathscr{K}(Y_0) \to I(R)$ ($I(R)$ the dense ideal in $E$ generated by $R$), which can be factored[2] through $\mathscr{K}(X)$ and $\mathscr{K}(Y)$, respectively, via the canonical injections $\mathscr{K}(X_0) \to \mathscr{K}(X)$ and $\mathscr{K}(Y_0) \to \mathscr{K}(Y)$. The set of restrictions $\mathsf{M}_0 := \mathsf{M} | X_0$ is vaguely compact in $\mathscr{M}(X_0)$ and, by the unicity of $P$ in Bauer's theorem (II.5.7), the unique minimal such set for which $i$ is a norm isomorphism of $(\mathscr{K}(X_0), p_{\mathsf{M}_0})$ into $E$; a corresponding assertion holds for $j$ and $\mathsf{N}_0$. From this it follows that the adjoint of $j^{-1} \circ i$ must carry $\mathsf{N}_0$ onto $\mathsf{M}_0$. Finally, it is clear that the pair $(X_0, \mathsf{M}_0)$ (as well as $(Y_0, \mathsf{N}_0)$) satisfies (i). □

Theorem 5.3 suggests the following definition where, as before, $\mathscr{K}(X)$ denotes the vector lattice of all continuous real functions on $X$ having compact support.

**5.4 Definition.** *Let $E$ be a Banach lattice. A locally compact space $X$ is called a* representation space *for $E$ if $\mathscr{K}(X)$ can be identified with a dense ideal in $E$. If, in addition, $X$ is a direct sum of compact spaces*[3], *$X$ is called a* strong representation space *for $E$.*

The existence of t.o.s. (Def. 5.1) in $E$ can now be characterized by the existence of certain types of representation spaces for $E$, as follows.

**5.5 Proposition.** *These properties of a Banach lattice $E$ are equivalent:*

(a) *There exists a t.o.s. of $E$.*

(b) *There exists a strong representation space for $E$.*

(c) *There exists a paracompact representation space for $E$.*

*Moreover, there exists a bijective correspondence between the set of all t.o.s. $R$ of $E$ and the set of equivalence classes (under homeomorphism) of strong representation spaces for $E$. This correspondence is given by $R \mapsto V_R$, where $V_R$ denotes the set of all valuations $\varphi$ of $E$ such that $\varphi(u)=1$ for some $u \in R$ under the topology of simple convergence*[4].

[2] As maps into $E$.

[3] That is, if $X$ is the union of a disjoint set of compact open subspaces.

[4] The coarsest topology making each of the maps $\varphi \mapsto \varphi(x)$, $x \in E$, of $V_R$ into $\overline{\mathbb{R}}$ continuous.

*Proof*. (a)⇒(b) is immediately clear from (5.3) (i).

(b)⇒(c): We recall that a (Hausdorff) topological space $X$ is called *paracompact* if each open cover of $X$ has an open locally finite refinement (cf. Bourbaki [1961], Top. générale, Chap. I, § 9, no. 10). Thus if $X$ is a direct topological sum of compact spaces, $X$ is evidently paracompact.

(c)⇒(a): A basic theorem of general topology (Bourbaki, l.c.) asserts that if (and only if) $X$ is locally compact and paracompact then $X$ is the direct topological sum, $X=\sum_{\alpha\in\mathsf{A}} X_\alpha$ say, of locally compact spaces $X_\alpha$ countable at infinity. Suppose, then, that $X$ is a representation space for $E$ given in this form. For each $\alpha\in\mathsf{A}$, there exists a sequence of functions $f_{\alpha,n}\in\mathscr{K}(X_\alpha)_+$ such that $X_\alpha=\bigcup_{n\in\mathbb{N}}\{t: f_{\alpha,n}(t)>0\}$. Let $u_{\alpha,n}\in E$ be the element corresponding to $f_{\alpha,n}\in\mathscr{K}(X)$ under the identification $\mathscr{K}(X)\to E$ which makes $X$ a representation space for $E$. We define

$$u_\alpha=\sum_{n\in\mathbb{N}} u_{\alpha,n}/2^n\|u_{\alpha,n}\| .$$

The sum converges, since $E$ is complete and, clearly, the elements $u_\alpha\in E$ are mutually orthogonal. Since the ideal generated by this orthogonal system contains the image of $\mathscr{K}(X)$ which is a dense ideal by hypothesis, it follows that $\{u_\alpha:\alpha\in\mathsf{A}\}$ is a t.o.s. of $E$.

The second assertion is easily intelligible after a close inspection of the first part of the proof of (i) in (5.3), which can be left to the reader. ☐

The set of all t.o.s. of a Banach lattice $E$ has a natural partial order under which it is a semi-lattice, and which is reflected in the relationship of the corresponding ideals and strong representation spaces. The interested reader is referred to Exerc. 12 (d).

**Examples.** 5. Kakutani's representation theorem (II.7.4) for $AM$-spaces with unit, upon which much of the preceding theory is built, and his representation theorem (II.8.5) for $AL$-spaces both emerge as special cases of (5.3). If $E$ is an $AM$-space with unit, there exists precisely one (to within homeomorphism) representation space, namely, the structure space $K(E)$ (Def. 4.2), and the set $\mathsf{M}$ of (5.3) (i) is necessarily the set of all Dirac measures on $K(E)$. If $E$ is an $AL$-space, a compact representation space exists iff $E_+$ contains a weak order unit (and in this case, any compact representation space is homeomorphic to $K(E)$, cf. (4.1)); but even if $E_+$ contains weak order units, it can be natural to consider non-compact representation spaces. For example, if $E=l^1(\mathbb{N})$ then $\mathbb{N}$ is a more convenient representation space, for most purposes, than the structure space $\beta\mathbb{N}$. Note also that for an $AL$-space, the set $\mathsf{M}$ of (5.3) (i) is always a singleton.

6. More generally, a Banach lattice $E$ possesses a compact representation space iff $E_+$ contains quasi-interior points (which is the case, in particular, whenever $E$ is separable (II.6.2)). If $E$ has a connected representation space $X$, then either $E$ has a compact representation space ($\cong K(E)$) or $E$ has no strong representation space (Exerc. 12, 13). The former or latter case presents itself accordingly as $X$ is, or is not, countable at infinity (cf. Example 4 above).

7. Suppose $E$ is any Banach lattice, $X$ a locally compact space, and that there exists an isomorphism $j$ of $\mathscr{K}(X)$ onto a dense vector sublattice of $E$.

Simple examples (such as $X=\mathbb{R}$, $\mu$ Lebesgue measure, $E=L^1(\mu)$) show that in general, $X$ is not a representation space for $E$ (Exerc. 13). *If for each* $f\in\mathscr{K}(X)$, $j([-f,f])$ *is closed in E, then X is a representation space for E.* In fact, the bipolar theorem [S, IV.1.5 p. 126] shows the closure of $j([-f,f])$ in $E$ to be $(-j(f)+E_+)\cap(j(f)-E_+)$ so that, under the hypothesis made, the range of $j$ is an ideal in $E$. A similar observation applies, of course, to the completion of $(\mathscr{K}(X),p)$ where $p$ is any lattice norm on $\mathscr{K}(X)$.

We conclude this section with a characterization of order completeness properties of Banach lattices possessing t.o.s., through topological properties of any strong representation space. This extends the characterization (II.7.7 Cor.) given earlier for spaces $C(K)$. Recall that a topological space is called Stonian (respectively, quasi-Stonian) if each open set (respectively, each open $F_\sigma$-set) has open closure. For direct sums of compact spaces, these properties are reflected by the corresponding property of each summand.

**5.6 Proposition.** *A Banach lattice E possessing a t.o.s. is order complete (countably order complete) if and only if any strong representation space for E is Stonian (quasi-Stonian).*

*Proof.* The condition is necessary. Assume that $E$ is order complete and denote by $X=\sum_{\alpha\in\mathrm{A}}K_\alpha$ any strong representation space for $E$ ($K_\alpha$ compact). Identifying $\mathscr{K}(X)$ with a dense ideal in $E$, we denote by $I_\alpha$ the principal ideal in $E$ generated by the characteristic function $u_\alpha$ of $K_\alpha$. Then $I_\alpha$ can be identified with $C(K_\alpha)$ (cf. discussion preceding 4.1); since $I_\alpha$ is order complete, $K_\alpha$ is a Stonian space by (II.7.7 Cor.). The proof for $E$ countably order complete is analogous.

The condition is sufficient. First we prove that a Banach lattice with quasi-interior positive elements and a quasi-Stonian structure space $K$, is countably order complete. Let $B\subset E_+$ be a countable, directed ($\leqq$) set majorized by $u\in E_+$; in view of Prop. II.6.2 we can assume that $u$ is quasi-interior to $E_+$. Since $E_u$ is isomorphic with $C(K)$ (Def. 4.2), it follows from (II.7.7) that $\sup B$ exists in $E_u$ and hence, since $E_u$ is an ideal in $E$, $\sup B$ exists in $E$. Assume now that $E$ is a Banach lattice possessing a t.o.s. and that $X=\sum_{\alpha\in\mathrm{A}}K_\alpha$ is a strong representation space for $E$ with each $K_\alpha$ compact quasi-Stonian. As before we denote by $\{u_\alpha\colon\alpha\in\mathrm{A}\}$ the t.o.s. of $E$ corresponding to the set of characteristic functions of $K_\alpha\subset X$, and by $E_\alpha$ the closed ideal in $E$ generated by $u_\alpha$. By the preceding, each $E_\alpha$ is countably order complete. Let $A\subset E_+$ be a countable, directed ($\leqq$) set majorized by $w\in E_+$. If $w=\sum_\alpha w_\alpha$ is the expansion of $w$ according to (3) of (5.2), the order interval $[0,w_\alpha]$ is countably order complete, since it is contained in $E_\alpha$. The image $A_\alpha$ of $A$ under the band projection $E\to E_\alpha$ (cf. 5.2) is countable, directed ($\leqq$) and majorized by $w_\alpha$. Hence

$$x_\alpha := \sup A_\alpha \in E_\alpha$$

exists and we have, for any $v\in A$, $v_\alpha\leqq x_\alpha\leqq w_\alpha$ ($\alpha\in\mathrm{A}$). Since $0\leqq x_\alpha-v_\alpha\leqq w_\alpha-v_\alpha$ and $v=\sum_\alpha v_\alpha$, $w-v=\sum_\alpha w_\alpha-v_\alpha$ are unconditionally convergent in $E$ by (5.2), $\sum_\alpha x_\alpha$ is unconditionally Cauchy and hence, unconditionally convergent in $E$.

It is straightforward that $x:=\sum_\alpha x_\alpha$ equals $\sup A$. Thus $E$ is countably order complete. The proof for the case where $X$ is a Stonian space and hence $E$ order complete will be omitted, since it contains no new idea. ☐

## § 6. The Stone Approximation Theorem

*Throughout this section, E will denote a Banach lattice possessing a topological orthogonal system* (or equivalently, a strong representation space). We are concerned with the characterization of closed ideals under a given representation $E \to \hat{E}$ (in the sense of 5.3), and with the representation of quotients and certain vector sublattices, thus extending results obtained in § 4 for Banach lattices with quasi-interior positive elements. This will lead naturally to an extension of Stone's approximation theorem (II.7.3).

We point out that, in general, a closed ideal $I \subset E$ does not inherit from $E$ the property of possessing a t.o.s. (In fact, the space $C_0(X)$ of § 5, Example 4 can be viewed as a maximal ideal in $C(\dot{X})$ where $\dot{X}$ denotes the one-point compactification of $X$.) On the other hand, quotients $E/I$ always do; as might be expected, there exists a bijective correspondence between $\boldsymbol{A}(E)$ (the lattice of closed ideals in $E$) and a lattice of closed subsets of $X$, for any strong representation space $X$.

From (5.5) we recall that without loss of generality a strong representation space for $E$ can be assumed to be of the form $V_S$, $S$ denoting a suitable t.o.s. of $E$. This notation will be employed throughout the present section.

**6.1 Proposition.** *Let $X:=V_S$ be a strong representation space for E. Then the mapping $I \mapsto X_I:=\{\varphi \in V_S: \varphi(I)=\{0\}\}$ is an anti-isomorphism of $\boldsymbol{A}(E)$ onto a sublattice of the lattice of closed subsets of $X$. Moreover, $X_I$ is a strong representation space for E/I.*

*Proof.* Let $S=\{u_\alpha: \alpha \in \mathsf{A}\}$, and let $x=\sum_\alpha x_\alpha$ according to (3) of (5.2). Since $I$ is a closed ideal in $E$, $x \in I$ is equivalent to the set of relations $x_\alpha \in I \cap E_\alpha$ $(\alpha \in \mathsf{A})$. Moreover, $X=\sum_\alpha K_\alpha$ where $K_\alpha$ is the compact space of valuations $\varphi$ of $E$ satisfying $\varphi(u_\alpha)=1$ (§ 4). Defining $X_{I,\alpha}:=\{\varphi \in K_\alpha: \varphi(I \cap E_\alpha)=\{0\}\}$, we recall from the proof of (4.3) that $x_\alpha \in I \cap E_\alpha$ if and only if $\varphi(x_\alpha)=0$ for all $\varphi \in X_{I,\alpha}$. Thus if we set $X_I=\bigcup_\alpha X_{I,\alpha}$ then $X_I$ is the unique closed subset of $X$ such that $x \in I$ is equivalent to $\varphi(x)=0$ for all $\varphi \in X_I$. Moreover, it is clear that $X_{I+J}=X_I \cap X_J$ for any pair $I,J$ of closed ideals; $X_{I \cap J}=X_I \cup X_J$ follows from (2.2) (*e*), since for each $\varphi \in V_S$, $\varphi^{-1}(0)$ is a prime ideal of $E$ (3.2). Hence, $I \mapsto X_I$ is an anti-isomorphism of lattices as asserted.

Furthermore, if $q: E \to E/I$ denotes the canonical map then $R=\{q(u): u \in S, u \notin I\}$ is a t.o.s. for $E/I$ (cf. Exerc. 12), and the set $V_R$ of valuations $\psi$ of $E/I$ such that $\psi(v)=1$ for some $v \in R$, corresponds bijectively to $X_I$ by virtue of $\psi \mapsto \varphi=\psi \circ q$. Clearly, $\psi \mapsto \varphi$ is continuous from $V_R$ onto $X_I$ (cf. 5.5) and hence, since $V_R$ and $X_I$ are sums of compact spaces, a homeomorphism. ☐

Thus if $X$ is a strong representation space for $E$ and $E \to \hat{E}$ is an associated representation, in the sense of (5.3), of $E$ by continuous numerical functions on $X$, then each closed ideal $I \subset E$ corresponds to the ideal of all $f \in \hat{E}$ vanishing on $X_I$. As has been pointed out earlier, in general not each closed subset of $X$ determines a closed ideal of $E$ in this manner (Exerc. 10, 15); note also that $\{X_I : I \in A(E)\}$ is a Boolean algebra of (open and) closed subsets of $X$ iff each $I \in A(E)$ is a projection band (Exerc. 10). Finally, if $(X, \mathsf{M})$ is a representation pair for $E$ in the sense of (5.3) and if $\mathsf{M}_I$ denotes the vaguely compact set in $\mathcal{M}(X_I)$ obtained by restricting each $\mu \in \mathsf{M}$ to $X_I$, then the norm of $[f] \in \hat{E}/\hat{I}$ is given by

$$\sup_{\nu \in \mathsf{M}_I} \int |f| \, d\nu \qquad (f \in [f]);$$

however, simple examples show that $\mathsf{M}_I$ need not be a *minimal* vaguely compact set with this property.

The following result is a simple generalization of (4.4) to the present case where $E$ is a Banach lattice possessing a t.o.s. which is not necessarily a singleton.

**6.2 Proposition.** *Let $H$ be a closed vector sublattice of $E$, denote by $j: H \to E$ the canonical injection, and suppose that $H$ contains a t.o.s. $S$ of $E$. Then the restriction map $\varphi \mapsto \varphi \circ j$ defines a continuous surjection of the strong representation space $V_S(E)$ for $E$ onto the strong representation space $V_S(H)$ for $H$.*

This is an immediate consequence of (4.4) and the direct sum decompositions $V_S(E) = \sum_{u \in S} K_u(E)$ and $V_S(H) = \sum_{u \in S} K_u(H)$, respectively.

The analogous extension of the corollary of (4.4) is equally valid, but we prefer to present this extension in a somewhat different form which makes its relationship to the classical Stone approximation theorem (II.7.3) more transparent. This transparency is enhanced by isolating the following lemma.

**6.3 Lemma.** *Let $X = \sum_{\alpha \in \mathsf{A}} K_\alpha$ be a direct sum of compact spaces, and let $H$ be a vector sublattice of $\mathcal{K}(X)$ which separates the points of $X$ and contains the characteristic function of each $K_\alpha$. Then $H$ is dense in $\mathcal{K}(X)$ for the order topology.*

*Proof.* Each $f \in \mathcal{K}(X)$ is contained in some $C(K_{\mathsf{H}})$, where $K_{\mathsf{H}} := \sum_{\alpha \in \mathsf{H}} K_\alpha$ and $\mathsf{H}$ is a suitable finite subset of $\mathsf{A}$. The assumption implies that $H \cap C(K_{\mathsf{H}})$ is a vector sublattice of $C(K_{\mathsf{H}})$ which separates the points of $K_{\mathsf{H}}$ and contains the constant functions; thus by (II.7.3) $H \cap C(K_{\mathsf{H}})$ is dense in $C(K_{\mathsf{H}})$ (for the usual norm topology). Since the order topology of $\mathcal{K}(X)$ is the finest locally convex topology under which each of the injections $C(K_{\mathsf{H}}) \to \mathcal{K}(X)$ is continuous (cf. discussion preceding 5.3), it follows that $H$ is dense in $\mathcal{K}(X)$. ◻

**6.4 Theorem.** *Let $E$ be a Banach lattice, and let $H$ be a vector sublattice whose closure contains a t.o.s. $S$ of $E$. If $H$ separates the points of $V_S(E)$ then $H$ is dense in $E$.*

*Proof.* Obviously we can assume $H$ closed hence, $S \subset H$. Identifying $\mathcal{K}(V_S(E))$ with a dense ideal in $E$ we see that $H \cap \mathcal{K}(V_S(E))$ satisfies the assumptions of (6.3) and so is dense in $\mathcal{K}(V_S(E))$ for the order topology and *a fortiori* in $E$, since the injection $\mathcal{K}(V_S(E)) \to E$ is continuous. ◻

*Remark.* The hypothesis made in (6.4) that $H$ separate the points of $V_S$, is unnecessarily strong. In fact, this condition implies that $H \cap \mathcal{K}(V_S)$ is dense in $\mathscr{K}(V_S)$ for the order topology which, unless $E$ is an $AM$-space, is strictly finer than the topology induced by $E$. For example, consider $E:=l^1(\mathbb{N})$. If the structure space of $E$ is identified with $\beta\mathbb{N}$ (the Stone-Čech compactification of $\mathbb{N}$) and $\mu$ denotes the measure on $\beta\mathbb{N}$ for which $\mu(\{n\})=2^{-n}$ $(n\in\mathbb{N})$ and $\mu(\beta\mathbb{N}\setminus\mathbb{N})=0$, then $E$ can be represented as the Banach lattice of all continuous, numerical functions on $\beta\mathbb{N}$ which are $\mu$-integrable. The sublattice of functions which are either constant or have their support contained in $\mathbb{N}$, is dense and contains the unit of $C(\beta\mathbb{N})$, but it does not separate the points of $\beta\mathbb{N}\setminus\mathbb{N}$. The reason for this phenomenon lies, of course, in the fact that the measure $\mu$ ignores $\beta\mathbb{N}\setminus\mathbb{N}$. A more appropriate criterion for the denseness of a vector sublattice $H$ must, therefore, take into account certain rare subsets of a strong representation space $X$ which are inessential for the set $\mathsf{M}\subset\mathscr{M}(X)_+$ defining the norm of $E$ (see 5.3). Even without explicit reference to $\mathsf{M}$, this can be done as follows.

Let $V_S$ be a strong representation space for $E$ (cf. 5.5). A set $N\subset V_S$ is called *E-negligible* if the ideal $\{x\in E: \varphi(x)=0 \text{ for all } \varphi\in N\}$ is dense in $E$. A subset $A$ of $E$ is said *to separate* $V_S$ *E-essentially* if for each pair $M_1, M_2$ of disjoint closed subsets of $V_S$, there exists an $E$-negligible set $N$ such that $A$ separates $M_1\setminus N$ from $M_2\setminus N$ (that is, for each $\varphi_1\in M_1\setminus N$ and $\varphi_2\in M_2\setminus N$, there exists $x\in A$ satisfying $\hat{x}(\varphi_1)\neq\hat{x}(\varphi_2)$, or equivalently, satisfying at least one of the relations $\varphi_1(x^+)\neq\varphi_2(x^+)$, $\varphi_1(x^-)\neq\varphi_2(x^-)$). Then the following theorem, which is an essential improvement of (6.4), can be shown to hold.

**Theorem.** *Let $E$ be a Banach lattice, and let $H$ be a vector sublattice containing a t.o.s. $S$ of $E$. $H$ is dense in $E$ if and only if $H$ separates $V_S$ E-essentially.*

For further details and an outline of the proof, the interested reader is referred to Exerc. 15.

## § 7. Mean Ergodic Semi-Groups of Operators

The earliest results in the now vast area called ergodic theory, and those with the greatest intuitive appeal, are concerned with a certain type of asymptotic behavior of semi-groups of linear operators on Banach lattices (cf. the introductory treatment given in Chap. I, Sections 3—5 and 9). This aspect of operator theory on Banach lattices is discussed in several sections of the present chapter. For a fruitful investigation we need several basic and non-elementary results which are independent of order structures, on multiplicative semi-groups of operators on a Banach space; these are given in the present section (cf. 7.6, 7.9, and 7.11). For our purposes, a convenient and sufficiently general approach is to consider semi-groups imbedded in a locally convex algebra.

A *locally convex algebra* $A$ is a (Hausdorff) locally convex space $A_0$ together with a bilinear map $A_0\times A_0\to A_0$, called *multiplication*, which is assumed to be separately continuous and denoted by $(a,b)\mapsto ab$ [S, Chap. III, § 5]. By a *semi-group* $S$ in $A$ we understand a non-void subset $S$ of $A$ such that $SS\subset S$. The notions *abelian, unit, zero* refer to the multiplicative structure of $S$ induced by that of $A$; we point out that if a semi-group $S$ in $A$ has a unit or zero, respectively, then this element will, in general, not be a unit or zero of $A$ (see examples below). Primarily we have in mind the case where $A=\mathscr{L}_s(E)$, the algebra of continuous endomorphisms of a locally convex space $E$, endowed with the topology of simple convergence (the strong operator topology, cf. [S, Chap. III, § 3]). Note that if $S$ is a semi-group in a locally convex algebra $A$, then so are the convex hull $\operatorname{co} S$ and the closed convex hull $\overline{\operatorname{co}}\, S$; both are abelian iff $S$ is.

**7.1 Definition.** *A semi-group $S$ in a locally convex algebra $A$ is called* mean ergodic *if the semi-group $\overline{\operatorname{co}}\, S$ has a zero element.*

In the particular case where $E$ is a Banach space and $A=\mathscr{L}(E)$, a semi-group $S\subset A$ is sometimes called *strongly mean ergodic* (respectively, *uniformly mean ergodic*) if $S$ is mean ergodic in $\mathscr{L}_s(E)$ (respectively, in $\mathscr{L}_b(E)$). Mean ergodicity of $S$ for the strong and weak operator topologies are equivalent (recall that the weak operator topology is $\sigma(\mathscr{L}(E), E\otimes E')$), since for both topologies the dual of $\mathscr{L}(E)$ can be identified with $E\otimes E'$ [S, IV.4.3 Cor. 4] and hence, $\overline{\operatorname{co}}\, S\subset\mathscr{L}_s(E)$ agrees with $\overline{\operatorname{co}}\, S\subset\mathscr{L}_\sigma(E)$.

**Examples.** 1. The semi-group $D_n\subset\mathscr{L}(\mathbb{R}^n)$ of doubly stochastic $n\times n$-matrices is mean ergodic with zero element $J_n=n^{-1}(e\otimes e)$ (Chap. I, § 5); in fact, $J_n A=A J_n=J_n$ for all $A\in D_n$. $D_n$ is convex and compact in $\mathscr{L}_s(\mathbb{R}^n)$ and by (I.5.2), the convex hull of the group $P_n$ of all $n\times n$ permutation matrices (cf. 7.9 below). By contrast, the semi-group $S_n$ of all stochastic $n\times n$-matrices is not mean ergodic but each abelian subsemi-group is (cf. 7.6 below).

2. Let $E$ be any Banach space, let $T\in\mathscr{L}(E)$, and denote by $S$ the cyclic semi-group $\{T^n\colon n\in\mathbb{N}\}$ generated by $T$. If $S$ is relatively compact in $\mathscr{L}_s(E)$ (that is, if for each $x\in E$ the orbit $Sx$ is relatively compact in $E$), then $S$ is mean ergodic in $\mathscr{L}_s(E)$. In fact, it is easy to extend the proof of (I.3.1) to show that the sequence of averages

$$M_k:=k^{-1}(I+T+\cdots+T^{k-1})\qquad(k\in\mathbb{N})$$

converges strongly to a projection $P\in\mathscr{L}(E)$ which acts as the zero of $\overline{\operatorname{co}}\, S$; if the sequence $(T^k)_{k\in\mathbb{N}}$ converges, then $P=\lim_k T^k$. A trivial case of mean ergodicity of $S$ presents itself if the spectral radius of $T$ is $<1$ for, in this case, $\lim_k\|T^k\|=0$. A less trivial case is at hand if $r(T)=1$, $S$ is bounded and some power of $T$ is compact; in this case, $P\neq 0$ if and only if 1 is an eigenvalue of $T$, and $S$ is uniformly mean ergodic.

3. The hypothesis of the preceding example can be considerably weakened, as follows. We assume that $S=\{T^k\colon k\in\mathbb{N}\}$ is a bounded cyclic semi-group in $\mathscr{L}(E)$ ($E$ a Banach space) and that for each $x\in E$, the sequence $(M_k x)$ has a weak cluster point $x_0$. Suppose that $\|T^k\|\leqq c$ for all $k\in\mathbb{N}$, then from $(I-T)M_k=k^{-1}(I-T^k)$ it follows that

$$(\star)\qquad \|(I-T)M_k\|\leqq k^{-1}(1+c)\qquad(k\in\mathbb{N}).$$

Since $T$ is weakly continuous and norm closed convex subsets of $E$ are weakly closed, $(\star)$ implies that $Tx_0 = x_0$. Since $I - M_k = k^{-1} \sum_{l=1}^{k} (I - T^l) = (I - T) R_k$ for some $R_k \in \mathscr{L}(E)$, $x - x_0$ is in the closure $N$ of the vector subspace $(I - T)E$. The estimate $(\star)$ and an equicontinuity argument show that $\lim_k \|M_k y\| = 0$ for each $y \in N$. Since $M_k x_0 = x_0$ and hence $M_k x = x_0 + M_k(x - x_0)$, it follows that

$$\lim_k \|M_k x - x_0\| = 0 .$$

Thus $(M_k)$ converges in $\mathscr{L}_s(E)$ to an operator $P$ which, in view of $(\star)$, satisfies $TP = PT = P$; this implies $M_k P = P$ for all $k$ and hence, $P^2 = P$. More generally, we have $RP = PR = P$ for each $R \in \overline{\text{co}}\, S$, so $S$ is strongly mean ergodic. The hypothesis made in this example is, in particular, satisfied whenever $S = \{T^k\}$ is bounded and some power of $T$ is weakly compact[1]; the latter condition is automatically satisfied if $E$ is a reflexive Banach space. See also Exerc. 16.

**7.2 Proposition.** *Let $E$ be any locally convex space, and let $S \subset \mathscr{L}_s(E)$ be a mean ergodic semi-group with zero element $P$. Then $P$ is a (continuous) projection onto the fixed space of $S$ with kernel the closed linear hull of the set $\{(I - T)x \colon T \in S,\ x \in E\}$.*

*Proof*. From the defining property of $P$ it follows that $RP = PR = P$ for all $R \in \overline{\text{co}}\, S \subset \mathscr{L}_s(E)$. In particular, $P^2 = P$ hence, $P$ is a (continuous) projection commuting with $S$. Since for each $x \in E$, the map $T \mapsto Tx$ from $\mathscr{L}_s(E)$ into $E$ is linear and continuous, it follows that $Px \in \overline{\text{co}}\, Sx$. Thus if $x \in E$ is fixed under $S$ we have $\overline{\text{co}}\, Sx = \{x\}$ hence, $Px = x$. Conversely, $Px = x$ implies $Tx = TPx = Px = x$ for each $T \in S$, so $x$ is fixed under $S$. Finally, if $N$ denotes the closure of $\{(I - T)x \colon T \in S,\ x \in E\}$, it is clear that $PN = \{0\}$; on the other hand, $Px = 0$ implies $x = (I - P)x$ hence, $x \in N$. ▯

We observe that if $S \subset \mathscr{L}_s(E)$ is mean ergodic then the adjoint semi-group $S' := \{T' \colon T \in S\} \subset \mathscr{L}(E'_\sigma)$, which is evidently anti-isomorphic to $S$, is mean ergodic with respect to the topology $\sigma(\mathscr{L}(E'_\sigma), E' \otimes E)$. Special cases where stronger assertions can be made, are (i) If $E$ is a reflexive locally convex space [S, Chap. IV, § 5], then $S'$ is mean ergodic in $\mathscr{L}_s(E'_\beta)$, and (ii) If $E$ is a Banach space and $S$ is uniformly mean ergodic, then $S'$ is uniformly mean ergodic in $\mathscr{L}(E'_\beta)$. More generally, if $E$ is a Banach space and $S$ is a bounded mean ergodic semi-group in $\mathscr{L}_s(E)$ with zero element $P \in \overline{\text{co}}\, S$, then the Banach-Steinhaus theorem [S, III.4.6 p. 86] shows that the zero $P'$ of $S'$ is adherent to co $S'$ in $\mathscr{L}_b(E'_\kappa)$ (topology of uniform convergence on the bounded subsets of $E'$, $E'$ being endowed with the topology $\kappa(E', E)$ of compact convergence; cf. [S, Chap. III, § 3]).

Let us note also that the dual of the topological vector subspace $PE$ of $E$ (i.e., the fixed space of $S$) can be identified with $P'E'$ (i.e., the fixed space of $S'$); in fact, $(PE)'$ is canonically isomorphic with $E'/(PE)^\circ$, and $(PE)^\circ = (P')^{-1}(0)$ is a topological supplement of $P'E'$ in $E'$. For later reference we record without proof the following simple result; $E$ denotes a locally convex space.

[1] A linear map $T \colon E \to F$ ($E, F$ locally convex spaces) is called *weakly compact* if for each bounded $B \subset E$, $T(B)$ is relatively $\sigma(F, F')$-compact.

**7.3 Proposition.** *If $S$ is a mean ergodic semi-group in $\mathscr{L}_s(E)$ and $F$ is a closed $S$-invariant vector subspace of $E$, then the operator semi-groups induced by $S$ on $F$ and $E/F$ are mean ergodic in $\mathscr{L}_s(F)$ and $\mathscr{L}_s(E/F)$, respectively.*

The examples considered so far suggest that compactness in one form or another is needed to prove mean ergodicity of a semi-group $S$; however, the semi-group $S_n$ of stochastic $n\times n$-matrices (which is not mean ergodic; cf. Chap. I, § 4 and Example 1 above) serves well to illustrate the need for some additional hypothesis. We proceed to show that compactness of $S$ suffices for the desired conclusion if $S$ is abelian or if $S$ is a topological group.

**7.4 Lemma.** *Let $A$ be a locally convex algebra, and let $S$ be a compact, convex subset of $A$ which is a semi-group with unit $u$. If $t\in S$ is invertible in $S$ then $t$ is an extreme point of $S$.*

*Proof.* By the Krein-Milman theorem [S, II.10.4] we have $S=\overline{\text{co}}\,\mathsf{E}(S)$, where $\mathsf{E}(S)$ denotes the set of extreme points of $S$. We show first that $u\in\mathsf{E}(S)$. Suppose $u=\lambda x+\mu y$ where $x,y\in S$ and where $\lambda,\mu>0$, $\lambda+\mu=1$. If $z\in\mathsf{E}(S)$ then $z=zu=\lambda zx+\mu zy$ implies that $zx=zy$. Thus $vx=vy$ for all $v\in\text{co}\,\mathsf{E}(S)$ and the separate continuity of multiplication shows that $ux=uy$, that is, $x=y$. Therefore, $u\in\mathsf{E}(S)$. Now let $t\in S$ have an inverse $t^{-1}\in S$, and suppose $t=\lambda x+\mu y$. Then $u=t^{-1}t=\lambda t^{-1}x+\mu t^{-1}y$ and, since $u\in\mathsf{E}(S)$, it follows that $t^{-1}x=u=t^{-1}y$. This implies $x=t=y$, and so $t\in\mathsf{E}(S)$. □

A well known theorem of the theory of abstract semi-groups (cf. Berglund-Hofmann [1967]) asserts that each compact semi-group $S$ contains a minimal two-sided ideal $K$; $K$ is closed, and called the *kernel* (or *Suskevič kernel*) of $S$. (A two-sided ideal is a non-void subset $J$ of $S$ satisfying $JS\cup SJ\subset J$.) If $S$ is abelian, $K$ is a group; we need the following specialized version of this result.

**7.5 Lemma.** *Let $A$ be a locally convex algebra, and let $S$ be a (non-empty) compact subset of $A$ which is an abelian semi-group. There exists a smallest closed semi-group ideal $K$ in $S$, and $K$ is a group.*

*Proof.* Consider the set $\boldsymbol{J}$ of all closed semi-group ideals in $S$. $\boldsymbol{J}$ is not empty (since $S\in\boldsymbol{J}$) and has the finite intersection property; in fact, if $J_1,\dots,J_n$ is a finite subset of $\boldsymbol{J}$ then $J_1\dots J_n\subset\bigcap_{i=1}^n J_i$, since $S$ is abelian. Thus $K:=\bigcap_{J\in\boldsymbol{J}}J$ is non-empty, since $S$ is compact, and evidently the smallest closed ideal in $S$.

To show that $K$ is a group note that for each $x\in S$, $xK$ is an ideal contained in $K$ and compact (hence closed) by the separate continuity of multiplication; thus $xK=K$, since $K$ is minimal. In particular, $xK=K$ for each $x\in K$, and it suffices to prove that $K$ possesses a unit $e$. To this end, fix $x_0\in K$; since $x_0K=Kx_0=K$, there exists $e\in K$ such that $ex_0=x_0$. If $x\in K$ is arbitrary, there exists $y\in K$ such that $x_0y=x$. Now $ex=ex_0y=x_0y=x$, so $e$ is the unit of $K$. □

**7.6 Theorem.** *Let $S$ be an abelian semi-group in the locally convex algebra $A$ and suppose the convex closure of $S$ to be compact. Then $S$ is mean ergodic.*

*Proof.* Without loss of generality we suppose $S$ to be convex and compact. By Lemma (7.5), the kernel $K$ of $S$ is a group. Let $x\in K$. Then $xS$ is compact by separate continuity of multiplication, convex, and a semi-group ideal contained in $K$. Hence $xS=K$ which shows $K$ to be convex. By Lemma (7.4), each $x\in K$ is an extreme point of $K$. Hence $K$ is a singleton $\{p\}$, and from $KS=K$ it follows that $pS=\{p\}$. Hence $p$ is the zero element of $S$. □

**Corollary.** *Let $E$ be a Banach space, and let $S\subset\mathscr{L}(E)$ be an abelian semi-group which is relatively compact for the weak operator topology. Then $S$ is strongly mean ergodic.*

*Proof.* Since $\mathscr{L}_s(E)$ and $\mathscr{L}_\sigma(E)$ ($\sigma$ the weak operator topology $\sigma(\mathscr{L}(E), E'\otimes E)$) have the same dual $E'\otimes E$, and since each closed bounded subset of $\mathscr{L}_s(E)$ is complete by virtue of the Banach-Steinhaus theorem [S, III.4.6], it follows from Krein's theorem [S, IV.11.5 p. 189] that $\overline{\text{co}}\,S$ is compact in $\mathscr{L}_\sigma(E)$. Thus (7.6) applies with $A=\mathscr{L}_\sigma(E)$. But $\overline{\text{co}}\,S$ is also the convex closure of $S$ in $\mathscr{L}_s(E)$ hence, $S$ is strongly mean ergodic. □

Our second result is analogous to (7.6) but the commutativity hypothesis is, roughly speaking, replaced by the assumption that $S$ be a compact group. Again, we need two lemmas.

**7.7 Lemma.** *Let $A$ be a locally convex algebra, and let $S\subset A$ be a semi-group which is convex and compact. For each $s\in S$, the mappings $x\mapsto sx$ and $x\mapsto xs$ $(x\in S)$ have a common fixed point in $S$.*

*Proof.* The closed cyclic subsemi-group $S_0$ of $S$ generated by $s$ is abelian. Thus by (7.6) $S_0$ has a zero element $p$, which is evidently a common fixed point of the mappings in question. □

**7.8 Lemma.** *Consider $\mathbb{R}^m$ under the Euclidean norm*

$$\|x\|=(\textstyle\sum_{j=1}^m |\xi_j|^2)^{1/2}.$$

*If $x=n^{-1}(x_1+x_2+\cdots+x_n)$ and $\|x\|=\|x_i\|$ $(i=1,\dots,n)$, then $x_i=x$ for all $i$.*

*Proof.* Expanding the inner product

$$(x+n^{-1}(\textstyle\sum_{i=1}^n x_i)\,|\,x+n^{-1}(\sum_{i=1}^n x_i))$$

and using Schwarz' inequality one obtains $(x_i|x)=(x|x)$ for all $i$ which implies $x_i=x$ $(i=1,\dots,n)$. □

**7.9 Theorem.** *Let $A$ be a locally convex algebra and let $G$ be a compact group in $A$. Suppose, moreover, that $S:=\overline{\text{co}}\,G$ is compact and that multiplication is jointly continuous in $S$. Then $S$ is mean ergodic.*

*Proof.* Denote by $\mathsf{A}$ any finite subset of $G$ and denote by $\mathsf{B}$ any finite set of continuous linear forms on $A$. Further, let $M(\mathsf{A},\mathsf{B})$ denote the set of all pairs $(x,g)\in S\times G$ such that $\langle xag,b\rangle=\langle xg,b\rangle$ for all $a\in\mathsf{A}$ and $b\in\mathsf{B}$. Suppose first that for each $(\mathsf{A},\mathsf{B})$, $M(\mathsf{A},\mathsf{B})$ is non-void. Evidently, $M(\mathsf{A},\mathsf{B})$ is closed in $S\times G$ (product topology) and $\mathsf{A}\subset\mathsf{A}_1$, $\mathsf{B}\subset\mathsf{B}_1$ implies $M(\mathsf{A}_1,\mathsf{B}_1)\subset M(\mathsf{A},\mathsf{B})$, so this family of sets has the finite intersection property. Since $S\times G$ is compact, this implies the existence of a pair $(x_0, g_0)$ belonging to each set $M(\mathsf{A},\mathsf{B})$. Since the locally convex space $A_0$ underlying $A$ is Hausdorff (by definition), the Hahn-Banach theorem shows that $x_0 a g_0 = x_0 g_0$ for all $a \in G$ and hence, $G$ being a group, it follows that $x_0 a = x_0$ for all $a \in G$. This implies evidently that $x_0$ is a left zero of $S$. In like manner it is shown that $S$ possesses a right zero $y_0$; consequently, $x_0 = y_0$ is the zero of $S$ so $S$ is mean ergodic.

It remains to show that $M(\mathsf{A},\mathsf{B})$ is always non-void. Let $\mathsf{A} \subset G$, $\mathsf{B} \subset A_0'$ be non-void finite sets so that $\mathsf{A} = \{a_1, \ldots, a_n\}$ and $\mathsf{B} = \{b_1, \ldots, b_m\}$, say. By (7.7) there exists $x \in S$ such that $x = n^{-1} \sum_{i=1}^n x a_i$. Further, denote by $F$ the continuous function $S \to \mathbb{R}^m$ defined by $z \mapsto F(z) := (\langle z, b_1 \rangle, \ldots, \langle z, b_m \rangle)$. Since $G$ is compact, there exists an element $h \in G$ for which the continuous function $g \mapsto \|F(xg)\|$ (Euclidean norm of $\mathbb{R}^m$) assumes its maximum. By the defining property of $x$ we have

$$F(xh) = n^{-1} \textstyle\sum_{i=1}^n F(x a_i h).$$

By the triangle inequality and the choice of $h$ we obtain $\|F(x a_i h)\| = \|F(xh)\|$ for all $i$, and so Lemma (7.8) implies $F(x a_i h) = F(xh)$ $(i = 1, \ldots, n)$. Therefore, $(x, h) \in M(\mathsf{A},\mathsf{B})$. □

**Corollary 1.** *If $E$ is a Banach space, every compact group $G \subset \mathscr{L}_s(E)$ is mean ergodic.*

*Proof.* Since the closed bounded subsets of $\mathscr{L}_s(E)$ are complete (theorem of Banach-Steinhaus), it follows that $S := \overline{\mathrm{co}}\, G$ is compact in $\mathscr{L}_s(E)$, cf. [S, II.4.3]. In order to apply (7.9) with $A = \mathscr{L}_s(E)$, we only need to verify that multiplication is jointly continuous in $S$. But $S$ is bounded in $\mathscr{L}_s(E)$, hence norm bounded in $\mathscr{L}(E)$, and the joint continuity of multiplication $(R, T) \mapsto RT$ for the strong operator topology on norm bounded (or equivalently, equicontinuous) subsets of $\mathscr{L}(E)$ is readily inferred from the identity

$$RT - R_0 T_0 = R(T - T_0) + (R - R_0) T_0$$

where $R, R_0, T, T_0 \in \mathscr{L}(E)$. □

*Remark.* The preceding corollary is valid under the formally weaker assumption that $G \subset \mathscr{L}(E)$ is a group compact for the weak operator topology; this entails the compactness of $G \subset \mathscr{L}_s(E)$ (cf. Berglund-Hofmann [1967]; see also § 11).

The following fixed point theorem due to Kakutani can be proved more directly (cf. [DS, V.10.8 p. 457]) but the proof given below exhibits its close relationship to (7.9).

**Corollary 2.** *Let $C$ be a non-void compact convex subset of a locally convex space*[2]*, and let $G$ be an equicontinuous group of affine maps of $C$ into $C$. There exists a point $x_0$ in $C$ which is fixed under $G$.*

*Proof.* If $A(C)$ denotes the ordered Banach space (canonical order and supremum norm) of all continuous, real-valued affine functions on $C$, then the evaluation

[2] Local convexity is dispensable if it is assumed that $A(C)$ (see subsequent proof) separates the points of $C$.

map $f \mapsto f(x)$ $(x \in C)$ defines an affine homeomorphism of $C$ into the weak dual of $A(C)$ so $C$ can be identified with a compact convex subset of $A(C)'$ (precisely, the set of all (continuous) positive linear forms of norm 1). On the other hand, $h \mapsto T_h$ where $T_h f := f \circ h$ $(f \in A(C),\ h \in G)$, is an isomorphism of $G$ onto a group $\hat{G} \subset \mathscr{L}(A(C))$. Since the mappings $h \in G$ are equicontinuous, $\hat{G}$ is compact in $\mathscr{L}_s(A(C))$ by the Arzela-Ascoli theorem. Thus by Cor. 1 there exists a zero element $P \in \overline{\mathrm{co}}\, \hat{G}$. If $C$ is identified with a subset of $A(C)'$ as above, then the restriction of the adjoint group $\hat{G}'$ to $C$ agrees with $G$ and it is clear that each element $x_0 = P'x$ $(x \in C)$ is fixed under $G$. □

Theorems (7.6) and (7.9) establish mean ergodicity with all essential assumptions placed on the semi-group $S$. (7.11) below is different as it waives all algebraic hypotheses on $S$ but places severe restrictions on the normed structure of the underlying space and its relation to $S$. (The theorem is true under more general hypotheses (Exerc. 18) but we restrict attention to Hilbert space, which is sufficient for our later applications.) For bounded operator semi-groups on Banach spaces, which form the majority of applications, the proof of mean ergodicity is often facilitated by the following proposition which shows equicontinuity to enter essentially, and which is a partial converse of (7.2).

**7.10 Proposition.** *Let $E$ be a locally convex space, and let $S \subset \mathscr{L}(E)$ be a semi-group which is equicontinuous. The following are equivalent assertions:*

(a) *$S$ is mean ergodic in $\mathscr{L}_s(E)$.*

(b) *There exists a projection $P \in \mathscr{L}(E)$ satisfying $TP = PT = P$ for all $T \in S$ and $Px \in \overline{\mathrm{co}}\, Sx$ for all $x \in E$.*

*Proof.* (a)⇒(b) is clear from the definition of mean ergodicity, and from (7.2).

(b)⇒(a): We have to show that $P \in \overline{\mathrm{co}}\, S$ in $\mathscr{L}_s(E)$, that is, given a 0-neighborhood $U$ and a finite set $\{x_1, \dots, x_n\}$ in $E$ there exists $R \in \mathrm{co}\, S$ satisfying $Rx_i - Px_i \in U$ $(i = 1, \dots, n)$. We prove this by induction on $n$; the case $n = 1$ is clear in view of (b). Since $S$ is equicontinuous, there exists a 0-neighborhood $V$ such that $T(V) \subset U$ for all $T \in S$ and hence, such that $R(V) \subset U$ for all $R \in \mathrm{co}\, S$; by the induction hypothesis, there exists $R_0 \in \mathrm{co}\, S$ such that $R_0 x_i - Px_i \in V$ for $i = 1, \dots, n-1$. Applying (b) to $x := R_0 x_n$ we can choose $R_1 \in \mathrm{co}\, S$ such that $R_1 x - Px \in U$. Define $R := R_1 R_0 \in \mathrm{co}\, S$. Using that $R_1 P x_i = P x_i$ $(i = 1, \dots, n-1)$ and that $PR_0 x_n = Px_n$, we obtain $Rx_i - Px_i \in U$ $(i = 1, \dots, n)$. □

A bounded linear operator $T$ on a Banach space $E$ is called *contractive*, or a *contraction*, if $\|T\| \leqq 1$; a semi-group $S \subset \mathscr{L}(E)$ is called *contractive* if each $T \in S$ is contractive. (Note that $\overline{\mathrm{co}}\, S \subset \mathscr{L}_s(E)$ is contractive iff $S$ is.)

**7.11 Theorem.** *Every contractive semi-group $S \subset \mathscr{L}(H)$, where $H$ is any Hilbert space, is mean ergodic. More precisely: The fixed space $F := \{x \in H : Tx = x$, all $T \in S\}$ and the subspace $N := \{y \in H : 0 \in \overline{\mathrm{co}}\, Sy\}$ are orthogonal supplements of each other, and the orthogonal projection $P: H \to F$ is the zero of $\overline{\mathrm{co}}\, S \subset \mathscr{L}_s(H)$.*

*Proof.* The assumption that $S$ be contractive implies that the fixed space $F$ of $S$ and the fixed space $F^*$ of the adjoint semi-group $S^* \subset \mathscr{L}(H)$ are identical. In fact, if $T \in S$ and $x \in F$ are arbitrary then, since $\|T\| = \|T^*\| \leqq 1$, the relation $(x|x) = (Tx|x) = (x|T^*x)$ implies

$$(x|x) \leqq \|x\| \, \|T^*x\| \leqq \|x\|^2 = (x|x).$$

Hence, equality holds throughout which implies that $x = T^*x$. Thus $F \subset F^*$ and, by symmetry, $F^* \subset F$. Now denote by $N_1$ the closed linear hull of the set $\{(I-T)x \colon T \in S,\ x \in H\}$. It is well known and easy to verify that $F^* = N_1^\circ$ where $N_1^\circ$ stands for the orthogonal supplement of $N_1$ in $H$. Since $F = F^*$, $H = F + N_1$ is an orthogonal sum. Moreover, since $F = F^*$ is trivially $S^*$-invariant, it follows that $N_1$ is $S$-invariant. Thus $H = F + N_1$ reduces each $T \in S$ and we have $TP = PT = P$ for all $T \in S$.

It remains to show that $N_1 = N$ and that $P \in \overline{\text{co}}\, S$ in $\mathscr{L}_s(H)$. By (b) of (7.10) it suffices to show that $Px \in \overline{\text{co}}\, Sx$ for each $x \in H$, and this is immediate once we know that $N_1 = N$. To prove this, let $x \in H$ be given. Then $\overline{\text{co}}\, Sx$ is a closed convex subset of $H$, so there exists a unique element $x_0 \in \overline{\text{co}}\, Sx$ with minimal norm (cf. Bourbaki [1967], Esp. vect. top. Chap. V, § 1). Since $\overline{\text{co}}\, Sx$ is $S$-invariant and $\|Tx_0\| \leqq \|x_0\|$, it follows that $Tx_0 = x_0$ for each $T \in S$. Thus $x_0 \in F$ and $x - x_0 \in N$ whence it follows that $H = F + N$. But it is clear that $N \subset N_1$ and, since $H = F + N_1$ is an orthogonal sum, we conclude that $N = N_1$. □

In conjunction with Example 3 above, we obtain this corollary.

**Corollary.** *If $T \in \mathscr{L}(H)$ is a contraction, then the averages $M_k = k^{-1}(I + T + \cdots + T^{k-1})$ $(k \in \mathbb{N})$ converge strongly to the orthogonal projection of $H$ onto the fixed space of $T$.*

We conclude this section with the Markov-Kakutani fixed point theorem which, although not a special case of the preceding ergodic theorems, is closely related to this circle of ideas (cf. Exerc. 19).

**7.12 Theorem.** *Let $C$ be a non-void, compact convex subset of a topological vector space, and let $S$ be an abelian semi-group of continuous affine maps of $C$ into $C$. There exists a point $x_0 \in C$ which is fixed under $S$.*

*Proof.* Consider the set $\boldsymbol{C}$ of all non-void, compact convex subsets of $C$ which are invariant under $S$. A routine application of Zorn's lemma shows that $\boldsymbol{C}$ contains minimal elements. If $C_0$ is such a minimal element of $\boldsymbol{C}$ then $R(C_0) = C_0$ for each $R \in \text{co}\, S$, since ($S$ being abelian) $R(C_0)$ is a member of $\boldsymbol{C}$ and contained in $C_0$. Let $x_0 \in C_0$, $T \in S$ be given and form $M_k := k^{-1}(I + T + \cdots + T^{k-1})$ $(k \in \mathbb{N})$. Since $M_k \in \text{co}\, S$, by the preceding there exists $x_k \in C_0$ satisfying $x_0 = M_k x_k$. This implies (cf. Example 3 above)

$$Tx_0 - x_0 = (TM_k - M_k)x_k = k^{-1}(T^{k-1} - I)x_k \in k^{-1}(C - C)$$

for each $k \in \mathbb{N}$. Thus $Tx_0 - x_0 \in \bigcap_1^\infty k^{-1}(C - C) = \{0\}$. □

*Remark.* It is clear that the convex closure $\overline{\text{co}}\, Sx$ $(x \in C)$ of each orbit contains a fixed point of $S$. However, in contrast with (7.9) Cor. 2 (we use the notation of that proof and suppose additionally that $A(C)$ separate the points of $C$) the semi-group $\hat{S}$ corresponding to $S$ is, in general, not mean ergodic in $\mathscr{L}_s(A(C))$ (Exerc. 19). Still, the amenability of $S$ (§ 11), itself a consequence of (7.12), implies the existence of a (not necessarily continuous, or unique) affine map $P: C \to C$ which acts as a zero of $S$ and satisfies $Px \in \overline{\text{co}}\, Sx$ $(x \in C)$. Cf. Exerc. 19 and Day [1969].

## § 8. Operator Invariant Ideals

We now return to our general setting and assume $E$ to be a Banach lattice. Let $S \subset \mathscr{L}(E)$ be a semi-group. Although in this and the subsequent sections we are primarily concerned with positive operators, the following definition (which generalizes Def. I.8.1) makes good sense without this restriction. As before (§ 1) we denote by $\boldsymbol{A}(E)$ the lattice of closed ideals in $E$.

**8.1 Definition.** *$I \in \boldsymbol{A}(E)$ is called an* $S$-ideal *if $T(I) \subset I$ for all $T \in S$. An $S$-ideal $I$ is called* maximal *if it is maximal (with respect to set inclusion) among the $S$-ideals $\neq E$; $I$ is called* minimal *if it is minimal (with respect to set inclusion) among the $S$-ideals $\neq \{0\}$. The semi-group $S$ is called* irreducible *if $\{0\}$ is a maximal $S$-ideal (equivalently, if $E$ is a minimal $S$-ideal).*

If $S$ is a cyclic semi-group with generator $T \in \mathscr{L}(E)$, $S$-ideals will also be called *$T$-ideals*; accordingly, $T \in \mathscr{L}(E)$ is called *irreducible* if the cyclic semi-group $\{T^n : n \in \mathbb{N}\}$ is irreducible.

Obviously the set $\boldsymbol{A}_S(E)$ of all $S$-ideals is, for a given semi-group $S$, a sublattice of $\boldsymbol{A}(E)$ which is anti-isomorphic, by virtue of $I \mapsto I^\circ$, with the lattice $\boldsymbol{W}_S(E')$ of all $\sigma(E',E)$-closed ideals in $E'$ invariant under the adjoint semi-group $S' \subset \mathscr{L}(E')$. The following result on the induced semi-group $S_I \subset \mathscr{L}(E/I)$ $(I \in \boldsymbol{A}_S(E))$ is an easy generalization of (1.3); we omit its proof.

**8.2 Proposition.** *If $I \in \boldsymbol{A}_S(E)$ and $q: E \to E/I$ is the canonical map, then $J \mapsto q(J)$ is a lattice homomorphism of $\boldsymbol{A}_S(E)$, and a bijection of the lattice of $S$-ideals containing $I$, onto the lattice of $S_I$-ideals in $E/I$.*

**Corollary.** *An $S$-ideal $J$ is maximal if and only if $S_J$ is irreducible.*

Before considering examples let us establish the following simple criterion of irreducibility for semi-groups of positive operators.

**8.3 Proposition.** *Let $E$ be a Banach lattice of (linear) dimension $>1$, and let $S \subset \mathscr{L}(E)_+$ be a semi-group. These assertions are equivalent:*

(a) *$S$ is irreducible.*

(b) *For each $0 \neq x \in E_+$, the ideal generated by the orbit $Sx$ is dense in $E$.*

(c) *For each $0 < x \in E$ and each $0 < x' \in E'$, there exists $T \in S$ such that $\langle Tx, x' \rangle > 0$.*

*Proof.* The equivalence (a)$\Leftrightarrow$(b) is clear, since the ideal generated by $Sx$ is invariant under $S$.

(b)⇒(c): Given $x_0>0$, $x'_0>0$ in $E$ and $E'$, respectively, the closed ideal $J:=\{x\in E: \langle|x|,x'_0\rangle=0\}$ is $\neq E$. By (b) $Sx_0$ cannot be contained in $J$ hence, $Tx_0\notin J$ for some $T\in S$.

(c)⇒(b): Suppose there exists $x_0>0$ and an ideal $J\in \boldsymbol{A}(E)$ containing $Sx_0$. $J\neq E$. By the Hahn-Banach theorem there exists $z\in J^\circ\subset E'$, $z\neq 0$. By (II.4.7) we have $x'_0:=|z|\in J^\circ$ and hence, $\langle Tx_0,x'_0\rangle=0$ for all $T\in S$ which contradicts (c). □

**Corollary.** *The following properties of a Banach lattice E are equivalent:*

(a) *There exists an irreducible semi-group of positive operators which is a separable subset of $\mathscr{L}_s(E)$.*

(b) *$E_+$ contains a quasi-interior point and $E'_+$ contains a strictly positive linear form.*

*Proof.* (a)⇒(b): Let $x>0$ be arbitrary and let $\{T_n: n\in\mathbb{N}\}$ be a countable dense subset of $S\subset\mathscr{L}_s(E)$, where $S$ is some irreducible semi-group satisfying (a). Define

$$y:=\textstyle\sum_{n=1}^{\infty}(2^n\|T_n x\|)^{-1}T_n x\,,$$

omitting in the sum any term for which $T_n x=0$. The principal ideal $E_y$ is dense in $E$, since the ideal generated by $Sx$ is dense in $E$ by the hypothesis and (8.3); therefore, $y$ is a quasi-interior element of $E_+$ (Chap. II, § 6). Similarly, if $x'>0$ is arbitrary in $E'$ then $y':=\sum_{n=1}^{\infty}(2^n\|T'_n x'\|)^{-1}T'_n x'$ is a (continuous) strictly positive linear form on $E$.

(b)⇒(a): Suppose $y$ is quasi-interior to $E_+$ and $y'$ is a strictly positive linear form on $E$. The cyclic semi-group $S$ generated by the positive operator of rank 1, $T:=y'\otimes y$, is clearly irreducible and separable. □

**Examples.** 1. Let $\pi$ be a permutation on $n$ letters and denote by $P_\pi$ the permutation matrix corresponding to $\pi$, viewed as a linear operator on $\mathbb{R}^n$ (cf. Chap. I, § 5). If $\pi=(c_1)\dots(c_k)$ is the decomposition of $\pi$ into $k$ independent cycles, denote by $J_\kappa$ the set of all vectors $x\in\mathbb{R}^n$ such that all coordinates of $x$ vanish whose indices are not in $(c_\kappa)$ $(\kappa=1,\dots,k)$. Obviously, $J_\kappa$ are precisely the minimal $P_\pi$-ideals, each $P_\pi$-ideal is a direct sum of some of these $J_\kappa$, and $J_\kappa^\perp$ are the maximal $P_\pi$-ideals $(\kappa=1,\dots,k)$. In particular, $P_\pi$ (or equivalently, the cyclic group $S$ generated by $P_\pi$) is irreducible iff $\pi$ is cyclic. The invariant ideals of a doubly stochastic matrix are determined similarly by those of the permutation matrices occurring in a barycentric representation of $A$ (cf. I.5.3). Moreover, it is clear that the semi-group $D_n$ of doubly stochastic $n\times n$-matrices (and, a fortiori, the semi-group $S_n$ of stochastic $n\times n$-matrices) is irreducible.

2. Let $K$ be a non-void compact space and $E:=C(K)$. Every semi-group $S\subset\mathscr{L}(E)$ possesses maximal invariant ideals. In fact, if $J\neq E$ is any $S$-ideal in $E$, the family of all $S$-ideals $\neq E$ and containing $J$ is inductively ordered. For, let $\{I_\alpha: \alpha\in\mathsf{A}\}$ be a totally ordered (with respect to set inclusion) subfamily; we have $I_\alpha=\{f\in E: f(K_\alpha)=\{0\}\}$ for $K_\alpha$ closed in $K$ (§ 1, Example 1), and the family $\{K_\alpha: \alpha\in\mathsf{A}\}$ is totally ordered by downward inclusion. Since $K$ is compact, $K_0:=\bigcap_\alpha K_\alpha$ is non-void, and evidently $I_0:=\{f\in E: f(K_0)=\{0\}\}$ is an $S$-ideal which is the least upper bound of $\{I_\alpha\}$. Hence, by Zorn's lemma

there exists a maximal $S$-ideal containing $J$. For some important types of semi-groups, a representation of the set of all maximal $S$-ideals will be discussed below. See also Exerc. 22.

3. In contrast to the case of an $AM$-space $E$ with unit, maximal $S$-ideals do not always exist in a Banach lattice $E$, even for the simplest semi-groups $S$. Thus if $E=L^1(\mu)$ for a diffuse measure $\mu$ (e.g., Lebesgue measure on $\mathbb{R}^n$) and if $S$ consists of the identity map $1_E$ alone, there exists no maximal or minimal $S$-ideal (cf. § 1, Example 2). Another example is this. Let $T$ be the Volterra operator on $C_0[0,1]$ (continuous real functions on the unit interval vanishing at 0) defined by

$$Tf(s)=\int_0^s f(t)dt, \quad s\in[0,1].$$

Then each $T$-ideal is of the form $I_a=\{f: f([0,a])=\{0\}\}$ for some $a\in[0,1]$. The proper $T$-ideals are obtained for $a>0$, but there exist no maximal or minimal $T$-ideals.

4. Let $\Gamma$ denote the circle group and fix $\alpha\in\Gamma$. Consider the cyclic semi-group in $\mathscr{L}(C(\Gamma))$ generated by the rotation operator $T$,

$$Tf(s)=f(\alpha s), \quad s\in\Gamma.$$

If $\alpha$ is a primitive $n$-th root of unity, each $T$-ideal is of the form $I_A=\{f: f(A)=\{0\}\}$ where $A$ is a closed subset of $\Gamma$ invariant under the rotation $s\mapsto\alpha s$. In particular, the maximal $T$-ideals are determined by sets $A$ of exactly $n$ points forming the vertices of a regular polygon. If $\alpha$ is not a root of unity (i.e., $\alpha/\pi$ irrational), then $T$ is irreducible. This latter case is typical for the abstract dynamical systems discussed in § 10 below.

We now turn our attention to mean ergodic semi-groups of positive operators in a Banach lattice $E$ (cf. § 7). If $S$ is such a semi-group it is clear that the associated projection $P$ (Def. 7.1) is positive. The following result is preparatory; recall that $T\in\mathscr{L}(E)$ is *strictly positive* (in symbols: $T\gg 0$) if $x>0$ implies $Tx>0$ (equivalently, if $T\geqq 0$ and the absolute kernel of $T$ is the zero ideal).

**8.4 Proposition.** *Let $S\subset\mathscr{L}_s(E)$ be a mean ergodic semi-group of positive operators, and denote by $P$ the associated projection.*

(i) *If $P\gg 0$ then the fixed space of $S$ is a sublattice of $E$.*

(ii) *If the ideal generated by $P(E_+)$ is dense in $E$, the fixed space of $S'\subset\mathscr{L}(E')$ is a sublattice of $E'$.*

(iii) *If the assumptions of both* (i) *and* (ii) *are satisfied, the restriction map $x'\mapsto x'_{|PE}$ is an isomorphism of the Banach lattice $P'E'$ onto $(PE)'$.*

*Proof.* (i) If $F:=\{x\in E: Tx=x, \text{ all } T\in S\}$ denotes the fixed space of $S$ then $F=PE$ by (7.2). Clearly $F$ is a vector subspace of $E$; if $x=Tx$ then $|x|\leqq T|x|$ and $P(T|x|-|x|)=0$, so $P\gg 0$ implies $|x|=T|x|$ for all $x\in F$, $T\in S$ and hence, $F$ is a vector sublattice of $E$.

(ii) Similarly, $T'\in S'$ and $x'\in F':=P'E'$ implies $|x'|\leqq T'|x'|$, since $T'\geqq 0$. But $\langle Px, T'|x'|-|x'|\rangle=0$ for all $x\in E_+$ and hence, $\langle y, T'|x'|-|x'|\rangle=0$ for all elements $y$ of the ideal generated by $P(E_+)$. Since the latter ideal is dense in $E$ by hypothesis, it follows that $|x'|=T'|x'|$ for all $x'\in F'$, $T'\in S'$.

(iii) Consider the topological direct sums

$$E = PE + P^{-1}(0),$$
$$E' = P'E' + (P')^{-1}(0).$$

By virtue of the Hahn-Banach theorem $(PE)'$ is canonically norm isomorphic to $E'/(PE)^\circ$ [S, Chap. IV, p. 161], and it is immediate that $(PE)^\circ = (P')^{-1}(0)$. Thus the restriction map $\psi: x' \mapsto x'_{|PE}$ is a norm isomorphism of $P'E'$ onto $(PE)'$. Clearly $\psi \geqq 0$ and it follows from (II.5.6) that $\psi[(P'E')_+] = (PE)'_+$. So $\psi$ is an isomorphism of Banach lattices. ☐

*Remark.* Simple counterexamples show that the assumptions made in the preceding statements (i)—(iii) are not dispensable. Cf. (11.5).

We can now characterize irreducible mean ergodic semi-groups by properties of the associated projection, as follows.

**8.5 Proposition.** *Let $E$ be a Banach lattice, $S \subset \mathscr{L}_s(E)$ a mean ergodic semi-group of positive operators, and suppose $S$ has non-zero fixed vectors. The following are equivalent:*

(a) *$S$ is irreducible.*

(b) *The projection $P$ associated with $S$ is strictly positive, with range a one-dimensional subspace of $E$ spanned by a quasi-interior element of $E_+$.*

(c) *The projection $P'$ associated with $S' \subset \mathscr{L}(E')$ is strictly positive, with range a one-dimensional subspace of $E'$ spanned by a strictly positive linear form.*

*Proof.* We note that $P \neq 0$ by hypothesis (cf. 7.2).

(a)$\Rightarrow$(b): The absolute kernel $R := \{x \in E : P|x| = 0\}$ of $P$ is an $S$-ideal. Hence, $S$ being irreducible, it follows that $R = \{0\}$, since $R = E$ would imply $P = 0$. Thus $P \gg 0$ and, since for each $S$-fixed element $u \in E_+$ the principal ideal $E_u$ is $S$-invariant, we conclude that $Px$ must be quasi-interior to $E_+$ for each $x > 0$. Moreover, by (8.4)(i) $PE$ is a sublattice of $E$. This implies that for each $y \in PE$, we must either have $y^+ = 0$ or else $y^- = 0$, since the orthogonal principal ideals $E_{y^+}$ and $E_{y^-}$ cannot both be dense in $E$. Thus $PE$ is totally ordered and, since $PE$ is clearly Archimedean, isomorphic to $\mathbb{R}$ (II.3.4). It follows that $P = x'_0 \otimes x_0$ where $x_0$ is quasi-interior to $E_+$ and $x'_0$ is a strictly positive linear form on $E$ satisfying $\langle x_0, x'_0 \rangle = 1$.

(b)$\Rightarrow$(c) is trivial.

(c)$\Rightarrow$(a): By hypothesis we have $P' = x_0 \otimes x'_0$ where $x'_0$ is a strictly positive linear form on $E$ and where $x_0 \in E$ defines a strictly positive linear form on $E'$; clearly, this latter property is equivalent to $x_0$ being a quasi-interior element of $E_+$. To show that $S$ is irreducible, assume that $J \neq \{0\}$ is an $S$-ideal in $E$. Then $PJ \subset J$ and, since $x'_0$ is strictly positive, we obtain $x_0 \in J$; thus $J = E$. Therefore, $S$ is irreducible. ☐

For the following corollary, recall that every compact group $G \subset \mathscr{L}_s(E)$ ($E$ a Banach space) is mean ergodic by (7.9) Cor. 1.

**Corollary.** *Suppose $E$ is a Banach lattice and $G\subset\mathscr{L}_s(E)$ is a compact group (with identity $1_E$) of positive operators. Then the associated projection $P$ is strictly positive, and $S$ is irreducible if and only if $P$ is irreducible and of rank* 1.

*Proof.* In view of (8.5) it suffices to prove that the group property of $G$ implies that $P\gg 0$. If $x>0$, the orbit $Gx\subset E_+$ is compact and evidently does not contain 0. Since for each $y\in Gx$ there exists $z'\in E'_+$ satisfying $\langle y,z'\rangle>0$, the compactness of $Gx$ implies the existence of $x'\in E'_+$ and $\delta>0$ such that $\langle Tx,x'\rangle\geqq\delta$ for all $T\in G$. Since $Px\in\overline{\text{co}}\,Gx$ (cf. 7.10) it follows that $\langle Px,x'\rangle\geqq\delta$ and, therefore, $Px>0$. ∎

From the preceding corollary we note that if $E$ is a Banach lattice without quasi-interior positive elements, there exists no irreducible compact group of positive operators in $\mathscr{L}_s(E)$; this will be further elucidated by Theorem 10.4 below.

**8.6 Lemma.** *Let $S\subset\mathscr{L}_s(E)$ be a positive mean ergodic semi-group whose fixed space is not contained in any closed ideal $\neq E$. Then the absolute kernel of the projection $P$ associated with $S$, is contained in the intersection of the set of all maximal $S$-ideals.*

*Proof.* Let $R:=\{x\colon P|x|=0\}$ be the absolute kernel of $P$, and let $I$ denote any maximal $S$-ideal. Now $J:=\{x\colon P|x|\in I\}$ is a closed ideal containing $R$. But $J$ also contains $I$, since $I$ is $S$- and hence $P$-invariant, and $J\neq E$, since $PE\not\subset I$ by hypothesis. Hence $J=I$ by the maximality of $I$, and so $R\subset I$. ∎

Note that Lemma 8.6 does not presuppose, or imply, the existence of maximal $S$-ideals; the same observation applies to (8.7) and (8.11) below. However, in the special case where $E$ is an $AM$-space with unit and $S$ is a mean ergodic semi-group of Markov operators (Def. 8.8), there always exist maximal $S$-ideals (Example 2 above) and $R$ equals the intersection of the set of all maximal $S$-ideals (Exerc. 21). The following theorem gives the representation of maximal $S$-ideals announced above; note that it contains (8.5) as a special case.

**8.7 Theorem.** *Let $S\subset\mathscr{L}_s(E)$ be a mean ergodic semi-group of positive operators on a Banach lattice $E$, denote by $S'$ the adjoint semi-group in $\mathscr{L}(E')$, and suppose that the fixed space of $S$ is not contained in any proper closed ideal of $E$. There exists a bijective correspondence between the set of all maximal $S$-ideals and the set of all extreme rays of the positive cone of the fixed space of $S'$. This correspondence is given by $I\mapsto I^\circ\cap P'(E'_+)$ ($P'$ the zero of $S'$).*

*Proof.* The proof is given in two stages. First, we assume additionally that $P\gg 0$ where, as before, $P$ is the zero of $S$; second, we remove this assumption with the aid of (8.6).

1. By (8.4) the fixed spaces $F:=PE$ and $F':=P'E'$ of $S$ and $S'$ are closed vector sublattices of $E$ and $E'$, respectively, and the map $x'\mapsto x'_{|F}$ is an isomorphism of the Banach lattice $F'$ onto the dual of $F$. Since $F$ is norm complete, by (II.5.5) the dual of $F$ agrees with the order dual $F^\star$. Hence (II.4.4) applies, so that $I\mapsto I^\circ\cap F'_+$ is a bijection of the set of maximal ideals in $F$ onto the set of

extreme rays of $F'_+$. (Note that the semigroup $S_{|F}$ consists of the identity map only.)

We claim that the map $k: I \mapsto k(I)$, where

$$k(I) := \{x \in E : P|x| \in I\},$$

is an injection of $\boldsymbol{A}(F)$ into $\boldsymbol{A}_S(E)$ such that $k(I) \cap F = I$ and $k(I)$ is the largest $S$-ideal intersecting $F$ in $I$. Clearly, $k(I)$ is an $S$-ideal because $k(I)$ is closed and $x \in k(I)$ implies $P|Tx| \leq PT|x| = P|x|$ hence, $Tx \in k(I)$ for all $T \in S$. Next we show that $k(I) \cap F = I$. The inclusion $I \subset k(I) \cap F$ is trivial; but also $k(I) \cap F \subset I$ for, $x \in k(I) \cap F$ implies $|x| \in k(I) \cap F$, since $F$ is a sublattice of $E$, and hence $|x| = P|x| \in I$ which implies $x \in I$. Thus $k$ is injective. Finally, if $L \in \boldsymbol{A}_S(E)$ satisfies $L \cap F = I$ $(I \in \boldsymbol{A}(F))$ then $x \in L$ implies $P|x| \in L \cap F = I$, since $L$ is $P$-invariant, and hence $x \in k(I)$. Therefore, $L \subset k(I)$.

It follows from the stated properties of $k$ that $k$ maps the maximal ideals in $F$ bijectively onto the set of maximal $S$-ideals in $E$. Moreover, if $I \in \boldsymbol{A}(F)$ is maximal and $I^\circ \cap F'_+$ is the extreme ray of the positive cone of $F'$ corresponding to $I$ by (II.4.4), then $I^\circ \cap F'_+ = I^\circ \cap P'(E'_+) = k(I)^\circ \cap P'(E'_+)$. This proves the theorem in case $P \gg 0$.

2. The general case is reduced to the preceding by the following device. Let $R := \{x \in E : P|x| = 0\}$ be the absolute kernel of $P$ and denote by $S_R$ the semigroup induced by $S$ on $E/R$. Then by (8.6) and (8.2), the quotient map $q: E \to E/R$ defines a bijection $J \mapsto q(J)$ of the set of all maximal $S$-ideals onto the set of all maximal $S_R$-ideals. The zero element $P_R$ of $S_R$, which satisfies $P_R \circ q = q \circ P$, is a strictly positive projection: If $\hat{x} = qx > 0$ then $|x| \notin R$ so $P|x| > 0$ which implies $P|x| \notin R$, since $P^2 = P$; that is, $P_R \hat{x} > 0$. On the other hand, the dual of $E/R$ can be identified with the ideal $R^\circ \subset E'$ and, since $P'E' \subset R^\circ$, the restriction of $P'$ to $R^\circ$ can be identified with $P'_R$. This reduces the proof to case 1., and it is now easily seen that the extreme ray of $F'_+$ corresponding to the maximal $S$-ideal $J \subset E$ is given by $q(J)^\circ \cap F'_+ = J^\circ \cap F'_+$. □

The remainder of this section is concerned with the theme of Theorem 8.7 in the important special cases where $E$ is an $AM$-space with unit, or an $AL$-space (Chap. II, §§ 7 et seq.). Many semi-groups occurring in these applications satisfy an additional condition covered by the following definition.

**8.8 Definition.** *A positive linear operator $T$ on an $AM$-space with unit $e$, is called a* Markov operator *if $Te = e$. Dually, a positive linear operator $T$ on an $AL$-space is called* stochastic *if $\|Tx\| = \|x\|$ for all $x \geqq 0$.*

Thus Markov operators can be viewed as generalizations of row stochastic matrices, while stochastic operators generalize column stochastic matrices (Def. I.4.1).

If $E = C(K)$ ($K$ compact), the cone $E'_+ = M(K)_+$ of positive Radon measures on $K$ has the set $\mathsf{P}$ of all probability measures on $K$ as a $\sigma(E', E)$-compact base, and it is easy to see that the extreme rays of $M(K)_+$ are generated by the extreme points of $\mathsf{P}$ (cf. § 1, Example 1). More generally, the extreme rays of any $\sigma(E', E)$-closed convex subcone $Q \subset M(K)_+$ are generated by the extreme points of

$Q \cap \mathsf{P}$; in view of this remark, the following theorem is an immediate consequence of (8.7) and requires no further proof. Note that in the present situation, the set of maximal $S$-ideals is non-void (Example 2 above).

**8.9 Theorem.** *Let $S$ be a mean ergodic semi-group of Markov operators on $C(K)$ ($K$ compact). Then $\varphi \mapsto I_\varphi := \{f : \varphi(|f|) = 0\}$ is a bijection of the extreme points of the set of all $S$-invariant probability measures onto the set of maximal $S$-ideals. In particular, $S$ is irreducible if and only if there exists a unique $S$-invariant probability measure on $K$.*

This result has extensions in several directions. If the assumption that $S$ be mean ergodic is dropped but $S$ is assumed abelian (or even amenable, cf. Day [1969]), then $\varphi \mapsto I_\varphi$ still maps a subset of the extreme $S$-invariant probability measures (which are sometimes called *ergodic measures* with respect to $S$) onto the set of all maximal $S$-ideals, but the mapping can fail to be injective. In particular, if $K$ is quasi-Stonian and $S$ monothetic it can be shown that the number of $S$-ergodic measures whose absolute kernel is a given maximal $S$-ideal, is either one or infinity (Exerc. 22). Here we confine ourselves to proving the following.

**8.10 Proposition.** *Let $S$ be an abelian semi-group of Markov operators on $C(K)$ ($K$ compact). Then each maximal $S$-ideal is the absolute kernel of some $S$-invariant probability measure.*

*Proof.* Let $I$ be a maximal $S$-ideal; $S_I$ is irreducible on $C(K)/I$ by the corollary of (8.2), and $C(K)/I$ can be identified with $C(K_0)$ for some closed subset $K_0 \subset K$ (§ 1, Example 1). The set $\mathsf{P}_0$ of probability measures on $K_0$ is invariant under the adjoint semi-group $S_I'$, and the restrictions to $\mathsf{P}_0$ of the elements of $S_I'$ constitute an abelian semi-group of affine, continuous transformations of the $\sigma(M(K_0), C(K_0))$-compact convex set $\mathsf{P}_0$. By the Markov-Kakutani theorem (7.12), there exists $\varphi_0 \in \mathsf{P}_0$ which is fixed under $S_I'$; since $S_I$ is irreducible and the absolute kernel of $\varphi_0$ is an $S_I$-ideal, it follows that $\varphi_0$ is a strictly positive Radon measure on $K_0$. If $q: C(K) \to C(K_0)$ denotes the canonical map, it is clear that $\varphi := \varphi_0 \circ q$ is an $S$-invariant probability measure on $K$ for which $I = \{f \in C(K) : \varphi(|f|) = 0\}$. □

If $S = \{T^n : n \in \mathbb{N}\}$ is cyclic then the assumption that $T$ be a Markov operator can be dropped also; that is, for each positive operator $T$ on $C(K)$, every maximal $T$-ideal is the absolute kernel of a suitable positive eigenvector of $T'$ (Exerc. 22).

We conclude this section with a result dual to (8.9), whose proof is rather elementary, and with applications of the foregoing theory.

**8.11 Proposition.** *Let $E$ be an AL-space and let $S$ be a mean ergodic semi-group of stochastic operators on $E$. Denote by $X$ the set of all positive normalized vectors fixed under $S$. There exist minimal $S$-ideals if and only if $X$ has extreme points; if so, the map $x \mapsto \bar{E}_x$ defines a bijection of the extreme points of $X$ onto the minimal $S$-ideals in $E$.*

*Remark.* It will be shown later (Chap. V, § 8) that a semi-group of positive contractions on an $AL$-space $E$ is mean ergodic whenever the set

$\{x \in E_+ : Tx \leq x, \text{ all } T \in S\}$ is not contained in any proper closed ideal of $E$. Thus if $S$ is a semi-group of stochastic operators, mean ergodicity is automatically present whenever $X$ is not contained in any closed ideal $\neq E$. Note also that (8.11) can be used, since it is a generalization of (I.8.4) Cor. to the infinite dimensional case, to discuss Markov chains with a countably infinite state space (§ 11, Example 2).

*Proof* of (8.11). We observe first that the fixed space $F$ of $S$ is a vector sublattice of $E$: In fact, $x \in F$ and $T \in S$ implies $|x| \leq T|x|$ but $|x|$ and $T|x|$ have the same norm, since $T$ is stochastic, and so $|x| = T|x|$; that is, $|x| \in F$. Denote by $\mathsf{E}(X)$ the (possibly empty) set of extreme points of $X$. If $x \in X$ then $\bar{E}_x$ (the closure of the principal ideal generated by $x$) is clearly an $S$-ideal. Suppose now $x \in \mathsf{E}(X)$ and let $I \subset \bar{E}_x$ be an $S$-ideal $\neq \{0\}$. Now if $P$ is the zero of $S$ (cf. 7.2), then $0 < y \in E_x$ implies $Py \in E_x$ and hence $Py = cx$ for some $0 < c \in \mathbb{R}$, since $x$ is extreme in $X$. Thus $PE_x$ and, therefore, $P\bar{E}_x$ is the one-dimensional subspace of $E$ spanned by $x$. It follows that $PI$ contains $x$ (note that, since $P$ is stochastic, $PI \neq \{0\}$); on the other hand, $PI \subset I$ since $I$ is $S$-invariant, and so $x \in I$. This shows $I = \bar{E}_x$, that is, $\bar{E}_x$ is a minimal $S$-ideal.

Conversely, let $J$ be a minimal $S$-ideal. Since $P$ is norm preserving on $E_+$ it follows that $J \cap X \neq \emptyset$ and by the preceding characterization (8.5) of irreducible semi-groups, $PJ$ is one-dimensional; that is, $J \cap X$ is a singleton $\{x\}$, and it is clear that $J = \bar{E}_x$. To show that $x \in \mathsf{E}(X)$, we observe that $x = \lambda y + \mu z$ ($\lambda > 0$, $\mu > 0$, $\lambda + \mu = 1$; $y, z \in X$) implies $y \in J \cap X$ and $z \in J \cap X$ hence, $x = y = z$. ☐

**Examples.** 5. Let $K$ denote a compact space and let $H$ denote a semi-group of continuous maps $h: K \to K$. If $E := C(K)$ then $\hat{H} := \{T_h : h \in H\}$, where $T_h f = f \circ h$ ($f \in E$), is a semi-group of Markov operators isomorphic with $H$. A closed ideal $I_A = \{f \in E : f(A) = \{0\}\}$, where $A \subset K$ is closed (cf. Example 2 above and § 1, Example 1), is an $\hat{H}$-ideal if and only if $H(A) \subset A$. Thus the lattice of $\hat{H}$-ideals is anti-isomorphic to the lattice of closed $H$-invariant subsets of $K$; in particular, an $\hat{H}$-ideal $I_A$ is maximal iff $A$ is a minimal (non-void) $H$-invariant closed subset of $K$. The *minimal transformation groups* $H$ of topological dynamics (cf. R. Ellis [1969]) are thus precisely those for which the operator group $\hat{H}$ is irreducible.

An extreme $H$-invariant positive Radon measure on $K$ with minimal support is frequently called *minimal ergodic* (with respect to $H$); hence by (8.9) and (8.10), if $\hat{H}$ is mean ergodic or abelian then ergodic measures always exist and their absolute kernels are precisely the maximal $\hat{H}$-ideals. In particular, if $\hat{H}$ is mean ergodic then $\hat{H}$ is irreducible iff there exists a unique ergodic measure with global support (Exerc. 22).

6. Let $G$ denote a compact topological group. Since the group $H$ of all left rotations $s \mapsto ts$ ($s, t \in G$) is uniformly equicontinuous (in fact, it defines the left uniformity of $G$), it follows from the Arzela-Ascoli theorem that the operator group $\hat{H}$ corresponding to $H$ is compact in $\mathscr{L}_s(C(G))$ (cf. proof of (7.9) Cor. 2) hence, mean ergodic by (7.9) Cor. 1; evidently $\hat{H}$ is irreducible. Therefore, (8.9) contains the assertion that there exists a unique positive, normalized measure on $G$ invariant under all left rotations which is called

*left Haar measure* on $G$; it is well known and easy to verify that this measure is right invariant also. It is, therefore, briefly called *Haar measure* on $G$.

7. Let $X$ be a topological space, and let $G$ be a group of homeomorphisms of $X$. Suppose that $\mu$ is a finite (or $\sigma$-finite) countably additive measure defined on a $\sigma$-field of subsets of $X$ and that each $g \in G$ defines a measure preserving transformation (cf. § 9, Example 1); then the *abstract dynamical system* $(X, G)$ is called *ergodic* with respect to $\mu$ (or $\mu$ an *ergodic invariant measure*) if whenever $X = X_1 \cup X_2$ is a partition of $X$ into measurable $G$-invariant sets (modulo null sets), then $\mu(X_1)=0$ or $\mu(X_2)=0$. (Note that as in the situation discussed in Example 5 above, this notion of ergodicity must not be confused with mean ergodicity.) In each of the spaces $L^p(\mu)$ $(1 \leqq p \leqq +\infty)$, $G$ induces a group $\hat{G} := \{T_g : g \in G\}$ of positive operators (§ 9, Example 1). If $p < +\infty$, then $(X, G)$ is ergodic iff $\hat{G}$ is irreducible; this is easily seen from the representation of closed ideals in $L^p(\mu)$ (§ 1, Example 2). A complete characterization of irreducible compact groups of positive operators on Banach lattices will be given below (§ 10).

8. Let $X \neq \emptyset$ be a set, $\Sigma$ a $\sigma$-algebra of subsets, and denote by $E$ the vector lattice of all real valued bounded, countably additive set functions on $\Sigma$ (Chap. II, § 5, Example 3). Under the norm $\mu \mapsto \|\mu\|$ ($\|\mu\|$ the total variation of $\mu$) $E$ is an $AL$-space. A *stochastic kernel* is a function $P: \Sigma \times X \to [0,1]$ such that for each $s \in X$, $B \mapsto P(B,s)$ is a probability measure on $\Sigma$, and such that for each $B \in \Sigma$, $s \mapsto P(B,s)$ is measurable $(\Sigma)$. The endomorphism $\nu \mapsto \mu$ of $E$ defined by

$$\mu(B) = \int P(B,s)\, d\nu(s) \qquad (B \in \Sigma)$$

is a stochastic operator (Def. 8.8); evidently, the set of these operators is a semi-group $S \subset \mathscr{L}(E)$. Let $S_0$ be a subsemi-group of $S$ which is mean ergodic and contains a weakly compact operator. Applying (II.9.9) Cor. 1 we see that the fixed space of $S_0$ is a finite dimensional sublattice of $E$, and hence the set $X$ of (8.11) is a finite dimensional simplex with extreme points $\mu_m$ $(m=1, \ldots, n)$, say. By virtue of the Radon-Nikodym theorem (II.8.7) Cor. the closed ideal $I_m \subset E$ generated by $\mu_m$ is the band of all $\mu_m$-continuous elements of $E$. Now (8.11) asserts that $I_m$ contains no $S_0$-invariant measures other than the scalar multiples of $\mu_m$, and the $n$-ideals $I_m$ are the only $S_0$-ideals in $E$ with this property.

## § 9. Homomorphisms of Vector Lattices

We recall (Def. II.2.4) that a *vector lattice homomorphism* (or briefly, *lattice homomorphism*) of the vector lattice $E$ into a vector lattice $F$ is a linear map preserving the lattice operations. In view of the translation invariance of the order in $E, F$ and in view of the linearity of $T$, a lattice homomorphism $T: E \to F$ is characterized by the property that $|Tx| = T|x|$ for all $x \in E$ (II.2.5). (This characterization serves well to extend the notion of lattice homomorphism to complex vector lattices, cf. Chap. II, § 11.) Thus lattice homomorphisms commute with the (finite) lattice operations; but it should be pointed out that, in general, a lattice homomorphism $T: E \to F$ does not commute with the formation of infinite suprema and infima (i.e., $T$ is not automatically order continuous), nor does the adjoint of $T$ in general define a lattice homomorphism of the order dual $F^\star$ into $E^\star$ (Exerc. 24). Examples will be considered below.

Now let $E, F$ be Banach lattices. A lattice homomorphism $T$ of $E$ into $F$ is necessarily continuous, since it is a positive linear map (II.5.3); we seek a characterization of $T$ in terms of the representation theory (whenever it applies) discussed in Sections 4 through 6 of this chapter. Guided by the finite dimensional case (I.4.4) we shall first characterize lattice homomorphisms between $AM$-spaces $E, F$ with respective units $u, v$ and consider positive linear maps $T: E \to F$ satisfying $Tu = v$; $E$ and $F$ can, of course, be assumed of the form $C(K)$ and $C(L)$ for suitable compact spaces $K$ and $L$ (Chap. II, § 7). The following theorem will then be the key for the discussion of the general case.

**9.1 Theorem.** *Let $K, L$ be compact spaces and denote by $e_K, e_L$ the respective units (constant-one functions) of $E := C(K)$ and $F := C(L)$. Define $H := \{T \in \mathscr{L}(E,F): T \geqq 0, Te_K = e_L\}$. The following assertions are equivalent:*

(a) *$T$ is an extreme point of $H$.*

(b) *$T$ is an algebra homomorphism satisfying $Te_K = e_L$.*

(c) *$T$ is a lattice homomorphism satisfying $Te_K = e_L$.*

(d) *There exists a (unique) continuous map $k: L \to K$ such that $Tf = f \circ k$ for all $f \in C(K)$.*

*Remark.* In accordance with the general assumption made in this chapter, we consider the *real* Banach lattices (respectively, algebras) $C(K)$, $C(L)$. However, the theorem is valid in the complex case with little change (Exerc. 25(a)), and the situation is similar with all subsequent results. Note also that lattice homomorphisms $T: E \to F$ not satisfying $Te_K = e_L$, can be characterized in a fashion similar to (d) (Exerc. 25(b)).

*Proof* of (9.1). (a)$\Rightarrow$(b): Let $f_0 \in C(K)$ be any fixed function satisfying $e_K \leqq f_0 \leqq 2e_K$. Define the linear maps $T_1, T_2 \in \mathscr{L}(E,F)$ by $T_1 f := T(f_0 f)/Tf_0$ and $T_2 f := 2Tf - T_1 f$ (juxtaposition of functions denoting pointwise multiplication). Since $Tf_0 \geqq Te_K = e_L$, these maps are well defined and it is straightforward to verify that $T_1 \in H$, $T_2 \in H$. Since $T$ is extreme in $H$ and $T = \frac{1}{2}(T_1 + T_2)$ it follows that $T = T_1$, that is, $T(f_0 f) = Tf_0\, Tf$ for all $f \in C(K)$. Now let $g \in C(K)$ be arbitrary. If $g = 0$ then, clearly, $T(gf) = Tg\, Tf$ for all $f$. If $g \neq 0$, define $f_0 := \frac{3}{2} e_K + g/2\|g\|$ then $e_K \leqq f_0 \leqq 2e_K$. So $T(f_0 f) = Tf_0\, Tf$ by the preceding argument, and a simple computation shows that $T(gf) = Tg\, Tf$ for all $f \in C(K)$. Hence, $T$ is a homomorphism of algebras.

(b)$\Rightarrow$(c): First we observe that $T \geqq 0$, since $T(f^2) = (Tf)^2 \geqq 0$ for all $f \in C(K)$ and since $f \mapsto f^2$ maps $E_+$ onto itself. Now the chain $(T|f|)^2 = T(|f|^2) = T(f^2) = (Tf)^2 = |Tf|^2$ shows that $|Tf| = T|f|$ for all $f \in C(K)$.

(c)$\Rightarrow$(d): The Dirac measures $\delta_s$ $(s \in K)$ are the only normalized scalar valued lattice homomorphisms on $C(K)$ (§ 1, Example 1). On the other hand, $T$ being a lattice homomorphism, for each $t \in L$ $T'\delta_t = \delta_t \circ T$ is a scalar valued lattice homomorphism on $C(K)$ which is normalized, since $\langle e_K, T'\delta_t\rangle = \langle Te_K, \delta_t\rangle = \langle e_L, \delta_t\rangle = 1$. Hence, $T'\delta_t = \delta_s$ for some $s \in K$ uniquely determined by $t$; moreover, the mapping $k: t \mapsto s$ of $L$ into $K$ is continuous, since the adjoint $T'$ is continuous from the weak dual of $C(L)$ into that of $C(K)$. Finally, for each $t \in L$ and $f \in C(K)$ we obtain

$$Tf(t) = \langle Tf, \delta_t\rangle = \langle f, T'\delta_t\rangle = \langle f, \delta_{k(t)}\rangle = f \circ k(t)$$

which is the desired conclusion.

(d)$\Rightarrow$(a): Clearly, (d) implies that $Te_K = e_L$ and $T \geqq 0$. To show that $T$ is an extreme point of $H$, suppose that $T = \lambda T_1 + \mu T_2$ where $T_1, T_2 \in H$ and $\lambda > 0$, $\mu > 0$, $\lambda + \mu = 1$. By hypothesis, for each $t \in L$ we have $\delta_{k(t)} = T'\delta_t = \lambda T_1'\delta_t + \mu T_2'\delta_t$. But $\delta_{k(t)}$ is an extreme point of the set of all (Radon) probability measures on $K$ and so it follows that $T'\delta_t = T_1'\delta_t = T_2'\delta_t$ for all $t \in L$, since $T_i'\delta_t$ $(i = 1, 2)$ are probability measures also. The last line of the preceding paragraph now shows that $T_1 = T_2$, that is, $T$ is an extreme point of $H$. □

The equivalences (a)$\Leftrightarrow$(c)$\Leftrightarrow$(d) of (9.1) are readily extended to the case where $E, F$ are Banach lattices possessing quasi-interior positive elements. If $E, F$ are Banach lattices with this property, by (4.5) there exist isomorphisms $U: E \to \hat{E}$ and $V: F \to \hat{F}$ onto Banach lattices of continuous numerical functions on the respective structure spaces $K(E)$ and $K(F)$ (Def. 4.2), and each $T \in \mathscr{L}(E, F)$ determines a linear map $\hat{T} \in \mathscr{L}(\hat{E}, \hat{F})$ by virtue of $\hat{T} \circ U = V \circ T$; $\hat{T}$ is called *conjugate* to $T$ (with respect to $U, V$). For convenience of expression, in the following proposition $K(E)$ and $K(F)$ are realized as the compact spaces $K_u$ and $K_v$ of $u$-normalized and $v$-normalized valuations of $E$ and $F$, respectively (§ 4). We point out that for each lattice homomorphism $T: E \to F$ mapping some quasi-interior point $u$ of $E_+$ onto a quasi-interior point $v$ of $F_+$, the choice of $u$ is immaterial (Exerc. 24(d)), and $T$ is characterized by a unique continuous map of $K(F)$ into $K(E)$.

**9.2 Proposition.** *Let $E, F$ be Banach lattices with respective quasi-interior elements $u \in E_+$, $v \in F_+$ and define $H := \{T \in L(E, F): T \geqq 0, Tu = v\}$. For $T \in H$, the following assertions are equivalent:*

(a) *$T$ is an extreme point of $H$.*

(b) *$T$ is a lattice homomorphism of $E$ into $F$.*

(c) *There exists a continuous map $k: K_v \to K_u$ such that $\hat{T}f = f \circ k$ for all $f \in \hat{E}$.*

*Proof.* In the spirit of the discussion preceding (4.1), the dense ideals $E_u \subset E$, $F_v \subset F$ will be identified with $C(K_u)$ and $C(K_v)$, respectively. Let $T_0$ denote the restriction to $E_u$ of $T \in H$, then $T_0$ can be viewed as a positive operator of $C(K_u)$ into $C(K_v)$. It is clear that $T_0$ is a lattice homomorphism iff $T$ is.

(a)$\Leftrightarrow$(b): By virtue of (9.1) it suffices to show that if and only if $T$ is extreme in $H$, then $T_0$ is an extreme point of $\{R \in L(E_u, F_v): R \geqq 0, Ru = v\} =: H_0$ (this set was denoted by $H$ in (9.1)). This follows from the fact that $T \mapsto T_0$ maps $H$ onto a face of $H_0$: In fact, let $T_0 = \lambda R + \mu S$ where $\lambda > 0$, $\mu > 0$, $\lambda + \mu = 1$ and $R, S \in H_0$; then $R \leqq \lambda^{-1} T_0$, $S \leqq \mu^{-1} T_0$. Now if $(x_n)$ is a null sequence in $E$ and contained in $E_u$, we obtain $|R(x_n)| \leqq R(|x_n|) \leqq \lambda^{-1} T_0(|x_n|)$ so $(R x_n)$ converges to 0 in $F$, that is, $R$ is the restriction to $E_u$ of some element of $H$, and similarly for $S$. Consequently, $T$ is extreme in $H$ iff $T_0$ is extreme in $H_0$ which proves the equivalence of (a) and (b).

(b)$\Rightarrow$(c): Since $T_0: E_u \to F_v$ is a lattice homomorphism mapping the constant-one function $e_u$ of $C(K_u)$ onto the constant-one function $e_v$ of $C(K_v)$, (9.1) implies the existence of a continuous map $k: K_v \to K_u$ such that $Tf = f \circ k$ for all $f \in C(K_u) \subset \hat{E}$. If $0 < f \in \hat{E}$ then $f = \lim_n (f \wedge n e_u)$ in norm, since $u$ is quasi-interior to $E_+$ (§ 5, Equation (2)), and thus $\hat{T}f = \lim_n \hat{T}(f \wedge n e_u)$ in the norm of $F$. This implies $\hat{T}f = \sup_n \hat{T}(f \wedge n e_u)$ and, since $\hat{T}(f \wedge n e_u)(s) = (f \circ k)(s) \wedge n$ for $s \in K_u$ by the preceding, it follows from the properties (*) of valuations (§ 3) that $\hat{T}f(s) = f \circ k(s)$ for all $s \in K_u$. Clearly, then, $\hat{T}f = f \circ k$ for all $f \in \hat{E}$.

(c)$\Rightarrow$(b): By (9.1) $T_0$ is a lattice homomorphism of $E_u$ into $F_v$ and hence, $T$ is a lattice homomorphism of $E$ into $F$. □

The preceding characterization of lattice homomorphisms can be extended to Banach lattices possessing topological orthogonal systems (cf. Exerc. 24(d)) and certain more general situations. Since our later applications are concerned exclusively with Banach lattices possessing quasi-interior positive elements, we will not pursue this subject further. On the other hand, before considering examples and specific applications, let us exhibit the contravariant behavior of the lattice homomorphisms studied in (9.2) and the associated continuous maps between the respective structure spaces. As explained before Proposition 9.2, the choice of $u$ is not essential, and reference to $u, v$ is made only for convenience.

**9.3 Proposition.** *Let $T: E \to F$ be a lattice homomorphism mapping the quasi-interior element $u \in E_+$ onto the quasi-interior element $v \in F_+$, and denote by $k$ the continuous map $K_v \to K_u$ associated with $T$* (cf. 9.2).

(i) *$T$ is injective if and only if $k$ is surjective.*

(ii) *$TE$ is a dense ideal in $F$ if and only if $k$ is injective.*

(iii) *If $k$ is open*[1] *then $T$ is order continuous.*

(iv) *If $T$ is order continuous and $E, F$ are order complete, then $k$ is open*[1].

*Proof.* (i) Since $\hat{T}f = 0$ $(f \in \hat{E})$ is equivalent to $f[k(K_v)] = \{0\}$, it is clear from (4.5) and an application of Urysohn's theorem that $T$ is injective if and only if $k$ is surjective.

(ii) $T^{-1}(0)$ is a closed ideal in $E$. By factoring $T$ through the quotient $E/T^{-1}(0)$ if necessary, we can suppose $T$ to be injective and hence, by (i), we can assume $k$ surjective. (Note that by (4.3) and (6.1) $k(K_v)$ is homeomorphic with the structure space of $E/T^{-1}(0)$.) Next, it is not difficult to verify that $TE_u = F_v$ if and only if $k$ is a homeomorphism of $K_v$ onto $K_u$. Thus if $TE$ is an ideal in $F$, $TE_u$ is an ideal in $F$ containing $v$ and contained in $F_v$, whence $TE_u = F_v$. So, $k$ is injective. Conversely, suppose $k$ to be injective; then $T$ maps $E_u$ isomorphically onto $F_v$. To see that $TE$ is an ideal in $F$, it suffices to show that $F_{|y|} \subset TE$ for each $y \in TE$. Let $y = Tx$ then $|y| = T|x|$ and $T(|x| \vee u) = |y| \vee v$. But $|x| \vee u$ and $|y| \vee v$ are quasi-interior points of $E_+$ and $F_+$, respectively, and by the corollary of (4.1) $E_u$ is canonically isomorphic to $E_{|x| \vee u}$ while $F_v$ is canonically isomorphic

[1] That is, for each open $V \subset K_v$, $k(V)$ is open in $K_u$.

to $F_{|y| \vee v}$. From the construction of these canonical isomorphisms and the fact that $T$ is a lattice monomorphism of $E$ into $F$ and an isomorphism of $E_u$ onto $F_v$, it can be seen that also $T$ is an isomorphism of $E_{|x| \vee u}$ onto $F_{|y| \vee v}$. Thus $F_{|y|} \subset TE$ and hence, $TE$ is an ideal in $F$.

(iii) Suppose $(f_\alpha)$ is a directed ($\geqq$) family in $\hat{E}$ with $\inf_\alpha f_\alpha = 0$. From (4.5) it is easily seen that one must have $\inf_\alpha f_\alpha(s) = 0$ for all $s \in K_u$ with the exception of a meager (first category) subset $M \subset K_u$. The assumption on $k$ implies that $N := k^{-1}(M)$ is meager in $K_v$. Since $Tf_\alpha(s) = f_\alpha \circ k(s)$ for all $\alpha$ and all $s \in K_v$, we have $\inf_\alpha Tf_\alpha(s) = 0$ for $s$ in the complement of $N$, that is, in a dense subset of $K_v$. Hence $\inf_\alpha Tf_\alpha = 0$ in $\hat{F}$ which shows $T$ to be order continuous.

(iv) Let us note first that a closed ideal $I$ in a Banach lattice $G$ (with quasi-interior elements in $G_+$) is a projection band (Chap. II, § 2) if and only if the structure space of $G/I$ can be canonically identified (cf. 4.3) with an open-and-closed subspace of $K(G)$ (Exerc. 10). Since $F$ is assumed order complete, $K_v$ is Stonian by (5.6). So in order to show that $k$ is open, it suffices to show that $k(V)$ is open in $K_u$ for each open-and-closed set $V \subset K_v$. Consider the subset $I := \{g \in \hat{F} : g(V) \subset \{0\}\}$; then $I$ is a closed ideal and, by the preceding remark, a projection band in $F$. On the other hand, for $f \in \hat{E}$ the relations $\hat{T}f \in I$ and $f[k(V)] = \{0\}$ are equivalent. Thus $\hat{T}^{-1}(I)$ is the ideal $J := \{f \in \hat{E} : f[k(V)] = \{0\}\}$ in $\hat{E}$ and, since $T$ is order continuous by hypothesis, $J$ is a band in $E$. Since $E$ is order complete, $J$ is a projection band in $E$ (II.2.10). From the above remark it follows that $k(V)$ is an open subset of $K_u$. ∎

**Examples.** 1. Let $(X, \Sigma, \mu)$ and $(Y, T, \nu)$ denote measure spaces. A transformation (mapping) $\varphi: Y \to X$ is called *measurable* whenever $B \in \Sigma$ implies $\varphi^{-1}(B) \in T$, *non-singular* if $\varphi$ is measurable and $\mu(B) = 0$ implies $\nu[\varphi^{-1}(B)] = 0$, *measure preserving* if $\varphi$ is measurable and $\nu[\varphi^{-1}(B)] = \mu(B)$ for all $B \in \Sigma$. Accordingly as $\varphi$ has one of these three properties in the stated order, $\varphi$ defines a lattice homomorphism $f \mapsto T_\varphi f := f \circ \varphi$ of $M(X, \Sigma)$ into $M(Y, T)$, of $L^\infty(\mu)$ into $L^\infty(\nu)$, and of $L^p(\mu)$ into $L^p(\nu)$ $(1 \leqq p \leqq +\infty)$, respectively. ($M(X, \Sigma)$ denotes the vector lattice of real valued $\Sigma$-measurable functions on $X$.) In the third case, for all $p$ considered $T_\varphi$ is an isometry which is surjective iff $\{\varphi^{-1}(B) : B \in \Sigma\}$ agrees with $T$ modulo null sets; in particular, if $X = Y$ and $p = 2$ then $T_\varphi$ is unitary. If $\varphi$ is non-singular, $T_\varphi$ is sequentially order continuous.

2. If in the preceding example $X$ and $Y$ are localizable (§ 1, Example 2) then $L^\infty(\mu)$ and $L^\infty(\nu)$ are order complete $AM$-spaces with unit, and isomorphic to $C(K)$ for suitable compact spaces $K$. But even if $T: L^\infty(\mu) \to L^\infty(\nu)$ is a lattice homomorphism satisfying $Te_X = e_Y$, it must not be inferred from (9.1) that $T = T_\varphi$ for some $\varphi: Y \to X$. For example, let $X = Y = \mathbb{N}$ and define $T: l^\infty \to l^\infty$ by $Tx = \delta(x)e$, where $e = (1, 1, \ldots)$ and $\delta(x) = \lim_U \xi_n$ $(x = (\xi_1, \xi_2, \ldots) \in l^\infty)$ for some free ultrafilter $U$ on $\mathbb{N}$ (that is, an ultrafilter not generated by some singleton $\{n\}$). Since every ultrafilter on a bounded interval in $\mathbb{R}$ converges, $\delta$ defines a real valued lattice homomorphism on $l^\infty$ satisfying $\delta(e) = 1$. Thus $T$ is a lattice homomorphism of $l^\infty$ satisfying $Te = e$, but $T$ is not induced by a transformation $\varphi$ of $\mathbb{N}$ into $\mathbb{N}$. (This situation is reconciled with (9.1) by the fact that $\delta$ is a Dirac measure on the Stone-Čech compactification

$\beta\mathbb{N}$ of $\mathbb{N}$ corresponding to a point of $\beta\mathbb{N}\setminus\mathbb{N}$, and $l^\infty \cong C(\beta\mathbb{N})$.) However, if $T$ is an order continuous lattice homomorphism of $l^\infty$ with $Te=e$, then $T=T_\varphi$ for some $\varphi: \mathbb{N}\to\mathbb{N}$. In fact, if $\delta_n(x)=\xi_n$ $(n\in\mathbb{N})$ then $\delta_n \circ T$ is an order continuous scalar lattice homomorphism of $l^\infty$ satisfying $\delta_n \circ T(e)=1$, hence of the form $\delta_m$ for some $m=\varphi(n)$ (Exerc. 10), and so $T=T_\varphi$. Since $\beta\mathbb{N}$ is the structure space of $l^\infty$, this situation corresponds precisely to the case where the continuous map $k: \beta\mathbb{N}\to\beta\mathbb{N}$ associated with $T$, is open (cf. 9.3).

3. Suppose $G$ is a compact topological group and suppose $H$ is a closed (not necessarily normal) subgroup. We consider the *homogeneous space* $G/H$ of all right cosets $gH$ $(g\in G)$; $G/H$ is compact under the quotient topology. Let $E:=C(G/H)$. By virtue of

$$T_g f(hH)=f(ghH) \qquad (f\in E, h\in G)$$

each $g\in G$ induces a lattice isomorphism $T_g$ of $E$; $T_g$ is called a *left quasi-rotation* on $G/H$ (respectively, a *left rotation* on the quotient group $G/H$ if $H$ is normal in $G$). The set of all left quasi-rotations is an irreducible compact group of positive operators in $\mathscr{L}_s(E)$ (and a homomorphic image of $G$), cf. § 8, Example 6. Below (§ 10) it will be shown that, roughly speaking, every irreducible compact group of positive operators on a Banach lattice is induced by a group of quasi-rotations. Note that, if $m$ denotes Haar measure on $G$ (§ 8, Example 6) and if $m_H$ is the image of $m$ under the quotient map $G\to G/H$, the group of all left quasi-rotations on $G/H$ defines (by continuous extension) an irreducible compact group of lattice isomorphisms in each of the spaces $L^p(m_H)$ $(1\leq p<+\infty)$.

If $(X,\Sigma,\tau)$ is a finite measure space, then the (class of the) constant-one function $e$ on $X$ is a common quasi-interior point of the positive cone of each of the Banach lattices $L^p(\tau)$; hence all of these spaces have the same structure space, that is, the Stone representation space $K$ of the Boolean $\sigma$-algebra $\Sigma/N$ ($N$ the ideal of $\tau$-null sets in $\Sigma$) (cf. § 5, Example 5 and Chap. II, § 8). $\tau$ defines a bounded, regular Borel measure $\mu$ on $K$ strictly positive on non-void open sets and vanishing on rare sets, such that (by 4.5) $L^p(\tau)$ can be identified with the vector lattice of all continuous numerical functions $f: K\to\overline{\mathbb{R}}$ for which $|f|^p$ is $\mu$-integrable, under the obvious norm. This simultaneous representation is the content of the following lemma. (Note that such a simultaneous representation is still possible if $\tau$ is only supposed to be $\sigma$-finite, but then the measure on $K$ needed to represent $L^p(\tau)$, can depend on $p$.)

**9.4 Lemma.** *Let $K$ be an extremally disconnected compact space, and let $\mu$ be a bounded, regular Borel measure on $K$ which is strictly positive on non-void open sets and vanishes on rare sets. Then $L^p(\mu)$ can be identified with the space of all continuous numerical functions $f$ on $K$ for which $|f|^p$ is $\mu$-integrable $(1\leq p<+\infty)$, while $L^\infty(\mu)$ can be identified with $C(K)$.*

*Proof.* We show that for each Borel set $A\subset K$, there exists an open-and-closed set $V\subset K$ such that $\mu(V\triangle A)=0$. By the regularity of $\mu$, there exist open sets $U_n\supset A$ satisfying $\mu(U_n\setminus A)\leq n^{-1}$ $(n\in\mathbb{N})$; since $\bar{U}_n\setminus U_n$ is rare, we conclude

$\mu(\bar{U}_n \setminus A) \leqq n^{-1}$. $V_0 := \bigcap_n \bar{U}_n$ is closed, contains $A$ and satisfies $\mu(V_0 \setminus A) = 0$. $W := K \setminus V_0$ is open, and again $\mu(\bar{W}) = \mu(W)$ by the hypothesis on $\mu$. Now $\bar{W}$ is open-and-closed, so $V := K \setminus \bar{W}$ is open-and-closed and satisfies $\mu(V \Delta A) = 0$. The characteristic function $\chi_V$ of $V$ is continuous and, since $\mu$ is strictly positive on non-void open sets, the unique continuous function in its equivalence class in $L^\infty(\mu)$. Therefore, the canonical imbedding $C(K) \to L^\infty(\mu)$ is a surjective isometry which proves the assertion for $p = +\infty$. Since $L^\infty(\mu)$ is a dense ideal in $L^p(\mu)$ $(1 \leqq p < +\infty)$, the remainder is a consequence of (4.5). □

*Remark.* The lemma actually establishes a strong lifting $L^\infty(\mu) \to \mathscr{L}^\infty(\mu)$ (cf. A. and C. Ionescu-Tulcea [1969]) which extends to $L^p(\mu) \to \mathscr{L}^p(\mu)$ with range consisting of continuous numerical functions (here, two functions in $\mathscr{L}^p(\mu)$ are to be identified whenever they agree on the dense subset of $K$ on which both are finite). From the preceding proof it is also clear that the lifting $L^\infty(\mu) \to \mathscr{L}^\infty(\mu)$ is a monomorphism of algebras.

**9.5 Proposition.** *Let $K, H$ be compact, extremally disconnected spaces and let $\mu, \nu$ be measures on $K$ and $H$, respectively, each satisfying the hypothesis of* (9.4). *Denote by $k: H \to K$ any continuous map, and by $T_k$ the associated lattice homomorphism of $C(K) \cong L^\infty(\mu)$ into $C(H) \cong L^\infty(\nu)$. The following assertions are equivalent:*

(a) *$k$ maps open subsets of $H$ onto open subsets of $K$.*
(b) *$T_k$ is order continuous.*
(c) *$\nu \circ k^{-1}$ is absolutely continuous with respect to $\mu$.*
(d) *There exists a bounded, regular Borel measure $\tilde{\nu}$ on $H$ which is equivalent to $\nu$ and such that $T_k$ has a continuous extension to $L^p(\mu)$ with values in $L^p(\tilde{\nu})$ $(1 \leqq p < +\infty)$.*

*Proof.* The equivalence (a)$\Leftrightarrow$(b) is implied by (iii), (iv) of (9.3).

(b)$\Rightarrow$(c): We have to show that for each Borel set $A \subset K$, $\mu(A) = 0$ implies $\nu[k^{-1}(A)] = 0$. By the regularity of $\mu$, there exists a decreasing sequence of open sets $U_n \supset A$ for which $\mu(U_n) \to 0$; since $\mu$ vanishes on rare sets and $K$ is extremally disconnected (Stonian), we can assume $U_n$ open and closed. The characteristic functions $\chi_n \in C(K)$ of $U_n$ order converge to 0 in $C(K)$; since $T_k$ is order continuous, the characteristic functions $T_k \chi_n$ of $k^{-1}(U_n)$ order converge to 0 in $C(H)$. Clearly this implies $\nu[k^{-1}(U_n)] \to 0$ hence, $\nu[k^{-1}(A)] = 0$.

(c)$\Rightarrow$(d): By (9.4) we identify $L^1(\mu)$ with the vector lattice of all continuous numerical functions on $K$ which are $\mu$-integrable. Since $\nu \circ k^{-1}$ is absolutely $\mu$-continuous by hypothesis, the Radon-Nikodym theorem asserts the existence of a unique $h \in L^1(\mu)$ satisfying $\nu \circ k^{-1} = h.\mu$. Since $\mu$ vanishes on rare sets and $\nu$ is strictly positive on non-void open sets, it follows that $k^{-1}(B)$ is rare in $H$ for rare subsets $B \subset K$. Now we define $h_0 := h \vee e_K$ ($e_K$ the constant-one function on $K$) and $g := h_0^{-1} \circ k$; then $0 \leqq g \leqq e_H$ and the set of zeros of $g$ is rare in $H$, since $h$ (and hence $h_0$) takes the value $+\infty$ only in a rare subset of $K$ (cf. 4.5). We define $\tilde{\nu} := g.\nu$; $\tilde{\nu}$ is bounded and, by the preceding, equivalent to $\nu$. Moreover, since $T_k$ is also a homomorphism of algebras (cf. 9.1), for $f \in C(K) \cong L^\infty(\mu)$ we have $(p \geqq 1)$

$$\int |T_k f|^p \, d\tilde{\nu} = \int T_k |f|^p \, d\tilde{\nu} = \int |f|^p h_0^{-1} \, d(\nu \circ k^{-1}) \leqq \int |f|^p \, d\mu\,,$$

which proves that $T_k$ has a continuous extension to $L^p(\mu)$ with values in $L^p(\tilde{\nu})$ $(p \geqq 1)$.

(d)$\Rightarrow$(b): Since for directed subsets of the spaces $L^p(\mu)$ order convergence and norm convergence are equivalent $(1 \leqq p < +\infty)$, the hypothesis implies that $T_k$ is order continuous. ☐

In a fashion analogous to Prop. 9.5, criteria can be formulated for the isomorphism of $L^p$-spaces constructed from distinct measures (Exerc. 27). Let us note also that, if $(X, \Sigma, \mu)$ and $(Y, T, \nu)$ are arbitrary finite measure spaces and $T$ is a (even order continuous) lattice homomorphism $L^\infty(\mu) \to L^\infty(\nu)$, it is not always possible to find a measurable transformation $\varphi: Y \to X$ such that $T = T_\varphi$ (A. and C. Ionescu-Tulcea [1969], § 10, and Exerc. 27). However, the answer is affirmative if an order continuous lifting $T_0: \mathscr{L}^\infty(\mu) \to \mathscr{L}^\infty(\nu)$ of $T$ exists (Exerc. 26).

## § 10. Irreducible Groups of Positive Operators. The Halmos-von Neumann Theorem

We have now gathered sufficient information for a successful study of irreducible compact groups of positive operators in $\mathscr{L}_s(E)$, where $E$ is an arbitrary Banach lattice. (Throughout this section, it is understood that an operator group is a multiplicative group $G \subset \mathscr{L}(E)$ with identity $1_E$.) By (7.9) Cor. 1 such a group is mean ergodic; in particular, if $G$ is an irreducible group (Def. 8.1) of positive operators then by the corollary of (8.5) the associated projection $P$ (the zero element of $\overline{\text{co}}\, G$) is irreducible and of rank 1 so that $E_+$ contains quasi-interior points (Chap. II, § 6). Moreover, since each $T \in G$ has a positive inverse, $G$ contains only lattice isomorphisms (II.2.6 Cor.). Our main result (10.4) shows that for some closed (in general, non-normal) subgroup $H \subset G$, $E$ can be viewed as a Banach lattice of $m_H$-integrable functions on $G/H$ ($m_H$ the measure induced by Haar measure $m$ on $G$) and that, under this identification, $G \subset \mathscr{L}(E)$ is induced by the group of all left (or right) quasi-rotations on $G/H$. We begin with the following result which is of independent interest, particularly in view of its corollary.

**10.1 Proposition.** *Let $S$ be an irreducible bounded semi-group of vector lattice homomorphisms of the Banach lattice $E$. If $S$ has a fixed vector $u \neq 0$, then each finite-dimensional $S$-invariant vector subspace of $E$ is contained in the principal ideal $E_u$.*

*Proof*. Since $Tu = u$ and $T \in S$ implies $T|u| = |Tu| = |u|$, we can assume that $u > 0$. Let $F$ denote any finite dimensional, $S$-invariant subspace $\neq \{0\}$ of $E$ and denote by $\{x_1, \ldots, x_n\}$ a Hamel basis of $F$. Since $S$ is bounded, there exists a constant $c > 0$ such that $x \in F$, $T \in S$, and $Tx = \sum_{i=1}^n \alpha_i x_i$ implies $\sup_i |\alpha_i| \leqq c\|x\|$. Defining $x_0 := \sum_{i=1}^n |x_i|$, we obtain $T|x| = |Tx| \leqq c\|x\| x_0$ for all $x \in F$, $T \in S$. Now if $(c x_0 - \xi u)^+$ were quasi-interior to $E_+$ for all real $\xi > 0$, we would obtain $(c x_0 - \xi u)^- = 0$ and hence $\xi u \leqq c x_0$ for all $\xi > 0$; but this is impossible, since $u > 0$ and $E$ is Archimedean ordered. Thus we can choose $\xi_0 > 0$ such that

$z := (c x_0 - \xi_0 u)^+$ is not quasi-interior to $E_+$. Since each $T \in S$ is a lattice homomorphism, we obtain for $x \in F$, $\|x\| \leqq 1$, and $T \in S$

$$T(|x| - \xi_0 u)^+ = (T|x| - \xi_0 u)^+ \leqq z\,.$$

But this means that for each $x \in F$, the orbit under $S$ of $(|x| - \xi_0 \|x\| u)^+$ is contained in the non-dense ideal $E_z$; by (8.3) this implies $(|x| - \xi_0 \|x\| u)^+ = 0$, or equivalently, $|x| \leqq \xi_0 \|x\| u$. Therefore, $F \subset E_u$. □

**Corollary.** *Let $(X, \Sigma, \mu)$ be a finite measure space, let $1 \leqq p < +\infty$ be fixed, and let $G$ be a group of positive contractions in $L^p(\mu)$. If the fixed space of $G$ consists precisely of the constant functions, then each finite dimensional $G$-invariant subspace of $L^p(\mu)$ is contained in $L^\infty(\mu)$.*

*Proof.* As usual, we identify functions with their equivalence classes modulo $\mu$-null functions, and we identify $L^\infty(\mu)$ with the ideal in $L^p(\mu)$ generated by the constant-one function $e$. Then (10.1) applies, provided that we can show $G$ to be irreducible. To this end, let $J$ denote a closed $G$-invariant ideal in $L^p(\mu)$. There exists a set $B \in \Sigma$ such that $J$ is the ideal of functions vanishing on $B$ (§ 1, Example 2). Let $e_1$ denote the characteristic function of $B$ and define $e_2 := e - e_1$. Since $G$ is assumed to be an operator group of positive contractions, it follows that each $T \in G$ is an isometric isomorphism of $L^p(\mu)$ onto itself. Now $Te = e$ and $TJ \subset J$ implies $Te_2 \leqq e_2 \wedge e = e_2$ and, since $T$ is an isometry, we have $Te_2 = e_2$ for each $T \in G$. From the hypothesis that each $G$-fixed function be constant it follows that $e_2 = 0$ or $e_2 = e$. Hence, $J = \{0\}$ or $J = L^p(\mu)$ which shows $G$ to be irreducible. □

The following special case of (10.4) is well known in topological dynamics (see § 8, Example 5 and R. Ellis [1969]); the concept of quasi-rotation has been defined above (§ 9, Example 3).

**10.2 Proposition.** *Let $K$ be a compact space, and let $G$ be an irreducible group of Markov operators on $C(K)$ which is compact in the strong operator topology. There exists a closed subgroup $H$ of $G$ and a homeomorphism of $K$ onto $G/H$ under which $G$ is conjugate to the group of all left quasi-rotations on $G/H$. If $G$ is abelian then $H = \{1_{C(K)}\}$ and $G$ is conjugate to the group of all left rotations in $C(G)$.*

*Note.* The homeomorphism $K \to G/H$ whose existence is asserted, induces an isomorphism of $C(G/H)$ onto $C(K)$ (cf. 9.1); it is with respect to this isomorphism that the conjugacy of $G$ with the operator group of all left quasi-rotations in $C(G/H)$ is claimed.

*Proof* of (10.2). Since $G$ is a group of lattice isomorphisms of $C(K)$ which leave the constant-one function fixed (and thus are even isometries), (9.1) and (9.3) implies that $G$ is isomorphic (as a group) to a group $G_0 := \{k_T : T \in G\}$ of homeomorphisms of $K$. We now fix $s_0 \in K$ and define a continuous map $\psi : G \to K$ by $\psi(T) := k_T s_0$. Clearly $\psi(G)$ is closed and the irreducibility of $G$ implies that $\psi$ is surjective; in fact, $K$ is a minimal (non-void) $G_0$-invariant subset of $K$ (§ 8, Example 5) and for each $s \in K$, the compact orbit $G_0 s$ equals $K$. The set

$H:=\{T\in G: k_T s_0 = s_0\}$ is a closed subgroup of $G$, and evidently $\psi(T_1)=\psi(T_2)$ is equivalent to $T_1 T_2^{-1}\in H$. So $\psi$ can be factored through the compact quotient $G/H$ and thus defines a continuous bijection $\psi_0: G/H\to K$. It follows that $\psi_0$ is a homeomorphism and hence defines an isomorphism $U$ of the Banach lattice $C(K)$ onto $C(G/H)$; the operator $\hat{T}:=UTU^{-1}\in\mathscr{L}(C(G/H))$ runs through the group of all left quasi-rotations on $G/H$ as $T$ runs through $G$. In fact, we have

$$\hat{T}f(SH)=Tf(k_S s_0)=f(k_T k_S s_0)=f(TSH) \qquad (S\in G)$$

with obvious notation, and it is easy to see that the conjugation $T\mapsto\hat{T}$ is an isomorphism of (topological) operator groups.

Finally, if $G$ is abelian then $T_0\in H$ implies $k_T s_0 = k_T k_{T_0} s_0 = k_{T_0} k_T s_0$ for all $T\in G$; since, as observed above, the orbit $\{k_T s_0: T\in G\}$ equals $K$, it follows that $k_{T_0}$ is the identity map of $K$, or $T_0=\{1_{C(K)}\}$. Hence in this case, $H$ is trivial as asserted. □

*Remark.* The subgroup $H$ constructed in the preceding proof is, in general, not normal (cf. Example 2 below); identifying $G_0$ with $G$, we have taken $H$ as the so-called *isotropy subgroup* $H(s_0)$ of $G$. Here the choice of $s_0\in K$ is immaterial; for distinct values of $s_0$ the corresponding subgroups are conjugate under some inner automorphism of $G$. Moreover, the above construction can be reversed in the following sense: If, for some closed subgroup $H\subset G$, the operator group $G$ is conjugate to the group of all left (or right) quasi-rotations in $C(G/H)$, then under $G\to G_0$ $H$ is isomorphic to some isotropy subgroup of $G$. In particular, if $G$ is abelian then necessarily $H=\{1\}$.

For an extension of (10.2) to arbitrary Banach lattices we need the following basic theorem from the theory of almost periodic (Banach space valued) functions (see Berglund-Hofmann [1967]).

**10.3 Lemma.** *Let $G$ be a compact group in $\mathscr{L}_s(E)$, where $E$ is any Banach space. Then the union of all finite dimensional $G$-invariant vector subspaces of $E$ is dense in $E$.*

The principal approximation theorem (Peter-Weyl Theorem) of the theory of $B$-space valued almost periodic functions on groups asserts that each such function can be uniformly approximated by functions whose orbit under all right and left rotations is contained in a finite dimensional subspace of the range space (see Glicksberg-de Leeuw [1961]). If $G$ is a compact topological group, every continuous function on $G$ into a Banach space is almost periodic (l.c.); in the case of a compact operator group $G\subset\mathscr{L}_s(E)$ which concerns us here, the functions in question are the functions $T\mapsto Tx$ $(x\in E)$. Moreover, if $G$ is abelian then the minimal $G$-invariant subspaces $\neq\{0\}$ of $E$ are of dimension at most two or one, accordingly as $E$ is a Banach space over the real or complex field.

**10.4 Theorem.** *Let $E$ be any Banach lattice, and let $G$ be an irreducible group (with identity $1_E$) of positive operators which is compact in $\mathscr{L}_s(E)$. There exists a minimal closed subgroup $H$ of $G$, which is unique to within conjugacy under inner* *automorphisms of G and which is the identity subgroup if G is abelian, such that the following holds:*

(i) *If $m_H$ denotes the image of normalized Haar measure m on G under the quotient map $G \to G/H$, the canonical injection $C(G/H) \to L^1(m_H)$ can be factored through E:*

$$C(G/H) \xrightarrow{J_1} E \xrightarrow{J_2} L^1(m_H)$$

*where $J_1$ and $J_2$ are continuous lattice monomorphisms with dense range;*

(ii) *If (under $J_2$) E is identified with a vector sublattice of $L^1(m_H)$, then the operator group of all left quasi-rotations in $L^1(m_H)$ induces the given group $G \subset \mathscr{L}(E)$.*

*Proof.* For greater transparency consider this diagram:

$$\begin{array}{ccc} F \xrightarrow{I_1} E \xrightarrow{I_2} & (E,q)^{\sim} \\ \uparrow U_1 & \downarrow U_2 \\ C(G/H) \xrightarrow{\quad I \quad} & L^1(m_H) \end{array}$$

where the symbols employed have the following meaning. $F$ is a Banach lattice algebraically identical with a dense sublattice of $E$; $I_1$ is the canonical injection, and continuous; $I_2$ is the canonical imbedding of $E$ into the completion $(E,q)^{\sim}$ with respect to a continuous $L$-norm $q$ on $E$ (Chap. II, § 8). Finally, $I$ is the canonical imbedding (which is injective, since $m_H$ is a strictly positive Radon measure on $G/H$), $U_1$ and $U_2$ are isomorphisms of Banach lattices, and the diagram commutes. The proof is divided into five steps: 1. Construction of $F$, 2. Denseness of $F$ in $E$, 3. Construction of $U_1$, 4. Construction of $U_2$, 5. Essential uniqueness of $H$. The maps $J_1, J_2$ are then obtained as $J_1 := I_1 \circ U_1$, $J_2 := U_2 \circ I_2$. Assertion (ii) will become clear from these steps; more precisely, the group $G_{|F}$ of restrictions $T_{|F}$ $(T \in G)$ will be shown to be conjugate (under $U_1$) to the group of all left quasi-rotations on $G/H$.

1. *Construction of F.* By the corollary of (8.5) there exists a unique normalized quasi-interior point $u \in E_+$ which is fixed under $G$. The principal ideal $E_u = \bigcup_{n=1}^{\infty} n[-u,u]$ is an $AM$-space with unit $u$ and unit ball the order interval $[-u,u]$ (Chap. II, § 7); clearly $E_u$ is invariant under $G$. We define $F$ to be the set of vectors $x \in E_u$ for which $Gx$ is relatively compact in the $AM$-space $E_u$. Since the elements $T \in G$ are lattice homomorphisms, $F$ is a vector sublattice of $E_u$, and since each $T \in G$ defines a Markov operator in $E_u$ an equicontinuity argument shows $F$ to be closed in $E_u$. But $u \in F$ and so $F$ is itself an $AM$-space with unit $u$. Clearly $F$ is $G$-invariant, and if we denote by $G_{|F}$ the group of restrictions $T_{|F}$ $(T \in G)$ then $G_{|F}$ is a group of Markov operators which is relatively compact in $\mathscr{L}_s(F)$.

2. *F is dense in E.* Since the order interval $[-u,u]$ is bounded in $E$, it is clear that the injection $I_1: F \to E$ is continuous. To show that $F$ is dense in $E$, let $F_0$ denote the set of all vectors $x \in E$ which are contained in some finite

dimensional $G$-invariant subspace of $E$; since $G \subset \mathscr{L}_s(E)$ is compact by hypothesis, it follows from (10.3) that $F_0$ is a dense vector subspace of $E$. By (10.1) we have $F_0 \subset E_u$; but since on finite dimensional subspaces of $E_u$ the topologies induced by $E$ and by $E_u$ agree we conclude that $F_0 \subset F$. Therefore, $F$ is dense in $E$.

3. *Construction of* $U_1$. We consider the group $G_{|F}$ (Step 1). We observe that $G_{|F}$ is closed in $\mathscr{L}_s(F)$. In fact, if $T_0 \in \mathscr{L}_s(F)$ is in the closure of $G_{|F}$ then, since $F$ is dense in $E$ and $G$ is bounded (hence equicontinuous) in $\mathscr{L}(E)$, $T_0$ has a continuous extension to $E$ which is in the closure of $G$ in $\mathscr{L}_s(E)$ and hence belongs to $G$. Consequently $G_{|F}$ is compact in $\mathscr{L}_s(F)$ (cf. Step 1). Now denote by $\boldsymbol{T}_F$ the topology on $G$ induced by $\mathscr{L}_s(F,E)$, then $\boldsymbol{T}_F$ agrees with the topology induced by $\mathscr{L}_s(E)$, since the latter is compact and $\boldsymbol{T}_F$ is a coarser Hausdorff topology. On the other hand, the map $G_{|F} \to (G, \boldsymbol{T}_F)$ is evidently continuous, since $I_1: F \to E$ is continuous, and hence a homeomorphism. It follows that the restriction map $G \to G_{|F}$ is an isomorphism of the compact group $G$ onto the compact group $G_{|F} \subset \mathscr{L}_s(F)$.

By the corollary of (8.5) the projection $P$ associated with $G$ is of the form $P = \mu \otimes u$, where $\mu$ is a strictly positive linear form on $E$ satisfying $\mu(u) = 1$. Since evidently $P_{|F}$ is the projection associated with $G_{|F}$, (8.5) shows $G_{|F}$ to be irreducible. Hence (10.2) can be applied to $G_{|F}$, $F$ being an $AM$-space with unit. This yields the existence of a closed subgroup $H \subset G$ ($H = \{1\}$ if $G$ is abelian) and an isomorphism $U_1$ of $C(G/H)$ onto $F$ under which $G_{|F}$ is conjugate to the group of all left quasi-rotations on $G/H$. This establishes the existence of $U_1$.

4. *Construction of* $U_2$. With the aid of (8.9) it is easy to see that the adjoint $U_1'$ of $U_1$ maps the unique $u$-normalized $G_{|F}$-invariant positive linear form $\mu$ (more precisely, the restriction to $F$ of the above mentioned $\mu \in E'$) onto $m_H$, the Radon measure on $G/H$ obtained from normalized Haar measure $m$ on $G$ under the quotient map $G \to G/H$. We consider the $L$-norms (Chap. II, § 8) $f \mapsto p(f) := m_H(|f|)$ on $C(G/H)$ and $x \mapsto q(x) := \mu(|x|)$ on $E$, respectively. Since $q$ is continuous on $E$, the denseness of $F$ in $E$ (Step 2) implies that $F$ is a dense vector sublattice of $(E, q)$. From the preceding it follows that $U_1: C(G/H, p) \to (F, q)$ is an isomorphism of normed vector lattices which extends uniquely to an isomorphism $V$ of $L^1(m_H)$ onto $(E,q)\tilde{}$. Then $V^{-1}$ defines the desired isomorphism $U_2$.

5. *Essential Uniqueness of* $H$. If $\tilde{H}$ is any closed subgroup of $G$ for which assertions (i) and (ii) of the theorem hold, there exists a continuous lattice monomorphism $C(G/\tilde{H}) \to E$ with range a dense sublattice $\tilde{F}$ of $E$; $\tilde{F}$ is an $AM$-space whose unit can, without loss of generality, assumed to be $u$ (cf. Step 1). Since by hypothesis $G_{|\tilde{F}}$ is conjugate to the group of all left quasi-rotations on $G/\tilde{H}$, it follows that for each $\tilde{x} \in \tilde{F}$, the orbit $G\tilde{x}$ is compact in $\tilde{F}$ and hence in $E_u$ (note that $\tilde{F}$ is a closed vector sublattice of $E_u$). Therefore, by construction of $F$ (Step 1) we have $\tilde{F} \subset F$. The subgroup $H$ of $G$ obtained in Step 3 for which $C(G/H)$ is isomorphic to $F$, is determined uniquely up to an inner automorphism of $G$ (Remark following 10.2). Let $K$, $\tilde{K}$ denote the compact spaces $G/H$ and $G/\tilde{H}$, respectively. The injection $\tilde{F} \to F$ defines a Markov operator $C(\tilde{K}) \to C(K)$ given by $\tilde{f} \mapsto \tilde{f} \circ \varphi$, where by (9.1) and (9.3) $\varphi$ is a continuous surjection $K \to \tilde{K}$. Now (cf. proof of 10.2) $G$ can be viewed as a transformation group $\{\tilde{k}_T : T \in G\}$ on $\tilde{K}$ and as a transformation group $\{k_T : T \in G\}$ on $K$, and we must have $\tilde{k}_T \circ \varphi = \varphi \circ k_T$ for all $T \in G$. Identifying $H$ and $\tilde{H}$ with isotropy subgroups of

the corresponding transformation groups corresponding to points $s_0 \in K$ and $\tilde{s}_0 \in \tilde{K}$, respectively, we can arrange (by applying an inner automorphism of $G$ to $H$, if necessary) that $\varphi(s_0) = \tilde{s}_0$. If $H'$ denotes the image of $H$ under such an automorphism, the relation $\tilde{k}_T \circ \varphi = \varphi \circ k_T$ $(T \in G)$ shows that $H' \subset \tilde{H}$. Thus if $\tilde{H}$ is a minimal closed subgroup of $G$ satisfying (i) and (ii) of the theorem, we must have $H' = \tilde{H}$. That is, $\tilde{H}$ is determined uniquely to within conjugacy under an inner automorphism of $G$.

This completes the proof of (10.4). □

**Examples.** 1. Important examples of irreducible groups of positive operators are the operator groups in $L^p(\mu)$ $(1 \leqq p < +\infty)$ induced by the minimal transformation groups (ergodic dynamical systems) of topological dynamics (§ 8, Example 7); here the compactness of the operator group is often verified through equicontinuity of the transformation group (cf. 7.9 Cor. 2), or by conditions on the point spectrum of the corresponding operators (see 10.5 below). If, more generally, $G$ is an irreducible compact group of positive operators in a space $L^p(\mu)$ then $L^p(\mu)$ can be "sandwiched" between $C(G/H)$ and $L^1(m_H)$ with $G$ induced by the group of all left (or right) quasi-rotations on $G/H$; if, in addition, $G$ is a group of isometries then $L^p(\mu)$ can be identified with $L^p(m_H)$ $(1 \leqq p < +\infty)$.

2. The following simple but instructive example shows why, in the non-abelian case, it is necessary to consider quotients $G/H$ over non-normal subgroups $H$ of $G$. Let $G$ denote the group of all $n \times n$-permutation matrices considered as a group of positive operators on $\mathbb{R}^n$ (Chap. I, § 4). $G$ is isomorphic to the symmetric group $\mathsf{S}_n$ of order $n!$ and by the construction given in the proof of (10.2) we can take $H$ as the subgroup of those $P_\pi \in G$ $(\pi \in \mathsf{S}_n)$ for which $\pi(1) = 1$. Then if $P_k$ corresponds to some permutation taking 1 into $k$ $(k = 1, \ldots, n)$, $G/H$ consists of the $n$-cosets $P_1 H, \ldots, P_n H$. Clearly $H$ (and any of the remaining $n-1$ choices) is not necessarily a normal subgroup of $G$. The example also illuminates the fact that a transitive, commutative subgroup of $\mathsf{S}_n$ must have order $n$.

3. Let $K$ be a compact space and denote by $T$ a Markov operator on $C(K)$ (Def. 8.8) which is irreducible. We suppose in addition that the canonical extension of $T$ (again denoted by $T$) to the complex Banach lattice $C(K)$ (Chap. II, § 11) possesses a set of unimodular eigenvalues whose eigenvectors form a total subset of $C(K)$. Let $G$ denote the closure in $\mathscr{L}_s(C(K))$ of the cyclic semi-group generated by $T$. It is not difficult to see that $G$ is compact and that $G$ must contain the identity map $1_{C(K)}$ (cf. proof of 10.5 (i) below). But then the smallest closed semi-group ideal (kernel) of $G$ must also contain $1_{C(K)}$ and hence, Lemma 7.5 shows $G$ to be a group. Therefore, $G$ is an abelian, compact, irreducible group of Markov operators in $\mathscr{L}_s(C(K))$.

By (10.2) $K$ is homeomorphic to $G$, and identifying $K$ with $G$ transforms $T$ into the translation operator $T_g$ on $C(G)$ defined by a generator $g$ of the monothetic group $G$. Since $T_g$ is an isometry, it is clear that each eigenvalue of $T$ must be unimodular; if $\alpha$ is such an eigenvalue and $f \in C(G)$ is a corresponding eigenvector normalized to satisfy $f(1) = 1$, then from $\alpha f(h) = f(gh)$ $(h \in G)$ we conclude $\alpha = f(g)$. Thus $f(gh) = f(g) f(h)$ for all

$h \in G$ and, since $g$ is a generator of $G$, the continuity of $f$ shows $f$ to be a character of $G$. Conversely, every continuous character of $G$ is an eigenvector of $T_g$ and the map $f \mapsto f(g)$ (which is injective, since $g$ generates $G$) is an isomorphism of the character group $G^*$ onto the point spectrum of $T_g$, considered as a subgroup of the circle group. (Cf. also § 8, Example 4.)

For an extension due to I. Glicksberg see Exerc. 28.

In the spirit of Theorem 10.4, the preceding example can be extended to arbitrary Banach lattices $E$. Since the hypothesis involves complex eigenvalues of an operator, we shall suppose here that $E$ is a *complex Banach lattice* (Chap. II, § 11). Also, since the group $G$ to be considered is abelian, we shall speak of *translations* rather than rotations.

**10.5 Theorem** (Halmos-von Neumann). *Let $E$ be any complex Banach lattice and let $T$ be a positive irreducible operator on $E$. Suppose that the cyclic semi-group $S := \{T^n : n \in \mathbb{N}\}$ is bounded in $\mathscr{L}(E)$ and that the eigenvectors pertaining to unimodular eigenvalues of $T$, span a dense vector subspace of $E$. Then the following holds:*

(i) *$T$ is a lattice isomorphism of $E$, and the closure $G$ of $S$ in $\mathscr{L}_s(E)$ is a compact group with identity $1_E$;*

(ii) *If $m$ denotes normalized Haar measure on $G$, the canonical injection $C(G) \to L^1(m)$ can be factored through $E$:*

$$C(G) \xrightarrow{J_1} E \xrightarrow{J_2} L^1(m)$$

*where $J_1$ and $J_2$ are continuous lattice monomorphisms with dense range such that under $J_2$, $G$ is induced by the group of all translation operators in $L^1(m)$;*

(iii) *The point spectrum of $T$ is a subgroup of the circle group which is isomorphic (as a discrete group) with the character group $G^*$ of $G$.*

*Proof.* Denote by $E_0$ the linear span of all eigenvectors of $T$ pertaining to unimodular eigenvalues; $E_0$ is dense in $E$ by hypothesis.

(i) Clearly, the orbit $Sx$ is relatively compact in $E$ for each $x \in E_0$; since $E_0$ is dense in $E$ and $S$ is bounded in $\mathscr{L}(E)$, the standard equicontinuity argument shows $S$ to be relatively compact in $\mathscr{L}_s(E)$. Thus $G$ is compact in $\mathscr{L}_s(E)$. Now let $\Phi = \{x_1, \dots, x_n\}$ be a finite set of eigenvectors pertaining to unimodular eigenvalues of $T$ so that $Tx_i = \alpha_i x_i$ and $|\alpha_i| = 1$ $(i = 1, \dots, n)$, say. By the compactness of $\Gamma^n$ ($\Gamma$ the circle group), there exists a sequence $(n_k)_{k \in \mathbb{N}}$ of positive integers such that $\lim_k \alpha_i^{n_k} = 1$ for $i = 1, \dots, n$. Hence, for each cluster point $R$ in $G$ of the sequence $(T^{n_k})$ we have $Rx_i = x_i$ $(i = 1, \dots, n)$. Thus if for each finite set $\Phi \subset E_0$ we denote by $P_\Phi$ the subset of $G$ whose elements leave each $x \in \Phi$ fixed, we have shown that $P_\Phi$ is non-void for all $\Phi$. Clearly, the family $(P_\Phi)$ is directed downward (with respect to the natural direction of finite subsets of $E_0$) and has the finite intersection property; since each $P_\Phi$ is closed in $G$ and $G$ is compact, it follows that $\bigcap_\Phi P_\Phi$ is non-void and hence equals $\{1_E\}$. Since multiplication in $G$ is jointly continuous (cf. proof of 7.9 Cor. 1), it is easy to see that the intersection $K$ of all closed semi-group ideals of $G$ must contain $1_E$; so $K = G$

and we infer from (7.5) that $G$ is a group (which is monothetic with generator $T$, and irreducible since $T$ is).

(ii) This assertion is an immediate consequence of (10.4).

(iii) From the construction of the $AM$-space $F$ (Steps 1 and 2 of the proof of 10.4) it follows that each eigenvector of $T$ is contained in $F$. On the other hand, $F$ is isomorphic with $C(G)$ (the isomorphism in question being the map $U_1$, cf. Step 3) in such a way that the group $G$ is conjugate (under $U_1$) to the group of all translation operators on $C(G)$. This reduces the present case to the situation considered in Example 3 above which completes the proof. □

The original version of the theorem (Halmos and von Neumann [1942]) refers to ergodic dynamical systems (cf. § 8, Example 7) with discrete spectrum. Thus a quadruple $(X,\Sigma,\mu,\varphi)$ is considered where $(X,\Sigma,\mu)$ is a finite measure space and $\varphi$ an invertible measure preserving transformation on $X$ admitting no non-trivial $\varphi$-invariant measurable disjunction (cf. § 9, Example 1) of $X$; the term *discrete spectrum* then refers to the hypothesis that the induced unitary operator $T_\varphi$ on $L^2(\mu)$ have pure point spectrum, or equivalently, that the eigenvectors of $T_\varphi$ form a total subset of $L^2(\mu)$. Thus the classical version of (10.5) reads as follows.

**Corollary.** *Each ergodic dynamical system with discrete spectrum is isomorphic to a system $(G,\Sigma,m,g)$, where $G$ is a compact monothetic group with generator $g$, $\Sigma$ the Borel field of $G$, and $m$ Haar measure on $G$.*

Here the term *isomorphism* is to be understood as an isomorphism of measure algebras (Exerc. 27).

## § 11. Positive Projections

The Riesz decomposition theorem (II. 2.10) asserts that each band $B$ of an order complete vector lattice $E$ is an order direct factor of $E$, that is, $E=B+B^\perp$. If, in addition, $E$ is a normed (or, more generally, a topological) vector lattice, then each band decomposition of $E$ is a topological direct sum (II. 5.2). It is natural to ask under what conditions on a Banach lattice $E$ it is true that each closed ideal, or each closed vector sublattice, of $E$ is the range of a positive projection. We first take up this question for ideals.

**11.1 Proposition.** *For any Banach lattice $E$, the following assertions are equivalent:*

(a) *Each closed ideal of $E$ is the range of a positive projection.*

(b) *Each directed ($\leqq$) majorized family in $E$ is norm convergent.*

(c) *For any directed ($\geqq$) family $(x_\alpha)$ in $E$, $\inf_\alpha x_\alpha=0$ implies $\lim_\alpha\|x_\alpha\|=0$ (i.e., $E$ has order continuous norm).*

*Proof.* (a) ⇒ (c): Let $I$ denote any closed ideal in $E$ and let $P\geqq 0$ be a (necessarily continuous, cf. II. 5.3) projection with range $PE=I$. If $x\in E_+$ then $Px-x\leqq Px$ hence, $(Px-x)^+\leqq Px$ and this implies $(Px-x)^+\in I$. If we can show that $(Px-x)^-=(x-Px)^+\in I^\perp$ then the identity

$$x=[Px-(Px-x)^+]+(Px-x)^-$$

shows that $E=I+I^{\perp}$ and hence, that $I$ is a projection band (Def. II. 1.8; note, however, that $P$ is not necessarily the associated band projection). To show that $(x-Px)^{+}\in I^{\perp}$ we show that $y\in I$ and $0\leqq y\leqq (x-Px)^{+}$ implies $y=0$. In fact, $0\leqq y\leqq (x-Px)^{+}$ implies $y\leqq x$ and hence we have $y=Py\leqq Px$, since $P\geqq 0$. Therefore, $y\leqq (x-Px)^{+}\leqq x-y$ which shows that $2y\leqq x$; thus $2y\leqq Px$ and so $y\leqq (x-Px)^{+}\leqq x-2y$. Proceeding inductively we obtain $ny\leqq x$ for all $n\in\mathbb{N}$ whence $y=0$, since $E$ is Archimedean.

Thus each closed ideal of $E$ is a projection band. Now suppose that $(x_\alpha)$ is a directed ($\geqq$) family in $E$ satisfying $\inf_\alpha x_\alpha=0$, and let $0<\varepsilon\in\mathbb{R}$ be preassigned. Clearly, we can assume that $x_\alpha\leqq x$ for some $x\in E_+$ and all $\alpha$. Define $z_\alpha:=(x_\alpha-\varepsilon x)^{+}$ and denote by $P_\alpha$ the band projection onto the closed ideal generated by the singleton $\{z_\alpha\}$. We have

$$(1)\qquad (1_E-P_\alpha)(x_\alpha-\varepsilon x)=-(x_\alpha-\varepsilon x)^{-}\leqq 0,$$

$$(2)\qquad P_\alpha(x_\alpha-\varepsilon x)=\phantom{-}(x_\alpha-\varepsilon x)^{+}\geqq 0.$$

Since $0\leqq P_\alpha\leqq 1_E$, (1) shows that $0\leqq (1_E-P_\alpha)x_\alpha\leqq \varepsilon(1_E-P_\alpha)x\leqq \varepsilon x$, whence $\|(1_E-P_\alpha)x_\alpha\|\leqq\varepsilon\|x\|$ for all $\alpha$.

On the other hand, (2) implies $\varepsilon P_\alpha x\leqq P_\alpha x_\alpha\leqq x_\alpha$ for all $\alpha$, and so $\inf_\alpha P_\alpha x=0$; letting $y_\alpha:=(1_E-P_\alpha)x$ we obtain $x=\sup_\alpha y_\alpha$. Since the closed ideal generated by the family $(y_\alpha)$ is a projection band, it contains $x$ and hence, there exists $0<c\in\mathbb{R}$ and an index $\alpha_0$ such that $\|x-z\|<\varepsilon$ for some $z$ satisfying $0\leqq z\leqq c y_{\alpha_0}$. Now $x\wedge z\leqq x\wedge c y_{\alpha_0}\leqq (1_E-P_{\alpha_0})x$ which implies $P_{\alpha_0}x\leqq x-x\wedge z\leqq |x-z|$. It follows that $\|P_{\alpha_0}x\|\leqq\|x-z\|<\varepsilon$ and, using the inequality obtained at the end of the preceding paragraph, we obtain $\|x_{\alpha_0}\|\leqq\|P_{\alpha_0}x_{\alpha_0}\|+\|(1_E-P_{\alpha_0})x_{\alpha_0}\|\leqq\varepsilon(1+\|x\|)$. This proves (c).

(c) $\Rightarrow$ (b): For $(y_\beta)$ a directed ($\leqq$) majorized family in $E$, denote by $(z_\gamma)$ the family of all upper bounds of $(y_\beta)$. Evidently $(u_{\gamma\beta})$, where $u_{\gamma\beta}=z_\gamma-y_\beta$, is a directed ($\geqq$) family with infimum 0 and hence, norm convergent to 0 by hypothesis. This implies $(y_\beta)$ to be a Cauchy family, and so $(y_\beta)$ norm converges (to its supremum) in $E$.

(b) $\Rightarrow$ (a): (b) implies that $E$ is order complete and, moreover, that each closed ideal in $E$ is a band. Thus by (II. 2.10) each closed ideal in $E$ is the range of a positive projection (namely, the band projection associated with it). □

**Corollary.** *If a Banach lattice E satisfies one (and hence all) of the equivalent assertions of* (11.1), *then so does each closed ideal I of E and each quotient E/I.*

It is to be expected that the class of Banach lattices in which each closed vector sublattice is the range of a positive projection, is even more restricted than the class characterized by (11.1). We shall show that every space $L^p(X,\Sigma,\mu)$, where $1\leqq p<+\infty$ and $(X,\Sigma,\mu)$ is an arbitrary measure space has the property in question (cf. Remark 1 below). The essential tools will be a characterization of certain vector sublattices of $L^p(\mu)$ for finite $\mu$, the Radon-Nikodym theorem, the Riesz convexity theorem, and some representation theory (§ 5). In the sequel measurable functions and their equivalence classes (modulo $\mu$-null functions) will be identified when permissible; to preclude inessential complications, a $\sigma$-subalgebra of $\Sigma$ will be assumed to contain all $\mu$-null sets in $\Sigma$.

**11.2 Proposition.** *Let $E := L^p(\mu)$, where $1 \leqq p < +\infty$ and where $(X, \Sigma, \mu)$ is a finite measure space. The relation*

$$H = \{f \in E : f \text{ is measurable } \Sigma'\}$$

*defines a bijective correspondence between the set of all closed vector sublattices $H$ of $E$ that contain the constant functions, and the set of all ($\mu$-complete) $\sigma$-subalgebras $\Sigma'$ of $\Sigma$.*

*Proof.* It will be convenient to restrict attention to finite measurable functions on $X$. Given a $\sigma$-subalgebra $\Sigma'$ of $\Sigma$, it is clear from standard arguments of measure theory [H, § 19] that the set $H$ of $\Sigma'$-measurable functions in $E$ is a vector sublattice of $E$. For any Cauchy sequence in $H \subset E$, there exists a subsequence whose representatives can be assumed to converge everywhere [H, § 22 Thm. D] to some function $f$ which is evidently $\Sigma'$-measurable. This shows $H$ to be closed in $E$.

Conversely, let $H$ be a closed vector sublattice of $E$ that contains the constant functions; then the $\sigma$-ring $\Sigma'$ generated by the sets $f^{-1}(B)$ ($f \in H$; $B$ Borel set in $\mathbb{R}$) and completed by adding all $\mu$-null sets, is a $\sigma$-subalgebra of $\Sigma$. We have to show that whenever $g \in E$ is $\Sigma'$-measurable, then $g \in H$.

To this end, denote by $e$ the constant-one function on $X$ and define, for $f \in H$ and real numbers $\alpha$, $\beta$ satisfying $\alpha > \beta$,

$$h_{\alpha\beta}(f) := (\alpha - \beta)^{-1}(f \wedge \alpha e - f \wedge \beta e).$$

A simple computation shows that $0 \leqq h_{\alpha\beta}(f) \leqq e$ and that, for every increasing sequence $(\beta_n)$ converging to $\alpha$, $(h_{\alpha\beta_n}(f))$ is a decreasing sequence converging everywhere to the characteristic function $\chi_{[f \geqq \alpha]}$ of the set $\{t \in X : f(t) \geqq \alpha\}$. Since $H$ is a closed vector sublattice of $E$, we have $\chi_{[f \geqq \alpha]} \in H$ for all $\alpha \in \mathbb{R}$, and it follows in the standard fashion that the characteristic function of each $f^{-1}(B)$ ($f \in H$, $B$ real Borel set) is in $H$. Now let $\Sigma_0$ denote the family of all sets $A \in \Sigma$ for which $\chi_A \in H$. Then (since $e \in H$) $\Sigma_0$ is a $\sigma$-subalgebra of $\Sigma$ which, by the preceding, contains $\Sigma'$ and is contained in $\Sigma'$. Therefore $\Sigma' = \Sigma_0$.

Now if $g \in E$ is $\Sigma'$-measurable, $g$ is the pointwise limit of a sequence $(g_n)$ of linear combinations of $\Sigma'$-measurable characteristic functions such that $|g_n| \leqq |g|$ for all $n$ [H, § 20, Thm. B]. Thus we have $g_n \in H$, $|g(t) - g_n(t)| \to 0$ $(t \in X)$ and $|g - g_n|^p \leqq 2^p |g|^p$. By Lebesgue's dominated convergence theorem we conclude $g \in H$, since $H$ is closed. ☐

**11.3 Lemma.** *Let $H$ be a closed vector sublattice of $L^p(\mu)$, where $1 \leqq p < +\infty$ and $\mu$ is a finite measure, and suppose that $H$ contains the constant functions. There exists a positive projection (with norm 1) of $L^p(\mu)$ onto $H$.*

*Proof.* First we consider the case $p=1$. By (11.2) there exists a (unique) $\sigma$-subalgebra $\Sigma'$ of $\Sigma$ such that $H$ is the set of all $\Sigma'$-measurable functions in $L^1(\mu)$. Denote by $\mu'$ the restriction of $\mu$ to $\Sigma'$ and fix $f\in L^1(\mu)$. Then

$$A\mapsto \nu(A):=\int_A f\,d\mu \qquad (A\in\Sigma')$$

is a measure on $\Sigma'$ which is (absolutely) $\mu'$-continuous. By the Radon-Nikodym theorem ([H, § 31, Thm. B] or (II. 8.7) Cor.) there exists a (essentially) unique $\Sigma'$-measurable function $f_0$ such that $\nu(A)=\int_A f_0\,d\mu'$ for all $A\in\Sigma'$ (and such that $\|\nu\|=\int|f_0|\,d\mu'$).

From the uniqueness of $f_0$ it follows that the map $P\colon f\mapsto f_0$ is linear. Evidently $P\geqq 0$ and $Pf=f$ if $f\in L^1(\mu)$ is $\Sigma'$-measurable; so $P$ is a positive projection of $L^1(\mu)$ onto $H$. To see that $\|P\|=1$, it suffices to observe that

$$\begin{aligned}\|Pf\|&=\int|f_0|\,d\mu=\int|f_0|\,d\mu'=\sup\textstyle\sum_i\left|\int_{A_i}f_0\,d\mu'\right|\\ &=\sup\textstyle\sum_i\left|\int_{A_i}f\,d\mu\right|\leqq\int|f|\,d\mu=\|f\|,\end{aligned}$$

where $(A_i)$ denotes any finite $\Sigma'$-measurable partition of $X$.

Now suppose that $p>1$. Since $\mu$ is finite, we can identify $L^p(\mu)$ with a dense vector sublattice of $L^1(\mu)$ containing $L^\infty(\mu)$ (cf. 9.4). The closure $\bar{H}$ of $H$ in $L^1(\mu)$ is a vector sublattice of $L^1(\mu)$ containing the constant functions. By the preceding there exists a positive projection $P$ (with norm 1) of $L^1(\mu)$ onto $\bar{H}$. Since $e\in H$, we have $Pe=e$ and so the restriction of $P$ to $L^\infty(\mu)$ is a projection, again of norm 1, in $L^\infty(\mu)$ (with range $H\cap L^\infty(\mu)$). By the M. Riesz convexity theorem (Chap. V, § 8), $P$ induces a bounded operator $P_0$ in $L^p(\mu)$, which is obviously a positive projection and also of norm 1. If $K_1$ denotes the kernel of $P$ in $L^1(\mu)$ then $H+K_1$ is dense in $L^1(\mu)$, and $P(H+K_1)=H$. Thus, since $P_0$ is continuous in $L^p(\mu)$ and $H$ is closed, we obtain $P_0(L^p(\mu))=H$ as desired. □

**Example.** 1. If $(X,\Sigma,\mu)$ is a probability space, the random variable $f_0$ introduced in the preceding proof, is called the *conditional expectation* of $f$ with respect to the class of events $\Sigma'$. If $(A_j)_{j\in\mathbb{N}}$ is a countable partition of $X$ into $\Sigma$-measurable sets (i. e., if $A_j\in\Sigma$) and $\Sigma'$ is the set of all unions of sets $A_j$ ($j\in\mathbb{N}$), the projection $P$ can be written down explicitly. (This special case is often used for an intuitive introduction of conditional expectation.) In fact, supposing that $\mu(A_j)>0$ for all $j$, we obtain $f_0=\sum_{j=1}^\infty(\mu(A_j)^{-1}\int_{A_j}f\,d\mu)\chi_{A_j}$ and hence,

$$P=\textstyle\sum_j\varphi_j\otimes\chi_{A_j}$$

where

$$\varphi_j(f):=\mu(A_j)^{-1}\int_{A_j}f\,d\mu \qquad (j\in\mathbb{N},\ f\in L^1(\mu)).$$

It is easy to see that the sum converges for the strong operator topology (that is, in $\mathscr{L}_s(L^1(\mu))$); in fact, letting $P_n:=\sum_{j=1}^n\varphi_j\otimes\chi_{A_j}$, for $f\geqq 0$ one obtains $\|Pf-P_nf\|\to 0$ employing the monotone convergence theorem. On the other hand, $\lim_n P_n=P$ in the uniform operator topology if and only if $P$ is compact, or equivalently, if and only if $\Sigma'$ is finite.

It can also be seen from this example that in general, the projection $P$ is not a lattice homomorphism.

**11.4 Theorem.** *Let $E$ be a Banach lattice $L^p(\mu)$, where $1 \leqq p < +\infty$ and $(X, \Sigma, \mu)$ is an arbitrary measure space. Every closed vector sublattice $H$ of $E$ is the range of a positive contractive projection.*

*Proof.* Let $S = (u_\alpha)_{\alpha \in A}$ be a maximal orthogonal system in the Banach lattice $H$, denote by $B$ the closed ideal in $E$ generated by $S$, and denote by $E_\alpha$ the closed ideal in $E$ generated by $u_\alpha$ $(\alpha \in A)$. Moreover, define $H_\alpha := H \cap E_\alpha$. $B$ and $E_\alpha$ are projection bands in $E$, and $S$ is a topological orthogonal system of $B$ (§ 5, Example 2). It is enough to prove the existence of a positive contractive projection $P$ of $B$ onto $H$, since $H \subset B$ and $E = B + B^\perp$.

By definition of $E_\alpha$, $u_\alpha$ is a quasi-interior positive element of $E_\alpha$; from (9.4) and the discussion immediately preceding it we conclude that $\mu$ defines a bounded, regular Borel measure $\mu_\alpha$ on the structure space $K_\alpha$ of $E_\alpha$, such that $E_\alpha$ is (isometrically) isomorphic to $L^p(\mu_\alpha)$. (In fact, $\sum_\alpha K_\alpha$ is a strong representation space for $B$.) By (11.3) there exists a contractive projection $P_\alpha \geqq 0$ of $E_\alpha$ onto $H_\alpha$, which extends continuously to $L^1(\mu_\alpha)$ (with range the closure $\bar{H}_\alpha$ in $L^1(\mu_\alpha)$).

Denoting by $Q_\alpha$ the band projection $B \to E_\alpha$, we claim that

$$Pf := \sum_\alpha P_\alpha Q_\alpha f \qquad (f \in B)$$

defines a projection of $B$ onto $H$ with required properties. Supposing for the moment that the right hand side converges for all $f \in B$, it is clear that $P \geqq 0$; moreover, we have $Pf \in H$ $(f \in B)$ and $Pf = f$ $(f \in H)$, the latter relation because $Q_\alpha H = H_\alpha$ and $f = \sum_\alpha Q_\alpha f$ by (5.2). Let $f_\alpha := Q_\alpha f$; the unconditional convergence (5.2) of the expansion $f = \sum_\alpha f_\alpha$ in $B$ is equivalent to the unconditional convergence of $|f|^p = \sum_\alpha |f_\alpha|^p$ in $L^1(\mu)$. Since $P_\alpha$ is a contractive projection of $E_\alpha \cong L^p(\mu_\alpha)$ onto $H_\alpha$, we obtain

$$\begin{aligned}\sum_\alpha \|P_\alpha f_\alpha\|^p &= \sum_\alpha \int |P_\alpha f_\alpha|^p \, d\mu_\alpha \leqq \sum_\alpha \int |f_\alpha|^p \, d\mu_\alpha \\ &= \int \left(\sum_\alpha |f_\alpha|^p\right) d\mu = \int |f|^p \, d\mu = \|f\|^p\end{aligned}$$

which shows that the series defining $P$ converges unconditionally for $f \in B$, and that $P$ is contractive. ☐

*Remarks.* 1. The preceding result, due to T. Ando [1969], also holds for Banach lattices $c_0(A)$ ("null sequences" on the discrete space $A$, under the supremum norm) (Exerc. 29). As shown by Ando (l.c.), this exhausts the class of Banach lattices for which (11.4) is valid: *If $E$ is a Banach lattice in which each closed vector sublattice is the range of a positive contractive projection, then $E$ is (isometrically) isomorphic to some $L^p(\mu)$ $(1 \leqq p < +\infty)$ or $c_0(A)$.*

In particular, (11.4) cannot be extended to the case $p = +\infty$; examples are furnished by the sublattices $c_0 \subset l^\infty$ and $C[0,1] \subset L^\infty(\mu)$ ($\mu$ Lebesgue measure on $[0,1]$, cf. (II. 10.4) Cor. 2).

2. L. Tzafriri [1969c], [1971] has shown that a countably order complete Banach lattice $E$ is isomorphic (as a topological vector lattice) to $c_0(A)$ or

some $L^p(\mu)$ $(1 \leqq p < +\infty)$ whenever each closed vector sublattice of $E$ is the range of a positive (not necessarily contractive) projection; this result even continues to hold without the positivity requirement, that is, for Banach lattices $E$ in which each closed vector sublattice is complemented (J. Lindenstrauss and L. Tzafriri [1971]). (Let us observe here that in the same paper, it is proved that a Banach space in which every closed subspace is complemented, is isomorphic to Hilbert space.)

3. Following earlier results by Grothendieck [1955b], Douglas [1965], and Ando [1966], Tzafriri [1969a] proves that if $P$ is any contractive (not necessarily positive) projection in $L^p(\mu)$ (where $1 \leqq p < +\infty$ and $(X, \Sigma, \mu)$ is any measure space), then the range $M$ of $P$ is isometric to $L^p(\tilde{\mu})$ for a suitable measure $\tilde{\mu}$. If, in addition, $p \neq 2$ then $M$ is isometric (by unitary multiplication) with a closed vector sublattice of $L^p(\mu)$ (Bernau-Lacey [1973]).

4. If $H$ is a closed vector sublattice of the Banach lattice $E$, then each positive contractive projection $P: E \to H$ defines a norm preserving positive extension $x' \mapsto x' \circ P$ of the continuous linear forms $x' \in H'$ to $E$, *simultaneously* for all $x' \in H'$. Thus the adjoint $P'$ of $P$ can be viewed as a "lifting" of $E'/H^\circ$ into $E'$. Such liftings exist if $E$ is some $L^p(\mu)$ $(1 \leqq p < +\infty)$ or a sublattice of $C(K)$ (Exerc. 29), and these Banach lattices are characterized by this property (Ando [1969]).

In view of the preceding remarks, it is worthwhile to find particular cases of interest where a sublattice of a Banach lattice, not necessarily an $L^p$-space $(1 \leqq p < +\infty)$ or $c_0(\mathsf{A})$, is the range of a positive projection. We shall discuss one such case which is of great interest in ergodic theory, and closely related to the existence of invariant means on semi-groups.

Let $S$ be an abstract semi-group (cf. § 7) and a topological space such that multiplication is separately continuous, and denote by $C_b(S)$ the Banach lattice of all real valued, bounded continuous functions on $S$ endowed with the supremum norm. A positive linear form $\mu$ on $C_b(S)$ satisfying $\mu(e)=1$ ($e$ the constant-one function) is called a *mean on* $S$. Denote by ${}_sT$ and $T_s$ the left and right translation operators on $C_b(S)$ defined by ${}_sTf(t) := f(st)$ and $T_s f(t) := f(ts)$ $(s, t \in S, f \in C_b(S))$, respectively. A mean $\mu$ on $S$ is called a *left, right,* or *two-sided invariant mean* on $S$ accordingly as $\mu \circ {}_sT = \mu$, $\mu \circ T_s = \mu$, or both relations hold for all $s \in S$. We shall call the semi-group $S$ *amenable* if there exists a two-sided invariant mean on $S$. (Frequently the term "amenable" is used with respect to the discrete topology of $S$, so that $C_b(S)$ becomes the Banach lattice $B(S)$ of all real valued bounded functions on $S$. An excellent survey of the related theory is found in Day [1969].) Important examples of amenable semi-groups are furnished by abelian semi-groups $S$ with separately continuous multiplication (the amenability of these semi-groups being a consequence of the Markov-Kakutani theorem (7.12)), and by compact topological groups $G$, on which normalized Haar measure is the unique invariant mean (§ 8, Example 6).

Suppose now that $S \subset \mathscr{L}(E)$ is a semi-group of operators on a Banach space $E$ (§ 7). We shall assume that $S$ is compact in the weak operator topology $\sigma(\mathscr{L}(E), E \otimes E')$ so that for each $x \in E$, the orbit $Sx$ is weakly compact in $E$. A vector $x \in E$ is called *S-reversible* if $z \in Sx$ implies $x \in Sz$, and $y \in E$ is called an *S-escape vector* if $0 \in Sy$. (Note that this corresponds precisely to the definition of non-transient and transient states of a Markov chain; cf. Chap. I, § 9 and Example 2 below.) If we define $E_r$ to be the set of $S$-reversible vectors and $E_0$ to be the set of $S$-escape vectors in $E$ it is clear that $E_r \cap E_0 = \{0\}$, and it can be shown that $E = E_r + E_0$ (cf. Jacobs [1963]); simple examples (Jacobs, l.c.) however show that in general, $E_r$ and $E_0$ are not vector subspaces of $E$. A fundamental result of Glicksberg and de Leeuw [1961, 1965] asserts that much more can be said for amenable semi-groups $S$. *Precisely, if (and only if) S is amenable, the following is true:* $E_r$ and $E_0$ are closed $S$-invariant vector subspaces of $E$ (and hence, topological supplements of each other). The projection $Q: E \to E_r$ vanishing on $E_0$ is the unique idempotent in the semi-group kernel of $S$ (i.e., in the intersection $K$ of all closed two-sided semi-group ideals, cf. 7.5), and $K$ is a group. Moreover, on $E_r$ $S$ acts as a group which is (isomorphic to $K$ and) compact in the strong operator topology.

To apply this theorem to semi-groups of positive operators in a Banach lattice, we need the following lemma on positive projections.

**11.5 Proposition.** *Let $P$ be a positive projection in $\mathscr{L}(E)$, where $E$ is any Banach lattice. The range $PE$ is a vector lattice under the order induced by $E$ and a Banach lattice under a norm equivalent to the norm induced by $E$. If $P$ is strictly positive, then $PE$ is a sublattice of $E$.*

*Proof.* Let $F := PE$ then the positive cone of $F$ for the order induced by $E$ is the set $F_+ = E_+ \cap PE$. It follows that $F_+ = PE_+$ hence, $F = F_+ - F_+$. To prove that $F$ is a vector lattice it suffices to show that for all $x, y \in F_+$, $\sup_F(x,y)$ exists. We claim that $P(x \vee y) = \sup_F(x,y)$. Clearly, $P(x \vee y) \geqq Px = x$ and, similarly, $P(x \vee y) \geqq y$. On the other hand, if $u \in F$ satisfies $u \geqq x$ and $u \geqq y$ then $u \geqq x \vee y$ in $E$, whence $u = Pu \geqq P(x \vee y)$.

To show that $F$ is a Banach lattice under some norm equivalent to the norm induced by $E$, it is enough to show that whenever $(x_n)$ is a null sequence in $F$ and $(y_n)$ is a sequence in $F$ such that $0 \leqq y_n \leqq |x_n|_F$ for all $n$, then $(y_n)$ is a null sequence in $F$ (Chap. II, Exerc. 13); but $|x_n|_F = P|x_n|$ by the preceding, and the continuity of $P$ implies the desired conclusion. Finally, $F$ is a vector sublattice of $E$ iff $|x| = P|x|$ for all $x \in F$. Now if $x \in F$ then $|x| = |Px| \leqq P|x|$ and $0 \leqq P|x| - |x| \in P^{-1}(0)$; thus $|x| = P|x|$ whenever $P$ is strictly positive. ▯

**11.6 Theorem.** *Let $E$ be a Banach lattice, $S$ a semi-group of positive operators on $E$ which is compact in the weak operator topology. If $S$ is amenable then the space $E_r$ of $S$-reversible vectors is (under the induced order and a suitable equivalent norm) a Banach lattice on which $S$ acts as a compact*[1] *group of lattice isomorphisms, and which contains the $S$-fixed vectors as a closed sublattice. In addition, if no $x > 0$ is an escape vector then $E_r$ is a sublattice of $E$.*

*Proof.* In view of the Glicksberg-de Leeuw theorem quoted above (and in view of (7.5) if $S$ is abelian) we need only observe that the unit $Q$ of the kernel $K$ of $S$ is a positive projection in $\mathscr{L}(E)$, and apply (11.5). It is, moreover, clear that the restrictions $T_{|E_r}$ $(T \in S)$ are lattice isomorphisms of $E_r$, since they are positive

[1] In the strong operator topology.

surjective maps with positive inverse. From the corollary of (8.5) it follows now that the fixed space $F$ of $S$, evidently contained in $E_r$, is the range of a strictly positive projection $P: E_r \to F$, and hence by (11.5) $F$ is a vector sublattice of $E_r$. Finally, if no $x>0$ is an $S$-escape vector then $Q$ is strictly positive, and again from (11.5) it follows that $E_r$ is a vector sublattice of $E$ (and hence, a Banach lattice under the norm induced by $E$). ▯

We note that under the hypothesis of (11.6), the semi-group $S$ is mean ergodic in $\mathscr{L}_s(E)$ (Exerc. 17); thus (8.7) applies to the group $S_{|E_r}$ and, if no $x>0$ is an escape vector, to the given semi-group $S$.

**Examples.** 2. Let $X$ be a non-empty set, $\Sigma$ a $\sigma$-algebra of subsets, and denote by $E$ the Banach lattice ($AL$-space) $M(X,\Sigma)$ of all real valued, bounded, countably additive set functions on $\Sigma$. Every stochastic kernel $P$ on $\Sigma \times X$ (§ 8, Example 8) can be viewed as the *transition kernel* of a homogeneous (stationary) Markov process with discrete parameter $n \in \mathbb{N}$ and state space $X$. We consider the closure $S$ of $\{P^n : n \in \mathbb{N}\}$ in $\mathscr{L}_s(E)$ and suppose that $S$ contains a compact map. (For this it suffices, for example, that $P$ satisfy Doeblin's condition (hypothesis D in Doob [1953, V, § 5]; cf. also Jacobs [1960, II, § 4]).) Then (11.6) applies and, since on $E_r$ $S$ acts as a group (with identity $1_{E_r}$) containing a compact map, it follows that $E_r$ is finite dimensional; on the other hand, the dimension of $E_r$ is at least 1, because the projection $Q$ (see proof of 11.6) must be a stochastic operator on $E$ (Def. 8.8). Moreover, $E_r$ is a sublattice of $E$ (and hence, an $AL$-space); in fact, if $x \in E_r$ then $x = Qx$, so $|x| \leq Q|x|$ and this shows $|x| = Q|x|$ or $|x| \in E_r$, since $Q$ is contractive.

If $\phi$ denotes the set of probability measures in $E_r$ then (since $\phi$ is a finite dimensional simplex) it is clear that the set $\mathsf{E}(\phi)$ of extreme points of $\phi$ is finite. But $P$ acts as an isometry on $\phi$, and hence it permutes $\mathsf{E}(\phi)$. If $(c_1)\dots(c_k)$ is the decomposition of the corresponding permutation into independent cycles, let $M_\kappa$ denote the subset of $\mathsf{E}(\phi)$ corresponding to $(c_\kappa)$. Moreover, since the elements of $\mathsf{E}(\phi)$ are pairwise (lattice) orthogonal, there exists a disjoint family of sets $A_\mu \in \Sigma$ each of which supports some $\mu \in \mathsf{E}(\phi)$. Then $A := \bigcup \{A_\mu : \mu \in \mathsf{E}(\phi)\}$ is called the set of *non-transient states* of the process, $A_0 := X \setminus A$ the set of *transient states*, and each of the sets $A_\kappa := \bigcup \{A_\mu : \mu \in M_\kappa\}$ $(\kappa = 1, \dots, k)$ is called an *ergodic class*. In this manner, the theory of stationary Markov processes with finite state space as presented in Chap. I, § 9 can be carried over to the present setting (Doob [1953]).

3. Let $E := L^2(\mu)$ where $(X,\Sigma,\mu)$ is an arbitrary measure space. If $S$ is *any* semi-group of contractions in $\mathscr{L}(E)$, the closure of $S$ in the weak operator topology is compact. Using methods similar to those employed in the proof of (7.11), it can be shown that $E = E_r + E_0$ is an orthogonal direct sum (Exerc. 23); however, simple examples (such as the group of unitary operators on $E$) show that $E_r = \{0\}$ in general. Assume now that $\mu$ is finite, and that $S$ is a semi-group of positive contractions of $L^2(\mu)$ which leaves the constant functions fixed. As in the preceding example, it can be verified directly that $E_r$ is a sublattice of $E$. Hence from (11.2) we obtain: *The space of all $S$-reversible vectors is the set of all $\Sigma'$-measurable functions in $L^2(\mu)$ for a suitable $\sigma$-subalgebra $\Sigma'$ of $\Sigma$.*

**Notes**

§ 1: A systematic study of the lattice of closed ideals in a Banach lattice was carried out by Lotz [1969]. The first part of (1.2) was proved by Lotz [1968] and, under somewhat more general assumptions, independently by Davies [1968]. See also Lotz [1973 b].

§ 2: This section contains only those results on prime ideals which are needed in the sequel. For further results and references, see Johnson-Kist [1962], to whom (2.2) is due, and Luxemburg-Zaanen [1971 b]. Coprefilters on abstract lattices are considered in Birkhoff [1967]; our terminology follows Bourbaki [1961].

§ 3: The concept of valuation (in the sense of Def. 3.1) and the results of this section are due to Lotz [1969].

§ 4: The representation theorem (4.5) was proved independently by Davies [1969], Lotz [1969], and Goullet de Rugy [1971]. While Davies uses Choquet theory, Lotz bases the proof on his ideal theory (Lotz [1969]). The present approach essentially follows Schaefer [1972 a].

§ 5: The theory developed in this section is found in Schaefer [1972 a] and based on the results of Section 4. Noteworthy features are the characterization of the normed structure of the Banach lattice $E$ and the essential unicity of a strong representation space (5.3); the latter unicity assertion appears to be new even for $AL$-spaces. Hackenbroch [1972], [1974] obtains, along similar lines, a representation theory for countably order complete vector lattices; see also Exerc. 14. For other approaches to the representation of vector lattices, see Vulikh [1967] and Luxemburg-Zaanen [1971 b].

§ 6: A Stone-Weierstrass type approximation theorem for Banach lattices with quasi-interior positive elements is proved by Lotz [1969]. (6.4) is due to Schaefer [1972 a], while the stronger theorem quoted at the end of the section is due to Nagel [1973 a].

§ 7: Mean ergodic theorems were obtained first by von Neumann [1932], Alaoglu-Birkhoff [1940], Eberlein [1949] and, later on, by many others; for a survey, see Day [1969]. The present approach, following in part unpublished lecture notes by H. P. Lotz, is based on the theory of topological semi-groups which affords greater clarity and many simplifications. The proofs of (7.6) and (7.9) are adapted from Berglund-Hofmann [1967]. (7.10) is due to Nagel [1973c].

§ 8: All results of this section are motivated by known results and problems of ergodic theory (see, e. g., Jacobs [1963]). For cyclic semi-groups of positive operators on an $AM$-space with unit, they were obtained by Schaefer [1967]; their present general form is largely due to Lotz (unpublished lecture notes). (See also Nagel [1973 c].) For a discussion of the case where the relation between $S$-ergodic measures and maximal $S$-ideals (see 8.9) is not bijective, see Raimi [1964] and Ando [1968]. (8.11) is due to Schaefer [1967].

§ 9: (9.1) is due to Phelps [1963] and Ellis [1964]. The remaining results in this section are essentially due to Nagel [1972].

§ 10: Representation theorems for monothetic operator groups (generated by an operator with discrete spectrum) were obtained by Halmos-von Neumann [1942] for operators on Hilbert space, and by Glicksberg [1965] and Lotz [1968] for operators on $AM$-spaces with unit. These theorems are all special cases of (10.5) which, in turn, is a considerable specialization of (10.4). Theorem 10.4 is due to Nagel-Wolff [1972].

§ 11: (11.1) and (11.4) were first proved by Ando [1969]. Further relevant results can be found in Tzafriri [1969 a], Lindenstrauss-Tzafriri [1971], Bernau [1973], and Bernau-Lacey [1973]. Finally, (11.6) is an application, due to the author, of a theorem of Glicksberg-de Leeuw [1961].

## Exercises

1. Suppose $E$ to be a topological vector lattice (Chap. II, § 5) and denote by $I, J$ closed ideals in $E$.

(a) If $E/I \cap J$ is complete for the quotient topology, then $I+J$ is closed in $E$. In particular, if $E$ is complete then $I+J$ is closed whenever $I \cap J = \{0\}$.

(b) If $E$ is a complete metrizable topological vector lattice, then $I+J$ is closed. (Use (a) and the fact that each quotient of a complete metrizable topological vector space over a closed vector subspace is complete [S, I. 6.3].)

(c) Give an example of a normed vector lattice $E$ and closed ideals $I, J$ in $E$ such that $I+J$ is not closed. (Consider the vector sublattice $E \subset C[0,1]$ of all continuous, piecewise affine real functions on $[0,1]$, and let $I := \{f \in E : f(n^{-1}) = 0$ *for n even*$\}$, $J := \{f \in E : f(n^{-1}) = 0$ *for n odd*$\}$.)

2. Let $E$ be a Banach lattice and let $I_i$ $(i=1,\dots,n)$ denote closed ideals in $E$. By virtue of $x \mapsto (q_1(x),\dots,q_n(x))$, the quotient mappings $q_i: E \to E/I_i$ define a canonical map $q$ of $E$ into $\prod_{i=1}^n E/I_i$.

(a) Show that $q$ is a topological lattice homomorphism (that is, $q$ preserves the lattice operations and maps open subsets of $E$ onto open subsets of $q(E)$). (Observe that the adjoint $q'$ has range $\sum_{i=1}^n I_i^\circ$ in $E'$, and apply (1.2) and [S, IV. 7.7].)

(b) Show that $q(E)$ consists of all vectors $(q_1(x_1),\dots,q_n(x_n))$ for which $x_i - x_j \in I_i + I_j$ $(i,j=1,\dots,n)$.

3. Let $\boldsymbol{A}(E)$ denote the lattice of closed ideals in the Banach lattice $E$. Then $\boldsymbol{A}(E)$ is a *Brouwerian lattice;* that is, for each pair $(I,J) \in \boldsymbol{A}(E) \times \boldsymbol{A}(E)$, there exists precisely one $K \in \boldsymbol{A}(E)$ such that $J \cap H \subset I \Leftrightarrow H \subset K$, for all $H \in \boldsymbol{A}(E)$. (Consider the ideal $q^{-1}(q(J)^\perp)$, where $q$ denotes the quotient map $E \to E/I$.)

4. A *Fréchet lattice* is a locally convex vector lattice (Chap. II, § 5) which is metrizable and (uniformly) complete. Extend the results of § 1 to Fréchet lattices, and find out to what extent the results of §§ 4—6 can be generalized.

5. Let $E$ denote a Banach lattice possessing quasi-interior positive elements (Chap. II, § 6).

(a) If $I \in A(E)$, $I \neq \{0\}$, there exists $J \in A(E)$, $J \neq E$, such that $I + J = E$.

(b) If $I, J \in A(E)$ and if $I$ is properly contained in $J$, there exists $K \in A(E)$ such that $I + K \neq J + K = E$ (*Wallman's separation property;* cf. Birkhoff [1967].)

6. Let $E$ be the set of continuous functions $f: [0,1] \to \bar{\mathbb{R}}$ ($\bar{\mathbb{R}}$ the two-point compactification of $\mathbb{R}$) which have the following property: For each $s \in [0,1]$ there exists a neighborhood $U$ of $s$ such that for all $t \in U$, $t \neq s$,

$$f(t) = \sum_{\nu=1}^{n} \frac{a_\nu}{|t-s|^\nu} + g(t),$$

where $U$, $n$, $a_\nu$, and $g \in C[0,1]$ depend on $f$ and $s$.

(a) Show that $E$ can be made into an Archimedean vector lattice in a natural way (§ 2, Example 3) and show that for each $s \in [0,1]$, the set $J_s := \{f \in E: f(s) = 0\}$ is a prime ideal in $E$ for which $E/J_s$ is isomorphic to the vector lattice of all finitely non-zero sequences under the lexicographic ordering (Chap. II, § 1, Example 7).

(b) There exist no maximal ideals of $E$.

7. Let $E$ be any vector lattice and fix $u \in E_+$.

(a) Denote by $S$ the set of all mappings $f: E_+ \to \bar{\mathbb{R}}_+$ satisfying (i) $f(x+y) = f(x) + f(y)$ $(x, y \in E_+)$ and (ii) $f(\lambda x) = \lambda f(x)$ $(x \in E_+, \lambda \in \mathbb{R}_+)$. (For the definition of addition and multiplication in $\bar{\mathbb{R}}_+$, see the comment preceding Proposition 3.1.) $S$ is an additive abelian semi-group over $\mathbb{R}_+$, and the set $\phi_u := \{f \in S: f(u) = 1\}$ is a convex subset of $S$. Show that $\varphi \in \phi_u$ is a valuation of $E$ iff $\varphi$ is an extreme point of $\phi_u$.

(b) Let $\varphi$ be any valuation of $E$ satisfying $\varphi(u) = 1$. Show that $\varphi$ can be constructed from its kernel $\varphi^{-1}(0)$ by the formula

$$\varphi(x) = \inf\{\lambda \in \mathbb{R}: E_{(x-\lambda u)^+} \subset \varphi^{-1}(0)\} = \sup\{\lambda \in \mathbb{R}: E_{(x-\lambda u)^-} \subset \varphi^{-1}(0)\},$$

valid for all $x \in E_+$ (cf. Theorem 3.6). (Use the fact that $\varphi^{-1}(0)$ is a prime ideal of $E$.)

(c) Infer from (b) that for each $u \in E_+$, there exists a bijective correspondence between the set of $u$-normalized valuations of $E$ and the set of prime ideals $I \subset E$ such that $q(u)$ generates a minimal ideal in $E/I$, where $q$ denotes the quotient map $E \to E/I$.

8. Let $E$ denote a Banach lattice possessing a t.o.s. $S$ (Def. 5.1), and suppose $\varphi$ is a valuation of $E$ such that $\varphi(u) = 1$ for some $u \in S$. The following assertions are equivalent:

($\alpha$) $\varphi$ is finite valued.

($\beta$) $\varphi$ is continuous (as a map of $E$ into the one-point compactification $\bar{\mathbb{R}}_+$ of $\mathbb{R}_+$).

($\gamma$) $\varphi$ is lower semi-continuous.

($\delta$) $\varphi^{-1}(0)$ is a closed ideal.

($\varepsilon$) $\varphi^{-1}(0)$ contains a maximal ideal of $E$.

($\zeta$) There exists a continuous linear form $f$ on $E$ such that $\varphi(x)=f(|x|)$, all $x\in E$.

(Reduce the problem to the case where $E_+$ possesses quasi-interior elements and use (3.2), (3.3), and (3.3) Cor.)

9. Let $E$ be a Banach lattice possessing quasi-interior positive elements and denote by $\mathfrak{B}$ the set of all ultra-coprefilters in $\boldsymbol{A}(E)$ (Def. 2.4).

(a) Show that each $V\in\mathfrak{B}$ is a prime ideal of the lattice $\boldsymbol{A}(E)$ (that is, $I,J\in\boldsymbol{A}(E)$ and $I\cap J\in V$ implies that $I\in V$ or $J\in V$).

(b) For each (not necessarily closed) ideal $I\in\boldsymbol{I}(E)$, define

$$\mathfrak{O}_I=\{V\in\mathfrak{B}\colon \textit{there exists } x\in I_+ \textit{ such that } \bar{E}_x\notin V\}.$$

Show that the family $(\mathfrak{O}_I)_{I\in\boldsymbol{I}(E)}$ of subsets of $\mathfrak{B}$ is stable under finite intersections and arbitrary unions, and hence defines a topology on $\mathfrak{B}$. (Observe that $\mathfrak{O}_I=\emptyset$ iff $I=\{0\}$, $\mathfrak{O}_I=\mathfrak{B}$ iff $I$ contains a quasi-interior point of $E_+$, and that $\bigcup_\alpha\mathfrak{O}_{I_\alpha}=\mathfrak{O}_{\Sigma_\alpha I_\alpha}$, $\bigcap_\alpha\mathfrak{O}_{I_\alpha}=\mathfrak{O}_{\bigcap_\alpha I_\alpha}$.)

(c) Under the hull-kernel topology defined in (b) and with respect to the mapping ($\star$) of (3.6), $\mathfrak{B}$ is homeomorphic to the compact space $K_u(E)$ defined in § 4 (cf. Lotz [1969]). Deduce from this another proof of (4.1).

(d) Interpret Propositions (4.3) and (4.4) in terms of the space $\mathfrak{B}$ and its topology defined in (b).

(e) Generalize the preceding to the case where $E$ is a Banach lattice possessing a t.o.s. (Def. 5.1).

10. We use the notation of (5.3) and (5.4). Let $E$ be a Banach lattice possessing a t.o.s., let $X$ denote any strong representation space for $E$, and let $E\to\hat{E}$ denote the associated representation of $E$ into $\bar{\mathbb{R}}^X$. Let $I$ be a closed ideal in $E$ and let $X_I$ be the closed subset of $X$ corresponding to $I$ by (6.1).

(a) $I$ is a projection band in $E$ (Def. II. 2.8) iff $X_I$ is open in $X$.

(b) Infer from (a) that the ideal $\{f\in\hat{E}\colon f(A)=\{0\}\}$, where $A$ is a closed subset of $X$, is not closed in general (cf. Exerc. 15).

(c) The set of order continuous, real valued lattice homomorphisms of $E$ corresponds (to within some normalization) bijectively to the set of isolated points of $X$.

11. Let $E=C[0,1]$ and denote by $H$ the closed vector sublattice of all $f\in E$ satisfying $f(\frac{1}{2})=0$. Show that the structure space of $H$ (Def. 4.2) is not a continuous image of the real unit interval $[0,1]$. (Observe that the structure space of $H$ is not connected, cf. Exerc. 10(a).)

12. Let $E$ denote a Banach lattice possessing a topological orthogonal system (Def. 5.1).

(a) If $F$ is a Banach lattice and $T\colon E\to F$ is a vector lattice homomorphism with dense range, then $F$ has a t.o.s. In particular, each quotient $E/I$ over a closed ideal $I\subset E$, possesses a t.o.s.

(b) Show that for each $x\in E$ the set of indices $\alpha$ for which $x_\alpha\neq 0$ in the expansion (3) of (5.2), is countable. Infer from this that under any representation

$E \to \hat{E}$ in the sense of (5.3), each $f \in \hat{E}$ has its support contained in a countable union of compact sets.

(c) $E$ has a compact representation space (Def. 5.4) if and only if $E_+$ has quasi-interior elements. If $E$ has a compact representation space $K$ then for any strong representation space $X$, $K$ is a compactification of $X$. Illustrate this by considering the spaces $l^p$ $(1 \leq p < +\infty)$, and explain why $\mathbb{N}$ is not a representation space for $l^\infty$.

(d) If $R, S$ are t.o.s. of $E$, define the relation $R \prec S$ to mean that for each $w \in R$ there exists $u \in S$ such that $w \in E_u$. Denoting by $I(R)$, $I(S)$ the ideals in $E$ generated by $R, S$, respectively, show that "$\prec$" defines a pre-order (i. e., a reflexive and transitive binary relation) on the set of all t.o.s. of $E$ such that $R \prec S$ is equivalent to $I(R) \subset I(S)$.

(e) Denote by $\boldsymbol{R}$ the set of equivalence classes of t.o.s. of $E$ according to "$R_1 \sim R_2$ iff $R_1 \prec R_2$ *and* $R_2 \prec R_1$". Show that the relation $\prec$ defines a partial order of $\boldsymbol{R}$ under which $\boldsymbol{R}$ is isomorphic to the set of dense ideals $\{I(R): R \in \boldsymbol{R}\}$ ordered by inclusion, and hence a semi-lattice. (Observe that if $S, T \in \boldsymbol{R}$ and $S \wedge T$ is the t.o.s. of $E$ (cf. 5.2) whose elements are the non-zero infima $u \wedge v$ $(u \in S, v \in T)$, then $I(S \wedge T) = I(S) \cap I(T)$.)

(f) If $E$ contains a t.o.s. $S$ such that the norm of $E$ is (sequentially) order continuous on each $E_u$ $(u \in S)$, then $E$ has (sequentially) order continuous norm.

(g) Generalize the concept of topological orthogonal system to locally convex vector lattices (Chap. II, § 5), and extend the results of §§ 5, 6 to this setting.

13. (a) If $E$ is a Banach lattice possessing a connected representation space (Def. 5.4), then either $E_+$ contains quasi-interior points (and hence, $E$ possesses a compact representation space, Exerc. 12 (c)) or else $E$ contains no t.o.s. (Using (5.2), prove that the presence of t.o.s. containing more than one element, forces any representation space for $E$ to be disconnected.)

(b) If $X$ is a locally compact space not countable at infinity there exists no continuous real function on $X$ which is strictly positive and vanishes at infinity. Equivalently, if $\dot{X}$ denotes the Alexandrov (one-point) compactification of $X$ and $\omega$ denotes the point at infinity, then the maximal ideal $\{f \in C(\dot{X}): f(\omega) = 0\}$ is not the closure of any principal ideal in $C(\dot{X})$.

(c) If $X$ is a locally compact space and if $\mu$ is a strictly positive Radon measure on $X$, then $\mathscr{K}(X)$ (continuous real functions with compact support) can be identified with a dense vector sublattice of $L^1(\mu)$. Show by examples that, in general, $X$ is not a representation space for $L^1(\mu)$.

14. Let $E$ be an Archimedean vector lattice and denote by $S$ any maximal orthogonal system of $E$ (Chap. II, § 1).

(a) Prove the existence of a locally compact, paracompact space $X$ with these properties: $E$ is isomorphic to a vector lattice $\hat{E}$ of continuous numerical functions which are finite except on rare subsets of $X$, where it is understood that the vector operations in $\hat{E}$ agree with the corresponding pointwise operations on sets of finiteness. In addition, if $E$ satisfies axiom (OS) (Def. II. 1.8) then $\mathscr{K}(X)$ is an ideal $J$ in $\hat{E}$ for which $J^\perp = \{0\}$ (that is, $J$ is a *foundation* of $E$ in the sense of Vulikh [1967]). (For each $u \in S$, consider the space $K_u(E)$ of

$u$-normalized valuations, topologized as in § 4, and define $X := \sum_{u \in S} K_u$. To establish the isomorphism $E \to \hat{E}$ proceed as in the proof of (4.5), using (II. 1.9).)

(b) Investigate analogs of (6.1) and (6.2), supposing $E$ to satisfy axiom (OS) (Def. II. 1.8).

(c) If the band $E^\star_{00}$ of order continuous linear forms (Chap. II, § 4) separates $E$, there exists a Hausdorff topology $\boldsymbol{T}$ on $E$ such that the completion of $(E, \boldsymbol{T})$ is a locally convex vector lattice in which $S$ is a t.o.s. (cf. Exerc. 12 (g)).

(d) Suppose $E$ is an order complete vector lattice which is separated by some solid $\sigma(E^\star, E)$-compact subset of $E^\star_{00}$. Show that $E$ can be identified with a dense ideal in a Banach lattice $F$ such that $S$ defines a t.o.s. of $F$. (Use II. 4.12, II. 5.9.)

15. ($E$-negligible Sets. R. Nagel [1973a].) Let $E$ denote a Banach lattice possessing quasi-interior positive elements and identify $E$, in the spirit of (4.5), with a Banach lattice of continuous numerical functions on the structure space $K$ of $E$ (Def. 4.2). Denote by $\mathsf{M}$ the minimal vaguely compact set of positive Radon (respectively, regular Borel) measures on $K$ determining the norm of $E$ (cf. 5.3).

(a) A subset $N \subset K$ is called *$E$-negligible* if the ideal $I_N := \{f \in E : f(N) \subset \{0\}\}$ is dense in $E$. Show that for a subset $A \subset K$ with closure $\bar{A}$, the following are equivalent:

($\alpha$) $A$ is $E$-negligible.

($\beta$) $\mu(\bar{A}) = 0$ for each $\mu \in \mathsf{M}$.

($\gamma$) For each $\varepsilon > 0$ there exists $f \in E_+$ such that $\|f\| < \varepsilon$ and $f(A) \subset \{1\}$.

($\delta$) There exists $f \in E$ such that $f(A) \subset \{+\infty\}$.

Deduce from this that the family $\boldsymbol{N}(E)$ of all $E$-negligible subsets of $K$ is a (hereditary) $\sigma$-ring of rare subsets of $K$.

(b) Since $I_N = I_{\bar{N}}$, the closure of each $E$-negligible set is $E$-negligible. For each closed $N \in \boldsymbol{N}(E)$, $K$ is homeomorphic to the Stone-Čech compactification of $K \setminus N$ (cf. II. 7.7). Infer that if $K$ is metrizable then $E = C(K)$.

(c) If $(f_n)$ is a convergent sequence in $E$ with limit $f$, there exists $N \in \boldsymbol{N}(E)$ and a subsequence $(g_n)$ such that $\lim_n g_n(s) = f(s)$ whenever $s \in K \setminus N$.

(d) $\boldsymbol{N}(E) = \{\emptyset\}$ iff $E$ is isomorphic to an $AM$-space with unit, and $\boldsymbol{N}(E)$ is the family of all rare subsets of $K$ iff each directed ($\geqq$) family $(x_\alpha)$ in $E$ with $\inf_\alpha x_\alpha = 0$, converges to 0 (cf. Theorem II. 5.10).

(e) Let $H$ be a vector sublattice of $E$ containing the constant-one function on $K$. Then $H$ is dense in $E$ if and only if for each pair $K_1$, $K_2$ of disjoint closed subsets of $K$, there exists $N \in \boldsymbol{N}(E)$ such that $H$ distinguishes the points of $K_1 \setminus N$ from those of $K_2 \setminus N$. (To prove the necessity of the condition, use (c). Sufficiency: If $K(\bar{H})$ denotes the structure space of the Banach lattice $\bar{H} \subset E$ then by (9.3), the canonical imbedding $\bar{H} \to E$ defines a surjective map $k: K(E) \to K(\bar{H})$ which is injective iff $\bar{H} = E$. Suppose that $\bar{H} \neq E$; there exist orthogonal linear forms $\mu, \nu \in E'_+$ whose restrictions to $\bar{H}$ agree and which satisfy $\mu(e) = \nu(e) = 1$ ($e$ the constant-one function on $K$). Thus there exist disjoint closed subsets $K_1, K_2$ of $K$ such that $\mu(K_1) > \frac{1}{2}$, $\nu(K_2) > \frac{1}{2}$. This implies that $k(K_1 \setminus N) \cap k(K_2 \setminus N) \neq \emptyset$ for all $E$-negligible Borel sets $N$, since otherwise one obtains the contradictory inequality $1 < \mu(K_1 \setminus N) + \nu(K_2 \setminus N) \leqq \mu \circ k^{-1}(k(K_1 \setminus N) \cup k(K_2 \setminus N))$.)

(f) Generalize the preceding results to Banach lattices possessing a topological orthogonal system (cf. 5.5).

16. Let $(T^n)_{n\in\mathbb{N}}$ be a bounded cyclic semi-group in $\mathscr{L}(E)$, where $E$ is any Banach space, let $M_k:=k^{-1}(1+T+\cdots+T^{k-1})$ $(k\in\mathbb{N})$, and denote by $T'$, $M_k'$ the respective adjoint maps in $\mathscr{L}(E')$. (We write 1 in place of $1_E$, $1_{E'}$ for the respective identity maps.)

(a) Each $\sigma(E',E)$-cluster point of the sequence $(M_k'x')$, $x'\in E'$, is a fixed point of $T'$. (Apply formula (*) of § 7, Example 3, noting that norm closed balls in $E'$ are $\sigma(E',E)$-closed.)

(b) If $(1-T')E'$ is norm closed in $E'$ then $(M_k')$ converges strongly to a (norm continuous) projection $Q$ of $E'$ onto the fixed space of $T'$. (Use the fact that under the present hypothesis, $(1-T')E'$ is $\sigma(E',E)$-closed in $E'$ [S, IV. 7.7 and IV. 7.9], and argue as in § 7, Example 3.)

17. (a) Every compact amenable (§ 11) semi-group $S\subset\mathscr{L}_\sigma(E)$, where $E$ is a Banach space and $\sigma$ denotes the weak operator topology, is mean ergodic in $\mathscr{L}_s(E)$ with $P=\int_S T\,dm(T)$ the zero element of $S$. (Here $m$ denotes an invariant mean on $S$; note that the integral exists in $\mathscr{L}_\sigma(E)$, since $\overline{\text{co}}\,S$ is compact in $\mathscr{L}_\sigma(E)$.)

(b) Obtain from (a) proofs of (7.6) Cor. and (7.9) Cor. 1, using the existence of Haar measure if $S$ is a group and using the Markov-Kakutani theorem (7.12) if $S$ is abelian.

(c) Conversely, it can be shown that each compact mean ergodic semi-group $S\subset\mathscr{L}_\sigma(E)$ is amenable (cf. Berglund-Hofmann [1967]).

18. The norm of a Banach space $E$ is said to be *uniformly convex* if for each $\varepsilon>0$, there exists $\delta>0$ such that $\|x\|=\|y\|=1$ and $\|x-y\|\geqq\varepsilon$ implies $\|\frac{1}{2}(x+y)\|\leqq 1-\delta$.

(a) Let $E$ be a Banach space with uniformly convex norm. Then $E$ is reflexive, and each non-void closed convex subset of $E$ contains a unique element of minimal norm. (Show that the unit ball of $E$ is weakly complete and hence, weakly compact; for the second assertion note that in any normed space, the norm is weakly lower semi-continuous.)

(b) Let $E$ be a Banach space such that the norm of $E$ as well as the dual norm of $E'$ are uniformly convex (for example, let $E=L^p(\mu)$ $(1<p<+\infty)$). Every contractive semi-group $S\subset\mathscr{L}(E)$ is mean ergodic. (Generalize the proof of (7.11), using (a) and the fact that the adjoint semi-group $S'$ is contractive.)

(c) Retaining the hypothesis of (b), let $S$ be a contractive semi-group in $\mathscr{L}(E)$, and denote by $\bar{S}$ its closure in the weak operator topology. Then $E$ is the direct sum of $E_r$ ($\bar{S}$-reversible vectors) and $E_0$ ($\bar{S}$-escape vectors) (see § 11), and $\bar{S}_{|E_r}$ is a compact group in $\mathscr{L}_s(E_r)$. (Cf. Jacobs [1963].)

19. Let $E$ be a Banach space and let $S$ be an amenable (§ 11) contractive semi-group in $\mathscr{L}_s(E)$. Denote by $S'$ the adjoint semi-group and by $F$, $F'$ the respective fixed spaces of $S$, $S'$.

(a) For each $x'\in E'$, $\overline{\text{co}}\,S'x'\subset E'_\sigma$ contains a fixed point of $S'$. (Cf. Exerc. 16 (a); for a proof, see Day [1969], § 6.)

(b) For $S$ to be mean ergodic in $\mathscr{L}_s(E)$, it is necessary and sufficient that $F$ separate the points of $F'$. (The necessity of the condition follows from the discussion preceding (7.3). For the sufficiency observe that if $F$ separates $F'$, there exists a *unique* $S'$-fixed point $y' \in \overline{\mathrm{co}}\, S'x' \subset E'_\sigma$; this implies ($F'$ being $\sigma(E',E)$-closed) that the projection $P': x' \mapsto y'$ of $E'$ onto $F'$ is $\sigma(E',E)$-continuous and hence, the adjoint of a projection $P \in \overline{\mathrm{co}}\, S \subset \mathscr{L}_s(E)$ (cf. 7.10) which turns out to be the zero element of $S$.)

20. Let $C$ be a non-void, compact convex subset of a locally convex vector space $E$ and let $S$ be a semi-group of continuous, affine maps $T: C \to C$ such that whenever $x \neq y$, 0 is not in the closure of $\{Tx - Ty: T \in S\}$. Then $S$ has a fixed point in $C$. (Theorem of Ryll-Nardzewski; for a proof, see Namioka-Asplund [1967].)

21. Let $S$ be a semi-group of positive operators in $\mathscr{L}(E)$, where $E$ is a Banach lattice. The intersection of the set of all maximal $S$-ideals (Def. 8.1) is called the *radical* $R$ of $S$ (respectively, of $T \in \mathscr{L}(E)$ if $S = (T^n)_{n \in \mathbb{N}}$ is cyclic); if $R = \{0\}$ then $S$ (respectively, $T$) is called *radical-free*. In (a)—(f) we suppose that $E = C(K)$, $K$ compact.

(a) The semi-group $S_R$ (induced by $S$ on $E/R$) is radical-free. (Use 8.2.)

(b) If $T$ is a Markov operator and $S := (T^n)_{n \in \mathbb{N}}$ is uniformly mean ergodic (§ 7), then $f \in R$ iff $\lim_n T^n |f| = 0$. (For the necessity of the condition, show that $T_{|R}$ has spectral radius $< 1$. Cf. Schaefer [1968].)

(c) If $S$ is a strongly mean ergodic semi-group of Markov operators with associated projection $P$, then $R = \{f: P|f| = 0\}$. (Apply 8.9.)

(d) Suppose that $T$ is a radical-free Markov operator such that some power of $T$ is compact. Then $K$ is the union of all minimal closed $T$-invariant subsets. (A closed subset $A$ of $K$ is called *T-invariant* if for each Radon measure $\mu$ with support in $A$, $\mu \circ T$ has its support contained in $A$.) Infer from this that if $K$ is connected, $T$ is irreducible.

(e) Show by an example that (d) fails (even for projections) when the compactness assumption is dropped.

(f) If $T$ is radical-free with spectral radius 0, then $T = 0$.

(g) Investigate to what extent the preceding results can be carried over to more general Banach lattices $E$.

22. Let $E := C(K)$, $K \neq \emptyset$ compact. In (a)—(d), we suppose $T$ to be a Markov operator on $E$ and let $S := (T^n)_{n \in \mathbb{N}}$.

(a) If the $T$-invariant probability measure whose existence is a consequence of (8.10) and § 8, Example 2 is unique, then $S$ is mean ergodic in $\mathscr{L}_s(E)$. (Show that the sequence $(M'_k)_{k \in \mathbb{N}}$ (notation of Exerc. 16) converges for $\sigma(\mathscr{L}(E'), E' \otimes E)$. This is a special case of Exerc. 19.)

(b) Suppose $T$ to be radical-free (Exerc. 21). If each maximal $T$-ideal $J$ is the absolute kernel of precisely one $T$-invariant probability measure $\varphi_J$ then $S'$ is strongly mean ergodic in the Banach space $E'$ $(= M(K))$ with associated projection

$$P' = \textstyle\sum_J \varphi_J \otimes \chi_J,$$

where $\chi_J$ is the characteristic function of the closed subset of $K$ on which each $f \in J$ vanishes (§ 1, Example 1).

(c) If $S$ is mean ergodic, irreducibility of $T$ is equivalent to the existence of a unique $T$-invariant probability measure with global support.

(d) If $T$ is irreducible and if $K$ is quasi-Stonian (Chap. II, § 7), the set of $T$-invariant probability measures is either a singleton or infinite. (Ando [1968].)

(e) Let $T$ be any positive operator on $E$. The set of maximal $T$-ideals is not empty (§ 8, Example 2), and each maximal $T$-ideal $J$ is the absolute kernel of a probability measure on $K$ which is an eigenvector of $T'$. (Schaefer [1967].)

23. Let $H$ be any Hilbert space and denote by $S$ an arbitrary semi-group of contractions of $H$. Then the closure $\bar{S}$ in the weak operator topology is compact and $H = H_r \oplus H_0$ is an orthogonal direct sum, where $H_r$, $H_0$ denote the closed vector subspaces of all $\bar{S}$-reversible and all $\bar{S}$-escape vectors, respectively (Exerc. 18 (c)). Show that $H_r = \{0\}$ is equivalent to $0 \in \bar{S}$, and show that $H_r = \{0\}$ can occur even if $S$ is an abelian group of unitary operators which are lattice isomorphisms for some ordering under which $H$ is a Banach lattice. (Consider the shift operator on $l^2(\mathbb{Z})$.)

24. Let $E, F$ be Archimedean vector lattices whose respective order duals $E^\star$, $F^\star$ (Chap. II, § 4) separate $E, F$.

(a) A positive linear map $T: E \to F$ is a lattice homomorphism iff for each $y', 0 < y' \in F^\star$, one has $T^\star[0, y'] = [0, T^\star y']$ where, as usual, square brackets denote order intervals and $T^\star$ denotes the *order adjoint* of $T$ (that is, the restriction of the algebraic adjoint to $F^\star$). (Use that $E^\star$, $F^\star$ are order complete (II. 4.2) and the fact that $T^\star[0, y']$ is $\sigma(E^\star, E)$-compact, since $[0, y']$ is $\sigma(F^\star, F)$-compact and $T^\star$ continuous for the corresponding weak topologies.)

(b) If $T: E \to F$ is a lattice homomorphism then so is its second order adjoint $T^{\star\star}$ of $E^{\star\star}$ into $F^{\star\star}$. In particular, if $E, F$ are Banach lattices then a linear map $T: E \to F$ is a lattice homomorphism iff its second adjoint $T'': E'' \to F''$ is.

(c) In general, the order adjoint $T^\star$ of a lattice homomorphism $T$ is not a lattice homomorphism (cf. Chap. I, Propositions 4.3 and 4.4). Prove that if $T: E \to F$ is a lattice homomorphism and $TE$ is an ideal in $F$ then $T^\star$ is a lattice homomorphism. (Show that $T^{\star\star}(E^{\star\star})$ is an ideal in $F^{\star\star}$, and apply (a). Cf. (9.3).)

(d) Suppose $E, F$ to be Banach lattices and let $T: E \to F$ be a lattice homomorphism. If $T$ maps some t.o.s. of $E$ (Def. 5.1) onto a t.o.s. of $F$ then $T$ maps each t.o.s. of $E$ onto a t.o.s. of $F$. In particular, if $E_+$ has quasi-interior points and $TE$ is dense in $F$ then $T$ maps the quasi-interior of $E_+$ into the quasi-interior of $F_+$. Use this to show that the map $k$ of (9.3) between the respective structure spaces is well defined, independently of the choice of $u \in E_+$.

(e) If $K$ is a compact space and $\delta_t$ the Dirac measure of a point $t \in K$ which is not isolated, then $T := \delta_t \otimes e$ ($e$ the constant-one function on $K$) defines a lattice homomorphism of rank 1 on $C(K)$ which is not order continuous (cf. Exerc. 10 (c)).

25. Let $K, L$ be compact spaces and denote by $E, F$ the *complex* Banach lattices $C(K)$, $C(L)$ respectively (cf. Chap. II, § 11).

(a) Show that (9.1) remains valid if a lattice homomorphism $T: E \to F$ is defined to be a linear map satisfying $|Tf| = T|f|$ $(f \in E)$, and if a homomorphism of complex algebras is understood to be a $*$-homomorphism (where the asterisk denotes complex conjugation).

(b) Each lattice homomorphism $T: E \to F$ is of the form $Tf(s) = g(s) f \circ k(s)$ ($f \in E, s \in L$), where $0 \leqq g \in F$ and $k: L \to K$ is a map continuous and determined by $T$ uniquely on the subset $\{s: g(s) > 0\}$ of $L$.

(c) A lattice homomorphism $E \to F$ which is a compact linear map, is necessarily of finite rank. (For a proof of this and further results, see M. Wolff [1969a].)

26. Let $(X, \Sigma, \mu)$ and $(Y, T, \nu)$ be localizable measure spaces (§ 1, Example 2); we suppose that $\Sigma$, $T$ are $\sigma$-algebras of subsets separating the points of $X$, $Y$ respectively, and containing no atoms of infinite measure. As usual, we denote by $\mathscr{L}^\infty(\mu)$ the vector lattice of all bounded, locally measurable real functions on $X$ and by $L^\infty(\mu)$ its quotient over the ideal of locally negligible functions; similarly for $\mathscr{L}^\infty(\nu)$, $L^\infty(\nu)$.

(a) Show the following assertions to be equivalent:

($\alpha$) $T: \mathscr{L}^\infty(\mu) \to \mathscr{L}^\infty(\nu)$ is a lattice homomorphism such that for each directed ($\geqq$) family $(f_\lambda)$ in $\mathscr{L}^\infty(\mu)$ satisfying $\lim f_\lambda(t) = 0$ locally a.e. ($\mu$), one has $\lim Tf_\lambda(s) = 0$ locally a.e. ($\nu$).

($\beta$) There exists $g \in \mathscr{L}^\infty(\nu)$ and a non-singular locally measurable transformation $k: Y \to X$ (§ 9, Example 1) such that $Tf(s) = g(s) f \circ k(s)$ locally a.e. ($\nu$) for each $f \in \mathscr{L}^\infty(\mu)$. (Consider $X$ as a dense subspace of the Stone representation space $\bar{X}$ of the Boolean algebra $\Sigma_1$ of all locally measurable sets, and similarly for $Y$, $\bar{Y}$; identify $\mathscr{L}^\infty(\mu)$ and $\mathscr{L}^\infty(\nu)$ with $C(\bar{X})$, $C(\bar{Y})$ respectively, and use Exerc. 25(b). Show that the resulting map $\bar{k}: \bar{Y} \to \bar{X}$ can be modified to yield the desired map $k$.)

(b) Extend (a) to operators $T: \mathscr{L}^p(\mu) \to \mathscr{L}^p(\nu)$ ($1 \leqq p < +\infty$) under suitable assumptions.

(c) A *lifting* $\varrho: L^\infty(\mu) \to \mathscr{L}^\infty(\mu)$ is a lattice homomorphism such that $\varrho(e) = 1$ and $\pi \circ \varrho$ is the identity map of $L^\infty(\mu)$, where $\pi$ denotes the canonical map $\mathscr{L}^\infty(\mu) \to L^\infty(\mu)$ (cf. A. and C. Ionescu-Tulcea [1969]). Suppose $\varrho'$ to be a lifting of $L^\infty(\nu)$ into $\mathscr{L}^\infty(\nu)$ and suppose that $T$ is a lattice homomorphism of $L^\infty(\mu)$ into $L^\infty(\nu)$. Define a lattice homomorphism $T_0: \mathscr{L}^\infty(\mu) \to \mathscr{L}^\infty(\nu)$ by "lifting" $T$, that is, define $T_0 := \varrho' \circ T \circ \pi$. If $T_0$ is order continuous, then $T$ is induced by a non-singular locally measurable transformation $k: Y \to X$. (Apply (a).)

27. Let $(X, \Sigma', \mu)$ and $(Y, T', \nu)$ be $\sigma$-finite measure spaces and denote by $\Sigma$ and $T$ the respective $\sigma$-complete Boolean algebras of measurable sets modulo null sets. A bijection $\psi: \Sigma \to T$ is called an isomorphism of measure algebras if $\psi$ and $\psi^{-1}$ commute with the algebraic operations of $\Sigma$ and $T$, respectively.

(a) Each measure preserving bijection $Y \to X$ (§ 9, Example 1) with measurable inverse induces an isomorphism of measure algebras $\Sigma \to T$ and a family of isometric lattice isomorphisms $L^p(\mu) \to L^p(\nu)$ ($1 \leqq p \leqq +\infty$). The converse of this assertion is false (for a classical example, see Halmos and v. Neumann [1942]; cf. Exerc. 26(c)).

(b) Suppose $\mu$, $\nu$ to be finite and let $U$ be a continuous linear map $L^1(\mu) \to L^1(\nu)$ satisfying $Ue_X = e_Y$, where $e_X$, $e_Y$ denote the (classes of the) respective constant-one functions. Show that $U$ is an isomorphism of Banach lattices if and only if $U$ is induced by a measure preserving isomorphism $\Sigma \to T$ of measure algebras.

(c) These assertions are equivalent:

($\alpha$) The measure algebra $\Sigma$ is isomorphic to $T$.

($\beta$) There exists $p$, $1 \leqq p \leqq +\infty$; such that $L^p(\mu)$ is isomorphic (as a vector lattice) to $L^p(\nu)$.

($\gamma$) The respective structure spaces of $L^p(\mu)$ and $L^p(\nu)$ are homeomorphic.

Observe that, in particular, $L^1(X,\Sigma',\mu)$ is isomorphic to $L^1(X,\Sigma',\mu')$ iff $\mu$ is equivalent to $\mu'$.

28. (Cf. I. Glicksberg [1969].) Let $K$ be a compact space and denote by $C(K)$ the (complex) Banach lattice of all continuous, complex valued functions on $K$. Suppose $T$ is a Markov operator on $C(K)$ (Def. 8.8) such that the union of the supports of all $T$-invariant Radon measures on $K$ (cf. 8.10) is dense in $K$.

(a) The closed linear span of all eigenfuctions of $T$ pertaining to unimodular eigenvalues, is a conjugation invariant subalgebra $A$ of $C(K)$. (Show that if $\alpha f = Tf$, $|\alpha| = 1$, then for each $s \in K$ one has $f(t) = \alpha f(s)$ for all $t$ in the support of the Radon measure $T' \delta_s$, $\delta_s$ denoting the Dirac measure of $s$.) Infer from this that $A$ is isomorphic (as a complex Banach lattice and algebra) to $C(K_0)$ for a suitable quotient of $K$.

(b) The restriction $T_0 := T_{|A}$ defines a radical-free Markov operator on $C(K_0)$ (Exerc. 21), and the closure $G$ of the cyclic semi-group $(T_0^n)_{n \in \mathbb{N}}$ in $\mathscr{L}_s(A)$ is a compact, monothetic group of operators with identity $1_A$. Conclude that $T_0$ is an automorphism of (the Banach lattice and Banach algebra) $C(K_0)$ so that $G$ induces a monothetic transformation group $\tilde{G}$ on $K_0$. (Consider quotients of $C(K_0)$ over maximal $T_0$-ideals, and apply Theorem 10.5.)

(c) Infer from (b) that each orbit $\tilde{G}(s)$ $(s \in K_0)$ is a minimal $T_0$-invariant subset of $K_0$ (supporting an ergodic measure, cf. § 8, Examples 5, 7), thus providing an example where $K_0$ is the union of all minimal closed $T_0$-invariant subsets (cf. Exerc. 21 (d)).

(d) Identifying the character group $G^*$ of $\tilde{G}$ with a subgroup of the circle group (cf. 10.5), show that the unimodular eigenvalues of $T$ are the elements of $\bigcup_{s \in K_0} \tilde{G}_s^\circ$ where $\tilde{G}_s^\circ$ denotes the annihilator of the isotropy subgroup $\tilde{G}_s$ whose elements leave $s \in K_0$ fixed. (See Glicksberg [1969].)

29. (Cf. T. Ando [1969].)

(a) Each closed vector sublattice of $c_0(\mathsf{A})$ is the range of a positive contractive projection. ($c_0(\mathsf{A})$ denotes the Banach lattice of continuous real functions on the discrete locally compact space $\mathsf{A}$ that vanish at infinity, endowed with the supremum norm.)

(b) If $H$ is a closed vector sublattice of the Banach lattice $C(K)$, where $K$ is compact, there exists a positive norm preserving linear map $P': H' \to E'$ such that for each $x' \in H'$, the restriction of $P'x'$ to $H$ agrees with $x'$. ("Lifting" of $E'/H^\circ$ into $E'$.)

30. Under the compactness hypothesis of § 11, Example 2, extend the results of Chap. I, § 9 to homogeneous Markov chains with countably infinite state space. (Cf. Doob [1953].)

Chapter IV

# Lattices of Operators

## Introduction

The present chapter is primarily concerned with vector lattices of linear operators between Banach lattices or, more precisely, with the problem of exhibiting significant classes of linear operators possessing a (linear) modulus. Up to the late 1960's, the available knowledge in this area appeared somewhat fragmentary and incoherent; above all, however, of little relevance to the mainstream of operator theory. The fact is that order bounded operators are very intimately related to operators of recognized significance, as is evidenced by Proposition 5.7 (iii); more specifically, a linear map from an $AM$-space into an $AL$-space is order bounded (equivalently, possesses a modulus) if and only if it is integral in the sense of Grothendieck [1955a]. Combined with a judicious use of $AM$- and $AL$-spaces as structural companions of general Banach lattices (cf. § 3), the systematic exploitation of those facts permits many interesting applications: For example, a fairly easy access to the theory of integral and absolutely summing maps, construction of Banach lattice tensor products with identification of their duals, clarification of the relations between kernel and nuclear operators, and others.

Section 1 collects elementary facts on the modulus of a linear operator and gives the (essentially sharp, cf. Exerc. 16) Theorem 1.5; then it proceeds to the case of complex scalars, thus extending the results of Chap. II, § 11. Section 2 is largely preparatory, establishing some basic results on the classical $\pi$- and $\varepsilon$-tensor products and the metric approximation property; it seemed appropriate to include (2.3) here even though its proof must borrow an early result of Section 5. The new characterization of $AM$- and $AL$-spaces found near the end of the section is indispensable for the sequel. Section 3 introduces two mutually dual classes of operators which (up to a technical condition on the range space) always possess a modulus. In view of the characterization of $AM$- and $AL$-spaces derived in Section 2, the earlier results of Section 4 (particularly, (4.3)) appear as the natural generalization of (1.5) to arbitrary Banach lattices; Theorem 4.6 is the basis for the construction of tensor products given in Section 7. Section 5 shows Grothendieck's integral maps to be intimately related to our subject (see, especially, (5.4) and (5.6)); the section gives an account of their theory, as well as of some related material, leading up to the theorem of Dvoretzky-Rogers. Section 6 is a treatment of Hilbert-Schmidt operators, including their basic classical theory as well as their remarkable order theoretic characterizations (6.9).

Section 7 constructs two types of tensor products of Banach lattices which are themselves Banach lattices under their natural orderings; these tensor products are isomorphic under transposition, but both types are retained because of the considerable clarity which operator theoretic applications gain by their distinction. The relation of these tensor products with certain $\varepsilon$-tensor products is the basis for the identification of their duals. The section is supplemented by results leading to the duality of vector-valued $L^p$-spaces, the latter being special cases of the tensor products considered. Section 8 then applies this theory to the construction of Banach lattices of compact operators, and to the relation of these Banach lattices with the Banach spaces of nuclear and (all) compact operators in important special cases; the section contains extensive applications to operators between $L^p$-spaces defined by measurable kernels.

Motivated by these applications, in Section 9 we pursue the characterization of kernel operators between very general types of Banach function lattices (Theorems 9.6 and 9.7); a characterization which is, in particular, useful in connection with representation of abstract Banach lattices as Banach function lattices (Chap. III, § 4—6). Finally, Section 10 gives internal conditions (with respect to the lattice of regular operators) under which a kernel operator is compact; this property is encountered much less frequently than might be expected.

## § 1. The Modulus of a Linear Operator

Let $E, F$ be vector lattices (over $\mathbb{R}$) and denote by $L(E,F)$ the vector space of all linear maps $E \to F$ (as in [S], we write $L(E)$ if $E=F$). Under its canonical ordering (the vector ordering defined by $T \geqq 0 \Leftrightarrow T(E_+) \subset F_+$; Chap. II, § 2) $L(E,F)$ is not a vector lattice in general; for example, it is not difficult to show that the algebraic dual $E^*$ is a vector lattice if and only if $E$ is the ordered direct sum of $d$ copies of $\mathbb{R}$, where $d$ is the linear dimension of $E$ (Exerc. 1). But even subspaces of $L(E,F)$, such as the space $\mathscr{L}(E,F)$ of continuous linear operators $E \to F$ when $E, F$ are Banach lattices, are not vector lattices in general for their canonical ordering (see Example 2 below). Thus it is natural to ask for subspaces of $L(E,F)$ which are lattices under the induced ordering, or more specifically: For what operators $T \in L(E,F)$ does the (linear) modulus $|T| := T \vee (-T)$ exist in $L(E,F)$?

In this first section, we will derive some elementary but important results along with their extensions to complex vector lattices (Chap. II, § 11).

**1.1 Definition.** *Let $E, F$ be ordered vector spaces. A linear map $T\colon E \to F$ is called* regular *if $T = T_1 - T_2$ for suitable positive linear maps $T_i\colon E \to F$ $(i=1,2)$.*

It is immediately clear from this definition that the set of all regular maps $E \to F$ forms a vector subspace $L^r(E,F)$ of $L(E,F)$. Moreover, for each $T \in L^r(E,F)$ the algebraic adjoint $T^*\colon F^* \to E^*$ is regular (cf. [S, Chap. IV, § 2]), provided the canonical orderings of $E^*$, $F^*$ are well defined.

**1.2 Proposition.** *Let $E, F$ be vector lattices. For $T \in L(E,F)$ consider these assertions:*

(a) *$|T| := T \vee (-T)$ exists in the canonical order of $L(E,F)$.*

(b) *$T$ is regular.*

(c) *$T$ maps order bounded sets onto order bounded sets.*

*Then* (a) $\Rightarrow$ (b) $\Rightarrow$ (c) *and, if $F$ is order complete,* (c) $\Rightarrow$ (a).

*Proof.* (a) $\Rightarrow$ (b): If $|T|$ exists then, clearly, $|T| \geqq 0$ and the identity $T = |T| - (|T| - T)$ shows $T$ to be regular.

(b) $\Rightarrow$ (c): Let $T = T_1 - T_2$ where $T_1, T_2 \geqq 0$. If $A$ is an order bounded subset of $E$, there exists $b \in E_+$ such that $|a| \leqq b$ for all $a \in A$; it follows that $|Ta| \leqq (T_1 + T_2)|a| \leqq (T_1 + T_2) b$ for all $a \in A$.

(c) $\Rightarrow$ (a): Suppose $F$ order complete, and let $S: E_+ \to F_+$ be defined by

$$Sx := \sup_{|z| \leqq x} |Tz| \qquad (x \in E_+);$$

in fact, this amounts to defining $Sx := \sup T[-x, x]$ and this definition makes sense, since $T$ is order bounded and $F$ order complete by hypothesis. Moreover, $S(\lambda x) = \lambda Sx$ for $\lambda \in \mathbb{R}_+$, and since $|z| \leqq x_1 + x_2$ $(x_1, x_2 \geqq 0)$ implies $z = z_1 + z_2$ for suitable elements $z_i \in [-x_i, x_i]$ $(i = 1, 2)$ by (II. 1.6), it follows that $S(x_1 + x_2) = Sx_1 + Sx_2$. Therefore, $S$ extends to a unique linear map $\bar{S}: E \to F$ (see discussion following (II. 2.4)) which has the property that $\pm T \leqq \bar{S}$, and that $\pm T \leqq R \in L(E,F)$ implies $\bar{S} \leqq R$; consequently, $\bar{S} = |T|$ in the canonical order of $L(E,F)$. □

**1.3 Proposition.** *If $E, F$ are vector lattices and $F$ is order complete, then $L^r(E,F)$ is an order complete vector lattice under its canonical ordering.*

*Proof.* From (1.2) it follows that $|T| = T \vee (-T)$ exists for each $T \in L^r(E,F)$. The equations $T = T^+ - T^-$, $|T| = T^+ + T^-$ then determine $T^+ := T \vee 0$ and $T^- := (-T) \vee 0$ (cf. remark following (II.1.4)); generally, for $S, T \in L^r(E,F)$ we have $S \vee T = S + (T - S)^+$ and $S \wedge T = S - (T - S)^-$ by the translation invariance of the ordering. Finally, if $(T_\alpha)$ is a directed $(\leqq)$ majorized subset of $L^r(E,F)$, the mapping $T \in L^r(E,F)$ given by

$$Tx := \sup_\alpha T_\alpha x \qquad (x \in E_+)$$

is readily seen to be $\sup_\alpha T_\alpha$ in $L^r(E,F)$. □

It should be noted that, in complete analogy with Formulae (1), (3), (4) of Chap. II, § 4, the following relations hold for all $S, T \in L^r(E,F)$ and all $x \in E_+$ ($E, F$ being vector lattices, $F$ order complete):

$$(T \vee S)x = \sup\{Ty + Sz : y \geqq 0, z \geqq 0, y + z = x\},$$
$$(T \wedge S)x = \inf\{Ty + Sz : y \geqq 0, z \geqq 0, y + z = x\}; \tag{1}$$

$$|T|x = \sup\{|Tz| : |z| \leqq x\}. \tag{2}$$

Moreover, if $(T_\alpha)_{\alpha\in \mathsf{A}}$ is a non-void majorized (not necessarily directed) subset of $L^r(E,F)$, then $T:=\sup_\alpha T_\alpha$ is given by

$$Tx = \sup(T_{\alpha_1}x_1 + \cdots + T_{\alpha_n}x_n) \tag{3}$$

where $\{\alpha_1,\dots,\alpha_n\}$ runs through all non-void finite subsets of $\mathsf{A}$ and for each such finite set, $\{x_1,\dots,x_n\}$ runs through all $n$-tuples of positive vectors satisfying $x_1+\cdots+x_n=x$.

Let us note the following corollary of the preceding results, whose simple proof is omitted.

**Corollary.** *If $E$ is an order complete vector lattice, the vector lattice $L^r(E)$ of regular linear maps $E\to E$ is a subalgebra of $L(E)$. Moreover, one has $|T_1T_2|\leqq|T_1|\,|T_2|$ for each pair $T_1,T_2\in L^r(E)$.*

Suppose now that $E,F$ are normed vector lattices (Chap. II, § 5) and denote, as usual, by $\mathscr{L}(E,F)$ the normed space of continuous linear maps $E\to F$ under the *operator norm* $\|T\|:=\sup\{\|Tx\|:\|x\|\leqq 1\}$. If $T\in\mathscr{L}(E,F)$ is a positive continuous map, the operator norm of $T$ satisfies

$$\|T\| = \sup\{\|Tx\|: x\geqq 0, \|x\|\leqq 1\} \qquad (T\geqq 0). \tag{4}$$

In fact, since $x\in E$ and $\|x\|\leqq 1$ implies $\||x|\|\leqq 1$, we have $\|Tx\|\leqq\|T|x|\|\leqq\|T\|$ whenever $\|x\|\leqq 1$ (cf. Chap. I, § 2, Formula (1)).

Supposing, in addition, $F$ to be order complete, we consider the space of regular maps $E\to F$ with continuous modulus.

**1.4 Proposition.** *Let $E,F$ be normed vector lattices, $F$ order complete, and denote by $\mathscr{L}^r(E,F)$ the ideal of those $T\in L^r(E,F)$ for which $|T|\in\mathscr{L}(E,F)$. Under the norm (called* the $r$-norm*)*

$$\|T\|_r := \|\,|T|\,\|, \tag{5}$$

*$\mathscr{L}^r(E,F)$ is an order complete normed vector lattice. $\mathscr{L}^r(E,F)$ is a Banach lattice whenever $F$ is.*

*Proof.* First, we observe that $S\in L(E,F)$, $T\in\mathscr{L}(E,F)$ and $0\leqq S\leqq T$ implies $S\in\mathscr{L}(E,F)$ (i.e., the continuity of $S$); thus, $\mathscr{L}^r(E,F)$ is an ideal in $L^r(E,F)$ and hence order complete by (1.3). Moreover, (4) shows that the $r$-norm defined by (5) is a lattice norm.

To show that $\mathscr{L}^r(E,F)$ is norm complete whenever $F$ is, let $(T_n)$ denote a Cauchy sequence in $\mathscr{L}^r(E,F)$; without loss of generality, we can assume that $\|T_{n+1}-T_n\|_r<2^{-(n+1)}$ for all $n\in\mathbb{N}$. Since $\|T\|\leqq\|T\|_r$ for each $T\in\mathscr{L}^r(E,F)$, the sequence converges to some $T\in\mathscr{L}(E,F)$ in the operator norm. For each $z\in E$ and $p\in\mathbb{N}$, we have $Tz-T_pz=\sum_{n\geqq p}(T_{n+1}-T_n)z$ and hence,

$$\sup_{|z|\leqq x}|(T-T_p)z|\leqq\textstyle\sum_{n\geqq p}|T_{n+1}-T_n|x$$

for every $x\in E_+$; this shows that $T\in\mathscr{L}^r(E,F)$ and that $\|T-T_p\|_r\leqq 2^{-p}$. □

If $E$ is a Banach lattice, then $\mathscr{L}^r(E,F)=L^r(E,F)$ by (II. 5.3). We also note that if $T\in\mathscr{L}^r(E,F)$ and $T'\colon F'\to E'$ is the adjoint operator then, since $E'$ is order complete (II. 5.5) and since $\pm T'\leqq |T|'$, we have $T'\in\mathscr{L}^r(F',E')$; however, $|T'|<|T|'$ in general (Exerc. 2) and hence, the map $T\mapsto T'$ possibly need not preserve the $r$-norm.

**Corollary.** *Let $E,F$ be normed vector lattices, $F$ order complete. The adjoint mapping $T\mapsto T'$ defines a norm decreasing injection of $\mathscr{L}^r(E,F)$ into $\mathscr{L}^r(F',E')$.*

Also, if $E,F$ are normed vector lattices but $F$ is not assumed order complete, a generalization of the $r$-norm can be defined as follows. Let $\mathscr{L}^r(E,F)$ denote the vector space of all linear maps $T\colon E\to F$ possessing a decomposition $T=T_1-T_2$ for suitable positive maps $T_1,T_2\in\mathscr{L}(E,F)$; then the norm

$$(5')\qquad \|T\|_r := \inf\|T_1+T_2\|,$$

where the infimum is taken over all of those decompositions of $T$, makes $\mathscr{L}^r(E,F)$ into an ordered normed space whose positive cone is normal and generating (Exerc. 4). Again, $\mathscr{L}^r(E,F)$ is a Banach space if $F$ is norm complete, and agrees with the set of all regular maps $E\to F$ whenever $E$ is norm complete.

**Examples.** 1. If $A=(\alpha_{ij})$ is an $m\times n$-matrix over $\mathbb{R}$, considered as a linear operator of the vector lattice $\mathbb{R}^n$ into $\mathbb{R}^m$ (canonical orderings), its modulus $|A|$ is given by the matrix $(|\alpha_{ij}|)$ (Chap. I, § 2).

In general, we have $\|A\|_r=\big\||A|\big\|>\|A\|$ for the matrix norm associated with given lattice norms on $\mathbb{R}^n$ and $\mathbb{R}^m$, respectively. More specifically, if $m=n=2^p$ $(p\in\mathbb{N})$ and $\mathbb{R}^n$ is given its Euclidean norm (Chap. I, § 1, Example 2), there exist orthogonal $n\times n$-matrices $A$ such that ($\|A\|=1$ and) $\|A\|_r=\sqrt{n}$.

In fact, if $n=2^p$ there exist $n$ orthogonal vectors $u_1,\dots,u_n$ each with coordinates $\pm 1$ only. This is clear for $p=1$; if $u_\nu$ $(\nu=1,\dots,n)$ have the required property for $n=2^p$, the $2n$-vectors $v_1=(u_1,u_1),\dots,v_n=(u_n,u_n)$; $v_{n+1}=(u_1,-u_1),\dots,v_{2n}=(u_n,-u_n)$ have the required property for $n=2^{p+1}$. Now if $A_n$ is the $n\times n$-matrix with rows $n^{-1/2}u_\nu$ $(\nu=1,\dots,n)$, $A_n$ is orthogonal and hence, $\|A_n\|=1$. On the other hand, $|A_n|$ is symmetric with largest eigenvalue $\sqrt{n}$ and hence, $\|A_n\|_r=\sqrt{n}$. (Cf. the corollary of (I. 2.3).)

2. If $E$ is an infinite dimensional order complete Banach lattice, it can happen that a continuous operator $T\in\mathscr{L}(E)$ possesses no modulus (that is, by (1.2), that $T$ is not "order bounded"), even if $T$ is compact; on the other hand, a compact operator can have a modulus which is not compact.

To construct examples, let $E_p$ denote $2^p$-dimensional Euclidean space under its canonical ordering, and let $A_p$ be an operator on $E_p$ for which $\|A_p\|=1$, $\|A_p\|_r=\sqrt{2^p}$ (Example 1). Let $E$ denote the vector space of all sequences $\boldsymbol{x}=(x_p)$ where $x_p\in E_p$ $(p\in\mathbb{N})$ and $\lim_p\|x_p\|=0$. Under the norm $\|\boldsymbol{x}\|:=\sup_p\|x_p\|$ and the canonical (coordinatewise) ordering, $E$ is an order complete Banach lattice with order continuous norm.

Now if $(k_p)$ is any bounded sequence of positive real numbers, the mapping $(x_p)\mapsto(k_pA_px_p)$ defines an operator $T\in\mathscr{L}(E)$; it is easily verified that $T$ is compact iff $(k_n)$ is a null sequence. On the other hand, the modulus

$|T|$ exists iff the sequence $(k_p \sqrt{2^p})$ is bounded. Therefore, if $k_p := 2^{-p/2}$ $(p \in \mathbb{N})$ then $T$ is compact with non-compact modulus; if $k_p := 2^{-p/3}$, $T$ is compact but possesses no modulus.

3. If $E$ is any Banach lattice, $F$ an $AM$-space (Def. II. 7.1), and if $T \in \mathscr{L}(E,F)$ is compact then $|T|$ exists and is compact. In fact, $F$ can be considered a closed vector sublattice of some $C(K)$ ($K$ compact) (for instance, of its bidual $F''$ which is an $AM$-space with unit (II. 9.1)); if $F$ is so represented, the image $B := T(U)$ of the unit ball $U$ of $E$ is a uniformly bounded, equicontinuous set of real functions on $K$. Thus (cf. (II. 7.6) and the discussion preceding it) the set $C$ of suprema of non-void finite subsets of $B$ is again equicontinuous with the same uniform bound as $B$, and hence $C$ is relatively compact in $F$; thus by (2), $|T|$ exists and is a compact linear map $E \to F$, and $\|T\|_r = \|T\|$.

We consider a special case. Let $X$, $Y$ denote compact spaces, let $E := C(X)$, $F := C(Y)$, and let $T: E \to F$ be given by

$$Tf(t) = \int K(s,t)\, f(s)\, d\mu(s) \qquad (t \in Y,\ f \in E)$$

where $K \in C(X \times Y)$ is continuous and $\mu$ is a positive Radon measure on $X$. Then $T$ is compact, and the compact modulus $|T|$ is given by the kernel $|K|$; that is,

$$|T|\, f(t) = \int |K(s,t)|\, f(s)\, d\mu(s) \qquad (t \in Y,\ f \in E).$$

It is clear that the operator defined by $|K|$ is $\geqq \pm T$; we have to show that $\pm T \leqq S \in \mathscr{L}(E,F)$ implies $\int |K|\, f\, d\mu \leqq Sf$ for all $f \in E_+$. But if $f \in E_+$ and $f = g + h$ is any decomposition of $f$ into positive summands, we have $Sf \geqq Tg - Th$ and hence, letting $\mu_t := K(\ ,t).\mu$ $(t \in Y)$,

$$|\mu_t|(f) = \sup_{f = g+h} \mu_t(g-h) \leqq Sf(t)$$

for all $t \in Y$. But it is well known and easy to see that $|\mu_t| = |K(\ ,t)|.\mu$, and this proves the assertion.

Thus Examples 1, 2 show that when $E, F$ are Banach lattices, the mapping (canonical imbedding) $\mathscr{L}^r(E,F) \to \mathscr{L}(E,F)$ is not surjective, in general, and the $r$-norm greater than the operator norm. Apart from the situation where $E$ or $F$ has finite dimension there are two notable exceptions, as follows.

**1.5 Theorem.** *Let $E, F$ be a Banach lattices. Then $\mathscr{L}^r(E,F) = \mathscr{L}(E,F)$ (or equivalently, $\mathscr{L}(E,F)$ is a Banach lattice under its canonical order and the operator norm) whenever at least one of the following conditions is satisfied:*

(i) *$F$ is an order complete $AM$-space with unit.*

(ii) *$E$ is an $AL$-space, and there exists a positive contractive projection $P: F'' \to F$.*[1]

[1] By means of evaluation, $F$ is considered a subspace of $F''$.

*Proof.* (i) Let a continuous $T: E \to F$ be given and let $e$ denote the unit of $F$. Since the unit ball of $F$ is the order interval $[-e,e]$, from Formula (2) it follows that $|T|$ exists and has operator norm $\leqq \|T\|$; thus $\|T\|_r = \|T\|$ for each $T \in \mathscr{L}(E,F)$.

(ii) By (II. 9.1), $E'$ is an order complete $AM$-space with unit. Hence if $T \in \mathscr{L}(E,F)$, by the preceding part (i) the adjoint $T': F' \to E'$ is regular and we have $\|T'\|_r = \|T'\| = \|T\|$. Thus the second adjoint $T''$ is regular which implies the existence of $|T''|$; moreover, if $Q: E \to E''$ denotes the evaluation map, we have $\pm T \leqq P \circ |T''| \circ Q$ which shows $T$ to be regular.

On the other hand, the existence of the projection $P$ implies $F$ to be order complete (see (4.2) below) (namely, for every directed ($\leqq$) majorized subset $D$ of $F$, one has $P(\sup_{F''} D) = \sup_F D$); consequently, $|T|$ exists in $\mathscr{L}(E,F)$. Finally, $|T| \leqq P \circ |T''| \circ Q$ implies $\|T\|_r \leqq \|T''\|_r$ while $\|T''\|_r \leqq \|T'\|_r$ by the corollary of (1.4); since $\|T'\|_r = \|T\|$ was established above, we obtain $\|T\|_r \leqq \|T\|$ and hence, $\|T\|_r = \|T\|$. □

Turning to the case of complex scalars, we now suppose that $E_{\mathbb{C}}, F_{\mathbb{C}}$ are complex vector lattices (Def. II. 11.1); by definition, this implies that the underlying (real) vector lattices $E, F$ satisfy Axiom (OS) (Def. I. 1.8). Recall that each linear map $T \in L(E_{\mathbb{C}}, F_{\mathbb{C}})$ has a unique representation $T = T_1 + iT_2$ where $T_j$ are real (Chap. II, § 11, Formula (3′)); in accordance with previous definitions (Chap. II, § 11 (iii)) we define $T \in L(E_{\mathbb{C}}, F_{\mathbb{C}})$ to be *regular* if (the restrictions to $E$ of) $T_1$ and $T_2$ are regular in the sense of Def. 1.1. It is clear that the regular maps $E_{\mathbb{C}} \to F_{\mathbb{C}}$ form a vector subspace of $L(E_{\mathbb{C}}, F_{\mathbb{C}})$; we denote this subspace by $L^r(E_{\mathbb{C}}, F_{\mathbb{C}})$.

A subset $A$ of a complex vector lattice $E_{\mathbb{C}}$ is called *absolutely majorized* if there exists $b \in E_+$ such that $|a| \leqq b$ for all $a \in A$.

**1.6 Proposition.** *Let $E_{\mathbb{C}}, F_{\mathbb{C}}$ be complex vector lattices and let $T \in L(E_{\mathbb{C}}, F_{\mathbb{C}})$. Consider the assertions:*

(a) *$T$ is regular.*

(b) *$T$ maps absolutely majorized subsets of $E_{\mathbb{C}}$ onto like subsets of $F_{\mathbb{C}}$.*

*Then* (a) $\Rightarrow$ (b) *and, if $F$ is order complete,* (b) $\Rightarrow$ (a).

*Proof.* (a) $\Rightarrow$ (b): If $T$ is regular, there exist positive maps $R_1, R_2$ and $S_1, S_2$ for which $T = (R_1 - R_2) + i(S_1 - S_2)$. Clearly, then, $|z| \leqq x$ $(z \in E_{\mathbb{C}}, x \in E_+)$ implies $|Tz| \leqq (R_1 + R_2 + S_1 + S_2)x$ (Chap. II, § 11, Formula (1)).

(b) $\Rightarrow$ (a): Suppose $F$ order complete. We define a map $|T|: E_+ \to F_+$ by

$$(2^\star) \qquad |T|x := \sup_{|z| \leqq x} |Tz| \qquad (x \in E_+).$$

It is clear that $|T|(\lambda x) = \lambda |T| x$ for all real $\lambda \geqq 0$, and from (II. 11.2) it follows that $|T|$ is additive on $E_+$. Thus $|T|$ extends uniquely to a positive linear map $E_{\mathbb{C}} \to F_{\mathbb{C}}$; this extension is again denoted by $|T|$. Now if $T = T_1 + iT_2$ is the canonical decomposition of $T$, it is clear from ($2^\star$) that $|T_j y| \leqq |T| x$ whenever $x, y \in E$ and $|y| \leqq x$ $(j = 1, 2)$; therefore, $T_1$ and $T_2$ (restricted to $E$) are regular and hence, $T$ is regular according to the definition above. □

The preceding proof motivates this definition.

**1.7 Definition.** *Let $E_{\mathbb{C}}$, $F_{\mathbb{C}}$ be complex vector lattices and suppose $F$ to be order complete. For each regular $T: E_{\mathbb{C}} \to F_{\mathbb{C}}$, the positive mapping $|T|: E_{\mathbb{C}} \to F_{\mathbb{C}}$ defined by*

$$|T|x := \sup_{|z| \leq x} |Tz| \qquad (x \in E_+),$$

*is called the* modulus *of $T$.*

Clearly, the modulus $|T|$ of $T \in L^r(E_{\mathbb{C}}, F_{\mathbb{C}})$ satisfies $|Tz| \leq |T|\,|z|$ for all $z \in E_{\mathbb{C}}$, and if $|Tz| \leq S|z|$ $(z \in E_{\mathbb{C}})$ then $|T| \leq S$; moreover, the modulus function $T \mapsto |T|$ satisfies Axioms (2) of Chap. II, § 11. In particular, if $E = F$ then $|T_1 \circ T_2| \leq |T_1|\,|T_2|$ for all pairs $T_1, T_2 \in L^r(E_{\mathbb{C}})$ (Exerc. 3). Also, if $E_{\mathbb{C}}$ and $F_{\mathbb{C}}$ are normed complex vector lattices (cf. II. 11.3) and $\mathscr{L}^r(E_{\mathbb{C}}, F_{\mathbb{C}})$ denotes the space of all regular maps $T: E_{\mathbb{C}} \to F_{\mathbb{C}}$ with continuous modulus $|T|$, then the $r$-norm

$$\|T\|_r := \big\| \, |T| \, \big\| \tag{5*}$$

has the property required of a complex lattice norm (Chap. II, § 11, Formula (5′)).

It is, therefore, natural to ask (supposing again $F$ order complete) if the space $L^r(E_{\mathbb{C}}, F_{\mathbb{C}})$ of complex regular maps, endowed with the modulus function $T \mapsto |T|$, is a complex vector lattice in the sense of Def. II. 11.3; likewise, supposing $E_{\mathbb{C}}$ and $F_{\mathbb{C}}$ to be complex normed vector lattices, is the same true of $\mathscr{L}^r(E_{\mathbb{C}}, F_{\mathbb{C}})$ when endowed with the modulus function $T \mapsto |T|$ and the $r$-norm (5*)? The answer is affirmative; so all results of Chap. II, § 11 apply to these spaces of operators.

**1.8 Theorem.** *Let $E_{\mathbb{C}}$, $F_{\mathbb{C}}$ be complex vector lattices and suppose $F$ to be order complete. For $T \in L^r(E_{\mathbb{C}}, F_{\mathbb{C}})$ the modulus $|T|$* (Def. 1.7) *satisfies*

$$|T| = \sup_{0 \leq \theta < 2\pi} |(\cos\theta)\, T_1 + (\sin\theta)\, T_2|$$

*where $T = T_1 + i\,T_2$ is the canonical decomposition of $T$. Hence $L^r(E_{\mathbb{C}}, F_{\mathbb{C}})$, endowed with the modulus function $T \mapsto |T|$, is isomorphic with the complexification of the vector lattice $L^r(E, F)$.*

*Proof.* Let $T = T_1 + i\,T_2 \in L^r(E_{\mathbb{C}}, F_{\mathbb{C}})$ be fixed. Since $L^r(E, F)$ is order complete by (1.3), it follows from Chap. II, § 11 (iii) that $S := \sup_\theta |(\cos\theta)\, T_1 + (\sin\theta)\, T_2|$ exists in $L(E, F)$ (we do not distinguish between elements of $L(E, F)$ and their canonical extensions $E_{\mathbb{C}} \to F_{\mathbb{C}}$).

(i) We show that $S \leq |T|$. Letting $T_\theta := (\cos\theta)\, T_1 + (\sin\theta)\, T_2$, from (3) above we conclude that for $x \in E_+$,

$$Sx = \sup \textstyle\sum_j T_{\theta_j} y_j$$

where the supremum is taken over all finite families $\{(\theta_j, y_j)\}$ such that $\theta_j \in \mathbb{R}$, $y_j \in E_+$ and $\sum_j y_j = x$. For any such family, let $u := \sum_j (\cos\theta_j)\, y_j$, $v := \sum_j (\sin\theta_j)\, y_j$, and $z := u - iv$; then $|z| \leq x$. Now $|T_1 u + T_2 v| = |\mathrm{Re}\, Tz| \leq |Tz|$ in $F_{\mathbb{C}}$; but, since $T_1 u + T_2 v = \sum_j T_{\theta_j} y_j$, it follows that $Sx \leq \sup_{|z| \leq x} |Tz|$ and hence, $S \leq |T|$.

(ii) To prove that $|T| \leq S$ we observe first that for each $x \in E_+$, the complex principal ideal $(E_{\mathbb{C}})_x = \bigcup_{n=1}^{\infty} n\{z : |z| \leq x\}$ can be identified with a complex $AM$-

space $C(K_x)$ ($K_x$ compact, cf. Chap. II, § 11 (i)), under an isomorphism of complex Banach lattices carrying $x$ onto the constant-one function. By considering suitable partitions of unity on $K_x$, it is readily seen that each $z \in (E_{\mathbb{C}})_x \cong C(K_x)$, $|z| \leqq x$, can be approximated, uniformly on $K_x$, by finite sums $\sum_j \alpha_j x_j$ where $\alpha_j \in \mathbb{C}$, $|\alpha_j| \leqq 1$, $x_j \geqq 0$ and $\sum_j x_j = x$.

Since $T$ induces a continuous linear map of $(E_{\mathbb{C}})_x$ into the ideal (and complex $AM$-space) $(F_{\mathbb{C}})_u$, where $u := |T|x$, it follows that

$$\sup_{|z| \leqq x} |Tz| = \sup |T(\textstyle\sum_j \alpha_j x_j)|$$

where the supremum on the right is taken over all finite families $\{(\alpha_j, x_j)\}$ satisfying $\alpha_j \in \mathbb{C}$, $|\alpha_j| \leqq 1$, $x_j \geqq 0$ and $\sum_j x_j = x$. By definition of the modulus in $F_{\mathbb{C}}$ (Chap. II, § 11 (1)) and by the definition of $S$ we have

$$|Tx_j| = \sup_\theta |(\cos\theta)\, T_1 x_j + (\sin\theta)\, T_2 x_j| \leqq S x_j$$

for each $x_j$. Therefore,

$$|T(\textstyle\sum_j \alpha_j x_j)| \leqq \sum_j |\alpha_j|\, |Tx_j| \leqq \sum_j |\alpha_j| S x_j = S(\sum_j |\alpha_j| x_j) \leqq Sx$$

which shows that $\sup_{|z| \leqq x} |Tz| \leqq Sx$ $(x \in E_+)$ and hence, $|T| \leqq S$. □

The following corollaries are now easy consequences of (1.4) and (1.5).

**Corollary 1.** *If $E_{\mathbb{C}}$ and $F_{\mathbb{C}}$ are complex Banach lattices ($F$ order complete), then $\mathscr{L}^r(E_{\mathbb{C}}, F_{\mathbb{C}})$, endowed with the r-norm (5*) and the modulus function $T \mapsto |T|$* (Def. 1.7), *is a complex Banach lattice isomorphic to the complexification of $\mathscr{L}^r(E, F)$.*

**Corollary 2.** *If $E, F$ are Banach lattices satisfying the hypothesis (either* (i) *or* (ii)*) of* (1.5) *then $\mathscr{L}(E_{\mathbb{C}}, F_{\mathbb{C}})$ is a complex Banach lattice with respect to the modulus function $T \mapsto |T|$ and the operator norm.*

We point out that Cor. 2 applies, in particular, whenever $E = F$ and $E$ is an $AL$-space or an order complete $AM$-space with unit (cf. Chacon-Krengel [1964]).

**Corollary 3.** *With respect to the modulus $f \mapsto |f|$ defined by*

$$|f|(x) := \sup_{|z| \leqq x} |f(z)| \qquad (x \in E_+),$$

*the dual of a complex Banach lattice is a complex Banach lattice.*

## § 2. Preliminaries on Tensor Products. New Characterization of $AM$- and $AL$-Spaces

We assume the reader to be familiar with the elementary algebraic properties of tensor products of vector spaces. We begin by recalling the basic properties of the so-called $\pi$- and $\varepsilon$-tensor norms; the scalar field considered can be either $\mathbb{R}$ or $\mathbb{C}$.[1]

Let $G, H$ be normed spaces. A norm on $G \otimes H$ is called a *cross-norm* if $\|x \otimes y\| = \|x\| \, \|y\|$ for all $x \in G$, $y \in H$. The $\pi$-norm on $G \otimes H$ is the cross-norm defined by

$$\|u\|_\pi := \inf \textstyle\sum_i \|x_i\| \, \|y_i\| \tag{1}$$

where the infimum is taken over all representations $u = \sum_i x_i \otimes y_i$ of the element $u \in G \otimes H$. Denoting by $U_1$, $U_2$ the respective unit balls of $G$ and $H$, the $\pi$-norm is the gauge function of the convex circled hull $\Gamma U_1 \otimes U_2$, so that the canonical bilinear map $\chi: G \times H \to G \otimes H$ has norm 1 (supposing $G, H \neq \{0\}$ and $G \times H$ normed by the gauge of $U_1 \times U_2$).

It is easy to determine the dual of the normed space $G \otimes_\pi H$; in fact, the continuous linear forms on $G \otimes_\pi H$ correspond, under the mapping $f \mapsto \varphi := f \circ \chi$, biunivocally to the continuous bilinear forms $\varphi$ on $G \times H$ [S, III. 6.2], and the dual norm is evidently given by $\|\varphi\| = \sup\{|\varphi(x,y)| : x \in U_1, y \in U_2\}$. Thus under this norm, the space $\mathscr{B}(G,H)$ of continuous bilinear forms on $G \times H$ is the Banach dual of $G \otimes_\pi H$ (and, of course, of the completion $G \tilde{\otimes}_\pi H$). In turn, the correspondence $\varphi \mapsto T_\varphi$ given by

$$\varphi(x,y) = \langle T_\varphi x, y \rangle \qquad (x \in G, \, y \in H) \tag{2}$$

is an isomorphism of $\mathscr{B}(G,H)$ onto the space $\mathscr{L}(G,H')$ of continuous linear operators $G \to H'$, endowed with the operator norm $T \mapsto \|T\| = \sup\{\|Tx\| : \|x\| \leq 1\}$. (Similarly, $\mathscr{B}(G,H)$ is isomorphic with $\mathscr{L}(H,G')$, cf. [S, Chap. IV, § 9].) Thus, by virtue of (2), $\mathscr{L}(G,H')$ can be identified with the Banach dual of $G \otimes_\pi H$.

On the other hand, each pair $(x',y') \in G' \times H'$ defines a continuous bilinear form on $G \times H$ by virtue of $(x,y) \mapsto \langle x,x' \rangle \langle y,y' \rangle$, and the tensor product $G' \otimes H'$ thus becomes a subspace of $\mathscr{B}(G,H)$; identifying, as above, $\mathscr{B}(G,H)$ with $\mathscr{L}(G,H')$, $G' \otimes H'$ becomes the vector subspace of $\mathscr{L}(G,H')$ containing precisely the continuous linear maps $G \to H'$ of finite rank. The duality between $G \otimes H$ and $G' \otimes H'$ so defined, is called *canonical* and denoted by $\langle G \otimes H, G' \otimes H' \rangle$; it is an easy consequence of the Hahn-Banach theorem that it is separated [S, Chap. IV, § 1].

Let $V_1 := U_1^\circ$, $V_2 := U_2^\circ$ denote the respective unit balls of the dual Banach spaces $G'$ and $H'$; then the polar $U_\varepsilon \subset G \otimes H$ of the convex circled hull $\Gamma V_1 \otimes V_2$ (with respect to the canonical duality), is called the *$\varepsilon$-unit ball* of $G \otimes H$; its gauge function is a cross-norm on $G \otimes H$ given by

$$\|u\|_\varepsilon = \sup \textstyle\sum_i \langle x_i, x' \rangle \langle y_i, y' \rangle \tag{3}$$

[1] A brief introduction to topological tensor products can be found in [S, Chap. III, § 6]; the notation employed there will be used in the sequel (with the exception of the completed $\varepsilon$-tensor product, here denoted by $G \tilde{\otimes}_\varepsilon H$).

where the supremum is taken over all pairs $(x',y')\in V_1\times V_2$ and (in contrast with $\sum\|x_i\|\,\|y_i\|$ used in (1)) the sum $\sum\langle x_i,x'\rangle\langle y_i,y'\rangle$ does not depend on the particular representation $u=\sum x_i\otimes y_i$ of $u\in G\otimes H$. If, as above, we identify the dual of $G'\otimes_\pi H'$ with $\mathscr{B}(G',H')$ or equivalently, with $\mathscr{L}(G',H'')$ so that $G\otimes H$ becomes the space of all $H$-valued weak* continuous linear maps $G'\to H''$ of finite rank, then the $\varepsilon$-norm (3) on $G\otimes H$ is the norm induced by the operator norm of $\mathscr{L}(G',H'')$.

**2.1 Definition.** *Let $G,H$ be normed vector spaces. Then $G\otimes H$, endowed with the norm (1) (respectively, (3)) is called the* $\pi$-tensor product *of $G$ and $H$ and denoted by $G\otimes_\pi H$ (respectively, the* $\varepsilon$-tensor product *of $G$ and $H$ and denoted by $G\otimes_\varepsilon H$). The respective completions are denoted by $G\tilde{\otimes}_\pi H$ and $G\tilde{\otimes}_\varepsilon H$.*

*Remark.* The $\pi$- and $\varepsilon$-tensor products of $G$ and $H$ are *symmetric* in the sense that $G\tilde{\otimes}_\pi H\cong H\tilde{\otimes}_\pi G$ *and* $G\tilde{\otimes}_\varepsilon H\cong H\tilde{\otimes}_\varepsilon G$, where the symbol $\cong$ indicates that the canonical isomorphism $G\otimes H\to H\otimes G$ extends to a norm isomorphism of the respective completions. This is not true for all cross-norms (see § 7 below); cf. also Exerc. 22.

**Examples.** 1. Let $K$ denote a compact space, $G$ any Banach space. The vector space of continuous functions $f\colon K\to G$ is a Banach space $C(K,G)$ under the norm

$$\|f\|:=\sup_{s\in K}\|f(s)\|\,.$$

$C(K,G)$ is isomorphic to $C(K)\tilde{\otimes}_\varepsilon G$. To prove this we first note that the linear map $j$ which assigns to each $f\otimes x\in C(K)\otimes G$ the function $s\mapsto f(s)x$ in $C(K,G)$, is an isometry. In fact, if $\sum_i f_i\otimes x_i\in C(K)\otimes G$ then

$$\begin{aligned}\|j(\textstyle\sum_i f_i\otimes x_i)\| &= \sup_{s\in K}\|\textstyle\sum_i f_i(s)x_i\| = \sup_{s\in K,\,x'\in U^\circ}\textstyle\sum_i\langle f_i,\delta_s\rangle\langle x_i,x'\rangle\\ &= \|\textstyle\sum_i f_i\otimes x_i\|_\varepsilon\end{aligned}$$

where $U^\circ$ is the unit ball of $G'$ and $\delta_s$ Dirac measure at $s\in K$ (for the last equality in the chain, note that the unit ball of $C(K)'$ is the weak* closed convex hull of the set $\{\pm\delta_s\colon s\in K\}$). Thus $C(K)\otimes_\varepsilon G$ can be identified with a normed vector subspace of $C(K,G)$; our assertion will be proved if this subspace can be shown to be dense. To this end, let $f\in C(K,G)$ and $\delta>0$ be given; by the uniform continuity of $f$, there exists an open cover $\{A_i\colon i=1,\dots,n\}$ of $K$ such that for all $i$, $(s,t)\in A_i\times A_i$ implies $\|f(s)-f(t)\|<\delta$. Now let $\{f_i\colon i=1,\dots,n\}$ be positive elements of $C(K)$ such that $\sum_i f_i(s)=1$ $(s\in K)$ and $f_i(s)=0$ for $s\notin A_i$ and all $i$ (i.e., a partition of unity subordinated to $\{A_i\}$); if $x_i\in f(A_i)$ $(i=1,\dots,n)$ are chosen arbitrarily, it is readily verified that

$$\|\textstyle\sum_i f_i\otimes x_i-f\|<\delta.$$

2. A closely related example is the following. Let $G$ be any Banach space and denote by $c_0(G)$ the vector space of all sequences $\boldsymbol{x}=(x_n)$ satisfying

$x_n \in G$ $(n \in \mathbb{N})$ and $\lim_n \|x_n\| = 0$. (Cf. Chap. II, Exerc. 30.) Then (if, as usual, $c_0$ is the space of scalar null sequences) $c_0(G)$ is a Banach space isomorphic to $c_0 \tilde{\otimes}_\varepsilon G$.

3. Let $(X, \Sigma, \mu)$ be a measure space, $G$ a Banach space. The Banach space $L^1_G(\mu)$ of $G$-valued integrable functions $f$ (see [DS, Chap. III, § 3]), under the norm $f \mapsto \int \|f\| \, d\mu$, is isomorphic to $L^1(\mu) \tilde{\otimes}_\pi G$. (For details, cf. [S, III. 6.5] and § 7, Example 4.)

4. The Banach spaces $l^1(G)$ of summable and $l^1[G]$ of absolutely summable sequences in $G$ ($G$ any Banach space), which will be discussed in some detail below, are isomorphic to the tensor products $l^1 \tilde{\otimes}_\varepsilon G$ and $l^1 \tilde{\otimes}_\pi G$, respectively. (Exerc. 6.)

5. Let $G, H$ be Banach spaces. If, as usual, $G' \otimes H$ is identified with the vector space of continuous linear maps $G \to H$ of finite rank; that is, if to $u = \sum_{i=1}^n x'_i \otimes y_i$ we let correspond the linear map

$$x \mapsto u(x) := \sum_{i=1}^n \langle x, x'_i \rangle y_i \qquad (x \in G),$$

then from (3) it follows that the $\varepsilon$-norm on $G' \otimes H$ agrees with the operator norm. Thus $G' \tilde{\otimes}_\varepsilon H$ can be identified with the closure of $G' \otimes H$ in $\mathscr{L}(G, H)$, which is contained in the space $\mathscr{K}(G, H)$ of all compact operators $G \to H$. The reader is encouraged to prove that if $G'$ or $H$ has the approximation property (Def. 2.2 below) then $G' \tilde{\otimes}_\varepsilon H = \mathscr{K}(G, H)$.

On the other hand, the imbedding $G' \otimes_\pi H \to \mathscr{L}(G, H)$ is clearly continuous and hence, extends to a continuous map $\tau \colon G' \tilde{\otimes}_\pi H \to \mathscr{L}(G, H)$; the operators in the range $\mathscr{N}(G, H)$ of $\tau$ are called *nuclear*. Sometimes the notation $G' \tilde{\otimes}_\pi H$ is used for the space of nuclear operators $G \to H$; more precisely, however, under its standard norm (again called the *nuclear* or $\pi$-norm) $\mathscr{N}(G, H)$ is isomorphic to the normed quotient $G' \tilde{\otimes}_\pi H / \tau^{-1}(0)$. We remark without proof that $\tau$ is injective (for all Banach spaces $G$) iff $H$ has the approximation property (cf. Grothendieck [1955], [S, Chap. IV, Exerc. 30]).

Let now $G, H$ be Banach spaces with duals $G', H'$. If $U_\varepsilon$, $U_\pi$ denote the $\varepsilon$- and $\pi$-unit balls of $G \otimes H$ and $V_\varepsilon$, $V_\pi$ denote the $\varepsilon$- and $\pi$-unit balls of $G' \otimes H'$, respectively, then from (3) it follows that $U_\varepsilon = V_\pi^\circ$ under the duality $\langle G \otimes H, G' \otimes H' \rangle$; however, the converse relation, $V_\varepsilon^\circ = U_\pi$, does not always hold but is related to a special property of $G$ (or $H$) which we recall briefly.

**2.2 Definition.** *A Banach space $G$ is said to have the* (metric) approximation property *if in the topology of compact convergence, the identity map $1_G$ can be approximated by continuous linear maps $G \to G$ of finite rank (and norm $\leq 1$).*

We note that the metric approximation property of $G$ is equivalent to denseness of the $\varepsilon$-unit ball of $G' \otimes G$ in the unit ball of $\mathscr{L}(G)$ for the topology of simple convergence, or even of simple convergence on a total subset of $G$ (the two topologies agreeing on bounded subsets of $\mathscr{L}(G)$, [S, III. 4.5]); this will be used below.

**2.3 Proposition.** *For any Banach space $G$, these assertions are equivalent:*

(a) *$G$ possesses the metric approximation property.*

(b) *For every Banach space $H$, the ε-unit ball $V_\varepsilon$ of $G'\otimes H'$ is dense in the unit ball of $\mathscr{B}(G,H)$ with respect to $\sigma(\mathscr{B}(G,H), G\otimes H)$.*

(c) *For every Banach space $H$, each linear map $H\to G$ of norm $\leqq 1$ can be approximated, uniformly on compact subsets of $H$, by maps of norm $\leqq 1$ contained in $H'\otimes G$.*

*Proof.* (a) ⇒ (b): As before, we identify $\mathscr{B}(G,H)$ with $\mathscr{L}(G,H')$ and suppose that $T\in\mathscr{L}(G,H')$, $\|T\|\leqq 1$, a finite set $A\subset G$, and a real number $\delta>0$ are preassigned. By (a) there exists $e_0\in G'\otimes G$, $\|e_0\|_\varepsilon\leqq 1$, such that $\|e_0(x)-x\|\leqq\delta$ whenever $x\in A$. Then $T_0:=T\circ e_0\in G'\otimes H'$ satisfies $\|T_0\|\leqq 1$ and $\|T_0x-Tx\|\leqq\delta$ for all $x\in A$. Translating back to $\mathscr{B}(G,H)$, we see that (b) holds.

(b) ⇒ (a): We apply (b) to $H=G'$; then it follows from (b) that the ε-unit ball $U''_\varepsilon$ of $G'\otimes G''$ is dense in the unit ball of $\mathscr{L}(G,G'')$ for the weak topology $\sigma(\mathscr{L}(G,G''), G\otimes G')$. But, considering $G'\otimes_\varepsilon G$ as a normed vector subspace of $G'\otimes_\varepsilon G''$, (5.4) Cor. 3 shows the ε-unit ball $U_\varepsilon$ of $G'\otimes G$ to be dense in $U''_\varepsilon$ for the weak topology $\sigma(G'\otimes G'', G\otimes G')$. Hence, $U_\varepsilon$ is dense in the unit ball of $\mathscr{L}(G)$ for the weak operator topology. But this implies the same assertion for the strong operator topology, since $U_\varepsilon$ is convex and the two topologies have the same dual $G\otimes G'$ [S, IV. 4.3 Cor. 4].

The proof of (a) ⇔ (c) is easy and will be omitted. □

**Corollary.** *Let $G, H$ be Banach spaces and denote by $U_\pi$ the π-unit ball of $G\otimes H$, by $V_\varepsilon$ the ε-unit ball of $G'\otimes H'$. Then $U_\pi=V_\varepsilon^\circ$ (or equivalently, $U_\pi=U_\pi^{\circ\circ}$) for all $H$ if and only if $G$ has the metric approximation property.*

It had long been an open question if every Banach space has the approximation property (Approximation Problem, see [S, Chap. III, § 9]). By a famous example of Enflo [1972] there exist reflexive, separable Banach spaces not having the approximation property (hence, in particular, not having the metric approximation property or a Schauder basis). It is still unknown if every Banach lattice has the approximation property; below, we show that certain frequently occurring Banach lattices have the metric approximation property, a result due to Grothendieck [1955]. It should be noted that for the Banach lattices for which the metric approximation property is established in (2.4), it is actually shown that the identity map $1_E$ can be approximated, uniformly on compact sets, by mappings $\varphi=\sum_{i=1}^n x'_i\otimes x_i$ such that $\|\varphi\|_\varepsilon\leqq 1$ and $x_i\in E_+$, $x'_i\in E'_+$ for $i=1,\dots,n$.

**2.4 Theorem.** *Every AM- and every AL-space has the metric approximation property. Moreover, if $(X,\Sigma,\mu)$ is any measure space, then $L^p(\mu)$ $(1\leqq p\leqq+\infty)$ has the metric approximation property.*

*Proof.* First, we consider the case of an $AM$-space $E$. By (II. 9.1) the bidual $E''$ is an $AM$-space with unit (that is, a space $C(K)$ for compact $K$ (II. 7.4)); if $E''$ has the metric approximation property then by (2.3)(c) the evaluation map $E\to E''$ can be approximated for $\sigma(\mathscr{L}(E,E''), E\otimes E')$ by elements of the ε-unit ball $U''_\varepsilon$ of $E'\otimes E''$ and hence, since the ε-unit ball $U_\varepsilon$ of $E'\otimes E$ is dense in $U''_\varepsilon$

for this topology by (5.4) Cor. 3, $1_E$ can be approximated by elements of $U_\varepsilon$. Thus it suffices to consider the case $E=C(K)$ ($K$ compact).

Let the finite set $H\subset E$ and the real number $\delta>0$ be given. Since continuous real functions on $K$ are uniformly continuous, there exists a finite open cover $\{A_i: i=1,\dots,n\}$ of $K$ such that $|f(s)-f(t)|<\delta$ whenever $f\in H$ and $(s,t)\in A_i\times A_i$ $(i=1,\dots,n)$. Let $\{e_i: i=1,\dots,n\}$ be a continuous partition of unity on $K$ subordinated to the cover $\{A_i\}$ (cf. Example 1), and choose $t_i\in A_i$ $(i=1,\dots,n$; we suppose, of course, $A_i\neq\emptyset$). Let $\varphi$ denote the mapping of finite rank $\sum_{i=1}^n \delta_{t_i}\otimes e_i\in C(K)'\otimes C(K)$; evidently, $\|\varphi\|_\varepsilon=1$ and

$$|f(s)-\varphi(f)(s)|\leqq \textstyle\sum_{i=1}^n e_i(s)\,|f(s)-f(t_i)|<\delta$$

for all $s\in K$ and $f\in H$; therefore, $\|f-\varphi(f)\|<\delta$ for all $f\in H$.

The proof for $E=L^p(\mu)$ $(1\leqq p<+\infty)$ uses the same method; it is enough to consider this case, since every $AL$-space is isomorphic to some $L^1(\mu)$ (II. 8.5) and since $L^\infty(\mu)$ is an $AM$-space. Given $\delta>0$ and a finite set $H\subset E$, there exists a set $A\in\Sigma$ of finite measure on which each $f\in H$ is bounded and such that $\|f\chi_0\|<\frac{1}{2}\delta$, where $\chi_0$ denotes the characteristic function of $X\setminus A$. If $\mu(A)=0$, we choose $\varphi=0$; so assume that $\mu(A)>0$. There exists a measurable, finite partition $\{A_i: i=1,\dots,n\}$ of $A$ such that $\mu(A_i)>0$ for all $i$, and such that $|f(s)-f(t)|\leqq\delta[2\mu(A)^{1/p}]^{-1}$ whenever $f\in H$ and $(s,t)\in A_i\times A_i$ $(i=1,\dots,n)$. We define the elements $\mu_i\in E'_+$ by

$$\mu_i(f):=\frac{1}{\mu(A_i)}\int_{A_i} f\,d\mu \qquad (f\in E)$$

and let $\chi_i$ denote the characteristic function of $A_i$. The mapping of finite rank $\varphi:=\sum_{i=1}^n \mu_i\otimes\chi_i$ has operator norm 1. From the identity $(f\in E, s\in A)$

$$f(s)-\varphi(f)(s)=\textstyle\sum_{i=1}^n \chi_i(s)\left[\dfrac{1}{\mu(A_i)}\displaystyle\int_{A_i}(f(s)-f(t))\,d\mu(t)\right]$$

it follows that for all $f\in H$, $|f-\varphi(f)|\chi_A\leqq\delta[2\mu(A)^{1/p}]^{-1}\chi_A$; hence, since $f-\varphi(f)=(f-\varphi(f))\chi_A+f\chi_0$, we obtain $\|f-\varphi(f)\|<\delta$ for all $f\in H$. □

For the anounced characterizations of $AM$- and $AL$-spaces, we need the concepts of a summable and absolutely summable sequence in a Banach space. We recall the definitions; for a brief discussion of the elementary properties, see [S, Chap. III, Exerc. 23]. A sequence $(x_n)_{n\in\mathbb{N}}$ in a Banach space $G$ is called *summable* if $\lim_{\mathsf{H}}\sum_{n\in\mathsf{H}} x_n$ exists in $G$, where $\mathsf{H}$ runs through the family of all finite subsets of $\mathbb{N}$ directed under inclusion; $(x_n)_{n\in\mathbb{N}}$ is called *absolutely summable* if $\sum_n\|x_n\|<+\infty$. For a summable sequence $(x_n)$ in $G$, the above limit $x\in G$ is called the *sum* of $(x_n)$ and written $x=\sum_n x_n$; also, the formal series $\sum_n x_n$ is called *unconditionally* (or *unordered*) *convergent* (cf. Day [1973]). By a famous theorem of Orlicz and Pettis, a sequence $(x_n)$ in a Banach space is summable if for each subsequence $(x_{n(k)})_{k\in\mathbb{N}}$, the series $\sum_{k=1}^\infty x_{n(k)}$ converges in $G$ for $\sigma(G,G')$ (Exerc. 6).

Now suppose $G$ to be a fixed Banach space. The set $S$ of all summable sequences $\boldsymbol{x}=(x_n)$ in $G$ (considered as a subset of the product $\prod_{n\in\mathbb{N}} G$) is a vector space on which

(4) $$\|\boldsymbol{x}\|_\varepsilon := \sup_{x'\in U^\circ} \sum_n |\langle x_n, x'\rangle|$$

($U^\circ$ the unit ball of $G'$) defines a norm called the $\varepsilon$-norm; under this norm, $S$ is a Banach space denoted by $l^1(G)$ (Exerc. 6). Similarly, the vector space $S_a \subset S$ of all absolutely summable sequences $\boldsymbol{x}=(x_n)$ in $G$ is a Banach space $l^1[G]$ under the norm

(5) $$\|\boldsymbol{x}\|_\pi := \sum_n \|x_n\|,$$

which is called the $\pi$-norm. The reason for this terminology is to be found in the relation of the norms (4) and (5) with the $\varepsilon$- and $\pi$-tensor norms. In fact, the bilinear map $l^1 \times G \to l^1[G] \subset l^1(G)$ given by $((\xi_n), x) \mapsto (\xi_n x)$ defines a linear injection $i: l^1 \otimes G \to l^1[G]$ which is called canonical; using Formulae (1), (3), (4), (5) it is not difficult to see that $i$ extends to an isomorphism of $l^1 \tilde{\otimes}_\pi G$ onto $l^1[G]$ and to an isomorphism of $l^1 \tilde{\otimes}_\varepsilon G$ onto $l^1(G)$, respectively [S, IV. 10.6 Cor. 1]. Let us note that if $G$ is one dimensional then $l^1(G)=l^1[G]$ is isomorphic to $l^1$; if $G$ is finite dimensional then $l^1[G] \to l^1(G)$ is a surjection but not, in general, an isometry; finally, if $G$ is infinite dimensional the canonical imbedding $l^1[G] \to l^1(G)$ is never surjective (and hence, the $\varepsilon$- and $\pi$-norm on $l^1 \otimes G$ never equivalent) (Theorem of Dvoretzky-Rogers; see (5.10) Cor. 4).

The following lemmas will be useful in the sequel.

**2.5 Lemma.** *Let $G$ denote a Banach space. If $\boldsymbol{x}=(x_n)$ is an absolutely summable sequence in $G$, then*

$$\|\boldsymbol{x}\|_\pi = \sup \sum_n \langle x_n, x'_n\rangle$$

*where the supremum is taken over all null sequences $\boldsymbol{x}'=(x'_n)$ in $G'$ satisfying $\|\boldsymbol{x}'\|_\infty := \sup_n \|x'_n\| \leqq 1$. If $G$ is a Banach lattice and $\boldsymbol{x}=(x_n)$ is a summable sequence in $G_+$, then*

$$\|\boldsymbol{x}\|_\varepsilon = \left\|\sum_n x_n\right\|.$$

*Proof.* For $(x_n)$ an absolutely summable sequence in $G$, it is clear that $\|\boldsymbol{x}\|_\pi \geqq \sum_n \langle x_n, x'_n\rangle$ for each $\boldsymbol{x}'=(x'_n)$ satisfying $\|\boldsymbol{x}'\|_\infty \leqq 1$. On the other hand, for each $n$ there exists $x'_n \in G'$, $\|x'_n\| \leqq 1$, such that $\langle x_n, x'_n\rangle = \|x_n\|$ and hence, the supremum on the right is $\geqq \sum_{n=1}^k \|x_n\|$ for each $k \in \mathbb{N}$ and thus, $\geqq \|\boldsymbol{x}\|_\pi$.

Second, if $G$ is a Banach lattice and $\boldsymbol{x}=(x_n)$ is summable with range in $G_+$, then $|\langle x_n, x'\rangle| \leqq \langle x_n, |x'|\rangle$ for all $n \in \mathbb{N}$ and $x' \in G'$. Since $\||x'|\| = \|x'\|$, it follows that

$$\|\boldsymbol{x}\|_\varepsilon = \sup_{x'\in U_+^\circ} \sum_n \langle x_n, x'\rangle = \sup_{x'\in U_+^\circ} \left\langle \sum_n x_n, x'\right\rangle = \left\|\sum_n x_n\right\|,$$

where $U_+^\circ$ denotes the positive part of the unit ball $U^\circ$ of $G'$. (Chap. II, § 5, Formula (2).) □

**2.6 Lemma.** *Let $G, H$ be Banach spaces and let $C$ be a cone*[2] *in $G$. If $T: G \to H$ is a linear map such that for each sequence $\boldsymbol{x}=(x_n)\in l^1(G)$ with range in $C$, the sequence $\boldsymbol{Tx}:=(Tx_n)$ is in $l^1[H]$, then there exists a constant $k\in\mathbb{R}$ such that*

$$\|\boldsymbol{Tx}\|_\pi \leqq k\|\boldsymbol{x}\|_\varepsilon$$

*for all sequences $\boldsymbol{x}\in l^1(G)$ having their range in $C$.*

*Proof.* If the assertion were false, there would exist finite sequences $\boldsymbol{x}_i=(x_n^{(i)})_{1\leqq n\leqq k_i}$ in $C$ for which $\|\boldsymbol{Tx}_i\|_\pi>i$ while $\|\boldsymbol{x}_i\|_\varepsilon<2^{-i}$. Now the ranges of these finite sequences $\boldsymbol{x}_i$ can be composed to form the range of a single (infinite) sequence $\boldsymbol{x}=(\boldsymbol{x}_1,\boldsymbol{x}_2,\ldots)$ in $C$ which is clearly summable in $G$; however, $\boldsymbol{Tx}$ is not absolutely summable in $H$, which is absurd. □

The method used in the proof of (2.6) is frequently useful and will be referred to as the *composition principle*. It should be noted that the conclusion of (2.6) does not depend on the completeness of $G$ or $H$ or related properties (such as category); rather, it is forced by the assumption that *every* summable sequence with range in $C$ be transformed by $T$ into an absolutely summable sequence in $H$.

The following characterization of $AM$- and $AL$-spaces (due to U. Schlotterbeck [1969]) is quite different from the well known characterizations of Krein-Kakutani (II. 7.4) and Kakutani (II. 8.5).

**2.7 Theorem.** *For any Banach lattice $E$, the following properties are equivalent:*

(a) *$E$ is isomorphic (as a topological vector lattice) to an $AL$-space.*

(b) *Every positive summable sequence in $E$ is absolutely summable.*

*Proof.* (a) ⇒ (b) is immediately clear, because the norm of an $AL$-space $E$ agrees on $E_+$ with a continuous linear form.

(b) ⇒ (a): For each $x\in E$, let

$$p(x):=\sup\{\textstyle\sum_n\|x_n\|:(x_n)\in l^1(E_+),\ \sum_n x_n=|x|\};$$

here $l^1(E_+)$ denotes the set of positive summable sequences in $E$. Since for each $\boldsymbol{x}=(x_n)\in l^1(E_+)$ we have $\|\boldsymbol{x}\|_\varepsilon=\|\sum_n x_n\|$ by (2.5), the hypothesis (b) and (2.6) imply the existence of a constant $l\in\mathbb{R}$ such that $p(x)\leqq l\|x\|$ for all $x\in E$. On the other hand, it is obvious that $\|x\|\leqq p(x)$; thus if we can show that $p$ is an $L$-norm on $E$ (Def. II. 8.1), it is an $L$-norm equivalent to the given norm of $E$, and the proof will be complete.

It is clear that $p(\lambda x)=|\lambda|p(|x|)$ for $\lambda\in\mathbb{R}$ and $x\in E$, and it is easy to see that $p(x+y)\geqq p(x)+p(y)$ for all $x,y\in E_+$. To show that $p(x+y)\leqq p(x)+p(y)$ for $x,y\in E_+$, let $z:=x+y$ and denote by $(z_n)$ any sequence in $l^1(E_+)$ satisfying $\sum_n z_n=z$. We define $w_n:=\sum_{\nu=1}^{n-1}z_\nu$, $u_n:=w_n\wedge x$, $v_n:=(w_n-x)^+$, and $x_n:=u_{n+1}-u_n$, $y_n:=v_{n+1}-v_n$. Then we have $x_n,y_n\in E_+$ for all $n$, $z_n=x_n+y_n$ and hence, $(x_n)$ and $(y_n)$ are sequences in $l^1(E_+)$ for which $\sum_n x_n=x$, $\sum_n y_n=y$. Therefore $\sum_n\|z_n\|\leqq\sum_n\|x_n\|+\sum_n\|y_n\|\leqq p(x)+p(y)$; since $(z_n)$ was an arbitrary, positive summable sequence with sum $x+y$, it follows that $p(x+y)\leqq p(x)+p(y)$. □

---

[2] A *cone* is a subset $C\subset G$ such that $\lambda C\subset C$ for all real $\lambda>0$.

By duality, this yields a new characterization of $AM$-spaces, as follows.

**2.8 Theorem.** *For any Banach lattice E, the following properties are equivalent:*
(a) *$E$ is isomorphic (as a topological vector lattice) to an $AM$-space.*
(b) *Every null sequence in $E$ is order bounded.*

*Proof.* (a) ⇒ (b): We can assume that the norm of $E$ is an $M$-norm (Def. II. 7.1). If $(x_n)$ is a given null sequence in $E$, let $y_0 := \sup\{|x_i| : \|x_i\| \geqq 1\}$ and let $y_m := \sup\{|x_i| : 2^{-m} \leqq \|x_i\| < 2^{-m+1}\}$ for $m \in \mathbb{N}$ $(m \geqq 1)$. Then, by the property of an $M$-norm, we obtain $\|y_m\| < 2^{-m+1}$ whenever $m \geqq 1$, and $y := \sum_0^\infty y_m$ is an element of $E$ majorizing the range of the sequence $(|x_n|)$.

(b) ⇒ (a): First, we observe that (b) implies the following: There exists a constant $k \in \mathbb{R}$ such that each null sequence contained in the unit ball $U$ of $E$, is majorized by some element in $E_+$ of norm $\leqq k$. (Otherwise, for each $n \in \mathbb{N}$ there exist null sequences in $U$ not majorized by any element of norm $\leqq n$, and a slight modification of the composition principle (see remark after (2.6)) permits the construction of a null sequence in $E$ which is not majorized; this is absurd.)

Now let $(x'_n)$ be a fixed, positive summable sequence in the Banach lattice $E'$; it is clear that

$$\textstyle\sum_{n=1}^m \|x'_n\| = \sup \sum_{n=1}^m \langle x_n, x'_n \rangle$$

for all $m \in \mathbb{N}$, where the supremum is taken over all positive null sequences $(x_n)$ contained in the unit ball of $E$. By the preceding, each such sequence is majorized by some element $x \in E_+$ satisfying $\|x\| \leqq k$; therefore, for all $m \in \mathbb{N}$

$$\begin{aligned} \textstyle\sum_{n=1}^m \|x'_n\| &\leqq \sup\{\textstyle\sum_{n=1}^\infty \langle x, x'_n \rangle : x \in E_+, \|x\| \leqq k\} \\ &= \sup\{\langle x, \textstyle\sum_{n=1}^\infty x'_n \rangle : x \in E_+, \|x\| \leqq k\} = k \left\| \textstyle\sum_{n=1}^\infty x'_n \right\|. \end{aligned}$$

This shows $(x'_n)$ to be absolutely summable. Hence (2.7) implies that (up to norm equivalence) $E'$ is an $AL$-space and $E''$ an $AM$-space (II. 9.1). Thus $E$ is isomorphic to an $AM$-space, since $E$ can be identified with a closed vector sublattice of $E''$ by (II. 5.5) Cor. 2. □

## § 3. Cone Absolutely Summing and Majorizing Maps

Throughout Chapters II and III, frequent use has been made of principal ideals $E_u$ generated by an element $u \in E_+$ of a vector lattice $E$; in particular, if $E$ is a Banach lattice then (for the norm which is the gauge of $[-u, u]$) $E_u$ is an $AM$-space with unit $u$ and the injection $E_u \to E$ is continuous (II. 7.2 Cor.). We will have need of a dual construction: If $E$ is a vector lattice and $x'$ is a positive linear form on $E$ with null ideal $N$, the completion of $E/N$ (with respect to the $L$-norm induced by $x \mapsto \langle |x|, x' \rangle$) is an $AL$-space $(E, x')$ (Chap. II, § 8, Example 1). The lattice homomorphism $E \to (E, x')$ is onto an ideal whenever $E$ is order complete and $x'$ is order continuous (cf. 9.3), and continuous if $E$ is a Banach lattice. It should be noted that, for $E$ a Banach lattice and $u \in E_+$, the mapping $E' \to (E', u)$ is the adjoint of $E_u \to E$ (Exerc. 9).

The characterization of $AL$- and $AM$-spaces given by Theorems (2.7) and (2.8), respectively, suggests the following definition; throughout this section, we denote by $E, F$ normed vector lattices and by $G, H$ normed vector spaces (over $\mathbb{R}$).

**3.1 Definition.** *A linear map $T: E \to H$ is called* cone absolutely summing *(abbreviated c.a.s.) if for every positive summable sequence $(x_n)$ in $E$, the sequence $(Tx_n)$ is absolutely summable in $H$.*

*A linear map $T: G \to F$ is called* majorizing *if for every null sequence $(x_n)$ in $G$, $(|Tx_n|)$ is a majorized sequence in $F$.*

It is easy to see that c.a.s. and majorizing linear maps are necessarily continuous; moreover, these elementary characterizations are valid.

**3.2 Lemma.** (i) *$T \in \mathscr{L}(E, H)$ is c.a.s. if and only if there exists a real constant $l \geqq 0$ such that*

$$\sum_n \|Tx_n\| \leqq l \sup_{x' \in U^\circ} \sum_n |\langle x_n, x' \rangle| \tag{1}$$

*for every positive summable sequence $(x_n)$ in $E$ ($U^\circ$ denoting the unit ball of $E'$).*

(ii) *$T \in \mathscr{L}(G, F)$ is majorizing if and only if there exists a real constant $m \geqq 0$ such that for each null sequence $(x_n)$ contained in the open unit ball of $G$, one has*

$$|Tx_n| \leqq y \tag{2}$$

*for all $n \in \mathbb{N}$ and some element $y \in F_+$, $\|y\| \leqq m$.*

While the proof of (i) is immediately contained in (2.6), the proof of (ii) is an easy application of the composition principle (remark after (2.6)). We proceed to characterize c.a.s. and majorizing maps by various equivalences, in particular, by factoring properties.

**3.3 Proposition.** *Let $E$ be a Banach lattice, $H$ a Banach space. For any $T \in \mathscr{L}(E, H)$, the following assertions are equivalent:*

(a) *$T$ is c.a.s. satisfying (1) with constant $l$.*

(b) *There exists $x' \in E'_+$, $\|x'\| \leqq l$, such that $\|Tx\| \leqq \langle |x|, x' \rangle$ for all $x \in E$.*

(c) *$T$ possesses a factoring $E \xrightarrow{T_1} L \xrightarrow{T_2} H$, where $L$ is an $AL$-space, $0 \leqq T_1 \in \mathscr{L}(E, L)$, $T_2 \in \mathscr{L}(L, H)$ and $\|T_1\| \leqq l$, $\|T_2\| \leqq 1$.*

(c′) *Same assertion as (c) but supposing, in addition, $T_1$ to be a lattice homomorphism with dense range.*

(d) *There exists a real constant $l \geqq 0$ such that*

$$\sum_i \|Tx_i\| \leqq l \left\| \sum_i x_i \right\| \tag{3}$$

*for all finite families $(x_i)$ in $E_+$.*

*Proof.* (a) $\Rightarrow$ (b): For each $x \in E$, define $p_T(x) := \sup\{\sum_n \|Tx_n\| : (x_n)$ summable in $E_+, \sum_n x_n = |x|\}$; by (2.5) and (3.2) we have $p_T(x) \leqq l \|x\|$, and it is seen precisely as in the proof of (2.7) that $p_T$ is a lattice semi-norm additive on $E_+$. Thus on

$E_+$ $p_T$ agrees with a positive linear form $x' \in E'$ of norm $\|x'\| \leqq l$; clearly, then, $\|Tx\| \leqq \langle |x|, x' \rangle$ for $x \in E$.

(b) $\Rightarrow$ (c′): The canonical map $E \to (E, x')$ (see beginning of this section) is a lattice homomorphism $T_1$ of $E$ into the $AL$-space $L := (E, x')$ with dense range and of norm $\|T_1\| \leqq l$; the linear map $T_2$ of $T(E) \subset L$ into $H$ defined by $T_2(T_1 x) := Tx$ $(x \in E)$ is continuous of norm $\leqq 1$, and its continuous extension $L \to H$ (again denoted by $T_2$) satisfies the requirement.

(c′) $\Rightarrow$ (c): Obvious.

(c) $\Rightarrow$ (d): Let $T = T_2 \circ T_1$ be factored as asserted in (c). For all summable sequences $(x_n)$ with range in $E_+$ we have $\sum_n \|T_1 x_n\| = \|\sum_n T_1 x_n\|$, since $T_1 \geqq 0$ and $L$ is an $AL$-space; and so, if $\mu$ denotes the norm functional of $L$, $\sum_n \|T_1 x_n\| = \sum_n \langle T_1 x_n, \mu \rangle = \langle \sum_n x_n, T_1' \mu \rangle \leqq l_1 \|\sum_n x_n\|$ where $l_1 := \|T_1' \mu\| \leqq l$. Therefore, since $\|T_2\| \leqq 1$, we obtain $\sum_n \|Tx_n\| \leqq l \|\sum_n x_n\|$ for all positive summable sequences in $E$.

(d) $\Rightarrow$ (a): For positive summable sequences in $E$ we have $\|\sum_n x_n\| = \sup\{\sum_n |\langle x_n, x' \rangle| : x' \in U^\circ\}$ by Lemma 2.5; thus (d) yields (a) by a simple continuity argument. □

Since every Banach lattice $E$ can be viewed as an ordered, normed vector subspace of the space of continuous functions $C(U_+^\circ)$ ($U_+^\circ$ the positive part of the dual unit ball, cf. (II. 5.7)), the equivalence (a) $\Leftrightarrow$ (b) of (3.3) and an application of the Hahn-Banach theorem yield the following corollary.

**Corollary.** *A linear map $T: E \to H$ is c.a.s. with constant $l$ if and only if there exists a positive Radon measure $\nu$ on the $\sigma(E', E)$-compact space $U_+^\circ$, of norm $\|\nu\| \leqq l$, such that*

$$\|Tx\| \leqq \int_{U_+^\circ} \langle |x|, x' \rangle \, d\nu(x') \qquad (x \in E).$$

This characterization shows the relationship of c.a.s. maps with absolutely summing linear maps $E \to H$ as characterized by Pietsch [1963b] (cf. (5.10) below).

**3.4 Proposition.** *Let $G$ be a Banach space, $F$ a Banach lattice. For any $T \in \mathscr{L}(G, F)$, the following assertions are equivalent:*

(a) *$T$ is majorizing, satisfying* (2) *with constant $m \geqq 0$.*

(b) *There exists $y'' \in F_+''$, $\|y''\| \leqq m$, such that $T$ maps the unit ball of $G$ into the order interval $[-y'', y'']$ of $F''$.*

(c) *$T$ possesses a factoring $G \xrightarrow{T_1} M \xrightarrow{T_2} F$ where $M$ is an $AM$-space, $T_1 \in \mathscr{L}(G, M)$, $0 \leqq T_2 \in \mathscr{L}(M, F)$ and $\|T_1\| \leqq 1$, $\|T_2\| \leqq m$.*

(c′) *Same assertion as* (c) *but supposing, in addition, $T_2$ to be a lattice isomorphism onto an ideal of $F$.*

(d) *There exists a real constant $m \geqq 0$ such that*

$$\|\sup_i |Tx_i|\| \leqq m \sup_i \|x_i\| \tag{4}$$

*for all finite families $(x_i)$ in $G$.*

*Proof.* (a) $\Rightarrow$ (d) is an immediate consequence of (3.2) (ii).

(d) $\Rightarrow$ (b): Let $y_{\mathsf{H}} := \sup_{x \in \mathsf{H}} |Tx|$ for every non-empty finite family $\mathsf{H}$ contained in the unit ball $U$ of $G$. Considering $F$ as a vector sublattice of the bidual $F''$ (II. 5.5 Cor. 2) and denoting by $V''$ the unit ball of $F''$, we see that $(y_{\mathsf{H}})$ is a directed $(\leqq)$ family contained in $mV''$. Since $mV''$ is $\sigma(F'',F')$-compact, $(y_{\mathsf{H}})$ converges for $\sigma(F'',F')$ to its supremum $y'' \in F''$ (cf. II. 5.8), and it is clear that $T(U) \subset [-y'', y'']$.

(b) $\Rightarrow$ (c'): The ideal $F''_{y''}$ is an $AM$-space with unit $y''$ and, since $F$ is norm closed in $F''$, the ideal $M := F \cap F''_{y''}$ of $F$ is an $AM$-space under the norm induced by the gauge of $[-y'', y'']$. It is clear that the mapping $T_1: G \to M$ defined by $T$, is of norm $\leqq 1$. Taking $T_2$ to be the imbedding map $M \to F$, we see that $T_1$, $T_2$ satisfy the requirements.

(c') $\Rightarrow$ (c): Obvious.

(c) $\Rightarrow$ (a): Let $(x_n)$ be a null sequence contained in the open unit ball of $G$, and let $\varepsilon := 1 - \sup_n \|T_1 x_n\|$. We define $y_k \in M$ $(k = 0, 1, 2, \ldots)$ by $y_0 := \sup\{|T_1 x_n| : \|T_1 x_n\| \geqq \varepsilon/2\}$, $y_k := \sup\{|T_1 x_n| : \varepsilon 2^{-k-1} \leqq \|T_1 x_n\| < \varepsilon 2^{-k}\}$ for all $k \geqq 1$. Since $M$ is an $AM$-space, we have $\|y_0\| \leqq 1 - \varepsilon$ and $\|y_k\| \leqq \varepsilon 2^{-k}$ for $k \geqq 1$; therefore, the sum $y := \sum_0^\infty y_k$ has norm $\|y\| \leqq 1$ and majorizes the sequence $(|T_1 x_n|)$. Now if $z := T_2 y$, then $\|z\| \leqq m$ and $|Tx_n| \leqq z$ for all $n$. ☐

Since every compact subset of a normed space is contained in the convex closure of a suitable null sequence (cf. [S, Chap. III, § 9, Lemma 1]), we obtain this corollary.

**Corollary.** *$T \in \mathscr{L}(G,F)$ is majorizing with constant $m$ if and only if each compact subset of the open unit ball of $G$ is mapped by $T$ into an order interval of $F$ whose elements have norm $\leqq m$.*

*Note.* Cone absolutely summing and majorizing maps can alternatively be characterized, as follows. Given vector spaces $G, H$ and a linear map $T: G \to H$, $T$ defines a linear map $\boldsymbol{T}$ of the space $\prod_{n \in \mathbb{N}} G$ of all sequences $\boldsymbol{x} = (x_n)$ in $G$, into $\prod_{n \in \mathbb{N}} H$ by virtue of $\boldsymbol{T}\boldsymbol{x} := (Tx_n)$. Now let $E, F$ denote Banach lattices and let $G, H$ denote Banach spaces.

If $l^1_+(E)$ denotes the vector space of all sequences $\boldsymbol{x} = (x_n)$ in $E$ for which the sequence $(|x_n|)$ is summable, then $l^1_+(E)$ is a Banach lattice under the norm

$$\|\boldsymbol{x}\| := \sup_{x' \in U_+^\circ} \sum_n \langle |x_n|, x' \rangle,$$

and $T \in \mathscr{L}(E,H)$ is c.a.s. with constant $l$ iff $\boldsymbol{T}$ maps $l^1_+(E)$ into $l^1[H]$ with norm $\leqq l$.

Similarly, let $c_0^+(F)$ denote the vector space of all null sequences $\boldsymbol{y} = (y_n)$ such that the sequence $(|y_n|)$ is majorized in $F$; to simplify, suppose there exists a positive contractive projection $F'' \to F$ (see Def. 4.2 below). Then $c_0^+(F)$ is a Banach lattice under the norm

$$\|\boldsymbol{y}\| := \|\sup_n |y_n|\|,$$

and $T \in \mathscr{L}(G,F)$ is majorizing with constant $m$ iff $T$ maps $c_0(G)$ (§ 2, Example 2) into $c_0^+(F)$ with norm $\leqq m$.

The detailed verification is left to the interested reader (Exerc. 6).

It is easy to see (for instance, from conditions (d) of (3.3) and (3.4), respectively) that the set $\mathscr{L}^l(E,H)$ of all c.a.s. maps $E \to H$ is a vector subspace of $\mathscr{L}(E,H)$, and that the set $\mathscr{L}^m(G,F)$ of all majorizing maps $G \to F$ is a vector subspace of $\mathscr{L}(G,F)$; obviously, the norm completeness of $E,F,G,H$ is inessential for this assertion. Moreover, the smallest constant $l$ (respectively, $m$) for which (3.2)(i) is valid for $T \in \mathscr{L}^l(E,H)$ (respectively, for which (3.2)(ii) is valid for $T \in \mathscr{L}^m(G,F)$) defines a norm on $\mathscr{L}^l(E,H)$ (respectively, $\mathscr{L}^m(G,F)$); these constants are given by the formulae

$$(3') \qquad \|T\|_l = \sup\{\textstyle\sum_i \|Tx_i\| : (x_i) \text{ finite } \subset E_+, \|\sum_i x_i\| = 1\}$$

and

$$(4') \qquad \|T\|_m = \sup\{\|\sup_i |Tx_i|\,\| : (x_i) \text{ finite}, \|x_i\| \leqq 1\}.$$

Therefore, we agree on this definition.

**3.5 Definition.** *Let $E, F$ denote normed vector lattices and let $G, H$ denote normed vector spaces. For each c.a.s. map $T: E \to H$, the number $\|T\|_l$ is called the* $l$-norm *of $T$; for each majorizing map $T: G \to F$, the number $\|T\|_m$ is called the* $m$-norm *of $T$.*

It is quickly seen from (3′) and (4′) that $\|T\| \leqq \|T\|_l$ whenever $T \in \mathscr{L}^l(E,H)$, and that $\|T\| \leqq \|T\|_m$ whenever $T \in \mathscr{L}^m(G,F)$. Moreover, if $E, F$ and $G, H$ are norm complete, then the $l$-norm is the smallest constant $l$ for which any one of the assertions of (3.3) is satisfied, and the $m$-norm is the smallest constant $m$ for which any one of the assertions of (3.4) is satisfied. We summarize and supplement this, as follows.

**3.6 Proposition.** *If $H$ is a Banach space, then $\mathscr{L}^l(E,H)$ is a Banach space under the $l$-norm; if $F$ is a Banach lattice, then $\mathscr{L}^m(G,F)$ is a Banach space under the $m$-norm. Moreover, the canonical injections $\mathscr{L}^l(E,H) \to \mathscr{L}(E,H)$ and $\mathscr{L}^m(G,F) \to \mathscr{L}(G,F)$ are continuous with norm $\leqq 1$.*

*Proof.* Suppose $H$ to be norm complete. To prove the norm-completeness of $\mathscr{L}^l(E,H)$ (for the $l$-norm) it suffices, since $\mathscr{L}(E,H)$ is norm complete (operator norm), to show that the $l$-unit ball $U_l$ of $\mathscr{L}^l(E,H)$ is closed in $\mathscr{L}(E,H)$ [S, I. 1.6]. Let $T$ be in the closure of $U_l$ in $\mathscr{L}(E,H)$ and let $\varepsilon > 0$ be preassigned. Given a finite family $(x_i)$ $(i=1,\dots,n)$ in $E_+$ satisfying $\|\sum_i x_i\| = 1$, there exists $T_0 \in U_l$ such that $\|T - T_0\| < \varepsilon/n$. Since $\sum_i \|x_i\| \leqq n$, we obtain

$$\textstyle\sum_i \|Tx_i\| \leqq \sum_i \|T_0 x_i\| + \|T - T_0\| \sum_i \|x_i\| \leqq 1 + \varepsilon.$$

Since $\varepsilon$ and the family $(x_i)$ were arbitrary, we conclude that $T \in \mathscr{L}^l(E,H)$ and $\|T\|_l \leqq 1$.

Similarly, if $T \in \mathscr{L}(G,F)$ is in the closure of the $m$-unit ball $U_m$ in $\mathscr{L}(G,F)$, then $T \in \mathscr{L}^m(G,F)$ and $\|T\|_m \leqq 1$. ∎

**Examples.** 1. If $E$ is an $AL$-space and $H$ is any Banach space, then $\mathscr{L}^l(E,H)=\mathscr{L}(E,H)$ and the $l$-norm agrees with the operator norm. Similarly, is $F$ is an $AM$-space and $G$ is any Banach space, then $\mathscr{L}^m(G,F)=\mathscr{L}(G,F)$ and the $m$-norm agrees with the operator norm. (Cf. Prop. 3.7 below.)

2. Generally, the continuous linear maps of finite rank $E\to H$ are c.a.s., and the continuous linear maps of finite rank $G\to F$ are majorizing (proof!); that is, $E'\otimes H\subset\mathscr{L}^l(E,H)$ and $G'\otimes F\subset\mathscr{L}^m(G,F)$. Moreover, the $l$-norm on $E'\otimes H$ and the $m$-norm on $G'\otimes H$ are both finer than the $\varepsilon$-norm and both coarser than the $\pi$-norm (§ 2), as can be seen from (3′) and (4′).

It can happen that the spaces $\mathscr{L}^l(E,H)$ and $\mathscr{L}^m(G,F)$ are not much larger than these tensor products; for instance, if $E=G=L^2(\mu)$ and $F=H=L^2(\nu)$ are Hilbert spaces of square integrable functions, endowed with their natural ordering, then $\mathscr{L}^l(E,F)=\mathscr{L}^m(E,F)$ is identical with the space of Hilbert-Schmidt operators under the Hilbert-Schmidt norm (see (6.10) below). More generally, if $E,F$ and $G,H$ are reflexive then it follows from the factoring properties (3.3) (c) and (3.4) (c) that every map in $\mathscr{L}^l(E,H)$ and in $\mathscr{L}^m(G,F)$ is compact, since $AL$- and $AM$-spaces have the Dunford-Pettis property (II. 9.8).

We can now extend the characterizations of $AL$-spaces and of $AM$-spaces given in the preceding section, as follows.

**3.7 Proposition.** *Let $E,F$ be Banach lattices and consider these propositions:*
(a) *$E$ is an $AL$-space.*
(b) *For every Banach space $H$, the canonical injection $\mathscr{L}^l(E,H)\to\mathscr{L}(E,H)$ is a surjective isometry.*
(c) *$F$ is an $AM$-space.*
(d) *For every Banach space $G$, the canonical injection $\mathscr{L}^m(G,F)\to\mathscr{L}(G,F)$ is a surjective isometry.*

*Then* (a)$\Leftrightarrow$(b) *and* (c)$\Leftrightarrow$(d).

*Proof.* (a)$\Rightarrow$(b): If $(x_n)$ is any positive summable sequence in $E$ and if $T\in\mathscr{L}(E,H)$, then $\sum_n\|Tx_n\|\leqq\|T\|\,\|\sum_n x_n\|$ because of $\|\sum_n x_n\|=\sum_n\|x_n\|$; hence, $T\in\mathscr{L}^l(E,H)$ and, moreover, $\|T\|_l\leqq\|T\|$; this shows $\|T\|_l=\|T\|$ by (3.6).

(b)$\Rightarrow$(a): Choosing $H=E$, we obtain $1_E\in\mathscr{L}^l(E,E)$ and hence, noting that $\|1_E\|_l=1$ by hypothesis, we observe that $\|x+y\|\geqq\|x\|+\|y\|$ for all $x,y\in E_+$; therefore, $E$ is an $AL$-space.

The proof of (c)$\Leftrightarrow$(d) is entirely analogous and will be omitted. □

We point out that the class of c.a.s. operators is a left operator ideal in the sense that if $T\in\mathscr{L}^l(E,H)$ and $R\colon H\to H_1$ is a continuous linear operator into a Banach space $H_1$, then $R\circ T\in\mathscr{L}^l(E,H_1)$; similarly, the class of majorizing operators is a right ideal. In particular, if $H=E$ and $F=G$ then $\mathscr{L}^l(E)$ and $\mathscr{L}^m(F)$ are left and right ideals in the Banach algebras $\mathscr{L}(E)$ and $\mathscr{L}(F)$, respectively. However, in general these ideals are not two-sided (cf. (5.11) below).

As the reader may have suspected, the classes of c.a.s. and majorizing maps are mutually dual with respect to the formation of adjoints; this is made precise in the following theorem.

**3.8 Theorem.** *Consider these propositions where $E, F$ are Banach lattices, $G, H$ are Banach spaces, where $T\in\mathscr{L}(E,H)$, $S\in\mathscr{L}(G,F)$, and where $T'\in\mathscr{L}(H',E')$, $S'\in\mathscr{L}(F',G')$ denote the corresponding adjoint operators:*

(a) *$T$ is c.a.s. with $\|T\|_l\leqq 1$.*

(b) *$T'$ is majorizing with $\|T'\|_m\leqq 1$.*

(c) *$S$ is majorizing with $\|S\|_m\leqq 1$.*

(d) *$S'$ is c.a.s. with $\|S'\|_l\leqq 1$.*

*Then* (a)$\Leftrightarrow$(b) *and* (c)$\Leftrightarrow$(d).

*Proof.* In view of (II. 9.1), (a)$\Rightarrow$(b) and (c)$\Rightarrow$(d) are immediately clear from Def. 3.5 and the characterizations (3.3) (c), (3.4) (c).

(b)$\Rightarrow$(a): By (c)$\Rightarrow$(d), the second adjoint $T''$ is c.a.s., so by (3.3) (b) there exists a positive linear form $z'\in E'''$, $\|z'\|\leqq 1$, such that $\|T''z\|\leqq\langle|z|,z'\rangle$ for all $z\in E''$; restricting $T''$ and $z'$ to $E$ yields the desired conclusion.

(d)$\Rightarrow$(c): By (a)$\Rightarrow$(b) and (3.4) (b) there exists $z\in F^{(\mathrm{iv})}$, $\|z\|\leqq 1$, such that $S''(W'')$ is contained in the order interval $[-z,z]$ of $F^{(\mathrm{iv})}$ where $W''$ denotes the the unit ball of $G''$. Now restricting linear forms on $F'''$ to $F'$ defines a positive contractive projection $P\colon F^{(\mathrm{iv})}\to F''$; so, if $W$ denotes the unit ball of $G$, we obtain $S(W)\subset[-Pz,Pz]$ which yields the desired conclusion by (3.4) (b). □

**Corollary.** *The canonical isometry $T\mapsto T'$ of $\mathscr{L}(E,H)$ into $\mathscr{L}(H',E')$ maps $\mathscr{L}^l(E,H)$ isometrically into $\mathscr{L}^m(H',E')$; a corresponding assertion is valid for $\mathscr{L}^m(G,F)$.*

We conclude this section by exhibiting remarkable extension properties of c.a.s. and majorizing operators, based on the factoring properties (3.3) (c) and (3.4) (c) and Theorems (II. 7.10) and (II. 8.9). Since every extension theorem yields a lifting theorem for the adjoint operator, it follows from (3.8) that c.a.s. and majorizing operators possess both lifting and extension properties; they share these properties with nuclear and integral operators (§ 5) without, however, presenting the difficulties caused by the latter operators through the dependence of their defining properties on the range space (see discussion preceding (5.8) below).

**3.9 Proposition.** *Let $E_0$ be a Banach sublattice of the Banach lattice $E$, and let $H$ be any Banach space. If $T_0\in\mathscr{L}^l(E_0,H)$ then $T_0$ possesses an extension $T\in\mathscr{L}^l(E,H)$ such that $\|T\|_l=\|T_0\|_l$.*

*Proof.* If $T_0=T_2\circ T_1$ is factored according to (3.3) (c), then by (II. 8.9) there exists a norm preserving positive extension $\tilde{T}_1\in\mathscr{L}(E,L)$ of $T_1$ which makes the diagram

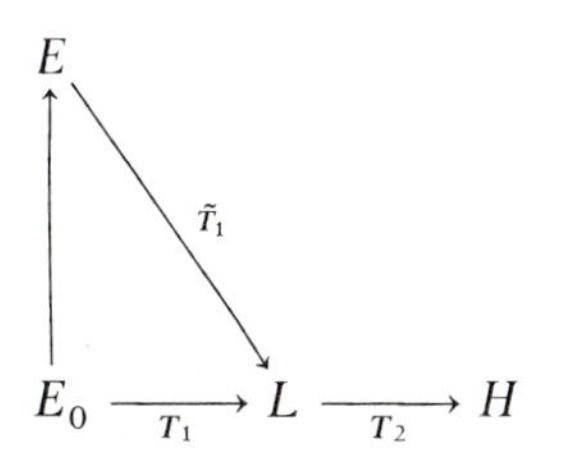

commute; $E_0 \to E$ is the canonical injection. The assertion follows from (3.3), letting $T := T_2 \circ \tilde{T}_1$. □

**3.10 Proposition.** *Let $G_0$ be a Banach subspace of the Banach space $G$, and let $F$ be a Banach lattice which is the range of a positive, contractive projection $P: F'' \to F$. If $T_0 \in \mathscr{L}^m(G_0, F)$ then $T_0$ possesses an extension $T \in \mathscr{L}^m(G, F)$ such that $\|T\|_m = \|T_0\|_m$.*

*Proof.* By (3.4)(c), $T_0$ has a factoring $G_0 \xrightarrow{T_1} M \xrightarrow{T_2} F$. The bidual $M''$ is an order complete $AM$-space with unit by (II. 9.1); thus, if $G_0 \to G$, $M \to M''$ denote the canonical injections, then from (II. 7.10) we conclude the existence of a norm preserving extension $\tilde{T}_1 \in \mathscr{L}(G, M'')$ of $T_1$ which makes the following diagram commute:

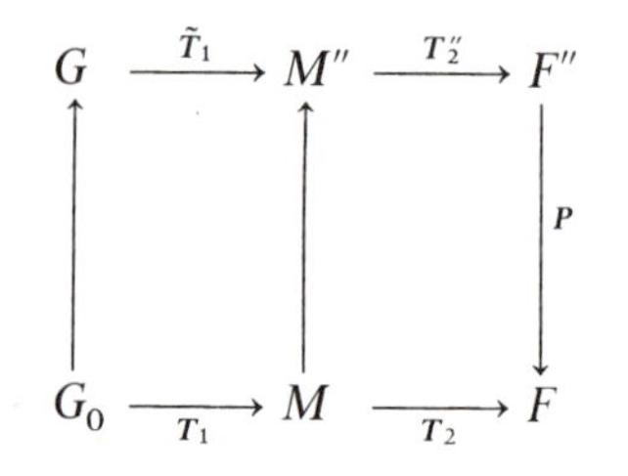


Now $P \circ T_2'' \circ \tilde{T}_1$ is an extension $T$ of $T_0$ which, by (3.4), has the desired properties. □

The reader will easily convince himself that in (3.9) and (3.10), provided that $G_0$, $G$, and $H$ are Banach lattices also and $T_0$ is positive, the extension $T$ can be chosen positive. (Note that in the proof of (3.10), (II. 7.10) must be substituted by (II. 7.10) Cor. 3.)

Finally, the preceding two extension theorems can be dualized to yield lifting theorems; the details are left to the interested reader (Exerc. 8).

## § 4. Banach Lattices of Operators

Let $E, F$ be Banach lattices and suppose $F$ to be order complete. Then by (1.4) the space $\mathscr{L}^r(E, F)$ of regular operators is a Banach lattice. However, we have seen (§ 1, Example 2) that in general $\mathscr{L}^r(E, F)$ does not even contain all compact linear maps $E \to F$; on the other hand, in certain special cases (see 1.5) $\mathscr{L}^r(E, F)$ agrees with $\mathscr{L}(E, F)$. This suggests that continuous linear operators with domain an $AL$-space or range an $AM$-space (with unit) show optimal behavior with respect to order boundedness; in conjunction with the factoring properties of c.a.s and majorizing maps (cf. (3.3) (c) and (3.4) (c)), Theorem 1.5 leads one to conjecture that the Banach spaces $\mathscr{L}^l(E, F)$ and $\mathscr{L}^m(E, F)$ are always Banach lattices under their natural ordering. We will show that, modulo an indispensable technical condition on $F$ (cf. Example 2 below), this is indeed true.

**4.1 Lemma.** *Let $E, F$ be Banach lattices and let $j: F \to F''$ denote the evaluation map. If $T: E \to F$ is c.a.s.* (Def. 3.1) *with* $\|T\|_l \leqq 1$ (Def. 3.5), *there exists a positive operator $S: E \to F''$ such that $\pm j \circ T \leqq S$ and $\|S\|_l \leqq 1$. Similarly, if $T: E \to F$ is majorizing with $\|T\|_m \leqq 1$, there exists $S: E \to F''$ such that $\pm j \circ T \leqq S$ and $\|S\|_m \leqq 1$.*

*Proof.* Let $\|T\|_l \leqq 1$ and consider a factoring $T: E \xrightarrow{T_1} L \xrightarrow{T_2} F$ as specified in (3.3) (c), so that $T_1 \geqq 0$ and $T_1, T_2$ are contractions. Because $j \circ T_2: L \to F''$ satisfies hypothesis (ii) of (1.5) (note that the restriction map $F^{(\text{iv})} \to F''$ is a positive contraction), (1.5) shows that the modulus $|j \circ T_2|$ exists and has norm $\leqq 1$. Defining $S := |j \circ T_2| \circ T_1$ we obtain $\pm j \circ T \leqq S$ and, using (3.3) (c) again, $\|S\|_l \leqq 1$.

The proof of the corresponding assertion for majorizing maps is similar and will be omitted. □

What is desired, of course, is to have the operator $S$ of the preceding lemma range in $F$; however, examples show that this is not possible, in general, even if $E$ is an $AL$-space and $F$ an order complete $AM$-space (without unit) (see Example 2 below). A condition on $F$ which remedies this situation, and which we have encountered before (cf. (1.5), Note after (3.4), and (3.10)), is the following.

**4.2 Definition.** *A Banach lattice $F$ is said to have* property $(P)$ *if there exists a positive, contractive projection $F'' \to F$, where $F$ (under evaluation) is identified with a vector sublattice of its bidual $F''$.*

Property $(P)$ implies that $F$ is order complete, but not necessarily an order complete sublattice of $F''$ (Exerc. 10). Clearly, every $KB$-space (Chap. II, § 5, Example 7) and hence every $AL$-space, has property $(P)$. Also, every order complete $AM$-space with unit has $(P)$ (II.7.10 Cor. 2 and subsequent remark); conversely, every $AM$-space which has $(P)$ is an order complete $AM$-space with unit (Exerc. 10).

**4.3 Theorem.** *Let $E, F$ be Banach lattices and suppose $F$ to have property $(P)$. Then $\mathscr{L}^l(E,F)$ and $\mathscr{L}^m(E,F)$ are Banach lattices, and ideals of $\mathscr{L}^r(E,F)$.*

*Proof.* Let $P: F'' \to F$ denote a positive contractive projection (Def. 4.2). If $T \in \mathscr{L}^l(E,F)$, by (4.1) there exists an operator $S$, $0 \leqq S \in \mathscr{L}^l(E,F'')$, such that $\|S\|_l = \|T\|_l$ and $\pm T \leqq P \circ S$. Since $F$ is order complete, $|T|$ exists and a glance at Def. 3.1 shows that $|T| \in \mathscr{L}^l(E,F)$; moreover, from (3′) we see that $\||T|\|_l \leqq \|P \circ S\|_l \leqq \|S\|_l = \|T\|_l$. This shows $\mathscr{L}^l(E,F)$ to be an ideal of $\mathscr{L}^r(E,F)$ and hence, by (1.4), to be order complete; furthermore, since $\|T\|_l \leqq \||T|\|_l$ (trivially) and since the $l$-norm is monotone on the positive cone of $\mathscr{L}^l(E,F)$, the $l$-norm is a lattice norm. Therefore, by (3.6) $\mathscr{L}^l(E,F)$ is a Banach lattice.

The proof concerning $\mathscr{L}^m(E,F)$ is entirely analogous and will be omitted. □

**Examples.** 1. If $E$ is an $AL$-space and $F$ has $(P)$ then, in view of (3.7), $\mathscr{L}^l(E,F) = \mathscr{L}^r(E,F) = \mathscr{L}(E,F)$ is a Banach lattice. Similarly, if $F$ is an $AM$-space having $(P)$ (cf. Exerc. 10) then $\mathscr{L}^m(E,F) = \mathscr{L}^r(E,F) = \mathscr{L}(E,F)$ is a Banach lattice. Thus (4.3) contains (1.5) as a special case.

2. The following example shows that property $(P)$ is not dispensable in (4.3). Let $E:=L^1(\mu)$, where $\mu$ denotes Lebesgue measure on $[0,1]$, and let $F:=c_0$. By (3.7), we have $\mathscr{L}^l(E,F)=\mathscr{L}^m(E,F)=\mathscr{L}(E,F)$.

Now let $(\varphi_n)_{n\in\mathbb{N}}$ be an orthonormal sequence in $L^2(\mu)$ which is contained and bounded in $L^\infty(\mu)$ and hence, order bounded in $E$ (for example, the Rademacher functions (cf. Chap. II, Exerc. 30) or the trigonometric functions); denote by $e_n:=(\delta_{nm})_{m\in\mathbb{N}}$ the $n$-th unit vector in $c_0$ $(=F)$. Since for each $f\in E$, $(\int f\varphi_n d\mu)$ is a null sequence, the equation

$$Tf:=\sum_{n=1}^{\infty}(\int f\varphi_n d\mu)e_n \qquad (f\in E)$$

defines an operator $T\in\mathscr{L}(E,F)$ which possesses no modulus. (In fact, if $[-u,u]\subset E$ is an order interval containing the sequence $(\varphi_n)$, $T[-u,u]$ contains the sequence $(e_n)$ which is not order bounded in $F$.)

The example shows that $\mathscr{L}^l(E,F)$ and $\mathscr{L}^m(E,F)$ are not vector lattices and, by virtue of (4.3), that $c_0$ does not have property $(P)$ (cf. Chap. II, § 10, Example 4).

3. Suppose $E$ is an $AM$-space and $F$ is an $AL$-space. Then $\mathscr{L}^l(E,F)=\mathscr{L}^m(E,F)$ $=\mathscr{L}^r(E,F)$ (see also 4.5 below); it will be shown in § 5 (cf. 5.6) that in the present circumstances, $\mathscr{L}^r(E,F)$ is the space of all integral linear maps in the sense of Grothendieck [1955a], and that the $r$-, $l$-, and $m$-norms all agree with the integral norm. If both $E$ and $F$ are infinite dimensional, it can be shown that there exist compact linear maps $E\to F$ not contained in $\mathscr{L}^r(E,F)$ (Exerc. 11).

It is natural to ask under what conditions on the Banach lattices $E$ and $F$, if any, $\mathscr{L}^l(E,F)$ (or $\mathscr{L}^m(E,F)$) is an $AL$- or $AM$-space; the following two propositions characterize these cases completely. It is interesting to note that in each case, $\mathscr{L}^l(E,F)$ and $\mathscr{L}^m(E,F)$ both coincide with $\mathscr{L}^r(E,F)$ (including the equality of norms).

**4.4 Proposition.** *Let $E,F$ be Banach lattices and suppose $F$ has property* $(P)$ (Def. 4.2). *The following assertions are equivalent:*

(a) *$\mathscr{L}^l(E,F)$ is an $AM$-space.*

(b) *$\mathscr{L}^m(E,F)$ is an $AM$-space.*

(c) *$E$ is an $AL$-space and $F$ is an $AM$-space.*

*Moreover, any one of these assertions implies that $\mathscr{L}^l(E,F)=\mathscr{L}^m(E,F)=\mathscr{L}^r(E,F)$ $=\mathscr{L}(E,F)$, and that $\mathscr{L}(E,F)$ is an order complete $AM$-space with unit.*

*Proof.* We observe first that both the $l$- and $m$-norms define cross-norms on $E'\otimes F\subset\mathscr{L}^l(E,F)\cap\mathscr{L}^m(E,F)$ (§ 3, Example 1): $\|x'\otimes y\|_l=\|x'\otimes y\|_m=\|x'\|\,\|y\|$ for all $x'\in E'$ and $y\in F$. This is immediately clear from § 3, Formulae (3′) and (4′).

(a)⇒(c): Consider the mappings $T_1:=x'_1\otimes y$, $T_2:=x'_2\otimes y$ in $\mathscr{L}^l(E,F)$, where $x'_1\in E'_+, x'_2\in E'_+, y\in F_+$ are chosen arbitrarily. It is clear that $T_1\vee T_2=(x'_1\vee x'_2)\otimes y$; hence, if $\mathscr{L}^l(E,F)$ is an $AM$-space, we have $\|T_1\vee T_2\|=\|T_1\|\vee\|T_2\|$ which implies $\|x'_1\vee x'_2\|=\|x'_1\|\vee\|x'_2\|$. Thus $E'$ is an $AM$-space and so, by (II.9.1) and (II.5.5) Cor. 2, $E$ is an $AL$-space.

Similarly, if $T_1:=x'\otimes y_1$ and $T_2:=x'\otimes y_2$ where $x'\in E'_+$ and $y_1\in F_+$, $y_2\in F_+$ are chosen arbitrarily, the hypothesis implies that $\|y_1\vee y_2\|=\|y_1\|\vee\|y_2\|$ and hence, that $F$ is an $AM$-space.

(c)⇒(a): Denote by $x'_0$ the linear form on $E$ defined by $\langle x, x'_0\rangle := \|x\|$ for $x \in E_+$, and denote by $e$ the unit of $F$ ($F$ has a unit by virtue of (P); see remarks after Def. 4.2). By (3.7) and (4.3), $\mathscr{L}^l(E,F)$ is a Banach lattice identical with $\mathscr{L}(E,F)$, and it is easily seen that for $T \in \mathscr{L}(E,F)$, $\|T\| \leq 1$ is equivalent to the relations $\pm T \leq x'_0 \otimes e$. Therefore, the unit ball of $\mathscr{L}^l(E,F) = \mathscr{L}(E,F)$ is the order interval $[-x'_0 \otimes e, x'_0 \otimes e]$ which shows $\mathscr{L}^l(E,F)$ to be an $AM$-space with unit $x'_0 \otimes e$.

The proof of the implications (b)⇒(c) and (c)⇒(b) is quite similar and is omitted. Finally, we note that by (3.7), (c) implies that the $l$- and $m$-norms both agree with the operator norm on $\mathscr{L}(E,F)$ (and, therefore, with the $r$-norm). □

**4.5 Proposition.** *Let $E, F$ be Banach lattices. The following assertions are equivalent:*

(a) *$\mathscr{L}^l(E,F)$ is an $AL$-space.*

(b) *$\mathscr{L}^m(E,F)$ is an $AL$-space.*

(c) *$E$ is an $AM$-space and $F$ is an $AL$-space.*

*Moreover, any one of these assertions implies that $\mathscr{L}^l(E,F) = \mathscr{L}^m(E,F) = \mathscr{L}^r(E,F)$.*[1]

*Proof.* The technique employed in the proof of (a)⇒(c) of (4.4) yields the implications (a)⇒(c) and (b)⇒(c).

Suppose (c) holds. Then from the definition of c.a.s. and majorizing maps (Def. 3.1) it follows that every positive $T \in \mathscr{L}(E,F)$ is both c.a.s. and majorizing; moreover, $\|T\|_l = \|T\|_m = \|T\|$ for $T \geq 0$ (§ 3, Formulae (3′) and (4′)). Since by (1.4) and (4.3), the $r$-, $l$-, and $m$-norms are lattice norms, these norms agree on $\mathscr{L}^r(E,F)$ and hence, $\mathscr{L}^r(E,F) = \mathscr{L}^l(E,F) = \mathscr{L}^m(E,F)$.

To prove (c)⇒(a) and (c)⇒(b), it now suffices to show that on the positive cone of $\mathscr{L}^r(E,F)$, the $r$-norm (that is, the operator norm) is additive; supposing $E$ to have a unit (which can be arranged, if necessary, by transition to $\mathscr{L}^r(E'', F'')$), this is an easy consequence of § 1, Formula (4). □

It is clear that whenever $E, F$ are Banach lattices and $G, H$ are Banach spaces, then $E' \otimes H \subset \mathscr{L}^l(E,H)$ and $G' \otimes F \subset \mathscr{L}^m(G,F)$ (§ 3, Example 2); in particular, $E' \otimes F \subset \mathscr{L}^l(E,F) \cap \mathscr{L}^m(E,F)$. We conclude this section with the following interesting theorem, where $H^l$ and $H^m$ denote the respective closures of $E' \otimes F$ in the Banach spaces $\mathscr{L}^l(E,F)$ and $\mathscr{L}^m(E,F)$; note that $F$ is not assumed to be order complete.

**4.6 Theorem.** *Let $E, F$ be Banach lattices. Then $H^l$ and $H^m$ are Banach lattices under their canonical ordering*[2] *and the $l$- and $m$-norms, respectively.*

*Proof.* We first consider the closure $H^m$ of $E' \otimes F$ in the Banach space $\mathscr{L}^m(E,F)$ and show that for each $u \in E' \otimes F$, the modulus $|u| \in \mathscr{L}(E,F)$ exists and is contained in $H^m$; from (4.1) and the obvious monotonicity of the $m$-norm on the positive cone of $H^m$ it then follows that $H^m$ is a Banach lattice as asserted.

Now let a continuous linear map of finite rank $u \in E' \otimes F$ be given; evidently, there exists $y \in F_+$ such that $|u(x)| \leq y$ for all $x \in U$ ($U$ the unit ball of $E$). Thus

---

[1] However, $\mathscr{L}^r(E,F) \neq \mathscr{L}(E,F)$ if $E, F$ are both infinite dimensional (Exerc. 11).

[2] I.e., the ordering induced by the canonical order of $\mathscr{L}(E,F)$.

$u$ can be viewed as a linear map $E \to F_y$; since this map is compact (in fact, of finite rank) and $F_y$ is an $AM$-space (II.7.2 Cor.), its modulus $|u| \in \mathscr{L}(E, F_y)$ exists and is compact (§ 1, Example 3). Thus $K := |u|(U)$ is relatively compact in $F_y$; so, if $\varepsilon > 0$ is given, there exists by (2.4) a mapping $\sum_{i=1}^n y_i' \otimes y_i \in (F_y)' \otimes F_y$ such that $\|y - \sum_{i=1}^n \langle y, y_i' \rangle y_i\|_0 < \varepsilon$ whenever $y \in K$. (Here the norm of $F_y$ is denoted by a subscript 0; from the proof of (2.4) we also note that $y_i$ and $y_i'$ can be chosen positive.) Letting $x_i' := |u|' y_i'$ $(i = 1, \dots, n)$, we obtain $\| \, |u|(x) - \sum_{i=1}^n \langle x, x_i' \rangle y_i\|_0 < \varepsilon$ for all $x \in U$. But since $F_y$ is an ideal in $F$, $|u|$ is the modulus of $u$ in $\mathscr{L}(E, F)$ as well; and, since by (3.7) the operator norm of $\mathscr{L}(E, F_y)$ is larger than the norm induced by the $m$-norm on the subspace $\mathscr{L}(E, F_y)$ of $\mathscr{L}^m(E, F)$, we conclude that

$$\| \, |u| - \textstyle\sum_{i=1}^n x_i' \otimes y_i \|_m \leq \varepsilon$$

for suitable elements $x_i' \in E'_+$, $y_i \in F_+$.

Next consider the closure $H^l$ of $E' \otimes F$ in $\mathscr{L}^l(E, F)$. Given $u \in E' \otimes F$, we conclude from (3.3) (b) that for a suitable $x' \in E'_+$, $u$ can be viewed as a continuous linear map of the $AL$-space $(E, x')$ into $F$. The dual of $(E, x')$ can be identified with the $AM$-space $(E')_{x'}$ (cf. beginning of § 3) so that the adjoint $u'$ of $u$ maps $F'$ into $(E')_{x'}$. Since $u'$ is of finite rank, the first part of the proof applies to $u'$ and hence, because $u''$ maps the dual of $(E')_{x'}$ into $F$, the modulus $|u'| \in \mathscr{L}(F', E')$ can be approximated in $\mathscr{L}^m(F', E')$ by elements of $F \otimes E'$ (in fact, by elements of the so-called projective cone $\operatorname{co} F_+ \otimes E'_+$). That is, there exists a sequence $(v_n)$ in $\operatorname{co} F_+ \otimes E'_+$ for which $|u'| = \lim_n v_n$ in $\mathscr{L}^m(F', E')$. Denote by $w_n \in E' \otimes F$ the pre-adjoint of $v_n$; by the corollary of (3.8), the sequence $(w_n)$ is a Cauchy sequence in $\mathscr{L}^l(E, F)$ and hence, convergent to some $w \in \mathscr{L}^l(E, F)$.

Clearly, $w' = |u'|$ and $\pm u \leq w$ so that $|u| \leq w$ in $\mathscr{L}^r(E, F)$. Hence, we have $|u|' \leq w' = |u'|$; but it is always true for adjoints that $|u'| \leq |u|'$ if the two moduli exist (cf. Exerc. 2), and so we obtain $|u|' = |u'| = w'$ which implies $|u| = w$. Therefore, $|u| \in H^l$. Finally, it is clear from (4.1) and the monotonicity of the $l$-norm on the positive cone of $H^l$ that $H^l$ is a Banach lattice. □

We will denote the Banach lattice $H^l$ by $E' \tilde{\otimes}_l F$ and the Banach lattice $H^m$ by $E' \tilde{\otimes}_m F$. Since the $l$- and $m$-norms are both larger than the $\varepsilon$-norm on $E' \otimes F$ (cf. 3.6), these Banach lattices contain only compact linear operators $E \to F$. We will return to the study of these tensor products below (§ 7); let us note, however, the following corollaries of (4.6).

**Corollary 1.** *Every positive operator in $E' \tilde{\otimes}_l F$ (respectively, $E' \tilde{\otimes}_m F$) can be approximated in the $l$-norm (respectively, $m$-norm) by elements of the projective cone $\operatorname{co} E'_+ \otimes F_+$.*

In general, however, it is unknown if every compact positive operator $E \to F$ ($E, F$ Banach lattices) can be so approximated in the operator norm.

**Corollary 2.** *If $E$ is an $AL$-space or if $F$ is an $AM$-space, the vector space of all compact linear operators $E \to F$ is a Banach lattice under its canonical ordering and the operator norm.*

*Proof.* If $E$ is an $AL$-space then by (3.7) the $l$-norm agrees on $E'\otimes F$ with the operator norm (and the $\varepsilon$-norm); therefore, $E'\tilde{\otimes}_l F = E'\tilde{\otimes}_\varepsilon F$. Since $E'$ is an $AM$-space and hence by (2.4) has the approximation property, it follows that $E'\tilde{\otimes}_\varepsilon F$ is the space of all compact operators in $\mathscr{L}(E,F)$ (cf. 2.3).

Similarly, if $F$ is an $AM$-space then by (3.7) the $m$-norm agrees on $E'\otimes F$ with the operator norm (and the $\varepsilon$-norm); therefore, $E'\tilde{\otimes}_m F = E'\tilde{\otimes}_\varepsilon F$. Again, since $F$ has the approximation property (2.4), it follows that $E'\tilde{\otimes}_\varepsilon F$ is the space of all compact operators in $\mathscr{L}(E,F)$. ▯

Thus we retrieve the result obtained in § 1, Example 3. Also, from Cor. 1 it follows that if $E$ is an $AL$-space or $F$ an $AM$-space, every positive compact operator $E\to F$ can be approximated (operator norm) by mappings $\sum_{i=1}^n x_i'\otimes y_i$ where $x_i'\in E'_+$, $y_i\in F_+$.

## § 5. Integral Linear Mappings

Let $G, H$ denote Banach spaces; we recall (§ 2) that the dual of $G\otimes_\pi H$ is the Banach space $\mathscr{B}(G,H)$ of continuous bilinear forms on $G\times H$, and canonically isomorphic to the Banach space $\mathscr{L}(G,H')$ (or $\mathscr{L}(H,G')$, cf. § 2, Formula (2)) under its operator norm. The dual of $G\otimes_\varepsilon H$ can thus be identified with a vector subspace $\mathscr{J}(G,H)$ of $\mathscr{B}(G,H)$; under the norm dual to the $\varepsilon$-norm, $\mathscr{J}(G,H)$ is a Banach space.

**5.1 Definition.** *Let $G,H$ be Banach spaces. A bilinear form $b$ on $G\times H$ is called* integral *if $b\in\mathscr{J}(G,H)$ (that is, if $b$ defines a continuous linear form on $G\otimes_\varepsilon H$); its norm in $\mathscr{J}(G,H)$ is called the* integral norm *of $b$.*

*A linear map $T\colon G\to H$ is called* integral *if the bilinear form $b_T\colon (x,y')\mapsto\langle Tx,y'\rangle$ on $G\times H'$ is integral; the* integral norm *$\|T\|_i$ is defined to be the integral norm of $b_T$.*

It is hence clear that under the integral norm, the vector space of all integral linear maps $G\to H$ is a Banach space $\mathscr{L}^i(G,H)$; since the $\varepsilon$-norm is coarser than the $\pi$-norm on $G\otimes H'$, it is also clear that the canonical imbedding $\mathscr{L}^i(G,H)\to\mathscr{L}(G,H)$ is of norm $\leqq 1$ (contractive). Moreover, it will be seen below (5.9) that under the correspondence of § 2, Formula (2), the Banach space $\mathscr{J}(G,H)$ is isomorphic to $\mathscr{L}^i(G,H')$.

**5.2 Proposition.** *If $G_0, G, H, H_1$ are Banach spaces, if $T\colon G\to H$ is integral, and if $R\colon G_0\to G$, $S\colon H\to H_1$ are continuous, then $S\circ T\circ R$ is integral and $\|S\circ T\circ R\|_i\leqq\|S\|\,\|T\|_i\|R\|$.*

*Proof.* It suffices to show that $T\circ R\colon G_0\to H$ is integral with integral norm $\|T\circ R\|_i\leqq\|T\|_i\|R\|$. The bilinear form $b$ on $G_0\times H'$ corresponding to $T\circ R$ is given by $b(x,y')=\langle TRx,y'\rangle$; since the linear mapping $R\otimes 1_{H'}$ of $G_0\otimes_\varepsilon H'$ into $G\otimes_\varepsilon H'$ is continuous of norm $\|R\|$ (§ 2, Formula (3); we suppose $H\neq\{0\}$), it follows that $b$ is continuous on $G_0\otimes_\varepsilon H'$ with integral norm $\leqq\|T\|_i\|R\|$. ▯

We next give an example which will turn out to be the prototype of all integral linear maps.

**5.3 Proposition.** *Let $K$ be a compact space and let $(X,\Sigma,\mu)$ be any measure space. Every positive linear map $T: C(K)\to L^1(\mu)$ is integral with integral norm $\|T\|_i=\|T\|$.*

*Proof.* Without loss of generality let us suppose that $\|T\|=1$. If $e_0$ denotes the unit (constant-one function) of $C(K)$, if $e:=Te_0$, and if $F:=\overline{F}_e$ denotes the closed ideal of $L^1(\mu)$ generated by $e$, then $T$ maps $C(K)$ into the $AM$-space $F_e$ and hence, possesses a factoring $C(K)\to F_e\to F\to L^1(\mu)$ where $j: F_e\to F$ and $j_1: F\to L^1(\mu)$ denote the canonical imbeddings; clearly, since $\|T\|=1$, all of the component maps have operator norm 1. By (5.2), it suffices to show that $j$ is integral with integral norm $\|j\|_i=1$.

By (II.8.5) we can identify $F$ with $L^1(\bar{\mu},K')$ for $K'$ a compact Stonian space and $\bar{\mu}$ a positive Radon measure on $K'$, in such a way that $F_e$ becomes $C(K')$ (Remark 2 after (8.5)) and that $\bar{\mu}(e)=1$. Now consider the bilinear form on $C(K')\times L^\infty(\bar{\mu})\cong C(K')\times C(K')$ canonically induced by $j$; this bilinear form is given by $\langle j(f),g\rangle=\int f(s)g(s)\,d\bar{\mu}(s)$ and hence, its associated linear form is the restriction to $C(K')\otimes C(K')$ of the Radon measure $h\mapsto v(h):=\int h(s,s)\,d\bar{\mu}(s)$ on $K'\times K'$. Since $\bar{\mu}(K')=1$ and the $\varepsilon$-norm on $C(K')\otimes C(K')$ is induced by the standard (supremum) norm of $C(K'\times K')$ (§ 2, Example 1), it follows from Def. 5.1 that $j$ is integral with integral norm $\|j\|_i=1$. □

**Examples.** 1. Every positive linear map $T$ of an $AM$-space $E$ into an $AL$-space $F$ is integral. In fact, if $E\to E''$ denotes the evaluation map, if $T'': E''\to F''$ the second adjoint of $T$ and $F''\to F$ is the band projection (II.8.3 (v)), then $T$ can be factored $E\to E''\to F''\to F$; since $T''$ is integral by (II.9.1) and (5.3), it follows that $T$ is integral with $\|T\|_i=\|T\|$. In particular, if $(X,\Sigma,\mu)$ is a finite measure space, then the canonical injection $L^\infty(\mu)\to L^1(\mu)$ is integral.

2. Let $\mu,v$ be $\sigma$-finite measures. Every majorizing linear map (Def. 3.1) $L^p(\mu)\to L^1(v)(1\leqq p\leqq+\infty)$ and every c.a.s. linear map $L^\infty(\mu)\to L^p(v)(1\leqq p\leqq+\infty)$ is integral by Example 1 and (3.3) (c), (3.4) (c) respectively.

3. More generally, let $G,H$ be Banach spaces and let $E$ be a Banach lattice. If $S: G\to E$ is majorizing and if $T: E\to H$ is c.a.s., then the composite map $T\circ S: G\to H$ is integral. In particular, every c.a.s. map with domain an $AM$-space and every majorizing map ranging in an $AL$-space, is integral.

The following characterization of integral maps explains the terminology and shows that Example 3 above comes close to describing the most general kind of integral mapping.

**5.4 Theorem.** *Let $G,H$ denote Banach spaces with respective unit balls $U,V$ and let $q: H\to H''$ denote the evaluation map. The following properties of an operator $T\in\mathscr{L}(G,H)$ are equivalent:*

(a) *$T$ is integral with $\|T\|_i\leqq 1$.*

(b) *On the $\sigma(G',G)\times\sigma(H'',H')$-compact product $U^\circ\times V^{\circ\circ}$, there exists a Radon measure $\mu$ of total variation $\|\mu\|\leqq 1$ such that $T=\int x'\otimes y''\,d\mu(x',y'')$.*

(c) *$q\circ T$ possesses a factoring $G\to M\to L\to H''$, where $M$ is an $AM$-space, $L$ an $AL$-space, $M\to L$ is positive, and all factors have operator norm $\leqq 1$.*

*Note.* The integral in (b) is a weak integral, i.e., it symbolizes that $\langle Tx,y'\rangle=\int\langle x,x'\rangle\langle y',y''\rangle\,d\mu(x',y'')$ for all $(x,y')\in G\times H'$. Moreover, in (c) $M$ can be supposed to be an order complete $AM$-space with unit (cf. proof of Cor. 1 below).

*Proof* of (5.4). (a)$\Rightarrow$(b): By means of evaluation, consider $G$ and $H'$ as Banach subspaces of $C(U^\circ)$ and $C(V^{\circ\circ})$, respectively; the definition of the $\varepsilon$-norm (§ 2, Formula (3)) shows that the imbedding $i: G\otimes_\varepsilon H' \to C(U^\circ)\tilde{\otimes}_\varepsilon C(V^{\circ\circ})$ is an isometry. But $C(U^\circ)\tilde{\otimes}_\varepsilon C(V^{\circ\circ}) \cong C(U^\circ, C(V^{\circ\circ})) \cong C(U^\circ \times V^{\circ\circ})$ (cf. § 2, Example 1) and hence, the adjoint $i'$ is a metric homomorphism of the dual $C(U^\circ \times V^{\circ\circ})' = M(U^\circ \times V^{\circ\circ})$ onto $\mathscr{J}(G,H')$. Therefore, if $b_T$ is the bilinear form on $G\times H'$ corresponding to $T$, we have $b_T = i'(\mu)$ for some Radon measure $\mu$ on $U^\circ \times V^{\circ\circ}$, $\|\mu\| \leqq 1$. That is, we have $b_T(x,y') = \int \langle x,x'\rangle \langle y',y''\rangle \, d\mu(x',y'')$ for all $(x,y')\in G\times H'$ as asserted.

(b)$\Rightarrow$(c): Let $K$ denote the compact product $U^\circ \times V^{\circ\circ}$ and let $M := C(K)$, $L := L^1(\mu, K)$. Define $R: G\to C(K)$ by $x\mapsto f_x$, where $f_x(x',y'') := \langle x,x'\rangle$, let $T_0$ denote the canonical imbedding $M\to L$, and define $S: L\to H''$ by the weak integral $Sf := \int f(x',y'')y''\,d\mu(x',y'')$. Then $q\circ T = S\circ T_0\circ R$ and, since $\|\mu\|\leqq 1$, it is clear that all factors have operator norm $\leqq 1$.

(c)$\Rightarrow$(a): By (5.2) and (5.3), $q\circ T$ is integral with integral norm $\leqq 1$; that is, $x\otimes z \mapsto \langle Tx, z\rangle$ ($x\in G$, $z\in H'''$) defines a continuous linear form of norm $\leqq 1$ on $G\otimes_\varepsilon H'''$. Since $G\otimes_\varepsilon H'$ can be identified with a normed vector subspace of $G\otimes_\varepsilon H'''$, it follows that $T$ is integral with $\|T\|_i \leqq 1$ (Def. 5.1). ▯

**Corollary 1.** *$T\in\mathscr{L}(G,H)$ is integral if and only if its adjoint $T'\in\mathscr{L}(H',G')$ is integral; in this case, $\|T\|_i = \|T'\|_i$.*

*Proof.* In view of (II.9.1), (5.4) (c) shows that $T\in\mathscr{L}^i(G,H)$ implies $T'\in\mathscr{L}^i(H',G')$ with integral norm not larger than $\|T\|_i$. Conversely, if $T'$ is integral then so is $T''$ by the preceding; thus, by (5.4) (c), $q''\circ T''$ has a factoring $G''\to M''\to L''\to H^{(\mathrm{iv})}$. Since the restriction map defines a contractive projection $H^{(\mathrm{iv})}\to H''$, it follows that $q\circ T$ has a factoring $G\to M''\to L''\to H''$, and so $T$ is integral. ▯

**Corollary 2.** *Every integral map is weakly compact.*

*Proof.* If $T: G\to H$ is integral, then by (5.4) (c) $q\circ T$ is weakly compact, since every positive linear map $M\to L$ is weakly compact (consider the second adjoint $M''\to L''$ and note that order intervals of an $AL$-space are weakly compact (II.8.8) Cor.). Clearly, then, $T$ is weakly compact. ▯

**Corollary 3.** *If $G, H$ are Banach spaces and $G\otimes H$ is canonically identified with a subspace of $G\otimes H''$, then the $\varepsilon$-unit ball $U_\varepsilon$ of $G\otimes H$ is $\sigma(G\otimes H'', G'\otimes H')$-dense in the $\varepsilon$-unit ball $U_\varepsilon''$ of $G\otimes H''$.*

*Proof.* Denoting by $j$ the evaluation map $H'\to H'''$, we consider $u\in G'\otimes H'$ as an integral map $G\to H'$ and $j\circ u$ as an integral map $G\to H'''$. Since there exists a contractive projection $H^{(\mathrm{v})}\to H'''$, (5.4) (c) implies that $\|u\|_i = \|j\circ u\|_i$. It follows that $u\in U_\varepsilon^\circ$ if and only if $u\in (U_\varepsilon'')^\circ$ for the dualities $\langle G\otimes H, G'\otimes H'\rangle$ and $\langle G\otimes H'', G'\otimes H'\rangle$, respectively. Hence the bipolar theorem shows that $U_\varepsilon'' = U_\varepsilon^{\circ\circ}$, that is, $U_\varepsilon$ is $\sigma(G\otimes H'', G'\otimes H')$-dense in $U_\varepsilon''$. ▯

To establish the relationship of integral maps with cone absolutely summing, majorizing, absolutely summing, and nuclear maps, we need the following important composition theorem (Grothendieck [1955a]). Recall that if $G, H$ are Banach spaces, a linear map $G \to H$ is called *nuclear* if it is of the form

$$x \mapsto Tx = \sum_{n=1}^{\infty} \langle x, x'_n \rangle y_n$$

where $(x'_n) \subset G'$ and $(y_n) \subset H$ are sequences such that $\sum_n \|x'_n\| \|y_n\| < +\infty$; the infimum of this sum (taken over all representations of $T$) is called the *nuclear* (or $\pi$-)*norm* of $T$. Under this norm, the vector space of all nuclear maps $G \to H$ is a Banach space $\mathcal{N}(G,H)$ (§ 2, Example 5).

**5.5 Theorem.** *Let $G, H, N$ denote Banach spaces. If $T: G \to H$ is integral and if $S: H \to N$ is weakly compact, then the composition $S \circ T: G \to N$ is nuclear.*

*Proof.* We show first that it suffices to consider the case of a separable Banach space $N$. Since $S$ is weakly compact, the second adjoint $S''$ maps $H''$ into $N$ and thus, arguing as in the proof of (5.4), (b)$\Rightarrow$(c), we find that $S \circ T = S'' \circ q \circ T$ ($q$ the evaluation map $H \to H''$) has a factoring $G \to C(K) \to L^1(\mu, K) \to H'' \to N$. Here the partial map $S_1: L^1(\mu, K) \to N$ is weakly compact and, since $L^1(\mu, K)$ has the Dunford-Pettis property (II.9.8), the partial map $C(K) \to N$ maps the unit ball of $C(K)$ onto a compact set $C \subset N$. Therefore, $S_1$ maps $L^1(\mu, K)$ into the closed linear hull $N_0$ of $C \subset N$, and $N_0$ is separable.

Thus it suffices to show that the mapping $R := S_1 \circ j$:

$$C(K) \to L^1(\mu, K) \to N_0$$

is nuclear, where $j: C(K) \to L^1(\mu, K)$ is the canonical imbedding. If $U_1$ and $V_1$ denote the respective unit balls of $C(K)$ and $L^1(\mu, K)$ then, since $j$ is integral by (5.3), by (5.4) there exists a Radon measure $\nu$ on $K_1 := U_1^\circ \times V_1^{\circ\circ}$ such that $jf = \int_{K_1} \langle f, x' \rangle y''\, d\nu(x', y'')$. Denoting by $\bar{\nu}$ the image of $\nu$ under the (weakly) continuous map $(x', y'') \mapsto (x', S_1'' y'')$ of $K_1$ onto $K_2 := U_1^\circ \times S_1''(V_1^{\circ\circ})$, we finally obtain the (weak) integral

$$Rf = \int_{K_2} \langle f, x' \rangle z\, d\bar{\nu}(x', z) \qquad (f \in C(K)).$$

Since $N_0$ is separable, the weakly continuous map $\psi: (x', z) \mapsto z$ of $K_2$ into the Banach space $N_0$ is $\bar{\nu}$-measurable [DS, III.6.11] and hence, since it is bounded, $\bar{\nu}$-integrable; that is, $\psi \in L^1_{N_0}(\bar{\nu}, K_2)$. But the latter space is isomorphic to $L^1(\bar{\nu}, K_2) \tilde{\otimes}_\pi N_0$ (§ 2, Example 3) and hence, $\psi$ is of the form $\sum_{n=1}^{\infty} f_n \otimes z_n$ where $f_n \in L^1(\bar{\nu}, K_2)$, $z_n \in N_0$ and $\sum_{n=1}^{\infty} \|f_n\| \|z_n\| < +\infty$ [S, III.6.4]. It follows that $R$ is given by

$$Rf = \sum_{n=1}^{\infty} \langle f, \lambda_n \rangle z_n \qquad (f \in C(K))$$

where $\langle f, \lambda_n \rangle := \int_{K_2} \langle f, x' \rangle f_n(x', z)\, d\bar{\nu}(x', z)$ for all $n \in \mathbb{N}$. Therefore, $R$ is nuclear. □

**Corollary 1.** *The composition of two integral linear maps is nuclear.*

**Corollary 2.** *If $H$ is reflexive, every integral linear map $G \to H$ is nuclear.*

The first corollary is clear from (5.5) and (5.4) Cor. 2, while the second follows from (5.5) and the fact that the identity map of a reflexive Banach space is weakly compact.

Let $E, F$ denote Banach lattices. If $E$ is an $AL$-space and $F$ an order complete $AM$-space with unit, we recall from (4.4) that $\mathscr{L}^r(E,F) = \mathscr{L}(E,F)$. If, conversely, $E$ is an $AM$-space and $F$ an $AL$-space, we can now show that $\mathscr{L}^r(E,F) = \mathscr{L}^i(E,F)$.

**5.6 Theorem.** *If $E$ is an $AM$-space and $F$ an $AL$-space, then $\mathscr{L}^r(E,F) = \mathscr{L}^i(E,F)$ and hence, the space of integral maps $E \to F$ is an $AL$-space under its canonical order and integral norm. Moreover, the space $\mathscr{N}(E,F)$ of nuclear operators $E \to F$ is the band of $\mathscr{L}^i(E,F)$ generated by $E' \otimes F$ and hence, an $AL$-space under its canonical order and nuclear norm.*

*Proof*. First, it is clear from (5.4) (c) that each $T \in \mathscr{L}^r(E,F)$ is integral. Conversely, suppose that $T \in \mathscr{L}^i(E,F)$. Because $F$ is a projection band in its bidual (II.8.3), $T$ can be factored $E \to E'' \to F'' \to F$ where the central map is $T''$. By (5.4) (c), $T''$ has a factoring $E'' \to M'' \to L'' \to F^{(iv)} \to F''$ where the map $M'' \to L''$ is positive; moreover, if $\|T\|_i = 1$ then all partial maps can be supposed to have operator norm 1. Thus $T$ itself possesses a factoring $T_3 \circ T_2 \circ T_1 : E \to M'' \to L'' \to F$ where $T_2 : M'' \to L''$ is positive and, supposing $\|T\|_i = 1$, where all factors have operator norm 1. From (1.5) it follows that $|T_1|$, $|T_3|$ exist with operator norm 1; evidently, $\pm T \leqq |T_3| \circ T_2 \circ |T_1|$. Therefore, $|T|$ exists, has operator norm $\leqq 1$, and is integral by our first remark; but from (5.4) (c) it follows that for positive operators from an $AM$-space into an $AL$-space the integral norm agrees with the operator norm; therefore, $\| |T| \|_i \leqq 1$. We conclude that $\mathscr{L}^r(E,F) = \mathscr{L}^i(E,F)$ and, since $\mathscr{L}^r(E,F)$ is an $AL$-space by (4.5), we must have $1 = \|T\|_i = \|T^+ - T^-\|_i \leqq \|T^+\|_i + \|T^-\|_i = \| |T| \|_i \leqq 1$. Thus the $r$-norm agrees with the integral norm.

Considering the duality $\langle E \otimes F', \mathscr{L}^i(E,F) \rangle$ we see from Def. 5.1 that the integral norm is dual to the $\varepsilon$-norm on $E \otimes F'$; thus if $E' \otimes F$ is considered a subspace of $\mathscr{L}^i(E,F)$ then, since $E$ has the metric approximation property (2.4), the corollary of (2.3) shows that the integral norm induces the $\pi$-norm on $E' \otimes F$. We will show that the ideal of $\mathscr{L}^r(E,F) = \mathscr{L}^i(E,F)$ generated by $E' \otimes F$ contains only nuclear operators; then, since $\mathscr{L}^i(E,F)$ is an $AL$-space, the closure of $E' \otimes F$ in $\mathscr{L}^i(E,F)$ agrees with the band generated by $E' \otimes F$ and, simultaneously, with $\mathscr{N}(E,F)$ $(\cong E' \tilde{\otimes}_\pi F)$.

Now the ideal of $\mathscr{L}^r(E,F)$ generated by $E' \otimes F$ consists of all regular maps $T: E \to F$ such that $0 \leqq |T| \leqq S$ for a suitable $S \in E' \otimes F$ where, evidently, $S$ can be assumed to be of rank 1, $S = x' \otimes y$. But $|T| \leqq x' \otimes y$ means that $|Tx| \leqq |T|\,|x| \leqq \langle |x|, x' \rangle\, y$ for all $x \in E$; hence, $T$ has a factoring

$$E \to (E, x') \to F_y \to F$$

(for notation, see § 3). Since the canonical mappings $E \to (E, x')$ and $F_y \to F$ are integral under the present assumptions on $E$ and $F$ (Example 1), it follows from (5.5) Cor. 1 that $T$ is nuclear. □

Let $E, F$ again be arbitrary Banach lattices. The classess of c.a.s., majorizing, and regular linear maps are closely related to integral maps: They can be characterized by the property that composition with the canonical maps $E_u \to E$ $(u \in E_+)$ and/or $F \to (F, v')$ $(v' \in F'_+)$ renders them integral, with appropriate relations between the corresponding norms. To simplify the notation, we let $i_u$ denote the continuous injection of the $AM$-space $E_u$ into $E$ $(u \in E_+)$, and we let $j_{v'}$ denote the continuous lattice homomorphism of $F$ into the $AL$-space $(F, v')$ $(v' \in F'_+)$. Moreover, for each $T \in \mathscr{L}(E, F)$ we write $T_u := T \circ i_u$, $T_{v'} := j_{v'} \circ T$, and $T_{u,v'} := j_{v'} \circ T \circ i_u$. The following result will be important in determining the duals of certain tensor products of Banach lattices (§ 7).

**5.7 Proposition.** *Let $E, F$ denote Banach lattices, and let $T \in \mathscr{L}(E, F)$.*

(i) *$T$ is c.a.s. with l-norm $\|T\|_l \leqq 1$ if and only if $T_u$ is integral with $\|T_u\|_i \leqq 1$ for all $u \in E_+$, $\|u\| \leqq 1$.*

(ii) *$T$ is majorizing with m-norm $\|T\|_m \leqq 1$ if and only if $T_{v'}$ is integral with $\|T_{v'}\|_i \leqq 1$ for all $v' \in F'_+$, $\|v'\| \leqq 1$.*

(iii) *Suppose $F$ has property (P)* (Def. 4.2). *$T$ is regular with r-norm $\|T\|_r \leqq 1$ if and only if $T_{u,v'}$ is integral with $\|T_{u,v'}\|_i \leqq 1$ for all $u \in E_+$, $\|u\| \leqq 1$ and all $v' \in F'_+$, $\|v'\| \leqq 1$.*

*Proof.* (i) If $\|T\|_l \leqq 1$ then from (3.3) (c) and (5.4) (c) it follows that $T_u \in \mathscr{L}^i(E_u, F)$ with $\|T_u\|_i \leqq 1$ for all $u \in E_+$, $\|u\| \leqq 1$.

Conversely, suppose that $(x_n)$ is a positive summable sequence in $E$ with $\varepsilon$-norm $\leqq 1$ (§ 2, Formula (4)); then by (2.5) we have $\|u\| \leqq 1$ for $u := \sum_n x_n$. Since by hypothesis $T_u$ is integral with integral norm $\leqq 1$, there exists a Radon measure $\mu$ of total variation $\|\mu\| \leqq 1$ on $K := W^\circ \times V^{\circ\circ}$ where $W, V$ denote the unit balls of $E_u$ and $F$, respectively, such that $T_u x = \int \langle x, x' \rangle y'' \, d\mu(x', y'')$ $(x \in E_u)$, by virtue of (5.4). Since $x_n \in E_u$ for all $n$, we obtain

$$\sum_{n=1}^k \|T x_n\| \leqq \int \sum_{n=1}^k \langle x_n, |x'| \rangle \, d|\mu|(x', y'') \leqq 1$$

for every $k \in \mathbb{N}$. It follows that $T$ is c.a.s. with $\|T\|_l \leqq 1$.

(ii) Noting that $(F')_{v'} \to F'$ is the adjoint of $F \to (F, v')$ (cf. Exerc. 9), we see that $(T_{v'})' = (T')_{v'}$ and hence, the assertion follows by duality from (i) in view of (3.8) and (5.4) Cor. 1.

(iii) If $T \in \mathscr{L}(E, F)$ is regular and $\|T\|_r \leqq 1$, it follows from (5.6) that $T_{u,v'}: E_u \to (F, v')$ is integral with integral norm $\leqq 1$ whenever $\|u\| \leqq 1$, $\|v'\| \leqq 1$.

Conversely, let $u \in E_+$ be given, $\|u\| \leqq 1$, and suppose $F$ has property $(P)$. If $T_{u,v'} = (T_u)_{v'}$ is integral with integral norm $\leqq 1$ whenever $\|v'\| \leqq 1$, then from (ii) we have $T_u \in \mathscr{L}^m(E_u, F)$ with $\|T_u\|_m \leqq 1$. Hence, by (3.4) (b), $T[-u, u]$ is contained in the unit ball of $F$ and majorized by an element $y'' \in F''$, $\|y''\| \leqq 1$; since $F$ has $(P)$, it follows that $T[-u, u] \subset [-y, y]$ for some $y \in F$, $\|y\| \leqq 1$.

Since $u$ was arbitrary, it follows that $T$ is regular with $r$-norm $\leqq 1$ (§ 1, Formulae (2), (5)). ▯

*Remark.* In the preceding characterization (i), $F$ can be replaced by any Banach space; similarly, in (ii) $E$ can be replaced by any Banach space. Finally, if in (iii) the condition that $F$ have property $(P)$ is omitted, a weaker characterization of the operator $T$ results (preregularity of $T$; cf. Exerc. 3).

It is a peculiar feature of integral maps that an integral map $G \to H_1$ with range in a Banach subspace $H$ of $H_1$ need not be integral when considered as a mapping $G \to H$ (Exerc. 14); the reason for this phenomenon lies in the fact that, in general, the natural map $G \otimes_\varepsilon H_1' \to G \otimes_\varepsilon H'$ is not open (and hence, the bilinear form corresponding to $T$ need not be continuous on $G \otimes_\varepsilon H'$). Conversely, it can happen that a non-integral map $G \to H$ can be made integral by right composition with (i.e., subsequent application of) an isometric injection $H \to H_1$; dually, certain non-integral maps $G \to H$ can be made integral by left composition with a metric homomorphism $G_0 \to G$ (that is, a linear surjection mapping the open unit ball of $G_0$ onto the open unit ball of $G$). We introduce these mappings, already studied by Grothendieck [1955a], because of their importance for the theory of summable and absolutely summable sequences (§ 2), and because they serve to clarify the relationship between integral maps and the classes of c.a.s. and majorizing maps studied above (§ 3).

**5.8 Definition.** *Let $G, H$ be Banach spaces, and let $T \in \mathscr{L}(G,H)$. $T$ is called* right semi-integral *if there exists a Banach space $H_1$ and an isometric injection $R: H \to H_1$ such that $R \circ T$ is integral. $T$ is called* left semi-integral *if there exists a Banach space $G_0$ and a metric homomorphism $S: G_0 \to G$ such that $T \circ S$ is integral.*

For the desired characterization of right and left semi-integral maps, we need this lemma which gives a simpler description of integral maps $G \to H'$ than Def. 5.1 (cf. [S, p. 169]).

**5.9 Lemma.** *Let $G, H$ be Banach spaces, and let $T \in \mathscr{L}(G,H')$. $T$ is integral if and only if the bilinear form $(x,y) \mapsto \langle Tx, y\rangle$ on $G \times H$ is integral.*

*Proof.* If $T$ is integral then, by Def. 5.1, the linear form $f$ defined by $x \otimes y'' \mapsto \langle Tx, y''\rangle$ is continuous on $G \otimes_\varepsilon H''$; since $G \otimes_\varepsilon H$ can be identified with a normed vector subspace of $G \otimes_\varepsilon H''$ (by virtue of § 2, Formula (3)), it follows that $f$ is continuous on $G \otimes_\varepsilon H$.

Conversely, if the linear form defined by $x \otimes y \mapsto \langle Tx, y\rangle$ is continuous on $G \otimes_\varepsilon H$, then $T$ has a factoring $G \to C(K) \to L^1(\mu, K) \to H'$ where the central map is positive, $K = U^\circ \times V^\circ$, and $\mu \in C(K)'_+$ (see proof of (5.4)); thus $T$ is integral by (5.4) (c). □

**5.10 Theorem.** *If $G, H$ are Banach spaces, then for $T \in \mathscr{L}(G,H)$ the following are equivalent assertions:*

(a) *$T$ is right semi-integral.*

(b) *There exists a positive Radon measure $\mu$ on the $\sigma(G', G)$-compact unit ball $U^\circ$ of $G'$ such that for all $x \in G$,*

$$\|Tx\| \leqq \int_{U^\circ} |\langle x, x'\rangle| \, d\mu(x').$$

(c) *$T$ maps summable sequences in $G$ onto absolutely summable sequences in $H$.*

(d) *For each continuous linear map $S: c_0 \to G$, $T \circ S$ is cone absolutely summing.*

(For a metric version of 5.10, see Exerc. 13).

*Proof.* (a)⇒(b): For a suitable isometric injection $R: H \to H_1$ the composition $R \circ T$ is integral; if $W$ denotes the unit ball of $H_1$, by (5.4) there exists a Radon measure $\nu$ on $U^\circ \times W^{\circ\circ}$ such that $R \circ T = \int x' \otimes y''\, d\nu(x', y'')$. Now let $\mu$ denote the positive Radon measure on $U^\circ$ which is the restriction of $|\nu|$ to the subspace $C(U^\circ)$ of $C(U^\circ \times W^{\circ\circ})$ ($C(U^\circ)$ can be identified with the space of continuous functions on $U^\circ \times W^{\circ\circ}$ "depending only on $x' \in U^\circ$"); then, since $R$ is an isometry, a simple estimate shows that $\|Tx\| \leq \int |\langle x, x' \rangle|\, d\mu(x')$ for $x \in G$.

(b)⇒(c): If $\boldsymbol{x} = (x_n)$ is a summable sequence in $G$ then $\sum_n |\langle x_n, x' \rangle|$ converges uniformly for $x' \in U^\circ$ and (with the notation of § 2) we obtain

$$\|T\boldsymbol{x}\|_\pi = \sum_n \|Tx_n\| \leq \int \sum |\langle x_n, x' \rangle|\, d\mu(x') \leq \|\mu\|\, \|\boldsymbol{x}\|_\varepsilon .$$

(c)⇒(d): Obvious.

(d)⇒(c): Let $e_n \in c_0$ denote the $n$-th unit vector $(\delta_{nm})_{m \in \mathbb{N}}$, let $(x_n)$ be a summable sequence in $G$, and define $S: c_0 \to G$ by $(\xi_n) \mapsto \sum_n \xi_n x_n$; $S$ is linear and continuous. If $z_p := (e_1, \ldots, e_p, 0, \ldots)$ then each $z_p$ ($p \in \mathbb{N}$) is a positive summable sequence in $c_0$ with $\varepsilon$-norm 1 (cf. 2.5); therefore, since $T \circ S$ is c.a.s. by hypothesis, we have $\|\boldsymbol{TS}\, z_p\|_\pi \leq \|T \circ S\|_l$; but $\|\boldsymbol{TS}\, z_p\|_\pi = \sum_{n=1}^p \|Tx_n\|$ and hence, $T$ has property (c).

(c)⇒(a): Choose an index set A such that there exists a metric homomorphism $\varphi: l^1(\mathsf{A}) \to H'$; then $\varphi': H'' \to l^\infty(\mathsf{A})$ is an isometric injection. Since $T$ transforms summable sequences in $G$ into absolutely summable sequences in $H$, the linear map $1_{l^1(\mathsf{A})} \otimes T: l^1(\mathsf{A}) \otimes_\varepsilon G \to l^1(\mathsf{A}) \otimes_\pi H$ is continuous. Now let $\tilde{T} := \varphi' \circ q \circ T$ ($q: H \to H''$ the evaluation map); then the bilinear form $(z, x) \mapsto \langle z, \tilde{T}x \rangle$ on $l^1(\mathsf{A}) \times G$ is integral and hence, $\tilde{T}: G \to l^\infty(\mathsf{A})$ is integral by (5.9). Since $\varphi' \circ q: H \to l^\infty(\mathsf{A})$ is an isometric injection, it follows that $T$ is right semi-integral. □

Because of the property expressed by (5.10) (c), right semi-integral maps are called *absolutely summing*; it is not difficult to show that under an appropriate norm, the vector space of all absolutely summing linear maps $G \to H$ is a Banach space (Exerc. 13). In the subsequent corollaries, we derive a number of important properties of absolutely summing maps.

**Corollary 1.** *Let $G, H$ be Banach spaces. Every absolutely summing linear map $T: G \to H$ permits a factoring*

$$G \to C(U^\circ) \to L^2(\mu, U^\circ) \to H$$

*where $\mu$ is a suitable positive Radon measure on the $\sigma(G', G)$-compact unit ball $U^\circ$ of $G'$. In particular, $T$ is weakly compact.*

*Proof.* If $\mu$ is a Radon measure as specified in (5.10) (b), then by means of the canonical imbedding $j: G \to C(U^\circ) \to L^2(\mu, U^\circ)$ (given by $x \mapsto f_x$, where $x \in G$ and $f_x(x') := \langle x, x' \rangle$, $x' \in U^\circ$), $G$ is mapped onto a vector subspace $N$ of $L^2(\mu, U^\circ)$. From the estimate of $\|Tx\|$ given in (5.10) (b), it follows that $T$ defines a continuous linear map $T_0: N \to H$ such that $T = T_0 \circ j$. Clearly, $T_0$ has a continuous extension to the closure $\bar{N} \subset L^2(\mu, U^\circ)$; since $L^2$ is a Hilbert space (and hence, $\bar{N}$ complemented), $T_0$ has a continuous extension $T_1: L^2(\mu, U^\circ) \to H$ so that $T = T_1 \circ j$. □

Note that in the factoring of Cor. 1, in general $L^2(\mu, U^\circ)$ cannot be replaced by $L^1(\mu, U^\circ)$; otherwise, by (5.3) every absolutely summing map would be integral which is false (cf. Example 4 below and Exerc. 14). However, the following is true.

**Corollary 2.** *Every absolutely summing map with domain or range an $AM$-space, is integral.*

*Proof.* By (5.4) Cor. 1 it is sufficient to show that the second adjoint of the mapping in question is integral; since the bidual of any $AM$-space is an order complete $AM$-space with unit (II.9.1), we can assume that the domain or range of the operator considered, is an order complete $C(K)$ ($K$ compact).

Suppose, then, that $T: C(K) \to H$ is absolutely summing, $H$ any Banach space. Since the compact space $K$ can be identified with the extreme boundary of the positive face of $U^\circ$ ($U$ the unit ball of $C(K)$; cf. (II.7.4) and its Corollary 2), the Radon measure $\mu$ of (5.10) (b) can be considered a Radon measure on $K$. The estimate of $\|Tx\|$ in (5.10) (b) then shows that $T$ is continuous (into $H$) on the dense normed subspace $C(K)$ of $L^1(\mu, K)$ and hence, $T$ has a factoring $C(K) \to L^1(\mu, K) \to H$; thus $T$ is integral by (5.3).

Second, let $T: G \to C(K)$ be absolutely summing, $G$ any Banach space. By (5.10) (b), $T$ has a factoring $G \to \bar{N} \to C(K)$ where $\bar{N}$ denotes the closure in $L^1(\mu, U^\circ)$ of the canonical image of $G$ (cf. proof of Cor. 1). But by (II.7.10), the mapping $\bar{N} \to C(K)$ has a continuous (even norm preserving) extension $L^1(\mu, U^\circ) \to C(K)$, since $C(K)$ is order complete (hence, $K$ Stonian (II.7.7) Cor.); therefore, $T$ is integral by (5.3). □

**Corollary 3.** *The composition of two absolutely summing linear maps is nuclear.*

*Proof.* If $T_1: G \to H$ and $T_2: H \to N$ are absolutely summing and $U, V$ denote the respective unit balls of $G, H$ then by Cor. 1, $T_2 \circ T_1$ has a factoring

$$G \to C(U^\circ) \to L^2(\mu, U^\circ) \to H \to C(V^\circ) \to L^2(v, V^\circ) \to N .$$

The partial map $G \to C(V^\circ)$ is absolutely summing, hence integral by Cor. 2; since the partial map $C(V^\circ) \to N$ (factoring through $L^2(v, V^\circ)$) is weakly compact, it follows from (5.5) that $T_2 \circ T_1$ is nuclear. □

**Corollary 4** (Dvoretzky-Rogers). *A Banach space in which every summable sequence is absolutely summable, is finite dimensional.*

*Proof.* In fact, by Cor. 3 the identity map of such a Banach space is nuclear, hence compact. □

It follows from the very definition of right and left semi-integral maps (Def. 5.8) and the mutual duality of isometric injections and metric homomorphisms of Banach spaces that a continuous linear map $T: G \to H$ is right (left) semi-integral if and only if its adjoint $T': H' \to G'$ is left (right) semi-integral. Thus from (3.8) and (5.10) the following characterization of left semi-integral maps can easily be derived; we omit its proof.

**5.11 Proposition.** *If $G, H$ are Banach spaces, then for $T \in \mathscr{L}(G,H)$ the following are equivalent assertions:*

(a) *$T$ is left semi-integral.*

(b) *For every Banach lattice $F$ and each continuous linear map $R: H \to F$, the composition $R \circ T$ is majorizing.*

(c) *For each continuous linear map $R: H \to l^1$, the composition $R \circ T$ is majorizing.*

*Moreover, any one of these assertions implies $T$ to be weakly compact.*

The preceding proposition suggests the following definition: A null sequence $(x_n)$ in a Banach space $H$ is called *hypermajorized* if for each continuous linear map $R: H \to l^1$, the sequence $(R x_n)$ is (absolutely) majorized in $l^1$. Accordingly, a linear map $T: G \to H$, where $G, H$ are Banach spaces, is called *hypermajorizing* if for each null sequence $(x_n)$ in $G$, the sequence $(T x_n)$ is hypermajorized in $H$. (For examples of hypermajorized sequences, cf. Exerc. 12).

It is well known that the adjoint of an absolutely summing map is not absolutely summing, in general (Exerc. 14); in other words, the concepts of right and left semi-integral maps are not coextensive. However, this dual characterization of absolutely summing linear maps results from (5.11).

**5.12 Proposition.** *Let $G, H$ be Banach spaces and for $T \in \mathscr{L}(G,H)$, denote by $T'$ the adjoint $H' \to G'$. Then $T$ is absolutely summing (respectively, hypermajorizing) if and only if $T'$ is hypermajorizing (respectively, absolutely summing).*

**Examples.** 4. Let $H_1, H_2$ be Hilbert spaces. It will be seen in (6.9) below that $T \in \mathscr{L}(H_1, H_2)$ is right semi-integral (absolutely summing) if and only if $T$ is a Hilbert-Schmidt operator. On the other hand, from (5.4) Cor. 2 it follows that $T$ is integral iff $T$ is nuclear; since Hilbert-Schmidt operators are not necessarily nuclear (see 6.2 Cor.), it follows that in general, absolutely summing operators are not integral (for noteworthy exceptions, cf. (5.10) Cor. 2). Thus by (5.4) Cor. 1 and by (5.12), hypermajorizing operators are not integral, in general, except when their domain or range is an $AL$-space.

5. As the identity maps of infinite dimensional $AL$- and $AM$-spaces show, cone absolutely summing maps are not absolutely summing and majorizing maps are not hypermajorizing, in general (cf. 5.10 Cor. 3). For the coalescence of these classes in special cases, see Exerc. 16.

6. If $E$ is an $AL$-space, every order bounded null sequence in $E$ is hypermajorized (since by (1.5), $\mathscr{L}(E, l^1)$ is a Banach lattice), and this property characterizes $AL$-spaces (Exerc. 16). Generally, every summable sequence in a Banach space is hypermajorized (Exerc. 12).

On the other hand, in every infinite dimensional Banach space $G$ there exist null sequences which are not hypermajorized; otherwise, the identity map $1_G$ would be left semi-integral by (5.12) and hence, by (5.10) Cor. 3, have a nuclear adjoint which is impossible.

Finally, if $F$ is a Banach lattice having property $(P)$ (Def. 4.2), then every hypermajorized sequence in $F$ is order bounded (Exerc. 12).

7. A very surprising result in operator theory asserts that *every continuous linear map of an $AL$-space into a Hilbert space is absolutely summing*; for a

discussion of this and related facts, the interested reader is referred to Exerc. 15. It is much easier to prove that the canonical map $l^1 \to l^2$ is absolutely summing (Exerc. 14 (e)).

Finally, if $E, F$ are Banach lattices and $F$ has property $(P)$, the following diagram (where vertically opposed properties are dual to each other) illustrates the implications among the properties of linear maps $E \to F$ considered in this section, with obvious abbreviations:

| | | abs. summing | ⇒ | cone abs. summing | | |
|---|---|---|---|---|---|---|
| integral | ⇗ ⇘ | | | | ⇘ ⇗ | regular |
| | | hypermaj. | ⇒ | majorizing | | |

We note that if $E$ is an $AM$-space and $F$ an $AL$-space, then by (5.6) all of these concepts are coextensive; in general, they are all distinct (see Examples above).

## § 6. Hilbert-Schmidt Operators and Hilbert Lattices

In the present section, the symbol $H$ will generally denote a Hilbert space with inner product (canonical sesquilinear form) $(x,y) \mapsto (x|y)$. It will be sufficient for our purposes to consider Hilbert spaces over $\mathbb{R}$; however, extension of all results of this section to complex scalars presents no difficulties (Exerc. 18). We recall that the Banach dual $H'$ of a Hilbert space $H$ is usually identified with $H$ by virtue of the fact that each element $f \in H'$ is given by a unique $x_f \in H$ through the identity $\langle x, f \rangle = (x|x_f)$ $(x \in H)$; the mapping $f \mapsto x_f$ is an isomorphism (a conjugate linear isometry if the scalar field is $\mathbb{C}$) of $H'$ onto $H$. In view of this identification, the *adjoint* $T^*$ of a continuous linear map $T: H_1 \to H_2$ is the mapping of $H_2$ into $H_1$ defined by the identity $(Tx|y) = (x|T^*y)$ for all $(x,y) \in H_1 \times H_2$. It is well known and easy to verify that $\|T \circ T^*\| = \|T^* \circ T\| = \|T\|^2$ for all $T \in \mathscr{L}(H_1, H_2)$. If $H_1 = H_2$, then an operator $T \in \mathscr{L}(H_1, H_2)$ is called *Hermitian* if $T = T^*$.

To characterize Hilbert-Schmidt and nuclear operators in $\mathscr{L}(H_1, H_2)$, we need the spectral theorem for compact Hermitian operators; for the convenience of the reader a condensed proof is included.

**6.1 Proposition.** *Let $T: H \to H$ be a non-zero compact Hermitian operator. There exists a finite or null sequence $(\lambda_i)$ of real numbers $\neq 0$ and a corresponding sequence $(P_i)$ of mutually annihilating Hermitian projections of finite rank, such that*

$$T = \sum_i \lambda_i P_i$$

*where the series converges in the operator norm. Moreover, $\|T\| = \sup_i |\lambda_i|$.*

*Proof.* Since $T$ is Hermitian and $H$ real, an easy computation shows that $(Tx|y) = \frac{1}{4}[(Tx+Ty|x+y) - (Tx-Ty|x-y)]$; hence, since $\|T\| = \sup\{\|Tx\| : \|x\| \leq 1\} = \sup\{(Tx|y) : \|x\| \leq 1,\ \|y\| \leq 1\}$, it follows that $\|T\| = \sup\{|(Tx|x)| : \|x\| \leq 1\}$.

Thus there exists a sequence $(x_n)$ in $H$, $\|x_n\|=1$, with the property that $\lim_n (Tx_n|x_n) = \varrho$ where $|\varrho|=\|T\|$. Since

$$0 \leqq \|Tx_n - \varrho x_n\|^2 \leqq \|T\|^2 + \varrho^2 - 2\varrho(Tx_n|x_n)$$

for all $n$, we conclude that $\lim_n \|Tx_n - \varrho x_n\| = 0$. Since $T$ is compact, $(Tx_n)$ contains a subsequence convergent to some $y \in H$, and since $\varrho \neq 0$, the corresponding subsequence of $(x_n)$ converges to $\varrho^{-1}y$. Therefore, we have $\varrho y = Ty$, $\|y\| = |\varrho|$, and it follows that at least one of the numbers $\|T\|$ and $-\|T\|$ is an eigenvalue of $T$.

Now let $H(\lambda) := \{x \in H: \lambda x = Tx\}$ for each $\lambda \in \mathbb{R}$. It is easy to see that $H(\lambda_1)$ is (inner product) orthogonal[1] to $H(\lambda_2)$ whenever $\lambda_1 \neq \lambda_2$; moreover, the compactness of $T$ implies $H(\lambda)$ to be finite dimensional for all $\lambda \neq 0$. Let $\Lambda$ denote the set of all $\lambda \in \mathbb{R}$ for which $\lambda \neq 0$ and $H(\lambda) \neq \{0\}$; by the preceding $\Lambda$ is non-void, and clearly bounded (note that $0 \neq x \in H(\lambda)$ implies $|\lambda| \leqq \|T\|$). Moreover, $\Lambda$ is either finite, or else countably infinite with 0 its only accumulation point: In fact, if $\lim_n \lambda_n \neq 0$ for some infinite sequence $(\lambda_n)$ of distinct elements of $\Lambda$, then there exists a bounded sequence of elements $x_n \in H(\lambda_n)$ satisfying $|\lambda_n|\,\|x_n\| = 1$ and, because of $\lambda_n x_n = Tx_n$, the sequence $(Tx_n)$ has no convergent subsequence. Therefore, $\Lambda$ is the range of a finite or null sequence $(\lambda_i)$ of distinct real numbers.

If $H_1 := \bigoplus_i H(\lambda_i)$ denotes the closure in $H$ of the vector subspace generated by the family $\{H(\lambda_i): \lambda_i \in \Lambda\}$ and if $H_0$ denotes the orthogonal complement of $H_1$, then $T$ leaves $H_0$ and $H_1$ invariant. By the first part of the proof and the construction of $H_1$, $T$ must vanish on $H_0$. Consequently, if $P_i$ denotes the Hermitian projection of $H$ onto $H(\lambda_i)$ $(\lambda_i \in \Lambda)$, then the $P_i$ are of finite (not necessarily uniformly bounded) rank and we obtain $T = \sum_i \lambda_i P_i$, where the series converges in the operator norm because $(\lambda_i)$ (if infinite) is a null sequence. Finally, it is clear that $\|T\| = \sup_i |\lambda_i|$. □

The representation $T = \sum_i \lambda_i P_i$ of (6.1) is unique if, as in the proof above, the eigenvalues $\lambda_i$ of $T$ are required to be distinct; if this requirement is not made, the projections belonging to one and the same eigenvalue $\lambda$ of $T$ can be chosen to yield any orthogonal decomposition of the eigenspace $H(\lambda)$. In particular, if $\{e_j: j \in \mathsf{H}_i\}$ is an orthonormal basis of $H(\lambda_i)$ and if $\mu_j := \lambda_i$ for $j \in \mathsf{H}_i$, the spectral representation of $T$ takes the form

$$(\star) \qquad T = \sum_j \mu_j (e_j \otimes e_j),$$

where $e_j \otimes e_j$ denotes the Hermitian projection of rank 1 given by $x \mapsto (x|e_j)e_j$ $(x \in H)$. In the sequel we will again write $(\lambda_i)$, not requiring these numbers to be necessarily distinct. Finally, the reader will note that $T$ is *Hermitian positive* (i.e., $(Tx|x) \geqq 0$ for all $x \in H$) iff all eigenvalues of $T$ are non-negative.

Now it is clear from the definition of the adjoint operator that for each continuous $T: H_1 \to H_2$, the compositions $T^* \circ T$ and $T \circ T^*$ are Hermitian positive operators on $H_1$ and $H_2$, respectively; in particular, these operators are compact

[1] Unless specific reference is made to a lattice structure, the terms *orthogonal*, *orthonormal* refer to inner products throughout this section.

if (and, as the proof of (6.2) will show, only if) $T$ is compact. This observation leads to the following useful representation of compact linear operators between Hilbert spaces.

**6.2 Proposition.** *Let $H_1, H_2$ denote Hilbert spaces, and let $T: H_1 \to H_2$ be any non-zero compact linear map. There exists a finite or null sequence $(\lambda_j)$ of real numbers $\neq 0$ and corresponding orthonormal sequences $(e_j)$ in $H_1$ and $(f_j)$ in $H_2$ such that $T$ has the representation*

$$T = \sum \lambda_j (e_j \otimes f_j),$$

*where the series converges in the operator norm. Moreover, $\|T\| = \sup_j |\lambda_j|$.*

*Proof.* $T^* \circ T \in \mathscr{L}(H_1)$ is Hermitian positive and compact; so, by (6.1), $T^* \circ T$ has a representation $\sum_j \mu_j (e_j \otimes e_j)$ (see (⋆) above) where $(e_j)$ is a (finite or infinite) orthonormal sequence in $H_1$, $\mu_j > 0$ for all $j$, and $(\mu_j)$—if infinite—is a null sequence. We define $\lambda_j := \sqrt{\mu_j} > 0$ and let $f_j := \lambda_j^{-1} T e_j$ for all $j$. An easy computation shows that $(f_j | f_k) = \lambda_j^{-1} \lambda_k^{-1} (T e_j | T e_k) = \delta_{jk}$, so $(f_j)$ is an orthonormal sequence in $H_2$. Moreover, if $x \in H_1$ is in the kernel of $T^* \circ T$ then $(T^* \circ T x | x) = \|Tx\|^2 = 0$; hence, $T$ vanishes on the kernel of $T^* \circ T$ (and conversely). Therefore, it follows that

$$Tx = T\left(\sum_j (x | e_j) e_j\right) = \sum_j \lambda_j (x | e_j) f_j$$

for all $x \in H_1$, which yields the desired representation of $T$. Moreover, the series $\sum_j \lambda_j (e_j \otimes f_j)$ converges in the operator norm because $(\lambda_j)$ (if infinite) is a null sequence, and it is clear from the final assertion of (6.1) and the relation $\|T^* \circ T\| = \|T\|^2$ that $\|T\| = \sup_j |\lambda_j|$. ☐

*Remark.* The above representation of $T$ is unique in the following sense: Whenever a compact operator $T: H_1 \to H_2$ is represented as specified in (6.2), the sequence $(\lambda_j^2)$ is the sequence of all non-zero eigenvalues of both $T^* \circ T$ and $T \circ T^*$, each occurring as frequently as the corresponding multiplicity (dimension of the eigenspace) indicates; and the sequence $(e_j)$ (respectively, $(f_j)$) is a complete orthonormal sequence of eigenvectors of $T^* \circ T$ (respectively, of $T \circ T^*$) pertaining to non-zero eigenvalues.

Conversely, if $(e_j)_{j \in \mathbb{N}}$ and $(f_j)_{j \in \mathbb{N}}$ are orthonormal sequences in $H_1$ and $H_2$, respectively, and if $(\lambda_j) \in c_0$, then the series $\sum_j \lambda_j (e_j \otimes f_j)$ converges in the operator norm to a compact operator $T \in \mathscr{L}(H_1, H_2)$; thus, if $H_1$ and $H_2$ are infinite dimensional, there exist Banach subspaces of $\mathscr{L}(H_1, H_2)$ isomorphic to $c_0$ (Exerc. 18). Through the representation (6.2), nuclear operators $H_1 \to H_2$ (cf. discussion preceding (5.5)) are characterized in the following elegant manner.

**Corollary.** *Let $T: H_1 \to H_2$ be a compact operator represented as specified in* (6.2). *$T$ is nuclear if and only if the sequence $(\lambda_j)$ is summable; in this case, $\sum_j |\lambda_j|$ is the nuclear norm of $T$.*

*Proof.* If $\sum_j |\lambda_j| < +\infty$ then, clearly, $T$ is nuclear. Conversely, if $T$ is nuclear then by definition (§ 5) there exist sequences $(x_n)$ in $H_1$ and $(y_n)$ in $H_2$ such that

$T=\sum_n x_n \otimes y_n$ and $\sum_n \|x_n\| \, \|y_n\| < +\infty$. Clearly, $T$ is compact; if $T$ is represented as in (6.2), it follows that

$$\sum_j |\lambda_j| = \sum_j |(Te_j | f_j)| \leqq \sum_{j,n} |(x_n | e_j)(y_n | f_j)| \leqq \sum_n \|x_n\| \, \|y_n\| .$$

Since the nuclear norm of $T$ is the infimum of $\sum_n \|x_n\| \, \|y_n\|$ over all of these representations $T=\sum_n x_n \otimes y_n$, it equals $\sum_j |\lambda_j|$ and the representation of $T$ given by (6.2) is optimal in this respect. □

A wider class of compact operators between Hilbert spaces than the nuclear ones are the so-called Hilbert-Schmidt operators; these operators, given by square integrable kernels when defined between spaces $L^2$, bear a much closer relationship to the Hilbert structure of their domain and range (cf. 6.4 and 6.5 below). Their appearance in the theory of linear integral equations goes back to the classical theories of D. Hilbert and E. Schmidt. The abstract definition of these operators is as follows.

**6.3 Definition.** *Let $H_1, H_2$ be Hilbert spaces. A linear map $T: H_1 \to H_2$ is called a* Hilbert-Schmidt operator *(abbreviated H.S. operator) if there exists an orthonormal basis $(e_\alpha)_{\alpha \in \mathrm{A}}$ of $H_1$ such that $\sum_{\alpha \in \mathrm{A}} \|Te_\alpha\|^2$ is finite.*

We observe first that the sum occurring in Def. 6.3 is independent of the particular orthonormal basis of $H_1$ chosen; in fact, if $(e'_\beta)_{\beta \in \mathrm{B}}$ is any orthonormal basis of $H_2$ then

$$\sum_\alpha \|Te_\alpha\|^2 = \sum_{\alpha,\beta} |(Te_\alpha | e'_\beta)|^2 = \sum_{\alpha,\beta} |(e_\alpha | T^* e'_\beta)|^2 = \sum_\beta \|T^* e'_\beta\|^2$$

by Parseval's equality; in particular, $T$ is H.S. if and only if $T^*$ is, and the respective sums of squares are equal. It is clear that the set of all H.S. operators $H_1 \to H_2$ is a vector subspace $\mathscr{H}(H_1, H_2)$ of $\mathscr{L}(H_1, H_2)$ (and, if $H_1 = H_2$, a self-adjoint subalgebra of $\mathscr{L}(H)$). Now let $(e_\alpha)_{\alpha \in \mathrm{A}}$ be any fixed orthonormal basis of $H_1$ and let $S, T \in \mathscr{H}(H_1, H_2)$. An application of the inequality of Schwarz-Buniakovski shows that the number

$$(S | T) := \sum_{\alpha \in \mathrm{A}} (Se_\alpha | Te_\alpha) \tag{1}$$

is well defined (and actually independent of the chosen basis $(e_\alpha)$ of $H_1$), and it is not difficult to see that the mapping $(S, T) \mapsto (S | T)$ thus defined is an inner product on $\mathscr{H}(H_1, H_2)$. Endowed with this inner product, $\mathscr{H}(H_1, H_2)$ is a Hilbert space in which $H_1 \otimes H_2$ is dense; the norm of $\mathscr{H}(H_1, H_2)$ derived from (1) is called the *Hilbert-Schmidt norm.*

For a characterization of Hilbert-Schmidt operators in connection with (6.2), we need the following simple but important result.

**6.4 Proposition.** *The composition of two Hilbert-Schmidt operators is nuclear.*

*Proof.* If $T: H_1 \to H_2$ and $T_2: H_2 \to H_3$ are continuous linear maps and $(e_\alpha)_{\alpha \in \mathrm{A}}$ is any orthonormal basis of $H_2$, then for each $x \in H_1$ the value of the composition $T_2 \circ T_1$ is given by

$$T_2(T_1 x) = \sum_\alpha (T_1 x | e_\alpha) T_2 e_\alpha = \sum_\alpha (x | T_1^* e_\alpha) T_2 e_\alpha ,$$

that is, the family $(T_1^* e_\alpha \otimes T_2 e_\alpha)_{\alpha\in A}$ is summable for the strong operator topology of $\mathscr{L}(H_1, H_3)$ with sum $T_2 \circ T_1$. Now if $T_1$ and $T_2$ are H.S. operators, this family is even absolutely summable for the norm topology of $\mathscr{L}(H_1, H_3)$: In fact, $\sum_\alpha \|T_1^* e_\alpha\| \|T_2 e_\alpha\|$ is finite ($\leqq$ the product of the respective H.S. norms of $T_1$ and $T_2$) and hence, $T_2 \circ T_1$ is nuclear. □

**6.5 Proposition.** *Let $H_1, H_2$ denote Hilbert spaces. For any $T \in \mathscr{L}(H_1, H_2)$, the following assertions are equivalent:*

(a) *$T$ is a Hilbert-Schmidt operator.*

(b) *There exist orthonormal sequences $(e_j)$ in $H_1$ and $(f_j)$ in $H_2$ such that $T = \sum_j \lambda_j (e_j \otimes f_j)$, where $(\lambda_j) \in l^2$ and the series converges in the Hilbert-Schmidt norm.*

(c) *If $H_1 = L^2(\mu)$ and $H_2 = L^2(\nu)$ over σ-finite measure spaces, $T$ is given by a kernel $K \in L^2(\mu \otimes \nu)$ through*

$$Tf(t) = \int K(s,t) f(s) d\mu(s),$$

*the formula holding a.e. $(\nu)$ for each $f \in L^2(\mu)$.*

*Proof.* First we note that each $T \in \mathscr{H}(H_1, H_2)$ is compact, which is easy to verify.

(a)⇒(b): Since $T$ is compact, $T$ has the desired representation by (6.2) and it is clear that $T^* \circ T = \sum_j \lambda_j^2 (e_j \otimes e_j)$. But by (6.4) $T^* \circ T$ is nuclear, hence we have $(\lambda_j) \in l^2$ by the corollary of (6.2). Moreover, if we let $T_n := \sum_{j \leqq n} \lambda_j (e_j \otimes f_j)$ it follows from the definitions that $T = \lim_n T_n$ in the Hilbert-Schmidt norm (in particular, $H_1 \otimes H_2$ is dense in the Hilbert space $\mathscr{H}(H_1, H_2)$).

(b)⇒(c): If $(e_j)$ and $(f_j)$ are orthonormal sequences in $L^2(\mu)$ and $L^2(\nu)$, respectively, then, since $(\lambda_j) \in l^2$, the series $\sum_j \lambda_j e_j(s) f_j(t)$ square mean converges in $L^2(\mu \otimes \nu)$ and hence, its sum defines $K \in L^2(\mu \otimes \nu)$ with the desired properties.

(c)⇒(a): Let $(e_\alpha)_{\alpha\in A}$ denote any orthonormal basis of $L^2(\mu)$ and let $(e'_\beta)_{\beta\in B}$ denote any orthonormal basis of $L^2(\nu)$. It is well known from measure theory that $(e_\alpha \otimes e'_\beta)_{(\alpha,\beta)\in A\times B}$ forms an orthonormal basis of $L^2(\mu \otimes \nu)$. Now if $T$ is given by a kernel $K$ as specified, then by Parseval's equality we have

$$\begin{aligned}\sum_\alpha \|Te_\alpha\|^2 &= \sum_{\alpha,\beta} |(Te_\alpha | e'_\beta)|^2 \\ &= \sum_{\alpha,\beta} \left|\int\int K(s,t) e_\alpha(s) e'_\beta(t) d\mu(s) d\nu(t)\right|^2 = \int K^2 d(\mu \otimes \nu).\end{aligned}$$

This proves (a). □

*Remark.* The unicity of the representation (b) is valid to the same extent as for compact operators $H_1 \to H_2$ (see Remark after (6.2)) and, of course, $(\sum_j \lambda_j^2)^{1/2}$ is the H.S. norm of $T$. In the more special situation of (c), (the class of) $K$ is determined uniquely by $T$ and, in fact, $L^2(\mu \otimes \nu)$ is isomorphic to $\mathscr{H}(H_1, H_2)$.

**Corollary.** *Every Hilbert-Schmidt operator is absolutely summing (equivalently, right semi-integral).*

*Proof.* Let $T$ be given as specified in (6.5) (b); without loss of generality we can assume that $(e_j)$ and $(f_j)$ are orthonormal bases of $H_1$ and $H_2$, respectively. Identifying each of $H_1$ and $H_2$ with $l^2$, the mapping $T$ takes the form $(\xi_j)\mapsto(\lambda_j\xi_j)$; however, since $(\lambda_j)\in l^2$, we actually have $(\lambda_j\xi_j)\in l^1$. Thus $T$ has a factoring $l^2\to l^1\to l^2$ where the mapping $l^1\to l^2$ is the canonical imbedding; since the latter is absolutely summing (Exerc. 14 (e)), the assertion follows. □

The main purpose of the present section is to investigate the relationship of the class of H.S. operators to the classes of linear maps introduced in Sections 3—5 above, in the case where the domain and/or range spaces are simultaneously Hilbert spaces and Banach lattices. To avoid misunderstanding, let us agree on this definition.

**6.6 Definition.** *By a* Hilbert lattice *we understand a Banach lattice whose underlying Banach space is a Hilbert space.*

Well known Hilbert lattices are the Banach lattices $l^2(\mathsf{A})$ ($\mathsf{A}$ any index set) and $L^2(\mu)$ ($(X,\Sigma,\mu)$ any measure space). As the example of a diffuse (atom-free) measure $\mu$ shows, a Banach lattice $L^2(\mu)$ is not necessarily isomorphic to any $l^2(\mathsf{A})$. It is thus a natural question to ask if there exist any types of Hilbert lattices other than $l^2(\mathsf{A})$, $L^2(\mu)$ ($\mu$ diffuse) and their products. The answer is negative (see the corollary below); this is not trivial since, in particular, it is not a priori clear if for a Hilbert lattice $H$, the canonical isomorphism of the dual $H'$ onto $H$ is an isomorphism of lattices. With the aid of the representation theory given in Chap. III, §§ 4—5, we can prove the following result. (To avoid ambiguity, we will exclusively use the term *disjoint* when referring to lattice orthogonality in the present context.)

**6.7 Theorem.** *For any Hilbert lattice $H$, there exists a locally compact space $X$ and a strictly positive Radon measure $\mu$ on $X$ such that $H$ is isomorphic to $L^2(\mu,X)$. Moreover, $X$ and $\mu$ can be chosen so that $L^2(\mu,X)$ can be identified with the vector lattice of all $\overline{\mathbb{R}}$-valued, continuous functions on $X$ which are square integrable with respect to $\mu$.*

*Proof.* We suppose first that the Banach lattice $H$ contains quasi-interior positive elements (Def. II.6.1). Then by (III.4.5), the underlying vector lattice of $H$ can be identified with a vector lattice $\hat{H}$ of continuous, $\overline{\mathbb{R}}$-valued functions on the structure space $K$ of $H$; the compact space $K$ is Stonian (III.5.6), since $H$ is order complete (II.5.11). Carrying the norm of $H$ over to $\hat{H}$, we define a set function $\mu$ on the Boolean algebra $\mathbf{Q}$ of open-and-closed subsets of $K$ by virtue of

$$\mu(U):=\|\chi_U\|^2 \qquad (U\in\mathbf{Q}),$$

where $\chi_U$ denotes the characteristic function of $U$. If $U,V\in\mathbf{Q}$ are disjoint then $|\chi_U+\chi_V|=|\chi_U-\chi_V|$ and hence, since the norm of $\hat{H}$ is a lattice norm, we have $\|\chi_U+\chi_V\|=\|\chi_U-\chi_V\|$. This implies $(\chi_U|\chi_V)=0$, and so $\|\chi_U+\chi_V\|^2=\|\chi_U\|^2+\|\chi_V\|^2$; that is, $\mu$ is additive on $\mathbf{Q}$. The linear form defined by $\mu$ on the linear hull $L$ of $\{\chi_U\colon U\in\mathbf{Q}\}$ is continuous for the uniform norm (note that $L$ is a dense vector sublattice of $C(K)$ containing the unit of $C(K)$); its unique continuous extension to $C(K)$ is a strictly positive Radon measure which we denote again by $\mu$.[1] Now each $f \in L$ is of the form $f = \sum_i \alpha_i \chi_{U_i}$ where the sets $U_i \in Q$ can be assumed to be disjoint; thus we obtain

$$\|f\|^2 = \|\sum_i \alpha_i \chi_{U_i}\|^2 = \sum_i \alpha_i^2 \|\chi_{U_i}\|^2 = \sum_i \alpha_i^2 \mu(U_i) = \int f^2 \, d\mu .$$

This shows that the norm of $\hat{H}$ carried over from $H$ induces on $L$ the norm of $L^2(\mu, K)$; consequently, since $C(K)$ is a dense ideal of $\hat{H}$, the elements of $\hat{H}$ are precisely the continuous numerical functions on $K$ square integrable for $\mu$. Thus $\hat{H}$ can be identified with $L^2(\mu, K)$ (and $C(K)$ with $L^\infty(\mu)$); in fact, since $\mu$ is strictly positive on $K$, each class $\in L^2(\mu)$ contains precisely one continuous numerical function.

To complete the proof, we observe that any maximal lattice disjoint family $(u_\alpha)_{\alpha \in \mathsf{A}}$ in $H$ is topological (Chap. III, § 5, Example 2). Let $H_\alpha$ denote the band of $H$ generated by $u_\alpha$ $(\alpha \in \mathsf{A})$ and let $K_\alpha$ denote the structure space of $H_\alpha$. Applying the preceding part of the proof to each $H_\alpha$ and employing (III.5.2), we obtain the desired representation of $H$ for $X$ the (locally compact) direct sum $\sum_\alpha K_\alpha$ and for $\mu$ the Radon measure on $X$ for which $\mu(U) = \|\chi_U\|^2$ whenever $U$ is a compact open subset of $X$. □

**Corollary.** *Let $H$ be any Hilbert lattice. There exists a discrete space $\mathsf{A}$, a locally compact space $X$ and a diffuse Radon measure $\mu$ on $X$ such that $H$ is isomorphic to the Hilbert and lattice direct sum $l^2(\mathsf{A}) \oplus L^2(\mu, X)$. If $X \cup \mathsf{A}$ is required to be a strong representation space for $H$* (Def. III.5.4), *this representation of $H$ is essentially unique.*

*Proof.* Denote by $H_a$ the band of $H$ generated by all atoms (cf. Chap. II, § 3 and Exerc. 7); since $H$ is order complete (II.5.11), $H = H_a + H_a^\perp$ is a direct sum of closed ideals. It is clear that $H_a$ is isomorphic to $l^2(\mathsf{A})$ where card $\mathsf{A}$ is unique (for example, $\mathsf{A}$ can be taken to be the set of all atoms of $H$); moreover, if (6.7) is applied to $H_a^\perp$ (note that $H_a$ and $H_a^\perp$ are inner product orthogonal) then a representation $H_a^\perp \cong L^2(\mu, X)$ results where $\mu$ is diffuse (non-atomic). Thus $H$ has the claimed representation, and if $X$ is a strong representation space for $H_a^\perp$ then $X$ is essentially unique (and hence $X \cup \mathsf{A}$ is essentially unique) in the sense specified in (III.5.3). □

We note in passing that an analogous representation can be obtained for any Banach lattice $E$ whose norm is $p$-additive $(1 \leqq p < +\infty$; cf. Chap. II, Exerc. 23).

The surprising fact is now that for any pair of Hilbert lattices $H_1, H_2$ the Hilbert-Schmidt operators $H_1 \to H_2$ coincide with the cone absolutely summing and majorizing (and hence, with the absolutely summing and hypermajorizing) linear maps from $H_1$ into $H_2$. From the proof of the main theorem we isolate this essential lemma.

**6.8 Lemma.** *Let $H$ be a Hilbert space and let $(Y, \mathsf{T}, \nu)$ be a measure space. Every majorizing linear map $T: H \to L^2(\nu)$ is a Hilbert-Schmidt operator and the m-norm of $T$* (§ 3, Formula (4')) *agrees with the Hilbert-Schmidt norm of $T$.*

[1] For an alternative approach using the Carathéodory extension of the set function $\mu$, see the Note after (II.8.5).

*Proof*. By (6.7) we can assume that $Y$ is locally compact, $\nu$ is a strictly positive Radon measure on $Y$, and that $L^2(\nu)$ is the space of all continuous functions $Y \to \overline{\mathbb{R}}$ square integrable with respect to $\nu$.

Since $T$ is majorizing, by (3.4) there exists $k \in L^2(\nu)$ such that $|Tx| \leqq k$ for all $x \in H$, $\|x\| \leqq 1$. Denoting by $(e_\alpha)_{\alpha \in \mathsf{A}}$ an orthonormal basis of $H$, we consider any finite subset $\mathsf{H} \subset \mathsf{A}$ and define $U := \{s \in Y : \sum_{\alpha \in \mathsf{H}} |Te_\alpha|(s) < +\infty\}$; since $\nu$ is strictly positive and each $Te_\alpha$ continuous, $U$ is a dense open subset of $Y$. For $s \in U$, we let $N(s) := (\sum_{\alpha \in \mathsf{H}} |Te_\alpha|(s)^2)^{1/2}$; $s \mapsto N(s)$ is continuous on $U$. Now for $t \in U$ and $\alpha \in \mathsf{H}$ we define $c_\alpha(t)$ by $c_\alpha(t) := Te_\alpha(t)/N(t)$ if $N(t) > 0$ and by $c_\alpha(t) := 0$ if $N(t) = 0$. Then for each $t \in U$, the element $x_t := \sum_{\alpha \in \mathsf{H}} c_\alpha(t) e_\alpha$ is in the unit ball of $H$; by the defining property of $k$ we have

$$|\textstyle\sum_{\alpha \in \mathsf{H}} Te_\alpha(t)\, Te_\alpha(s)| \leqq k(s)\, N(t)$$

whenever $(s,t) \in U \times U$. Letting $s = t$ we obtain $\sum_{\alpha \in \mathsf{H}} |Te_\alpha|(s)^2 \leqq k(s)^2$ for all $s \in U$, and by continuity this holds throughout $Y$. Integration of the latter inequality with respect to $\nu$ yields

$$\textstyle\sum_{\alpha \in \mathsf{H}} \|Te_\alpha\|^2 \leqq \|k\|^2 . \tag{2}$$

Since $\mathsf{H}$ was an arbitrary finite subset of $\mathsf{A}$, it follows that $T$ is a Hilbert-Schmidt operator.

It remains to prove that the $m$-norm $\|T\|_m$ agrees with the H.S. norm of $T$. Since $k$ was an arbitrary majorant of $\{Tx : \|x\| \leqq 1\}$ (cf. 3.4), (2) shows the H.S. norm of $T$ to be $\leqq \|T\|_m$. To prove the converse inequality, we identify $H$ with a suitable $L^2(\mu)$; since $T$ is compact, the domain and range of $T$ can be assumed to be separable and hence, $\mu$ and $\nu$ $\sigma$-finite. Using the representation (6.5)(c) of $T$ and defining $k \in L^2(\nu)$ by

$$k(s) := (\textstyle\int K(t,s)^2\, d\mu(t))^{1/2} \qquad (s \in Y),$$

we obtain $|Tf| \leqq k\|f\|$ for all $f \in L^2(\mu)$ so that $\|T\|_m \leqq \|k\|$. But $\|k\|$ is the Hilbert-Schmidt norm of $T$ (see Remark after (6.5)), and this completes the proof. □

**6.9 Theorem.** *Let $H_1, H_2$ be Hilbert spaces and let $T \in \mathscr{L}(H_1, H_2)$. In addition, consider any order structures on $H_1$ and $H_2$ under which both are Hilbert lattices. The following assertions are equivalent:*

(a) *$T$ is a Hilbert-Schmidt operator.*
(b) *The modulus $|T|$ exists (in the canonical order of $\mathscr{L}(H_1, H_2)$) and is a Hilbert-Schmidt operator.*
(c) *$T$ is cone absolutely summing.*
(c′) *$T$ is majorizing.*
(d) *$T$ is absolutely summing.*
(d′) *$T$ is hypermajorizing.*

*Proof*. Since $T$ is a H.S. operator if and only if its adjoint $T^*$ is (see above), the equivalences (c)⇔(c′) and (d)⇔(d′) are clear from (3.8) and (5.12), respectively, once we have shown that (a) is equivalent to (c) and (d). But (a)⇒(d) is (6.5) Cor., (d)⇒(c) is trivial, and (c)⇒(a) follows via (3.8) and (6.8) which shows that $T^*$ is a H.S. operator.

Finally, since $H_2$ is reflexive (and hence, as a Banach lattice has property $(P)$ of Def. 4.2) it follows from (4.3) and the implication (c)⇒(a) already proved, that (c)⇒(b). Similarly, (b)⇒(c) which ends the proof. □

It should be noted that for fixed Hilbert spaces $H_1, H_2$ and a fixed $T: H_1 \to H_2$, the modulus $|T|$ (whenever it exists) depends on the orderings of $H_1$ and $H_2$ considered. Thus if $T$ is a H.S. operator represented according to (6.5)(b) with all $\lambda_i > 0$, and if the orderings of $H_1$ and $H_2$ are such that $(e_j)$ and $(f_j)$ are positive sequences, then $T \geqq 0$; however, $T$ is not positive for other (Hilbert lattice) orderings of $H_1$ and $H_2$.

Since a H.S. operator $T$ and its adjoint $T^*$ have the same Hilbert-Schmidt norm, it follows from (3.8) and (6.8) that the H.S. norm of $T$ agrees with $\|T\|_m$ and $\|T\|_l$. Thus, in conjunction with (4.3), we obtain from (6.9) this result.

**6.10 Proposition.** *Let $H_1, H_2$ be Hilbert lattices. Then the Banach lattices $\mathscr{L}^l(H_1, H_2)$ and $\mathscr{L}^m(H_1, H_2)$ are identical with $\mathscr{H}(H_1, H_2)$ which is a Hilbert lattice under its canonical order and the Hilbert-Schmidt norm; in particular, $\mathscr{H}(H_1, H_2)$ is an ideal of $\mathscr{L}^r(H_1, H_2)$.*

Finally, since every nuclear $T: H_1 \to H_2$ is a H.S. operator, (6.9) implies that $|T|$ exists and is a H.S. operator; it does not seem to be known if $|T|$ is necessarily nuclear. However, it can be shown that in general, the nuclear operators $H_1 \to H_2$ do not constitute a (lattice) ideal of $\mathscr{H}(H_1, H_2)$ (Exerc. 19).

## § 7. Tensor Products of Banach Lattices

Let $E, F$ denote Banach lattices and let $G, H$ denote Banach spaces. We identify the tensor product $E \otimes H$ with a vector subspace of $\mathscr{L}(E', H)$ (cf. § 2, Example 5); under this identification, the elements of $E \otimes H$ are precisely the linear maps $E' \to H$ of finite rank which are continuous for the weak topologies $\sigma(E', E)$ and $\sigma(H, H')$, respectively. Clearly, $E \otimes H$ is contained in $\mathscr{L}^l(E', H)$ (§ 3, Example 2) and, as we have observed in § 4, the $l$-norm (§ 3, Formula (3′)) is a cross-norm on $E \otimes H$; that is, $\|x \otimes y\|_l = \|x\| \|y\|$ for each pair $(x, y) \in E \times H$. Similarly, identifying $G \otimes F$ with a vector subspace of $\mathscr{L}(G', F)$ we see that $G \otimes F \subset \mathscr{L}^m(G', F)$ and that the $m$-norm (§ 3, Formula (4′)) is a cross-norm on $G \otimes F$.

**7.1 Definition.** *For Banach lattices $E, F$ and Banach spaces $G, H$ we denote by $E \tilde{\otimes}_l H$ (respectively, by $G \tilde{\otimes}_m F$) the completion of $E \otimes H$ with respect to the $l$-norm (respectively, the completion of $G \otimes F$ with respect to the $m$-norm).*

Thus the Banach space $E \tilde{\otimes}_l H$ can be identified with a Banach subspace of $\mathscr{L}^l(E', H)$ and, similarly, $G \tilde{\otimes}_m F$ can be identified with a Banach subspace of $\mathscr{L}^m(G', F)$ (cf. 3.6); these (canonical) identifications will often be made tacitly in the sequel. From (3.8) it follows that the canonical algebraic isomorphism $E \otimes H \to H \otimes E$ is a norm isomorphism $E \otimes_l H \to H \otimes_m E$ and hence, extends to a norm isomorphism of the respective completions. It might thus appear that the introduction of both of these tensor products is unnecessary; however, we feel that the applications to spaces and lattices of linear maps (see § 8) gain considerably in clarity by the retention of both concepts. We point out that the $l$- and $m$-norms are not tensor norms in the general sense introduced by Grothendieck [1955 a]; however, they have all properties desirable in the order theoretic context considered here (Exerc. 22).

If $E, F$ are ordered vector spaces with respective positive cones $E_+, F_+$ (Def. II.1.2) then the set $C_p := \operatorname{co} E_+ \otimes F_+$ is a (proper) convex cone in $E \otimes F$ (Exerc. 22) and hence, defines an ordering satisfying Axioms $(\mathrm{LO})_1$ and $(\mathrm{LO})_2$ (Chap. II, § 1); $C_p$ is called the *projective* (positive) *cone* of $E \otimes F$. Now if $E, F$ are Banach lattices then, in general, the classical tensor products $E \tilde{\otimes}_\pi F$ and $E \tilde{\otimes}_\varepsilon F$ (§ 2) are not vector lattices (and much less Banach lattices) for the natural orderings defined by the respective closures of $C_p$; the fact that the tensor products $E \tilde{\otimes}_l F$ and $E \tilde{\otimes}_m F$ are Banach lattices under this ordering whenever $E, F$ are, warrants their interest for the theory considered here.

**7.2 Theorem.** *If $E$ and $F$ are Banach lattices, the tensor products $E \tilde{\otimes}_l F$ and $E \tilde{\otimes}_m F$ are Banach lattices under the ordering whose positive cone is the respective closure of the projective cone $C_p = \operatorname{co} E_+ \otimes F_+$. Moreover, the canonical algebraic isomorphism $E \otimes F \to F \otimes E$ extends to an isomorphism of Banach lattices $E \tilde{\otimes}_l F \to F \tilde{\otimes}_m E$.*

*Proof.* In view of Def. 7.1, the first assertion is an almost direct consequence of the proof of (4.6). Under the canonical identification of $E \otimes F$ with a subspace of $\mathscr{L}(E', F)$ and of $F \otimes E$ with a subspace of $\mathscr{L}(F', E)$, the canonical isomorphism $\varrho: E \otimes F \to F \otimes E$ corresponds to the formation of adjoints $u \mapsto u'$; thus, since $\mathscr{L}^m(F', E)$ can be considered a Banach subspace of $\mathscr{L}^m(F', E'')$ (cf. 3.4(d)), it follows from (3.8) that the map $\varrho$ is a norm isomorphism of $E \otimes_l F$ onto $F \otimes_m E$. Since $\varrho$ obviously maps the projective cone of $E \otimes F$ onto the projective cone of $F \otimes E$, it is clear that $\varrho$ extends to an isomorphism of Banach lattices. □

**Examples.** 1. Let $E$ be an $AM$-space, $F$ any Banach lattice. $E'$ is an $AL$-space (II.9.1) so by (3.7), $\mathscr{L}^l(E', F) = \mathscr{L}(E', F)$ and the $l$-norm agrees with the operator norm. It follows that $E \tilde{\otimes}_l F = F \tilde{\otimes}_m E = E \tilde{\otimes}_\varepsilon F$ and hence, $E \tilde{\otimes}_\varepsilon F$ is a Banach lattice; in particular, if $E = C(K)$ is an $AM$-space with unit, then $C(K) \tilde{\otimes}_l F = F \tilde{\otimes}_m C(K)$ is isomorphic to $C(K, F)$ (§ 2, Example 1).

If both $E$ and $F$ are $AM$-spaces then by (4.4), $E \tilde{\otimes}_\varepsilon F$ is an $AM$-space possessing a unit whenever $E$ or $F$ does. (Note, however, that in general $L^\infty(\mu) \tilde{\otimes}_\varepsilon L^\infty(\nu)$ is not isomorphic to $L^\infty(\mu \otimes \nu)$ even if $\mu, \nu$ are finite measures.)

2. Let $E, F$ both be $AL$-spaces. $E'$ is an $AM$-space (II.9.1) so by (5.6), $\mathscr{L}^r(E', F)$ is the Banach lattice of all integral maps $E' \to F$, the integral norm (which agrees with the $l$-, $m$-, and $r$-norms on $\mathscr{L}^r(E', F)$) inducing the $\pi$-norm on $E \otimes F$. Therefore, $E \tilde{\otimes}_l F = E \tilde{\otimes}_m F = E \tilde{\otimes}_\pi F$ and $E \tilde{\otimes}_\pi F$ is an $AL$-space.

In particular, if $E = L^1(\mu)$ where $(X, \Sigma, \mu)$ is any measure space then the Banach space $L^1_F(\mu)$ (which is isomorphic to $L^1(\mu) \tilde{\otimes}_\pi F$; § 2, Example 3) is an $AL$-space under its canonical ordering.

More generally, one has $L^1(\mu)\tilde{\otimes}_l H = H\tilde{\otimes}_m L^1(\mu) = L^1(\mu)\tilde{\otimes}_\pi H$ for any Banach space $H$ (see Example 4 below); in view of (II.8.5), then, it follows that $E\tilde{\otimes}_l H = E\tilde{\otimes}_\pi H$ for any $AL$-space $E$ and any Banach lattice or Banach space $H$.

3. Let $H_1, H_2$ be Hilbert spaces. The bi-sesquilinear form $((x,y),(\bar{x},\bar{y})) \mapsto (x|\bar{x})(y|\bar{y})$ on $(H_1\times H_2)\times(H_1\times H_2)$ induces a positive definite Hermitian form on $H_1\otimes H_2$; the corresponding completion of $H_1\otimes H_2$ is a Hilbert space denoted by $H_1\tilde{\otimes} H_2$. If $H_1\otimes H_2$ is canonically identified with a vector subspace of $\mathscr{L}(H_1, H_2)$, it is easy to verify from (6.5)(b) that $H_1\tilde{\otimes} H_2$ becomes the space of Hilbert-Schmidt operators $H_1\to H_2$ under the Hilbert-Schmidt norm. In particular, if $H_1$ and $H_2$ are Hilbert lattices (Def. 6.6) then by (6.9), (6.10) $H_1\tilde{\otimes} H_2$ is a Hilbert lattice isomorphic to $\mathscr{L}^l(H_1, H_2)$ and to $\mathscr{L}^m(H_1, H_2)$. Consequently, $H_1\tilde{\otimes}_l H_2 = H_1\tilde{\otimes}_m H_2 = H_1\tilde{\otimes} H_2$.

4. Let $(X,\Sigma,\mu)$ denote a measure space and let $H$ denote any Banach space. The vector space $L^p_H(\mu)$ $(1\leq p<+\infty)$ of all (equivalence classes mod $\mu$-null functions of) functions $f: X\to H$ which are $\mu$-measurable[1] and for which $\int\|f\|^p d\mu$ is finite, is a Banach space under the norm $f\mapsto\|f\|_p := (\int\|f\|^p d\mu)^{1/p}$ (see [DS], III.6.6) in which the $\mu$-simple functions are dense.

Let $\varphi$ denote the canonical imbedding $L^p(\mu)\otimes H\to L^p_H(\mu)$, that is, the mapping given by $\varphi(\sum_i f_i\otimes y_i)(s) = \sum_i f_i(s) y_i$ $(s\in X)$; we are going to show that $\varphi$ extends to an isomorphism of Banach spaces $L^p(\mu)\tilde{\otimes}_l H\to L^p_H(\mu)$. To this end it suffices to show that $\|u\|_l = \|\varphi(u)\|_p$ for elements $u=\sum_i \chi_i\otimes y_i$ where the $\chi_i$ are disjoint $\mu$-integrable characteristic functions, since the set of these elements is dense in $L^p(\mu)\tilde{\otimes}_l H$ and its image under $\varphi$ is dense in $L^p_H(\mu)$. Considering $u$ as a linear map $L^q(\mu)\to H$, where $q=p/(p-1)$ $(q=+\infty$ if $p=1)$ is the exponent conjugate to $p$, then by (3.8) the $l$-norm $\|u\|_l$ equals the $m$-norm of the adjoint $H'\to L^p(\mu)$ given by $y'\mapsto\sum_i\langle y_i, y'\rangle\chi_i$; therefore, § 3 Formula (4′) shows that

$$\|u\|_l = \Big\| \sup_{\|y'\|\leq 1} |\textstyle\sum_i\langle y_i, y'\rangle\chi_i| \Big\|.$$

But $|\sum_i\langle y_i, y'\rangle\chi_i| = \sum_i|\langle y_i, y'\rangle|\chi_i$, since the $\chi_i$ were supposed to be lattice disjoint; it is hence clear that $\|u\|_l = \|\sum_i\|y_i\|\chi_i\|_p = \|\varphi(u)\|_p$.

Finally, if $H$ is a Banach lattice $F$ and $L^p_F(\mu)$ is endowed with its canonical order (Chap. II, § 1, Example 3) then $\varphi$ maps the projective cone of $L^p(\mu)\otimes F$ onto a dense subcone of the positive cone of $L^p_F(\mu)$; therefore, $L^p_F(\mu)$ is a Banach lattice and $\varphi$ extends to an isomorphism of Banach lattices $L^p(\mu)\tilde{\otimes}_l F\to L^p_F(\mu)$ $(1\leq p<+\infty)$.

For $p=+\infty$, it is natural to define $L^\infty_H(\mu)$ as the vector space (of equivalence classes mod $\mu$-null functions) of measurable functions $\boldsymbol{g}: X\to H$ uniformly bounded a.e. $(\mu)$, endowed with the norm $\|\boldsymbol{g}\|_\infty := \sup\operatorname{ess}_{s\in X}\|\boldsymbol{g}(s)\|$. $L^\infty_H(\mu)$ is a Banach space and, if $H$ is a Banach lattice, a Banach lattice under

[1] I.e., which are $\mu$-essentially separably valued on integrable sets and such that $y'\circ f$ is measurable for each $y'\in H'$ [DS, III.6.11].

its canonical ordering. However, in general $L^\infty(\mu)\tilde{\otimes}_l H = L^\infty(\mu)\tilde{\otimes}_\varepsilon H$ (Example 1) is a proper closed subspace of $L^\infty_H(\mu)$ (Exerc. 20).

To determine the duals of the Banach spaces $E\tilde{\otimes}_l H$ and $G\tilde{\otimes}_m F$, we need the following representation of these tensor products; to abbreviate we denote by $U_+, V_+$ the positive parts of the respective unit balls $U\subset E$, $V\subset F$. As always, $E_x$ and $F_y$ denote the $AM$-spaces with unit given by principal ideals of $E$ and $F$ (II.7.2 Cor.).

**7.3 Proposition.** *Let $E,F$ denote Banach lattices and $G,H$ Banach spaces; then*

$$E\tilde{\otimes}_l H = \bigcup_{x\in E_+} E_x\tilde{\otimes}_\varepsilon H, \qquad G\tilde{\otimes}_m F = \bigcup_{y\in F_+} G\tilde{\otimes}_\varepsilon F_y .$$

*More precisely: If $U_x$ $(x\in E_+)$ denotes the unit ball of $E_x\tilde{\otimes}_\varepsilon H$, then $\bigcup_{x\in U_+} U$ is contained in the unit ball of $E\tilde{\otimes}_l H$ and contains its interior; an analogous assertion holding for $G\tilde{\otimes}_m F$.*

*Proof.* It is enough, by virtue of (7.2), to prove the assertion concerning $G\tilde{\otimes}_m F$. From (3.4) it is clear that for each $y\in F_+$, $G\tilde{\otimes}_\varepsilon F_y$ (§ 2) is canonically isomorphic to a vector subspace of $\mathscr{L}^m(G',F)$ contained in the $m$-closure of $G\otimes F$, and that $\|u\|_m \leqq 1$ whenever $u$ is in the unit ball of $G\tilde{\otimes}_\varepsilon F_y$ for some $y\in V_+$.

Conversely, suppose $u\in G\tilde{\otimes}_m F$ satisfies $\|u\|_m<1$; we show the existence of $y\in V_+$ such that $u$ is in the unit ball of $G\tilde{\otimes}_\varepsilon F_y$. There exists a $\delta>0$ and a sequence $(u_n)_{n\in\mathbb{N}}$ in $G\otimes F$ such that $(u_n)$ converges to $u$ in the $m$-norm, that $\|u_1\|_m<1-\delta$, and that $\|u_{n+1}-u_n\|_m<2^{-2n}\delta$ for all $n\in\mathbb{N}$. Let $v_n := u_{n+1}-u_n$; considering $v_n$ as a majorizing linear map $G'\to F$, there exists $y_n\in F_+$ such that $\|y_n\|\leqq 2^{-n}\delta$ and $|v_n(x')|\leqq 2^{-n}y_n$ for all $n\in\mathbb{N}$ and $x'\in W^\circ$ ($W^\circ$ the unit ball of $G'$). Moreover, there exists $y_0\in F_+$, $\|y_0\|\leqq 1-\delta$ such that $|u_1(x')|\leqq y_0$ $(x'\in W^\circ)$. Let $y := \sum_{n=0}^\infty y_n$, then $\|y\|\leqq 1$ and the previous estimates imply that

$$|u(x')-u_n(x')| \leqq \textstyle\sum_{k\geqq n} 2^{-k} y_n \leqq \left(\sum_{k\geqq n} 2^{-k}\right) y$$

for all $x'\in W^\circ$. This shows that $\lim_n u_n = u$ in $\mathscr{L}(G',F_y)$ for the uniform operator topology and hence, $u$ is contained in the unit ball of $G\tilde{\otimes}_\varepsilon F_y$. ☐

From (4.6) Cor. 1 we obtain this supplement.

**Corollary.** *If $E,F$ are Banach lattices then $E\tilde{\otimes}_l F = \bigcup_{x\in E_+} E_x\tilde{\otimes}_\varepsilon F$ and the positive part of the $l$-unit ball is the closure of the union of the positive parts of the unit balls of $E_x\tilde{\otimes}_\varepsilon F$, where $x\in U_+$; an analogous assertion holding for $E\tilde{\otimes}_m F$. Moreover, $E_x\tilde{\otimes}_\varepsilon F$ (respectively, $E\tilde{\otimes}_\varepsilon F_y$) can be identified with a vector sublattice of $E\tilde{\otimes}_l F$ (respectively, of $E\tilde{\otimes}_m F$).*

To identify the duals of the $l$- and $m$-tensor products we employ the canonical identification of bilinear forms and linear maps (§ 2); more specifically, if $G,H$ are vector spaces then each linear form $\varphi$ on $G\otimes H$ (or bilinear form $\varphi$ on $G\times H$) defines a linear mapping $T_\varphi: G\to H^*$ by virtue of $\langle x\otimes y,\varphi\rangle = \varphi(x,y) = \langle T_\varphi x, y\rangle$, and conversely.

**7.4 Theorem.** *For arbitrary Banach lattices $E, F$ and Banach spaces $G, H$, the normed dual of $E\tilde{\otimes}_l H$ (respectively, of $G\tilde{\otimes}_m F$) is canonically isomorphic to the Banach space $\mathscr{L}^l(E, H')$ (respectively, to $\mathscr{L}^m(G, F')$).*

*Proof.* Let $\varphi \in (E\tilde{\otimes}_l H)'$ be of norm $\leqq 1$. If $x \in E_+$ and $\|x\| \leqq 1$, then by (7.3) the restriction $\varphi_x$ of $\varphi$ to $E_x \tilde{\otimes}_\varepsilon H$ is continuous and of norm $\leqq 1$. On the other hand, the operator $T_x: E_x \to H'$ defined by $\varphi_x$ is evidently the composition $T_\varphi \circ i_x$: $E_x \to E \to H'$ (cf. notational convention preceding (5.7)); so $T_x$ is integral with integral norm $\leqq 1$ (Def. 5.1). Since this holds for all $x \in E_+$, $\|x\| \leqq 1$, it follows from (5.7)(i) that $T_\varphi \in \mathscr{L}^l(E, H')$ and that $\|T_\varphi\|_l \leqq 1$.

Conversely, if $T \in \mathscr{L}^l(E, H')$ has $l$-norm $\leqq 1$ then by virtue of (5.7)(i) and (7.3), $(x, y) \mapsto \langle Tx, y \rangle$ defines a linear form $\varphi_T$ on $E\tilde{\otimes}_l H$ which is bounded by 1 on the union $\bigcup_{x \in U_+} U_x$ of the respective unit balls $U_x \subset E_x \tilde{\otimes}_\varepsilon H$; since by (7.3) this union contains the interior of the unit ball of $E\tilde{\otimes}_l H$, it follows that $\varphi_T$ is continuous on $E\tilde{\otimes}_l H$ and of norm $\leqq 1$. Since $G\tilde{\otimes}_m F$ is isomorphic to $F\tilde{\otimes}_l G$ (7.2), the assertion concerning $G\tilde{\otimes}_m F$ is obtained noting that by virtue of (3.8), transposition $T \mapsto T'$ is an isomorphism of $\mathscr{L}^l(F, G')$ onto $\mathscr{L}^m(G, F')$. ☐

**Corollary.** *The $l$- and $m$-norms are self-dual with respect to the dualities $\langle E \otimes H, E' \otimes H' \rangle$ and $\langle G \otimes F, G' \otimes F' \rangle$, respectively.*

Since every dual Banach lattice has property (P) (Def. 4.2), $\mathscr{L}^l(E, F')$ and $\mathscr{L}^m(E, F')$ are Banach lattices by (4.3); it is now readily derived from (7.4) that these Banach lattices are duals themselves.

**7.5 Proposition.** *If $E, F$ are Banach lattices, the dual Banach lattices $(E\tilde{\otimes}_l F)'$ and $(E\tilde{\otimes}_m F)'$ are canonically isomorphic to $\mathscr{L}^l(E, F')$ and $\mathscr{L}^m(E, F')$, respectively.*

*Proof.* We restrict attention to the $l$-tensor product. By (7.4), it remains only to prove that the canonical isomorphism $(E\tilde{\otimes}_l F)' \to \mathscr{L}^l(E, F')$ is an order isomorphism (i.e., bi-positive); by virtue of (7.2), we have to show that $T \geqq 0$ in the canonical order of $\mathscr{L}^l(E, F')$ if and only if the linear form $\varphi$ given by $\langle x \otimes y, \varphi \rangle = \langle Tx, y \rangle$ is $\geqq 0$ on the projective cone $C_p$. But by definition of the canonical order of $\mathscr{L}(E, F')$, an operator $T: E \to F'$ is positive if and only if $\langle Tx, y \rangle \geqq 0$ for all $x \in E_+$, $y \in F_+$. ☐

In particular, we note that $\mathscr{L}^l(E, F')$ is isomorphic to $\mathscr{L}^m(F, E')$ under transposition $T \mapsto T'$.

**Example.** 5. Let $(X, \Sigma, \mu)$ be any measure space, let $1 \leqq p < +\infty$, and let $H$ be any Banach space. By Example 4 and by (7.4), the dual $(L_H^p(\mu))'$ is isomorphic to $\mathscr{L}^l(L^p(\mu), H')$, and this isomorphism is an isomorphism of Banach lattices whenever $H$ is a Banach lattice.

In the case of a separable Banach space (or Banach lattice) $H$ (and particularly if $H$ has a separable dual $H'$), the representation of the dual of $L_H^p(\mu)$ $(1 \leqq p < +\infty)$ can be further refined to take an unexpectedly familiar form, by virtue of the following theorem (cf. [DS, VI.8.6]). To make its formulation less cumbersome we introduce the following notation.

For a Banach space $H$ and a measure space $(X,\Sigma,\mu)$, we denote by $\overline{L}^p_{H'}(\mu)$ the Banach space (of equivalence classes mod $\mu$-null functions) of functions $\boldsymbol{g}: X \to H'$ which are $\sigma(H',H)$-measurable (i.e., for which $s \mapsto \langle y, \boldsymbol{g}(s)\rangle$ is measurable whenever $y \in H$) and for which $s \mapsto \|\boldsymbol{g}(s)\|^p$ is $\mu$-integrable (respectively, if $p=+\infty$, for which $\boldsymbol{g}$ is uniformly bounded a.e. $(\mu)$), under the obvious norms (cf. Example 4). Whenever $H$ is a Banach lattice, these spaces are Banach lattices (Exerc. 20) under their canonical orderings.

**7.6 Theorem.** *Let $(X,\Sigma,\mu)$ be a $\sigma$-finite measure space and let $H$ be a separable Banach space. Then $\mathscr{L}(L^1(\mu),H')$ is isomorphic to $\overline{L}^\infty_{H'}(\mu)$ under the correspondence $T \mapsto \boldsymbol{g}$ given by the identity of bilinear forms*

$$\langle Tf,y\rangle = \int f(s)\langle y,\boldsymbol{g}(s)\rangle \, d\mu(s)$$

*on $L^1(\mu)\times H$. This correspondence is an isomorphism of Banach lattices if $H$ is a Banach lattice.*

*Proof*. First, we suppose $L^1(\mu)$ to be separable and $\mu$ finite. Observing that $L^1(\mu)\tilde{\otimes}_\pi H$ is separable, we conclude from [S, IV.1.7] that the unit ball of $\mathscr{B}(L^1(\mu),H)$ is weak* metrizable. Let $T\in\mathscr{L}(L^1(\mu),H')$ be given; we can suppose that $\|T\|=1$. By (2.3) and (2.4) there exists a sequence $(\varphi_n)$ contained in the unit ball of $L^\infty(\mu)\otimes_\varepsilon H'$ such that $\lim_n \varphi_n(f,y)=\langle Tf,y\rangle$ for all $f\in L^1(\mu)$ and $y\in H$. The bilinear forms $\varphi_n$ can be chosen as $\mu$-simple functions $\boldsymbol{h}_n: X\to H'$ with (finite) range in the unit ball $V^\circ$ of $H'$, so that $\varphi_n(f,y)=\int f(s)\langle y,\boldsymbol{h}_n(s)\rangle\, d\mu(s)$.

Considering each $\boldsymbol{h}_n$ as a linear map $H\to L^\infty(\mu)$ of norm $\leqq 1$, the sequence $(\boldsymbol{h}_n)$ converges to the adjoint $T': H\to L^\infty(\mu)\subset L^1(\mu)$ for the weak operator topology of $\mathscr{L}(H,L^1(\mu))$. Since the unit ball of $\mathscr{L}(H,L^1(\mu))$ is metrizable for the strong operator topology [S, III.4.7], a suitable sequence of convex combinations $(\boldsymbol{k}_n)$ of the $\boldsymbol{h}_n$ converges to $T'$ for the strong operator topology of $\mathscr{L}(H,L^1(\mu))$; in particular, for each $y\in H$ the sequence $(\boldsymbol{k}_n(y))$ converges in the norm topology of $L^1(\mu)$, hence in measure. Now let $(y_m)$ denote a sequence dense in the unit ball of $H$; it is clear that there exists a subsequence $(\boldsymbol{g}_n)$ of $(\boldsymbol{k}_n)$ such that for each $m\in\mathbb{N}$, the sequence $(\langle y_m,\boldsymbol{g}_n(s)\rangle)_{n\in\mathbb{N}}$ converges for $\mu$-almost all $s\in X$. This means that for $\mu$-almost all $s\in X$, the $H'$-valued sequence $(\boldsymbol{g}_n(s))$ converges for $\sigma(H',H)$ to an element $\boldsymbol{g}(s)\in V^\circ$. (In fact, on $V^\circ$ $\sigma(H',H)$ agrees with the weak topology generated by the semi-norms $y'\mapsto|\langle y_m,y'\rangle|$ $(m\in\mathbb{N})$.) Clearly, this defines a function (class) $\boldsymbol{g}: X\to V^\circ$ such that $s\mapsto\langle y,\boldsymbol{g}(s)\rangle$ is measurable for each $y\in H$; it follows that $\boldsymbol{g}\in\overline{L}^\infty_{H'}(\mu)$, $\|\boldsymbol{g}\|_\infty=1$, and $\langle Tf,y\rangle=\int f(s)\langle y,\boldsymbol{g}(s)\rangle\, d\mu(s)$ on $L^1(\mu)\times H$.

Second, suppose $L^1(\mu)$ is not separable and let $U$ denote the unit ball of $L^1(\mu)$; we still suppose $\mu$ finite. Let $T\in\mathscr{L}(L^1(\mu),H')$ have norm 1. Since $V^\circ$ is compact metrizable for $\sigma(H',H)$, $T(U)\subset V^\circ$ is $\sigma(H',H)$-separable; let $(f_k)$ denote a sequence in $U$ such that the sequence $(Tf_k)$ is $\sigma(H',H)$-dense in $T(U)$. The closed vector sublattice of $L^1(\mu)$ generated by $(f_k)$ and the constant-one function is separable and, by (III.11.2), isomorphic to $L^1(X,\Sigma_0,\mu_0)$ where $\Sigma_0$ is a suitable $\sigma$-subalgebra of $\Sigma$ and where $\mu_0$ denotes the restriction of $\mu$ to $\Sigma_0$. By the preceding part of the proof there exists $\boldsymbol{g}\in\overline{L}^\infty_{H'}(\mu_0)$ such that $\langle Tf_k,y\rangle=\int f_k(s)\langle y,\boldsymbol{g}(s)\rangle d\mu_0(s)$ for all $k\in\mathbb{N}$ and $y\in H$; evidently, $\boldsymbol{g}\in\overline{L}^\infty_{H'}(\mu)$ and $\|\boldsymbol{g}\|_\infty=1$, and $\mu_0$ can be replaced by $\mu$. Now each $f \in U$ is in the closure of the sequence $(f_k)$ for the topology of simple convergence on the set $T'(V) \subset L^\infty(\mu)$ ($V$ the unit ball of $H$), whence it follows that $\langle Tf, y \rangle = \int f \langle y, \boldsymbol{g} \rangle d\mu$ holds identically on $L^1(\mu) \times H$. The unicity of $\boldsymbol{g}$ is clear, since $\langle y_m, \boldsymbol{g} \rangle = T' y_m \in L^\infty(\mu)$ for each element $y_m$ of a sequence dense in $V$.

Finally, it is clear that each $\boldsymbol{g} \in \overline{L}^\infty_{H'}(\mu)$ defines an operator $T \in \mathscr{L}(L^1(\mu), H')$ in this manner, so the isometry $T \mapsto \boldsymbol{g}$ just established is surjective. We can leave it to the reader to remove the assumption of a finite $\mu$ in favor of $\sigma$-finiteness, and to verify that the isomorphism is an isomorphism of Banach lattices whenever $H$ is a Banach lattice. □

**Corollary.** *Let $(X, \Sigma, \mu)$ be a $\sigma$-finite measure space and let $H$ be a separable Banach space. For $1 \leqq p < +\infty$ and $p' = p/(p-1)$, the formula of (7.6) establishes an isomorphism $\mathscr{L}^l(L^p(\mu), H') \to \overline{L}^{p'}_{H'}(\mu)$ which is an isomorphism of Banach lattices if $H$ is a Banach lattice.*

*Proof.* Let $T \in \mathscr{L}^l(L^p(\mu), H')$ have $l$-norm 1. Letting $E := L^p(\mu)$, by (3.3)(b) there exists a factoring $T: E \to (E, x') \to H'$ where $x' \in E'_+$ has norm 1. Now $x' = k . \mu$ (cf. Chap. II, Exerc. 23) for some $k \in L^{p'}(\mu)$, $\|k\|_{p'} = 1$. Defining $\tilde{\mu} = k . \mu$ we can identify $(E, x')$ with $L^1(\tilde{\mu})$; then by (7.6), the partial map $\tilde{T}: L^1(\tilde{\mu}) \to H'$ is given by the bilinear form $(f, y) \mapsto \langle \tilde{T} f, y \rangle = \int f \langle y, \tilde{\boldsymbol{g}} \rangle d\tilde{\mu}$ where $\tilde{\boldsymbol{g}} \in \overline{L}^\infty_{H'}(\tilde{\mu})$ has norm 1. Now $\boldsymbol{g}: X \to H'$, defined by $s \mapsto \boldsymbol{g}(s) = k(s) \tilde{\boldsymbol{g}}(s)$, is an element of $\overline{L}^{p'}_{H'}(\mu)$ of norm $\leqq 1$ such that $\langle Tf, y \rangle = \int f \langle y, \boldsymbol{g} \rangle d\mu$ holds identically on $L^p(\mu) \times H$; certainly, $\boldsymbol{g}$ is unique and we must have $\|\boldsymbol{g}\|_{p'} = 1$.

Conversely, it is clear that each $\boldsymbol{g} \in \overline{L}^{p'}_{H'}(\mu)$ defines, by virtue of $\langle Tf, y \rangle := \int f \langle y, \boldsymbol{g} \rangle d\mu$, an operator $T \in \mathscr{L}(L^p(\mu), H')$ whose adjoint $T': H \to L^{p'}(\mu)$ is given by $y \mapsto \langle y, \boldsymbol{g} \rangle$. Since $\boldsymbol{g} = k \tilde{\boldsymbol{g}}$, where $k(s) = \|\boldsymbol{g}(s)\|$ for $\mu$-almost all $s \in X$ and where $\tilde{\boldsymbol{g}} \in \overline{L}^\infty_{H'}(\mu)$, $T'$ can be factored through $L^\infty(\mu)$ and hence, is majorizing. Therefore, $T \in \mathscr{L}^l(L^p(\mu), H')$ by (3.8). □

Now if $H$ is a Banach space (or lattice) with (norm) separable dual $H'$, we have $\overline{L}^p_{H'}(\mu) = L^p_{H'}(\mu)$ (Example 4) for each $p$, $1 \leqq p \leqq +\infty$, since each function $X \to H'$ is automatically separably valued and hence measurable [DS, III.6.11]. Thus we obtain the following result which applies, in particular, to every separable reflexive Banach space (or lattice) $H$.

**7.7 Theorem.** *Let $(X, \Sigma, \mu)$ denote a $\sigma$-finite measure space, and let $H$ denote a Banach space with separable dual $H'$. For $1 \leqq p < +\infty$ and $p' = p/(p-1)$, the dual of $L^p_H(\mu)$ is isomorphic to $L^{p'}_{H'}(\mu)$, the canonical bilinear form being given by*

$$\langle \boldsymbol{f}, \boldsymbol{g} \rangle = \int \langle \boldsymbol{f}(s), \boldsymbol{g}(s) \rangle d\mu(s).$$

*In particular, this correspondence is an isomorphism of Banach lattices if $H$ is a Banach lattice, and $L^p_H(\mu)$ is reflexive whenever $H$ is reflexive and $1 < p < +\infty$.*

With the aid of Example 4, the proof is readily obtained from (7.6) and its corollary.

## § 8. Banach Lattices of Compact Maps. Examples

We turn to the discussion of Banach spaces and lattices of compact linear maps that are representable by the (completed) $l$- and $m$-tensor products. If for Banach spaces $G, H$ we denote by $\mathscr{K}(G,H)$ the Banach space of all compact linear maps $G \to H$ (operator norm) and by $\mathscr{N}(G,H)$ the Banach space of all nuclear linear maps $G \to H$ (nuclear norm), then $G' \tilde{\otimes}_\varepsilon H \to \mathscr{K}(G,H)$ is an isometric injection and $G' \tilde{\otimes}_\pi H \to \mathscr{N}(G,H)$ is a metric homomorphism (§ 2, Example 5); both mappings are bijections if $G'$ or $H$ has the approximation property.

Suppose $E, F$ are Banach lattices. It has been observed above (and follows readily from § 3, Formulae (3'), (4')) that the $l$-norm on $E' \otimes H$ (and, accordingly, the $m$-norm on $G' \otimes F$) is coarser than the $\pi$-norm and finer than the $\varepsilon$-norm; therefore, the imbeddings $\mathscr{N}(E,H) \to E' \tilde{\otimes}_l H \to E' \tilde{\otimes}_\varepsilon H$ and $\mathscr{N}(G,F) \to G' \tilde{\otimes}_m F \to G' \tilde{\otimes}_\varepsilon F$ are contractions and $E' \tilde{\otimes}_l H$, $G' \tilde{\otimes}_m F$ represent Banach spaces of compact linear maps. In particular, $E' \tilde{\otimes}_l F$ and $E' \tilde{\otimes}_m F$ are Banach lattices of compact operators, and vector lattices for the canonical order induced by $\mathscr{L}(E,F)$.

For the convenience of the reader, we collect in the following proposition the relevant results obtained so far (and contained in (4.6) Cor. 2, (5.6), (6.10), and (7.4), respectively).

**8.1 Proposition.** *Let $E, F$ denote Banach lattices.*

(i) *If $E$ is an AL-space or $F$ an AM-space, then $\mathscr{K}(E,F)$ is a Banach lattice whose dual is canonically isomorphic to $\mathscr{L}^i(E',F')$ (integral norm and canonical order).*

(ii) *If $E$ is an AM-space and $F$ an AL-space, then $\mathscr{N}(E,F)$ is a Banach lattice whose dual is canonically isomorphic to $\mathscr{L}(E',F')$.*

(iii) *If $E$ and $F$ are Hilbert lattices, then $\mathscr{H}(E,F)$ is a Hilbert lattice under its canonical order and the Hilbert-Schmidt norm.*

We next give an explicit characterization of the operators contained in $E' \tilde{\otimes}_l H$ and $G' \tilde{\otimes}_m F$, respectively; as usual, $E, F$ denote Banach lattices and $G, H$ denote Banach spaces.

**8.2 Proposition.** (i) *Let $T \in \mathscr{L}(E,H)$. Then $T \in E' \tilde{\otimes}_l H$ if and only if there exists an AL-space $L$ and a factoring $T = T_2 \circ T_1 : E \to L \to H$ such that $T_1 : E \to L$ is positive and $T_2 : L \to H$ is compact; if so, then $\|T\|_l = \inf \|T_1\| \|T_2\|$ where the infimum is taken over all factorings of this type.*

(ii) *Let $T \in \mathscr{L}(G,F)$. Then $T \in G' \tilde{\otimes}_m H$ if and only if there exists an AM-space $M$ and a factoring $T = T_2 \circ T_1 : G \to M \to F$ such that $T_1 : G \to M$ is compact and $T_2 : M \to F$ is positive; if so, then $\|T\|_m = \inf \|T_1\| \|T_2\|$ where the infimum is taken over all factorings of this type.*

*Proof.* It is sufficient to prove (i); (ii) can then be proved analogously, or by duality.

The condition is sufficient. In fact, if $T$ has a factoring of the type indicated, then the norm functional of $L$ defines (via the positive adjoint $L' \to E'$) a positive linear form $x' \in E'$ such that $T$ has a factoring $E \to (E,x') \to L \to H$, where the partial map $\tilde{T}_2 : (E,x') \to L \to H$ is compact. Since by (2.4), the dual $(E')_{x'}$ of the

$AL$-space $(E,x')$ (see § 3) has the approximation property, it follows (cf. 2.3) that $\tilde{T}_2 \in (E')_{x'} \tilde{\otimes}_\varepsilon H$ and hence, by (7.3), that $T \in E' \tilde{\otimes}_l H$. It is clear from (3.3) that $\|T\|_l \leqq \inf \|T_1\| \|T_2\|$, and it will be seen immediately that the infimum equals the $l$-norm of $T$.

The condition is necessary. In fact, let $T \in E' \tilde{\otimes}_l H$ and let the number $\varepsilon > 0$ be given. By (7.3), there exists $x' \in E'_+$, $\|x'\| < \|T\|_l + \varepsilon$, such that $T$ is contained in the unit ball of $(E')_{x'} \tilde{\otimes}_\varepsilon H$ and hence defines a compact linear map $T_2 : (E,x') \to H$ of (operator) norm $\leqq 1$. Since the canonical map $T_1 : E \to (E,x')$ has norm $\|x'\|$, it follows that $T = T_2 \circ T_1$ is a factoring of the desired type satisfying $\|T_1\| \|T_2\| < \|T\|_l + \varepsilon$. □

The reader may have noticed an asymmetry between (8.1)(i) and (8.1)(ii); this asymmetry is remedied by the following theorem which can be viewed as the full counterpart of (8.1)(i). (Also, this result is implicitly contained in § 7, Example 2.)

**8.3 Theorem.** *If $E$ is an $AM$-space or if $F$ is an $AL$-space, then $\mathcal{N}(E,F)$ is a Banach lattice whose dual is canonically isomorphic to $\mathcal{L}(E',F')$. Precisely: $\mathcal{N}(E,F) = E' \tilde{\otimes}_l F$ if $E$ is an $AM$-space, and $\mathcal{N}(E,F) = E' \tilde{\otimes}_m F$ if $F$ is an $AL$-space.*

*Proof*. Suppose that $E$ is an $AM$-space. Given $T \in E' \tilde{\otimes}_l F$ we have to show that $T$ is nuclear with nuclear norm $\|T\|_l$. Let $T = T_2 \circ T_1$ be factored as specified in (8.2)(i); we can arrange that $\|T_2\| \leqq 1$ and that $\|T_1\| < \|T\|_l + \varepsilon$ for a given $\varepsilon > 0$. Since $T_1$ is positive and $T_2$ compact, (5.6) shows $T_1$ to be integral and hence, (5.5) implies $T$ to be nuclear; moreover, $\|T\|_\pi = \|T\|_i \leqq \|T_1\| < \|T\|_l + \varepsilon$ (cf. 5.2). Thus we have $\mathcal{N}(E,F) = E' \tilde{\otimes}_l F$ $(= E' \tilde{\otimes}_\pi F)$ as asserted.

If $F$ is an $AL$-space, the proof of the isomorphy $\mathcal{N}(E,F) = E' \tilde{\otimes}_m F$ (using (8.2)(ii)) is entirely analogous and will be omitted. □

**Examples.** Throughout the following examples, we let $E := L^p(\mu)$ and $F := L^q(\nu)$, where $(X, \Sigma, \mu)$ and $(Y, \mathrm{T}, \nu)$ are $\sigma$-finite measure spaces and $p, q \in [1, +\infty]$; the conjugate exponents $p', q' \in [1, +\infty]$ being determined by the relations $1/p + 1/p' = 1$, $1/q + 1/q' = 1$ respectively. Recall that the dual of $L^p(\mu)$ can be identified with $L^{p'}(\mu)$ (Chap. II, Exerc. 23); conversely, (under evaluation) $L^p(\mu)$ is isomorphic to the band of all order continuous linear forms on $L^{p'}(\mu)$ (cf. II.8.7).

A linear operator $T : E \to F$ is called an *operator with absolute kernel* (briefly, a *kernel operator*) if there exists a $(\Sigma \times \mathrm{T})$-measurable function $K$ on $X \times Y$ such that for each $f \in E$, $s \mapsto K(s,t) f(s)$ is $\mu$-integrable for $\nu$-almost all $t \in Y$ and $t \mapsto \int |K(s,t) f(s)| \, d\mu(s)$ is in $F$, and

$$Tf(t) = \int K(s,t) f(s) \, d\mu(s) \tag{1}$$

holds a.e. $(\nu)$.

1. Clearly, each kernel operator $E \to F$ is regular, and the set $\mathscr{A}_{pq}$ of all kernel operators $L^p(\mu) \to L^q(\nu)$ is a vector subspace of $\mathcal{L}^r(E,F)$ (cf. 1.4). Moreover:

*If $T \in \mathscr{A}_{pq}$, the modulus $|T|$ is the kernel operator with kernel $|K|$, and $\mathscr{A}_{pq}$ is a Banach sublattice of $\mathscr{L}^r(E,F)$.*

For the proof we consider first the case where $T$ is given by a kernel $K$ which is bounded and vanishes outside a set of finite $(\mu \otimes \nu)$-measure. Then $(f,g) \mapsto \int\int K(s,t) f(s) g(t) d\mu(s) d\nu(t)$ is a continuous bilinear form on $L^1(\mu) \times L^1(\nu)$, hence defines a continuous linear form on the Banach lattice $L^1(\mu) \tilde{\otimes}_\pi L^1(\nu) \cong L^1(\mu \otimes \nu)$. But the dual of $L^1(\mu \otimes \nu)$ is isomorphic to $L^\infty(\mu \otimes \nu)$ (cf. II.8.7) and hence, since the modulus of $K \in L^\infty(\mu \otimes \nu)$ is given by $|K|$, the modulus of the above bilinear form is the form $(f,g) \mapsto \int |K| f g \, d(\mu \otimes \nu)$. Thus by virtue of the canonical isomorphism of Banach lattices $\mathscr{B}(L^1(\mu), L^1(\nu)) \to \mathscr{L}(L^1(\mu), L^\infty(\nu))$ (cf. Exerc. 5) the modulus $|T|$, $T$ being considered an operator $L^1(\mu) \to L^\infty(\nu)$, is a kernel operator with kernel $|K|$; clearly, then, the same is true of the modulus $|T|: E \to F$. (Note that for finite measures $\mu$ and $\nu$, $\mathscr{L}(L^1(\mu), L^\infty(\nu))$ can be identified with an ideal of $\mathscr{L}^r(L^p(\mu), L^q(\nu))$.)

In the general case let $(C_n)$ denote an increasing sequence of subsets of $X \times Y$, each of finite $(\mu \otimes \nu)$-measure and with union $X \times Y$. Let $\chi_n$ denote the characteristic function of $C_n$, and let $T_n^+$ (respectively, $T_n^-$) be the operator $E \to F$ with kernel $K^+ \wedge n\chi_n$ (respectively, $K^- \wedge n\chi_n$). The sequence $(T_n^+)$ converges for the weak operator topology to an operator $S_1$ with kernel $K^+$, and $S_1 = \sup_n T_n^+$ in $\mathscr{L}^r(E,F)$ by (II.5.8); similarly, $(T_n^-)$ converges to $S_2$ and $S_2 = \sup_n T_n^-$ has the kernel $K^-$. So the sequence $(T_n^+ - T_n^-)$ order converges to $T$ and hence, $(|T_n^+ - T_n^-|)$ order converges to $|T| \in \mathscr{L}^r(E,F)$. But $|T_n^+ - T_n^-| = T_n^+ + T_n^-$ by the first part of the proof, whence it follows that $|T|$ is a kernel operator with kernel $|K|$.

Finally, it is not difficult to verify that $\mathscr{A}_{pq}$ is closed in $\mathscr{L}^r(E,F)$ and hence, a Banach lattice under the $r$-norm.

2. We define $\mathscr{H}_{pq} := \mathscr{A}_{pq} \cap \mathscr{L}^m(E,F)$ and $\mathscr{J}_{pq} := \mathscr{A}_{pq} \cap \mathscr{L}^l(E,F)$. Since the $m$-norm is larger than the $r$-norm on $\mathscr{L}^m(E,F)$ and since $\mathscr{A}_{pq}$ is a Banach lattice under the $r$-norm (Example 1), the space $\mathscr{H}_{pq}$ of majorizing kernel operators $E \to F$ is a Banach lattice under the $m$-norm. It is tedious but not hard to verify that an operator with kernel $K$ is in $\mathscr{H}_{pq}$ if and only if

$$k(t) := \left(\int |K(s,t)|^{p'} d\mu(s)\right)^{1/p'}$$

(respectively, $k(t) := \mu\text{-sup ess}_s |K(s,t)|$ if $p=1$) is finite a.e. $(\nu)$ and defines a function $k \in F$; moreover, $\|K\|_m = \|k\|$ in this case if we use the symbol $K$ for the operator as well. Thus if $p>1$, $q<+\infty$ the norm of $K \in \mathscr{H}_{pq}$ is explicitly given by

$$\|K\|_m = \left[\int\left(\int |K(s,t)|^{p'} d\mu(s)\right)^{q/p'} d\nu(t)\right]^{1/q} \tag{2}$$

with a corresponding formula holding if $p=1$ and/or $q=+\infty$; the elements of $\mathscr{H}_{pq}$ are called *Hille-Tamarkin operators* (cf. Hille-Tamarkin [1934], Jörgens [1970]).

Dually, since the $l$-norm is larger than the $r$-norm on $\mathscr{L}^l(E,F)$ and $\mathscr{A}_{pq}$ is a Banach lattice under the $r$-norm, the space $\mathscr{J}_{pq}$ of cone absolutely summing kernel operators is a Banach lattice under the norm

$$\|K\|_l = \left[\int\left(\int |K(s,t)|^q \, dv(t)\right)^{p'/q} d\mu(s)\right]^{1/p'} \tag{2'}$$

provided that $p>1$, $q<+\infty$; a corresponding formula holding if $p=1$ and/or $q=+\infty$. This follows from (2) and (3.8) in view of the fact that for a kernel operator $K: E \to F$ the adjoint $K'$, at least when restricted to the order continuous dual $F_{00}^{\star}$ (Def. II.4.1), is given by the transposed kernel $(s,t) \mapsto K(t,s)$. Moreover, if $q=p'<+\infty$ then from (2) or (2') we obtain that $\mathscr{H}_{pp'}$ and $\mathscr{J}_{pp'}$ are both isomorphic to $L^q(\mu \otimes v)$. We summarize:

*The spaces $\mathscr{H}_{pq}$ of majorizing and $\mathscr{J}_{pq}$ of c.a.s. kernel operators are Banach lattices; for each pair $(p,q)$, $\mathscr{H}_{pq}$ is isomorphic to $\mathscr{J}_{q'p'}$ by transposition of kernels. If $q=p'<+\infty$, $\mathscr{H}_{pp'}=\mathscr{J}_{pp'} \cong L^q(\mu \otimes v)$.*

For $p=q=2$, the space of Hilbert-Schmidt operators (see (6.5)(c)) emerges as a special case.

3. We consider the special case $1<p,q<+\infty$. If $F$ is separable then $\mathscr{L}^l(E,F)=\mathscr{L}^l(L^p(\mu),F)$ is isomorphic to $L_F^{p'}(\mu)$ (7.6 Cor., 7.7) which is in turn isomorphic to $L^{p'}(\mu) \tilde{\otimes}_l F = E' \tilde{\otimes}_l F$ (§ 7, Example 4); therefore, $\mathscr{L}^l(E,F) = E' \tilde{\otimes}_l F$ in this case. If $F$ is not separable and $T \in \mathscr{L}^l(E,F)$ then, since $T$ is compact (§ 3, Example 2), $T$ has its range in a separable closed vector sublattice $F_0$ of $F$ (cf. Chap. II, Exerc. 5(e)), and so $T \in E' \tilde{\otimes}_l F_0$ by the preceding. But $E' \tilde{\otimes}_l F_0$ can be identified with a Banach sublattice of $E' \tilde{\otimes}_l F$ and hence, we have $\mathscr{L}^l(E,F) = E' \tilde{\otimes}_l F$ whether $F$ is separable or not. By (7.4) this implies that $E' \tilde{\otimes}_l F = \mathscr{J}_{pq}$ is reflexive with dual $E \tilde{\otimes}_l F' = \mathscr{J}_{p'q'}$. Since under transposition $\mathscr{L}^m(E,F)$ is isomorphic to $\mathscr{L}^l(F',E')$ by (3.8) and the reflexivity of $E$ and $F$, it follows that likewise, $\mathscr{L}^m(E,F) = E' \tilde{\otimes}_m F = \mathscr{H}_{pq}$ is reflexive. Thus we have:

*If $1<p,q<+\infty$ then $\mathscr{H}_{pq} = E' \tilde{\otimes}_m F$ and $\mathscr{J}_{pq} = E' \tilde{\otimes}_l F$ are reflexive Banach lattices with duals $\mathscr{H}_{p'q'} = E \tilde{\otimes}_m F'$ and $\mathscr{J}_{p'q'} = E \tilde{\otimes}_l F'$, respectively.*

*In particular, every majorizing (respectively, cone absolutely summing) linear operator $L^p(\mu) \to L^q(v)$ is a compact kernel operator and can be approximated in the m-norm (2) (respectively, in the l-norm (2')) by operators of finite rank.* (in m-case, $\|\cdot\|_m \geq \|\cdot\|_r \Rightarrow$ they're r-compact)

If $1<p<+\infty$ and $q=1$ then by (5.5) and (5.6) every operator in $\mathscr{L}^m(E,F)$ is nuclear and hence, $\mathscr{L}^m(E,F) = \mathscr{H}_{p1} = E' \tilde{\otimes}_\pi F$ by (8.3); similarly, if $p=+\infty$ and $1<q<+\infty$ then we have $\mathscr{L}^l(E,F) = E' \tilde{\otimes}_\pi F$; however, $\mathscr{J}_{\infty q} = E_{00}^{\star} \tilde{\otimes}_\pi F = L^1(\mu) \tilde{\otimes}_\pi L^q(v)$ in this case.

On the other hand, an operator in $\mathscr{H}_{\infty q}$ or $\mathscr{J}_{p1}$ $(1<p,q<+\infty)$ need not be compact (Exerc. 21) but one always has $\mathscr{H}_{\infty q} = \mathscr{A}_{\infty q}$ and $\mathscr{J}_{p1} = \mathscr{A}_{p1}$ in these cases.

4. A situation of particular interest (cf. 9.1 below) arises for $p=+\infty$, $q=1$. By Formulae (2), (2') above we have $\mathscr{A}_{\infty 1}=\mathscr{H}_{\infty 1}=\mathscr{J}_{\infty 1}\cong L^1(\mu\otimes\nu)\cong E_{00}^{\star}\tilde{\otimes}_{\pi}F$, noting that $L^1(\mu)$ is the order continuous dual of $L^\infty(\mu)$. Hence:

*Every kernel operator* $K: L^\infty(\mu)\to L^1(\nu)$ *is nuclear.*

5. Finally, we consider the case $p=1$, $q=+\infty$ excluded in Examples 3 and 4. Since $\mathscr{L}(L^1(\mu),L^\infty(\nu))$ is canonically isomorphic with $\mathscr{B}(L^1(\mu),L^1(\nu))$ (§ 2, Formula (2)) and the latter is the dual of $L^1(\mu)\tilde{\otimes}_{\pi}L^1(\nu)\cong L^1(\mu\otimes\nu)$, $\mathscr{L}(L^1(\mu),L^\infty(\nu))$ is Banach lattice isomorphic to $L^\infty(\mu\otimes\nu)$, the isomorphism being given by the formula $\langle Tf,g\rangle=\iint K(s,t)f(s)g(t)\,d\mu(s)\,d\nu(t)$ $(f\in L^1(\mu)$, $g\in L^1(\nu))$. This is a special case of (4.4); we have $\mathscr{H}_{1\infty}=\mathscr{J}_{1\infty}=\mathscr{A}_{1\infty}=\mathscr{L}(E,F)\cong L^\infty(\mu\otimes\nu)$. We point out that in the present circumstances it can happen that an operator $T\in\mathscr{L}(E,F)$ which is not even weakly compact, has a modulus $|T|$ of rank 1 (Exerc. 21).

The reader may find it worthwhile to examine the somewhat more general cases $p=1$, $q<+\infty$ and $p<+\infty$, $q=1$ not covered by the foregoing examples. (Cf. Exerc. 23.)

In conclusion, we point out that the preceding Examples can be extended, with minor changes, to localizable measure spaces (Chap. III, § 1, Example 2); in particular, to measure spaces defined by Radon measures on locally compact spaces.

## § 9. Operators Defined by Measurable Kernels

Again, let $E, F$ denote arbitrary Banach lattices. The preceding study of tensor products (§§7,8) was naturally brought about by the problem to find vector sublattices of $\mathscr{L}^r(E,F)$ generated, in some sense, by the linear mappings of finite rank $E'\otimes F$ (cf. Theorem 4.6). The abstract theme of the present section is the characterization of the band of $\mathscr{L}^r(E,F)$ ($F$ being supposed order complete) generated by $E'\otimes F$. It will turn out that this is closely related to the study of operators defined by measurable kernels (§ 8, Examples 1—5). An operator $T$ between spaces of measurable functions and given by a kernel $K$ (as in § 8, Formula (1)) will be briefly referred to as a *kernel operator*. However, in this section we will consider primarily locally compact spaces $X, Y$ and Radon measures on $X, Y$ respectively; this causes a slight inconvenience in applications to more general measure spaces but is compensated for by the applicability of the present results to abstract Banach lattices possessing a topological orthogonal system (Def. III.5.1) via their representation theory (Chap. III, §§ 3—6).

We begin with a special result that supplements (5.6) and the theorem of § 8, Example 4.

**9.1 Proposition.** *Let* $E:=C(X)$ *and* $F:=L^1(\nu,Y)$ *where* $X$ *is a compact space,* $Y$ *a locally compact space, and* $\nu$ *a positive Radon measure on* $Y$. *For an operator* $T\in\mathscr{L}(E,F)$ *the following assertions are equivalent:*

(a) *$T$ belongs to the band of $\mathscr{L}^r(E,F)$ generated by $E'\otimes F$.*
(b) *$T$ is nuclear.*
(c) *There exists a positive Radon measure $\mu$ on $X$ and $K\in L^1(\mu\otimes\nu)$ such that for each $f\in E$,*

$$(\star)\qquad Tf(t)=\int K(s,t)f(s)\,d\mu(s)$$

*holds locally a.e. $(\nu)$.*

*Proof.* (a)$\Leftrightarrow$(b) is contained in (5.6); note that in the present circumstances, $\mathscr{L}^r(E,F)$ is the space of all integral maps $E\to F$ under the integral norm.

(b)$\Rightarrow$(c): If $T$ is nuclear then $T=\sum_n x'_n\otimes y_n$, where $(x'_n)\subset C(X)'$ and $(y_n)\subset L^1(\nu,Y)$ are sequences satisfying $\sum_n\|x'_n\|\,\|y_n\|<+\infty$. Considering $\nu$ as a linear form on $F$, we define $\mu:=|T'|\nu$; then $T'(F')$ is contained in the principal band of $E'$ generated by $\mu$ and, by the Radon-Nikodym theorem (II.8.7) Cor., this band is isomorphic to $L^1(\mu,X)$. If $P$ denotes the band projection $E'\to L^1(\mu)$, it is clear that $T=\sum_n Px'_n\otimes y_n$; but $Px'_n=f_n\cdot\mu$ for functions $f_n\in L^1(\mu)$ satisfying $\|f_n\|\leqq\|x'_n\|$ for all $n$. Therefore, the series $\sum_n f_n\otimes y_n$ converges in $L^1(\mu)\tilde{\otimes}_\pi L^1(\nu)\cong L^1(\mu\otimes\nu)$ and hence, its sum defines a kernel $K\in L^1(\mu\otimes\nu)$ which evidently satisfies $(\star)$.

(c)$\Rightarrow$(b): Since $K\in L^1(\mu\otimes\nu)\cong L^1(\mu)\tilde{\otimes}_\pi L^1(\nu)$ and the latter is (isomorphic with) a normed subspace of $E'\tilde{\otimes}_\pi F$, the operator $T$ defined by $(\star)$ is nuclear. ☐

The argument used in proving (b)$\Rightarrow$(c) yields the following corollary.

**Corollary 1.** *Let $E$ be an AM-space with unit $e$, and let $F$ be an AL-space with norm functional $\nu$. If $T: E\to F$ is nuclear then $T$ has a representation $T=\sum_n\lambda_n x'_n\otimes y_n$ where $(\lambda_n)\in l^1$, $(x'_n)$ is a normalized sequence contained in the band of $E'$ generated by $|T'|\nu$, and where $(y_n)$ is a normalized sequence contained in the band of $F$ generated by $|T|e$.*

The second corollary is a slight extension of (9.1); we suppose $X$ locally compact and replace $E$ by $\mathscr{K}(X)$, the vector lattice of all continuous functions $X\to\mathbb{R}$ with compact support. For each compact subset $X_0\subset X$, let $\mathscr{K}(X_0)$ denote the ideal of $\mathscr{K}(X)$ containing all functions with support in $X_0$; under the supremum norm, $\mathscr{K}(X_0)$ is an $AM$-space (in general, without unit). Let $i_{X_0}$ denote the canonical injection $\mathscr{K}(X_0)\to\mathscr{K}(X)$.

**Corollary 2.** *Let $X, Y$ denote locally compact spaces, let $\nu$ denote a positive Radon measure on $Y$, and let $T$ denote a regular linear map $\mathscr{K}(X)\to L^1(\nu)$. The following are equivalent:*
(a) *For each compact subset $X_0\subset X$, the composition $T\circ i_{X_0}$ is nuclear.*
(b) *There exists a positive Radon measure $\mu$ on $X$ and an $X$-locally $(\mu\otimes\nu)$-integrable[1] real function $K$ on $X\times Y$ such that for each $f\in\mathscr{K}(X)$,*

$$Tf(t)=\int K(s,t)f(s)\,d\mu(s)$$

*holds locally a.e. $(\nu)$.*

[1] I.e., $K\chi_{X_0\times Y}\in L^1(\mu\otimes\nu)$ for each compact $X_0\subset X$.

*Proof*. The implication (b)$\Rightarrow$(a) is clear from (9.1).

(a)$\Rightarrow$(b): Since $T$ is regular, the order adjoint $T^\star$ (Chap. III, Exerc. 24) maps the order dual $L^\infty(\nu)$ of $L^1(\nu)$ into the order dual $\mathcal{M}(X)$ of $\mathcal{K}(X)$; we define $\mu:=|T^\star|e$ ($e$ the unit of $L^\infty(\nu)$). Let $X_0\subset X$ be compact and denote by $\mu_0$ the restriction of $\mu$ to $X_0$; since $T_0:=T\circ i_{X_0}$ is nuclear and since $T_0'(L^\infty(\nu))\subset L^1(\mu_0)$ $\subset\mathcal{M}(X)$, we have (cf. proof of (9.1), (b)$\Rightarrow$(c)) $T_0=\sum_n(f_n\cdot\mu_0)\otimes y_n$ where $(f_n)$ is a sequence in $L^1(\mu_0)$, $(y_n)$ is a sequence in $L^1(\nu)$, and $\sum_n\|f_n\|\,\|y_n\|<+\infty$. Hence, $T_0$ is a kernel operator with kernel $K_0\in L^1(\mu_0\otimes\nu, X_0\times Y)$; evidently, $K_0$ is $(\mu_0\otimes\nu)$-essentially unique. Since the measure $\mu\otimes\nu$ on $X\times Y$ is localizable (cf. Chap. III, § 1, Example 2), there exists an $X$-locally $(\mu\otimes\nu)$-integrable function $K$ on $X\times Y$ such that for each compact $X_0\subset X$, the restriction of $K$ to $X_0\times Y$ represents $T\circ i_{X_0}$. □

**9.2 Proposition.** *Let $E:=C(X)$ and $F:=L^1(\nu,Y)$, where $X,Y,\nu$ satisfy the hypothesis of (9.1). For each $\mu\in E'_+$, the mapping $\varphi: L^1(\mu\otimes\nu)\to\mathcal{L}^r(E,F)$ defined by ($\star$) of (9.1), is an isomorphism of Banach lattices with range the band of $\mathcal{L}^r(E,F)$ generated by $\mu\otimes F$.*

*Proof*. Since by (5.6), $E'\tilde{\otimes}_\pi F$ is the band of $\mathcal{L}^r(E,F)$ generated by $E'\otimes F$, every $T\in\mathcal{L}^r(E,F)$ contained in the band generated by $\mu\otimes F:=\{\mu\otimes y: y\in F\}$ is nuclear, and clearly $T'(F')\subset L^1(\mu,X)\subset E'$; as in the proof of (9.1), (b)$\Rightarrow$(c), we conclude that $T=\varphi(K)$ for some $K\in L^1(\mu\otimes\nu)$. Moreover, $\varphi$ is an isometry, since $L^1(\mu\otimes\nu)$ is isomorphic (under ($\star$)) to the Banach sublattice $L^1(\mu)\tilde{\otimes}_\pi L^1(\nu)$ of $E'\tilde{\otimes}_\pi F$, and the latter is isometrically imbedded in $\mathcal{L}^r(E,F)$ by (5.6). Clearly, $\varphi(K)\geqq 0$ if and only if $K\geqq 0$ and so $\varphi$ is an isomorphism of Banach lattices. □

**Corollary.** *If $\nu$ is a finite measure and $e$ denotes the constant-one function on $Y$, then $\varphi$ maps $L^1(\mu\otimes\nu)$ $(\mu\in E'_+)$ isomorphically onto the principal band of $\mathcal{L}^r(E,F)$ generated by $\mu\otimes e$.*

We now consider arbitrary Banach lattices $E,F$ and suppose $F$ to be order complete; then $\mathcal{L}^r(E,F)$ is order complete by (1.4). Our aim is to characterize the band $(E'\otimes F)^{\perp\perp}$ of $\mathcal{L}^r(E,F)$ generated by $E'\otimes F$; this characterization will be modeled on the special case considered in Theorem (5.6). To this end we consider the canonical maps $i_x: E_x\to E$ $(x\in E_+)$ and $j_{y'}: F\to(F,y')$ $(y'\in F'_+)$ (see § 3); for each $T\in\mathcal{L}^r(E,F)$, the bicomposition $T_{x,y'}=j_{y'}\circ T\circ i_x$ is an element of $\mathcal{L}^r(E_x,(F,y'))$ (cf. 5.7). Recall that $F^\star_{00}\subset F'$ denotes the band of order continuous linear forms on $F$ (Def. II.4.1).

**9.3 Lemma.** *If $0\leqq y'\in F^\star_{00}$ then the canonical image of $F$ in $(F,y')$ is an ideal.*

*Proof*. If $N$ denotes the absolute kernel of $y'$, then the canonical image $F/N$ of $F$ in $(F,y')$ is an ideal by (II.4.12). (It is clear that the assertion holds for any vector lattice $F$; conversely, if $y'$ is strictly positive and $F$ is an ideal in $(F,y')$, then $y'\in F^\star_{00}$ since $y'$ is order continuous on the $AL$-space $(F,y')$.) □

**9.4 Proposition.** *Let $E,F$ denote Banach lattices ($F$ order complete) and let $x\in E'_+$, $y'\in(F^\star_{00})_+$. The mapping $\psi_{x,y'}:\mathcal{L}^r(E,F)\to\mathcal{L}^r(E_x,(F,y'))$ defined by bicomposition $T\mapsto T_{x,y'}$, is an order continuous lattice homomorphism with kernel $\{T:\langle|T|x,y'\rangle=0\}$.*

*Proof*. Since the canonical image of $F$ in $(F,y')$ is an ideal by (9.3), it is readily seen that $|T|_{x,y'}=|T_{x,y'}|$ for each $T\in\mathscr{L}^r(E,F)$; thus $\psi_{x,y'}$ is a lattice homomorphism. Moreover, if $A$ is a directed $(\leqq)$ subset of $\mathscr{L}^r(E,F)$ and if $T=\sup A$ then, since $(F,y')$ is an $AL$-space, $\psi_{x,y'}(A)$ converges to $T_{x,y'}$ for the strong operator topology of $\mathscr{L}^r(E_x,(F,y'))$. Since the cone of positive linear maps $E_x\to(F,y')$ is closed for this topology, (II.5.8) shows that $T_{x,y'}=\sup\psi_{x,y'}(A)$; thus $\psi_{x,y'}$ is order continuous. The final assertion is clear. □

To prove the main theorem (9.6), we need another property of the mappings $\psi_{x,y'}$ just defined.

**9.5 Proposition.** *Let $x\in E_+$, $y'\in(F^\star_{00})_+$. The subset $J_{x,y'}:=\{T: T_{x,y'}$ is nuclear$\}$ of $\mathscr{L}^r(E,F)$ is a band, and the kernel of $\psi_{x,y'}$ contains the band $J_{x,y'}\cap(E'\otimes F)^\perp$. Moreover, $\psi_{x,y'}$ maps $J_{x,y'}$ onto an ideal of $E'_x\tilde{\otimes}_\pi(F,y')$.*

*Proof*. By (5.6) the nuclear maps $E_x\to(F,y')$ form a band in $\mathscr{L}^r(E_x,(F,y'))$, and $J_{x,y'}$ is the inverse image of this band under $\psi_{x,y'}$; since $\psi_{x,y'}$ is an order continuous lattice homomorphism by (9.4), $J_{x,y'}$ is a band which evidently contains $E'\otimes F$.

Next we show that $\psi_{x,y'}(J_{x,y'})$ is an ideal of $\mathscr{L}^r(E_x,(F,y'))$. First, let $T_0\in\mathscr{L}^r(E_x,(F,y'))$ be absolutely majorized by some element of $\psi_{x,y'}(E'\otimes F)$; that is, suppose there exist $x'\in E'_+$ and $y\in F_+$ such that $|T_0z|\leqq\langle|z|,x'\rangle y$ for all $z\in E_x$. By (5.6) $T_0$ is nuclear, and it is clear from (3.9) (or 3.10) that $T_0$ can be extended to a regular mapping $T: E\to F$; clearly, $T_0=\psi_{x,y'}(T)$ and so $T_0\in\psi_{x,y'}(J_{x,y'})$. Therefore, $\psi_{x,y'}(J_{x,y'})$ contains the ideal $I$ generated by $\psi_{x,y'}(E'\otimes F)$, and it follows from (9.1) Cor. 1 that $I$ is dense in $\psi_{x,y'}(J_{x,y'})$. Since the norm of $\mathscr{L}^r(E_x,(F,y'))$ is order continuous (5.6) and since $\psi_{x,y'}$ is an order continuous lattice homomorphism, it now follows that $\psi_{x,y'}(J_{x,y'})$ is an ideal of $\mathscr{L}^r(E_x,(F,y'))$ and, in fact, equals $\psi_{x,y'}((E'\otimes F)^{\perp\perp})$.

Clearly, this shows that the kernel of $\psi_{x,y'}$ contains $(E'\otimes F)^\perp\cap J_{x,y'}$. □

**9.6 Theorem.** *Let $E$ be a Banach lattice and let $F$ be an order complete Banach lattice separated by its order continuous dual $F^\star_{00}$. The band of $\mathscr{L}^r(E,F)$ generated by $E'\otimes F$ consists of those operators $T$ for which all bicompositions $T_{x,y'}$ $(x\in E_+$, $y'\in(F^\star_{00})_+)$ are nuclear.*

*Proof*. Define $J(E,F):=\bigcap_{(x,y')}J_{x,y'}$ where $(x,y')$ ranges over all pairs in $E_+\times(F^\star_{00})_+$. By (9.5) $J(E,F)$ is a band of $\mathscr{L}^r(E,F)$ which clearly contains $E'\otimes F$. Now let $T\in J(E,F)\cap(E'\otimes F)^\perp$. Then by (9.5) $T$ is in the kernel of each of the mappings $\psi_{x,y'}$ and hence, by (9.4) we have $\langle|T|x,y'\rangle=0$ for each $x\in E_+$ and $y'\in(F^\star_{00})_+$. Since $F^\star_{00}$ separates $F$, it follows that $T=0$. Therefore, $J(E,F)=(E'\otimes F)^{\perp\perp}$. □

To combine the results of (9.1) and (9.6), we now assume that $E, F$ are Banach lattices of extended real valued functions defined on locally compact spaces $X, Y$ respectively; more precisely, we assume that each element of $E$ is an equivalence class of universally measurable $\overline{\mathbb{R}}$-valued functions $f$, modulo some fixed equivalence relation disregarding the values of $f$ on a certain $\sigma$-ring of meagre subsets of $X$ on which these functions are possibly infinite; and similarly for $F$ (cf.

Chap. II, Exerc. 14 and Dieudonné [1951]). In addition, we assume $E$ to contain $\mathscr{K}(X)$ (the continuous real functions on $X$ with compact support, under their natural ordering) as a dense vector sublattice, and we suppose $F$ to be order complete with separating order continuous dual $F_{00}^{\star}$; finally, we suppose $F$ to contain $\mathscr{K}(Y)$ as a vector sublattice dense with respect to $\sigma(F, F_{00}^{\star})$. In this situation, the dual $E'$ can be identified with an ideal $\mathscr{M}_E(X)$ of $\mathscr{M}(X)$ (precisely, with the ideal of those Radon measures $\mu$ on $X$ for which each $f \in E$ is $|\mu|$-integrable; Exerc. 24). Similarly, $F_{00}^{\star}$ can be identified with an ideal $\mathscr{M}_{F_{00}^{\star}}(Y)$ of $\mathscr{M}(Y)$.

**9.7 Theorem.** *Let $E, F$ be Banach function lattices satisfying the preceding assumptions. The following properties of a regular linear map $T: E \to F$ are equivalent:*

(a) *$T$ belongs to the band of $\mathscr{L}^r(E, F)$ generated by $E' \otimes F$.*

(b) *For each positive $\nu \in \mathscr{M}_{F_{00}^{\star}}(Y)$ there exist a positive $\mu \in \mathscr{M}_E(X)$ and an $X$-locally $(\mu \otimes \nu)$-integrable kernel $K$ on $X \times Y$ such that for each $f \in E$,*

$$(\star) \qquad Tf(t) = \int K(s,t) f(s) \, d\mu(s)$$

*holds locally a.e. $(\nu)$ on $Y$.*

*Moreover, if an operator $T \in \mathscr{L}^r(E, F)$ is represented by a family $\{(K, \mu, \nu)\}$ of tripels as specified in* (b), *then $|T|$ is represented by the family $\{(|K|, \mu, \nu)\}$.*

*Note.* With the aid of (9.4), (9.5), and (II.2.5) it is not difficult to show that for each pair of positive Radon measures $(\mu, \nu) \in \mathscr{M}_E(X) \times \mathscr{M}_{F_{00}^{\star}}(Y)$, $(\star)$ defines a lattice homomorphism of the band of $\mathscr{L}^r(E, F)$ generated by $\mu \otimes F$ onto an ideal of $X$-locally $(\mu \otimes \nu)$-integrable functions $K$ on $X \times Y$.

*Proof* of (9.7). Before proving the equivalence of (a) and (b), we show that if the restriction $T_0$ to $\mathscr{K}(X)$ of an operator $T \in \mathscr{L}^r(E, F)$ is represented by a family $\{(K, \mu, \nu)\}$ as specified in (b), then the restriction $|T|_0$ to $\mathscr{K}(X)$ of $|T|$ is represented by the family $\{(|K|, \mu, \nu)\}$; from the continuity of $T$, $|T|$, and $\mu$ on $E$ and standard measure theoretic arguments it then follows that $|T| f(t) = \int |K(s,t)| f(s) \, d\mu(s)$ locally a.e. $(\nu)$ for each $f \in E$ and tripel $(K, \mu, \nu)$. Consequently, $(\star)$ holds for all $f \in E$ and $(K, \mu, \nu)$.

In fact, if $j_\nu \circ T_0: \mathscr{K}(X) \to (F, \nu) \cong L^1(\nu)$ is given by $(\star)$ then (9.2) shows that $(j_\nu \circ |T_0|) f(t) = \int |K(s,t)| f(s) \, d\mu(s)$ locally a.e. $(\nu)$ for each $f \in \mathscr{K}(X)$. On the other hand, for $f \in \mathscr{K}(X)_+$ we have $|T_0| f = \sup\{|Tg|: |g| \leq f, g \in \mathscr{K}(X)\}$ (§ 1, Formula (2)) and $|T| f = \sup\{|Tg|: |g| \leq f, g \in E\}$; since $\mathscr{K}(X)$ is a dense vector sublattice of $E$, the order interval $\{g \in \mathscr{K}(X): |g| \leq f\}$ of $\mathscr{K}(X)$ is dense in the order interval $[-f, f]$ of $E$. Hence, since $T$ is continuous, it follows that $|T_0| f = |T| f$ for all $f \in \mathscr{K}(X)_+$ which is the desired assertion.

(a)$\Rightarrow$(b): Let $T \in (E' \otimes F)^{\perp\perp}$. By (9.6), for each $f \in E_+$ and each $\nu \in (F_{00}^{\star})_+$ the bicomposition $T_{f,\nu}$ is nuclear; this implies that for each $\nu \in (F_{00}^{\star})_+$, assertion (a) of (9.1) Cor. 2 holds for the restriction $T_0$ of $T$ to $\mathscr{K}(X)$ (note that $(F, \nu) \cong L^1(\nu, Y)$). Hence, letting $\mu := |T'| \nu$ (cf. proof of (9.1) Cor. 2) we have $\mu \in \mathscr{M}_E(X)$, and there exists an $X$-locally $(\mu \otimes \nu)$-integrable kernel $K$ (i.e., a kernel $K$ satisfying $K \chi_{X_0 \times Y} \in L^1(\mu \otimes \nu)$ for each compact $X_0 \subset X$) on $X \times Y$ such that $(\star)$ holds for all $f \in \mathscr{K}(X)$. From the preceding it follows that $(\star)$ holds for all $f \in E$.

(b)$\Rightarrow$(a): $T$ being regular by assumption, by (9.6) we have to show that all bicompositions $T_{f,\nu}$ $(f \in E_+, \nu \in (F^\star_{00})_+)$ are nuclear. Let $\nu$ (and hence, $\mu$ and $K$) be given; we can assume that $T \geqq 0$.

To show that $T_{f,\nu}$ is nuclear, we first consider $f \in \mathscr{K}(X)_+$. The ideal of $E$ and $AM$-space $E_f$ is continuously imbedded in $L^\infty(\mu)$, so (*) implies that the bicomposition $T_{f,\nu}: E_f \to (F,\nu)$ is nuclear (§ 8, Example 4). Similarly, $T_{f,\nu}$ is nuclear for each $f \geqq 0$ contained in the ideal $I$ of $E$ generated by $\mathscr{K}(X)$. Now let $f$ be any element of $E_+$; since the ideal $I$ is dense in $E$, there exists an increasing sequence $(f_n)$ in $I$ which converges to $f$. The canonical isomorphism $E_f \to C(H)$ ($H$ compact) of (II.7.4) maps $(f_n)$ onto a sequence $(e_n)$ in $C(H)$ for which $\sup_n e_n = e$ ($e$ the unit of $C(H)$), and maps the $AM$-spaces $E_{f_n}$ onto principal ideals $I_n$ of $C(H)$. The mapping $S_n: C(H) \to I_n$ defined by pointwise multiplication with $e_n$, is continuous even if $I_n$ is given the norm topology (with unit ball $[-e_n, e_n]$) carried over from $E_{f_n}$; and the mappings $R_n := T_{f_n,\nu} \circ S_n$ from $C(H) \cong E_f$ into $(F,\nu)$ are nuclear, since $T_{f_n,\nu}$ are nuclear by the preceding. Clearly, $T_{f,\nu} = \sup_n R_n$ in $\mathscr{L}^r(E_f, (F,\nu))$ and so by (5.6), $T_{f,\nu}$ is nuclear. □

The preceding result applies, via the representation theorem (III.5.3), to all Banach lattices $E, F$ possessing a topological orthogonal system (Def. III.5.1); in particular, to Banach lattices with order continuous norm (Chap. III, § 5, Example 2). We single out the following special case.

**Corollary.** *Suppose $E, F$ are Banach lattices each containing quasi-interior positive elements; assume, in addition, that $F$ is order complete and admits a strictly positive order continuous linear form $\nu$. Identifying $E, F$ with their representations on the respective (compact) structure spaces $K, H$* (cf. III.4.5), *a regular operator $T: E \to F$ is in $(E' \otimes F)^{\perp\perp} \subset \mathscr{L}^r(E,F)$ if and only if there exists $\mu \in E'_+$ and $K \in L^1(\mu \otimes \nu)$ such that $T$ is given by Formula* (*) *of* (9.7).

*If the lattice $E$ has order continuous norm, $\mu$ can be chosen independently of $T$, and for each weak order unit $\mu$ of $E'$,* (*) *defines an isomorphism (of vector lattices) of $(E' \otimes F)^{\perp\perp}$ onto an ideal of $L^1(\mu \otimes \nu)$ containing $L^\infty(\mu \otimes \nu)$.*

*Proof*. The first assertion is clear from (9.7) and (III.4.5). Using the notation of (9.4), (9.5) and denoting by $e$ a quasi-interior element $e \in E_+$ (Def. III.6.1), we observe that $J_{e,\nu} = (E' \otimes F)^{\perp\perp}$ and $\psi_{e,\nu}$ is an isomorphism of $(E' \otimes F)^{\perp\perp}$ into the nuclear band of $\mathscr{L}^r(E_e, (F,\nu))$. Moreover, $(E' \otimes F)^{\perp\perp}$ is the union of the principal bands $J_\mu$ generated by $\mu \otimes f$, where $f$ denotes a fixed quasi-interior point of $F_+$.

Now if the lattice $E$ has order continuous norm, then by (II.6.6) $E'$ contains weak order units $\mu$, and $J_\mu = (E' \otimes F)^{\perp\perp}$ for each weak order unit $\mu$ of $E'$. The remainder is clear from (9.2). □

The preceding corollary applies, in particular, to Banach lattices $E = L^p(\mu)$ and $F = L^q(\nu)$ $(1 \leqq p, q \leqq +\infty)$ over $\sigma$-finite measure spaces $(X, \Sigma, \mu)$ and $(Y, \mathrm{T}, \nu)$, respectively. However, since in general $X$ and $Y$ are not the respective structure spaces of $E$ and $F$, some caution is necessary in adapting (9.7) Cor. to this setting: in fact, an operator $T$ in $(E' \otimes F)^{\perp\perp}$ need not be given (at least if $p = +\infty$) by a measurable kernel on $X \times Y$ (§ 8, Example 1), since an operator $T \in \mathscr{A}_{pq}$ (notation of that example) is necessarily order continuous and hence, its adjoint $T'$ maps

$F_{00}^{\star}$ ($\cong L^{q'}(\nu)$) into $E_{00}^{\star}$ ($\cong L^{p'}(\mu)$). Keeping this in mind, it is not difficult to prove the following variant of (9.7); we omit the details.

**9.8 Proposition.** *Let $(X, \Sigma, \mu)$ and $(Y, \mathrm{T}, \nu)$ be $\sigma$-finite measure spaces, and let $E := L^p(\mu)$, $F := L^q(\nu)$ $(1 \leqq p, q \leqq +\infty)$. For a regular operator $T: E \to F$, the following properties are equivalent:*

(a) *$T$ belongs to the band of $\mathscr{L}^r(E, F)$ generated by the order continuous linear operators of finite rank.*

(b) *There exists a $(\Sigma \times \mathrm{T})$-measurable kernel $K$ such that $T$ is given by Formula $(\star)$ of (9.7).*

*More precisely: The mapping $T \mapsto K$ given by $(\star)$ is an isomorphism of $(E_{00}^{\star} \otimes F)^{\perp\perp}$ onto an ideal of $(\Sigma \times \mathrm{T})$-measurable functions containing $L^{\infty}(\mu \otimes \nu)$ (and into $L^1(\mu \otimes \nu)$ if $\mu, \nu$ are finite). Moreover, $(E_{00}^{\star} \otimes F)^{\perp\perp} = \mathscr{A}_{pq}$* (§ 8, Example 1).

We conclude this study by a list of sufficient conditions for an operator $T \in \mathscr{L}^r(E, F)$ to belong to the band generated by $E' \otimes F$.

**9.9 Proposition.** *Let $E, F$ be Banach lattices and suppose $F$ order complete and separated by its order continuous dual. For $T \in \mathscr{L}^r(E, F)$, each of the following conditions on $E, F$ and/or $T$ ensures that $T \in (E' \otimes F)^{\perp\perp}$:*

(i) *$E$ is an AL-space, $T$ weakly compact.*

(ii) *$F$ is an AM-space, $T$ weakly compact.*

(iii) *$E$ is an AL-space, $F$ an AM-space.*

(iv) *$E$ is reflexive*[2], *$T$ majorizing* (Def. 3.1).

(v) *$F$ is reflexive*[2], *$T$ cone absolutely summing* (Def. 3.1).

*Proof.* By (9.6), in each case it suffices to show that all bicompositions $T_{x,y'}: E_x \to E \to F \to (F, y')$ $(x \in E_+, y' \in (F_{00}^{\star})_+)$ are nuclear. In cases (i) and (ii), this follows from (5.5) and (5.6). In case (iii), it follows from (5.5) Cor. 1 and (5.6). The remaining cases (iv) and (v) can be reduced to the preceding in view of the factoring properties of majorizing maps (3.4) and of cone absolutely summing maps (3.3), respectively. □

We note that (iv) includes the case where $T$ is hypermajorizing (equivalently, has an absolutely summing adjoint (5.12)) and that (v) includes the case where $T$ is absolutely summing. Finally, it should be pointed out that, in general, the conditions on $T$ given in (9.9) are not necessary for $T \in (E' \otimes F)^{\perp\perp}$ (Exerc. 21).

## § 10. Compactness of Kernel Operators

A glance at (9.9) shows that an operator $T \in \mathscr{L}^r(E, F)$ contained in the band generated by $E' \otimes F$, is not compact in general; on the other hand, in the special situation of Theorem 5.6 every such operator is nuclear. We devote this section to a search for sufficient and necessary, internal conditions for an operator $T \in (E' \otimes F)^{\perp\perp}$ to be compact. To facilitate the discussion we consider exclusively Banach lattices $E, F$ containing quasi-interior positive elements $e, f$ (Def. II.6.1).

[2] It suffices that $E$ (resp. $F'$) have order continuous norm.

**10.1 Proposition.** *Let $E, F$ denote Banach lattices containing quasi-interior positive elements, and suppose that $F$ is order complete and admits a strictly positive, order continuous linear form $v$. If $T \in (E' \otimes F)^{\perp\perp}$ and if the r-norm on the principal band of $\mathscr{L}^r(E,F)$ generated by $|T'| v \otimes f$ is order continuous, then $T$ is compact.*

*Proof.* Representing, as in the corollary of (9.7), $E$ and $F$ as Banach lattices of continuous numerical functions on their respective structure spaces $K_E$ and $K_F$ (III.4.5) and letting $\mu := |T'| v$ for a given $T \in (E' \otimes F)^{\perp\perp}$, we conclude from (9.7) Cor. that $T$ is a kernel operator with kernel $K \in L^1(\mu \otimes v)$. Clearly, then, $T$ and $|T|$ are contained in the band $J_\mu$ of $\mathscr{L}^r(E,F)$ generated by $\mu \otimes v$ (notation as in the proof of (9.7) Cor.) and, if $e$ is a quasi-interior point of $E_+$, the mapping $\psi_{e,v}$ induces an order isomorphism of $J_\mu$ onto an ideal of $L^1(\mu \otimes v, K_E \times K_F)$.

Thus to prove that $T$ is compact we can assume that $T$, and hence the corresponding kernel $K$, is $\geqq 0$. If $e_1$ denotes the constant-one function on $K_E \times K_F$ we have $K = \sup_n (K \wedge n e_1)$. Since $K_n := K \wedge n e_1$ is bounded for each $n \in \mathbb{N}$ $K_n$ is the limit a.e. $(\mu \otimes v)$ of a sequence $(K_{nm})_{m \in \mathbb{N}}$ of uniformly bounded continuous functions on $K_E \times K_F$; consequently, for each $n \in \mathbb{N}$ the sequence $(K_{nm})_{m \in \mathbb{N}}$ order converges to $K_n$. Since the operators $T_n$ and $T_{nm}$ from $E$ into $F$ given by the kernels $K_n$ and $K_{nm}$, respectively, are all in $J_\mu$, it follows from the assumed order continuity of the $r$-norm on $J_\mu$ that that $T$ is the $r$-norm limit of a sequence $(T_{n,m(n)})$ of operators defined by continuous kernels. Thus if we can show that an operator $T_0$ defined by a continuous kernel on $K_E \times K_F$ is compact, it follows that $T$ is compact.

But such an operator $T_0$ possesses a factoring $E \to L^1(\mu, K_E) \to C(K_F) \to F$, where the central map is a kernel operator with the same kernel as $T_0$ and is clearly compact, since $C(K_E \times K_F) \cong C(K_E) \tilde{\otimes}_\varepsilon C(K_F)$ is isomorphically imbedded in $L^\infty(\mu, K_E) \tilde{\otimes}_\varepsilon C(K_F) \cong L^1(\mu)' \tilde{\otimes}_\varepsilon C(K_F)$. Thus $T_0$ is compact (cf. Exerc. 25). ☐

**Corollary.** *If $E = C(K)$ ($K$ compact) and $F$ has order continuous norm, or if $E'$ has order continuous norm and $F$ is an AL-space; then each operator $T$ in the band of $\mathscr{L}^r(E,F)$ generated by $E' \otimes F$, is compact.*

*Proof.* In the first case we have $\|T\|_r = \||T|\| = \||T|e\|$ (§ 1, Formula (4)) where $e$ denotes the unit of $E$; since the norm of $F$ is order continuous by hypothesis, it follows that the norm of $\mathscr{L}^r(E,F)$ is order continuous. In the second case, it follows from the preceding that the norm of $\mathscr{L}^r(F', E')$ is order continuous, $F'$ being an $AM$-space with unit; thus to show that for a decreasing sequence $\inf_n T_n = 0$ in $\mathscr{L}^r(E,F)$ implies $\lim_n \|T_n\| = 0$, it suffices to prove that $\inf_n T_n = 0$ implies $\inf_n T_n' = 0$ in $\mathscr{L}^r(F', E')$. But every positive linear form $y'$ on an $AL$-space is order continuous and hence, $\inf_n T_n = 0$ implies $\lim_n T_n' y' = 0$ for $\sigma(E', E)$ and so $\inf_n T_n' y' = 0$ for each $y' \in F'_+$; therefore, $\inf_n T_n' = 0$.

So, under both pairs of assumptions on $E, F$ the Banach lattice $\mathscr{L}^r(E,F)$ has order continuous norm and (10.1) applies. ☐

**10.2 Proposition.** *Let $E, F$ denote Banach lattices containing quasi-interior positive elements and suppose $E'$ and $F$ to have order continuous norm. If $J$ is an ideal of $(E' \otimes F)^{\perp\perp} \subset \mathscr{L}^r(E,F)$ generated by a family of compact positive operators, then each operator in $J$ is compact.*

*Proof*. If $S \in J$, there exists a compact operator $T \in J$ such that $0 \leqq S^+ \leqq T$ and $0 \leqq S^- \leqq T$; therefore, it suffices to show that $0 \leqq S \leqq T$, $T \in J$ compact, implies $S$ to be compact. Also, by (II.6.6) the hypotheses made on $F$ imply the existence of a strictly positive linear form $v \in F' = F^\star_{00}$.

To show that $S$ is compact, let $(x_n)$ denote any bounded sequence in $E_+$ we prove that $(Sx_n)$ contains a convergent subsequence. First, there exists a subsequence $(\bar{x}_n)$ of $(x_n)$ for which $y := \lim_n T\bar{x}_n \in F$ exists and, by selecting another subsequence if necessary, we can even suppose that $y_1 := y + \sum_{n=1}^\infty |y - T\bar{x}_n|$ exists in $F$. Now consider the compositions $T_v = j_v \circ T$, $S_v = j_v \circ S$ where $j_v$ denotes the canonical map $F \to (F, v)$; since $j_v(F)$ is an ideal of $(F, v)$ by (9.3), $T_v$ is in the band of $\mathscr{L}^r(E, (F, v))$ generated by $E' \otimes (F, v)$ and it follows from (10.1) Cor. that $S_v$ is compact. Thus, there exists a subsequence $(\bar{\bar{x}}_n)$ of $(\bar{x}_n)$ such that $(S_v \bar{\bar{x}}_n)$ converges in $(F, v)$; moreover, since $0 \leqq S_v \bar{\bar{x}}_n \leqq T_v \bar{\bar{x}}_n = T\bar{\bar{x}}_n \leqq y_1 \in F$, we can assume that $(S_v \bar{\bar{x}}_n)$ order converges. But $j_v(F)$ is an ideal in $(F, v)$ and the sequence $(S_v \bar{\bar{x}}_n)$ is majorized by the element $y_1 \in F$; therefore, $(S\bar{\bar{x}}_n)$ order converges in $F$ and thus norm converges in $F$, because $F$ has order continuous norm. ▯

Finally, we are prepared to prove the following internal characterization of operators $T \in (E' \otimes F)^{\perp\perp}$ with compact modulus.

**10.3 Theorem.** *Let $E, F$ denote Banach lattices containing quasi-interior positive elements, and let $E'$ and $F$ have order continuous norm. For an operator $T \in (E' \otimes F)^{\perp\perp} \subset \mathscr{L}^r(E, F)$, the following assertions are equivalent:*

(a) *The modulus $|T|$ is compact.*

(b) *The r-norm is order continuous on the principal ideal of $\mathscr{L}^r(E, F)$ generated by $T$.*

*Proof*. (a)$\Rightarrow$(b): First, we note that $F$ is order complete (II.5.10) and hence. $\mathscr{L}^r(E, F)$ is order complete (1.4).

Now let $T \in (E' \otimes F)^{\perp\perp}$, let $J$ be the ideal generated by $|T|$; to show that the $r$-norm is order continuous on $J$ it suffices, by (II.5.10)(d) and the Remark following the proof of (II.5.10), to prove that $\lim_n \|T_n\| = 0$ whenever $(T_n)$ is a decreasing sequence such that $0 \leqq T_n \leqq |T|$ and $\inf_n T_n = 0$. To this end, we denote by $U_+^{\circ\circ}$ (respectively, $V_+^\circ$) the positive part of the unit ball of $E''$ (respectively, of $F'$) and consider the restriction $\varphi_n$ of the bilinear form $(x'', y') \mapsto \langle T_n'' x'', y' \rangle$ to the $\sigma(E'', E') \times \sigma(F', F)$-compact product $U_+^{\circ\circ} \times V_+^\circ$.

We note first that $\lim_n \varphi_n(x'', y') = 0$ pointwise. In fact, $\inf_n T_n = 0$ implies $\inf_n T_n' y' = 0$ for each $y' \in V_+^\circ$, since every positive linear form on $F$ is order continuous (II.5.10); thus, since $E'$ has order continuous norm, $\lim_n \varphi_n(x'', y') = 0$ uniformly on $U_+^{\circ\circ}$ for each $y' \in V_+^\circ$. Since $|T|$ is compact by hypothesis, it follows from (10.2) that each $T_n$ is a compact operator; so $T_n''(U_+^{\circ\circ})$ is a compact subset of $F$. Because $\sigma(F', F)$ and the topology of compact convergence agree on $V_+^\circ$ (theorem of Banach-Steinhaus, [S, III.4.6]), we conclude that each $\varphi_n$ is a (jointly) continuous function on $U_+^{\circ\circ} \times V_+^\circ$. Therefore, by Dini's theorem the monotone sequence $(\varphi_n)$ converges to 0 uniformly on $U_+^{\circ\circ} \times V_+^\circ$. This implies $\lim_n \|T_n\| = 0$, since $\|T_n\| = \|T_n''\| = \sup\{\langle T_n'' x'', y' \rangle : (x'', y') \in U_+^{\circ\circ} \times V_+^\circ\}$ (§ 1, Formula (4)).

(b)$\Rightarrow$(a): In the proof of (10.1) it has been established that for $T \in (E' \otimes F)^{\perp\perp} \subset \mathscr{L}^r(E, F)$, there exists an increasing sequence $(T_n)$ of positive operators (each represented by a bounded kernel on $K_E \times K_F$) such that $|T| = \sup_n T_n$, and such that each $T_n$ is the order limit (in $(E' \otimes F)^{\perp\perp}$) of a sequence $(T_{nm})_{m\in\mathbb{N}}$ of compact positive operators (each represented by a continuous kernel). These operators $T_{nm}$ need not be contained in the ideal $J$ generated by $T$; however, if we define $\tilde{T}_{nm} := T_{nm} \wedge T_n$ for all $n$ and $m$, then the operators $\tilde{T}_{nm}$ are compact by (10.2) and contained in $J$ while $T_n$ is still the order limit of the sequence $(\tilde{T}_{nm})_{m\in\mathbb{N}}$. Thus, since the $r$-norm (and, a fortiori, the operator norm) is order continuous on $J$ by hypothesis, each $T_n$ is a compact operator $E \to F$ and hence, so ist $|T|$. □

**Corollary 1.** *If, under the hypotheses of* (10.3), *$T$ is an operator in $(E' \otimes F)^{\perp\perp}$ with compact modulus $|T|$, then $T$ is compact.*

This follows at once from (10.3) and (10.2); it should be noted that under weaker assumptions on $E$ and $F$, the assertion of the corollary fails in general (Exerc. 21).

**Corollary 2.** *Under the hypotheses of* (10.3), *each positive compact operator $E \to F$ which is contained in $(E' \otimes F)^{\perp\perp}$, can be approximated in the r-norm by operators of finite rank. Consequently, the closure of $E' \otimes F$ in $\mathscr{L}^r(E,F)$ consists of all operators in $(E' \otimes F)^{\perp\perp}$ possessing a compact modulus.*

*Proof*. We use the notation of the proof of (10.1). If $T$ is positive and compact then by (10.3)(b), the operators $T_{nm}$ converge to $T_n$ $(m \to \infty)$ in the $r$-norm and the same is true for the convergence $T_n \to T$. On the other hand, each $T_{nm}$ is defined by a kernel $K_{nm}$ which is continuous on $K_E \times K_F$ and hence, contained in the $r$-norm closure of $E' \otimes F$. □

**Examples.** 1. The equivalence (a)$\Leftrightarrow$(b) of (10.3) is no longer valid if in (b), the ideal $J$ generated by $T$ is replaced by the band generated by $T$. For example, if $E = F = l^2$ under its canonical ordering and if $e_n$ denotes the $n$-th unit vector $(\delta_{nm})_{m\in\mathbb{N}}$ in $l^2$, then $T := \sum_n n^{-1}(e_n \otimes e_n)$ is a compact positive operator in $(E' \otimes F)^{\perp\perp}$. (In fact, $T$ is a Hilbert-Schmidt operator by (6.5).) Clearly, the identity map $\sum_n (e_n \otimes e_n)$ is in the band of $\mathscr{L}^r(E,F)$ generated by $T$, but not compact. Moreover, if we let $P_n := \sum_{k=1}^n (e_k \otimes e_k)$ for $n \in \mathbb{N}$ it is quickly verified that $T = \sup_n T \wedge P_n$ for each positive operator $T: E \to F$, so we have $(E' \otimes F)^{\perp\perp} = \mathscr{L}^r(E,F)$ in the present case.

2. Let $E$ be an $AL$-space with weak order unit and let $F := L^2(\nu)$, where $(Y, \mathrm{T}, \nu)$ is any measure space. By (1.5) every continuous linear map $T: E \to F$ is regular, and from (9.9)(i) it follows that $\mathscr{L}(E,F) = (E' \otimes F)^{\perp\perp}$ (cf. Exerc. 23). Thus, an operator $T \in (E' \otimes F)^{\perp\perp}$ is not compact in general.

To characterize the compact operators in $\mathscr{L}(E,F)$, we employ the representation of $E$ on its compact structure space $K_E$ (cf. II.8.5 and III.4.5); that is, we identify $E$ with $L^1(\mu)$ ($\mu$ an order continuous Radon measure on $K_E$) under an isomorphism which maps $E_e$ ($e$ a fixed weak order unit of $E$) onto $C(K_E) \cong L^\infty(\mu)$. Then the Banach lattice $E' \tilde{\otimes}_\varepsilon F$ of compact operators $E \to F$ (cf. 4.6 Cor. 2) becomes identical with $C(K_E) \tilde{\otimes}_\varepsilon L^2(\nu)$ which, in turn, is isomorphic to $C(K_E, L^2(\nu))$ (§ 2, Example 1). Under this isomorphism, each $T \in E' \tilde{\otimes}_\varepsilon F$ corresponds to a continuous function $s \mapsto \boldsymbol{g}(s)$ of $K_E$ into $L^2(\nu)$; on the other hand, by (9.8) $T$ is a kernel operator with $((\mu \otimes \nu)$-essentially unique) kernel $K \in L^1(\mu \otimes \nu)$. (Obviously, the $\sigma$-finiteness of $\nu$ assumed in (9.8) is inessential here.) Thus if $T$ is given by

$$(*) \qquad Tf(t) = \int K(s,t) f(s) d\mu(s)$$

for $f \in E$, we must have $\boldsymbol{g}(s) = K(s, \ ) \in L^2(\nu)$ for each $s \in K_E$. Since $\boldsymbol{g}$ is continuous, we have

$$(**) \qquad \lim_{s \to s_0} \int |K(s,t) - K(s_0,t)|^2 \, d\nu(t) = 0$$

for each $s_0 \in K_E$ (*square mean continuity* of $K$ with respect to $s \in K_E$). Hence, *$T: E \to F$ is compact if and only if its kernel $K$ satisfies* (**) *for all $s_0 \in K_E$.*

If, more generally, $E$ is given in the form $L^1(\mu, X)$ where $\mu$ is a positive Radon measure on some compact space $X$, then again by (9.8) each $T \in \mathscr{L}(E,F)$ is a kernel operator defined by a measurable kernel $K$ on $X \times Y$, and condition (**) is sufficient (but not necessary) for $T$ to be compact. In fact, in this case the kernel $K$ satisfies (**) iff the corresponding operator $T$ is contained in the Banach sublattice $C(X) \tilde{\otimes}_\varepsilon L^2(\nu)$ of $L^\infty(\mu) \tilde{\otimes}_\varepsilon L^2(\nu) \cong E' \tilde{\otimes}_\varepsilon F$. (Cf. Exerc. 25.)

3. Let $X$ be a compact space and let $(Y, \mathsf{T}, \nu)$ be a measure space; let $E := C(X)$, $F := L^p(\nu)$ where $1 \leqq p < +\infty$. Then (10.1) Cor. applies, so $(E' \otimes F)^{\perp\perp}$ is a band of compact operators in $\mathscr{L}^r(E,F)$. By (9.7), we obtain:

*Every kernel operator $C(X) \to L^p(\nu)$ $(1 \leqq p < +\infty)$ is compact.*

On the other hand, $(E' \otimes F)^{\perp\perp} \neq \mathscr{L}^r(E,F)$ in general. For example, if $Y = X$ is compact and $\nu = \mu$ is a diffuse strictly positive Radon measure on $X$ then the canonical injection $C(X) \to L^1(\mu, X)$ is not nuclear while, by (9.6), it would have to be if the canonical injection $C(X) \to L^p(\mu, X)$ were contained in $(E' \otimes F)^{\perp\perp}$.

## Notes

§1: The modulus of a linear operator and regular operators were already considered and their basic properties established by Kantorovič [1937], [1940]; see also Kantorovič-Vulikh-Pinsker [1950] and Vulikh [1967]. (1.5) is found in Krengel [1964a], (1.8) is due to the author. See also Luxemburg-Zaanen [1971a].

§2: The results of the first half of the section (through 2.4) are due to Grothendieck [1955a]; for summable sequences and the related technicalities, Pietsch [1965] is the standard reference. (2.5) through (2.8) are due to Schlotterbeck [1969]; Pietsch [1963a] supplements (2.7).

§3: The material of this section is essentially taken from Schlotterbeck [1969]. Kantorovič [1940] considers special cases of majorizing and cone absolutely summing mappings; further relevant references include Levin [1969b] and Nielsen [1973]. (3.9) is based on Lotz' theorem (II.8.9) (Lotz [1974]).

§ 4: Again, the results of this section (with the exception of (4.6) which was proved by Schaefer [1972b]) are due to Schlotterbeck [1969]. Ellis [1966] and Wickstead [1974] treat problems related to (4.4) for more general vector lattices $E, F$. For an extension of various ideas basic to Sections 3, 4 to ordered topological vector spaces which are not necessarily lattice ordered, see Walsh [1973].

§ 5: Integral linear maps (Def. 5.1) were defined and extensively studied by Grothendieck [1955a], where additional information on the properties of these mappings can be found. The relation between integral and absolutely summing maps (cf. 5.10) is also due to Grothendieck [1955a] and is treated, along with related phenomena, systematically in Grothendieck [1956]. Pietsch [1963b] was the first to consider absolutely summing maps outside the framework of topological tensor products. The present proofs of (5.3) and (5.5) were suggested by U. Schlotterbeck (oral communication), who also found the important characterization (5.7). The equivalence (b)$\Leftrightarrow$(c) of (5.10) is due to Pietsch [1963b]; this result was later generalized by Pietsch [1967] to characterize absolutely $p$-summing operators. The equivalence (a)$\Leftrightarrow$(d) of (5.10) as well as (5.11) are due to Schlotterbeck [1969]. For a proof of the Dvoretzky-Rogers theorem (5.10) Cor. 4 in its more precise, original form see Day [1973].

§ 6: The first part of the section (through 6.5) contains standard material; the unicity of the representation (6.2) is rarely mentioned. (6.7) and the proof of (6.8) are due to Schaefer [1974c]. Part of (6.9) is related to a more general result of Grothendieck [1956] (cf. also Lindenstrauss-Pelczynski [1968]). Other results in this direction are found in Pietsch [1967], Pelczynski [1967], and Schlotterbeck [1973].

§ 7: The material of this section is taken from Schaefer [1972b]; independently, Levin [1969] and Chaney [1972] obtained similar results. Special cases of (7.6) and (7.7) are due to Dunford-Pettis [1940] (see also [DS]); the present proof of (7.6) is due to the author. Chaney [1972] proved a slightly more general version of (7.7).

§ 8: (8.2) and (8.3) are taken from Schaefer [1972b]; the examples given in this section are found in Schaefer-Schlotterbeck [1974]. Nielsen [1973] obtained further results on the approximation (in the $m$-norm) of majorizing maps by maps of finite rank.

§ 9: The literature on kernel operators is quite extensive; for additional information and further references, we refer to [DS] and to Jörgens [1970]. However, the problem of determining under what conditions the modulus $|T|$ of a kernel operator $T$ is given by the modulus of the kernel defining $T$, has apparently been ignored until recently (cf. Luxemburg-Zaanen [1971a]). The idea behind our presentation is due to Nagel-Schlotterbeck [1972] where, in particular, the results (9.6), (9.7) Cor., and (9.9) are proved; in the present form, (9.7) is due to the author. Nakano apparently first recognized the relations between kernel operators and operators contained in the band of $\mathscr{L}^r(E,F)$ generated by $E' \otimes F$ (see Nakano [1953]).

§ 10: Compactness criteria related to (10.3) were obtained by Ando [1957] and Luxemburg-Zaanen [1963]. The results presented here are due to Nagel-Schlotterbeck [1973], [1974].

## Exercises

1. For a vector lattice $E$, the following assertions are equivalent:
   ($\alpha$) Every linear form on $E$ is order bounded (Def. II.4.1).
   ($\beta$) $E$ is Archimedean and every principal ideal of $E$ is finite dimensional.
   ($\gamma$) $E$ is isomorphic to the ordered direct sum (Chap. II, § 1, Example 6) of $d$ copies of $\mathbb{R}$, where $d$ is the linear dimension of $E$.
   ($\delta$) For every vector lattice $F$ satisfying Axiom (OS) (Def. II.1.8), the vector space $L(E,F)$ of all linear maps $E \to F$ is an order complete vector lattice under its canonical ordering.

(For ($\alpha$)$\Rightarrow$($\beta$), observe that ($\alpha$) implies the finest locally convex topology on $E$ for which order intervals are bounded, to agree with the finest locally convex topology on $E$ (cf. [S, Chap. II, Exerc. 7]. For ($\gamma$)$\Rightarrow$($\delta$), note that ($\gamma$) implies that for each $T \in L(E,F)$ and each $x \in E_+$, the restriction of $T$ to $E_x$ is of finite rank.)

2. Let $E, F$ denote Banach lattices, and let $T: E \to F$ denote a continuous linear map with first and second adjoints $T': F' \to E'$, $T'': E'' \to F''$.

(a) If $T \in E' \otimes F$ (i.e., if $T$ is of finite rank), then $|T|$ exists and $|T|' = |T'|$, but $|T| \notin E' \otimes F$ in general. (Cf. proof of (4.6).)

(b) If $T$ is nuclear, then $|T|$ exists and is compact. Is $|T|$ necessarily nuclear?

(c) If $E'$ and $F$ have order continuous norm and $T$ is regular, then $|T'| = |T|'$ and $|T''| = |T|''$. (Use the fact that under evaluation, $E'$ and $F$ are ideals in their respective biduals (II.5.10).)

(d) Whenever $|T|$ exists then $|T'| \leqq |T|'$, and equality holds iff $|T'|$ is continuous with respect to $\sigma(F', F)$ and $\sigma(E', E)$.

(e) Let $E = F = C[0,1]$. There exist regular operators $T$ such that $|T|$ exists and $|T''| < |T|''$. (Consider the operator $T$ given by $Tf(s) = \frac{1}{2}(f(s) - f(s^2))$ for all $f \in E$, $s \in [0,1]$.) Infer from (c) that one must necessarily have $|T'| < |T|'$ for such an operator $T$.

3. Let $E, F$ denote normed vector lattices.

(a) Denoting by $\mathscr{L}^r(E,F)$ the vector space of linear maps $T: E \to F$ possessing a decomposition $T = T_1 - T_2$ where $T_j$ are positive and continuous, show that Formula (5′) of § 1 defines a norm on $\mathscr{L}^r(E,F)$ for which the positive cone is a normal ***B***-cone [S, Chap. V, § 3], and for which $\mathscr{L}^r(E,F)$ is a Banach space whenever $F$ is norm complete. Moreover, $\mathscr{L}^r(E,F) = L^r(E,F)$ whenever $E$ is norm complete.

(b) If $T \in \mathscr{L}^r(E,F)$ (see (a)) and $T'$ denotes the adjoint map $F' \to E'$, then $T' \in \mathscr{L}^r(F', E')$ and $\|T'\|_r \leqq \|T\|_r$. Does equality necessarily hold? Show that the answer is affirmative if $F$ is a Banach lattice with property ($P$) (Def. 4.2).

(c) Show that even if $E, F$ are Banach lattices but $F$ is not order complete, a regular operator $T \in \mathscr{L}(E,F)$ need not possess a modulus. (Modify the construction of § 4, Example 2 by letting $F := C[0,1]$ and by letting $e_n$ $(n \in \mathbb{N})$ denote a function of norm 1 in $F_+$ such that $e_n(t) > 0$ if and only if $2^{-(n+1)} < t - \frac{1}{2} < 2^{-n}$.)

(d) Let $E, F$ denote Banach lattices. A linear operator $T \in \mathscr{L}(E,F)$ is called *preregular* if $q \circ T: E \to F''$ is regular, where $q: F \to F''$ denotes the evaluation map. Proceeding as in the proof of (5.7)(iii), show that $T$ is preregular if and only if all bicompositions $T_{u,v'}$ $(u \in E_+, v' \in F'_+)$ are integral.

4. Let $E, F, G$ be vector lattices and let $T_1: E \to F$, $T_2: F \to G$ denote linear mappings.

(a) If $T_1$ and $T_2$ are regular then so is $T_2 \circ T_1$. Moreover, if $|T_1|$, $|T_2|$, and $|T_2 \circ T_1|$ exist (in particular, if $F$ and $G$ are order complete), then $|T_2 \circ T_1| \leqq |T_2| \circ |T_1|$.

(b) If $E, F, G$ are normed vector lattices and if $T_1 \in \mathscr{L}^r(E,F)$, $T_2 \in \mathscr{L}^r(F,G)$ (Exerc. 3(a)) then $T_2 \circ T_1 \in \mathscr{L}^r(E,G)$ and $\|T_2 \circ T_1\|_r \leqq \|T_2\|_r \|T_1\|_r$.

(c) Suppose $E$ to be an order complete Banach lattice. Then, with multiplication defined by composition of mappings, $\mathscr{L}^r(E)$ is a *Banach lattice algebra* (i.e., a Banach algebra and a Banach lattice such that $|ST| \leqq |S||T|$ for all pairs of elements $S, T$). (Use (a), (b).)

Show that an analogous assertion is true for the complex Banach algebra $\mathscr{L}^r(E_{\mathbb{C}})$ (cf. (1.8) Cor. 1).

5. (Bilinear Forms.) Let $E, F$ denote vector lattices.

(a) A bilinear form $\varphi: E \times F \to \mathbb{R}$ is called *positive* if $\varphi(x,y) \geqq 0$ whenever $x \in E_+$ and $y \in F_+$; $\varphi$ is called *regular* if $\varphi = \varphi_1 - \varphi_2$ where $\varphi_j$ are positive. Under the canonical ordering (defined by "$\varphi \geqq \psi$ *iff* $\varphi - \psi \geqq 0$") the vector space $B^r(E,F)$ of all regular bilinear forms on $E \times F$ is an order complete vector lattice isomorphic to $L^r(E, F^\star)$ and to $L^r(F, E^\star)$. (Show that Formula (2) of § 2 establishes the desired isomorphism.)

(b) Suppose $E, F$ to be normed vector lattices and denote by $\mathscr{B}^r(E,F)$ the vector space of all regular bilinear forms $\varphi$ on $E \times F$ for which $|\varphi|$ is continuous. Show that $\mathscr{B}^r(E,F)$ is isomorphic to $\mathscr{L}^r(E,F')$ (cf. 1.4), and infer from this that $\mathscr{B}^r(E,F)$ is an ideal of $B^r(E,F)$, and an order complete Banach lattice under the $r$-norm $\|\varphi\|_r := \sup\{|\varphi|(x,y): x \in U_+, y \in V_+\}$, where $U_+$ and $V_+$ denote the positive parts of the unit balls of $E$ and $F$, respectively.

*If $E$ and $F$ are Banach lattices, every positive (and hence, every regular) bilinear form on $E \times F$ is continuous.* (Use (a) and (II.5.3).)

(c) Let $K, H$ be compact spaces and let $E \subset C(K)$, $F \subset C(H)$ denote vector sublattices which are dense and contain the constant functions. Then each $\varphi \in B^r(E,F)$ has a unique extension $\bar{\varphi} \in B^r(C(K), C(H))$. (Identify $B^r(E,F)$ with $L^r(E,F')$ and observe that each regular $T: E \to F'$ is continuous.)

Since every regular bilinear form on $C(K) \times C(H)$ is integral (§ 5, Example 1), $B^r(E,F)$ is isomorphic to the Banach lattice $M(K \times H)$ of Radon measures on $K \times H$ under the correspondence $\varphi \mapsto \mu$ given by $\varphi(f,g) = \int f(s) g(t) d\mu(s,t)$ $(f \in E, g \in F)$.

Deduce from this that whenever $\varphi$ is a positive bilinear form on $E\times F$, then $\varphi(u)\geqq 0$ for every $u\in E\otimes F$ contained in the closure of the projective cone $C_p=\operatorname{co}E_+\otimes F_+$ for the $\varepsilon$-topology (cf. § 2, Formula (3)).

(d) A bilinear form $\varphi$ on $E\times F$ is called a *lattice bimorphism* (cf. Fremlin [1972]) if for each $x\in E_+$ and each $y\in F_+$, the partial maps $y\mapsto\varphi(x,y)$ and $x\mapsto\varphi(x,y)$ are lattice homomorphisms of $F$ and $E$ into $\mathbb{R}$, respectively.

Under the assumptions made on $E$ and $F$ in (c) above, show that the following properties of $\varphi\in B(E,F)_+$ are equivalent:

($\alpha$) $\varphi$ is a lattice bimorphism.

($\beta$) There exist points $s\in K$, $t\in H$ and a number $c\geqq 0$ such that $\varphi(f,g)=cf(s)g(t)$ for all $f\in E$, $g\in F$.

($\gamma$) $\varphi$ is an extreme point of the ($\sigma(B(E,F),E\otimes F)$-compact) set $\{\psi\in B(E,F)_+:\psi(e_1,e_2)=\|\varphi\|\}$, where $e_1,e_2$ denote the constant-one functions on $K,H$, respectively.

(For ($\alpha$)$\Rightarrow$($\beta$), use Chap. III, § 1, Example 1. For ($\beta$)$\Leftrightarrow$($\gamma$), use (c).)

(e) Suppose $E$ to be a reflexive Banach lattice. Then the isomorphism $\mathscr{B}^r(E,E')\cong\mathscr{L}^r(E)$ (see (a)) induces on $\mathscr{B}^r(E,E')$ the structure of a Banach lattice algebra (Exerc. 4(c)).

6. (Summable Sequences.) Let $G$ be a topological vector space, and let $(x_n)_{n\in\mathbb{N}}$ be a sequence in $G$. Denote by $x_{\mathsf{H}}$ the sum $\sum_{n\in\mathsf{H}}x_n$ for each finite subset $\mathsf{H}\subset\mathbb{N}$. $(x_n)$ is called *summable* (in $G$) if $\lim_{\mathsf{H}}x_{\mathsf{H}}$ exists in $G$.

(a) The following assertions are equivalent:

($\alpha$) The set $\{x_{\mathsf{H}}:\mathsf{H}\text{ finite}\subset\mathbb{N}\}$ is relatively compact in $G$.

($\beta$) Every subsequence $(x_{n(k)})_{k\in\mathbb{N}}$ is summable.

If the closed bounded subsets of $G$ are complete, then ($\alpha$) and ($\beta$) are equivalent to ($\gamma$): $(x_n)$ is summable.

(b) *If $G$ is a Banach space and every subsequence of $(x_n)$ is summable for $\sigma(G,G')$, then $(x_n)$ is summable in $G$* (theorem of Orlicz-Pettis). (With the aid of (a), show that $(\xi_n)\mapsto\sum_\nu\xi_n x_n:=\sigma(G,G')\text{-}\lim_n\sum_{\nu=1}^n\xi_\nu x_\nu$ defines a weakly compact linear map $l^\infty\to G$, and use the fact that $l^\infty$ has the strict Dunford-Pettis property (Chap. II, Exerc. 26).)

(c) Suppose $G$ to be a Banach lattice and consider the Banach spaces $l^1(G)$ and $l^1[G]$ (§ 2, Formulae (4), (5)) under their canonical ordering defined by "$\boldsymbol{x}\geqq 0$ *iff* $x_n\geqq 0$ $\forall n$".

(i) $l^1[G]$ is a Banach lattice.

(ii) If $G$ is an $AM$-space, $l^1(G)$ is a Banach lattice.

(iii) In general, $l^1(G)$ is not a Banach lattice. (Consider an infinite dimensional $AL$-space and employ the theorem of Dvoretzky-Rogers (5.10) Cor. 4.)

(iv) The condition "$(x_n)$ *summable* $\Rightarrow(|x_n|)$ *summable*" characterizes $AM$-spaces within the class of Banach lattices (Exerc. 16(b)).

(d) Verify the statements of the Note after (3.4) Cor.

7. Let $F$ denote any Banach lattice. Show the following assertions on $F$ to be equivalent:

($\alpha$) The identity map $1_F$ can be approximated, uniformly on compact subsets of $F$, by positive linear maps $\sum_i y'_i\otimes y_i$ ($y'_i\in F'_+, y_i\in F_+$) of finite rank and norm $\leqq 1$.

($\beta$) For every Banach lattice $E$, each positive linear map $E \to F$ of norm $\leqq 1$ can be approximated, uniformly on compact subsets of $E$, by mappings of norm $\leqq 1$ contained in co $E'_+ \otimes F_+$.

Moreover, if the norm conditions in ($\alpha$) and ($\beta$) are omitted, the resulting weaker assertions ($\alpha'$) and ($\beta'$) are equivalent to

($\gamma'$) For every Banach lattice $E$, each compact positive linear map $E \to F$ can be approximated in the operator norm by mappings contained in co $E'_+ \otimes F_+$.

Problem: Does every Banach lattice $F$ satisfy ($\alpha$), or at least ($\gamma'$)?

8. Derive lifting theorems for cone absolutely summing and majorizing linear maps by dualizing Propositions (3.9) and (3.10), respectively. (Cf. the corollary of (II.7.11).)

9. Let $E$ be a Banach lattice and let $x \in E_+$, $x' \in E'_+$. (For the notation used below, see § 3.)

(a) The dual of $(E, x')$ is isomorphic to $(E')_{x'}$. On the other hand, $(E', x)$ (where $x$ is identified with its canonical image in $E''$) is the band in the dual of $E_x$ of all linear forms continuous for the topology induced on $E_x$ by $E$.

(b) Under the isomorphisms established in (a), the adjoint of $E_x \to E$ is the mapping $E' \to (E', x)$, and the adjoint of $E \to (E, x')$ is the mapping $(E')_{x'} \to E'$.

(c) The following propositions are equivalent:

($\alpha$) The interval $[-x, x] \subset E$ is weakly compact ($\beta$) $E_x$ is the dual of $(E', x)$ ($\gamma$) $(E', x)$ is the order continuous dual of $E_x$. (To prove ($\gamma$)$\Rightarrow$($\alpha$), proceed as in the proof of (II.9.3).)

10. (Property ($P$); Def. 4.2.)

(a) Every Banach lattice $F$ possessing ($P$) is order complete. (Show that $\sup_F A = P(\sup_{F''} A)$ for every directed ($\leqq$) majorized subset $A$ of $F$.) Give an example where $F$ has ($P$) but $F$ is not an order complete sublattice of $F''$ (equivalently, where the evaluation map $q: F \to F''$ is not order continuous).

(b) Every dual Banach lattice $F'$ has ($P$). (Consider the restriction map $F''' \to F'$.)

(c) An $AM$-space $F$ has property ($P$) if and only if $F$ is an order complete $AM$-space with unit. (To prove sufficiency apply (II.7.10) Cor. 2.)

Infer from this that an infinite dimensional, separable $AM$-space never has ($P$) (cf. (II.10.4) Cor. 2).

11. If $E$ is an $AM$-space and $F$ is an $AL$-space, and if every compact linear map $E \to F$ possesses a modulus, then $\min(\dim E, \dim F) < \infty$.

(Suppose that $\dim F = \infty$. Observing that by (2.4), $E$ and $F$ as well as their successive duals possess the approximation property, show that the assumption implies the $r$-norm to be equivalent to the $\varepsilon$-norm on $E' \otimes F$; hence, since the $r$-norm induces the $\pi$-norm on $E' \otimes F$ by (5.6), it follows that $E' \otimes_\pi F = E' \otimes_\varepsilon F$ (equivalence of norms). By (2.3) Cor., this implies $E \otimes_\varepsilon F' = E \otimes_\pi F'$ and hence, $E' \otimes_\pi F'' = E' \otimes_\varepsilon F''$. Since $F''$ contains $l^1$ as a band (and hence, as a complemented subspace), the identity map of $E' \otimes l^1$ permits a continuous factoring $E' \otimes_\varepsilon l^1 \to E' \otimes_\varepsilon F'' \to E' \otimes_\pi F'' \to E' \otimes_\pi l^1$ which implies $E'$ to be nuclear [S, IV.10.7] and hence, $E$ to be finite dimensional.)

12. (Hypermajorized Sequences.) Let $H$ denote a Banach space.

(a) If $(x_n)$ is a null sequence in $H$, $(x_n)$ is hypermajorized iff the map $\varphi: l^1 \to H$ given by $(\lambda_n) \mapsto \sum_{n=1}^{\infty} \lambda_n x_n$ is integral.

(b) Each summable sequence in $H$ is hypermajorized. (Show that whenever $(x_n)$ is summable in $H$, the map $\varphi$ of (a) can be factored $l^1 \to l^2 \to c_0 \to H$, and use the fact that the canonical imbedding $l^1 \to l^2$ is absolutely summing (Exerc. 14 (e)).)

(c) Show that if $F$ is a Banach lattice possessing Property $(P)$ (Def. 4.2), each hypermajorized (null) sequence in $F$ is order bounded.

13. Let $G, H$ be Banach spaces.

(a) The Banach space $l^1(G)$ (§ 2) is canonically isomorphic to $l^1 \tilde{\otimes}_\varepsilon G$, and the Banach space $l^1[G]$ is canonically isomorphic to $l^1 \tilde{\otimes}_\pi G$. (Cf. [S, IV.10.6].)

(b) For $T \in \mathscr{L}(G, H)$ the following are equivalent:

($\alpha$) $T$ is right semi-integral, and there exists a Banach space $H_1$ and an isometric injection $R: H \to H_1$ such that $R \circ T$ is integral with integral norm $\leqq 1$.

($\beta$) There exists a positive Radon measure $\mu$ on the $\sigma(G', G)$-compact unit ball $U^\circ$, of total variation $\|\mu\| \leqq 1$, such that $\|Tx\| \leqq \int |\langle x, x' \rangle| \, d\mu(x')$ for all $x \in G$.

($\gamma$) $T$ is absolutely summing and the induced map $\boldsymbol{T}: l^1(G) \to l^1[G]$ has norm $\leqq 1$ (for notation, see Note preceding (3.5)).

($\delta$) For each contraction $S: c_0 \to G$, $T \circ S$ is cone absolutely summing with $l$-norm $\leqq 1$.

(c) For each absolutely summing $T: G \to H$, let $\|T\|_{as}$ denote the minimal norm of a measure $\mu$ satisfying ($\beta$) of (b). On the vector space $\mathscr{L}^{as}(G, H)$ of absolutely summing maps $G \to H$, $T \mapsto \|T\|_{as}$ is a norm which makes $\mathscr{L}^{as}(G, H)$ into a Banach space. (Use the equivalence ($\beta$)$\Leftrightarrow$($\gamma$) of (b).)

(d) The vector space $\mathscr{L}^{hm}(G, H)$ of all hypermajorizing maps $G \to H$ is a Banach space under the norm $T \mapsto \|T\|_{hm} := \|T'\|_{as}$. (Use 5.12.)

14. The Rademacher function $\varphi_n$ $(n = 0, 1, 2, \ldots)$ is defined as follows: Divide $[0,1]$ into $2^n$ intervals of equal length and let $\varphi_n(t) = 0$ if $t$ is an endpoint, $\varphi_n(t) = (-1)^{k-1}$ if $t$ is an interior point of the $k$-th consecutive subinterval $(k = 1, \ldots, 2^n)$.

(a) (Khintchin's Inequality.) If $\xi_n$ $(n = 1, \ldots, p)$ are complex numbers then

$$\left(\sum_{n=1}^{p} |\xi_n|^2\right)^{1/2} \leqq \sqrt{3} \int_0^1 \left|\sum_{n=1}^{p} \xi_n \varphi_n(t)\right| dt .$$

(For each $p \in \mathbb{N}$, let $\varrho_p := \sum_{n=1}^{p} |\xi_n|^2$ and show that $\int_0^1 |\sum_{n=1}^{p} \xi_n \varphi_n(t)|^4 \, dt \leqq 3 \varrho_p^2$. Apply Hölder's inequality with conjugate exponents $\frac{3}{2}$ and 3 to $\varrho_p = \int_0^1 |\sum_{n=1}^{p} \xi_n \varphi_n(t)|^2 \, dt$.)

(b) Denote by $\mu$ Lebesgue measure on $[0,1]$ and let $F$ denote the closed vector subspace of $L^1(\mu)$ spanned by the set $\{\varphi_n : n = 0, 1, \ldots\}$. Using (a) show that as a topological vector space, $F$ is isomorphic to $l^2$.

(c) Let $B[0,1]$ denote the $AM$-space with unit of all bounded Borel functions on $[0,1]$ under its canonical order and the supremum norm. If $R[0,1]$ denotes the closed vector subspace of $B[0,1]$ spanned by the sequence $(\varphi_n)_{n \geqq 0}$, then $R[0,1]$ is an ordered $B$-space with unit under the induced order and norn.

(d) Show that $R[0,1]$ cannot be a vector sublattice of $B[0,1]$. (If $R[0,1]$ were an $AM$-space, then by (5.6) the imbedding $R[0,1] \to F$ would be absolutely summing and hence, by (5.5) Cor. 2 and (5.10) Cor. 2, nuclear which is absurd, since $(\varphi_n)$ is a weak but not a strong null sequence in $L^1(\mu)$.)

(e) The canonical imbedding $l^1 \to l^2$ is absolutely summing. (Cf. Exerc. 15 below.) (For any finite set $\mathsf{H} \subset \mathbb{N}$ and each $t$, $0 \leqq t \leqq 1$, consider the linear form $\varphi_{\mathsf{H}}(t): (\lambda_n) \mapsto \sum_{n \in \mathsf{H}} \lambda_n \varphi_n(t)$ on $l^1$ and apply (a).) Infer that there exist absolutely summing maps $T$ whose adjoint $T'$ fails to be absolutely summing.

15. (Grothendieck's Inequality. See Grothendieck [1956] or Lindenstrauss-Pelczynski [1968].)

(a) Let $(a_j)_{j \in \mathbb{N}}$ be a summable sequence in $l^1$, $a_j = (\alpha_{ij})_{i \in \mathbb{N}}$, and let $(x_l)$, $(y_k)$ be arbitrary sequences in $l^2$. Then for each $n \in \mathbb{N}$,

$$\left|\sum_{i,j=1}^{n} \alpha_{ij}(x_i \,|\, y_j)\right| \leqq K_G \, \|(a_j)\|_\varepsilon \sup_l \|x_l\| \sup_k \|y_k\|$$

where $K_G$ is an absolute constant $\leqq \frac{1}{2}(\exp \pi/2 - \exp(-\pi/2))$.

(b) Every continuous linear map $T$ from an $AL$-space $L$ into a Hilbert space $H$ is absolutely summing. (Use (a) to show that for any null sequence $(x_n)$ in $L$, the map $(\alpha_n) \mapsto \sum_n \alpha_n T x_n$ of $l^1$ into $H$ is absolutely summing. Employ this and Exerc. 12 to show that any composition of $T$ with a continuous linear map of $H$ into an $AM$-space is hypermajorizing and hence, integral.)

(c) Let $H_1, H_2$ be Hilbert spaces and let $T \in \mathscr{L}(H_1, H_2)$. $T$ is a Hilbert-Schmidt operator if and only if $T$ can be factored through an $AM$-space or through an $AL$-space (cf. 6.9).

(d) Let $G$ and $H$ denote Banach spaces. A linear map $T: G \to H$ is called *p-absolutely summing* $(1 \leqq p < +\infty)$ if there exists a constant $C \geqq 0$ such that for any finite subset $\{x_1, \ldots, x_n\}$ of $G$, one has

$$\left(\sum_{i=1}^{n} \|T x_i\|^p\right)^{1/p} \leqq C \sup_{\|x'\| \leqq 1} \left(\sum_{i=1}^{n} |\langle x_i, x' \rangle|^p\right)^{1/p}.$$

Any 2-absolutely summing map can be factored through a suitable Hilbert space (Pietsch [1967]). Show that any continuous linear map of an $AM$-space into an $AL$-space is 2-absolutely summing.

(e) Let $M$ denote an $AM$-space, $L$ an $AL$-space, and $H$ a Hilbert space. For any choice of the operators $R \in \mathscr{L}(H, M)$, $S \in \mathscr{L}(L, H)$, and $T \in \mathscr{L}(M, L)$, the compositions $S \circ T$ and $T \circ R$ are nuclear.

(f) For a Banach space $G$, the following assertions are equivalent:

($\alpha$) $G$ is isomorphic (as a topological vector space) to a Hilbert space.

($\beta$) Every c.a.s. map of a Banach lattice into $G$ is absolutely summing.

($\gamma$) Every majorizing map of $G$ into a Banach lattice is hypermajorizing.

(This characterization of Hilbert space is due to Schlotterbeck [1969].)

16. (Characterization of $AL$- and $AM$-spaces. For the following, cf. Lotz [1973a], Fremlin [1974b], Wickstead [1974].) Let $E$ denote a Banach lattice.

(a) These assertions are equivalent:

($\alpha$) As a topological vector lattice, $E$ is isomorphic to an $AL$-space.

($\beta$) Every continuous linear map of $E$ into any Banach lattice is preregular (cf. Exerc. 3).

($\gamma$) Each order bounded null sequence in $E$ is hypermajorized.

($\delta$) Every majorizing map of any Banach space into $E$ is hypermajorizing.

($\varepsilon$) Each positive linear map $c_0 \to E$ is hypermajorizing.

(Establish the implication ($\varepsilon$)$\Rightarrow$($\alpha$) through the following steps (Lotz [1973a]): 1. Every composition $S \circ T$, where $0 \leqq S \in \mathscr{L}(E, l^2)$ and $0 \leqq T \in \mathscr{L}(c_0, E)$, is nuclear (Exerc. 15). 2. There exists a constant $k \in \mathbb{R}_+$ such that $\|S \circ T\|_l \leqq k \|S\| \, \|T\|$ (Def. 3.5) for all $0 \leqq S \in \mathscr{L}(E, l^2)$ and $0 \leqq T \in \mathscr{L}(c_0, E)$. 3. No closed vector sublattice of $E$ is isomorphic to $c_0$ and hence, by (II.5.13), $E$ has order continuous norm and thus satisfies the assertions of (II.5.10). 4. If $(P_n)$ is a sequence of disjoint band projections on $E$ then for each $x \in E$, $(P_n x)$ is a summable sequence in $E$. Let $(\alpha_n) \in l^2$ and let $(x'_n)$ be a sequence in $E'_+$ satisfying $\|x'_n\| \leqq 1$ for all $n$. Then for each $u \in E_+$, the mapping $(\lambda_n) \mapsto (\langle \alpha_n \lambda_n P_n u, x'_n \rangle)_{n \in \mathbb{N}}$ of $c_0$ into $l^2$ is of $l$-norm $\leqq k \|(\alpha_n)\|_2 \|u\|$. Deduce from this that $(\sum_n \|P_n x\|^2)^{1/2} \leqq k \|x\|$ for all $x \in E$. 5. If $(x_n)$ is a positive, summable sequence of disjoint elements of $E$ then $\sum_n \|x_n\| \leqq k^2 \|\sum_n x_n\|$. 6. For $x \in E$, define

$$q(x) := \sup \{ \textstyle\sum_n \|x_n\| : x_n \in E_+, \sum_n x_n = |x|, \, x_n \wedge x_m = 0 \; (m \neq n) \} .$$

Show that $q$ is a lattice norm on $E$ equivalent to the given norm and such that $q(x+y) = q(x) + q(y)$ whenever $x, y \in E_+$ and $x \wedge y = 0$. This implies $q$ to be an $L$-norm (Def. II.8.1; cf. Bernau [1973]) and proves ($\varepsilon$)$\Rightarrow$($\alpha$).)

(b) Using the duality of $AL$- and $AM$-spaces and (a), show the following to be equivalent:

($\alpha$) As a topological vector lattice, $E$ is isomorphic to an $AM$-space.

($\beta$) Every continuous linear map of any Banach lattice into $E$ is preregular.

($\gamma$) Each summable sequence in $E$ is the difference of two positive summable sequences.

($\delta$) Every c.a.s. mapping of any Banach lattice into $E$ is absolutely summing.

($\varepsilon$) Each positive linear map $E \to l^1$ is absolutely summing.

(c) As a topological vector lattice, $E$ is isomorphic to an $AM$-space iff each compact linear map $c_0 \to E$ has a compact modulus. Dually, $E$ is (t.v.l.) isomorphic to an $AL$-space iff each compact linear map $E \to l^1$ has a compact modulus. (Use (b), assertion ($\gamma$).)

17. Let $E$ denote a Banach lattice.

(a) $E$ is isomorphic (as a topological vector lattice) to an $AL$-space if (and only if) each continuous linear map $E \to l^2$ is absolutely summing. (If $E$ contains a closed vector sublattice isomorphic (as a topological vector lattice) to $c_0$, then by (II.7.9) any continuous linear map $c_0 \to l^\infty$ can be factored through $E$; on the other hand, the mapping $(\lambda_n) \mapsto (n^{-1} \lambda_n)$ of $c_0$ into $l^2$ can be factored $c_0 \to l^\infty \to l^2$ but fails to be absolutely summing. Thus the assumption implies that $E$ does not contain a closed vector sublattice isomorphic to $c_0$; from here proceed as in Exerc. 16 (a), ($\varepsilon$)$\Rightarrow$($\alpha$).)

(b) Deduce from (a) that any $\mathscr{L}_1$-space (in the sense of Lindenstrauss-Pelczynski [1968]) which is a Banach lattice, is isomorphic (as a topological vector lattice) to an $AL$-space. Similarly, any $\mathscr{L}_\infty$-space which is a Banach lattice is isomorphic to an $AM$-space.

18. (Hilbert Lattices.)

(a) Let $H_1, H_2$ be Hilbert lattices (Def. 6.6) each isomorphic to $l^2$. The Banach lattice $\mathscr{L}^r(H_1, H_2)$ (cf. 1.4) contains Banach sublattices isomorphic to $l^\infty$ (and hence, to $c_0$). (Denoting by $e_n := (\delta_{nm})_{m\in\mathbb{N}}$ the *n-th* unit vector in $l^2$ ($n\in\mathbb{N}$), show that for each $(\lambda_n)\in l^\infty$, $\sum_n \lambda_n(e_n \otimes e_n)$ defines an operator $T: H_1 \to H_2$ such that $(\lambda_n)\mapsto T$ is the desired isomorphism $l^\infty \to \mathscr{L}^r(H_1, H_2)$.)

(b) Every infinite dimensional Hilbert lattice contains a Hilbert sublattice isomorphic to $l^2$. (Consider an infinite dimensional principal band $H_0$ of $H$ and represent $H_0$, as in the proof of (6.7), as $L^2(\mu, X_0)$ where $X_0$ is compact Stonian. There exists a sequence $(X_n)_{n\in\mathbb{N}}$ of disjoint, non-void, open-and-closed subsets of $X$ whose union is dense in $X$; the set of all $f\in H_0$ which are constant on each $X_n$ (cf. II.7.8), is a Hilbert sublattice of $H_0$ isomorphic to $l^2$.)

(c) If $H_1, H_2$ are infinite-dimensional Hilbert lattices, there always exist compact linear maps $H_1 \to H_2$ possessing no modulus. (Using (b), reduce the problem to the special case where $H_1 = H_2 = l^2$. Consider $l^2$ to be the Hilbert direct sum $\bigoplus_{p=1}^\infty l_{2^p}^2$ and adapt the method of § 1, Example 2.)

(d) Let $H_1, H_2$ be infinite dimensional Hilbert lattices. Then the Banach lattice $\mathscr{L}^r(H_1, H_2)$ never has order continuous norm; moreover, the ideal $\mathscr{H}(H_1, H_2)$ of Hilbert-Schmidt operators is properly contained in $\mathscr{L}^r(H_1, H_2)$. (Use (a), (b), (II.5.14), and (6.10). Note that if $(e_n)$ is a positive, orthonormal basis of a Hilbert sublattice of $H_1$ isomorphic to $l^2$ and if $(f_n)$ is a positive orthonormal sequence in $H_2$, then for $(\lambda_n)\in l^\infty$, $T = \sum_n \lambda_n(e_n \otimes f_n)$ is a regular operator from $H_1$ into $H_2$ with modulus $|T| = \sum_n |\lambda_n|(e_n \otimes f_n)$.)

(e) If $H_1, H_2$ are Hilbert lattices represented (as in (6.7)) as $L^2(\mu, X)$, $L^2(\nu, Y)$ for Radon measures $\mu, \nu$ on locally compact spaces $X$, $Y$, respectively, then $\mathscr{H}(H_1, H_2) \cong L^2(\mu \otimes \nu)$ (cf. 6.5).

(f) Extend all results of § 6 to the case of complex scalars, using the method of complexification explained in Chap. II, § 11 and the results of § 1 of the present chapter.

19. Let $H_1, H_2$ be Hilbert spaces and let $T\in\mathscr{L}(H_1, H_2)$.

(a) If $T$ is a Hilbert-Schmidt operator, its Hilbert-Schmidt norm $\|T\|_\sigma$ is given by $(\sum_j \lambda_j)^{1/2}$, where $\lambda_j$ ($>0$) runs through the sequence of non-zero eigenvalues of $T^* \circ T$ (or equivalently, of $T \circ T^*$), counted according to their multiplicity. If $T$ is nuclear, its nuclear norm $\|T\|_\pi$ equals $\sum_j \sqrt{\lambda_j}$. (Use (6.5) and its corollary.)

(b) If $H_1, H_2$ are Hilbert lattices and $T$ is Hilbert-Schmidt, then $\| |T| \|_\sigma = (\sum_j \lambda_j)^{1/2}$ (see (a)). If $T$ is nuclear and $(\varrho_k)$ denotes the sequence of non-zero eigenvalues of $|T|^* \circ |T|$, counted according to their multiplicity, then the modulus $|T|$ is nuclear iff $\sum_k \sqrt{\varrho_k}$ is finite and, in this case, $\| |T| \|_\pi = \sum_k \sqrt{\varrho_k}$. Note that $\sum_k \varrho_k = \sum_j \lambda_j$.

(c) Suppose $H_1, H_2$ to be infinite dimensional Hilbert lattices. There exist nuclear operators $T: H_1 \to H_2$ such that $\|T\|_\pi = 1$ while $\| |T| \|_\pi$ is arbitrarily small. (Use Exerc. 18 (b) and consider the orthogonal matrix $A_n$ ($n = 2^p$) constructed in § 1, Example 1. Using (a), (b) show that $\|A_n\|_\pi = 2^p$ while $\| |A_n| \|_\pi = 2^{p/2}$.)

(d) Conclude from (c) that whenever $H_1, H_2$ are infinite dimensional Hilbert lattices, the space $\mathscr{N}(H_1, H_2)$ is not an ideal of $\mathscr{L}^r(H_1, H_2)$. Moreover, $\mathscr{N}(H_1, H_2)$ is not a topological vector lattice under its canonical order and ($\pi-$) norm topology. (Observe that the dual of $\mathcal{N}(H_1, H_2)$ can be identified with $\mathcal{L}(H_1, H_2)$, and apply Exerc. 18 (c).)

Do the nuclear operators form a sublattice of $\mathcal{L}^r(H_1, H_2)$?

20. (Vector Valued $L^p$-Spaces.)

Let $(X, \Sigma, \mu)$ denote a measure space, $H$ a Banach space, and denote by $M = M(X, \Sigma, \mu)$ the vector space of all $\mathbb{R}$-valued $\Sigma$-measurable functions with identification of functions equal a.e. $(\mu)$. As is customary, we do not distinguish between a function and its equivalence class unless confusion is likely to result.

(a) A function $f \in M$ is called *simple* if $f = \sum_{i=1}^n \alpha_i \chi_i$ where $\alpha_i \in \mathbb{R}$ and $\chi_i$ is the characteristic function of a set $S_i \in \Sigma$ of finite $\mu$-measure $(i = 1, \dots, n)$; the set $M_0$ of simple functions is a vector subspace (and sublattice) of $M$. A function $f: X \to H$ is called *simple* (sometimes, *$\mu$-simple*) if $f(s) = \sum_{i=1}^n \chi_i(s) y_i$ where $y_i \in H$ and the $\chi_i$ are characteristic functions of sets $S_i$ such that $\mu(S_i) < +\infty$ $(i = 1, \dots, n)$. The $H$-valued simple functions form a vector space which is canonically isomorphic to $M_0 \otimes H$.

(b) A function $f: X \to H$ is called $\mu$-measurable (or, if $\mu$ is understood, measurable) if for each set $S \in \Sigma$ such that $\mu(S) < +\infty$, there exists a sequence $(f_n)$ in $M_0 \otimes H$ which converges to $f \chi_S$ in ($\mu$-)measure, that is, for which $\lim_n \mu\{s \in X : \|f_n(s) - f \chi_S(s)\| \geqq \varepsilon\} = 0$ for all $\varepsilon > 0$. Show that $f: X \to H$ is measurable if and only if (i) Each $S \in \Sigma$ with $\mu(S) < +\infty$ contains a $\mu$-null set $S_0$ such that $f(S \setminus S_0)$ is a separable subset of $H$, and (ii) For each $y' \in H'$, the real function $s \mapsto \langle f(s), y' \rangle$ is measurable. (In other words, $f$ is measurable iff $f$ is weakly measurable and $\mu$-essentially separably valued on integrable sets. Cf. [DS, III.6.11].)

(c) Let $1 \leqq p \leqq +\infty$. Show that the spaces $L^p_H(\mu)$ (§ 7, Example 4) are Banach spaces, and Banach lattices if $H$ is a Banach lattice. Similarly, the spaces $\overline{L}^p_{H'}(\mu)$ (see definition preceding (7.6)) are Banach spaces (respectively, Banach lattices if $H$ is a Banach lattice); moreover, $L^p_{H'}(\mu)$ can be identified with a Banach subspace (respectively, Banach sublattice) of $\overline{L}^p_{H'}(\mu)$ with identity occurring for separable $H'$. Finally, show that $L^\infty_H(\mu) \neq L^\infty(\mu) \tilde{\otimes}_\varepsilon H$ in general.

(d) We use the notation of § 8, Examples 1,2; in particular, let $(X, \Sigma, \mu)$ and $(Y, \mathrm{T}, \nu)$ be $\sigma$-finite measure spaces, and let $E := L^p(\mu)$, $F := L^q(\nu)$ where $1 \leqq p, q \leqq +\infty$ and $p', q'$ denote the conjugate exponents. Let $\overline{L}^q_{E'}(\nu)$ be defined as in the discussion preceding (7.6), and let $\overline{L}^{p'}_F(\mu)$ be the Banach lattice of $\sigma(L^q(\nu), L^{q'}(\nu))$-measurable functions $f: X \to F = L^q(\nu)$ for which $s \mapsto \|f(s)\|^{p'}$ is $\mu$-integrable if $p > 1$ (respectively, for which $s \mapsto \|f(s)\|$ is uniformly bounded a.e. $(\mu)$ if $p = 1$), under the obvious norm and canonical order.

Then (§ 8, Example 2, Formulae (2), (2')) $\mathcal{H}_{pq} \subset \overline{L}^q_{E'}(\nu)$ and $\mathcal{J}_{pq} \subset \overline{L}^{p'}_F(\mu)$, where "$\subset$" denotes an isomorphism onto a Banach sublattice. The first of these isomorphisms is surjective whenever $E$ is separable, and the second is surjective whenever $F$ is separable (or, in case $q = +\infty$, whenever $L^1(\nu)$ is separable).

However, if $1 < p, q < +\infty$ then $\mathcal{H}_{pq} \cong L^q_{E'}(\nu)$ and $\mathcal{J}_{pq} \cong L^{p'}_F(\mu)$ without any separability assumptions on $E$ or $F$ (§ 8, Example 3).

(e) For $\sigma$-finite measures $\mu, \nu$ and $F := L^1(\nu)$, one has the isomorphisms $L^1_F(\mu) \cong L^1(\mu) \tilde{\otimes}_\pi L^1(\nu) \cong L^1(\mu \otimes \nu)$ (§ 7, Examples 2,4). This is the counterpart, for the $\pi$-norm, to the isomorphisms $C(K, C(H)) \cong C(K) \tilde{\otimes}_\varepsilon C(H) \cong C(K \times H)$ where $K, H$ are compact spaces (§ 2, Example 1).

21. (Counterexamples.)

(a) Let $E, F$ be Banach lattices. Then, even if $F$ is supposed to be order complete or to have property ($P$) (Def. 4.2). $E' \tilde{\otimes}_\varepsilon F$ is not a vector sublattice of $\mathscr{L}^r(E,F)$ in general. (Use § 1, Example 2 or Exerc. 11.) It is unknown if the nuclear operators $E \to F$ constitute a sublattice of $\mathscr{L}^r(E,F)$ in general, even though each nuclear map is regular.

(b) Denote by $\mu$ Lebesgue measure on $[0,1]$ and let $E := L^1(\mu)$, $F := L^\infty(\mu)$. There exists an operator $T \in \mathscr{L}(E,F)$ which is not weakly compact but possesses a modulus $|T|$ of rank 1. (Let $\chi_n (n=0,1,2,\ldots)$ denote the characteristic function of the left open interval $(2^{-(n+1)}, 2^{-n}]$ and denote by $\varphi_n$ the $n$-th Rademacher function (see Exerc. 14 above). Consider the operator $T$ given by $Tf = \sum_{n=0}^\infty (\int f \varphi_n \, d\mu) \chi_n$.)

(c) An operator in $\mathscr{H}_{\infty q}$ or $\mathscr{J}_{p1}$ ($1 < p, q < +\infty$; § 8, Example 2) is not necessarily compact. (Consider the operator $T: L^\infty(\mu) \to L^2(\mu)$ given by $Tf = \sum_n (\int f \chi_n \, d\mu) \varphi_n$, where $\varphi_n, \chi_n, \mu$ have the same meaning as in (b). Then $T \in \mathscr{H}_{\infty 2}$ and $T' \in \mathscr{J}_{21}$ are not compact.)

(d) In each of the cases (i), (ii), (iv), (v) of (9.9), give an example of an operator $T \in (E' \otimes F)^{\perp\perp}$ so that $E$ and $F$ satisfy the respective hypotheses of (9.9) but $T$ does not. (Cf. § 10, Example 1.)

22. ($l$- and $m$-Tensor Products.)

(a) If $E$, $F$ are ordered vector spaces with respective positive cones $E_+, F_+$ (Def. II.1.2), then the convex hull $C_p$ of $E_+ \otimes F_+$ (the projective cone) is a proper cone (that is, satisfies $C_p \cap (-C_p) = \{0\}$). (Cf. Dermenjian-Saint Raymond [1970].)

(b) Suppose $E, E_1, F, F_1$ to be Banach lattices and let $S: E \to E_1$, $T: F \to F_1$ be regular (Def. 1.1). The tensor product $S \otimes T$ is continuous as a mapping $E \otimes_l F \to E_1 \otimes_l F_1$ (and as a mapping $E \otimes_m F \to E_1 \otimes_m F_1$). Moreover, $\|T\|_r \leqq 1$, $\|S\|_r \leqq 1$ implies $\|S \otimes T\|_r \leqq 1$ (for the extension to the respective completions) in either case.

(c) The assertion of (b) can fail if $S$ or $T$ is not regular. (Let $E$ be an infinite dimensional $AM$-space, $E_1$ an infinite dimensional $AL$-space, and let $S: E \to E_1$ be a continuous linear map which is not regular (Exerc. 11). Let $F = F_1 = l^1$ and let $T$ be the identity map of $l^1$. Since $E \otimes_l l^1 = E \otimes_\varepsilon l^1$ and $E_1 \otimes_l l^1 = E_1 \otimes_\pi l^1$ (§ 7, Examples 1, 2), the continuity of $S \otimes T: E \otimes_l F \to E_1 \otimes_l F_1$ would imply $S$ to be absolutely summing (§ 2, Example 4) and hence, $S$ would be regular by (4.3).)

(d) If $E, F$ are Banach lattices and we employ the notation of Theorem 7.2, then for every closed vector sublattice $F_0$ of $F$, $E \tilde{\otimes}_l F_0$ can be identified with a closed vector sublattice of $E \tilde{\otimes}_l F$ (respectively, $E \tilde{\otimes}_m F_0$ can be identified with a closed vector sublattice of $E \tilde{\otimes}_m F$). Using (7.5), deduce from this another proof of (II.8.9).

23. (Kernel Operators on $L^p$-Spaces.) We use the notation of § 8, Examples 1—5. $(X, \Sigma, \mu)$ and $(Y, \mathrm{T}, \nu)$ are supposed to be $\sigma$-finite measure spaces.

(a) A linear operator $T: L^\infty(\mu) \to L^1(\nu)$ is a kernel operator iff $T$ is nuclear and order continuous. (Cf. § 8, Example 4 and observe that every kernel operator $L^p(\mu) \to L^q(\nu)$ is order continuous.)

(b) Give a detailed proof of Prop. 9.8. (Use (a) and (9.6), observing that the canonical maps $E_x \to E$ $(x \in E_+)$ are order continuous.)

(c) Generalize Example 5 of § 8 as follows: If $p=1$, $1<q<+\infty$ or if $1<p<+\infty$, $q=+\infty$ then every continuous linear operator $L^p(\mu) \to L^q(\nu)$ is a kernel operator. (Use (1.5), (9.8), and (9.9).)

(d) If $1 \leqq p < +\infty$ then the integral maps (Def. 5.1) from $L^p(\mu)$ into $L^1(\nu)$ form a Banach lattice identical to $\mathscr{K}_{p1}$. Dually, if $1 < q \leqq +\infty$ then $\mathscr{J}_{\infty q}$ is the Banach lattice of all integral maps $L^\infty(\mu) \to L^q(\nu)$.

24. Let $E, F$ be Banach function lattices satisfying the assumptions made in Theorem 9.7.

(a) Let $j$ denote the lattice isomorphism (injection) $\mathscr{K}(X) \to E$ with dense range. The order adjoint $j^\star$ (Chap. III, Exerc. 24) is an injection $E' \to \mathscr{M}(X)$ which maps the Šilov boundary of $E_+$ (see II.5.7) onto a vaguely (i.e., $\sigma(\mathscr{M}(X), \mathscr{K}(X))$-) compact set $\mathsf{M}$ of positive Radon measures on $X$. Hence, the norm of an element $f \in E$ is given by $\|f\| = \sup_{\mu \in \mathsf{M}} \int |f| \, d\mu$, and a universally measurable function $f: X \to \overline{\mathbb{R}}$ belongs to (an equivalence class of) $E$ iff for each $\varepsilon > 0$, there exists $g \in \mathscr{K}(X)$ such that $\int |f-g| \, d\mu < \varepsilon$ uniformly for $\mu \in \mathsf{M}$.

(b) Show that $j^\star: E' \to \mathscr{M}(X)$ is a lattice isomorphism onto an ideal $\mathscr{M}_E(X)$ of $\mathscr{M}(X)$. Infer from this that a Radon measure $\mu$ on $X$ is in $\mathscr{M}_E(X)$ iff for each $f \in E_+$, $\int f \, d|\mu| < +\infty$. (Observe that for each $f \in \mathscr{K}(X)_+$, $j([-f, f])$ is dense in the order interval $[-j(f), j(f)]$ of $E$ and hence, $j^\star(|x'|) = |j^\star(x')|$ for $x' \in E'$. Moreover, if $0 \leqq \mu \leqq j^\star(x')$ for some $x' \in E'_+$, then $\mu$ is continuous on $\mathscr{K}(X)$ for the topology induced by $E$.)

(c) Let $k$ denote the lattice isomorphism (injection) $\mathscr{K}(Y) \to F$ with range dense for $\sigma(F, F_{00}^\star)$. The order adjoint $k^\star$ maps $F_{00}^\star$ onto an ideal $\mathscr{M}_{F_{00}^\star}(Y)$ of $\mathscr{M}(Y)$. $\mathscr{M}_{F_{00}^\star}(Y)$ is an ideal of order continuous (or normal) Radon measures on $Y$ iff $k$ is order continuous. (Remark that $k(\mathscr{K}(Y))$ is a dense vector sublattice of the locally convex vector lattice $(F, o(F, F_{00}^\star))$ (cf. Chap. II, Exerc. 28).)

25. (Kernel Operators $C(X) \to C(Y)$.) Let $E := C(X)$, $F := C(Y)$ where $X, Y$ are compact spaces. *For the purposes of this exercise only*, a continuous linear operator $T: E \to F$ is called a *kernel operator* if there exist a real function $K$ on $X \times Y$ and a positive Radon measure $\mu$ on $X$ with the following properties: The function $t \mapsto K(\ ,t)$ is continuous from $Y$ into $L^1(\mu)$ for the topology $\sigma(L^1(\mu), C(X))$, and $T$ is defined by $Tf(t) = \int K(s,t) f(s) \, d\mu(s)$ $(f \in E, t \in Y)$.

(a) $\mathscr{L}(E,F)$ (operator norm) is isomorphic with the Banach space $C(Y, E'_\sigma)$ of continuous functions $g: Y \to (E', \sigma(E', E))$ under the norm $\|g\| = \sup_{t \in Y} \|g(t)\|$. Letting $\mu_t := g(t)$, this isomorphism $T \mapsto g$ is given explicitly by the identity $Tf(t) = \int f(s) \, d\mu_t(s)$ $(f \in E, t \in Y)$. The function $g$ corresponding to $T$ will be denoted by $g_T$. (Identifying $Y$ with the $\sigma(F', F)$-compact set of all Dirac measures on $Y$, $g_T$ is nothing but the restriction of the adjoint $T': F' \to E'$ to $Y$.)

(b) $T \in \mathscr{L}(E,F)$ is weakly compact (respectively, compact) iff $g_T$ (see (a)) is continuous from $Y$ into $(E', \sigma(E', E''))$ (respectively, into the normed dual $E'$). (Cf. II.9.4.)

(c) $T \in \mathscr{L}(E,F)$ is a kernel operator if and only if $g_T$ maps $Y$ into a principal band of $E' = M(X)$. Moreover, a kernel operator given by the kernel $K$ and Radon measure $\mu$ is compact iff

$$(**) \qquad \lim_{t \to t_0} \int |K(s,t) - K(s,t_0)| \, d\mu(s) = 0$$

for each $t_0 \in Y$. (Observe that each principal band of $E' = M(X)$ is isomorphic to $L^1(\mu, X)$ for some positive Radon measure $\mu$ on $X$ (and conversely), and use (b).)

(d) Suppose $Y$ to be metrizable (or, equivalently, $C(Y)$ separable (II.7.5).) Every weakly compact $T \in \mathscr{L}(E,F)$ is a kernel operator; in particular, if in addition $X$ is quasi-Stonian, every $T \in \mathscr{L}(E,F)$ is a kernel operator. (Use (c), observing that $T'(Y)$ is contained in a separable subspace of $M(X)$. For the second assertion, employ (II.10.4) Cor. 1.)

(e) If $T \in \mathscr{L}(E,F)$ is represented by the function $g_T\colon Y \to E'_\sigma$, and if the function $t \mapsto |g_T(t)|$ is continuous from $Y$ into $E'_\sigma = (E', \sigma(E',E))$, then $|T|$ exists and is represented by $|g_T|$. In particular, if $T$ is a compact kernel operator given by the pair $(K, \mu)$, then $|T|$ exists and is a compact kernel operator given by the pair $(|K|, \mu)$. (A special case of this is considered in § 1, Example 3.)

Generally, the function $|g_T|\colon t \mapsto |\mu_t|$ is lower semi-continuous from $Y$ into $E'_\sigma$; thus if $Tf(t) = \int f(s) \, d\mu_t(s)$, the operator $S\colon E \to F''$ defined by $Sf(t) = \int f(s) \, d|\mu_t|(s)$ maps $E_+$ into the cone of positive, lower semi-continuous real functions on $Y$. Deduce from this that if $T$ is a kernel operator given by the kernel $K$ then, even if the modulus $|T| \in \mathscr{L}(E,F)$ exists, $|T|$ is not necessarily a kernel operator given by the kernel $|K|$. (Let $X = Y$ be the closed multiplicative subsemigroup of $[0,1]$ generated by some $s \in (0,1)$, and consider the operator $Tf(t) = f(t) - f(t^2)$; cf. Exerc. 2 (e).)

Chapter V

# Applications

## Introduction

This final chapter deals with applications of the theory presented in this book to various areas of analysis; especially to approximation theory, spectral theory, and ergodic theory. Extensive use is made of the results and techniques developed in Chapters II—IV, so the reader may judge their efficiency and usefulness. We proceed to give a more detailed survey.

Section 1 discusses an imbedding procedure (essentially the method of ultraproducts familiar from mathematical logic) which proves extremely useful in later sections (§§ 2, 4—7). Section 2 gives an account of the chapter of approximation theory known as Korovkin theory—the accent being on a complete and simple characterization of Banach lattices possessing a (finite) Korovkin family. Section 3 characterizes the cyclic Banach spaces of general spectral theory as Banach lattices with quasi-interior positive elements and order continuous norm. The relationship is so close that the theory of Banach lattices can be employed rather comfortably to establish many results not easily obtained otherwise; this application is a good, and perhaps the most beautiful, example of the relevance of order theory for an area of analysis upon which it would not immediately be suspected to bear. It is true that this investigation is far from complete, but it should serve well as the application it is meant to be.

Sections 4 and 5 derive most of the known and interesting results of the spectral theory of a single positive operator defined on a (complex) Banach lattice; Section 4 concentrates on the general case while Section 5 concentrates on irreducible operators (which, from other angles, were investigated in Chap. III, §§ 8, 10). Section 6 takes up the problem of finding criteria guaranteeing that an irreducible positive operator is not topologically nilpotent (or quasi-nilpotent). In a sense, these three sections close the circle of this book, taking up many topics of Chap. I in the infinite dimensional case and thus exhibiting the rather long distance that had to be covered to obtain the results now available.

The last two sections attempt to show how the modulus of a linear operator, treated in detail in Chapter IV, can be advantageously employed to extend the techniques previously developed for positive operators, to suitable classes of non-positive operators on Banach lattices. Section 7 pursues this goal in the spectral theory of a single operator while Section 8 does the same for mean ergodic theory. Again, these results—even though coherent—are intended to exemplify possible approaches rather than supply a final account. This seems only appropriate for areas of research that are far from exhausted.

## § 1. An Imbedding Procedure

Let $G$ denote a complex Banach space and let $\boldsymbol{F}$ denote a filter on $\mathbb{N}$ which is finer than the Fréchet filter $\boldsymbol{F}_0$. (Recall that $\boldsymbol{F}_0$ is the filter of all subsets of $\mathbb{N}$ with finite complement.) Then $m(G)$, the vector space of all bounded $G$-valued sequences $(x_n)$, is a Banach space under the norm $\|(x_n)\|:=\sup_n \|x_n\|$, and the space of all sequences $(x_n)\in m(G)$ for which $\lim_{\boldsymbol{F}} \|x_n\|=0$, is a Banach subspace $c_{\boldsymbol{F}}(G)$.

**1.1 Definition.** *By the* $\boldsymbol{F}$-product $\hat{G}_{\boldsymbol{F}}$ *(or briefly, by* $\hat{G}$ *if* $\boldsymbol{F}$ *is understood) we understand the quotient Banach space* $m(G)/c_{\boldsymbol{F}}(G)$. *Moreover, for a bounded operator* $T\in\mathscr{L}(G)$ *we denote by* $\hat{T}_{\boldsymbol{F}}$ *(or briefly, by* $\hat{T}$ *if* $\boldsymbol{F}$ *is understood) the operator in* $\mathscr{L}(\hat{G}_{\boldsymbol{F}})$ *defined by* $\hat{T}_{\boldsymbol{F}}((x_n)+c_{\boldsymbol{F}}(G)):=((Tx_n)+c_{\boldsymbol{F}}(G))$.

It is clear that by virtue of $x\mapsto((x,x,\ldots)+c_{\boldsymbol{F}}(G))$, each $x\in G$ defines an element $\hat{x}$ of $\hat{G}_{\boldsymbol{F}}$. This (linear) map, as well as the operator map $T\mapsto\hat{T}_{\boldsymbol{F}}$, are called *canonical;* they have the following properties.

**1.2 Proposition.** *Let* $G$ *be any Banach space, and let* $\hat{G}$ *be any fixed* $\boldsymbol{F}$*-product of* $G$. *The canonical map* $G\to\hat{G}$ *is an isometric injection, and the canonical map* $\mathscr{L}(G)\to\mathscr{L}(\hat{G})$ *is an isometric homomorphism of Banach algebras which maps the unit of* $\mathscr{L}(G)$ *onto the unit of* $\mathscr{L}(\hat{G})$. *Moreover, the norm of* $\hat{x}=((x_n)+c_{\boldsymbol{F}}(G))$ *is given by* $\|\hat{x}\|=\overline{\lim}_{\boldsymbol{F}}\|x_n\|$ *(and by* $\|\hat{x}\|=\lim_{\boldsymbol{U}}\|x_n\|$ *whenever* $\boldsymbol{F}$ *is an ultrafilter* $\boldsymbol{U}$*).*

*Proof.* We prove the last assertion first. Since the norm of $(x_n)\in m(G)$ is $\sup_n\|x_n\|$, we have $\|\hat{x}\|=\inf\sup_n\|x_n+y_n\|=\overline{\lim}_{\boldsymbol{F}}\|x_n\|$ where, by definition of the quotient norm, the infimum is to be taken over all elements $(y_n)\in c_{\boldsymbol{F}}(G)$. Thus if $x\in G$ the canonical image $\hat{x}$ has norm $\|\hat{x}\|=\|x\|$, and it is clear that $x\mapsto\hat{x}$ is linear; hence $G\to\hat{G}$ is a linear isometry.

Second, it is easily verified that $T\mapsto\hat{T}$ is a homomorphism of algebras and that $(1_G)\hat{}=1_{\hat{G}}$. Moreover, for $\hat{x}\in\hat{G}$ we have $\|\hat{T}\hat{x}\|=\overline{\lim}_{\boldsymbol{F}}\|Tx_n\|\leq\|T\|\,\|\hat{x}\|$ by the preceding, whence it follows that $\|\hat{T}\|\leq\|T\|$. On the other hand, $\hat{T}\hat{x}=(Tx)\hat{}$ for $x\in G$ and so $\|\hat{T}\|=\|T\|$. ☐

We are interested in the special case where $G=E$ is a complex Banach lattice (Chap. II, § 11). It is immediate that $m(E)$ is a Banach lattice under its canonical (coordinatewise) ordering (Chap. II, § 1, Example 6) and that $c_{\boldsymbol{F}}(E)$ is a closed ideal of $m(E)$; thus $\hat{E}_{\boldsymbol{F}}$ is a Banach lattice (II. 5.4). We record this supplement of (1.2).

**1.3 Proposition.** *Let* $E$ *denote a Banach lattice. Then any* $\boldsymbol{F}$*-product* $\hat{E}_{\boldsymbol{F}}$ *is a Banach lattice, the canonical map* $E\to\hat{E}_{\boldsymbol{F}}$ *is a lattice isomorphism, and the associated injection* $\mathscr{L}(E)\to\mathscr{L}(\hat{E}_{\boldsymbol{F}})$ *is an order isomorphism*[1]. *Moreover,* $T\in\mathscr{L}(E)$ *is a lattice homomorphism if and only if* $\hat{T}$ *is.*

[1] For the respective canonical orderings of $\mathscr{L}(E)$ and $\mathscr{L}(\hat{E}_{\boldsymbol{F}})$.

*Proof*. It is obvious that the canonical injection $E \to m(E)$ is a lattice isomorphism and hence, so is $E \to \hat{E}_F$ by (II. 2.6). Moreover, it is clear that $T \geqq 0$ implies $\hat{T} \geqq 0$. Conversely, suppose that $\hat{T} \geqq 0$ and let $U$ denote an ultrafilter finer than $F$. For each $\varphi \in E'_+$, $\langle \hat{x}, \hat{\varphi} \rangle := \lim_U \langle x_n, \varphi \rangle$ defines a positive linear form $\hat{\varphi}$ on $\hat{E}_F$; thus for any $x \in E_+$ we obtain $0 \leqq \langle \hat{T}\hat{x}, \hat{\varphi} \rangle = \langle Tx, \varphi \rangle$ which shows that $T \geqq 0$. Verification of the final assertion is left to the reader. □

**Examples.** 1. Let $E = C(K)$ ($K$ compact) be an $AM$-space with unit (Chap. II, § 7), and let $H := \sum_{n=1}^{\infty} K_n$ denote the topological sum of countably many copies of $K$. Then $m(E)$ can be identified with the Banach lattice of all bounded, continuous complex functions on $H$ and hence, with $C(\beta H)$ ($\beta H$ denoting the Stone-Čech compactification of $H$; cf. Chap. II, § 7, Examples 1 and 4 (a)). If $F$ is the Frechet filter $F_0$ then $c_F(E)$ becomes the ideal of all functions in $C(\beta H)$ vanishing on $\beta H \setminus H$, and thus $\hat{E}_F$ is the $AM$-space $C(\beta H \setminus H)$. If, more generally, $F$ is any filter finer than $F_0$ then $\hat{E}_F = C(H_F)$, $H_F$ denoting the subset of $\beta H \setminus H$ of all cluster points of the filter on $H$ which is the inverse image of $F$ under the mapping $H \to \mathbb{N}$ defined by $t \mapsto n$ ($t \in K_n$, $n \in \mathbb{N}$).

2. Let $E$ be a Banach lattice whose norm is $p$-additive ($1 \leqq p < +\infty$); more concretely, a Banach lattice $L^p(\mu)$ (Chap. II, Exerc. 23). If $U$ is a free ultrafilter on $\mathbb{N}$, it follows that for orthogonal elements $\hat{x}, \hat{y} \in \hat{E}_U$ one has $\|\hat{x} + \hat{y}\|^p = \lim_U \|x_n + y_n\|^p = \lim_U \|x_n\|^p + \lim_U \|y_n\|^p = \|\hat{x}\|^p + \|\hat{y}\|^p$ (where, of course, $(x_n) \in \hat{x}$ and $(y_n) \in \hat{y}$). Therefore, $E_U$ has $p$-additive norm and $\hat{E}_U = L^p(\hat{\mu})$ for a suitable measure $\hat{\mu}$ (cf. Exerc. 1).

One of the very useful features of the imbedding $G \to \hat{G}$ is the fact that the associated operator mapping $T \mapsto \hat{T}$ transforms the approximate point spectrum of $T$ into point spectrum of $\hat{T}$. (For the notions *spectrum*, *resolvent set*, *resolvent* etc. of a bounded operator on a Banach space we refer to [DS]; a brief discussion can also be found in [S, Appendix, § 1].) We recall specifically that the *approximate point spectrum* $A\sigma(T)$ of an operator $T \in \mathscr{L}(G)$ ($G$ a complex Banach space $\neq \{0\}$) is the set of all $\lambda \in \mathbb{C}$ for which $\lambda - T$ is not a (topological vector) isomorphism onto some Banach subspace of $G$; $A\sigma(T)$ contains, of course, the *point spectrum* $P\sigma(T)$ (the set of all $\lambda \in \mathbb{C}$ for which $\lambda - T$ fails to be injective). The *residual spectrum* $R\sigma(T)$ is the set of all $\lambda \in \mathbb{C}$ for which $\lambda - T$ is a (topological vector) isomorphism onto a closed subspace distinct from $G$. The *spectrum* $\sigma(T)$ is the disjoint union $A\sigma(T) \cup R\sigma(T)$, and the resolvent equation (cf. Exerc. 5) implies that the topological boundary of $\sigma(T)$ is contained in $A\sigma(T)$; hence $R\sigma(T)$ is open (but, in general, empty). Finally, $\varrho(T) := \mathbb{C} \setminus \sigma(T)$ is called the *resolvent set* of $T$.

**1.4 Theorem.** *Let $G$ be a complex Banach space $\neq \{0\}$, and let $\hat{G}$ be any $F$-product of $G$. For each $T \in \mathscr{L}(G)$ and its canonical image $\hat{T} \in \mathscr{L}(\hat{G})$, the following are valid:*

(i) $\sigma(T) = \sigma(\hat{T})$.

(ii) $A\sigma(T) = A\sigma(\hat{T}) = P\sigma(\hat{T})$.

(iii) $[(\lambda - T)^{-1}]\hat{} = (\lambda - \hat{T})^{-1}$ *for all* $\lambda \in \varrho(T) = \varrho(\hat{T})$.

*Proof*. For $\lambda \in \varrho(T)$, we use the standard notation $R(\lambda) := (\lambda - T)^{-1}$. Since $\hat{1}_G = 1_{\hat{G}}$, $(\lambda - T)R(\lambda) = R(\lambda)(\lambda - T) = 1_G$ implies $(\lambda - \hat{T})\hat{R}(\lambda) = \hat{R}(\lambda)(\lambda - \hat{T}) = 1_{\hat{G}}$ by

(1.2); therefore $\varrho(T) \subset \varrho(\hat{T})$. On the other hand, if $\lambda \in \sigma(T)$ then either $\lambda \in A\sigma(T)$ or $\lambda \in R\sigma(T)$. If $\lambda \in A\sigma(T)$ there exists a normalized sequence $(x_n)$ in $G$ such that $\lim_n(\lambda x_n - Tx_n) = 0$ and, a fortiori, $\lim_{\mathfrak{F}}(\lambda x_n - Tx_n) = 0$; thus $\hat{x} := (x_n) + c_{\mathfrak{F}}(G)$ has norm 1 and satisfies $\lambda \hat{x} = \hat{T}\hat{x}$. Also, if $\lambda \in A\sigma(\hat{T})$ it is readily seen that $\lambda \in A\sigma(T)$ and hence (ii) follows. If $\lambda \in R\sigma(T)$, there exists $y \in G$ such that $\|(\lambda - T)x - y\| \geqq 1$ for all $x \in G$ and hence, letting $\hat{y} := (y, y, \ldots) + c_{\mathfrak{F}}(G)$, we obtain $\|(\lambda - \hat{T})\hat{x} - \hat{y}\| \geqq 1$ for all $\hat{x} \in \hat{G}$; therefore, $\lambda - \hat{T}$ is not surjective and so $\lambda \in \sigma(\hat{T})$ as well. Consequently, $\sigma(T) = \sigma(\hat{T})$ and $\varrho(T) = \varrho(\hat{T})$. ∎

**Corollary.** *A number $\lambda_0 \in \sigma(T)$ is an (isolated) essential singularity (respectively, a pole of order $k$) of the resolvent $R(\lambda, T)$ if and only if the same is true of the resolvent $R(\lambda, \hat{T})$.*

*Proof.* In fact, $\lambda_0$ is an isolated point of $\sigma(T)$ if and only if it is an isolated point of $\sigma(\hat{T})$; it is then clear that corresponding coefficients of the Laurent expansion near $\lambda_0$ of $R(\lambda, T)$ and $R(\lambda, \hat{T})$, respectively, are related to each other by the canonical map $S \mapsto \hat{S}$. ∎

## § 2. Approximation of Lattice Homomorphisms (Korovkin Theory)

Let $f_i$ $(i = 0, 1, 2)$ denote the continuous real functions on $[0, 1]$ defined by $s \mapsto 1$, $s \mapsto s$, $s \mapsto s^2$, respectively. *If $(T_n)_{n \in \mathbb{N}}$ is a sequence of positive operators on $C[0, 1]$ satisfying* $\lim_n T_n f_i = f_i$ for $i = 0, 1, 2$ *then* $\lim_n T_n f = f$ *for every* $f \in C[0, 1]$. This classical result, due to Korovkin [1953], is surprising because it asserts that the sequence $(T_n)$ converges to the identity operator in $\mathscr{L}_s(C[0, 1])$ once convergence is known on a certain finite subset of $C[0, 1]$.

In the present section we make a systematic investigation of this phenomenon, culminating in the result (due to Wolff [1973b]) that a Banach lattice $E$ contains a finite subset with the above property if and only if $E$ is finitely generated (Theorem 2.9). We restrict attention to real Banach lattices, noting that extension to complex scalars requires no new ideas and provides no additional insight; we begin with two definitions.

**2.1 Definition.** *Let $E$ denote a Banach lattice. A subset $M \subset E$ is called a* Korovkin family *if for every Banach lattice $F$, for every lattice homomorphism $S: E \to F$, and for every equicontinuous sequence $(T_n)_{n \in \mathbb{N}}$ of positive operators $E \to F$, the following holds true:* $\lim_n T_n x = Sx$ *for all* $x \in M$ *implies* $\lim_n T_n x = Sx$ *for all* $x \in E$.

In particular, we note from this definition that $T \geqq 0$ and $Tx = Sx$ for $x \in M$ implies $T = S$. Also, we observe that in the Korovkin theorem quoted above, equicontinuity of the sequence $(T_n)$ is automatically present, since $\|T\| = \|Tf_0\|$ for every positive $T \in \mathscr{L}(C[0, 1])$ (Chap. IV, § 1, Formula (4)).

**2.2 Definition.** *A subset $M$ of a Banach lattice $E$ is called* generating *if $E$ equals the smallest closed vector sublattice containing $M$. $E$ is called* finitely generated *if $E$ contains a generating subset which is finite.*

**Example.** 1. Let $X$ be a compact subset of $\mathbb{R}^n$, where $n$ is any natural number. By the Stone-Weierstrass theorem (II. 7.3), the vector sublattice of $C(X)$ generated by the constant-one function $e$ and the $n$ coordinate functions $f_\nu: s \mapsto s_\nu$ $(s \in X; \nu = 1, \ldots, n)$ is dense; hence, $C(X)$ is finitely generated.

Similarly, if $\mu$ is a positive Radon measure on $X$ and if $1 \leqq p < +\infty$, the family $\{e, f_1, \ldots, f_n\}$ generates $L^p(\mu)$, since $C(X)$ is dense in $L^p(\mu)$.

**2.3 Proposition.** *Every Korovkin family of a Banach lattice E is generating.*

*Proof.* Let $M$ be a Korovkin family of $E$ and denote by $H$ the closed vector sublattice of $E$ generated by $M$. Suppose that $H \neq E$; there exists a non-zero $x' \in E'$ vanishing on $H$. Now we consider the canonical map $S: E \to (E, |x'|)$ (for notation, see Chap. IV, § 3) and observe that $S(H)$ cannot be dense in $(E, |x'|)$: In fact, there exists $z \in E$ such that $\langle z, (x')^+ \rangle \neq \langle z, (x')^- \rangle$, whence it follows that $Sz \notin \overline{S(H)}$. Since $(E, |x'|)$ is an $AL$-space and hence so is $\overline{S(H)}$, by (II. 8.9) Cor. 1 there exists a positive projection $P: (E, |x'|) \to \overline{S(H)}$. Now $S$ is a lattice homomorphism and $T := P \circ S$ is a positive operator such that $Tx = Sx$ for all $x \in M$ but $T \neq S$ which contradicts the defining property of the Korovkin family $M$ (see remark following Def. 2.1). ▯

The converse of (2.3) is not quite true, as the Korovkin theorem quoted at the beginning of the section suggests (cf. Exerc. 2); however, the relationship between families of generators and Korovkin families is quite close for $AM$-spaces with unit. This relationship will be exhibited, via the concept of Choquet boundary, in Propositions (2.5) through (2.8) below; the general case of a finitely generated Banach lattice $E$ then becomes accessible by finding suitable dense sublattices $C(K_0) \subset E$ ($K_0$ compact).

**2.4 Definition.** *Let $K$ be a compact space and let $N \subset C(K)$. The set $\{s \in K : 0 \leqq \mu \in C(K)'$ and $\mu(f) = \delta_s(f)$ for all $f \in N$ implies $\mu = \delta_s\}$ is called the* Choquet boundary *of $N$ and denoted by $\partial_c N$.*

Usually the Choquet boundary is introduced for vector subspaces $N$ of $C(K)$; however, from our definition it is clear that $\partial_c N = \partial_c [N]^-$ where $[N]^-$ denotes the closed linear hull of $N$. The following consequence of Def. 2.4 is immediate:

**2.5 Proposition.** *If $K$ is compact and if $M$ is a Korovkin family of $C(K)$, then $\partial_c M = K$.*

*Proof.* In fact, each Dirac measure $\delta_s$ $(s \in K)$ is a lattice homomorphism $C(K) \to \mathbb{R}$ and each $\mu \in C(K)'_+$ is a positive operator $C(K) \to \mathbb{R}$; thus the result is clear from the remark following (2.1). ▯

**2.6 Proposition.** *Let $K$ be compact and suppose $A \subset C(K)$ is a subset separating the points of $K$ and containing the unit $e$ of $C(K)$. The set $M := A \cup \{f^2 : f \in A\}$ has Choquet boundary $\partial_c M = K$.*

*Proof.* Let $s_0 \in K$, let $0 \leqq \mu \in C(K)'$ and suppose that $\mu(f) = \delta_{s_0}(f)$ for all $f \in M$. For each $f \in A$, we consider the function $h_f := (f(s_0)e - f)^2$ which is contained in the linear hull of $M$. Since $\mu(h_f) = \delta_{s_0}(h_f)$ by hypothesis, we have $\mu(h_f) = 0$

for each $f \in A$. But $\mu \geqq 0$ and $h_f \geqq 0$, and so support $\mu \subset \bigcap_{f \in A} h_f^{-1}(0) = \{s_0\}$ because $A$ separates the points of $K$. It follows that $\mu$ is a scalar multiple of $\delta_{s_0}$; but $e \in A$ and so $\mu = \delta_{s_0}$. Since $s_0$ was arbitrary, we conclude that $\partial_c M = K$. ▯

The use of the imbedding procedure (§ 1) now permits us to reduce the convergence problem for sequences of positive operators to a "stationary" problem; we collect the relevant facts in this lemma.

**2.7 Lemma.** *Let $K$ be a compact space, $F$ any Banach lattice, $S: C(K) \to F$ a lattice homomorphism, and define $v := S e_K$. Suppose $(T_n)_{n \in \mathbb{N}}$ is a sequence of positive operators $C(K) \to F$, denote by $\hat{F}$ the $\mathbf{F}_0$-product*[1] *of $F$ (Def. 1.1) and by $j: F \to \hat{F}$ the canonical map. For the mappings $\hat{S} := j \circ S$, and $\hat{T}: C(K) \to \hat{F}$ defined by $\hat{T} f := (T_n f) + c_{\mathbf{F}_0}(F)$, the following assertions are valid:*

(i) *For any $f \in C(K)$, $\lim T_n f = Sf \Leftrightarrow \hat{T} f = \hat{S} f$.*

(ii) *$\hat{S}$ is a lattice homomorphism.*

(iii) *$\hat{T} e_K = \hat{S} e_K \Rightarrow \hat{T}(C(K)) \subset (\hat{F})_{\hat{v}}$.*

*Proof.* (ii) is clear, and (i) follows from the construction of the $\boldsymbol{F}_0$-product. (iii): $\hat{T} e_K = \hat{S} e_K$, which by (i) is equivalent to $\lim_n T_n e_K = S e_K = v$, shows that for each $f \in C(K)$ we have $|T_n f| \leqq \|f\| \, T_n e_K = \|f\| \, v + y_n$ $(n \in \mathbb{N})$ where $(y_n)$ is a null sequence in $F$ depending on $f$; therefore, $|\hat{T} f| \leqq \|f\| \, \hat{v}$ which proves the assertion. ▯

**2.8 Theorem.** *Let $K$ be a compact space, and let $M \subset C(K)$ be a subset satisfying $e_K \in M$ and $\partial_c M = K$. Then $M$ is a Korovkin family of $C(K)$.*

*Proof.* Suppose $F$ is any Banach lattice, $S: C(K) \to F$ is a lattice homomorphism, and that $(T_n)$ is a sequence of positive operators $C(K) \to F$ such that $\lim_n T_n f = Sf$ for each $f \in M$. We have to show that $\lim_n T_n f = Sf$ for each $f \in C(K)$ or equivalently, by Lemma 2.7, that $\hat{T} = \hat{S}$. Since $e_K \in M$, we observe that $\hat{T}$ (as well as $\hat{S}$) maps $C(K)$ into the $AM$-space $(\hat{F})_{\hat{v}}$ where $\hat{v} = j(v) = \hat{S} e_K$ as in (2.7); hence, it suffices to show that $\hat{S}$ and $\hat{T}$ are equal as mappings into the $AM$-space $(\hat{F})_{\hat{v}}$ with unit $\hat{v}$. Identifying $(\hat{F})_{\hat{v}}$ with $C(H)$ for a suitable compact $H$ and noting that $\hat{S}$ is a lattice homomorphism satisfying $\hat{S} e_K = e_H$, we infer from (III. 9.1) the existence of a continuous map $k: H \to K$ such that $\hat{S} f = f \circ k$ $(f \in C(K))$, that is, $\hat{S}' \delta_t = \delta_{k(t)}$ for all $t \in H$. Now $\hat{T} f = \hat{S} f$ for $f \in M$ implies that the restrictions of $\delta_{k(t)}$ and $\hat{T}' \delta_t$ to $M$ are identical; since $\partial_c M = K$ by hypothesis and $\hat{T}' \delta_t$ $(t \in H)$ is a positive Radon measure on $K$, we obtain $\delta_{k(t)} = \hat{S}' \delta_t = \hat{T}' \delta_t$ $(t \in H)$ by the defining property of $\partial_c M$. Thus $\hat{S}' = \hat{T}'$ and hence $\hat{S} = \hat{T}$ which completes the proof. ▯

The following corollary, now immediate from (2.6), contains the classical Korovkin theorem as a very special case.

**Corollary.** *Let $A$ be a subset of $C(K)$ containing $e_K$ and separating the points of $K$. Then $A \cup A^2$ is a Korovkin family of $C(K)$.*

Our principal result is now as follows (Wolff [1973b]).

---

[1] $\boldsymbol{F}_0$ denotes the Fréchet filter on $\mathbb{N}$ (§ 1), $e_K$ the unit of $C(K)$, $\hat{v}$ the canonical image of $v$.

**2.9 Theorem.** *For any Banach lattice E, these assertions are equivalent:*
(a) *E is finitely generated.*
(b) *E possesses a finite Korovkin family.*

*Proof.* (b) ⇒ (a) is clear from (2.3).
(a) ⇒ (b): Let $A$ denote a finite family of generators of $E$, and let $u := \sum_{x \in A} |x|$. It is clear that $u$ is a quasi-interior element of $E_+$ (Def. II. 6.1). Consider the principal ideal ($AM$-space) $E_u$ and denote by $E_0$ the closed vector sublattice of $E_u \cong C(K)$ generated by $\{A, u\}$. $E_0$ is an $AM$-space $C(K_0)$ which is dense in $E$. (Precisely, $K_0$ is the quotient of $K$ over the equivalence relation on $K$ defined by "$s \sim t$ iff $f(s) = f(t)$ for all $f \in A$".) Clearly, $A$ is a generating subset of $E_0$; hence, by the corollary of (2.8), $\{A, u\} \cup A^2$ is a Korovkin family of $E_0$. Now since $E_0$ is dense in $E$, the equicontinuity requirement of Def. 2.1 guarantees that $\{A, u\} \cup A^2$ is a (finite) Korovkin family of $E$. □

In view of (2.9) it is natural to ask for some useful characterization of finitely generated Banach lattices. In the proof of (2.9), (a) ⇒ (b), it has been observed that a finitely generated Banach lattice $E$ necessarily contains quasi-interior positive elements (a fact also clear from (II. 6.2), since $E$ is separable; cf. Chap. II, Exerc. 5). Let us say that a Banach lattice $E$ with quasi-interior point $u \in E_+$ has a *system of n generators* $x_\nu$ ($\nu = 1, \ldots, n$) if $E$ equals the smallest closed vector sublattice containing the set $\{u; x_1, \ldots, x_n\}$; this usage closely parallels the terminology customary in the theory of Banach algebras.

Concerning the term *Banach function lattice*, we use the terminology previously employed in the formulation of Theorem IV. 9.7 (cf. also Chap. IV, Exerc. 24). Specifically, if $X \subset \mathbb{R}^n$ is compact and $M$ is a vaguely compact set of positive Radon measures on $X$ whose supports have a union dense in $X$, then the vector lattice $B(X, M)$ of all finite Borel functions $f$ on $K$ for which $p_M(f) := \sup_{\mu \in M} \int |f| \, d\mu$ is finite, is complete under the lattice semi-norm $p_M$. The closure $\overline{C(X)}$ in $B(X, M)$ is complete as well, and the quotient $\overline{C(X)}/p_M^{-1}(0)$ is a Banach function lattice denoted by $E(X, M)$.

**2.10 Theorem.** *Let E be a Banach lattice possessing quasi-interior positive elements. The following are equivalent:*
(a) *E has a system of n generators* ($n \in \mathbb{N}$).
(b) *E is isomorphic to a Banach function lattice* $E(X, M)$ *where X is a compact subset of* $\mathbb{R}^n$.

*Proof.* (b) ⇒ (a): By the definition of $E(X, M)$ (see above), $C(X)$ can be identified with a dense vector sublattice of $E(X, M)$, and it is clear that the constant-one function $e_X$ is a quasi-interior point of $E(X, M)_+$. Since $X$ is compact in $\mathbb{R}^n$, by (II. 7.3) the $n$ coordinate functions $s \mapsto s_\nu$ ($s \in X$; $\nu = 1, \ldots, n$) constitute a family of generators for $C(X)$ and hence, for $E(X, M)$.
(a) ⇒ (b): Let $u$ denote a quasi-interior element of $E_+$ and let $\{x_1, \ldots, x_n\}$ be a family of generators of $E$. Then $v := u + \sum_1^n |x_\nu|$ is quasi-interior to $E_+$, and the dense ideal $E_v$ is an $AM$-space $C(K)$ (where the compact space $K$ is, in fact, the structure space of $E$; cf. (III. 4.5)). Identifying the elements $x_\nu$ ($\nu = 1, \ldots, n$) as elements of $C(K)$, we find that $k: s \mapsto (x_1(s), \ldots, x_n(s))$ maps $K$ continuously onto

a compact subset $X$ of $\mathbb{R}^n$. Thus the mapping $j: f \mapsto f \circ k$ is an injective lattice homomorphism (cf. III. 9.3) of $C(X)$ into $C(K) \subset E$ whose range is dense in $E$, since $v \in j(C(X))$ and since $\{x_1, \ldots, x_n\}$ is a family of generators of $E$. Therefore (cf. Chap. IV, Exerc. 24) the adjoint $j'$ is a lattice isomorphism of $E'$ onto an ideal of $C(X)'$, and $j'$ maps the Šilov-boundary of $E_+$ (see II. 5.7) onto a vaguely compact set $M$ of positive Radon measures on $X$. Clearly, if $C(X)$ is identified with its image under $j$, the norm on $C(X)$ induced by $E$ is given by $\|f\| = \sup_{\mu \in M} \int |f| \, d\mu$. Consequently, $E$ is isomorphic to the Banach function lattice $E(X, M)$. □

Thus the most general, finitely generated Banach lattice is very similar to the Banach lattices considered in Example 1 above; on the other hand, spaces such as an infinite dimensional $L^\infty(\mu)$ ($\mu$ a positive Radon measure on some compact set $X \subset \mathbb{R}^n$), or $C(K)$ where $K = [0,1]^{\mathbb{N}}$, are not finitely generated and so do not contain any finite Korovkin family.

## § 3. Banach Lattices and Cyclic Banach Spaces

Let $G$ denote a (real or complex) Banach space. If $\boldsymbol{P}$ is a Boolean algebra of projections in $\mathscr{L}(G)$ (see Def. 3.1 below) and if $x_0$ is a fixed vector in $G$, the closed linear hull $G_0$ of $\{P x_0 : P \in \boldsymbol{P}\}$ is called the *cyclic subspace of G generated by* $\boldsymbol{P}$ *and* $x_0$ or, briefly, a *cyclic Banach space*. Simple examples show that condition (i) of the following definition does not even imply $\boldsymbol{P}$ to be bounded in $\mathscr{L}(G)$; on the other hand, (ii) entails the presence of a surprisingly rich structure on a cyclic Banach space. For a comparison of the results of this section to what can be obtained under weaker assumptions, the interested reader is referred to Exerc. 4.

**3.1 Definition.** *Let G be any Banach space. A set* $\boldsymbol{P} \subset \mathscr{L}(G)$ *will be called a* σ-complete Boolean algebra of projections *(σ-complete B.a.p.) if these axioms are satisfied:*

(i) $\boldsymbol{P}$ *is a Boolean algebra under the operations* $P \vee Q := P + Q - PQ$, $P \wedge Q := PQ$ *with* 0 *the zero operator and* 1 *the identity operator on G.*

(ii) *If* $(P_n)_{n \in \mathbb{N}}$ *is an increasing sequence in* $\boldsymbol{P}$, *then* $P := \sup_n P_n$ *exists in* $\boldsymbol{P}$ *and* $\lim_n \langle P_n x, x' \rangle = \langle P x, x' \rangle$ *for all* $x \in G$, $x' \in G'$.

The cyclic Banach space generated by $\boldsymbol{P}$ and $x_0$ will be denoted by $(G, \boldsymbol{P}, x_0)$ or, if $\boldsymbol{P}$ and $x_0$ are understood, simply by $G_0$. $\boldsymbol{P}$ is called *reduced* with respect to $x_0$ if $P x_0 = 0$ implies $P = 0$. It is clear that the subset $\boldsymbol{J} := \{P \in \boldsymbol{P} : P x_0 = 0\}$ is a σ-ideal of $\boldsymbol{P}$, and that $\boldsymbol{P}/\boldsymbol{J}$ defines a σ-complete B.a.p. on $G_0$ which is reduced with respect to $x_0$.

**Examples.** 1. Let $E$ denote a Banach lattice with order continuous norm (Def. II. 5.12; cf. II. 5.10). The Boolean algebra $\boldsymbol{P} := \boldsymbol{P}(E)$ of all band projections of $E$ (cf. II. 2.9) is complete (as an abstract Boolean algebra), since $E$ is order complete, and it satisfies (ii) of (3.1) (even for arbitrary directed families), since $E$ has order continuous norm. If $x_0 \in E_+$ and if $E_0$ denotes the band of $E$ generated by $x_0$, then $x_0$ is a quasi-interior point of $(E_0)_+$ (II. 6.5). By (II. 2.12) and (II. 2.13), the $x_0$-reduced Boolean algebra $\boldsymbol{P}_0$ is isomorphic to the Boolean algebra $B_{x_0} = \{Px_0 : P \in \boldsymbol{P}\}$ of all extreme points of the order interval $[0, x_0]$. Since $[0, x_0]$ is weakly compact (II. 5.10) and a total subset of $E_0$, $E_0$ is the closed linear hull of $B_{x_0}$ and hence, $E_0 = (E, \boldsymbol{P}, x_0)$ is a cyclic Banach space.

More specially, if $(X, \Sigma, \mu)$ is a $\sigma$-finite measure space and if $1 \leqq p < +\infty$, then $L^p(\mu)$ is a cyclic Banach space generated by the Boolean algebra of all band projections and any function $f_0$ which is $>0$ a.e. $(\mu)$ (cf. Chap. II, § 6, Example 1).

2. If $(X, \Sigma, \mu)$ is $\sigma$-finite and $E := L^\infty(\mu)$ is infinite dimensional, the Boolean algebra $\boldsymbol{P}(E)$ of band projections is complete (as an abstract Boolean algebra) but does not satisfy (ii) of Def. 3.1. Cf. Exerc. 4.

The surprising fact is now that Example 1, although it looks rather special, already furnishes the most general example of a cyclic Banach space; this will be the main result of this section (see 3.6 below).

From now on through the proof of (3.6), we consider a fixed Banach space $G$ and a $\sigma$-complete B.a.p. $\boldsymbol{P}$ on $G$. Let $K$ denote the Stone representation space of $\boldsymbol{P}$ (Chap. II, Exerc. 1); since $\boldsymbol{P}$ is $\sigma$-complete, $K$ is quasi-Stonian and, by construction of $K$, the Boolean algebra $\boldsymbol{Q}$ of open-and-closed subsets of $K$ is isomorphic to $\boldsymbol{P}$ (the isomorphism being denoted by $U \mapsto P_U$, $U \in \boldsymbol{Q}$). Let $\boldsymbol{D}$ denote the $\sigma$-ring of subsets of $K$ generated by all closed, rare $G_\delta$-sets, and let $\Sigma$ denote the $\sigma$-algebra generated by $\boldsymbol{Q} \cup \boldsymbol{D}$; it is tedious but not difficult to verify that for each $S \in \Sigma$, there exists a unique $U \in \boldsymbol{Q}$ such that the symmetric difference $S \triangle U$ is in $\boldsymbol{D}$. (Incidentally, $\Sigma$ is the Baire field of $K$.) If for each $S \in \Sigma$ we define $\mu(S) := P_U$ (where $S \triangle U \in \boldsymbol{D}$) then, by virtue of the axioms of Def. 3.1, $\mu : \Sigma \to \mathscr{L}(G)$ is a set function which is countably additive for the weak operator topology of $\mathscr{L}(G)$, and multiplicative ($\mu(S_1 \cap S_2) = \mu(S_1)\mu(S_2)$ for all $S_1, S_2 \in \Sigma$); that is, $\mu$ is a *spectral (Baire) measure* on $K$.

**3.2 Proposition.** *The B.a.p. $\boldsymbol{P}$ is a bounded subset of $\mathscr{L}(G)$.*

*Proof.* Let $\mu$ be the spectral measure on $\Sigma$ associated with $\boldsymbol{P}$; for fixed $x \in G$ and $x' \in G'$, the mapping $S \mapsto \mu_{x,x'}(S) := \langle \mu(S)x, x' \rangle$ is a scalar valued measure on $\Sigma$ and hence bounded [DS, III. 4.4]; the principle of uniform boundedness shows $\boldsymbol{P}$ to be bounded. □

**3.3 Proposition.** *The representation of $\boldsymbol{P}$ on its Stone space $K$ induces a continuous homomorphism of Banach algebras $m : C(K) \to \mathscr{L}(G)$. Moreover, the norm on $G$ defined by*

$$(\star) \qquad \|x\|_1 := \sup_{\|f\| \leqq 1} \|m(f)x\| \qquad (x \in G)$$

*is equivalent to the original norm, and with respect to the norm $(\star)$ $m$ is an isomorphism of the group of unimodular functions in $C(K)$ onto a group of isometries of $G$. Finally, $m$ is contractive for the operator norm on $\mathscr{L}(G)$ arising from $(\star)$.*

*Proof.* Let $L$ denote the linear hull of all characteristic functions $\chi_U$, $U \in \boldsymbol{Q}$. By (II. 7.3), $L$ is a dense sublattice and subalgebra of $C(K)$. Each non-zero $f \in L$ can be written $f = \sum_i \alpha_i \chi_{U_i}$ where $U_i \in \boldsymbol{Q}$ are mutually disjoint and non-void; we have $\|f\| = \sup_i |\alpha_i|$. The definition $m(f) := \sum_i \alpha_i P_{U_i}$ is easily seen to be unambiguous, and $m: L \to \mathscr{L}(G)$ is a homomorphism of algebras. To show that $m$ is continuous for the sup-norm on $L$ and the operator norm on $\mathscr{L}(G)$, respectively, we select $x \in G$, $x' \in G'$ arbitrarily and let $M$ denote a uniform bound of $\boldsymbol{P}$ (3.2). If $f \in L$, $0 < \|f\| \leqq 1$, is represented as above we obtain

$$|\langle m(f)x, x'\rangle| \leqq \textstyle\sum_i |\langle m(U_i)x, x'\rangle| \leqq \sum_i |\mathrm{Re}\langle m(U_i)x, x'\rangle| + \sum_i |\mathrm{Im}\langle m(U_i)x, x'\rangle|.$$

But

$$\textstyle\sum_i |\mathrm{Re}\langle m(U_i)x, x'\rangle| = \sum_i^+ \mathrm{Re}\langle m(U_i)x, x'\rangle - \sum_i^- \mathrm{Re}\langle m(U_i)x, x'\rangle$$

where $\sum^+$ is extended over those subscripts $i$ for which $\mathrm{Re}\langle m(U_i)x, x'\rangle \geqq 0$ and $\sum^-$ over those for which $\mathrm{Re}\langle m(U_i)x, x'\rangle < 0$. Now if $U := \bigcup^+ U_i$ and $V := \bigcup^- U_i$ in corresponding notation, we obtain $\sum_i^+ \mathrm{Re}\langle m(U_i)x, x'\rangle = \mathrm{Re}\langle m(U)x, x'\rangle \leqq M\|x\|\,\|x'\|$ and accordingly, $\left|\sum_i^- \mathrm{Re}\langle m(U_i)x, x'\rangle\right| = |\mathrm{Re}\langle m(V)x, x'\rangle| \leqq M\|x\|\,\|x'\|$. Estimating the sum over the imaginary parts similarly, we finally obtain $|\langle m(f)x, x'\rangle| \leqq 4M\|x\|\,\|x'\|$ whenever $\|f\| \leqq 1$. It follows that $m$ is continuous of norm $\leqq 4M$, and its continuous extension to $C(K)$ (again denoted by $m$) is clearly a homomorphism of Banach algebras.

Let $B$ denote the unit ball of $C(K)$. Since $1_G \in m(B)$ and since $m$ is continuous, it is clear that (*) is equivalent to the original norm of $G$. Moreover, if $f \in C(K)$ is unimodular then $fB = B$ and $\|m(f)x\|_1 = \sup_{g \in B} \|m(g)m(f)x\| = \sup_{g \in B} \|m(g)x\| = \|x\|_1$, since $m$ is multiplicative; hence, $m(f)$ is an isometry of $G$ for the norm (*). Finally, if $\|f\| \leqq 1$ then $fB \subset B$ and, by the same argument, $\|m(f)x\|_1 \leqq \|x\|_1$; so $\|m(f)\| \leqq 1$ for the operator norm on $\mathscr{L}(G)$ arising from (*). ▯

*Note.* The preceding proof applies verbatim when $L$ is replaced by the linear hull $L_\Sigma$ of all characteristic functions of Baire sets $S \in \Sigma$, and yields an extension of the homomorphism $L \to \mathscr{L}(G)$ induced by the spectral measure $\mu$ (see above) to a homomorphism $\bar{m}: B(\Sigma) \to \mathscr{L}(G)$ of Banach algebras, where $B(\Sigma)$ denotes the Banach algebra of all bounded Baire functions on $K$. For $f \in B(\Sigma)$, $\bar{m}(f) \in \mathscr{L}(G)$ is frequently denoted by $\int f\,d\mu$ and called the integral of $f$ with respect to the spectral measure $\mu$.

**3.4 Proposition.** *Let $x_0 \in G$ and let $G_0 = (G, \boldsymbol{P}, x_0)$. If $\boldsymbol{P}$ is reduced with respect to $x_0$, the mapping $m_0: C(K) \to G_0$ defined by $m_0(f) := m(f)x_0$, is injective.*

*Proof.* Suppose $m(f)x_0 = 0$, where $0 \neq f \in C(K)$. There exists $\delta > 0$ such that the set $\{s \in K: |f(s)| > \delta\}$ is non-void; since the latter is an $F_\sigma$-set and $K$ is quasi-Stonian, its closure $U$ is open. Hence, the function $f_0$ defined by $f_0(s) = 1/f(s)$ if $s \in U$ and by $f_0(s) = 0$ if $s \notin U$, is in $C(K)$, and $f_0 f = \chi_U$. Thus $0 = m(f)x_0 = m(f_0)m(f)x_0 = m(\chi_U)x_0 = P_U x_0$ which contradicts the assumption that $\boldsymbol{P}$ be reduced with respect to $x_0$. ▯

From the main theorem (3.6) we isolate this crucial lemma.

**3.5 Lemma.** *Let* $x_0 \in G$ *and let* $(f_n)$ *be a sequence in the unit ball* $B$ *of* $C(K)$ *such that* $(m(f_n)x_0)_{n\in\mathbb{N}}$ *is a Cauchy sequence in* $G$. *There exists* $f \in B$ *such that* $m(f)x_0 = \lim_n m(f_n)x_0$.

*Proof.* By (II. 7.8) Cor. 1, there exists $g_n \in B$ such that $|f_{n+1}-f_n| = g_n(f_{n+1}-f_n)$ $(n\in\mathbb{N})$. (Consider the closure $U$ of $\{s\in K: |f_{n+1}-f_n|(s)>0\}$ and extend $|f_{n+1}-f_n|/(f_{n+1}-f_n)$ continuously to $K$.) Since $m$ is multiplicative we obtain $m(|f_{n+1}-f_n|)x_0 = m(g_n)m(f_{n+1}-f_n)x_0$ and hence, we obtain $\|m(|f_{n+1}-f_n|)x_0\| \leqq 4M\|m(f_{n+1}-f_n)x_0\|$.

Second, letting $W^\circ$ denote the unit ball of $G'$, we consider the Baire extension $\mu_{x_0,x'}$ of the Radon measure $f \mapsto \langle m(f)x_0, x'\rangle$ $(x'\in W^\circ)$ (this notation agrees with that used in the proof of (3.2)). $\mu_{x_0,x'}$ are complex Baire measures on $K$ but, considering real and imaginary parts separately if necessary, we can assume that the functions $f_n$ and the measures $\mu_{x_0,x'}$ are real valued. For fixed $x'\in W^\circ$ let $K = S_1 \cup S_2$ denote a Jordan decomposition for $\mu_{x_0,x'}$; since $\mu_{x_0,x'}$ vanishes on $\boldsymbol{D}$, we can suppose $S_1, S_2 \in \boldsymbol{Q}$ and disjoint (cf. discussion preceding (3.2)). Letting $y_1' = m(\chi_{S_1})'x'$ and $y_2' = -m(\chi_{S_2})'x'$ and assuming that $\mu_{x_0,x'}$ is positive on $S_1$ and negative on $S_2$, we obtain $|\mu_{x_0,x'}| = \mu_{x_0,y_1'} + \mu_{x_0,y_2'}$ where $y_i' \in 4M\,W^\circ$ $(i=1,2)$. Thus it suffices to show that for some $f\in B$, we have $\lim_n \langle m(f_n - f)x_0, y'\rangle = 0$ for those $y'\in W^\circ$ for which $\mu_{x_0,y'} \geqq 0$; let us denote this subset of $W^\circ$ by $W_+^\circ$.

To this end, we select a subsequence (again denoted by $(f_n)$) satisfying $\|m(|f_{n+1}-f_n|)x_0\| \leqq (4M)^{-1}2^{-n}$ for all $n\in\mathbb{N}$; then we have $\int |f_{n+1}-f_n|\,d\mu_{x_0,y'} \leqq 2^{-n}$ uniformly for $y'\in W_+^\circ$. The series $\sum_{n=1}^\infty |f_{n+1}(s)-f_n(s)|$ converges to a Baire function on $K$ which is $\mu_{x_0,y'}$-integrable for all $y'\in W_+^\circ$ and hence, finite outside a Baire set $N$ which is null for each of these measures (B. Levi's theorem). Since $(f_n)\subset B$ there exists a Baire function $h$, absolutely bounded by 1, such that $h(s) = \lim_n f_n(s)$ whenever $s\in K\setminus N$. On the other hand, since $(f_n)$ is order bounded and $K$ quasi-Stonian, $g_k := \sup_{n\geqq k} f_n$ $(k\in\mathbb{N})$ and $f := \inf_k g_k$ exist in $C(K)$; we claim that the subset $[h\neq f]$ of $K$ (for notation, see (II. 7.7) et seq.) is a null set for each $\mu_{x_0,y'}$ $(y'\in W_+^\circ)$. In fact, if $\tilde{g}_k$ denotes the pointwise supremum $\tilde{g}_k(s) := \sup_p(f_k(s)\vee\ldots\vee f_{k+p}(s))$ then the set $D_k := [\tilde{g}_k - g_k < 0]$ is in $\boldsymbol{D}$ (see discussion preceding (3.2)) because of $D_k = \bigcup_{n=1}^\infty B_{k,n}$, where $B_{k,n} := [\tilde{g}_k - g_k \leqq -n^{-1}]$ is a closed rare $G_\delta$-set. (Note that $B_{k,n}$ is closed, since $\tilde{g}_k - g_k$ is upper semicontinuous, and that $B_{k,n} = \bigcap_{p=1}^\infty \Big[\sup_{k\leqq m\leqq k+p} f_m \leqq -n^{-1} + g_k\Big]$ where each of the intersectands is a $G_\delta$.) Thus, if $N_1\in\boldsymbol{D}$ denotes the subset of $K$ where $f$ differs from the pointwise infimum of the sequence $(g_k)$, we have $[h\neq f] \subset N\cup N_1\cup(\bigcup_1^\infty D_k)\in\boldsymbol{D}$ and so $[h\neq f]$ is $\mu_{x_0,y'}$-null for each $y'\in W_+^\circ$. Therefore,

$$|\langle m(f_n)x_0 - m(f)x_0, y'\rangle| \leqq \int |f_n - f|\,d\mu_{x_0,y'} \leqq 2^{-n}$$

uniformly for $y'$ in $W_+^\circ$ and the assertion is proved. ☐

**3.6 Theorem.** *Let* $G_0 = (G, \boldsymbol{P}, x_0)$ *be a cyclic Banach space. There exists an order structure on* $G_0$ *such that under the equivalent norm* (*) *of* (3.3), $G_0$ *becomes a Banach lattice* $E$ *with order continuous norm and quasi-interior positive element*

*$x_0$, and on which $\boldsymbol{P}$ induces the Boolean algebra $\boldsymbol{P}(E)$ of all band projections. Moreover, if $\boldsymbol{P}$ is reduced with respect to $x_0$, the Stone space of $\boldsymbol{P}$ is (homeomorphic with) the structure space of $E$.*[1]

*Proof.* Without loss of generality we assume $\boldsymbol{P}$ reduced with respect to $x_0$. By (3.4) we can identify $C(K)$ algebraically with its image under $m_0$; we show that the norm $(\star)$ of (3.3) induces a lattice norm (Def. II. 5.1) on $C(K)$. For this it suffices to prove that $|g| \leqq |h|$ implies $\|m(g)x_0\|_1 \leqq \|m(h)x_0\|_1$ whenever $g, h \in L$ (i.e., whenever $g, h$ are continuous functions on $K$ with finite range); for, $L$ is clearly dense in $C(K)$ (even for the supremum norm of $C(K)$, $K$ being quasi-Stonian). But if $g, h \in L$ and $|g| \leqq |h|$, there exists $p \in L$ such that $g = ph$; this implies $gB \subset hB$, $B$ denoting the unit ball of $C(K)$. Since $m$ is multiplicative, we obtain

$$\|m(g)x_0\|_1 = \sup_{f \in B} \|m(gf)x_0\| \leqq \sup_{f \in B} \|m(hf)x_0\| = \|m(h)x_0\|_1,$$

which is the desired inequality.

It follows (II.5.2) that the lattice operations (of $C(K)$) are uniformly continuous on the dense subspace $C(K)$ of $G_0$; their continuous extension to $G_0$ defines on $G_0$, with respect to the norm $(\star)$, the structure of a Banach lattice $E$. Moreover, from (3.5) we conclude that the unit ball $B$ of $C(K)$ is complete, hence closed in $G_0$; this shows $C(K)$ to be an ideal of $E$. In fact, if $x \in C(K)$, $y \in E$, and if $|y| \leqq |x|$ and $(y_n)$ is a sequence in $C(K)$ converging to $y$, then $z_n := y_n^+ \wedge |x|$ is a sequence in $C(K)$ which is order bounded and Cauchy in the norm $(\star)$, and which converges to $y^+ \in E$; therefore, $y^+ \in C(K)$ and, similarly, $y^- \in C(K)$. (Cf. Chap. III, § 5, Example 7.) The fact that $C(K)$ is a dense ideal of $E$ now shows $K$ to be the structure space of $E$ (III. 4.5); in particular, $E$ is countably order complete (III. 5.6). To see that $\boldsymbol{P}$ induces the Boolean algebra $\boldsymbol{P}(E)$ of all band projections on $E$, we observe that $\boldsymbol{P}(E)$ is in natural one-to-one correspondence with the Boolean algebra $\boldsymbol{Q}$ of all open-and-closed subsets of $K$ (cf. II. 2.12).

Finally, we have to show that $E$ has order continuous norm. Since $E$ is countably order complete, by (II. 5.10) it is enough to show that for every sequence $(x_n)$ in $E$, $x_n \downarrow 0$ implies $\lim_n \|x_n\| = 0$. Here it can even be assumed that $(x_n)$ is a sequence $(f_n)$ in $C(K)$ (for, $C(K)$ being a dense ideal of $E$, for each sequence $x_n \downarrow 0$ in $E$ there exists a sequence $f_n \downarrow 0$ in $C(K)$ such that $\|f_n - x_n\| < n^{-1}$ $(n \in \mathbb{N})$). Now if $f_n \downarrow 0$ in $C(K)$ then $\lim_n f_n(s) = 0$ for $s \in K \setminus N$, where $N \in \boldsymbol{D}$ (the $\sigma$-ring of subsets of $K$ generated by the rare, closed $G_\delta$-sets). Consequently, by Axiom (ii) of Def. 3.1, $\lim_n \langle m(f_n)x_0, x' \rangle = 0$ for each $x' \in E'_+$; therefore, (II. 5.9) implies that $\lim_n \|m(f_n)x_0\|_1 = 0$.

This ends the proof of (3.6). □

**Corollary 1.** *The following properties of a Banach space $G$ are equivalent:*

(a) *$G$ is a cyclic Banach space with respect to some $\sigma$-complete Boolean algebra of projections* (Def. 3.1).

---

[1] Cf. Chap. III, § 4. The subsequent proof carries over without difficulty if $G$ is a complex Banach space (cf. Chap. II, § 11).

(b) *G is capable of an equivalent norm and an order structure under which it is a Banach lattice with order continuous norm and quasi-interior positive elements.*

*Proof.* (a) ⇒ (b) is contained in (3.6) while (b) ⇒ (a) is contained in Example 1.

As an application it follows that a cyclic Banach space is reflexive iff it does not contain any Banach subspace isomorphic to $c_0$ or $l^1$ (Tzafriri [1969]; see Exerc. 3).

**Corollary 2.** *Every reduced σ-complete B.a.p.* $\boldsymbol{P}$ *is complete and each orthogonal subset of* $\boldsymbol{P}$ *is countable* [2].

*Proof.* By (3.6) and (II. 2.12), the $x_0$-reduced B.a.p. $\boldsymbol{P}$ is isomorphic to the Boolean algebra $B_{x_0}:=\{x\in E: x\wedge(x_0-x)=0\}$; since $E$ is order complete (II. 5.10), $B_{x_0}$ and hence $\boldsymbol{P}$ is complete. Moreover, under the isomorphism $\boldsymbol{P}\to B_{x_0}$, each orthogonal subset of $\boldsymbol{P}$ corresponds to an order bounded orthogonal subset of $E$; since $E$ has order continuous norm, such a set must be countable. □

As before (cf. discussion preceding (3.2)) we denote by $\mu$ the spectral (Baire) measure associated with $\boldsymbol{P}$. The linear form $x_0'$ of the following corollary is sometimes called a "Badé functional" (cf. Badé [1959]).

**Corollary 3.** *If* $G_0=(G,\boldsymbol{P},x_0)$ *is a cyclic Banach space and* $\mu\colon \Sigma\to\mathscr{L}(G)$ *is the spectral measure associated with* $\boldsymbol{P}$*, there exists a linear form* $x_0'\in G_0'$ *such that* $\langle\mu(S)x_0,x_0'\rangle=0$ *implies* $\mu(S)_{|G_0}=0$.

*Proof.* According to (3.6) we provide $G_0$ with the structure of a Banach lattice $E$ with order continuous norm and quasi-interior element $x_0\in E_+$. Since $\mu(S)\geqq 0$ (all $S\in\Sigma$) for the canonical order of $\mathscr{L}(E)$, and since by (II. 6.6) there exists a weak order unit $x_0'\in E'$, it follows that $\langle\mu(S)x_0,x_0'\rangle=0$ implies $\mu(S)x_0=0$ and hence $\mu(S)_{|G_0}=0$, because $x_0$ is quasi-interior to $E_+$. □

Before considering examples, we point out that under assumptions on $\boldsymbol{P}$ weaker than those of Def. 3.1, a weakened form of Theorem 3.6 can be obtained (Exerc. 4).

**Examples.** 3. Let $G$ be a Banach space with unconditional basis $(e_n)_{n\in\mathbb{N}}$. If $f_n$ $(n\in\mathbb{N})$ denotes the $n$-th coefficient form with respect to this basis, the projections $P_J\colon x\mapsto\sum_{n\in J}f_n(x)e_n$ ($J$ any subset of $\mathbb{N}$) form a B.a.p. $\boldsymbol{P}$ isomorphic to $2^{\mathbb{N}}$. In particular, the Stone space $K$ of $\boldsymbol{P}$ is $\beta\mathbb{N}$, the spectral measure $\mu$ is purely atomic, and if $e=\sum_n n^{-2}e_n$ then $G=(G,\boldsymbol{P},e)$ is cyclic. Under the norm and ordering introduced in (3.6), $G$ becomes a Banach lattice $E$ of real (or complex) sequences, the isomorphism $G\to E$ being given by $x\mapsto(f_1(x), f_2(x),\dots)$ $(x\in G)$. Moreover, $E$ is an ideal of $\mathbb{R}^{\mathbb{N}}$ (respectively, $\mathbb{C}^{\mathbb{N}}$).

4. Let $H$ be any Hilbert space, $A\subset\mathscr{L}(H)$ a commuting family of Hermitian operators containing the identity operator $1_H$. The strongly closed algebra (in the complex case, $W^*$-algebra) $C$ generated by $A$ is isomorphic to $C(K)$ for some compact $K$ (the space of pure states of $C$). (A short proof is obtained

[2] I.e., $\boldsymbol{P}$ is of countable type (Chap. II, Exerc. 4).

by considering $A$ as an algebra ordered by the cone of Hermitian-positive elements and applying Stone's algebra theorem [S, Chap. V, Exerc. 24].) If $\boldsymbol{P}$ denotes the set of Hermitian projections in $C$, then $\boldsymbol{P}$ is a B.a.p. on $H$ isomorphic to the Boolean algebra of extreme points of $[0, 1_H] \subset C$ (II. 2.13) and, of course, $K$ is the Stone representation space of $\boldsymbol{P}$.

Let $x_0 \in H$ be a fixed vector; the cyclic subspace $H_0 := (H, \boldsymbol{P}, x_0)$ is a Hilbert lattice (Def. IV. 6.5) under the ordering introduced in (3.6) and the original norm. In fact, the norm (*) of (3.3) equals the original norm on $H$ in the present case: If $f = \sum_i \alpha_i \chi_{U_i}$ (where $U_i$ runs through a finite family of disjoint, open-and-closed subsets of $K$) and $\|f\| \leqq 1$, then $m(f) = \sum_i \alpha_i P_i$ where $|\alpha_i| \leqq 1$ and $P_i$ are orthogonal Hermitian projections. Thus

$$\|m(f)x\|^2 = (m(f)x \mid m(f^*)x) = \sum_i |\alpha_i|^2 \, \|P_i x\|^2 \leqq \|x\|^2$$

for each $f \in B \cap L$ (for notation, see proof of (3.3)). Consequently, $\|x\| = \|x\|_1$ for all $x \in H$.

Next, we prove that the conclusion of (3.6) Cor. 2 subsists without requiring $\boldsymbol{P}$ to be reduced whenever $G$ is separable; this implies, in particular, that $C(K)$ ($K$ the Stone space of $\boldsymbol{P}$) is super Dedekind complete (Chap. II, § 4).

**3.7 Proposition.** *Let $G$ be a separable Banach space. Every $\sigma$-complete B.a.p. on $G$ is complete and of countable type.*

*Proof.* By (II. 4.9) it suffices to prove the existence of a strictly positive linear form $\varphi$ on $C(K)$. Let $\{x_n : n \in \mathbb{N}\}$ denote a countable total subset of $G$ and let $G_n := (G, \boldsymbol{P}, x_n)$. If $\boldsymbol{J}_n := \{P \in \boldsymbol{P} : P x_n = 0\}$ $(n \in \mathbb{N})$ and $K_n$ is the Stone space of the $x_n$-reduced Boolean algebra $\boldsymbol{P}/\boldsymbol{J}_n$ then $C(K_n) \cong C(K)/J_n$, $J_n$ denoting the $\sigma$-ideal $\{f : m(f) x_n = 0\}$ of $C(K)$. By (3.6) Cor. 3 there exists a strictly positive linear form $\varphi_n$ on $C(K_n)$; on the other hand, $K_n$ can be identified with the closed subset $\{s : f(s) = 0 \text{ for all } f \in J_n\}$ of $K$. Since, clearly, $\bigcap_n J_n = \{0\}$ it follows that $\bigcup_n K_n$ is dense in $K$; therefore, $\varphi := \sum_n \varphi_n / 2^n \|\varphi_n\|$ defines a strictly positive linear form on $C(K)$. ▯

The presence of quasi-interior positive elements in the Banach lattice $E$ corresponding, according to (3.6), to a cyclic Banach space $G_0$ suggests the application of the representation theory (Chap. III, §§ 4—6); the proof of the following proposition is an immediate consequence of (III. 4.5).

**3.8 Proposition.** *Every cyclic Banach space $G_0 = (G, \boldsymbol{P}, x_0)$, where $\boldsymbol{P}$ is assumed reduced with respect to $x_0$, is isomorphic (as a topological vector space) to a Banach lattice of continuous numerical functions on the Stone space $K$ of $\boldsymbol{P}$.*

Our present interest in this representation is based on its usefulness in the construction of closed, densely defined linear operators in a cyclic Banach space $G_0$ that map the cyclic vector $x_0$ onto a given $x \in G_0$. Again, we consider the Banach lattice $E$ with quasi-interior positive element $x_0$ constructed from $(G, \boldsymbol{P}, x_0)$ in (3.6); we may also assume for simplicity of presentation that $G$ is a real Banach space, since extension to complex scalars causes no difficulty. Let $\hat{E}$ denote the Banach lattice of continuous functions $K \to \overline{\mathbb{R}}$ representing $E$ by

(III. 4.5); for each $f \in \hat{E}$ and $n \in \mathbb{N}$, we let $f_n := f \chi_n$ where $\chi_n$ is the characteristic function of the (open) closure of the set $\{s: |f(s)| < n\}$ in $K$. Clearly, then, $\chi_n \in C(K)$ and $f_n \in C(K)$. Now we define the operator $T_f: D_f \to E$ by $T_f(x) := \lim_n m(f_n)x$, where $D_f$ consists precisely of those $x \in E$ for which the limit exists (in norm). $D_f$ is a vector subspace of $E$ which is dense because the set of all $x$ for which the sequence $(m(f_n)x)_{n \in \mathbb{N}}$ is eventually stationary, is dense (in fact, this latter set contains the dense subset $\{x \in E: m(\chi_n)x = x$ for some $n\}$ of $E$); moreover, it is not difficult to show that $T_f$ is closed. Of course, $T_f = m(f) \in \mathscr{L}(E)$ whenever $f \in C(K)$.

**3.9 Proposition.** *Let $G_0 = (G, \boldsymbol{P}, x_0)$ be a cyclic Banach space. For each $x \in G_0$ there exists a closed, densely defined linear operator $T_x$ in $G_0$ such that $x = T_x(x_0)$, and such that the domain of each $T_x$ contains a fixed dense subspace of $G_0$.*

*Proof.* If $E$ is the Banach lattice constructed from $G_0$ according to (3.6), then the mapping $m_0$ of (3.4) maps $C(K)$ onto a dense ideal of $E$; without loss of generality we suppose $\boldsymbol{P}$ to be $x_0$-reduced so that $m_0$ is injective. Thus the representation (III. 4.5) can be thought of as an extension $\tilde{m}_0: \hat{E} \to E$ of $m_0$. Now if $f \in \hat{E}$ and if $f_n \in C(K)$ is defined by $f_n(s) := f(s)$ for $s \in [|f| < n]^-$ and by $f(s) := 0$ otherwise (see above), it follows that $\|f_n - f\| \to 0$, since $E$ has order continuous norm. Therefore, if $x \in E$ is given and $x = \tilde{m}_0(f)$, then we have $x = \lim_n m(f_n)x_0 = T_f(x_0)$ and $T_x := T_f$ is an operator of the desired kind. Finally, if $x \in m_0(C(K))$, $x = m_0(h)$ say, then $\lim_n m(f_n)x = \lim_n m(f_n h)x_0$ exists, since $\|f_n h - fh\| \to 0$ in $\hat{E}$; so the dense subspace $m_0(C(K))$ belongs to the domain of each $T_f$ $(f \in \hat{E})$. ☐

*Note.* In terms of the spectral (Baire) measure $\mu$ associated with $\boldsymbol{P}$ (see Note preceding (3.4)), the operator $T_f$ is usually written $\int f \, d\mu$ and defined (as a closed linear operator with dense domain) for every Baire function $f$ on $K$ which is finite a.e. ($\mu$). The particular feature afforded by the representation of $E$ on its structure space $K$ is that the operators $T_x$ of (3.9) can be obtained by integrating *continuous* numerical functions on $K$ with respect to $\mu$; precisely, the functions $f$ contained in the representation $\hat{E}$ of $E$.

On the other hand, if $E$ is identified with $\hat{E}$, the operator $T_f$ $(f \in \hat{E})$ is simply multiplication by $f$: $T_f(g) = fg$, the domain $D_f$ $(\supset C(K))$ being the set of all $g \in \hat{E}$ for which $fg \in \hat{E}$ as well. (In fact, it follows from (II. 7.8) and its corollaries that $C_\infty(K)$ is an algebra, multiplication of functions in $C_\infty(K)$ being defined pointwise on sets of finiteness. Moreover, $\hat{E}$ is a lattice ideal of $C_\infty(K)$.)

## § 4. The Peripheral Spectrum of Positive Operators

The spectral properties of positive square matrices (Chap. I, §§ 2, 4—7) motivate the investigation of spectral properties of positive operators defined on more or less special types of (complex) Banach lattices. We begin our discussion by recalling a few standard facts of the spectral theory of a bounded operator $T \in \mathscr{L}(G)$, where $G$ is a complex Banach space $\neq \{0\}$ (cf. [DS] and the discussion preceding (1.4)).

The *spectrum* $\sigma(T)$ is defined to be the set $\{\lambda \in \mathbb{C}: \lambda - T$ is not a bijection of $G\}$;[1] $\sigma(T)$ is compact, non-void and the disjoint union $A\sigma(T) \cup R\sigma(T)$ of the approximate point spectrum and the residual spectrum of $T$ (§ 1); clearly, $R\sigma(T)$ is contained in the point spectrum $P\sigma(T')$, $T'$ denoting the adjoint $T' \in \mathscr{L}(G')$. $\varrho(T) := \mathbb{C} \setminus \sigma(T)$ is the resolvent set of $T$, and for $\lambda \in \varrho(T)$ we write $(\lambda - T)^{-1} =: R(\lambda, T)$ (or $R(\lambda)$ if $T$ is understood). The number $r(T) := \sup\{|\lambda| : \lambda \in \sigma(T)\}$ is called the *spectral radius* of $T$, and the set $\{\lambda : |\lambda| = r(T)\} \cap \sigma(T)$ (which is a closed non-void subset of $\sigma(T)$) is called the *peripheral spectrum* of $T$. Finally, if $G_0$ is a $T$-invariant Banach subspace of $G$ then $\sigma(T) \subset \sigma(T_{|G_0}) \cup \sigma(T_{G_0})$ where $T_{G_0}$ is the operator on $G/G_0$ induced by $T$; moreover, if $\lambda$ is in the unbounded connected component of $\varrho(T)$ then $R(\lambda, T_{|G_0}) = R(\lambda, T)_{|G_0}$ and, $q$ denoting the canonical map $G \to G/G_0$, $R(\lambda, T_{G_0}) \circ q = q \circ R(\lambda, T)$ (Exerc. 5).

Now let $E$ denote a complex Banach lattice; accordingly, the terms ideal, vector sublattice etc. are to be understood in the complex sense (Chap. II, § 11). If $T \in \mathscr{L}(E)$ is a positive operator, from C. Neumann's series

$$R(\lambda, T) = \sum_{n=0}^{\infty} \lambda^{-(n+1)} T^n \qquad (|\lambda| > r(T)) \tag{1}$$

it follows that $\lambda > r(T)$ implies $R(\lambda, T) \geqq 0$ (the converse being equally valid if $E \neq \{0\}$; Exerc. 5); moreover, $\lambda \mapsto R(\lambda, T)$ is increasing (and unbounded, see (4.1) below) as $\lambda \downarrow r(T)$. Also, (1) shows that

$$0 \leqq TR(\lambda, T) \leqq \lambda R(\lambda, T) \qquad (\lambda > r(T)) \tag{2}$$

for positive $T$; in particular, if $x \in E_+$ and $z := R(\lambda, T)x$ for some $\lambda > r(T)$ then the principal ideal $E_z$ of $E$ is $T$-invariant. Dually, if $0 \leqq x' \in E'$ and if $T'(E'_{x'}) \subset E'_{x'}$ then $T$ has a continuous extension $\tilde{T}$ to the $AL$-space $(E, x')$ (for notation, see Chap. IV, § 3).

Our first result holds for any positive operator and is proved exactly as in the finite dimensional case (cf. I. 2.3 and Exerc. 7).

**4.1 Proposition.** $0 < T \in \mathscr{L}(E)$ *implies* $r(T) \in \sigma(T)$.

*Proof.* By definition of $r(T)$, the family of operators $(R(\lambda, T))_{|\lambda| > r(T)}$ cannot be uniformly bounded in $\mathscr{L}(E)$; hence, there exists a sequence $(\lambda_n)$ such that $|\lambda_n| \downarrow r(T)$ and $\lim_n \|R(\lambda_n, T)x\| = +\infty$ for a suitable $x \in E$. Now

$$|R(\lambda_n, T)x| \leqq \sum_{n=0}^{\infty} |\lambda_n|^{-(n+1)} T^n |x| = R(|\lambda_n|, T)|x|$$

shows that $\lim_n \|R(|\lambda_n|, T)|x|\| = +\infty$; consequently, $r(T) \in \sigma(T)$. □

We have seen that the peripheral spectrum of every positive square matrix is cyclic (I. 2.7); it turns out that the corresponding problem for arbitrary positive operators on Banach lattices is not so easily settled (and, in fact, as yet

[1] We write $\lambda - T$ in place of $\lambda 1_G - T$.

unsolved in full generality). The main objective of this section is to give an affirmative answer for a large class of positive operators (Theorems 4.4 and 4.9 below), in addition to establishing some interesting auxiliary results concerning the peripheral point spectrum.

If $E=C(K)$ ($K\neq\emptyset$ compact) is an $AM$-space with unit $e$ (Def. II.7.1), a positive operator $T$ on $E$ is called a *Markov operator* if $Te=e$ (Def. III.8.8). By a *unimodular eigenfunction* of $T$ we understand a function $f\in C(K)$ satisfying $|f|=e$ and $\alpha f=Tf$ for some $\alpha\in\mathbb{C}$, $|\alpha|=1$.

**4.2 Proposition.** *If $f$ is a unimodular eigenfunction of the Markov operator $T$ on $C(K)$ pertaining to the eigenvalue $\alpha=\alpha(f)$ and if $\mu_s:=T'\delta_s$ ($\delta_s$ Dirac measure at $s\in K$), then for each $s\in K$ one has $\alpha f(s)=f(t)$ for all $t$ in the support of $\mu_s$. Moreover, the set $G$ of all unimodular eigenfunctions of $T$ is a group $G$ under pointwise multiplication, and $f\mapsto\alpha(f)$ is a character of $G$.*

*Proof.* In fact, since $Tf(s)=\int f(t)\,d\mu_s(t)$ for all $s\in K$, we obtain $1=\int\alpha^{-1}f(s)^{-1}f(t)\,d\mu_s(t)$. Now for fixed $s$, the integrand is a unimodular function of $t$ and $\mu_s$ is a probability measure on $K$, since $T$ is Markov; therefore, the integrand must be identically $=1$ on the support of $\mu_s$. From this it is immediate that if $g$ is another unimodular eigenfunction of $T$, $\beta g=Tg$, then $\beta^* g^*=T(g^*)$ (*denoting complex conjugation) and

$$(\alpha\beta^*)\,f(s)g^*(s)=\int f(t)g^*(t)\,d\mu_s(t)=T(fg^*)(s)$$

for each $s\in K$; this proves the remaining assertions. ▯

**Corollary 1.** *Let $F$ denote the closed vector sublattice of $C(K)$ generated by $G$. Then $F$ is an $AM$-space with unit $C(K_0)$ and the restriction of $T$ to $F$ is a lattice homomorphism.*

*Proof.* It is clear that $F$ is a (complex) $AM$-space with unit; for, $T$ being a Markov operator, $F$ contains the unit $e$ of $C(K)$. Moreover, $F$ is a conjugation invariant subalgebra of $C(K)$ and it is easily verified that $T_{|F}$ is multiplicative; thus from the complex version of (III.9.1) it follows that $T_{|F}$ is a lattice homomorphism (Chap. II, § 11 (iii)). ▯

We recall (Def. I.2.5) that a subset $A\subset\mathbb{C}$ is said to be *cyclic* if $\alpha\in A$, $\alpha=|\alpha|\gamma$ implies $|\alpha|\gamma^k\in A$ for all $k\in\mathbb{Z}$.

**Corollary 2.** *Let $E$ be any Banach lattice and let $T$ denote a lattice homomorphism of $E$. The point spectrum $P\sigma(T)$ is cyclic.*

*Proof.* Let $\lambda x=Tx$ where $\lambda\neq 0$ and $x\neq 0$. Since $T$ is a lattice homomorphism, we have $|\lambda||x|=T|x|$ and hence, $|\lambda|^{-1}T$ induces a Markov operator on the $AM$-space $E_{|x|}\cong C(K)$. Since $|x|$ corresponds to the unit of $C(K)$, $x$ defines a unimodular eigenfunction of this Markov operator and the assertion follows from (4.2). ▯

Our next objective is to prove that the entire spectrum of any lattice homomorphism is cyclic; to this end we need the following lemma.

**4.3 Lemma.** *Let $T\in\mathscr{L}(E)$, $E$ any Banach lattice, be a lattice homomorphism and suppose that $\alpha\in P\sigma(T')$, $\alpha=|\alpha|\gamma\neq 0$. At least one of these assertions is true:*
(i) $|\alpha|\gamma^k\in P\sigma(T')$ *for all* $k\in\mathbb{Z}$,
(ii) $\{\lambda\in\mathbb{C}: |\lambda|<|\alpha|\}\subset P\sigma(T')$.

*Proof*. Suppose that $\alpha x'=T'x'$ for some $x'\in E'$, $x'\neq 0$. We define $r_1:=r(T)+1$ and $z':=R(r_1, T')|x'|$. From Formula (2) above we find that $T'z'\leqq r_1 z'$ so $T'$ leaves the ideal $(E')_{z'}$ of $E'$ invariant. As observed above this implies that $T$ induces a positive operator $\tilde{T}$ on the $AL$-space $(E,z')$. On the other hand, we have $|\alpha||x'|\leqq T'|x'|$ and operating with $R(r_1,T')$ on this inequality yields $|\alpha|z'\leqq T'z'$, since $T'$ commutes with its resolvent. Now for any $x\in E$, we have

$$|\alpha|\langle|x|,z'\rangle\leqq\langle|x|,T'z'\rangle=\langle T|x|,z'\rangle=\langle|Tx|,z'\rangle,$$

because $T$ is a lattice homomorphism of $E$. Consequently, if $p$ denotes the norm of the $AL$-space $(E,z')$, we obtain $|\alpha|p(\tilde{x})\leqq p(\tilde{T}\tilde{x})$ for all $\tilde{x}\in(E,z')$. Clearly, this implies that $A\sigma(\tilde{T})\cap\{\lambda: |\lambda|<|\alpha|\}=\emptyset$. We show that (i) or (ii) holds, accordingly as $0\notin\sigma(\tilde{T})$ or $0\in\sigma(\tilde{T})$.

(i) If $0\notin\sigma(\tilde{T})$ then $\tilde{T}$ is a lattice isomorphism of $(E,z')$. In fact, $\tilde{T}$ is bijective, and if $\tilde{T}\tilde{x}\geqq 0$ then $\tilde{T}\tilde{x}=|\tilde{T}\tilde{x}|=\tilde{T}|\tilde{x}|$ which shows that $\tilde{x}=|\tilde{x}|\geqq 0$; so $\tilde{T}$ has a positive inverse. But then $\tilde{T}'$, which agrees with the restriction of $T'$ to the ideal $(E')_{z'}$, is a lattice isomorphism as well and assertion (i) holds by (4.2) Cor. 2.

(ii) If $0\in\sigma(\tilde{T})$ then, since $A\sigma(\tilde{T})\cap\{\lambda: |\lambda|<|\alpha|\}=\emptyset$, the entire set $\{\lambda: |\lambda|<|\alpha|\}$ must belong to the residual spectrum $R\sigma(\tilde{T})$. Thus $\{\lambda: |\lambda|<|\alpha|\}\subset P\sigma(\tilde{T}')\subset P\sigma(T')$ which shows that (ii) occurs. ▯

**4.4 Theorem.** *The spectrum of every lattice homomorphism $T$ of a Banach lattice $E$ is cyclic.*

*Proof*. By imbedding $E$ into some $\boldsymbol{F}$-product $\hat{E}$ (1.3) and applying (1.4), we can assume that each $\lambda\in\sigma(T)$ is an element of either the point or the residual spectrum of $T$; that is, we can assume that $\sigma(T)\subset P\sigma(T)\cup P\sigma(T')$. The assertion now follows from (4.2) Cor. 2 and (4.3), respectively. ▯

Stronger assertions can be made in special cases; for example, if $E=C(K)$ and $T$ is a lattice homomorphism and Markov operator on $E$, then either $\sigma(T)$ is a cyclic subset of the circle group $\Gamma$, or else $\sigma(T)=\{\lambda\in\mathbb{C}: |\lambda|\leqq 1\}$ (Exerc. 6; cf. (I.4.5)). Also, each compact cyclic subset of $\mathbb{C}$ is the spectrum of some lattice homomorphism defined on a suitable Banach lattice $E$ (Exerc. 6).

We now turn to our second main objective in this section, the fact that each $(G)$-solvable positive operator (Def. 4.7 below) has cyclic peripheral spectrum. The method of proof will again be a reduction, via several technical devices, of the general case to the case of a Markov operator on an $AM$-space (cf. 4.2). To formulate an intermediate concept (Def. 4.5) which is of some independent interest, we need the concept of a Banach function lattice that we have encountered in varying generality before (Theorems (III.4.5), (III.5.3), (IV.9.7), (V.2.10)).

For the present purpose, by a *Banach function lattice* we understand a (complex) vector lattice $E$ of equivalence classes of complex valued functions $f$ defined on a set $X$ (two functions being equivalent if they agree on $X\setminus N$, where $N$ is a member of some fixed $\sigma$-ideal of subsets of $X$), endowed with some lattice norm under which $E$ is complete. In addition, we require that for each fixed $f \in E$, the set $B_f$ of all bounded functions $g: X \to \mathbb{C}$ such that $|f| g \in E$, be a (complex) $AM$-space with unit under its canonical order and the supremum norm.

**4.5 Definition.** *Let $E$ denote a Banach function lattice in the sense just specified; let $0 \leqq T \in \mathscr{L}(E)$ and let $r(T)=1$. The peripheral point spectrum of $T$ is called* fully cyclic[2] *if $\alpha f = Tf$, $|\alpha|=1$, and $f=|f|g$ imply that $\alpha^k |f| g^k = T(|f| g^k)$ for all $k \in \mathbb{Z}$.*

The following generalization of (4.2) is the concrete basis on which Theorem 4.9 eventually rests.

**4.6 Proposition.** *Let $0 \leqq T \in \mathscr{L}(E)$, where $E$ is a Banach function lattice as defined above, let $r(T)=1$, and suppose that $T'\varphi \leqq \varphi$ for some strictly positive linear form $\varphi$ on $E$. The peripheral point spectrum of $T$ is fully cyclic.*

*Proof.* Let $\alpha f = Tf$, where $f \neq 0$ and $|\alpha|=1$. Then $|f| \leqq T|f|$, and since $T'\varphi \leqq \varphi$ we have $0 < \langle |f|, \varphi\rangle \leqq \langle T|f|, \varphi\rangle = \langle |f|, T'\varphi\rangle \leqq \langle |f|, \varphi\rangle$; consequently, $\langle T|f| - |f|, \varphi\rangle = 0$ and so $|f| = T|f|$, since $\varphi$ is strictly positive. Now since $B_f := \{g \text{ bounded}: |f| g \in E\}$ is an $AM$-space by hypothesis (see above), $T$ induces a Markov operator $U$ on $B_f$ for which $\alpha g_0 = U g_0$, where $f = |f| g_0$. Hence by (4.2), $\alpha^k g_0^k = U(g_0^k)$ for all $k \in \mathbb{Z}$ which proves the assertion. □

**Example.** 1. Let $E = L^1(\mu)$ be an $AL$-space (cf. II.8.5), and suppose $T$ is a positive contraction on $E$ satisfying $r(T)=1$. By (4.6) the peripheral point spectrum of $T$ is fully cyclic, since $\|T\| \leqq 1$ means that $\int Tf\,d\mu \leqq \int f\,d\mu$ for all $f \in E_+$ and hence, that $T'\mu \leqq \mu$.

More generally, the peripheral spectrum of $T$ is cyclic, since by (1.4) it agrees with the peripheral point spectrum of $\hat{T}$ on the ultraproduct $\hat{E}_U$, and $\hat{T}$ is a positive contraction on the $AL$-space $\hat{E}_U$ (§ 1, Example 2).

**4.7 Definition.** *A positive operator $T$ on a Banach lattice $E$ is said to satisfy the* growth condition (G) *if the family of operators $(\lambda - r(T))R(\lambda, T)$ is uniformly bounded for $\lambda > r(T)$.*

*$T$ is said to be* (G)-solvable *if there exists a chain $(E_k)_{0 \leqq k \leqq n}$ $(n \in \mathbb{N})$ of closed $T$-invariant ideals such that $\{0\} = E_0 \subset E_1 \subset \cdots \subset E_n = E$ and such that the operator $T_k$ induced by $T$ on $E_k/E_{k-1}$ satisfies* (G) *for each $k$, $1 \leqq k \leqq n$.*

**Examples.** 2. If $r(T)=1$ and the sequence $(T^n)_{n \in \mathbb{N}}$ is bounded in $\mathscr{L}(E)$, an easy estimate using the Neumann series (1) shows $T$ to satisfy (G).

3. If $r = r(T)$ is a pole of $R(\lambda, T)$ of order 1 then $T$ satisfies (G), since $P := \lim_{\lambda \to r} (\lambda - r) R(\lambda, T)$ exists in $\mathscr{L}(E)$. (We observe that if $P$ is of finite rank, then $(T^n)_{n \in \mathbb{N}}$ is necessarily bounded; see Exerc. 7).

4. More generally, suppose that $r(T)=1$ is a pole of $R(\lambda, T)$ of arbitrary order; we claim that $T$ is (G)-solvable. (This occurs, in particular, whenever some power of $T$ is compact.) The proof is by induction over the order $n$ of the pole $\lambda = 1$; for $n=1$, the assertion is clear (Example 3) so assume the

[2] Cf. Def. I.2.5.

result to be valid for poles of order $<n$ $(n \geqq 2)$. Let $Q := \lim_{\lambda \to 1} (\lambda - 1)^n R(\lambda, T)$ denote the highest coefficient in the Laurent expansion of $R(\lambda, T)$ at $\lambda = 1$; then $QT = Q$ and $J := \{x \in E; Q|x| = 0\}$ is a $T$-ideal (Def. III.8.1). It is clear that the resolvent of $T_{|J}$ has a pole at $\lambda = 1$ of order $<n$ and hence, $T_{|J}$ is $(G)$-solvable by hypothesis. On the other hand, if $q: E \to E/J$ denotes the quotient map and $S_{-k} = (T-1)^{k-1} P$ ($P$ the residuum of $R(\lambda, T)$ at $\lambda = 1$) is the $k$-th negative coefficient of $R(\lambda, T)$, then $\hat{S}_{-k} \circ q = q \circ S_{-k}$ determines the corresponding coefficient in the expansion of $\hat{R}(\lambda) := R(\lambda, T_J)$ at $\lambda = 1$. Since $S_{-k} = 0$ for $k > n$, it follows that $S_{-k}(E) \subset J$ whenever $k > 1$; therefore, $\hat{S}_{-k} = 0$ which shows $\hat{R}(\lambda)$ to have a pole of order 1 at $\lambda = 1$ (Lotz [1968]).

The growth condition $(G)$ ensures the existence of $T$-invariant linear forms that can play a role similar to that played by $\varphi$ in (4.6).

**4.8 Lemma.** *Let $E$ be any Banach lattice and let $0 < T \in \mathscr{L}(E)$ satisfy $r(T) = 1$ and condition $(G)$ (Def. 4.7). If $x_0 \leqq T x_0$ for some $x_0 > 0$, there exists a linear form $x_0' \in E_+'$ such that $x_0' = T' x_0'$ and $\langle x_0, x_0' \rangle > 0$.*

*Proof.* Since $x_0 \leqq T x_0$ implies $x_0 \leqq T^k x_0$ for all $k \in \mathbb{N}$, a glance at (1) shows that we have $R(\lambda, T) x_0 \geqq (\lambda - 1)^{-1} x_0$ for all $\lambda > 1$. On the other hand, $(G)$ implies that for each $x' \in E'$, the family $x_\lambda' := (\lambda - 1) R(\lambda, T)' x'$ has a $\sigma(E', E)$-limit point $x_0'$ as $\lambda \to 1$, and it is clear that $T' x_0' = x_0'$, because $T' R(\lambda, T)' = \lambda R(\lambda, T)' - 1_{E'}$ and because $T'$ is $\sigma(E', E)$-continuous. Now if $x' \in E_+'$ is chosen so that $\langle x_0, x' \rangle > 0$, we obtain $\langle x_0, x_0' \rangle \geqq \varliminf_{\lambda \to 1} (\lambda - 1) \langle R(\lambda, T) x_0, x' \rangle \geqq \langle x_0, x' \rangle > 0$. □

**4.9 Theorem.** *Let $E$ be any Banach lattice. Every $(G)$-solvable positive operator on $E$ has cyclic peripheral spectrum.*

*Proof.* It is no loss of generality to assume that the operator $T$, $0 < T \in \mathscr{L}(E)$, satisfies $r(T) = 1$. First, suppose that $T$ satisfies $(G)$ (Def. 4.7). By imbedding $E$ into some $\boldsymbol{F}$-product (Def. 1.1) and applying (1.3) and (1.4), we can assume that the peripheral spectrum is point spectrum of $T$ (notice that by (1.2), the transition $T \mapsto \hat{T}$ leaves $(G)$ intact). Now let $\alpha x = Tx$ where $x \neq 0$, $|\alpha| = 1$. Then $|x| \leqq T|x|$ and by (4.8), there exists a linear form $x' > 0$ on $E$ such that $x' = T' x'$ and $\langle |x|, x' \rangle > 0$. Let $N := \{x \in E: \langle |x|, x' \rangle = 0\}$ denote the null ideal of $x'$; then $N$ is a $T$-ideal, and $x \notin N$. If $\hat{x}$ denotes the canonical image of $x$ in $E/N$, we have $\alpha \hat{x} = T_N \hat{x}$ and $|\hat{x}| = T_N |\hat{x}|$. Now $T_N$ induces a Markov operator $U$ on the ideal $(E/N)_{|\hat{x}|}$ of $E/N$ and by (4.6) $U$ has cyclic peripheral point spectrum. Hence, $T_N$ has each $\alpha^k$ $(k \in \mathbb{Z})$ as an eigenvalue, and because of $|\alpha| = 1 = r(T_N)$ it follows that $\{\alpha^k : k \in \mathbb{Z}\} \subset \sigma(T)$.

If, more generally, $T$ is $(G)$-solvable the assertion follows by induction over the chain length $n$ (cf. Def. 4.7). For $n = 1$, the result has just been proved; so we assume that it holds for all $(G)$-solvable operators with an ideal chain of length $<n$ $(n \geqq 2)$. Suppose $\{0\} = E_0 \subset E_1 \subset \cdots \subset E_{n-1} \subset E_n = E$ is a chain of $T$-ideals and suppose that $\alpha \in \sigma(T)$, $|\alpha| = 1$. If $T_n$ denotes the operator induced by $T$ on $E/E_{n-1}$ then, as remarked above, at least one of the relations $\alpha \in \sigma(T_{|E_{n-1}})$ or $\alpha \in \sigma(T_n)$ holds. If $\alpha \in \sigma(T_{|E_{n-1}})$ then $\alpha^k \in \sigma(T_{|E_{n-1}})$ for all $k \in \mathbb{Z}$ by the induction hypothesis; since all $\alpha^k$ are in the approximate point spectrum of $T_{|E_{n-1}}$, it

follows that they are elements of $\sigma(T)$ as well. On the other hand, if $\alpha\in\sigma(T_n)$ then $\alpha^k\in\sigma(T_n)$ (all $k\in\mathbb{Z}$) by the first part of the proof, since $T_n$ satisfies $(G)$ by definition of $(G)$-solvability. Again, this implies $\{\alpha^k: k\in\mathbb{Z}\}\subset\sigma(T)$, and the proof is complete. □

**Corollary.** *Let $T$ be a positive operator on $E$ with spectral radius $r(T)>0$. If $r(T)$ is a pole of the resolvent (in particular, if some power of $T$ is compact), then $T$ has cyclic peripheral spectrum.*

Since every operator satisfying these conditions is $(G)$-solvable (Example 4), the corollary is an immediate consequence of (4.9). For other conditions entailing the cyclicity of the peripheral spectrum, see Exerc. 6.

## § 5. The Peripheral Point Spectrum of Irreducible Positive Operators

As in the finite dimensional case (Chap. I, § 6), considerably stronger results than those of Section 4 are available for irreducible positive operators, whose definition has been given earlier (Def. III.8.1) and which arise naturally in ergodic theory (Chap. III, § 8, Examples 5—7). We will exclusively be concerned with the peripheral point spectrum for which the state of affairs is largely clarified. Since the results of this (as well as the preceding) section are interesting only for operators $T$ with $r(T)>0$, we will now generally assume that $r(T)=1$; the question of when irreducibility of $T$ implies $r(T)>0$, will be studied in detail below (§ 6). Throughout this section, the field of scalars is $\mathbb{C}$.

**5.1 Proposition.** *Let $E$ denote a Banach lattice and let $0<T\in\mathscr{L}(E)$ be irreducible with $r(T)=1$; suppose, in addition, that there exists a non-zero linear form $\varphi\in E'_+$ which is $T$-subinvariant (i.e., for which $T'\varphi\leqq\varphi$). If $\alpha$, $|\alpha|=1$, is an eigenvalue of $T$, there exists a surjective isometry $V: E\to E$ such that $\alpha T=V^{-1}TV$.*

*Proof.* $T'\varphi\leqq\varphi$ $(\varphi>0)$ implies that the null ideal $N_\varphi$ is a $T$-ideal (Def. III.8.1); hence, since $T$ is irreducible, $N_\varphi=\{0\}$. Now let $\alpha x=Tx$ where $x\neq 0$; then $|x|\leqq T|x|$ and it follows that $|x|=T|x|$, since $\varphi\gg 0$ and $\varphi(T|x|-|x|)=0$. (It follows similarly that $T'\varphi=\varphi$, since $|x|$ is quasi-interior to $E_+$.) The principal ideal $E_{|x|}$ is a complex $AM$-space $C(K)$ ($K$ compact) on which $T$ induces a Markov operator $T_0$ while $x$ defines a unimodular eigenfunction $f$ of $T_0$. As in (4.2), we let $\mu_s:=T_0'\delta_s$ $(s\in K)$ and obtain $\alpha f(s)=f(t)$ for all $t$ in the support of $\mu_s$. Hence if $h\in C(K)$, we have $\alpha f(s)h(t)=f(t)h(t)$ for all $t\in\operatorname{Supp}\mu_s$ and so $T_0(fh)=\alpha f\,T_0(h)$. If $V_0$ denotes the multiplication operator $h\mapsto fh$ of $C(K)$, we obtain $\alpha T_0=V_0^{-1}T_0V_0$. Since $f$ is unimodular, one has $|V_0h|=|h|$ for all $h\in C(K)=E_{|x|}$ and hence, $\|V_0h\|=\|h\|$ for the norm induced by $E$; but $E_{|x|}$ is dense in $E$, since $T$ is irreducible, and so $V_0$ is an isometry of a dense ideal of $E$. Clearly, the continuous extension $V$ of $V_0$ to $E$ is an isometry satisfying $\alpha T=V^{-1}TV$. □

**Corollary.** *Under the assumptions on $T$ made in* (5.1), *each pole of $R(\lambda,T)$ on the spectral circle $\{\lambda:|\lambda|=1\}$ is of order one.*

*Proof.* If $\alpha$, $|\alpha|=1$, is a pole of $R(\lambda,T)$ then from $\alpha T=V^{-1}TV$ it follows that $\lambda=1$ is a pole of the same order as $\alpha$; it suffices, therefore, to show that the pole

$\lambda=1$ of $R(\lambda,T)$ has order 1. Now if $\lambda=1$ is a pole of order $k$, then $Q:=\lim_{\lambda\to 1}(\lambda-1)^k R(\lambda,T)$ is a positive operator, and $Q^2=0$ whenever $k>1$ (cf. § 4, Example 4). But $QT=Q$ implies that $J:=\{x\in E: Q|x|=0\}$ is a $T$-ideal; since $J\neq E$ and $T$ is irreducible, it follows that $J=\{0\}$. But $J=\{0\}$ is incompatible with $Q^2=0$, and so $k=1$. ∎

The following theorem is closely related to the Halmos-von Neumann theorem (III.10.5); precisely: If $T$ satisfies the hypotheses of (5.2) and if $E_0$ denotes the (complex) closed vector sublattice of $E$ generated by the eigenvectors of $T$ pertaining to unimodular eigenvalues, then the restriction of $T$ to $E_0$ generates a compact monothetic group in $\mathscr{L}_s(E_0)$. Compare also (III.11.6).

**5.2 Theorem.** *Let $0<T\in\mathscr{L}(E)$, $r(T)=1$, be an irreducible operator which has non-void peripheral point spectrum and possesses an invariant form $\varphi=T'\varphi$ $(0<\varphi\in E'_+)$.*[1] *The following assertions are true:*

(i) *The fixed space $F$ of $T$ is one dimensional and spanned by a (unique normalized) quasi-interior point of $E_+$.*

(ii) *The peripheral point spectrum of $T$ is a subgroup of the circle group.*

(iii) *Each peripheral eigenvalue $\alpha$ of $T$ is simple, and $\sigma(T)=\alpha\sigma(T)$.*

(iv) *1 is the unique eigenvalue of $T$ with a positive eigenvector.*

*Proof.* (i) First, we observe that $F\neq\{0\}$. In fact, by hypothesis there exist $\alpha$, $|\alpha|=1$, and $x\neq 0$ such that $\alpha x=Tx$. This implies $|x|\leqq T|x|$ and here equality must hold because of $\varphi=T'\varphi,\varphi$ being strictly positive (cf. proof of (5.1)). Therefore, $F\neq\{0\}$ and $x=Tx$ implies $|x|=T|x|$ by the same argument so $F$ is a vector sublattice of $E$. Since $T$ is irreducible, each non-zero $x\in F_+$ is a quasi-interior point of $E_+$ and hence, by (II.6.3) Cor. 2, of $F_+$; thus, the underlying real vector lattice of $F$ is simple and it follows from (II.3.4) that $\dim F=1$.

(ii) If $\alpha,\beta$ are eigenvalues of $T$ on the spectral circle $\{\lambda:|\lambda|=1\}$ then the complex conjugate $\beta^*$ is an eigenvalue of $T$, since $T$ is positive and hence, a real operator on $E$ (Chap. II, § 11). By (5.1) there exist isometries $U,V$ of $E$ such that $\beta^*T=U^{-1}TU$, $\alpha T=V^{-1}TV$. Consequently, $\alpha\beta^*T=(VU)^{-1}T(VU)$ which shows $\alpha\beta^*$ to be an eigenvalue of $T$.

(iii) This is clear from (5.1), since $\sigma(V^{-1}TV)=\sigma(T)$ and $\sigma(\alpha T)=\alpha\sigma(T)$.

(iv) Suppose $\lambda x=Tx$ where $x>0$; since $\varphi\gg 0$, we obtain $\lambda\langle x,\varphi\rangle=\langle Tx,\varphi\rangle$ $=\langle x,\varphi\rangle>0$ whence $\lambda=1$. ∎

We remark that (iv) of (5.2) can be proved under somewhat more general assumptions on $T$ (Exerc. 8). Also, (5.2) applies in the following special cases.

**Corollary.** *If $0<T\in\mathscr{L}(E)$ is irreducible, the remaining hypotheses of (5.2) are automatically fulfilled in the following cases:*

(i) *$E=C(K)$ ($K$ compact) and $T$ is a Markov operator.*

(ii) *$r(T)=1$ is a pole of the resolvent of $T$.*

[1] By (4.8) this is true if $T$ satisfies $(G)$.

*Proof*. In case (i), the existence of a $T$-invariant form $\varphi \gg 0$ is implied by (III.8.10). (ii): The pole $\lambda=1$ has order 1 by the proof of (5.1) Cor. and $P=\lim_{\lambda\to 1}(\lambda-1)R(\lambda,T)$ is a strictly positive projection of rank 1, $P=\varphi\otimes u$ (cf. III.8.5). □

We note that condition (ii) of the corollary is satisfied by operators $T$ having a compact power $T^k$; especially, by weakly compact operators on $AM$- and $AL$-spaces (II.9.9) Cor. 1.

Generally, if $\lambda=1$ is a pole of $R(\lambda,T)$ then by (5.1) each eigenvalue of $T$ on the spectral circle is a (first order) pole of $R(\lambda,T)$; consequently, the peripheral point spectrum of $T$ is a finite group of roots of unity. Under these circumstances, can there exists elements $\lambda\in\sigma(T)$, $|\lambda|=1$, which are not eigenvalues of $T$ (and hence, not poles of $R(\lambda,T)$)? The answer is negative; this constitutes a rather deep theorem which was first proved in full generality by Niiro and Sawashima [1966]. The proof given below follows Lotz and Schaefer [1968]; first, we need this lemma.

**5.3 Lemma.** *Let $E$ be a Banach lattice, let $T\in\mathscr{L}(E)$ and let $J$ denote a closed $T$-invariant ideal of $E$; by $T_1:=T_J$ and $T_2:=T_{|J}$ we denote the induced operators on $E/J$ and $J$, respectively. If $\lambda_0\in\mathbb{C}$ is a pole of $R(\lambda,T_1)$ of order $k_1\geq 0$, and if $\lambda_0$ is a pole of $R(\lambda,T_2)$ of order $k_2\geq 0$, then $\lambda_0$ is a pole of $R(\lambda,T)$ whose order $k$ satisfies $\sup(k_1,k_2)\leq k\leq k_1+k_2$.*

*Proof*. By (II.5.5) Cor. 1, the polar $J^\circ$ is a band in the order complete dual $E'$ and hence, by (II.2.12), $E'=J^\circ+(J^\circ)^\perp$. With respect to this direct sum, the adjoint $T'$ has a matrix representation $T'=\begin{pmatrix} T_2' & S \\ 0 & T_1'\end{pmatrix}$ where $J^\circ$ and $(J^\circ)^\perp$ are identified with the duals of $E/J$ and $J$, respectively, and where $S\in\mathscr{L}((J^\circ)^\perp,J^\circ)$. It follows that near $\lambda_0$ the resolvent of $T'$ has the matrix representation

$$R(\lambda,T')=\begin{pmatrix} R(\lambda,T_2') & R(\lambda,T_2')SR(\lambda,T_1') \\ 0 & R(\lambda,T_1')\end{pmatrix}$$

which shows that $\lambda_0$ is a pole of $R(\lambda,T')$ of order $k$, where $\sup(k_1,k_2)\leq k\leq k_1+k_2$. Since $R(\lambda,T)'=R(\lambda,T')$, the lemma is proved. □

**5.4 Theorem.** *Let $E$ be a Banach lattice, and let $0<T\in\mathscr{L}(E)$ be irreducible with $r(T)=1$. If the peripheral spectrum of $T$ contains 1 as a pole of $R(\lambda,T)$, then it is a group of roots of unity consisting entirely of first order poles of $R(\lambda,T)$.*

*Proof*. The main idea of the proof consists in applying the imbedding procedure (§1) in order to deal with point spectrum only, and then to consider an ideal of a suitable quotient to retrieve irreducibility. By (5.1) and (5.2), using the fact that $\lambda=1$ is a pole of $R(\lambda,T')$, we have only to show that each $\alpha\in\sigma(T)$, $|\alpha|=1$, is a pole of $R(\lambda,T)$. The proof is divided into four steps.

1. We select a free ultrafilter $\boldsymbol{U}$ on $\mathbb{N}$ and imbed $E$ into the $\boldsymbol{U}$-product $\hat{E}$ (1.1). Since by (5.1) and the corollary of (5.2), $\lambda=1$ is a first order pole of $R(\lambda,T)$ with residue $P=\varphi\otimes u$ of rank 1, (1.2) shows that the resolvent $R(\lambda,\hat{T})$ of the induced operator $\hat{T}$ has a first order pole at $\lambda=1$ with residue $\hat{P}=\hat{\varphi}\otimes\hat{u}$, where $\hat{u}\in\hat{E}$

is the canonical image of $u$ and where $\hat{\varphi}\in\hat{E}'$ is given by $\hat{\varphi}(\hat{x})=\lim_U \varphi(x_n)$ whenever $(x_n)\in\hat{x}$. The null ideal $R:=\{\hat{x}:\hat{\varphi}(|\hat{x}|)=0\}$ is a $\hat{T}$-ideal, and the operator $\tilde{T}$ defined by $\hat{T}$ on $\tilde{E}:=\hat{E}/R$ has a resolvent $R(\lambda,\tilde{T})$ with like properties as $R(\lambda,\hat{T})$, except that the residue $\tilde{P}$ at the pole $\lambda=1$ is strictly positive. Finally, let $J$ denote the closure of the principal ideal $\tilde{E}_{\tilde{u}}$, where $\tilde{u}\in\tilde{E}$ is the canonical image of $\hat{u}$. Since $\tilde{T}\tilde{u}=\tilde{u}$, $J$ is a $\tilde{T}$-ideal; we define $S:=\tilde{T}_{|J}$.

2. It is clear from the preceding that $R(\lambda,S)$ has a first order pole at $\lambda=1$ with residue the restriction of $\tilde{P}$ to $J$. But $S$ is irreducible: In fact, $\tilde{P}$ leaves every $S$-ideal of $J$ invariant, and is strictly positive and of rank 1; hence, every $S$-ideal $G\neq\{0\}$ contains $\tilde{P}(G)=\{\varrho\tilde{u}:\varrho\in\mathbb{C}\}$ and so $G=(\tilde{E}_{\tilde{u}})^{-}=J$.

3. Let $\alpha\in\sigma(T)$, $|\alpha|=1$. Then $\alpha\hat{x}=\hat{T}\hat{x}$ for some $0\neq\hat{x}\in\hat{E}$ by (1.4); since $|\hat{x}|\leqq\hat{T}|\hat{x}|$, we have $0<|\hat{x}|\leqq\hat{P}|\hat{x}|$ which shows that $\hat{x}\notin R$. Thus if $\tilde{x}\in\tilde{E}$ is the canonical image of $\hat{x}$, we have $|\tilde{x}|=\tilde{P}|\tilde{x}|\neq 0$, since $\tilde{P}$ is strictly positive. So $0\neq\tilde{x}\in J$ and $\alpha\tilde{x}=S\tilde{x}$; therefore, (5.1) applies and $\alpha$ is a first order pole of $R(\lambda,S)$.

4. Now we apply (5.3) twice to return to $\hat{E}$. First, $\tilde{P}(\tilde{E})\subset J$ shows (in conjunction with (4.1)) that the operator $\tilde{T}_J$ on $\tilde{E}/J$ has spectral radius $<1$; hence by (5.3), $\alpha$ is a first order pole of $R(\lambda,\tilde{T})$. Second, $\hat{P}(R)=\{0\}$ shows that the restriction $\hat{T}_{|R}$ has spectral radius $<1$; hence by (5.3), $\alpha$ is a first order pole of $R(\lambda,\hat{T})$. It is now clear from the corollary of (1.4) that $\alpha$ is a first order pole of $R(\lambda,T)$, and the proof is complete. ∎

**Corollary.** *Under the assumptions of* (5.4), *$T$ has a representation*

$$T=\sum_{\mu=0}^{m-1}\zeta^{\mu}P_{\mu}+V,$$

*where $m\in\mathbb{N}$, $\zeta$ is a primitive $m$-th root of unity, $P_\mu$ are projections of rank* 1 *commuting with $T$, and $V$ is an operator commuting with $T$ and spectral radius $r(V)<1$.*

The proof of this corollary can be left to the reader (cf. [DS, VII.3]).

It is a natural question now if irreducibility of $T$ is an indispensable assumption in (5.4). For example, if $E$ is a finite direct sum of minimal $T$-ideals (Def. III.8.1) then, clearly, the conclusion of (5.4) (with the exception of the group property) obtains. The following result is an extension of this type, and the subsequent example will show that the hypothesis of (5.5) cannot be further relaxed.

**5.5 Theorem.** *Let $E$ be a Banach lattice, let $0<T\in\mathscr{L}(E)$, and suppose that $r(T)=1$ is a pole of $R(\lambda,T)$. If the residuum $P$ of $R(\lambda,T)$ at $\lambda=1$ is of finite rank, then the peripheral spectrum of $T$ consists entirely of poles of $R(\lambda,T)$.*

> *Note.* It is not difficult to prove that the pole at $\lambda=1$ is of maximal order on the spectral circle (Exerc. 7); moreover, from (4.9) it follows that the peripheral spectrum of $T$ is a finite union of finite groups of roots of unity, and a representation of $T$ analogous to that of (5.4) Cor. is valid.

*Proof* of (5.5). 1. First, we assume that $\lambda=1$ is a first order pole of $R(\lambda,T)$ and that $P$ is strictly positive, $P\gg 0$. If $E_0$ denotes the underlying real vector lattice of $E$ (so that $E=E_0+iE_0$), it follows from (III.11.5) that $P(E_0)$ is a finite dimensional vector sublattice of $E_0$ and hence, $P(E_0)\cong\mathbb{R}^n$ for some $n\in\mathbb{N}$ by (II.3.9) Cor. 1. Therefore, $P(E)$ can be identified with $\mathbb{C}^n$. Now if $\{e_1,\dots,e_n\}$ denotes the standard basis of $\mathbb{C}^n$ and if $J_k$ denotes the closed ideal of $E$ generated by $e_k$ $(k=1,\dots,n)$ we conclude from (III.8.5) that each $T_k := T_{|J_k}$ is irreducible (note that the cyclic semi-group $(T^n)_{n\in\mathbb{N}}$ is mean ergodic). Hence by (5.4), the peripheral spectrum if each $T_k$ consists entirely of poles of $R(\lambda, T_k)$, and the same is obviously true of $T_{|J}$ where $J := \sum_k J_k$ is a closed ideal by (III.1.2); evidently, $J$ is a $T$-ideal of $E$. Since $P(E) \subset J$, the operator $T_J$ induced by $T$ on $E/J$ has spectral radius $<1$ (cf. 4.1); thus (5.3) shows the peripheral spectrum of $T$ to consist entirely of poles of $R(\lambda, T)$.

2. Next, if $\lambda=1$ is a first order pole of $R(\lambda, T)$ but $P$ is not strictly positive, the operator $T_R$ induced on $E/R$, where $R := \{x : P|x| = 0\}$ is the null ideal of $P$, satisfies the hypothesis of step 1. On the other hand, $r(T_{|R})<1$ by a now familiar conclusion, and (5.3) proves the assertion.

3. Finally, let $\lambda=1$ be a pole of $R(\lambda, T)$ of order $k>1$. We consider the operator $Q := \lim_{\lambda\to 1}(\lambda-1)^k R(\lambda, T) > 0$ and we have $Q^2=0$ (cf. § 4, Example 4). If $J := \{x\in E : Q|x|=0\}$, then $J$ is a $T$-ideal and $R(\lambda, T)_{|J} = R(\lambda, T_{|J})$ has at $\lambda=1$ a pole of order $<k$ while on $E/J$, $R(\lambda, T_J)$ has at $\lambda=1$ a pole of order 1. Therefore, the present case can be reduced to 2. by employing (5.3) and induction over $k$. □

**Example.** 1. Let $A_n$ denote a $(2\times 2)$ positive matrix with eigenvalues $1, -1+n^{-1}$ $(n\in\mathbb{N})$. If $E$ is the Banach lattice of all sequences $x=(x_n)$ $(x_n\in\mathbb{C}^2)$ satisfying $\sum_n \|x_n\|^2 < +\infty$, then $E$ is isomorphic to $l^2$ and the positive operator $A : E\to E$ defined by $Ax := (A_n x_n)$ (supposing the norms of $A_n$ to be uniformly bounded, which can be achieved by a suitable choice of each matrix $A_n$), has the eigenvalues 1 and $-1+n^{-1}$ $(n\in\mathbb{N})$. Thus 1 is a first order pole of $R(\lambda, A)$ but $-1\in\sigma(A)$ is not a pole; clearly, the residuum $P$ of $R(\lambda, A)$ at $\lambda=1$ is not of finite rank. This shows that the central assumption of (5.5) cannot be relaxed.

We conclude this section with a condition on $T$ (sometimes called *strong positivity*) which ensures that the peripheral point spectrum contains at most $r(T)$; of course, this condition implies $T$ to be irreducible.

**5.6 Proposition.** *Let $0<T\in\mathscr{L}(E)$, $r(T)=1$, be an operator such that for each $x>0$, there exists $n\in\mathbb{N}$ so that $T^n x$ is quasi-interior to $E_+$; moreover, suppose that $\varphi = T'\varphi$ for some non-zero $\varphi\in E'_+$. The peripheral point spectrum of $T$ contains at most the number* 1.

*Proof.* Assume that $\alpha x = Tx$ where $x\neq 0$ and $|\alpha|=1$, $\alpha\neq 1$. Then $|x|\leqq T|x|$ and equality must hold, since $\varphi$ is strictly positive ($T$ being necessarily irreducible). On the $AM$-space $E_{|x|}\cong C(K)$, $T$ induces a Markov operator $T_0$. By (4.2) Cor. 1, the closed vector sublattice $F$ of $C(K)$ generated by the group of unimodular eigenfunctions of $T_0$, is an $AM$-space and $T_{0|F}$ is a lattice homomorphism. Let $x, y\in F$ satisfy $x\wedge y=0$ and $x>0$; by the hypothesis on $T$, there exists $n\in\mathbb{N}$ such that $T^n x$ is quasi-interior to $E_+$ and hence, a weak order unit of $F$. But $0 = T^n(x\wedge y) = T^n x\wedge T^n y$, and so $T^n y = 0$. But this is incompatible with the assumption on $T$ unless $y=0$. Therefore, $\dim F = 1$ and this is contradictory, since $x\in F$ and $|x|\in F$ are linearly independent as eigenvectors of $T$ pertaining to distinct eigenvalues. □

We point out that if $\psi$ is a measure preserving transformation of a measure space (Chap. III, § 9, Example 1), the associated operator $T_\psi$ on $L^p(\mu)$ $(1 \leqq p < +\infty)$ has peripheral point spectrum $\{1\}$ iff $\psi$ is weakly mixing (Exerc. 14). See also Exerc. 8.

## § 6. Topological Nilpotency of Irreducible Positive Operators

Throughout the preceding section we have considered irreducible operators $T$, $0 < T \in \mathscr{L}(E)$, satisfying $r(T)=1$; that is, up to a simple normalization, we have assumed that $T$ have a spectral radius $r(T)>0$. Under what conditions does irreducibility of $T>0$ imply $r(T)>0$? It is easy to see that an irreducible operator $T>0$ on a Banach lattice $E$ cannot be nilpotent; in fact, for no $n \in \mathbb{N}$ and $x>0$ can it occur that $T^n x=0$. For, if $x>0$ is given and $n$ is the smallest integer such that $T^n x=0$, then the ideal $E_y$ generated by $y:=T^{n-1}x$ is $T$-invariant, and $T(E_y)=\{0\}$; but $E_y$ is dense in $E$ by irreducibility of $T$ and so $T=0$, a contradiction. In particular, each irreducible $T>0$ on a finite dimensional Banach lattice has $r(T)>0$ (cf. I.6.3); on the other hand, in the general case of infinite dimensional $E$ this question is much more difficult to answer. On certain Banach lattices $E$, there exist irreducible operators $T>0$ which are *topologically nilpotent (quasi-nilpotent)*, that is, for which $r(T)=0$ (Exerc. 9); it is the purpose of this section to show that for a large class of irreducible operators $T$ and/or domain spaces $E$, one does have $r(T)>0$. It should be observed that for compact operators $T$, the assertion $r(T)>0$ implies the existence of eigenvalues (cf. Theorem 6.6 below).

We begin our discussion by observing that an operator $T$, where $0 < T \in \mathscr{L}(E)$ and $E$ is any Banach lattice, is irreducible if and only if for each $\lambda > r(T)$ and $x>0$, $R(\lambda, T)x$ is a quasi-interior point of $E_+$; this is clear, since the principal ideal of $E$ generated by $R(\lambda, T)x$ is invariant under $T$ (§ 4, Formula (2)). (In particular, irreducible positive operators can have for their domain only Banach lattices $E$ possessing quasi-interior positive elements, (III.8.3) Cor.) Moreover, $T$ is irreducible if and only if its adjoint $T'$ possesses no $\sigma(E', E)$-closed invariant bands other than $\{0\}, E'$ (cf. (II.5.5) Cor. 1). Thus if $T'$ is irreducible then so is $T$, but not conversely as the following examples show.

**Examples.** 1. If $E=C(K)$, where $K$ is a compact space and the set $K$ is uncountable, the dual Banach lattice $E'$ possesses no quasi-interior positive elements (Chap. II, § 6, Example 4); hence, no operator $S>0$ on $E'$ (and, in particular, no adjoint $T'$ of some $T \in \mathscr{L}(E)$) is irreducible. On $E$, there exist irreducible positive operators if and only if $K$ admits a strictly positive Radon measure (cf. III.8.10).

2. Let $E=L^1(\mu)$, where $(X, \Sigma, \mu)$ is a $\sigma$-finite measure space, and let $T \geqq 0$ be an operator on $E$ such that the semi-group $(T^n)_{n \in \mathbb{N}}$ is mean ergodic (Def. III.7.1) with zero element (projection) $P \geqq 0$. $T$ is irreducible iff $P=\varphi \otimes u$ where $u \in E_+$ is quasi-interior, $\varphi$ is a strictly positive linear form on $E$, and $Tu=u$, $T'\varphi=\varphi$ (III.8.5). Here we have $\varphi=f.\ \mu$ where $f \in L^\infty(\mu)$ is a function

such that $f(s)>0$ a.e. $(\mu)$. If $T$ is irreducible, the adjoint $T'$ on $E'$ is irreducible if and only if $f(s)\geqq\delta$ a.e. $(\mu)$ for some $\delta>0$ (Chap. II, § 6, Example 1).

3. If $E$ is a reflexive Banach lattice and $0\leqq T\in\mathscr{L}(E)$, then $T$ is irreducible if and only if $T'$ is.

4. Let $(X,\Sigma,\mu)$ be a $\sigma$-finite measure space, and let $E:=L^p(\mu)$ $(1\leqq p<+\infty)$. Assume that $T\in\mathscr{L}(E)$ is a kernel operator given by the $(\Sigma\times\Sigma)$-measurable kernel $K\geqq 0$ on $X\times X$, so that

$$Tf(t)=\int K(s,t)f(s)\,d\mu(s)$$

holds a.e. $(\mu)$ for each $f\in E$ (Chap. IV, § 8, Example 1). *$T$ is irreducible if and only if*

$$(\star)\qquad \int_{X\setminus S}\left[\int_S K(s,t)\,d\mu(s)\right]d\mu(t)>0$$

*for each $S\in\Sigma$ such that $\mu(S)>0$ and $\mu(X\setminus S)>0$.* In fact, $(\star)$ means that for some $f\in E_+$ living on $S$, the function $Tf$ is $>0$ on a set of (strictly) positive measure disjoint from $S$; a condition clearly equivalent to the non-existence of $T$-ideals $\neq\{0\},E$ (Chap. III, § 1, Example 2). In particular, $T$ is irreducible whenever $K(s,t)>0$ a.e. $(\mu\otimes\mu)$, but this latter condition is much more stringent than $(\star)$ (cf. 5.6). If $p=+\infty$, $(\star)$ is not sufficient for $T$ to be irreducible (Example 2).

Throughout this section, $E$ denotes a (real or complex) Banach lattice.

**6.1 Proposition.** *Let $0<T\in\mathscr{L}(E)$ be irreducible. Each of the following assumptions implies that $r(T)>0$:*

(i) *The positive cone $E_+$ contains an extreme ray.*

(ii) *$E$ is a closed ideal of an AM-space with unit.*

*Proof.* Under either of these conditions, we will establish the existence of elements $0<x_0\in E_+$ and $0<\varepsilon\in\mathbb{R}$ satisfying the relation $Sx_0\geqq\varepsilon x_0$, where $S:=TR(r+1,T)$ ($R(\lambda,T)$ the resolvent of $T$) and $r:=r(T)$. If $\psi$ is the holomorphic function on $\{z\in\mathbb{C}:|z|<r+1\}$ defined by $\psi(z):=\sum_{n=1}^{\infty}(r+1)^{-n}z^n$, $\psi$ is holomorphic on $\sigma(T)$ and we have $S=\psi(T)$. On the other hand, $\sigma(S)=\psi[\sigma(T)]$ by the spectral mapping theorem [DS, VII.3.11]. But $Sx_0\geqq\varepsilon x_0$ clearly implies that $r(S)\geqq\varepsilon$ (cf. § 4, Formula (1)) and hence, since $\psi(0)=0$, we must have $r(T)>0$.

(i) Let $x_0>0$ generate an extreme ray of $E_+$ and denote by $J_0$ the (one dimensional) ideal generated by $x_0$. Since $TS\leqq(r+1)S$ (§ 4, Formula (2)), the ideal $J$ of $E$ generated by $Sx_0$ is $T$-invariant and, since $T$ is irreducible, dense in $E$. Therefore, $J\supset J_0$ and it follows that $Sx_0\geqq\varepsilon x_0$ for some real number $\varepsilon>0$.

(ii) If $E$ is a (non-zero) closed ideal of an $AM$-space with unit $C(K)$, then $E=\{f\in C(K): f(A)\subset\{0\}\}$ for some closed subset $A\subsetneqq K$ (Chap. III, § 1, Example 1). Hence, $E$ is isomorphic to $C_0(U)$ (the continuous functions vanishing at infinity) of the locally compact space $U:=K\setminus A$. Let $K_0$ be any compact subset of $U$ containing an interior point, and let $0<f_0\in C_0(U)$ be a function supported by $K_0$. Since $T$ is irreducible, $Sf_0$ is a strictly positive function on $U$; in particular, $\inf_{s\in K_0} Sf_0(s)>0$. Therefore, $Sf_0\geqq\varepsilon f_0$ for some real number $\varepsilon>0$. □

In view of Examples 1, 2 above, the following result is very useful; it can be considered a generalization of the fact that if $E$ is reflexive, the adjoint $T'$ of an irreducible operator $T \in \mathscr{L}(E)$ is irreducible.

**6.2 Proposition.** *Let $0 < T \in \mathscr{L}(E)$ be irreducible and weakly compact. There exists a (norm closed) $\sigma(E', E)$-dense minimal $T'$-ideal of $E'$.*

*Proof.* We consider those $T''$-ideals of the bidual $E''$ which are $\sigma(E'', E')$-closed (that is, which are polars of $T'$-ideals of $E'$ with respect to the duality $\langle E', E'' \rangle$). If $J$ is such an ideal, we have either $J \cap E = \{0\}$ or else $J \supset E$, since $T$ is irreducible; hence, in the latter case we have $J = E''$. Thus if $J$ is proper, $J \cap E = \{0\}$ and hence, $T''(J) = \{0\}$ because of $T''(E'') \subset E$ by weak compactness of $T$ (II.9.4). Therefore, the null ideal $J_0 := \{x \in E'' : T''|x| = 0\}$ is a maximal (proper) $\sigma(E'', E')$-closed $T''$-ideal. (The $\sigma(E'', E')$-closure $\bar{J}_0$ is a $T''$-ideal, since $\bar{J}_0$ is the bipolar of $J_0$ with respect to $\langle E', E'' \rangle$; but $T''$ is $\sigma(E'', E')$-continuous and hence, $J_0 = \bar{J}_0$.) It follows that $(J_0)^\circ \subset E'$ is a minimal $T'$-ideal of $E'$ which is $\sigma(E', E)$-dense, since $T$ is irreducible. □

**6.3 Theorem.** *If $E$ is an AM- or AL-space, every weakly compact*[1] *irreducible operator $T > 0$ on $E$ has spectral radius $r(T) > 0$.*

*Proof.* First, if $E$ is an $AL$-space then $E'$ is an $AM$-space with unit (II.9.1), and by (6.2) there exists a minimal $T'$-ideal $I$ of $E'$. By (6.1), the restriction $T'_{|I}$ has spectral radius $>0$; a fortiori, $r(T') = r(T) > 0$.

Second, if $E$ is an $AM$-space (note that the case of an $AM$-space without unit is not covered by (6.1)) then again by (6.2), there exists a minimal $T'$-ideal $I$ of $E'$. Since $T'$ and hence $T'_{|I}$ is weakly compact, it follows from the first part of the proof that $r(T'_{|I}) > 0$; therefore, $r(T') = r(T) > 0$. □

Since compactness and topological nilpotency of operators on Banach spaces are independent properties, it is generally impossible to assert that a compact operator has non-void point spectrum; the notable exception being compact normal operators on Hilbert space. It is not yet known if a compact, irreducible positive operator on a Banach lattice has non-void point spectrum; therefore, the following corollary of (6.3) (which is supplemented by (5.4) after a simple normalization) appears noteworthy.

**Corollary.** *Every weakly compact, irreducible positive operator on an AL- or AM-space has non-void point spectrum.*

*Proof.* From (6.3) we conclude that $r(T) > 0$. Since $T^2$ is compact by (II.9.9) Cor. 1, it follows that $r(T)^2$ is a pole of $R(\lambda, T^2)$ and hence, $r(T)$ is a pole of $R(\lambda, T)$. □

We have seen that (6.3) rests on the two preceding propositions; in turn, together with the abstract characterization of kernel operators (Thm. IV. 9.7), (6.3) is the key to the rather deep theorem of Ando-Krieger (6.5 below) via the following lemma, which is of some independent interest.

---

[1] It suffices to assume that some power of $T$ be weakly compact (Exerc. 9).

**6.4 Lemma.** *Let $T \geqq 0$ be a (not necessarily irreducible) operator on E. There exists $z \in E_+$ such that $T(E_z) \subset E_z$ and the restriction $T_0 := T_{|E_z}$ to the AM-space $E_z$ has spectral radius $r(T_0) = r(T)$.*

*Proof.* Supposing that $E \neq \{0\}$, we select $x \in E_+$ and a strictly decreasing sequence $(\lambda_n)$ in $\mathbb{R}$ such that $\lim_n \lambda_n = r(T)$ and the sequence $(R(\lambda_n, T)x)_{n \in \mathbb{N}}$ is unbounded in $E$; this choice of $x$ is possible by (4.1). Define $z_n := \varrho_n R(\lambda_n, T)x$ where $\varrho_n^{-1} := 2^n \|R(\lambda_n, T)x\|$ $(n \in \mathbb{N})$. Clearly, the sum $z := \sum_{p=1}^{\infty} z_k$ exists in norm; for each $n \in \mathbb{N}$, we define $y_n := \sum_{k \geqq n} z_k$. Since $Tz \leqq \lambda_1 z$ (cf. §4, Formula (2)), the principal ideal $E_z$ is invariant under $T$; moreover, since $R(\lambda_n, T)x \geqq R(\lambda_m, T)x$ whenever $n \geqq m$, each $y_n$ is an order unit of $E_z$. Because of $Ty_n \leqq \lambda_n y_n$ we have $r(T_0) \leqq \lambda_n$ for each $n \in \mathbb{N}$ and hence, $r(T_0) \leqq r(T)$. On the other hand, $x \in E_z$ and the sequence $(R(\lambda_n, T)x)_{n \in \mathbb{N}}$ is unbounded in $E$ by the choice of $x$; a fortiori, the sequence $(R(\lambda_n, T_0)x)_{n \in \mathbb{N}}$ is unbounded in $E_z$ which shows that $r(T_0) \geqq r(T)$. □

**6.5 Theorem.** *Let E be a Banach lattice with order continuous norm. If $T > 0$ is an irreducible operator contained in the band $(E' \otimes E)^{\perp\perp}$ of $\mathscr{L}^r(E)$,[2] then $r(T) > 0$.*

*Proof.* We reduce the proof to the case of an irreducible kernel operator on a space $L^1(\mu)$ (Example 4 above), where $\mu$ denotes a positive Radon measure on the compact structure space $X$ of $E$ (cf. III.4.5). By (6.4), there exists a $T$-invariant ideal ($AM$-space) $E_z$ of $E$ such that $T_0 := T_{|E_z}$ has spectral radius $r(T_0) = r(T)$; since $T$ is irreducible (and hence, $E_z$ dense in $E$), $E_z$ is isomorphic to $C(X)$. Moreover, since $E$ has order continuous norm, $E_z$ can be identified with the dual of the $AL$-space $(E', z)$ (Chap. IV, Exerc. 9) and $T_0 = S'$, where $S$ is the continuous extension of $T'$ to $(E', z)$. (Note that by (IV.9.3), $E'$ is a dense ideal of $(E', z)$.)

We show that $S$ is irreducible. In fact, since $E'$ is a dense ideal of $(E', z)$, for any closed ideal $J$ of $(E', z)$ the intersection $J \cap E'$ is a $\sigma(E', E)$-closed ideal of $E'$; moreover, by (III.1.1) $J \cap E' = \{0\}$ implies $J = \{0\}$ and $J \cap E' = E'$ implies $J = (E', z)$. Since by irreducibility of $T$ there exists no $\sigma(E', E)$-closed $T'$-invariant ideal of $E'$, $S$ must be irreducible.

Now if the $AL$-space $(E', z)$ (which, by irreducibility of $S$, contains weak order units) is represented by $L^1(\mu, X_0)$ in such a way that $C(X_0) \cong L^\infty(\mu)$ ($X_0$ compact; see II.8.5) then we have $X_0$ homeomorphic to $X$ because of the isomorphisms $L^\infty(\mu) \cong (E', z)' \cong E_z$. Since $T \in (E' \otimes E)^{\perp\perp}$ by hypothesis, the corollary of (IV.9.7) shows that $T$ (and hence, $T_0$) is given by a measurable kernel $K \in L^1(\mu \otimes \mu)$. Clearly, then, $S$ is given by the transposed kernel $K'$ which satisfies condition (*) of Example 4 above. But so does the truncated kernel $K'_0 := K' \wedge e$ ($e$ the constant-one function on $X \times X$) and hence, the operator $S_0 \leqq S$ defined by $K'_0$ is irreducible on $L^1(\mu)$. The point of considering $K'_0$ is that $S_0$ can be factored $L^1(\mu) \to L^\infty(\mu) \to L^1(\mu)$ where the second map is the weakly compact canonical injection; hence, $S_0$ is weakly compact and so $r(S_0) > 0$ by by (6.3). Clearly, now, $r(S_0) \leqq r(S) = r(T_0) = r(T)$ which proves that $r(T) > 0$. □

We note that by virtue of (IV.9.9), the assertion of (6.3) concerning $AL$-spaces is contained in (6.5). In addition, from (IV.9.9) we obtain this corollary (see also Exerc. 13).

[2] Cf. Def. IV.1.1 and Prop. IV.1.4.

**Corollary.** *Let $0 < T \in \mathscr{L}(E)$, E denoting a reflexive Banach lattice. If T is irreducible and majorizing (or irreducible and cone absolutely summing,* Def. IV.3.1) *then* $r(T) > 0$.

As an application of the results of this and earlier sections, we spell out in detail the following general version of a classical theorem on kernel operators (Theorem of Jentzsch).

**6.6 Theorem.** *Let $E := L^p(\mu)$, where $1 \leqq p \leqq +\infty$ and $(X, \Sigma, \mu)$ is a $\sigma$-finite measure space. Suppose $T \in \mathscr{L}(E)$ is an operator given by a $(\Sigma \times \Sigma)$-measurable kernel $K \geqq 0$, satisfying these two assumptions:*

(i) *Some power of T is compact.*

(ii) *$S \in \Sigma$ and $\mu(S) > 0$, $\mu(X \setminus S) > 0$ implies*

$$(\star) \qquad \int_{X \setminus S} \int_S K(s,t)\, d\mu(s)\, d\mu(t) > 0.$$

*Then $r(T) > 0$ is an eigenvalue of T with a unique normalized eigenfunction f satisfying $f(s) > 0$ a.e. $(\mu)$. Moreover, if $K(s,t) > 0$ a.e. $(\mu \otimes \mu)$ then every other eigenvalue $\lambda$ of T has modulus $|\lambda| < r(T)$.*

*Proof.* By Example 4 above, $(\star)$ signifies irreducibility of $T$ if $p < +\infty$; in case $p = +\infty$, $(\star)$ signifies irreducibility of the operator (preadjoint of $T$) on $L^1(\mu)$ defined by the transposed kernel $K'(s,t) := K(t,s)$. Thus we can restrict attention to the case $p < +\infty$. (6.5) then shows, in conjunction with (IV.9.8), that $r(T) > 0$. Condition (i) now implies that $r(T)$ is an eigenvalue of $T$ (in fact, a first order pole of the resolvent), and (5.2) proves the unicity of $f$. Finally, if $K(s,t) > 0$ *a.e.* $(\mu \otimes \mu)$ then $Tg$ is a quasi-interior point of $L^p(\mu)_+$ $(p < +\infty)$ for each $g > 0$, and the last assertion follows from (5.6). □

**Example.** 5. It may not be superfluous to indicate some instances in which Condition (i) of (6.6) is fulfilled. This will be so in the following cases:

($\alpha$) $p = 1$ or $p = +\infty$, and some power of $T$ is weakly compact (cf. II.9.9 Cor. 1). In particular, $p = 1$, $\mu$ is finite and $T^k$ maps $L^1(\mu)$ into $L^q(\mu)$ for some integer $k \geqq 1$ and some $q > 1$; or $p = +\infty$, $\mu$ is finite, and $T^k$ has a continuous extension $L^q(\mu) \to L^\infty(\mu)$ for some integer $k \geqq 1$ and some $q < +\infty$.

($\beta$) $1 < p < +\infty$, $\mu$ is finite, and some power $T^k$ maps $L^p(\mu)$ into $L^\infty(\mu)$. Then $T^k$ has a factoring: $L^p(\mu) \to L^\infty(\mu) \to L^p(\mu)$ and so is compact, because $L^\infty(\mu)$ has the Dunford-Pettis property (II.9.9).

In conclusion, we point out again that there exist irreducible operators $T > 0$ which are quasi-nilpotent. For an example where $T$ is an operator on $L^1(\mu, \Gamma)$ ($\mu$ Haar measure on the circle group $\Gamma$) which induces an irreducible quasi-nilpotent operator on each $L^p(\mu)$ $(1 \leqq p < +\infty)$ and even leaves $C(\Gamma)$ invariant, the interested reader is referred to Exerc. 9.

## § 7. Application to Non-Positive Operators

The present section is intended to show how the methods and (mutatis mutandis) results of Sections 4,5 on spectral properties of positive operators can be extended to certain classes of non-positive operators possessing a modulus.

Let $E_0$ denote an ordered vector space over $\mathbb{R}$ whose order is Archimedean (i.e., satisfies (A) of Def. II.1.8). It is well known that if $E_0$ is generated by its positive cone $(E_0)_+$ then $E_0$ can be imbedded in an order complete vector lattice $\bar{E}_0$ (Dedekind completion; Chap. II, Exerc. 9) under preservation of arbitrary suprema and infima.

The complexification $E:=E_0+iE_0$ (Chap. II, § 11) is considered ordered with positive cone $E_+:=(E_0)_+$ (cf. [S, V, § 2]). We recall that each $x \in E$ has a canonical decomposition $x=x_1+ix_2$, where $x_1, x_2 \in E_0$, and each linear operator $T: E \to E$ has a canonical decomposition $T=T_1+iT_2$, where $T_1, T_2$ are real operators on $E$ (that is, operators leaving the real subspace $E_0$ of $E$ invariant; cf. Chap. II, § 11).

**7.1 Definition.** *Let $E=E_0+iE_0$ denote the complexification of an Archimedean ordered real vector space $E_0$, and let $r$ denote an integer $\geqq 1$. An element $x \in E$ is called* $r$-cyclic *if there exist a primitive $r$-th root of unity $\zeta \in \mathbb{C}$ and elements $y_j \in E_+$ $(j=0,1,\ldots,r-1)$ satisfying $y_j \wedge y_k=0$ $(j \neq k)$, such that $x=\sum_{j=0}^{r-1} \zeta^i y_j$.*

**Examples.** 1. If $E$ is a complex vector lattice (Def. II.11.1), an element $x \in E$ is 1-cyclic iff $x \geqq 0$, and 2-cyclic iff $x$ is a real element of $E$ ($x=x^+-x^-$ being a representation as required by Def. 7.1).

2. Let $X$ be a set and let $\mathbb{C}^X$ denote the complex algebra and vector lattice of all functions $f: X \to \mathbb{C}$. An element $f \in \mathbb{C}^X$ is $r$-cyclic iff $f^r \geqq 0$; in fact, if $\zeta$ is any primitive $r$-th root of unity and if $X_j:=\{s \in X: f(s)=\zeta^j|f(s)| \neq 0\}$ $(j=0,\ldots,r-1)$, then the characteristic functions $\chi_j$ of $X_j$ are mutually orthogonal and $f=\sum_j \zeta^j |f| \chi_j$.

3. By the preceding example, an $(n \times n)$-matrix $A=(\alpha_{ik})$ (where $n \in \mathbb{N}$ and $A$ is considered an element of $\mathbb{C}^{n^2}$ under its canonical order; Chap. I, § 2) is $r$-cyclic iff $(\alpha_{ik})^r \geqq 0$ for all $i,k=1,\ldots,n$. More generally, if $(X,\Sigma,\mu)$ and $(Y,T,\nu)$ are $\sigma$-finite measure spaces, an operator $L^p(\mu) \to L^q(\nu)$ $(1 \leqq p,q \leqq +\infty)$ given by a $(\Sigma \times T)$-measurable complex kernel $K$ (Chap. IV, § 8, Example 1) is $r$-cyclic iff $(K(s,t))^r \geqq 0$ a.e. $(\mu \otimes \nu)$. (Note that by virtue of (IV.9.8), the kernel operators $L^p \to L^q$ form an ideal of regular operators.)

We note without proof that the representation $x=\sum_{j=0}^{r-1} \zeta^j y_j$ (Def. 7.1) of an $r$-cyclic element $x \in E$ is unique (Exerc. 10). Also, if $E$ is a countably order complete complex vector lattice, the Approximation Lemma II.11.5 shows that the totality of $r$-cyclic elements ($r \in \mathbb{N}$) is a dense subset of $E$ with respect to relative uniform convergence.

If $E_0$ is an ordered vector space over $\mathbb{R}$ and if $x \in E_0$ is an element such that $x \vee (-x)$ exists, it is customary to say that the modulus $|x|$ of $x$ exists and to define $|x|:=x \vee (-x)$. Accordingly, if $E=E_0+iE_0$ is the complexification of

$E_0$ and $x=x_1+ix_2$ $(x_1,x_2\in E_0)$ is an element of $E$ such that $\sup_{\theta\in\mathbb{R}}[(\cos\theta)x_1+(\sin\theta)x_2]$ exists (in $E_0$), then we say the *modulus* $|x|$ exists and define it to be this supremum. (This definition is in accordance with the definition of the modulus function in a complex vector lattice, cf. (II.11.1).) Likewise, $x=x_1+ix_2$ is called a *regular* element of $E$ if both $x_1$ and $x_2$ are regular elements of $E_0$ (Def. IV.1.1).

**7.2 Proposition.** *Let $E=E_0+iE_0$ denote the complexification of an Archimedean ordered real vector space $E_0$. Given a regular element $x=x_1+ix_2\in E$, consider these assertions:*

(a) *$x$ is $r$-cyclic.*

(b) *$|x|$ exists and $x$ has a representation $x=\sum_{j=0}^{r-1}\zeta^j y_j$, where $\zeta$ is a primitive $r$-th root of unity and where $y_j\geqq 0$, such that $|x|=\sum_{j=0}^{r-1}y_j$.*

(c) *$|x|$ exists and $|x|=\sup_j[(\cos\theta_j)x_1+(\sin\theta_j)x_2]$, where $\theta_j:=2\pi j/r$ $(j=0,\dots,r-1)$.*

*Then* (a)$\Leftrightarrow$(b)$\Rightarrow$(c) *and, if $E_0$ is a vector lattice satisfying Axiom* (OS) (Def. II.1.8), (c)$\Rightarrow$(a).

*Proof.* Since $x$ is regular, we can assume $E_0$ to be generated by its positive cone and so, $E_0$ being Archimedean, we consider $E_0$ imbedded in an order complete vector lattice $\bar{E}_0$ under a suprema-preserving linear map (Chap. II, Exerc. 9); accordingly, $E$ is considered imbedded in the complexification $\bar{E}:=\bar{E}_0+i\bar{E}_0$.

(a)$\Rightarrow$(b): Let $y\in E_0$ satisfy $y\geqq(\cos\theta)x_1+(\sin\theta)x_2=\mathrm{Re}(e^{-i\theta}x)$ for all $\theta\in\mathbb{R}$. Let $x=\sum_{j=0}^{r-1}\zeta^j y_j$ be a defining representation (so that $y_j\wedge y_k=0$ for $j\neq k$, Def. 7.1) and denote by $P_j$ the band projection of $\bar{E}_0$ onto the band generated by $y_j$; if $\theta_j:=2\pi j/r$, then from $y\geqq\mathrm{Re}(e^{-i\theta_j}x)$ it follows that $y\geqq P_j y\geqq P_j\mathrm{Re}(e^{-i\theta_j}x)=y_j$ $(j=0,\dots,r-1)$. Thus $y\geqq y_j$ for all $j$; since $y_j$ are orthogonal, we have $\sup_j y_j=\sum_j y_j$ and so $y\geqq\sum_j y_j$. On the other hand, it is clear that $(\cos\theta)x_1+(\sin\theta)x_2$ $=\mathrm{Re}(e^{-i\theta}x)\leqq\sum_j y_j$ for all $\theta\in\mathbb{R}$ and hence, $|x|=\sum_j y_j$.

(a)$\Rightarrow$(c): From the preceding it follows that $y\geqq(\cos\theta_j)x_1+(\sin\theta_j)x_2$ $(\theta_j=2\pi j/r; j=0,\dots,r-1)$ implies $y\geqq y_j$ and hence, $y\geqq\sum_j y_j=|x|$; therefore, $|x|=\sup_j[(\cos\theta_j)x_1+(\sin\theta_j)x_2]$.

(b)$\Rightarrow$(a): We consider the principal ideal (complex $AM$-space with unit) $\bar{E}_{|x|}\cong C(K)$ of $\bar{E}$ generated by $|x|$. Under this identification, $x$ and $y_j\geqq 0$ are continuous functions on the compact space $K$. Now $x(s)=\sum_j\zeta^j y_j(s)$ and $|x(s)|=\sum_j y_j(s)$ show that for each $s\in K$, all non-zero summands $\zeta^j y_j(s)$ must be complex numbers with the same argument. Hence, since $\zeta^j$ $(j=0,\dots,r-1)$ all have distinct arguments, it follows that $x(s)$ assumes precisely the values $\zeta^j|x(s)|$ so that $\{s\in K: y_j(s)>0\}=\{s\in K: x(s)=\zeta^j|x(s)|\neq 0\}$ for each $j$. Therefore, $y_j\wedge y_k=0$ in $E_0$ whenever $j\neq k$ and $x$ is $r$-cyclic.

(c)$\Rightarrow$(a): If $E_0$ is a vector lattice satisfying (OS), then the ideal $E_{|x|}$ of $E$ is a complex $AM$-space with unit $C(K)$ (cf. (II.7.2) and Chap. II, § 11 (vii)); therefore $x, x_1, x_2$ can be considered to be continuous functions on $K$ (in particular, $x$ is unimodular, since $|x|$ corresponds to the unit of $C(K)$). The assumption implies that for each $s\in K$ there exists $j$, $0\leqq j\leqq r-1$, such that $|x(s)|=(\cos\theta_j)x_1(s)$ $+(\sin\theta_j)x_2(s)$; hence, $j$ is unique and $x(s)=\zeta^j$. It is clear that each set $K_j:=\{s\in K: x(s)=\zeta^j\}$ is open and closed and hence, its characteristic function $y_j$ continuous on $K$; therefore, $y_j\in E_0$ and $x=\sum_j\zeta^j y_j$ is $r$-cyclic. $\square$

**Corollary.** *Let $E$ denote a complex vector lattice, and let $T \in L(E)$ be a linear map with canonical decomposition $T = T_1 + i T_2$. $T$ is $r$-cyclic (with respect to the canonical order of $L(E)$) if and only if* (i) $|T| := \sup_{\theta \in \mathbb{R}} [(\cos\theta) T_1 + (\sin\theta) T_2]$ *exists in $L(E)$ and* (ii) *there exists a representation $T = \sum_{j=0}^{r-1} \zeta^j S_j$ (where $S_j \geqq 0$ and $\zeta$ is a primitive $r$-th root of unity) such that $|T| = \sum_{j=0}^{r-1} S_j$.*

The proof is clear from (a)$\Leftrightarrow$(b) of (7.2), since the canonical order of $L(E_0)$ is Archimedean. (See the discussion after (II.2.4) and cf. Chap. II, § 11; note that by Def. II.11.1, the underlying real vector lattice $E_0$ of $E$ is required to satisfy (OS).)

It is not immediately clear that the operator modulus

$$|T| := \sup_{\theta \in \mathbb{R}} [(\cos\theta) T_1 + (\sin\theta) T_2]$$

—whenever it exists—satisfies Def. IV.1.7, because (IV.1.8) is not immediately applicable unless $E_0$ is order complete. However, an inspection of the proof of (IV.1.8) shows that the answer is affirmative; we state this explicitly as a lemma.

**7.3 Lemma.** *Let $E$ be a complex vector lattice and let $T = T_1 + i T_2 \in L(E)$ ($T_1, T_2$ real). The modulus $|T| := \sup_{\theta \in \mathbb{R}} [(\cos\theta) T_1 + (\sin\theta) T_2]$ exists in $L(E)$ if and only if for each $x \in E_+$, $\sup_{|z| \leqq x} |Tz|$ exists in $E$; in this case, one has $|T| x = \sup_{|z| \leqq x} |Tz|$.*

The following proposition is related to (4.2), and its role in the subsequent proofs is similar to the role played by (4.2) in the proofs of Section 4.

**7.4 Proposition.** *Let $T$ be an operator on the complex Banach lattice $C(K)$ ($K$ compact) such that $|T|$ exists and is a Markov operator. If $f$ is a unimodular eigenfunction of $T$ corresponding to the eigenvalue $\alpha$, $|\alpha| = 1$, then multiplication by $f$ is an isometry $V$ of $C(K)$ satisfying $T = \alpha V |T| V^{-1}$.*

*If, in addition, $T$ is $r$-cyclic then $\alpha^{rk+1} f^{rk+1} = T(f^{rk+1})$ and $\alpha^{rk} f^{rk} = |T|(f^{rk})$ for all $k \in \mathbb{Z}$.*

*Proof.* Let $R := \alpha^* V^{-1} T V$. If $e$ denotes the unit (constant-one function) of $C(K)$, it is clear that $Re = e$; moreover, it is easy to see from (7.3) that $|R| = |T|$. Now let $R = R_1 + i R_2$ be the canonical decomposition of $R$; since $Re = R_1 e = |T| e = e$ and $|T| - R_1 \geqq 0$, it follows that $|T| = R_1$. In turn, this implies $|Rh| = |R_1 h + i R_2 h| \leqq R_1 h$ for all $h$, $0 \leqq h \in C(K)$; therefore, $R_2 = 0$ and $|T| = R$. This proves that $T = \alpha V |T| V^{-1}$.

Now suppose $T$ to be $r$-cyclic. If $T = \sum_{j=0}^{r-1} \zeta^j S_j$ is a decomposition of $T$ as in the corollary of (7.2) and if we define $Q_j := \zeta^j \alpha^* V^{-1} S_j V$ ($j = 0, \dots, r-1$), we obtain $|Q_j| = S_j$ and $R = \sum_j Q_j = |T| = \sum_j S_j$. Clearly, this implies $Q_j = S_j$ for all $j$ by an argument similar to the preceding one. By definition of $Q_j$ we obtain $S_j = \zeta^j \alpha^* V^{-1} S_j V$, and iteration shows that $S_j = \zeta^{kj} (\alpha^*)^k V^{-k} S_j V^k$ for all $k \in \mathbb{Z}$. Replacing $k$ by $rk$ this yields $S_j = (\alpha^*)^{rk} V^{-rk} S_j V^{rk}$, because $\zeta^r = 1$.

Now $e=|T|e=\sum_j S_j e=\sum_j(\alpha^*)^{rk}V^{-rk}S_jV^{rk}e$ shows that $\alpha^{rk}f^{rk}=|T|(f^{rk})$, in view of the definition of $V$. Because of $T=\alpha V|T|V^{-1}$ we finally obtain $\alpha^{rk+1}f^{rk+1}=T(f^{rk+1})$ for all $k\in\mathbb{Z}$. □

In accordance with Def. III.8.1, a (not necessarily positive) operator $T$ on a real or complex Banach lattice $E$ is called *irreducible* if $T$ possesses no closed invariant ideals other than $\{0\}$, $E$.

**Corollary.** *Let $T$ be an irreducible, r-cyclic operator on $C(K)$ ($K$ compact) whose modulus $|T|$ is a Markov operator. If $T$ has a unimodular eigenvalue $\alpha$, then the peripheral point spectrum of $T$ is the coset $\alpha H$ where $H$, the peripheral point spectrum of $|T|$, is a subgroup of the circle group.*

*Proof.* First, if $T$ is irreducible then so is $|T|$; thus by (5.2) every normalized eigenfunction $f$ of $T$ corresponding to $\alpha$ must be unimodular: In fact, $\alpha f=Tf$ implies $|f|\leqq|T||f|$; since $\varphi=|T|'\varphi$ for some $\varphi\gg 0$ by (III.8.10), we have $|f|=|T||f|$ and so $|f|=e$, $|T|$ being a Markov operator by hypothesis. The remainder is now clear from (5.2) and the relation $T=\alpha V|T|V^{-1}$ of (7.4). □

**7.5 Theorem.** *Let $E$ denote a Banach lattice, and let $T\in\mathscr{L}(E)$ be an r-cyclic operator such that both $T$ and $|T|$ have spectral radius* 1.

(i) *If $|T|$ satisfies* (G) (Def. 4.7), *then $\alpha^{rk+1}\in\sigma(T)$ and $\alpha^{rk}\in\sigma(|T|)$ for each $\alpha\in\sigma(T)$, $|\alpha|=1$, and all $k\in\mathbb{Z}$.*

(ii) *If there exists a strictly positive linear form $\varphi$ on $E$ such that $|T|'\varphi\leqq\varphi$, then an assertion analogous to* (i) *holds for the peripheral point spectrum of $T$ and $|T|$, respectively.*

*Proof.* We prove (ii) first. Let $\alpha x=Tx$, $|\alpha|=1$, $x\neq 0$. Then $|x|\leqq|T||x|$, and this implies $|x|=|T||x|$, since $\varphi(|T||x|-|x|)=0$. Considering the $|T|$-ideal $E_{|x|}\cong C(K)$, which is clearly invariant under $T$, we can apply (7.4) provided we can show that the restriction $T_0$ of $T$ to $E_{|x|}$ is $r$-cyclic. But since $E_{|x|}$ is an ideal of $E$, it follows from (7.3) that $|T_0|$ exists and equals the restriction of $|T|$ to $E_{|x|}$; therefore, $T_0$ is $r$-cyclic by the corollary of (7.2).

We reduce the proof of (i) to the case of peripheral point spectrum by imbedding $E$ into an arbitrary $\boldsymbol{F}$-product $\hat{E}$ (Def. 1.1); let $\hat{T},\hat{S}$ denote the operators on $\hat{E}$ corresponding to $T,|T|$ respectively. Let $\alpha\hat{x}=\hat{T}\hat{x}$ where $|\alpha|=1$ and $\hat{x}\neq 0$; then $|\hat{x}|\leqq\hat{S}|\hat{x}|$ and, since the transition $|T|\mapsto\hat{S}$ leaves (G) intact, by (4.8) there exists a linear form $\varphi=\hat{S}'\varphi>0$ on $\hat{E}$ such that $\varphi(|\hat{x}|)>0$. Now we imbed $\hat{E}$ into its bidual $\hat{E}''$ (which is order complete, and on which $\varphi$ defines an order continuous linear form), and we consider the null ideal $N_\varphi\subset\hat{E}''$. If $T=\sum_{j=0}^{r-1}\zeta^j S_j$ is an $r$-cyclic decomposition of $T$ (Def. 7.1) so that $|T|=\sum_{j=0}^{r-1}S_j$ by (7.2), we have $\hat{S}=\sum_j\hat{S}_j$ and it is easy to see that $N_\varphi$ is invariant under the second adjoints of the operators $\hat{T},\hat{S},\hat{S}_j$ $(0\leqq j\leqq r-1)$, respectively. But, $\varphi$ being order continuous on $\hat{E}''$, $N_\varphi$ is a projection band of $\hat{E}''$ and so the quotient $\tilde{E}:=\hat{E}''/N_\varphi$ is order complete, since it is isomorphic to $N_\varphi^\perp$. If we denote the elements and operators in $\tilde{E}$ corresponding to $\hat{x},|\hat{x}|,\hat{S},\hat{T}$, and $\hat{S}_j$ by a tilde $\tilde{}$, we obtain $\alpha\tilde{x}=\tilde{T}\tilde{x}$ and $|\tilde{x}|=\tilde{S}|\tilde{x}|$, since $\varphi$ defines a strictly positive linear form on $\tilde{E}$ which is fixed under $\tilde{S}'$. Finally, let $F$ denote the band of $\tilde{E}$ generated by $|\tilde{x}|$; if $T_0$ denotes the restriction of $\tilde{T}$ to $F$, then $|\tilde{x}|\leqq|T_0||\tilde{x}|\leqq\tilde{S}|\tilde{x}|$ and this shows that $\tilde{S}_{|F}=|T_0|$ and hence,

$|T_0| = \sum_j (\tilde{S}_j)_0$ where $(\tilde{S}_j)_0$ is the restriction of $\tilde{S}_j$ to $F$. Therefore, $T_0$ is $r$-cyclic by (7.2) Cor. and (ii) applies. Since the point spectrum of $T_0$ and $|T_0|$ is contained in the spectrum of $T$ and $|T|$, respectively, the proof is complete. □

**Corollary 1.** *If $T$ satisfies the assumptions of* (7.5)(i), *the peripheral spectrum of $T$ is either finite or the entire spectral circle. Similary, if $T$ satisfies the assumptions of* (7.5)(ii), *the peripheral point spectrum of $T$ is either finite or dense in the spectral circle.*

Of the many possible specializations of (7.5) (cf. Exerc. 12), we single out the following which seems to be the most striking.

**Corollary 2.** *Let $T$ denote a real, contractive operator on an $AL$-space. If $\alpha$, $|\alpha|=1$, is in $\sigma(T)$ then $\alpha^{2k+1}\in\sigma(T)$ and $\alpha^{2k}\in\sigma(|T|)$ for all $k\in\mathbb{Z}$; a corresponding assertion holds for the peripheral point spectrum.*

*Proof.* It is clear that $T$ is 2-cyclic by (IV.1.5). Since for the spectral radii of $T$ and $|T|$ we have $r(T)\leqq r(|T|)\leqq \| |T| \| = \|T\|$, the assumption $\|T\|\leqq 1$ and the presence of a unimodular element $\alpha\in\sigma(T)$ imply that $r(T)=r(|T|)=1$. Moreover, $|T|$ satisfies $(G)$ and $|T|'\varphi\leqq\varphi$, $\varphi$ denoting the norm functional of its domain. □

We conclude this section by an application of Theorems (5.2) and (5.4) to irreducible operators which are not necessarily positive.

**7.6 Proposition.** *Let $E$ denote a Banach lattice, and let $T\in\mathscr{L}(E)$ be an irreducible operator possessing a modulus $|T|$ with spectral radius* 1; *in addition, suppose that $\varphi\geqq|T|'\varphi$ for some linear form $\varphi>0$ on $E$.*[1]

*If $\alpha$, $|\alpha|=1$, is an eigenvalue of $T$ then the peripheral point spectrum of $T$ is the coset $\alpha H$, where $H$ is the peripheral point spectrum of $|T|$ (and hence, a subgroup of the circle group).*

*In particular, if $\alpha$ is a pole of the resolvent $R(\lambda, T)$, then the peripheral spectrum of $T$ equals $\alpha H$, $H$ is the group of all $m$-th roots of unity for some $m\in\mathbb{N}$, and each $\varepsilon\in\alpha H$ is a first order pole of $R(\lambda. T)$ and a simple eigenvalue of $T$.*

*Proof.* Clearly, if $T$ is irreducible then so is $|T|$; hence, $\varphi\gg 0$. Let $\alpha x = Tx$, $x\neq 0$; then $|x|\leqq|T||x|$, and $\varphi\,(|T||x|-|x|)=0$ shows that $|x|=|T||x|$. Moreover, since $|T|$ is irreducible, the $T$-invariant ideal $E_{|x|}$ (complex $AM$-space with unit $|x|$) is dense in $E$. As usual, we identify $E_{|x|}$ with $C(K)$, $K$ denoting the structure space of $E$.

On $C(K)$, $T$ induces an operator $T_0$ whose modulus $|T_0|$ is the restriction of $|T|$ to $E_{|x|}$; moreover, $x$ defines a unimodular eigenfunction of $T_0$. By (7.4) there exists an isometry $V$ of $C(K)$ satisfying $T_0=\alpha V|T_0|V^{-1}$ and $|Vh|=|h|$ for all $h\in C(K)$; evidently, $V$ has a (unique) continuous extension $\bar{V}$ to $E$ which is an isometry satisfying $T=\alpha\bar{V}|T|\bar{V}^{-1}$. The assertions are now immediate consequences of (5.2) and (5.4), respectively. □

**Examples.** 4. The assertion of (7.5) (ii) holds for all $r$-cyclic complex $(n\times n)$-matrices $A$ (Example 3) such that $A$ and $|A|$ have the same spectral radius

[1] Cf. Lemma 4.8.

$(=1)$. In fact, if $|A|$ is irreducible (Def. I.6.1) then the existence of a linear form $\varphi > 0$ satisfying $\varphi = {}^t|A|\varphi$ is assured by (I.6.2); in the general case, the assertion is obtained by considering irreducible square blocks along the main diagonal of $A$ and employing induction on $n$.

5. Let $E := L^1(\mu)$, $(X, \Sigma, \mu)$ being a $\sigma$-finite measure space. Suppose $K$ is a kernel operator on $E$ (Chap. IV, § 8, Example 1) given by a $(\Sigma \times \Sigma)$-measurable real kernel satisfying $K(s,t) \neq 0$ a.e. $(\mu \otimes \mu)$.

*If $K$ has an eigenvalue $\lambda_0$ of modulus $\|K\|$, then either $\lambda_0 = \|K\|$ or $\lambda_0 = -\|K\|$ and $\lambda_0$ is a simple eigenvalue with an eigenfunction $f_0$ satisfying $f_0(s) \neq 0$ a.e. $(\mu)$.*

Without loss of generality we can suppose that $\|K\| = 1$. Then the first assertion of (7.6) holds for $K$, and by (5.6) $H$ is the trivial group $\{1\}$. From (7.5) (ii) it follows that $\lambda_0^2 = 1$. By the proof of (7.6), there exists an isometry $V$ of $E$ such that $|Vf| = |f|$ $(f \in E)$ and $K = \lambda_0 V|K|V^{-1}$; thus if $f_0$ is an eigenfunction of $K$ for $\lambda_0$, $V^{-1}f_0$ is fixed under $|K|$. Hence by (5.2), $|V^{-1}f_0| = |f_0|$ is fixed under $|K|$ and $>0$ a.e. $(\mu)$. Finally, it is clear that $f_0$ can be chosen to be real valued.

## § 8. Mean Ergodicity of Order Contractive Semi-Groups. The Little Riesz Theorem

This section is mainly concerned with supplementing the criteria for mean ergodicity of operators and semi-groups of operators given in Chapter III (§§ 7, 8), and some related material; essential use is made of the representation theory of Banach lattices with quasi-interior positive elements (Chap. III, § 4) and the modulus of linear operators (Chap. IV, § 1). It will thus be possible to extend Theorem III.7.11 to a large class of Banach lattices and semi-groups of (not necessarily positive) operators. We recall that a semi-group $S \subset \mathscr{L}(E)$, $E$ any Banach space, is *contractive* if $\|T\| \leq 1$ for all $T \in S$. The Banach spaces and lattices considered can be either real or complex (cf. Chap. II, § 11).

**8.1 Definition.** *Let $E$ be a Banach lattice, and let $S \subset \mathscr{L}(E)$ be an equicontinuous semi-group of operators* (Chap. III, § 7) *each possessing a modulus. $S$ is called* order contractive *if there exist a quasi-interior element $u \in E_+$ and a strictly positive linear form $\mu$ on $E$ such that $|T|u \leq u$ and $|T|'\mu \leq \mu$ for all $T \in S$. An operator $T \in \mathscr{L}(E)$ is called* order contractive (*or an* order contraction) *if the cyclic semi-group $(T^n)_{n \in \mathbb{N}}$ is order contractive.*

We note that if $S \subset \mathscr{L}(E)$ is order contractive and the set $\{|T| : T \in S\}$ is equicontinuous, then the semi-group generated by the moduli $|T|$ $(T \in S)$ is order contractive.

**Examples.** 1. If $E$ is an $AL$-space and $S \subset \mathscr{L}(E)$ is a contractive semi-group such that $|T|u \leq u$ for some quasi-interior $u \in E_+$ and all $T \in S$, then $S$ (and even the semi-group generated by $\{|T| : T \in S\}$) is order contractive.

2. If $K$ is compact and $S$ is an irreducible, abelian semi-group of (positive) Markov operators on $C(K)$, then $S$ is order contractive by (III.8.10).

3. The Dunford-Schwartz operators of Ionescu-Tulcea [1969] are order contractive.

4. Let $S$ be an irreducible semi-group of positive contractions on a Banach lattice $E$. If $S$ is mean ergodic with associated projection $P\neq 0$, then $S$ is order contractive (III.8.5).

The following result, which we call the *Little Riesz Theorem* and which is one of the most frequently used forms of the M. Riesz convexity theorem [DS, VI.10.12], can be proved in a very direct manner by using the representation of $AL$-spaces given in (II.8.5).

**8.2 Proposition.** *Let $(X,\Sigma,\mu)$ be a finite measure space, and let $T$: $L^1(\mu)\to L^1(\mu)$ be a linear operator leaving the ideal $L^\infty(\mu)$ invariant. If $\|T\|_1\leqq 1$ and if $\|T\|_\infty\leqq 1$, then $|T|$ (and, a fortiori, $T$) induces a contraction and an order contraction on each of the spaces $L^p(\mu)$ $(1\leqq p\leqq +\infty)$.*

*Proof.* By (IV.1.8) Cor. 2, the modulus $|T|$ exists and has $L^1$-norm and $L^\infty$-norm $\leqq 1$; hence, $|T|e\leqq e$ and $|T|'\mu\leqq\mu$ where $e$ denotes the constant-one function on $X$. Since for each $p$, $1\leqq p\leqq +\infty$, $e$ is a quasi-interior point of $L^p(\mu)_+$ and $\mu$ defines a strictly positive linear form on $L^p(\mu)$, it suffices to show that the restriction of $|T|$ to $L^\infty(\mu)$ is a contraction with respect to the norm $f\mapsto\|f\|_p:=(\int|f|^p\,d\mu)^{1/p}$.

To this end, we identify $L^\infty(\mu)$ with $C(K)$, where $K$ is the Stone representation space of the Boolean algebra $\Sigma/N$ (Chap. II, § 7, Example 5). Defining $\mu_s:=|T|'\delta_s$ $(s\in K)$ we obtain $\mu_s\geqq 0$ and $\mu_s(K)\leqq 1$; thus by Hölder's inequality we obtain

$$|T|(f^p)(s)=\int f^p(t)\,d\mu_s(t)\geqq\left(\int f(t)\,d\mu_s(t)\right)^p=(|T|\,f(s))^p$$

for all $0<f\in C(K)$ (and $p<+\infty$); since $\|\,|T|\,\|_1\leqq 1$,

$$\int(|T|\,f)^p\,d\mu\leqq\int|T|(f^p)\,d\mu\leqq\int f^p\,d\mu$$

and it follows that $|T|$ is contractive on $L^p(\mu)$. □

**8.3 Lemma.** *Let $K$ be a compact space, and let $E,F$ be Banach lattices with order continuous norm each containing $C(K)$ as a dense ideal*[1]. *On each order bounded subset of $C(K)$, the weak topology $\sigma(E,E')$ agrees with $\sigma(F,F')$ and the norm topology $\beta(E,E')$ agrees with $\beta(F,F')$.*

*Proof.* It is enough to prove the assertions for the unit ball $U:=\{z\in C(K)\colon |z|\leqq e\}$ of $C(K)$. By (II.6.6), $E'_+$ and $F'_+$ contain weak order units $\mu,\nu$ respectively, which can be viewed as strictly positive, order continuous Radon measures on $K$; if $\lambda:=\mu\wedge\nu$ then $\lambda$ is still order continuous and strictly positive. Now by (II.5.10), $U$ is $\sigma(E,E')$- compact and $\sigma(F,F')$-compact, and $\sigma(L^1(\lambda),L^\infty(\lambda))$-compact by (II.8.8). Since the latter topology is coarser on $C(K)$ than both $\sigma(E,E')$ and $\sigma(F,F')$, it follows that all three topologies agree on $U$.

Second, suppose $(x_n)$ to be a $\beta(E,E')$-null sequence in $U$; clearly, it suffices to show that some subsequence $(y_n)$ of $(x_n)$ is a null sequence for $\beta(F,F')$. In fact, if

[1] I.e., $E$ and $F$ are supposed to have the same structure space $K$ (cf. III.4.5).

$(y_n)$ is chosen to satisfy $\sum_n \|y_n\|_E < +\infty$ then $y_0 := \sup_n \sum_{m=1}^n |y_m|$ exists in $E$. Now let $z_n := \sup_{m \geqq n} |y_m|$; then we obtain $\inf_n z_n = 0$. But $z_n \in U$ for all $n$ and hence $\lim_n \|y_n\|_F = 0$, since $|y_n| \leqq z_n$ and since $F$ has order continuous norm. □

Theorem III.7.11 and its extension to uniformly convex Banach spaces (Chap. III, Exerc. 18) appear to be the only mean ergodic theorems placing no restriction on the algebraic structure (such as commutativity, or amenability) of the semi-group $S$ considered. It is, therefore, remarkable that for arbitrary order contractive semi-groups $S$ of operators on Banach lattices with order continuos norm, the validity of (III.7.11) forces mean ergodicity of $S$.

**8.4 Theorem.** *Let $E$ be a Banach lattice with order continuous norm. Every order contractive semi-group $S \subset \mathscr{L}(E)$ is mean ergodic in $\mathscr{L}_s(E)$.*

*Proof.* By Def. 8.1 there exist a quasi-interior $u \in E_+$ and a strictly positive linear form $\mu \in E'$ such that $|T|u \leqq u$ and $|T'|\mu \leqq \mu$ for all $T \in S$. Consider the $AM$-space with unit $u$, $E_u \cong C(K)$ and the $AL$-space $(E, \mu) \cong L^1(\mu, K)$ (for notation, see Chap. IV, § 3), where the compact space $K$ is the structure space of $E$. Since $E$ has order continuous norm, $\mu$ is order continuous on $E$ and hence by (IV.9.3), $E$ is an ideal of $L^1(\mu, K)$ and so $C(K)$ can be identified with $L^\infty(\mu)$. The hypothesis on $S$ (Def. 8.1) implies that $S$ defines contractive semi-groups $S_1$ and $S_\infty$ on $L^1(\mu)$ and $L^\infty(\mu)$, respectively; therefore, the Riesz theorem (8.2) shows that $S$ induces a contractive semi-group $S_p$ on each $L^p(\mu)$ $(1 < p < +\infty)$.

The semi-group $S_2$ is mean ergodic (Def. III.7.1) in $\mathscr{L}_s(L^2(\mu))$ by (III.7.11); let $P_2$ denote the corresponding projection (zero element of $S_2$). If $x$ denotes any element of the unit ball $U$ of $L^\infty(\mu)$ then by (III.7.10), $P_2 x \in \overline{\mathrm{co}}\, Sx$ where the closure is taken for the weak (or norm) topology induced on $U$ by $L^2(\mu)$; by (8.3), this topology agrees on $U$ with the weak (or norm) topology induced by $E$. Therefore, $\|P_2 x\|_E \leqq c\|x\|_E$ for all $x \in U$ and some $c > 0$ independent of $x$, and it follows that $P_2$ defines a projection $P \in \mathscr{L}(E)$ satisfying $TP = PT = P$ for all $T \in S$. In addition, it follows that $P \in \overline{\mathrm{co}}\, S$ in $\mathscr{L}_s(E)$, since $S$ is equicontinuous (Def. 8.1) and hence, the topologies of simple convergence on the dense ideal $C(K) \subset E$ and on $E$, respectively, agree on $S$ [S. III.4.5]. Therefore, $S \subset \mathscr{L}_s(E)$ is mean ergodic by (III.7.10). □

**Corollary 1.** *Let $(X, \Sigma, \mu)$ be a $\sigma$-finite measure space. If $S$ is a contractive semi-group on $L^1(\mu)$ such that $|T|f \leqq f$ for all $T \in S$ and some $f$ satisfying $f(s) > 0$ a.e. $(\mu)$, then $S$ is mean ergodic.*

**Corollary 2.** *If $E$ has order continuous norm, if $S$ is an order contractive semi-group in $\mathscr{L}(E)$, and if the semi-group $S_a$ generated by the moduli $|T|\,(T \in S)$ is equicontinuous, then $S_a$ is mean ergodic.*

The method, due to Nagel [1974], applied in the proof of (8.4) proves equally fruitful for the extension of Hilbert space results to Banach lattices in other cases; an example is given below (8.5 and its corollary). This method is symbolized by the following diagram:

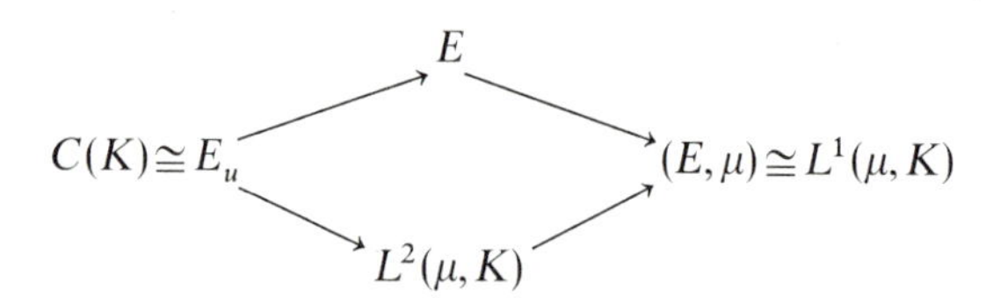


Here $u, \mu$ denote the elements designed into the definition of order contractive semi-groups (Def. 8.1); if $E$ is a Banach lattice with order continuous norm, each arrow in the diagram represents a continuous lattice isomorphism with range a dense ideal (so that $K$ is the common structure space of $E$, $L^1(\mu)$, and $L^2(\mu)$). By (8.2), $S$ induces a contractive semi-group $S_2$ in $L^2(\mu)$, and (8.3) ensures that the topological properties of $S_2$ are valid for the restriction of $S$ to the dense ideal $C(K)$, whence they carry over to $E$ by equicontinuity of $S$.

The following results (8.5) and (8.7) have their roots in the investigation of mixing properties of measure preserving transformations (Chap. III, § 9, Example 1; cf. Exerc. 14).

**8.5 Proposition.** *Let $H$ denote a Hilbert space, and let $(T^n)_{n\in\mathbb{N}}$ be a contractive, cyclic semi-group in $\mathscr{L}(H)$ with associated projection $P$.*[2] *The following are equivalent:*

(a) *The sequence $(T^n)_{n\in\mathbb{N}}$ converges to $P$ in the weak operator topology of $\mathscr{L}(H)$.*

(b) *For all infinite subsequences $(n_i)_{i\in\mathbb{N}}$ of $\mathbb{N}$, the averages*

$$\frac{1}{N}\sum_{i=1}^{N} T^{n_i}$$

*converge strongly to $P$.*

*Proof.* The equivalence (a)$\Leftrightarrow$(b) is trivial for the restriction of $T$ to its fixed space $P(H)$; hence, it suffices to consider the complementary subspace $(1_H - P)(H)$ or, equivalently, to consider the case $P=0$.

(a)$\Rightarrow$(b): Let $x\in H$; by assumption, $\lim_n T^n x = 0$ weakly. If $\lim_n \|T^n x\| = 0$ as well then (b) clearly holds; if not, we can assume that $\lim_n \|T^n x\| = 1$ keeping in mind that, by virtue of $\|T\|\leqq 1$, $(\|T^n x\|)_{n\in\mathbb{N}}$ is a decreasing sequence and hence the limit exists. We have to show that for any sequence $(n_i)$ of positive integers,

$$\left\|\frac{1}{N}\sum_{i=1}^{N} T^{n_i}x\right\|^2 = \frac{1}{N^2}\sum_{1\leqq i\leqq j\leqq N} 2\,\mathrm{Re}\,(T^{n_i}x\,|\,T^{n_j}x) \tag{1}$$

converges to 0 for $N\to\infty$. Let $\varepsilon>0$ be preassigned; by hypothesis, there exists an integer $m_0$ such that $1\leqq \|T^{m_0+l}x\|^2 < 1+\varepsilon$ for all $l\geqq 0$, and an integer $k_0$ such that $|(T^{m_0}x\,|\,T^{m_0+k}x)|<\varepsilon$ for all $k\geqq k_0$. This implies

$$\begin{aligned} 2\,\mathrm{Re}(T^{m_0+l}x\,|\,T^{m_0+k+l}x) &= \|T^{m_0+l}x + T^{m_0+k+l}x\|^2 - \|T^{m_0+l}x\|^2 - \|T^{m_0+k+l}x\|^2 \\ &\leqq \|T^{m_0}x + T^{m_0+k}x\|^2 - 2 < 4\varepsilon \end{aligned} \tag{2}$$

[2] By (III.7.11), $(T^n)$ is mean ergodic in $\mathscr{L}_s(H)$.

for all $l \geqq 0$ and $k \geqq k_0$. We now subdivide the set of all pairs $\{(i,j): i \leqq j; i,j=1,\dots,N\}$ into four subsets $A,B,C,D$ by specifying $A:=\{(i,j): i=j\}$, $B:=\{(i,j): i<j, i<m_0\}$, $C:=\{(i,j): m_0 \leqq i, i<j<k_0+i\}$, and $D:=\{(i,j): m_0 \leqq i, k_0+i \leqq j\}$. The point of this partition is that the number of elements in each of $A,B,C$ is bounded by $cN$ where $c>0$ depends only on $m_0$ and $k_0$ (and hence, on $\varepsilon$ and $x$). On the other hand, $|\mathrm{Re}(T^{n_i}x|T^{n_j}x)| \leqq \|x\|^2$ for all $(i,j)$ and $\mathrm{Re}(T^{n_i}x|T^{n_j}x)<2\varepsilon$ for $(i,j)\in D$ by (2); therefore, the right hand side of (1) is $<5\varepsilon$ for all sufficiently large $N$.

(b)⇒(a): (b) implies that for each pair of elements $x,y\in H$, $\lim_N N^{-1}\sum_{i=1}^N (T^{n_i}x|y)=0$ for any subsequence $(n_i)_{i\in\mathbb{N}}$ of $\mathbb{N}$. Since $((T^n x|y))_{n\in\mathbb{N}}$ is a bounded sequence, it follows from a standard result on Cesaro convergence that $\lim_n (T^n x|y)=0$. Hence, $(T^n)_{n\in\mathbb{N}}$ converges to 0 for the weak operator topology. □

The following corollary is easily obtained by the method explained in the discussion preceding (8.5).

**Corollary.** *Let $E$ denote a Banach lattice with order continuous norm, and let $(T^n)_{n\in\mathbb{N}}$ be an order contractive, cyclic semi-group in $\mathscr{L}(E)$ with associated projection $P$.*[3] *Then assertions* (a) *and* (b) *of* (8.5) *(with $H$ replaced by $E$) are equivalent.*

This corollary can be strengthened for $AL$-spaces $E$ so as to apply to all (metric) contractions $T$ on $E$; we include this result in view of the importance of $AL$-spaces for ergodic theory. For its proof we need a lemma which is of independent interest (and which makes strong use of the structure of an $AL$-space).

**8.6 Lemma.** *Let $E$ be an $AL$-space, and suppose that $T\in\mathscr{L}(E)$ is a contraction such that $|T|$ has no fixed points $\neq 0$. For each $x\in E$ whose orbit $\{T^n x: n\in\mathbb{N}\}$ is relatively $\sigma(E,E')$-compact, one has $\lim_n \|T^n x\|=0$.*

*Proof.* If $\{T^n x: n\in\mathbb{N}\}$ is relatively $\sigma(E,E')$-compact, then by (II.8.8) Cor. so is the set of moduli $A:=\{|T^n x|: n\in\mathbb{N}\}$. Denote by $B\subset E_+$ the set of all weak cluster points of the sequence $(|T^n x|)_{n\in\mathbb{N}}$, and let $B_0\subset B$ be the subset of those $z\in B$ for which $\|z\|=\inf_{y\in B}\|y\|$; $B_0$ is weakly compact. We claim that $|T|B_0\subset B_0$. In fact, if $z\in B_0$ then $z$ is the $\sigma(E,E')$-limit of some subsequence $(|T^{n_k}x|)_{k\in\mathbb{N}}$; if $y\in B$ is a cluster point of the sequence $(|T^{n_k+1}x|)_{k\in\mathbb{N}}$ then $y\leqq|T|z$. Hence, since by (IV.1.5) $|T|$ is a contraction, we have $y\in B_0$ and it follows that $y=|T|z$; therefore, $|T|B_0\subset B_0$. Consequently, $|T|C\subset C$ where $C$ denotes the convex closure of $B_0$; but $C$ is $\sigma(E,E')$-compact by Krein's theorem [S, IV.11.4] and so by the Markov-Kakutani theorem (III.7.12) $|T|$ has a fixed point $z\in C$. Now $z=0$ by hypothesis; but, $E$ being an $AL$-space, all elements of $C$ have equal norm and so $B_0=\{0\}$. Thus some subsequence of $(|T^n x|)_{n\in\mathbb{N}}$ converges to 0 weakly and hence, since the sequence is positive, in norm; since $T$ is contractive, we have $\|T^{n+1}x\|\leqq\|T^n x\|$ for all $n\in\mathbb{N}$ and the lemma is proved. □

**Corollary.** *Let $T$ be a contraction in an $AL$-space $E$ such that $\{T^n: n\in\mathbb{N}\}$ is order bounded in $\mathscr{L}(E)$. If the modulus $|T|$ has no fixed vectors $\neq 0$, then $\lim_n\|T^n x\|=0$ for all $x\in E$.*

---

[3] $(T^n)$ is mean ergodic by (8.4).

The proof is clear from (8.6) in view of the fact that order intervals in $E$ are weakly compact (II.8.8) Cor.

The following is now the announced theorem of Akcoglu-Sucheston [1972] and Sato [1973].

**8.7 Theorem.** *Let $E$ be an AL-space. For any contraction $T\in\mathscr{L}(E)$, the following assertions are equivalent:*

(a) *The sequence $(T^n)_{n\in\mathbb{N}}$ converges to a projection $P$ in the weak operator topology of $\mathscr{L}(E)$.*

(b) *For all infinite subsequences $(n_i)_{i\in\mathbb{N}}$ of $\mathbb{N}$, the averages $N^{-1}\sum_{i=1}^{N} T^{n_i}$ converge strongly to $P\in\mathscr{L}(E)$ as $N\to\infty$.*

*Proof.* Since, clearly, (b) implies $T$ to be mean ergodic (Def. III.7.1), the implication (b)⇒(a) results from the same argument as the implication (b)⇒(a) of (8.5).

(a)⇒(b): Let $x\in E_+$ be fixed; by considering the closed ($T$-invariant) ideal of $E$ generated by the orbit $\{|T|^n x: n\in\mathbb{N}\}$, we can and will assume that $E$ contains weak order units (or equivalently, quasi-interior positive elements (II.6.5)).

Let $\boldsymbol{J}$ denote the set of all (closed) $T$-ideals of $E$ such that on each $J\in\boldsymbol{J}$, $T$ induces an order contraction (Def. 8.1). For each $J\in\boldsymbol{J}$, let $u_J\in E_+$ be a weak order unit of $J$ satisfying $|T|u_J\leqq u_J$; moreover, let $v$ denote a weak order unit of $E$ and let $v_J$ denote the band component of $v$ in $J$. Since $|T|(u_{J_1}+u_{J_2})\leqq u_{J_1}+u_{J_2}$ and $u_{J_1}+u_{J_2}$ is a weak order unit of $J_1+J_2$ for any pair $J_1,J_2\in\boldsymbol{J}$, it follows that $\{v_J: J\in\boldsymbol{J}\}$ is a directed ($\leqq$) subset of $[0,v]$. Define $v_0:=\sup_J v_J$; since $E$ is super Dedekind complete by (II.5.10) Cor. 1, there exists a sequence $(J_n)_{n\in\mathbb{N}}$ in $\boldsymbol{J}$ such that $v_0=\sup_n v_{J_n}$. Letting $u_0:=\sum_{n\in\mathbb{N}}\varrho_n u_{J_n}$ (where $\varrho_n^{-1}:=2^n\|u_{J_n}\|$), it is clear that $|T|u_0\leqq u_0$ and that $u_0$ is a weak order unit of the closed ideal $J_0$ generated by $v_0$; therefore, $J_0$ is the largest ideal of $E$ on which $T$ induces an order contraction.

We consider the band decomposition $E=J_0+J_0^\perp$ and let $Q$ denote the band projection $E\to J_0^\perp$. For any $x\in J_0$, the norm convergence of $N^{-1}\sum_{i=1}^{N} T^{n_i}x$ as $N\to\infty$ follows from the corollary of (8.5), $(n_i)_{i\in\mathbb{N}}$ being an arbitrary subsequence of $\mathbb{N}$. It remains to show that the sequence $N^{-1}(\sum_{i=1}^{N} T^{n_i}x)$ converges for each $x\in J_0^\perp$.

To this end we consider the contraction $T_0:=QTQ\in\mathscr{L}(E)$. Then $|T_0|=Q|T|Q$ and it is easily seen that $z=|T_0|z$ implies $z\in J_0^\perp$ and $z=|T|z$, whence $z=0$ by the construction of $J_0$; therefore, (8.6) is applicable to $T_0$. Moreover, since $J_0$ is $T$-invariant, it follows that $T_0^n=QT^nQ$ for all $n\in\mathbb{N}$; hence, if $x\in J_0^\perp$ and $\varepsilon>0$ are preassigned, by (8.6) there exists an integer $m_0$ such that $\|QT^{m_0}x\|<\varepsilon$. From the corollary of (8.5) and the construction of $J_0$ it follows that the sequence with general term

$$\frac{1}{N}\sum_{k=1}^{N} T^{m_k}(1_E-Q)T^{m_0}x \qquad (N\in\mathbb{N}) \tag{3}$$

norm converges, where $(m_k)_{k\in\mathbb{N}}$ denotes the sequence of all positive integers $n_i-m_0$ ($i\in\mathbb{N}$) and $(n_i)$ is a given subsequence of $\mathbb{N}$. It is clear that $\lim_N N^{-1}\|\sum_{i=1}^{N} T^{n_i}-\sum_{k=1}^{N} T^{m_k}T^{m_0}\|=0$; on the other hand,

$$N^{-1}\|\sum_{k=1}^{N} T^{m_k} T^{m_0} x - \sum_{k=1}^{N} T^{m_k}(1_E - Q) T^{m_0} x\| < \varepsilon$$

for all $N \in \mathbb{N}$, since $\|Q T^{m_0} x\| < \varepsilon$. Therefore, from the norm convergence of the sequence (3) it follows that $N^{-1}(\sum_{i=1}^{N} T^{n_i} x)$ converges in norm as $N \to \infty$, completing the proof. □

Of course, the equivalence (a)⇔(b) of (8.7) cannot be expected to hold in arbitrary Banach lattices (Exerc. 15). We conclude these investigations with a result due to Choquet-Foias [1973]. When compared with (8.7), it shows once again the contrasting behavior of $AL$- and $AM$-spaces.

**8.8 Proposition.** *Let $E$ be an $AM$-space with unit and let $0 \leqq T \in \mathscr{L}(E)$. If $\lim_n T^n = 0$ in the weak operator topology then $\lim_n \|T^n\| = 0$.*

*Proof.* We consider $E$ as a space $C(K)$ (II.7.4) and denote by $e$ the constant-one function on $K$; since $\lim_n T^n e = 0$ weakly (hence, pointwise on $K$), there exists $r \in \mathbb{N}$ such that $\inf_{1 \leqq \varrho \leqq r} T^{\varrho} e(t) < 1$ for all $t \in K$. For reasons of continuity we must still have $\hat{g}(t) := \inf_{1 \leqq \varrho \leqq r} \hat{T}^{\varrho} e(t) < 1$ for all $t$, where $\hat{T} := T + \theta 1_E$ and $0 < \theta < 1$ is suitably chosen. Then we obtain $\hat{g} \leqq \inf_{0 \leqq \varrho \leqq r-1} \hat{T}^{\varrho} e$ and $\hat{T}\hat{g} \leqq \inf_{1 \leqq \varrho \leqq r} \hat{T}^{\varrho} e = \hat{g}$, since $\hat{T} \geqq 0$. It follows that $T\hat{g} \leqq (1-\theta)\hat{g}$, whence $T^n \hat{g} \leqq (1-\theta)^n \hat{g}$ for all $n \in \mathbb{N}$; on the other hand, $\hat{g} \geqq \theta^r e$ and so $T^n e \leqq (1-\theta)^n \theta^{-r} \hat{g}$. Since $\|T^n\| = \|T^n e\|$ for all $n$ (Chap. IV, § 1, Formula (4)), we conclude that $\lim_n \|T^n\| = 0$. □

## Notes

§ 1: Apparently the imbedding procedure was first applied to spectral theory by Berberian [1962]; its application to positive operators on Banach lattices is due to Lotz [1968].

§ 2: The arrangement of the material in this section is due to M. Wolff, as are the major results (2.8) through (2.10). For further results and references, see Wolff [1973 a]—[1973 c]. (2.7) and its application are due to Scheffold [1973 a]. Other relevant references include Korovkin [1953], Šaškin [1966], Wulbert [1968], Scheffold [1973 b], Bernau [1974]; see also the monograph of Lorentz [1967].

§ 3: Boolean algebras of projections on Banach spaces were studied extensively by Badé (cf. Badé [1959]); for a thorough discussion and references, see Dunford-Schwartz [1971]. Following an approach initiated by Schaefer [1962 a], Walsh [1965] first recognized the importance of the natural ordering induced on the cyclic subspaces (of a locally convex space) generated by a spectral measure; this ordering is also used by Tzafriri [1969 b]. Theorem 3.6 is due to Schaefer [1974 b].

§ 4: Spectral properties of positive operators, under more or less special assumptions on the operator and/or type of ordering of the domain space, were discovered by Krein-Rutman [1948], Ando [1957], Karlin [1959], Schaefer [1960c], and others. Results valid in general or special Banach lattices were subsequently established by Rota [1961], Schaefer [1963b]—[1965], Lotz [1968], Wolff [1969a], and Scheffold [1971b]. Krieger [1969] gives a systematic and comprehensive account (including his own contributions), to which we refer for further details and references. The presentation of the material in this section is selective and largely based on Lotz [1968] (to whom (4.9) is due) and Scheffold [1971b] (to whom (4.4) is due).

§ 5: The concept of an irreducible operator seems to have been employed first by Ando [1957] and, independently, by Schaefer [1960c], [1963b]. Further results are due to Schaefer [1966b]—[1968] and to Lotz [1968]. In full generality, Theorems 5.4 and 5.5 were first proved by Niiro-Sawashima [1966]; the proofs given here are due to Lotz-Schaefer [1968]. See also Krieger [1969].

§ 6: With the exception of (6.5) and (6.6), the material of this section is taken from Schaefer [1970b]. Theorem 6.5 was first established for compact operators by Ando [1957], then improved by Krieger [1969] who removed the compactness assumption. (Note that by virtue of (II.6.6), the hypothesis of (6.5) is equivalent to the hypothesis made by Krieger [1969, 4.2.1].) (6.6) and the present proof of (6.5) are due to the author.

§ 7: The application of the methods and techniques developed for positive operators, to non-positive operators is of recent origin; see Schaefer [1971b], Wolff [1974]. Special cases of (7.5) (particularly, (7.5) Cor. 2) were obtained by the author, then extended by Wolff [1974] to whom the concept of $r$-cyclic element is due.

§ 8: The arrangement and most of the results of this section (especially, Theorem 8.4) are due to Nagel [1974], who also introduced the concept of order contraction (Def. 8.1). (8.7) is due to Akcoglu-Sucheston [1972] and Sato [1973]. (8.8) is due to Choquet and Foias [1973].

## Exercises

1. We use the terminology and notation of § 1.

(a) Let $E$ denote a Banach lattice with quasi-interior positive elements and structure space $K$ (cf. III. 4.5), and let $\hat{E}_{\boldsymbol{F}}$ denote any $\boldsymbol{F}$-product of $E$ (Def. 1.1). Show that the Banach lattice $l^p(E)$ of all sequences $x=(x_n)$, such that $x_n \in E$ $(n \in \mathbb{N})$ and $\|x\|:=(\sum_n \|x_n\|^p)^{1/p} < +\infty$, has the structure space $\beta(\mathbb{N} \times K)$ ($\mathbb{N}$ discrete) whenever $1 \leqq p < +\infty$. On the other hand, $m(E)$ contains $C(\beta(\mathbb{N} \times K))$ as an order dense (but not, unless $E$ is an $AM$-space with unit, as a topologically dense) ideal; hence, $m(E)$ contains weak order units but, in general, no quasi-interior positive elements.

(b) Show that if $E$ is a Banach lattice with $p$-additive norm ($1 \leqq p < +\infty$; cf. Chap. II, Exerc. 23) and $\boldsymbol{U}$ is a free ultrafilter on $\mathbb{N}$, then $\hat{E}_{\boldsymbol{U}}$ has $p$-additive norm. Using the representation theory (Chap. III, §§ 4—6), determine the dual of $\hat{E}_{\boldsymbol{U}}$ and its relationship to $E'$ when $E = L^p(\mu)$ for a $\sigma$-finite measure $\mu$.

(c) Let $G, H$ be Banach spaces and let $\hat{H}_{\boldsymbol{F}}$ denote any $\boldsymbol{F}$-product of $H$. If $\mathscr{L}(G,H)$ denotes the Banach space of bounded linear operators $G \to H$ under its natural norm, show that there exists a natural, norm decreasing map of $(\mathscr{L}(G,H))\hat{}_{\boldsymbol{F}}$ into $\mathscr{L}(G, \hat{H}_{\boldsymbol{F}})$, and that $\mathscr{L}(G,H)$ can be identified with a Banach subspace of $\mathscr{L}(G,\hat{H}_{\boldsymbol{F}})$ by virtue of the composition $\mathscr{L}(G,H) \to (\mathscr{L}(G,H))\hat{}_{\boldsymbol{F}} \to \mathscr{L}(G,\hat{H}_{\boldsymbol{F}})$.

2. (Korovkin Families.)

(a) Let $E := C[0,1]$ and let $f_i$ $(i=0,1,2)$ denote the functions $t \mapsto t^i$ $(t \in [0,1])$. The family $\{f_0, f_1\}$ is generating but not a Korovkin family. (Find a Radon measure $\mu \geqq 0$ on $[0,1]$ such that $\mu(f_0)=1$, $\mu(f_1)=f_1(s)$ for some $s \in [0,1]$ but $\mu \neq \delta_s$, and use (2.5). Note, however, that $\{f_0, f_1\}$ is a Korovkin family in $L^p[0,1]$ in a restricted sense for $1 \leqq p < +\infty$. Cf. Bernau [1974].)

(b) For $t \in [0,1]$ and $f \in C[0,1]$, define $B_n f(t) := \sum_{k=0}^{n} \binom{n}{k} f\left(\frac{k}{n}\right) t^k (1-t)^{n-k}$ (Bernstein polynomials of $f$) and show that $\lim_n B_n f = f$ uniformly on compact subsets of $C[0,1]$. (Use the fact that by (2.6) and (2.8), $\{f_0, f_1, f_2\}$ (see (a)) constitutes a Korovkin family of $C[0,1]$.)

3. Let $G$ be a cyclic Banach space (§ 3). Show that $G$ is reflexive iff $G$ does not contain a Banach subspace isomorphic (as a topological vector space) to $c_0$ or $l^1$. (Use (3.6) and Chap. II, Exerc. 17. Cf. Tzafriri [1969].)

4. (Boolean Algebras of Projections.) Let $G$ be a Banach space and suppose there exists a Banach subspace $H$ of $G'$ whose unit ball $V$ is $\sigma(G', G)$-dense in the unit ball $W^\circ$ of $G'$. (Examples are furnished by dual Banach spaces $G = M'$ with $H$ a Banach subspace of $G'$ containing $M$.) For the purpose of this exercise only, we understand by $\boldsymbol{P}$ a Boolean algebra of projections on $G$ which satisfies Condition (i) of Def. 3.1, while (ii) is weakened as follows:

(ii)′ *If* $(P_n)_{n \in \mathbb{N}}$ *is an increasing sequence in* $\boldsymbol{P}$, $P := \sup_n P_n$ *exists in* $\boldsymbol{P}$ *and* $\lim_n \langle P_n x, y' \rangle = \langle Px, y' \rangle$ *for all* $x \in G$, $y' \in H$.

(a) If $K$ denotes the Stone representation space of $\boldsymbol{P}$, there exists a continuous homomorphism of $B$-algebras $m: C(K) \to \mathscr{L}(G)$, which has a continuous extension $\bar{m}: B(K) \to \mathscr{L}(G)$ ($B(K)$ the Banach algebra of bounded Baire functions on $K$) such that on the Baire field $\Sigma$ of $K$, $\bar{m}$ induces a spectral measure $\mu: \Sigma \to \mathscr{L}(G)$ which is countably additive for the topology $\sigma(\mathscr{L}(G), G \otimes H)$. Observe, in particular, that $\mu(\Sigma) = \boldsymbol{P}$, and that $m$ and $\bar{m}$ have identical ranges in $\mathscr{L}(G)$.

(b) Assume $\boldsymbol{P}$ to be reduced with respect to a vector $x_0 \in G$. If $G_0 := (G, \boldsymbol{P}, x_0)$ is the cyclic subspace of $G$ generated by $\boldsymbol{P}$ and $x_0$, then $G_0$ is capable of an ordering and equivalent norm under which it becomes a Banach lattice $E$ containing $x_0$ as a quasi-interior positive element and $C(K)$ as a dense ideal; in particular, $K$ is (homeomorphic with) the structure space of $E$ (cf. III.4.5). Moreover, $E$ is countably order complete and $\boldsymbol{P}$ induces on $E$ the Boolean algebra of all band projections.

(Prove (a) and (b) by adapting the proofs given in § 3.)

(c) Thm. 3.6 and its corollaries must be modified, as follows: In general, $\boldsymbol{P}$ is not complete; even if $\boldsymbol{P}$ is complete, it is not necessarily of countable type (cf. Chap. II, Exerc. 4; as before, $\boldsymbol{P}$ is assumed to be reduced with respect to $x_0$).

In general, the norm of $E$ is not order continuous and no Badé functional exists on $E$. (As counterexamples, consider the Boolean algebra $\boldsymbol{P}(E)$ of all band projections, where $E=B(X)$ (bounded Baire functions on some compact space $X$) or $E=L^\infty(\mu)$, where $\mu$ is a positive Radon measure on a suitable locally compact space.)

(d) Prop. 3.8 remains valid; find out how (3.9) must be modified.

5. (Spectrum and Resolvent.) Let $G\neq\{0\}$ denote a complex Banach space and let $T\in\mathscr{L}(G)$; generally, we use the notation introduced in § 4.

(a) Suppose $G_0$ is a $T$-invariant Banach subspace of $G$ and let $T_1:=T_{|G_0}$ (restriction of $T$ to $G_0$), $T_2:=T_{G_0}$ (map induced on $G/G_0$). Then $\sigma(T)\subset\sigma(T_1)\cup\sigma(T_2)$.

(b) Denote by $\varrho_\infty(T)$ the unbounded connected component of $\varrho(T)$. If $q\colon G\to G/G_0$ is the canonical map and if $\lambda\in\varrho_\infty(T)$, then $\lambda\in\varrho(T_2)$ and $R(\lambda,T_2)\circ q=q\circ R(\lambda,T)$. Similarly, if $\lambda\in\varrho_\infty(T)$ then $\lambda\in\varrho(T_1)$ and $R(\lambda,T_1)=R(\lambda,T)_{|G_0}$.

(c) Deduce from (a) and (b) that whenever $\varrho(T)$ is connected (for example, if $T$ is compact or $\sigma(T)$ real) then $\sigma(T)=\sigma(T_1)\cup\sigma(T_2)$ (cf. Chap. I, Exerc. 1).

(d) Show that whenever there exists $n\in\mathbb{N}$ such that $T^n$ is compact, then each non-zero element $\lambda\in\sigma(T)$ is a pole of $R(\lambda,T)$. (Show that if $0\neq\lambda\in\sigma(T)$ and $\lambda^n$ is a pole of $R(\mu,T^n)$ then $\lambda$ is a pole of $R(\lambda,T)$.)

(e) Suppose $G$ to be a complex Banach lattice (Def. II.11.3). The C. Neumann series (§ 4, Formula (1)) implies that whenever $T\geqq 0$ and $\lambda_1>\lambda_2>r(T)$, then $R(\lambda_2,T)>R(\lambda_1,T)>0$. Using the resolvent equation

$$R(\lambda,T)-R(\mu,T)=-(\lambda-\mu)R(\lambda,T)R(\mu,T)$$

$(\lambda,\mu\in\varrho(T))$, show that $R(\lambda,T)\geqq 0\Leftrightarrow(\lambda\in\mathbb{R}$ and $\lambda>r(T))$ for all $T\geqq 0$.

6. (a) Let $K$ be a compact space and suppose $T\colon C(K)\to C(K)$ is a Markov operator such that for each $f\in C(K)$, $f\geqq 0$, $T[0,f]$ is dense in $[0,Tf]$. Then either $\sigma(T)$ is a cyclic subset of the circle group $\Gamma$, or else $\sigma(T)=\{\lambda\in\mathbb{C}\colon|\lambda|\leqq 1\}$. (Observe that the adjoint operator $T'$ is a lattice homomorphism and an isometry; hence, the topological boundary of $\sigma(T)$ is contained in $\Gamma$. Use (4.4).)

(b) For each compact, cyclic subset $A$ of $\mathbb{C}$, there exists a Banach lattice $E$ and a lattice homomorphism $T\in\mathscr{L}(E)$ such that $\sigma(T)=A$. (Cf. Scheffold [1971 b].)

(c) Let $E$ be a Banach lattice and let $0<T\in\mathscr{L}(E)$ have spectral radius $r(T)=1$. Suppose, in addition, that 1 is an isolated singularity of $R(\lambda,T)$ with residuum $P$ of finite rank. The peripheral spectrum of $T$ is cyclic and finite. (Observe that $\lambda=1$ is necessarily a pole of $R(\lambda,T)$ and apply (5.5).)

*Note.* H.J. Krieger [1969] has shown that if $0<T\in\mathscr{L}(E)$, if $r(T)$ is an isolated singularity of $R(\lambda,T)$ and if $[0,r(T))\subset\varrho(T)$, then $T$ has cyclic peripheral spectrum.

7. Let $E$ denote a Banach lattice, let $0<T\in\mathscr{L}(E)$, and suppose that $r(T)=1$ is a pole of the resolvent $R(\lambda,T)$.

(a) If $\alpha$, $|\alpha|=1$, is a pole of $R(\lambda,T)$ then the order of the pole $\alpha$ is not greater than the order of the pole $\lambda=1$. (Cf. [S, Appendix, (2.4)].)

(b) $T$ possesses a fixed vector $x>0$.

(c) If $\lambda=1$ is a first order pole of $R(\lambda,T)$ with residuum of finite rank, then the semi-group $(T^n)_{n\in\mathbb{N}}$ is bounded and uniformly mean ergodic in $\mathscr{L}(E)$ (Chap. III, § 7). (Use (5.5) and (a).)

8. Let $E$ denote a Banach lattice, and let $T$ be a positive operator on $E$ with spectral radius $r(T)=1$.

(a) Supply a detailed proof of the corollary of (5.4).

(b) If $T$ is irreducible and if the fixed space $\{x\in E: Tx=x\}$ is a vector sublattice $\neq\{0\}$ of $E$, then $\lambda=1$ is the only eigenvalue of $T$ with an eigenvector $>0$.

(c) Suppose that $T$ satisfies Condition (G) (Def. 4.7) and that for each $x\in E_+\setminus\{0\}$, there exists $n\in\mathbb{N}$ such that $T^n x$ is a weak order unit of $E$. Then the peripheral point spectrum of $T$ contains at most $\lambda=1$. (Observe that by virtue of (4.8), the proof of (5.6) can be adapted to the present hypothesis.)

9. (Topological Nilpotency of Irreducible Operators.)

(a) If $0<T\in\mathscr{L}(E)$, $E$ any Banach lattice, is an irreducible operator such that some power $T^p$ ($p\in\mathbb{N}$) is weakly compact, there exists a (norm closed) $\sigma(E',E)$-dense minimal $T'$-ideal of $E'$. (Adapt the proof of (6.2), observing that an irreducible $T>0$ cannot be nilpotent.)

(b) If $E$ is an $AM$- or $AL$-space and if $T>0$ is an irreducible operator on $E$ such that some power $T^p$ is weakly compact, then $r(T)>0$. (Use (a) and proceed as in the proof of (6.3).)

(c) Let $E:=L^1(\mu,\Gamma)$ where $\mu$ is Haar measure on the circle group $\Gamma$. There exists an irreducible operator $T>0$ on $E$ such that $r(T)=0$. (Consider the operator $T$ defined by $Tf(z):=g(z)f(\alpha z)$ $(z\in\Gamma)$, where $\alpha\in\Gamma$ is not a root of unity while $g\in C(\Gamma)$ satisfies $0<g(z)\leqq 1$ $(z\neq 1)$ and vanishes sufficiently fast as $z\to 1$. Cf. Schaefer [1970b].)

10. ($r$-Cyclic Elements.) Let $E$ denote an ordered complex vector space in the sense of Def. 7.1.

(a) If $x\in E$ is $r$-cyclic, show that $x$ is $s$-cyclic for every integral multiple $s>0$ of $r$. The least integer $r_0\in\mathbb{N}$ such that $x\in E$ is $r_0$-cyclic, is called the *cyclic rank* of $x$.

(b) If $x$ is an $r$-cyclic element of $E$, the representation $x=\sum_{j=0}^{r-1}\zeta^j y_j$ (where $\zeta$ is a primitive $r$-th root of unity and $y_j\wedge y_k=0$ whenever $j\neq k$), is unique (even though $\zeta$ is not unique if $r\geqq 3$). (Imbed $E_0$ in an order complete vector lattice $\bar{E}_0$ and, for two cyclic representations $x=\sum_{j=0}^{r-1}\zeta^j y_j=\sum_{j=0}^{r-1}\zeta^j z_j$, consider the band projections of $\bar{E}_0$ onto the principal bands generated by $y_j$ and $z_j$ $(j=0,\dots,r-1)$, respectively.)

(c) If $E$ is a complex vector lattice (Def. II.11.1), an element $x\in E$ is real iff $x$ is $r$-cyclic of cyclic rank $\leqq 2$.

11. (a) Let $T$ be an $r$-cyclic operator on $C(K)$ ($K$ compact) whose modulus $|T|$ is a Markov operator, and suppose $T$ has a unimodular eigenvalue with corresponding unimodular eigenfunction. Then either $T$ is a scalar multiple of a positive operator, or else no component $T_k\geqq 0$ in a cyclic representation $T=\sum_{j=0}^{r-1}\zeta^j T_j$ $(r\geqq 2)$ is irreducible. (Consider the closed vector sublattice $F$ of $C(K)$ generated by the group of unimodular eigenfunctions of $T$, and use (4.2)

to conclude that $|T|$ induces a lattice homomorphism of $F$. Employing (7.4) and (III.9.1), infer that $T_j e \wedge T_k e = 0$ $(j \neq k)$ where $e$ is the unit of $C(K)$.)

(b) Show by examples that, in general, the $r$-th power of an $r$-cyclic operator is not positive.

12. (a) Let $E$ be any Banach lattice, and suppose that $T \in \mathscr{L}(E)$ is majorizing with $m$-norm $\|T\|_m = 1$ or that $T$ is c.a.s. with $l$-norm $\|T\|_l = 1$ (Def. IV.3.1). If $T$ is $r$-cyclic, then $|T|$ exists and satisfies $(G)$ (Def. 4.7); in particular, assertion (i) of (7.5) holds for the spectrum of $T$. Accordingly, (7.5) (ii) applies to $T$ (provided that $|T|' \varphi \leqq \varphi$ for some strictly positive linear form $\varphi$ on $E$).

(b) Suppose $E$ to be order complete and assume that $0 < T \in \mathscr{L}(E)$ is irreducible and majorizing (or irreducible and c.a.s.; Def. IV.3.1). If $r(T) = \|T\|_m = 1$ (respectively, if $r(T) = \|T\|_l = 1$), then the first part of the assertions of (7.6) is valid for $T$. Investigate if the second part carries over as well.

13. The corollary of (6.5) can be strengthened, as follows:

*Let $E$ be any Banach lattice, and let $0 < T \in \mathscr{L}(E)$ be majorizing (or* c.a.s.; Def. IV.3.1*) and weakly compact. If $T$ is irreducible, then $r(T) > 0$ and each non-zero element $\lambda \in \sigma(T)$ is a pole of $R(\lambda, T)$ (so that, in particular, (5.4) applies).*

(Supposing $T$ to be majorizing, observe that by (IV.3.4) $T$ has a factoring $E \xrightarrow{T_1} M \xrightarrow{T_2} E$ where $M$ is an $AM$-space. By a suitable choice of $M$ (namely, so that $M$ is an ideal of $E$, $T_2: M \to E$ the canonical injection, and no proper closed ideal of the $AM$-space $M$ contains $T_1(E)$), it can be arranged that $S := T_1 \circ T_2 \in \mathscr{L}(M)$ is irreducible. Since $S^2$ is weakly compact, conclude that $r(S) > 0$ (Exerc. 9) and so $r(T) > 0$; finally, observe that some power of $T$ is compact and use Exerc. 5(d). A similar argument applies if $T$ is c.a.s.)

14. (Mixing Properties.) Let $E$ denote a Banach lattice with order continuous norm and let $0 < T \in \mathscr{L}(E)$ denote an operator satisfying $Tu = u$ and $T'\mu = \mu$, where $u$ and $\mu$ denote a quasi-interior positive element of $E$ and a strictly positive linear form on $E$, respectively; suppose, in addition, that $\{T^n : n \in \mathbb{N}\}$ is equicontinuous. Then by (8.4), the cyclic semi-group $(T^n)_{n \in \mathbb{N}}$ is mean ergodic in $\mathscr{L}_s(E)$ with associated projection $P \in \overline{\text{co}}\{T^n : n \in \mathbb{N}\}$.

(a) $T$ is irreducible if and only if $P = \mu \otimes u$.

(b) The following are equivalent:

($\alpha$) $P = \mu \otimes u$ and $P$ is in the closure of $\{T^n : n \in \mathbb{N}\}$ in $\mathscr{L}_\sigma(E)$.

($\beta$) $T$ is irreducible and the peripheral point spectrum of $T$ equals $\{1\}$.

($\gamma$) For all $x \in E$ and $x' \in E'$ one has

$$\lim_n n^{-1} \sum_{m=1}^n |\langle T^m x, x' \rangle - \langle x, \mu \rangle \langle u, x' \rangle| = 0 .$$

(c) The following are equivalent:

($\alpha$) $P = \mu \otimes u$ and $\lim_n T^n = P$ in the weak operator topology.

($\beta$) $P = \mu \otimes u$ and for all subsequences $(n_i)_{i \in \mathbb{N}}$ of $\mathbb{N}$, $\lim_N N^{-1} \sum_{i=1}^N T^{n_i} = P$ weakly.

($\gamma$) $P = \mu \otimes u$ and for all subsequences $(n_i)_{i \in \mathbb{N}}$ of $\mathbb{N}$, $\lim_N N^{-1} \sum_{i=1}^N T^{n_i} = P$ strongly.

(d) For the equivalent properties of $T$ listed under (a), (b), (c), respectively, one clearly has (c)$\Rightarrow$(b)$\Rightarrow$(a); however, none of the reverse implications holds.

(e) Let $\varphi$ be a measure preserving transformation (Chap. III, § 9, Example 1) of a probability space $(X, \Sigma, \mu)$, and let $T_\varphi$ be the operator on $L^p(\mu)$ $(1 \leqq p < +\infty)$ induced by $\varphi$. Show that $T_\varphi$ has the equivalent properties listed under (a) (respectively, (b) or (c)) if and only if the transformation $\varphi$ is ergodic (respectively, mixing or strongly mixing) in the sense customary in ergodic theory (cf. Jacobs [1963]).

15. Show that the equivalence (a)$\Leftrightarrow$(b) of (8.7) fails for operators $T$ on $AM$-spaces, even if $T$ is supposed to be a Markov operator and lattice homomorphism. (Construct a compact space $K$ and a continuous map $k: K \to K$ with the following properties: For some $s_0 \in K$ and all $s \in K$, $\lim_n k^n(s) = s_0$; there exists a closed subset $K_0$ not containing $s_0$ and such that for preassigned $n_0 \in \mathbb{N}$, $K_0$ contains elements $t$ satisfying $k^n(t) \in K_0$ whenever $n \leqq n_0$. The operator $T: f \mapsto f \circ k$ on $C(K)$ will yield the desired counterexample.)

# Bibliography

*Note.* In the text frequent reference is made to the monographs of Dunford-Schwartz [1958], Halmos [1950], Schaefer [1966a]; these are abbreviated by [DS], [H], [S] respectively.

Akcoglu, M. and L. Sucheston

[1972] On operator convergence in Hilbert space and in Lebesgue space. Periodica Math. Hung. **2**, 235—244 (1972).

Alaoglu, L. and G. Birkhoff

[1940] General ergodic theorems. Ann. of Math. **41**, 293—309 (1940).

Alfsen, E. M.

[1971] Compact Convex Sets and Boundary Integrals. Berlin-Heidelberg-New York: Springer 1971.

Ando, T.

[1957] Positive operators in semi-ordered linear spaces. J. Fac. Sci. Hokkaido Univ., Ser. I., **13**, 214—228 (1957).

[1960] Linear functionals on Orlicz spaces. Nieuw Archief voor Wiskunde (3), **8**, 1—16 (1960).

[1961] Convergent sequences of finitely additive measures. Pacific J. Math. **11**, 395—404 (1961).

[1962] On compactness of integral operators. Indag. Math. **24**, 235—239 (1962).

[1965] Extensions of linear functionals on Riesz spaces. Indag. Math. **27**, 388—395 (1965).

[1966] Contractive projections in $L_p$-spaces. Pacific J. Math. **17**, 391—405 (1966).

[1968] Invariante Masse positiver Kontraktionen in $C(X)$. Studia Math. **31**, 173—187 (1968).

[1969] Banachverbände und positive Projektionen. Math. Z. **109**, 121—130 (1969).

Asplund, E. and I. Namioka

[1967] A geometric proof of Ryll-Nardzewski's fixed point theorem. Bull. Amer. Math. Soc. **73**, 443—445 (1967).

Badé, W. G.

[1959] A multiplicity theory for Boolean algebras of projections in Banach spaces. Trans. Amer. Math. Soc. **92**, 508—530 (1959).

Bartle, R. G.

[1963] Spectral localization of operators in Banach spaces. Math. Ann. **153**, 261—269 (1963).

[1970] Spectral decomposition of operators in Banach spaces. Proc. London Math. Soc. (3) **20**, 438—450 (1970).

Bauer, F. L.

[1965] An elementary proof of the Hopf inequality for positive operators. Numer. Math. **7**, 331—337 (1965).

Bauer, F. L., Deutsch, E. und J. Stoer

[1969] Abschätzungen für die Eigenwerte positiver linearer Operatoren. Linear Algebra and Appl. **2**, 275—301 (1969).

Bauer, H.

[1957] Sur le prolongement des formes linéaires positives dans un espace vectoriel ordonné. C. R. Acad. Sci. Paris **244**, 289—292 (1957).

[1961] Šilovscher Rand und Dirichletsches Problem. Ann. Inst. Fourier (Grenoble) **11**, 89—136 (1961).

Berberian, S. K.
[1962] Approximate proper vectors. Proc. Amer. Math. Soc. **13**, 111—114 (1962).

Berglund, J. E. and K. H. Hofmann
[1967] Compact Semitopological Semigroups and Weakly Almost Periodic Functions. Lecture Notes in Math. 42, Berlin-Heidelberg-New York: Springer 1967.

Bernau, S. J.
[1973] A note on $L_p$-spaces. Math. Ann. **200**, 281—286 (1973).
[1974] Theorems of Korovkin type for $L_p$-spaces. (Preprint).

Bernau, S. J. and H. E. Lacey
[1973] The range of a contractive projection on an $L_p$-space. (Preprint).

Birkhoff, G. (See also Alaoglu, L. and G. Birkhoff)
[1946] Three observations on linear algebra. Universidad Nacional de Tucuman, Revista Ser. A, **5**, 147—151 (1946).
[1967] Lattice Theory. 3rd ed., Amer. Math. Soc. Colloquium Publications, Vol. 25, 1967.

Blum, J. R. and D. L. Hanson
[1960] On the mean ergodic theorem for subsequences. Bull. Amer. Math. Soc. **66**, 308—311 (1960)

Bonsall, F. F. and G. E. H. Reuter
[1956] A fixed-point theorem for transition operators in an $(L)$-space. Quart. J. Math. Oxford Ser. **7**, 244—248 (1956).

Bourbaki, N.
[1961] Topologie Générale, chap. 1 et 2, 3e ed. Paris: Hermann 1961.
[1965] Intégration, chap. 1, 2, 3 et 4, 2e ed. Paris: Hermann 1965.
[1967] Espaces Vectoriels Topologiques, chap. 3, 4 et 5, 2e ed. Paris: Hermann 1967.

Brace, J. W.
[1953] Transformations on Banach spaces. Dissertation, Cornell University 1953.

Brauer, A.
[1957] The theorems of Ledermann and Ostrowski on positive matrices. Duke Math. J. **24**, 265—274 (1957).
[1961] On the characteristic roots of a power-positive matrix. Duke Math. J. **28**, 439—445 (1961).
[1964] On the characteristic roots of non-negative matrices. In: Schneider [1964], pp. 3—38.
[1971] A remark on the paper of H. H. Schaefer. Numer. Math. **17**, 163—165 (1971).

Braunschweiger, C. C. (See Fullerton, R. E. and C. C. Braunschweiger).

Caubet, J. P.
[1965] Semi-groupes généralisés de matrices positives. Ann. Inst. H. Poincaré, Sect. B, **1**, 239—310 (1964/65).

Chacon, R. V. and U. Krengel
[1964] Linear modulus of a linear operator. Proc. Amer. Math. Soc. **15**, 553—559 (1964).

Chaney, J.
[1972] Banach lattices of compact maps. Math. Z. **129**, 1—19 (1972).

Choquet, G. et C. Foias
[1973] Solution d'un problème sur les itérés d'un opérateur sur $C(K)$, et propriétés ergodiques associées. (Preprint).

Chung, K. L.
[1960] Markov Chains with Stationary Transition Probabilities. Berlin-Heidelberg-New York: Springer 1960.

Davies, E. B.
[1968] The structure and ideal theory of the pre-dual of a Banach lattice. Trans. Amer. Math. Soc. **131**, 544—555 (1968).
[1969] The Choquet theory and the representation theory of ordered Banach spaces. Illinois J. Math. **13**, 176—187 (1969).

Day, M. M.
[1969] Semigroups and amenability. In: Folley, K. W.: Semigroups. New York-London: Academic Press 1969.
[1973] Normed Linear Spaces, 2nd edition. Berlin-Heidelberg-New York: Springer 1973.

Dean, D. W.
[1965] Direct factors of $(AL)$-spaces. Bull. Amer. Math. Soc. **71**, 368—371 (1965).

Dermenjian, Y. et J. Saint-Raymond
[1970] Produit tensoriel de deux cônes convexes saillants. Séminaire Choquet, 9e année, 1969/70, n° 20, 6 p.

Deutsch, E. (See Bauer, F., Deutsch, E. und J. Stoer).

Dieudonné, J.
[1951] Sur les espaces de Köthe. J. Analyse Math. **1**, 81—115 (1951).

Dixmier, J.
[1951] Sur certains espaces considérés par M. M. Stone. Summa Brasil. Math. **2**, 151—182 (1951).

Doob, J. L.
[1953] Stochastic Processes. New York-London: Wiley 1953.

Douglas, R. G.
[1965] Contractive projections on an $L^1$-space. Pacific J. Math. **15**, 443—462 (1965).

Dunford, N. and J. Pettis
[1940] Linear operations on summable functions. Trans. Amer. Math. Soc. **47**, 323—392 (1940).

Dunford, N. and J. T. Schwartz
[1958] Linear Operators. Part I: General Theory. New York: Wiley 1958. (Abbreviated [DS]).
[1963] Linear Operators. Part II: Spectral Theory. New York: Wiley 1963.
[1971] Linear Operators. Part III: Spectral Operators. New York: Wiley 1971.

Dvoretzky, A. and C. A. Rogers
[1950] Absolute and unconditional convergence in normed linear spaces. Proc. Nat. Acad. Sci. U.S.A. **36**, 192—197 (1950).

Eberlein, W. F.
[1949] Abstract ergodic theorems and weakly almost periodic functions. Trans. Amer. Math. Soc. **67**, 217—240 (1949).

Ellis, A. J.
[1964] Extreme positive operators. Quart. J. Math. Oxford Ser. **15**, 342—344 (1964).
[1966] Linear operators in partially ordered normed vector spaces. J. London Math. Soc. **41**, 323—332 (1966).

Ellis, R.
[1969] Lectures on Topological Dynamics. New York: W. A. Benjamin 1969.

Enflo, P.
[1973] A counterexample to the approximation problem in Banach spaces. Acta Math. **130**, 309—317 (1973).

Fan, Ky
[1958] Topological proofs for certain theorems on matrices with non-negative elements. Math. Monatshefte **62**, 219—237 (1958).

Farahat, H.K.
[1966] The semigroup of doubly stochastic matrices. Proc. Glasgow Math. Assoc. **7**, 178—183 (1966).

Foias, C. (See Choquet, G. et C. Foias).

Fremlin, D.H.
[1972] Tensor products of Archimedean vector lattices. Amer. J. Math. **44**, 777—798 (1972).
[1974a] Topological Riesz Spaces and Measure Theory. Cambridge Univ. Press 1974.
[1974b] A characterization of *L*-spaces. (Preprint).

Freudenthal, H.
[1936] Teilweise geordnete Moduln. Proc. Acad. Sci. Amsterdam **39**, 641—651 (1936).

Frobenius, G.
[1909] Über Matrizen aus positiven Elementen. Sitz.-Berichte Kgl. Preuß. Akad. Wiss. Berlin, 471—476 (1908), 514—518 (1909).
[1912] Über Matrizen aus nicht-negativen Elementen. Sitz.-Berichte Kgl. Preuß. Akad. Wiss. Berlin, 456—477 (1912).

Fuchs, L.
[1966] Teilweise geordnete algebraische Strukturen. Göttingen: Vandenhoeck u. Rupprecht 1966.

Fullerton, R.E.
[1957] Quasi-interior points of cones in a linear space. ASTICA Doc., No. AD-120406 (1957).

Fullerton, R.E. and C.C. Braunschweiger
[1966] Quasi-interior points and the extension of linear functionals. Math. Ann. **162**, 214—224 (1966).

Gantmacher, F.R.
[1970] Matrizenrechnung I. Berlin: Deutscher Verlag der Wissenschaften 1970.
[1971] Matrizenrechnung II. Berlin: Deutscher Verlag der Wissenschaften 1971.

Gillespie, T.A. and T.T. West
[1972] Operators generating weakly compact groups. Indiana Univ. Math. J. **21**, 671—688 (1972).

Glicksberg, I.
[1965] A remark on almost periodic transition operators. Israel J. Math. **3**, 71—74 (1965).
[1969] A remark on eigenvalues of certain positive operators. Israel J. Math. **7**, 147—150 (1969).

Glicksberg, I. and K. de Leeuw
[1961] Applications of almost periodic compactifications. Acta Math. **105**, 63—97 (1961).
[1965] The decomposition of certain group representations. J. Analyse Math. **15**, 135—192 (1965).

Goullet de Rugy, A.
[1971] La théorie des cônes biréticulés. Ann. Inst. Fourier (Grenoble) 21, n° 4, 1—18 (1971).
[1972] La structure idéale des *M*-espaces. J. Math. Pures Appl. IX. Sér. **51**, 331—373 (1972).
[1974] Representation of Banach lattices. In: Foundations of Quantum Mechanics and Ordered Linear Spaces. Lecture Notes in Physics 28, pp. 41—46. Berlin-Heidelberg-New York: Springer 1974.

Gordon, H.
[1964] Relative uniform convergene. Math. Ann. **153**, 418—427 (1964).

Grothendieck, A.
[1953] Sur les applications linéaires faiblement compactes d'espaces du type *C*(*K*). Canad. J. Math. **5**, 129—173 (1953).
[1954] Résultats nouveaux dans la théorie des opérateurs linéaires. C. R. Acad. Sci. Paris **239**, 577—579, 607—609 (1954).
[1955a] Produits tensoriels topologiques et espaces nucléaires. Mem. Amer. Math. Soc. **16** (1955).

[1955b] Une charactérisation vectorielle-métrique des espaces $L^1$. Canad. J. Math. **7**, 552—561 (1955).
[1956] Résumé de la théorie métrique des produits tensoriels topologiques. Bol. Soc. Mat. Sao Paulo **8**, 1—79 (1956).

Hackenbroch, W.
[1972] Zur Darstellungstheorie $\sigma$-vollständiger Vektorverbände. Math. Z. **128**, 115—128 (1972).
[1973] Zum Radon-Nikodymschen Satz für positive Vektormaße. Math. Ann. **206**, 63—65 (1973).
[1974] Eindeutigkeit des Darstellungsraumes von Vektorverbänden. Math. Z. **135**, 285—288 (1974).

Haeuslein, G.K.
[1972] Skew product ergodic flows on the torus. (Preprint).

Halmos, P.R.
[1970] Measure Theory. New York-Toronto-London: D. van Nostrand 1950 (Abbreviated [H]).
[1956] Lectures on Ergodic Theorie. New York: Chelsea Publ. Comp. 1956.

Halmos, P.R. and J. von Neumann
[1942] Operator methods in classical mechanics, II. Ann. of Math. **43**, 332—350 (1942).

Hanson, D. L. (See Blum, J. R. and D. L. Hanson).

Hardy, G. H., Littlewood, J. E., and G. Polya
[1952] Inequalities. Cambridge University Press 1952.

Hille, E. and J. D. Tamarkin
[1930] On the theory of linear integral equations, I. Ann. of Math. **31**, 479—528 (1930).
[1934] On the theory of linear integral equations, II. Ann. of Math. **35**, 445—455 (1934).

Hoffman, A. J. and H. Wielandt
[1953] The variation of the spectrum of a normal matrix. Duke Math. J. **20**, 37—40 (1953).

Hofmann, K.H. (See Berglund, J.E. and K.H. Hofmann).

Hopf, E.
[1963] An inequality for positive linear integral operators. J. Math. Mech. **12**, 683—692 (1963).

Ionescu-Tulcea, A. and C. Ionescu-Tulcea
[1969] Topics in the Theory of Lifting. Berlin-Heidelberg-New York: Springer 1969.

Ito, T.
[1967] On some properties of a vector lattice. Sci. Papers College General Educ. Univ. Tokyo **17**, 161—172 (1967).

Jacobs, K.
[1960] Neuere Methoden und Ergebnisse der Ergodentheorie. Berlin-Heidelberg-New York: Springer 1960.
[1963] Lectures on Ergodic Theory. Aarhus 1962/63.

Jameson, G.
[1970] Ordered Linear Spaces. Lecture Notes in Math. 141. Berlin-Heidelberg-New York: Springer 1970.

Jentzsch, P.
[1912] Über Integralgleichungen mit positivem Kern. J. Reine Angew. Math. **141**, 235—244 (1912).

Jörgens, K.
[1970] Lineare Integraloperatoren. Stuttgart: B. G. Teubner 1970.

Johnson, D. G. and J. E. Kist
[1962] Prime ideals in vector lattices. Canad. J. Math. **14**, 517—528 (1962).

Jones, L. K. and V. Kuftinec
[1971] A note on the Blum-Hanson theorem. Proc. Amer. Math. Soc. **30**, 202—203 (1971).

Judin, A. J.
[1939] Solution of two problems on the theory of partially ordered spaces. Dokl. Akad. Nauk SSSR **23**, 418—422 (1939).

Kaashoek, M. A. and T. T. West
[1974] Locally Compact Semi-Algebras, With Applications to Spectral Theory of Positive Operators. Amsterdam: North-Holland 1974.

Kakutani, S.
[1941a] Concrete representation of abstract $L$-spaces and the mean ergodic theorem. Ann. of Math. **42**, 523—537 (1941).
[1941b] Concrete representation of abstract $M$-spaces. Ann. of Math. **42**, 994—1024 (1941).

Kantorovič, L. V.
[1937] Lineare halbgeordnete Räume. Recueil Math. (Moscow) **2**, 121—168 (1937).
[1940] Linear operators in semi-ordered spaces. Mat. Sbornik **7**, 209—284 (1940).

Kantorovič, L. V., Vulikh, B. C., and A. G. Pinsker
[1950] Functional Analysis in Partially Ordered Spaces. (Russian). Moscow-Leningrad 1950.

Karlin, S.
[1959] Positive operators. J. Math. Mech. **8**, 907—937 (1959).

Kemeny, J. G., Snell, J. L., and A. W. Knapp
[1966] Denumerable Markov Chains. Princeton: D. van Nostrand 1966.

Kist, J. E. (See Johnson, D. G. and J. E. Kist).

Knapp, A. W. (See Kemeny, J. G., Snell, J. L., and A. W. Knapp).

Korovkin, P. P.
[1953] On the convergence of linear positive operators in the space of continuous functions. Dokl. Akad. SSSR (N. S.) **90**, 961—964 (1953).

Krasnosels'kiĭ, M. A. and Ya. B. Rutickiĭ
[1961] Convex Functions and Orlicz Spaces. (Translated from the Russian). Groningen: Noordhoff 1961.

Krein, M. and S. Krein
[1940] On an inner characteristic of the set of all continuous functions defined on a bicompact Hausdorff space. Dokl. Akad. Nauk SSSR **27**, 427—430 (1940).

Krein, M. G. and M. A. Rutman
[1948] Linear operators leaving invariant a cone in a Banach space. Uspehi Mat. Nauk (N. S.) **3**, no. 1 (23), 3—95 (1948). (Russian). Also Amer. Math. Soc. Transl. no. 26 (1950).

Krengel, U. (See also Chacon, R. V. and U. Krengel)
[1964a] Über den Absolutbetrag stetiger linearer Operatoren und seine Anwendung auf ergodische Zerlegungen. Math. Scand. **13**, 151—187 (1964).
[1964b] Ein Approximationslemma in komplexen Banachräumen mit positivem Kegel. Math. Scand. **15**, 69—78 (1964).
[1966] Remark on the modulus of compact operators. Bull. Amer. Math. Soc. **72**, 132—133 (1966).

Krieger, H. J.
[1969] Beiträge zur Theorie positiver Operatoren. Schriftenreihe der Institute für Math., Reihe A, Heft 6. Berlin: Akademie-Verlag 1969.

Kuftinec, V. (See Jones, L. K. and V. Kuftinec).

Lacey, H. E. (See Bernau, S. J. and H. E. Lacey).

Ledermann, W.
[1950] Bounds for the greatest latent root of a matrix. J. London Math. Soc. **25**, 265—268 (1950).

de Leeuw, K. (See Glicksberg, I. and K. de Leeuw).

Levin, V. L.
[1969a] Tensor products and functors in categories of Banach spaces defined by *KB*-lineals. Trudy Moscov Mat. Obšč. **20**, 43—82 (1969). (Russian).
[1969b] Two classes of linear mappings which operate between Banach spaces and Banach lattices. Sibirsk Mat. Ž. **10**, 903—909 (1969). (Russian).

Lindenstrauss, J.
[1970] Some aspects of the theory of Banach spaces. Advances in Math. **5**, 159—180 (1970).

Lindenstrauss, J. and A. Pelczynski
[1968] Absolutely summing operators in $\mathscr{L}_p$-spaces and their applications. Studia Math. **29**, 275—326 (1968).

Lindenstrauss, J. and L. Tzafriri
[1971] On the complemented subspaces problem. Israel J. Math. **9**, 263—269 (1971).
[1973] Classical Banach Spaces. Lecture Notes in Math. 338. Berlin-Heidelberg-New York: Springer 1973.

Lindenstrauss, J. and D. E. Wulbert
[1969] On the classification of the Banach spaces whose duals are $L_1$-spaces. J. Functional Analysis **4**, 332—349 (1969).

Littlewood, J. E. (See Hardy, G. H., Littlewood, J. E., and G. Polya).

Lloyd, S. P.
[1963] On finitely additive set functions. Proc. Amer. Math. Soc. **14**, 701—704 (1963).

Loomis, L. H.
[1953] An Introduction to Abstract Harmonic Analysis. New York-Toronto-London: D. van Nostrand 1953.

Lorentz, G. G.
[1967] Approximation of Functions. New York: Holt, Rinehart and Winston 1967.

Lotz, H. P.
[1968] Über das Spektrum positiver Operatoren. Math. Z. **108**, 15—32 (1968).
[1969] Zur Idealstruktur in Banachverbänden. Habilitationsschrift. Tübingen 1969.
[1973a] Minimal and reflexive Banach lattices. Math. Ann. **209**, 117—126 (1974).
[1973b] A note on the sum of two closed lattice ideals. Proc. Amer. Math. Soc. **44**, 389—390 (1974).
[1974] Extensions and liftings of positive linear mappings on Banach lattices. To appear in Trans. Amer. Math. Soc.
[1975] Grothendieck ideals of operators in Banach spaces. To appear in Lecture Notes in Math. Berlin-Heidelberg-New York: Springer.

Lotz, H. P. und H. H. Schaefer
[1968] Über einen Satz von F. Niiro und I. Sawashima. Math. Z. **108**, 33—36 (1968).

Lozanovskii, G. Ja.
[1967] On Banach lattices and bases. Funkcional'. Analiz Priloženija **1**, 92 (1967).
[1968] Some topological properties of Banach lattices and reflexivity conditions of them. Soviet Math., Doklady **9**, 1415—1418 (1968).
[1970] On Banach lattices with unit element. Izvestija Vysš. Zaved. Matematika **1**, 65—69 (1970). (Russian).

Luxemburg, W. A. J.
[1965a] Notes on Banach function spaces XIV. Indag. Math. **27**, 229—248 (1965).
[1965b] Notes on Banach function spaces XV. Indag. Math. **27**, 415—446 (1965).
[1965c] Notes on Banach function spaces XVI. Indag. Math. **27**, 646—667 (1965).
[1967] Rearrangement-invariant Banach function spaces. Proc. Symp. Analysis, Queen's University Kingston, Ontario, 83—144 (1967).

Luxemburg, W. A. J. and A. C. Zaanen
[1963a] Notes on Banach function spaces I, II. Indag. Math. **25**, 135—153 (1963).
[1963b] Notes on Banach function spaces III, IV. Indag. Math. **25**, 239—263 (1963).
[1963c] Notes on Banach function spaces V. Indag. Math. **25**, 496—504 (1963).
[1963d] Notes on Banach function spaces VI, VII. Indag. Math. **25**, 655—681 (1963).
[1963e] Compactness of integral operators in Banach function spaces. Math. Ann. **149**, 150—180 (1963).
[1964a] Notes on Banach function spaces VIII. Indag. Math. **26**, 104—119 (1964).
[1964b] Notes on Banach function spaces IX. Indag. Math. **26**, 360—376 (1964).
[1964c] Notes on Banach function spaces X, XI, XII, XIII. Indag. Math. **26**, 493—543 (1964).
[1971a] The linear modulus of an order-bounded linear transformation. Indag. Math. **33**, 422—447 (1971).
[1971b] Riesz Spaces I. Amsterdam: North Holland 1971.

Marcus, M. and H. Minc
[1964] A Survey of Matrix Theory and Matrix Inequalities. Boston: Allyn and Bacon 1964.

Marcus, M., Minc, H., and B. Moyls
[1961] Some results on non-negative matrices. J. Res. Nat. Bur. Standards, Sect. B, **65**, 205—209 (1961).

Marek, I.
[1966] Spektrale Eigenschaften der $\mathcal{K}$-positiven Operatoren und Einschließungssätze für den Spektralradius. Czechoslovak Math. J. **16** (91), 493—517 (1966).
[1970] Frobenius theory of positive operators: comparison theorems and applications. SIAM J. Appl. Math. **19**, 607—628 (1970).

Marti, J. T.
[1970] Topological representation of abstract $L_p$-spaces. Math. Ann. **185**, 315—321 (1970).

Meyer-Nieberg, P.
[1973a] Charakterisierung einiger topologischer und ordnungstheoretischer Eigenschaften von Banachverbänden mit Hilfe disjunkter Folgen. Arch. Math. **24**, 640—647 (1973).
[1973b] Zur schwachen Kompaktheit in Banachverbänden. Math. Z. **134**, 303—315 (1973).

Minc, H. (See Marcus, M. and H. Minc; Marcus, M., Minc, H., and B. Moyls).

Mirsky, L.
[1963] Results and problems in the theory of doubly-stochastic matrices. Z. Wahrscheinlichkeitstheorie und Verw. Gebiete **1**, 319—334 (1963).

Mittelmeyer, G. und M. Wolff
[1974] Über ein einheitliches Axiomensystem für Vektorverbände über $\mathbb{R}$ oder $\mathbb{C}$. Math. Z. **137**, 87—92 (1974).

Moyls, B. (See Marcus, M., Minc, H., and B. Moyls).

Nagel, R. J.
[1972] Darstellung von Verbandsoperatoren auf Banachverbänden. Revista Acad. Ciencias Zaragossa, II. Ser., **27**, 281—288 (1972).
[1973a] A Stone-Weierstrass theorem for Banach lattices. Studia Math. **47**, 75—82 (1973).
[1973b] Ordnungsstetigkeit in Banachverbänden. Manuscripta Math. **9**, 9—27 (1973).
[1973c] Mittelergodische Halbgruppen linearer Operatoren. Ann. Inst. Fourier (Grenoble) **23**, 75—87 (1973).

[1974] Ergodic and mixing properties of linear operators. To appear in Proc. Royal Irish Acad.

Nagel, R.J. und U. Schlotterbeck
[1972] Integraldarstellung regulärer Operatoren auf Banachverbänden. Math. Z. **127**, 293—300 (1972).
[1973] Kompaktheit von Integraloperatoren auf Banachverbänden. Math. Ann. **202**, 301—306 (1973).
[1974] Zur Approximation kompakter Operatoren durch Operatoren endlichen Ranges. To appear in Arch. Math.

Nagel, R.J. and M. Wolff
[1972] Abstract dynamical systems with an application to operators with discrete spectrum. Arch. Math. **23**, 170—176 (1972).

Nakano, H.
[1941] Über das System aller stetigen Funktionen auf einem topologischen Raum. Proc. Imp. Acad. Tokyo **17**, 308—310 (1941).
[1950] Modulared Semi-Ordered Linear Spaces. Tokyo: Maruzen 1950.
[1953] Product spaces of semi-ordered linear spaces. J. Fac. Sci. Hokkaido Univ. Ser. I, **12**, 163—210 (1953).
[1955] Semi-Ordered Linear Spaces. Tokyo: Society for the Promotion of Science 1955.
[1966] Linear Lattices. Detroit: Wayne State Univ. Press 1966.

Namioka, I. (See also Asplund, E. and I. Namioka)
[1957] Partially ordered linear topological spaces. Mem. Amer. Math. Soc. **24** (1957).

von Neumann, J. (See also Halmos, P. R. and J. von Neumann)
[1932] Zur Operatorenmethode in der klassischen Mechanik. Ann. of Math. **33**, 587—648, 789—791 (1932).

Ng, K.F. (See Wong, Y.C. and K.F. Ng).

Nielsen, N.J.
[1973] On Banach ideals determined by Banach lattices and their applications. Diss. Math. CIX.

Niiro, F.
[1964a] On indecomposable operators in $l^p_{1<p<\infty}$ and a problem of H. H. Schaefer. Sci. Papers College General Educ. Univ. Tokyo **14**, 165—179 (1964).
[1964b] On indecomposable operators in $L^p_{1<p<\infty}$ and a problem of H. H. Schaefer. Sci. Papers College General Educ. Univ. Tokyo **15**, 53—64 (1964).

Niiro, F. and I. Sawashima
[1966] On spectral properties of positive irreducible operators in an arbitrary Banach-lattice and problems of H. H. Schaefer. Sci. Papers College General Educ. Univ. Tokyo **16**, 145—183 (1966).
[1973a] On a set including the spectra of positive operators. Sci. Papers College General Educ. Univ. Tokyo **23**, 93—98 (1973).
[1973b] Reduction of a sub-Markov operator to its irreducible components. Nat. Sci. Report Ochanomizu Univ. **24**, 35—59 (1973).

Ogasawara, T.
[1948] Vector Lattices. Tokyo 1948 (Japanese).

Ostrowski, A.W.
[1952] Bounds for the greatest latent root of a positive matrix. J. London Math. Soc. **27**, 253—256 (1962).
[1964] Positive matrices and functional analysis. In: Schneider [1964], pp. 81—101.

Pelczynski, A. (See also Lindenstrauss, J. and A. Pelczynski)
[1967] A characterization of Hilbert-Schmidt operators. Studia Math. **28**, 355—360 (1967).
[1968] On Banach spaces containing $L_1(\mu)$. Studia Math. **30**, 231—246 (1968).

Peressini, A. L.
[1963] Concerning the order structure of Köthe sequence spaces. Michigan Math. J. **10**, 409—415 (1963).
[1967] Ordered Topological Vector Spaces. New York-Evanston-London: Harper and Row 1967.

Peressini, A. L. and D. R. Sherbert
[1969] Ordered topological tensor products. Proc. London Math. Soc. (3) **19**, 177—190 (1969).

Perron, O.
[1907] Zur Theorie der Matrices. Math. Ann. **64**, 248—263 (1907).

Pettis, J. (See Dunford, N. and J. Pettis).

Phelps, R. R.
[1963] Extremal operators and homomorphisms. Trans. Amer. Math. Soc. **108**, 265—274 (1963).

Phillips, R. S.
[1940] On linear transformations. Trans. Amer. Math. Soc. **48**, 516—541 (1940).
[1962] Semi-groups of positive contractive operators. Czechoslovak Math. J. **12**, 294—312 (1962).

Pietsch, A.
[1963a] Absolute Summierbarkeit in Vektorverbänden. Math. Nachr. **26**, 15—32 (1963).
[1963b] Absolut summierende Abbildungen in lokalkonvexen Räumen. Math. Nachr. **27**, 77—103 (1963).
[1965] Nukleare lokalkonvexe Räume. Schriftenreihe der Institute für Math., Reihe A, Heft 1. Berlin: Akademie-Verlag 1965.
[1967] Absolut $p$-summierende Abbildungen in normierten Räumen. Studia Math. **28**, 333—353 (1967).

Pinsker, A. G. (See Kantorovič, L. V., Vulikh, B. C., and A. G. Pinsker).

Polya, G. (See Hardy, G. H., Littlewood, J. E., and G. Polya).

Popa, N.
[1968] Produits tensoriels ordonnés. Revue Roumaine Math. Pur. Appl. **8**, 235—246 (1968).

Raimi, R. A.
[1964] Minimal sets and ergodic measures in $\beta\mathbb{N}\setminus\mathbb{N}$. Bull. Amer. Math. Soc. **70**, 711—712 (1964).

Reuter, G. E. H. (See Bonsall, F. F. and G. E. H. Reuter).

Riesz, F.
[1930] Sur la décomposition des opérations fonctionelles linéaires. Atti Congr. Internaz. Mat. Bologna **128**, 3, 143—148 (1930).
[1940] Sur quelques notions fondamentales dans la théorie générale des opérations linéaires. Ann. of Math. **41**, 174—206 (1940).

Rogers, C. A. (See Dvoretzky, A. and C. A. Rogers).

Romanovsky, V.
[1936] Recherches sur les chaines de Markoff. Acta Math. **66**, 147—251 (1936).

Rosenthal, H. P.
[1973] On subspaces of $L^p$. Ann. of Math. **97**, 344—373 (1973).

Rota, G.-C.
[1961] On the eigenvalues of positive operators. Bull. Amer. Math. Soc. **67**, 556—558 (1961).
[1962] An „Alternierende Verfahren" for general positive operators. Bull. Amer. Math. Soc. **68**, 95—102 (1962).

Rutickiĭ, Ya. B. (See Krasnosels'kiĭ, M. A. and Ya. B. Rutickiĭ).

Rutman, M. A. (See Krein, M. G. and M. A. Rutman).

Ryff, J.V.
[1963] On the representation of doubly stochastic operators. Pacific J. Math. **13**, 1379—1386 (1963).
[1965] Orbits of $L^1$-functions under doubly stochastic transformations. Trans. Amer. Math. Soc. **117**, 92—100 (1965).
[1967] Extreme points of some convex subsets of $L^1(0,1)$. Proc. Amer. Math. Soc. **18**, 1026—1034 (1967).

Saint-Raymond, J. (See Dermenjian, Y. et J. Saint-Raymond).

Šaškin, Ju. A.
[1966] Korovkin systems in spaces of continuous functions. Izw. Akad. Nauk SSSR Ser. Mat. **26**, 495—512 (1962). English Transl.: Amer. Math. Soc. Transl. II. Ser. **54**, 125—144 (1966).

Sato, R.
[1973] On Akcoglu and Sucheston's operator convergence theorem in Lebesgue space. Proc. Amer. Math. Soc. **40**, 513—516 (1973).

Sawashima, I. (See also Niiro, F. and I. Sawashima)
[1964] On spectral properties of some positive operators. Nat. Sci. Report Ochanomizu Univ. **15**, 53—64 (1964).

Schaefer, H.H. (See also Lotz, H.P. und H.H. Schaefer)
[1955] Positive Transformationen in halbgeordneten lokalkonvexen Vektorräumen. Math. Ann. **129**, 323—329 (1955).
[1958] Halbgeordnete lokalkonvexe Vektorräume. Math. Ann. **135**, 115—141 (1958).
[1959a] On non-linear positive operators. Pacific J. Math. **9**, 847—860 (1959).
[1959b] Über algebraische Integralgleichungen mit nichtnegativen Koeffizienten. Math. Ann. **137**, 385—391 (1959).
[1959c] Halbgeordnete lokalkonvexe Vektorräume. II. Math. Ann. **138**, 259—286 (1959).
[1960a] On the singularities of an analytic function with values in a Banach space. Arch. Math. **11**, 40—43 (1960).
[1960b] Halbgeordnete lokalkonvexe Vektorräume. III. Math. Ann. **141**, 113—142 (1960).
[1960c] Some spectral properties of positive linear operators. Pacific J. Math. **10**, 1009—1019 (1960).
[1960d] On the completeness of topological vector lattices. Michigan Math. J. **7**, 303—309 (1960).
[1962a] Spectral measures in locally convex algebras. Acta Math. **107**, 125—173 (1962).
[1962b] A generalized moment problem. Math. Ann. **146**, 326—330 (1962).
[1962c] Some non-linear eigenvalue problems. Proc. Symposium on Non-linear Problems. Univ. of Wisconsin Press, 117—137, Madison 1962.
[1963a] Convex cones and spectral theory. Proc. of Symposia in Pure Math., Convexity, 451—471. Providence, R.I. 1963.
[1963b] Spektraleigenschaften positiver linearer Operatoren. Math. Z. **82**, 303—313 (1963).
[1964] On the point spectrum of positive operators. Proc. Amer. Math. Soc. **15**, 56—60 (1964).
[1965] Über das Randspektrum positiver Operatoren. Math. Ann. **162**, 289—293 (1965).
[1966a] Topological Vector Spaces. 1st and 2nd print, New York-London: Macmillan 1966 and 1967. 3rd print, Berlin-Heidelberg-New York: Springer 1971. (Abbreviated [S]).
[1966b] Eine Klasse irreduzibler positiver Operatoren. Math. Ann. **165**, 26—30 (1966).
[1967] Invariant ideals of positive operators in $C(X)$. I. Illinois J. Math. **11**, 703—715 (1967).
[1968] Invariant ideals of positive operators in $C(X)$. II. Illinois J. Math. **12**, 525—538 (1968).
[1970a] Eine Abschätzung der nichttrivialen Eigenwerte stochastischer Matrizen. Numer. Math. **15**, 219—223 (1970).
[1970b] Topologische Nilpotenz irreduzibler Operatoren. Math. Z. **117**, 135—140 (1970).
[1971a] Weak convergence of measures. Math. Ann. **193**, 57—64 (1971).
[1971b] On the characteristic roots of real matrices. Proc. Amer. Math. Soc. **28**, 91—92 (1971).
[1972a] On the representation of Banach lattices by continuous numerical functions. Math. Z. **125**, 215—232 (1972).
[1972b] Normed tensor products of Banach lattices. Israel J. Math. **13**, 400—515 (1972).

[1974a] The Šilov boundary of a convex cone. In: Foundations of Quantum Mechanics and Ordered Linear Spaces, pp. 342—344. Lecture Notes in Physics 28. Berlin-Heidelberg-New York: Springer 1974.
[1974b] Banach lattices and cyclic Banach spaces. Proc. Symposium on Spectral Theory, Royal Irish Acad. 1974 (to appear).
[1974c] Reticoli di Hilbert ed Operatori Hilbert-Schmidt. Boll. Un. Mat. Ital. (to appear).

Schaefer, H. H. and U. Schlotterbeck
[1974] On the approximation of kernel operators by operators of finite rank. J. Approximation Theory.

Scheffold, E.
[1971a] Die erzeugte Algebra eines Markov-Verbandsoperators in $C(X)$. Math. Ann. **193**, 76—82 (1971).
[1971b] Das Spektrum von Verbandsoperatoren in Banachverbänden. Math. Z. **123**, 177—190 (1971).
[1973a] Ein allgemeiner Korovkin-Satz für lokalkonvexe Vektorverbände. Math. Z. **132**, 209—214 (1973).
[1973b] Über die punktweise Konvergenz von Operatoren in $C(X)$. Revista Acad. Ciencias Zaragossa, II. Ser. **28**, 5—12 (1973).

Schlotterbeck, U. (See also Nagel, R. J. und U. Schlotterbeck; Schaefer, H. H. and U. Schlotterbeck)
[1969] Über Klassen majorisierbarer Operatoren auf Banachverbänden. Dissertation. Tübingen 1969.
[1973] Order-theoretic characterization of Hilbert-Schmidt operators. Arch. Math. **24**, 67—70 (1973).

Schneider, H.
[1964] Recent Advances in Matrix Theory. Univ. of Wisconsin Press 1964.

Schwartz, J. T. (See Dunford, N. and J. T. Schwartz).

Seever, G. L.
[1968] Measures on $F$-spaces. Trans. Amer. Math. Soc. **133**, 267—280 (1968).

Semadeni, Z.
[1964] On weak convergence of measures and $\sigma$-complete Boolean algebras. Colloq. Math. **12**, 229—233 (1964).
[1971] Banach Spaces of Continuous Functions. Warsaw: Polish Scientific Publishers 1971.

Sherbert, D. R. (See Peressini, A. L. and D. R. Sherbert).

Snell, J. L. (See Kemeny, J. G., Snell, J. L., and A. W. Knapp).

Soloviev, V. A.
[1966] Extension of a monotone norm from a normed lattice to its Dedekind completion. Sibirsk. Mat. Ž. **7**, 1360—1369 (1966). (Russian).

Stecenko, V. Ja.
[1967] On a spectral property of an indecomposable operator. Uspehi Mat. Nauk. **22**, 242—244 (1967).

Stoer, J. (See Bauer, F. L., Deutsch, E. und J. Stoer).

Stone, M. H.
[1937] Applications of the theory of Boolean rings to general topology. Trans. Amer. Math. Soc. **41**, 375—481 (1937).
[1940] A general theory of spectra. I. Proc. Nat. Acad. Sci. U.S.A. **26**, 280—283 (1940).
[1941] A general theory of spectra. II. Proc. Nat. Acad. Sci. U.S.A. **27**, 83—87 (1941).

Sucheston, L. (See Akcoglu, M. and L. Sucheston).

Tamarkin, J. D. (See Hille, E. and J. D. Tamarkin).

Taylor, J. C.
[1968] The Šilov boundary for a lattice-ordered semi-group. Pacific J. Math. **25**, 185—191 (1968).

Thoma, E.
[1956] Darstellung von vollständigen Vektorverbänden durch vollständige Funktionenverbände. Arch. Math. **7**, 11—22 (1956).

Tzafriri, L. (See also Lindenstrauss, J. and L. Tzafriri)
[1969a] Remarks on contractive projections in $L_p$-spaces. Israel J. Math. **7**, 9—15 (1969).
[1969b] Reflexivity of cyclic Banach spaces. Proc. Amer. Math. Soc. **22**, 61—68 (1969).
[1969c] An isomorphic characterization of $L_p$ and $c_0$-spaces. I. Studia Math. **32**, 295—304 (1969).
[1971] An isomorphic characterization of $L_p$ and $c_0$-spaces. II. Michigan Math. J. **18**, 21—31 (1971).
[1972] Reflexivity in Banach lattices and their subspaces. J. Functional Analysis **10**, 1—18 (1972).

Vulikh, B. C. (See also Kantorovič, L. V., Pinsker, A. G., and B. C. Vulikh)
[1967] Introduction to the Theory of Partially Ordered Vector Spaces. Groningen: Wolters-Noordhoff 1967.

Walk, H.
[1970] Approximation unbeschränkter Funktionen durch lineare positive Operatoren. Habilitationsschrift. Stuttgart 1970.

Walsh, B.
[1965] Structure of spectral measures on locally convex spaces. Trans. Amer. Math. Soc. **120**, 295—326 (1965).
[1973] Ordered vector sequence spaces and related classes of linear operators. Math. Ann. **206**, 89—138 (1973).

West, T. T. (See Gillespie, T. A. and T. T. West; Kaashoek, M.A. and T. T. West).

Wickstead, A. W.
[1974] Spaces of linear operators between partially ordered Banach spaces. Proc. London Math. Soc. **28**, 141—158 (1974).

Wielandt, H. (See also Hoffman, A. J. and H. Wielandt)
[1950] Unzerlegbare, nicht-negative Matrizen. Math. Z. **52**, 642—648 (1950).

Wolff, M. (See also Mittelmeyer, G. und M. Wolff; Nagel, R. J. and M. Wolff)
[1969a] Über das Spektrum von Verbandshomomorphismen in $C(X)$. Math. Ann. **182**, 161—169 (1969).
[1969b] Über das Spektrum von Verbandshomomorphismen in Banachverbänden. Math. Ann. **184**, 49—55 (1969).
[1970] Über das Spektrum konvexer Kombinationen von Operatoren. Arch. Math. **21**, 72—81 (1970).
[1971] Vektorwertige invariante Masse von rechtsamenablen Halbgruppen positiver Operatoren. Math. Z. **120**, 265—276 (1971).
[1973a] Darstellung von Banachverbänden und Sätze vom Korovkin-Typ. Math. Ann. **200**, 47—67 (1973).
[1973b] Über Korovkin-Sätze in lokalkonvexen Vektorverbänden. Math. Ann. **204**, 49—56 (1973).
[1973c] A general theorem of Korovkin type for vector lattices. J. Approximation Theory **9**, 517—521 (1973).
[1974] Über das Randspektrum komplexer ordnungsbeschränkter Operatoren in Banachverbänden. Math. Z. **135**, 293—302 (1974).

Wong, Y. C. and K. F. Ng
[1973] Partially Ordered Topological Vector Spaces. Oxford: Clarendon Press 1973.

Wulbert, D. E. (See also Lindenstrauss, J. and D. E. Wulbert)
[1968] Convergence of operators and Korovkin's theorem. J. Approximation Theory **1**, 381—390 (1968).

Zaanen, A. C. (See also Luxemburg, W. A. J. and A. C. Zaanen)
[1967] Integration. Amsterdam: North Holland 1967.

# Index of Symbols

Chapter IV:



Chapter V:

# Subject Index

**Die Grundlehren der mathematischen Wissenschaften in Einzeldarstellungen mit besonderer Berücksichtigung der Anwendungsgebiete**

*Eine Auswahl*

---

23. Pasch: Vorlesungen über neuere Geometrie. In Vorbereitung
41. Steinitz: Vorlesungen über die Theorie der Polyeder. In Vorbereitung
45. Alexandroff/Hopf: Topologie. Band 1
46. Nevanlinna: Eindeutige analytische Funktionen
63. Eichler: Quadratische Formen und orthogonale Gruppen
102. Nevanlinna/Nevanlinna: Absolute Analysis
114. Mac Lane: Homology
123. Yosida: Functional Analysis
127. Hermes: Enumerability, Decidability, Computability
131. Hirzebruch: Topological Methods in Algebraic Geometry
135. Handbook for Automatic Computation. Vol. 1/Part a: Rutishauser: Description of ALGOL 60
136. Greub: Multilinear Algebra
137. Handbook for Automatic Computation. Vol. 1/Part b: Grau/Hill/Langmaack: Translation of ALGOL 60
138. Hahn: Stability of Motion
139. Mathematische Hilfsmittel des Ingenieurs. 1. Teil
140. Mathematische Hilfsmittel des Ingenieurs. 2. Teil
141. Mathematische Hilfsmittel des Ingenieurs. 3. Teil
142. Mathematische Hilfsmittel des Ingenieurs. 4. Teil
143. Schur/Grunsky: Vorlesungen über Invariantentheorie
144. Weil: Basic Number Theory
145. Butzer/Berens: Semi-Groups of Operators and Approximation
146. Treves: Locally Convex Spaces and Linear Partial Differential Equations
147. Lamotke: Semisimpliziale algebraische Topologie
148. Chandrasekharan: Introduction to Analytic Number Theory
149. Sario/Oikawa: Capacity Functions
150. Iosifescu/Theodorescu: Random Processes and Learning
151. Mandl: Analytical Treatment of One-dimensional Markov Processes
152. Hewitt/Ross: Abstract Harmonic Analysis. Vol. 2: Structure and Analysis for Compact Groups. Analysis on Locally Compact Abelian Groups
153. Federer: Geometric Measure Theory
154. Singer: Bases in Banach Spaces I
155. Müller: Foundations of the Mathematical Theory of Electromagnetic Waves
156. van der Waerden: Mathematical Statistics
157. Prohorov/Rozanov: Probability Theory. Basic Concepts. Limit Theorems. Random Processes
158. Constantinescu/Cornea: Potential Theory on Harmonic Spaces
159. Köthe: Topological Vector Spaces I
160. Agrest/Maksimov: Theory of Incomplete Cylindrical Functions and their Applications
161. Bhatia/Szegö: Stability Theory of Dynamical Systems
162. Nevanlinna: Analytic Functions
163. Stoer/Witzgall: Convexity and Optimization in Finite Dimensions I
164. Sario/Nakai: Classification Theory of Riemann Surfaces
165. Mitrinović/Vasić: Analytic Inequalities
166. Grothendieck/Dieudonné: Eléments de Géométrie Algébrique I
167. Chandrasekharan: Arithmetical Functions
168. Palamodov: Linear Differential Operators with Constant Coefficients
169. Rademacher: Topics in Analytic Number Theory
170. Lions: Optimal Control of Systems Governed by Partial Differential Equations
171. Singer: Best Approximation in Normed Linear Spaces by Elements of Linear Subspaces

172. Bühlmann: Mathematical Methods in Risk Theory
173. Maeda/Maeda: Theory of Symmetric Lattices
174. Stiefel/Scheifele: Linear and Regular Celestial Mechanics. Perturbed Two-body Motion—Numerical Methods—Canonical Theory
175. Larsen: An Introduction to the Theory of Multipliers
176. Grauert/Remmert: Analytische Stellenalgebren
177. Flügge: Practical Quantum Mechanics I
178. Flügge: Practical Quantum Mechanics II
179. Giraud: Cohomologie non abélienne
180. Landkof: Foundations of Modern Potential Theory
181. Lions/Magenes: Non-Homogeneous Boundary Value Problems and Applications I
182. Lions/Magenes: Non-Homogeneous Boundary Value Problems and Applications II
183. Lions/Magenes: Non-Homogeneous Boundary Value Problems and Applications III
184. Rosenblatt: Markov Processes. Structure and Asymptotic Behavior
185. Rubinowicz: Sommerfeldsche Polynommethode
186. Handbook for Automatic Computation. Vol. 2. Wilkinson/Reinsch: Linear Algebra
187. Siegel/Moser: Lectures on Celestial Mechanics
188. Warner: Harmonic Analysis on Semi-Simple Lie Groups I
189. Warner: Harmonic Analysis on Semi-Simple Lie Groups II
190. Faith: Algebra: Rings, Modules, and Categories I
192. Mal'cev: Algebraic Systems
193. Pólya/Szegö: Problems and Theorems in Analysis I
194. Igusa: Theta Functions
195. Berberian: Baer *-Rings
196. Athreya/Ney: Branching Processes
197. Benz: Vorlesungen über Geometrie der Algebren
198. Gaal: Linear Analysis and Representation Theory
199. Nitsche: Vorlesungen über Minimalflächen
200. Dold: Lectures on Algebraic Topology
201. Beck: Continuous Flows in the Plane
202. Schmetterer: Introduction to Mathematical Statistics
203. Schoeneberg: Elliptic Modular Functions
204. Popov: Hyperstability of Control Systems
205. Nikol'skii: Approximation of Functions of Several Variables and Imbedding Theorems
206. André: Homologie des Algèbres Commutatives
207. Donoghue: Monotone Matrix Functions and Analytic Continuation
208. Lacey: The Isometric Theory of Classical Banach Spaces
209. Ringel: Map Color Theorem
210. Gihman/Skorohod: The Theory of Stochastic Processes I
211. Comfort/Negrepontis: The Theory of Ultrafilters
212. Switzer: Algebraic Topology—Homotopy and Homology
213. Shafrevich: Basic Algebraic Geometry
214. van der Waerden: Group Theory and Quatum Mechanics
215. Schaefer: Banach Lattices and Positive Operators
216. Pólya/Szegő: Problems and Theorems in Analysis II. In preparation